7th International High Performance Buildings Conference at Purdue 2022

West Lafayette, Indiana, USA
10-14 July 2022

ISBN: 978-1-7138-5930-7

Printed from e-media with permission by:

Curran Associates, Inc.
57 Morehouse Lane
Red Hook, NY 12571

CURRAN ASSOCIATES INC. proceedings.com

Some format issues inherent in the e-media version may also appear in this print version.

Ray W. Herrick Laboratories
School of Mechanical Engineering
Purdue University
177 S. Russell Street
West Lafayette, IN 47907-2099, USA.

Phone: +1 (765) 494-2132
Fax: +1 (765) 494-0787

rhlab@ecn.purdue.edu

Additional copies of this publication are available from:

Curran Associates, Inc.
57 Morehouse Lane
Red Hook, NY 12571 USA
Phone: 845-758-0400
Fax: 845-758-2633
Email: curran@proceedings.com
Web: www.proceedings.com

TABLE OF CONTENTS

Data Analytics for Analyzing Campus Energy Use

Whitman Jerman, Soowon Chang, William Hutzel

Purdue Polytechnic Institute

wjerman@purdue.edu, chang776@purdue.edu, hutzelw@purdue.edu

PURDUE UNIVERSITY | Polytechnic Institute

AeL Applied Energy Laboratory

Introduction

- Cogeneration plant supplies utilities to campus
- Chilled Water
- Steam
- Electricity
- Domestic Water

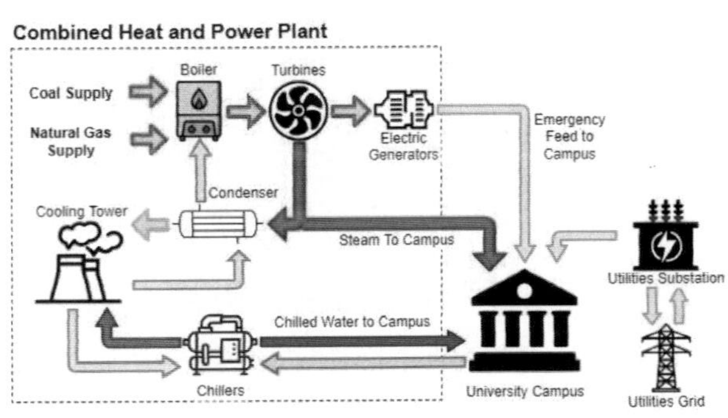

Problem

- Production and energy distribution metered at CHP plant and at buildings
- Data recorded every 5 – 60 seconds
- Improvements in how data is used can result in more efficient asset management and improve demand prediction

Methodology

- Bayesian multilevel additive modeling used to analyze significant variables and their impacts on supply and demand
- Population and group effects considered
- Data Collection
- R Programming Language
- Predictor and response variables
- Trending relationships
- Predictive modeling

$$y_{i,j} = \sum_{j=1}^{J} \alpha_j u_j + \sum_{k=1}^{K} \beta_k x_{ij,k} + \sum_{q=1}^{Q} f_q(z_{ij,q}) + \varepsilon_{ij}$$

Data Collection

- Data collected from 2018 to 2021
- 1 hour averages

(a) 2020 Electricity Production

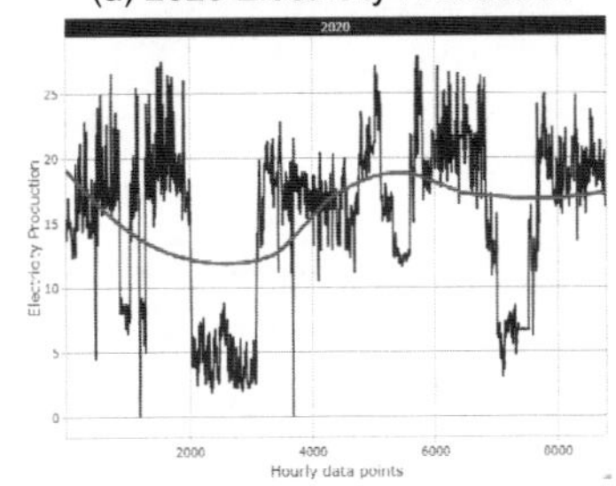

(b) 2020 Chilled Water Production

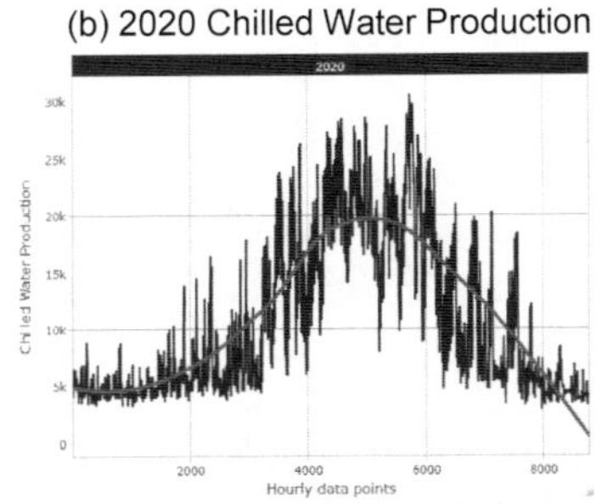

(c) 2020 Steam Demand

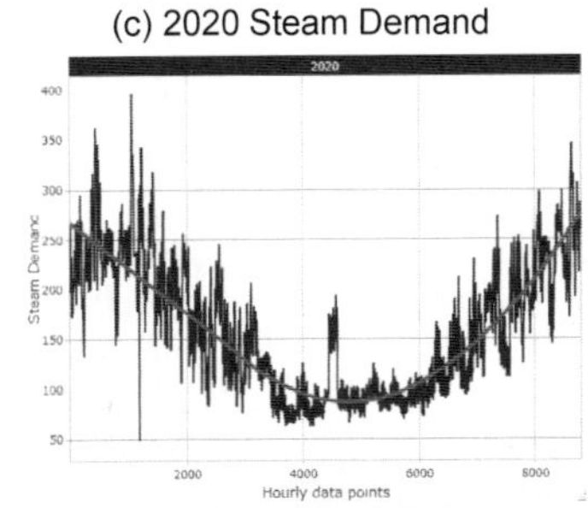

Variable Definitions

Demand- The energy required by all campus buildings

Production- The energy generated at the CHP plant

Linear Effects – variables with a direct correlation to the response variables

Non-Linear Effects – variables with a lower order impact on the response variables

Group Effects – variables that impact all response variables

Relationships Between Variables

Response variables / Predictor variables		Linear population effects	Non-linear population effects	Group effects
1. Demand (Building)	Steam	•Dry bulb air temperature	NA	•Campus size (gsf or gsm) •Time of Day (ToD) •Month
	Domestic Water	•Dry bulb air temperature	NA	
	Electricity	•Wet bulb temperature	NA	
2. Production (CHP)	Steam	•Steam use (building) •Electricity use (building)	•Dry bulb air temperature	
	Chilled Water	NA	•Wet bulb temperature •Domestic Water	
	Electricity	•Steam use (building) •Electricity use (building)	•Dry bulb air temperature	
3. Electricity Purchase		•Steam demand (building) •Electricity demand (building) •Electricity production •Chilled water production	NA	

Steam and Chilled Water Data

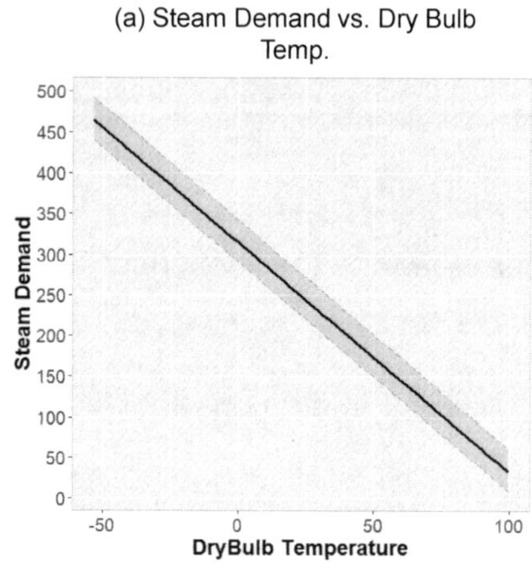

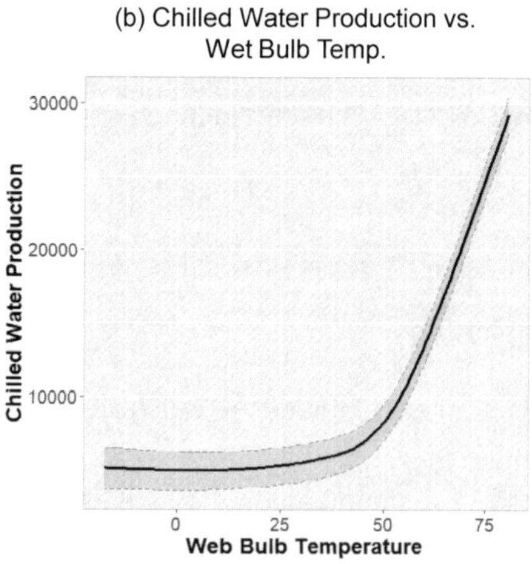

Electricity and Domestic Water Data

(c) Electricity Production vs. Dry Bulb Temp.

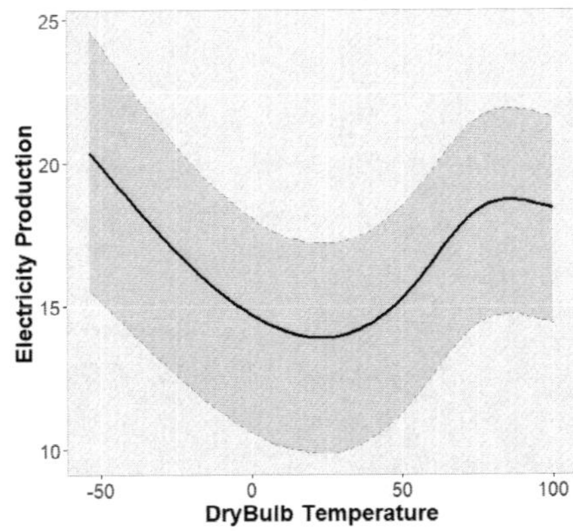

(d) Domestic Water vs. Dry Bulb Temp.

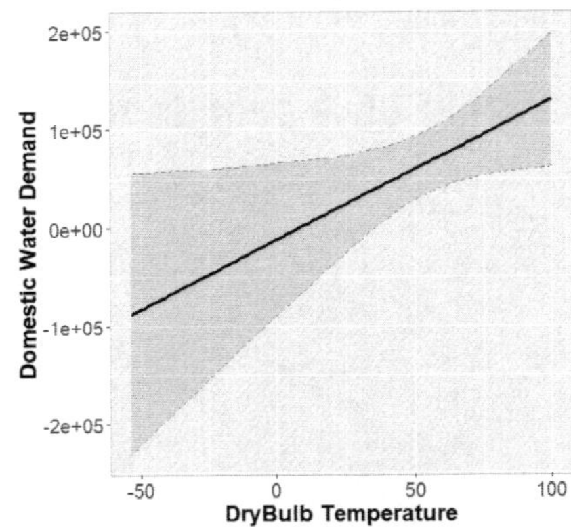

Posterior Predictive Checking

- ❑ Output of predictive model compared to observed data

(a) Steam Production

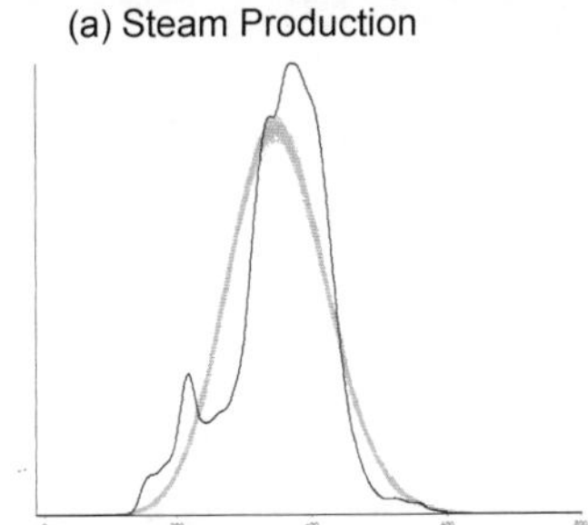

(b) Chilled Water Production

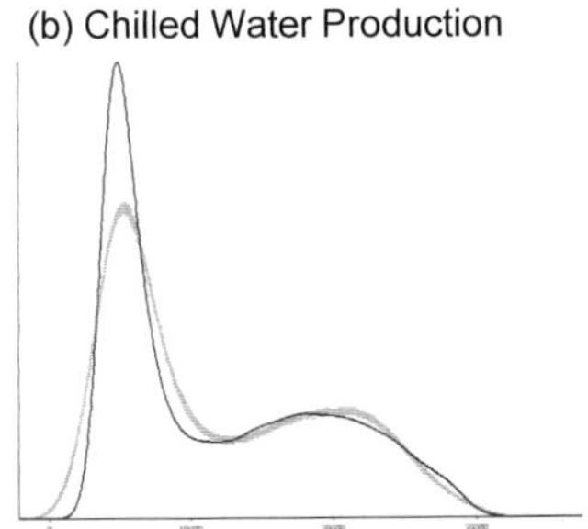

(c) Electricity Generation

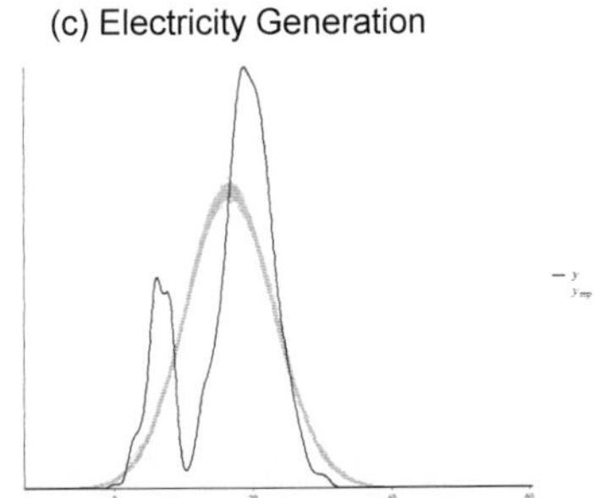

Predictions

- Sampled data points compared to outputs from models
- Within 95% confidence interval

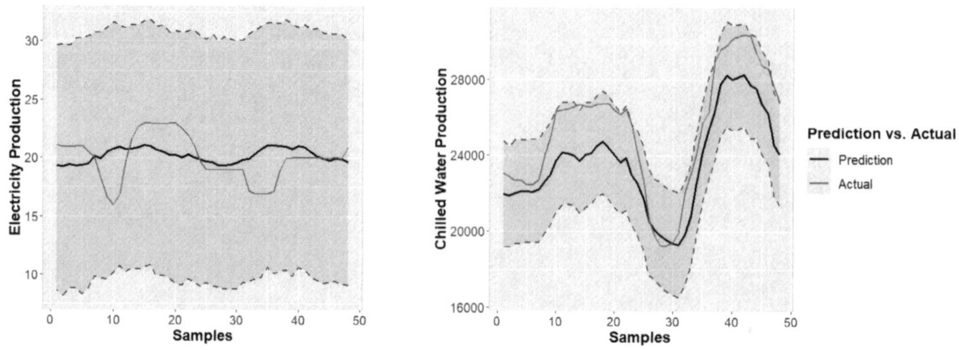

Conclusions

- Bayesian models are useful for identifying the relationship between variables such as air temperature and energy demand
- Models can be used to predict energy demand based on conditions such as weather forecasts
- The accuracy of models can be further improved through exploration of additional predictor variables

References

Amasyali, K. & El-Gohary, N. (2018): A Review of Data-Driven Building Energy Consumption Prediction Studies. Renewable and Sustainable Energy Reviews., 81(2018), 1192-1205. https://doi.org/10.1016/j.rser.2017.04.095

Bürkner, P.-C. (2017). brms: An R Package for Bayesian Multilevel Models using Stan. *Journal of Statistical Software*, *80*(1), 1–28. https://doi.org/10.18637/jss.v080.i01

Bürkner, P.-C. (2018). Advanced Bayesian Multilevel Modeling with the R Package brms. *The R Journal*, *10*(1), 395–411. https://journal.r-project.org/archive/2018/RJ-2018-017/RJ-2018-017.pdf

Chang, S., Castro-Lacouture, D., & Yamagata, Y. (2020). Estimating Building Electricity Performance Gaps with Internet of Things Data Using Bayesian Multilevel Additive Modeling. *Journal of Construction Engineering and Management*, *146*(12), 05020017. https://doi.org/10.1061/(asce)co.1943-7862.0001930

Chang, S., Yoshida, T., Castro-Lacouture, D., & Yamagata, Y. (2020). Block-level Building Transformation Strategies for Improving Energy Efficiency, Thermal Comfort and Visibility Performance Using Bayesian Multilevel Additive Modeling. *Journal of Architectural Engineering*, *27*(3). https://doi.org/10.1061/(ASCE)AE.1943-5568.0000491

Granderson, J., Piette, M. A., & Ghatikar, G. (2011). Building energy information systems: User case studies. *Energy Efficiency*, *4*(1), 17–30. https://doi.org/10.1007/S12053-010-9084-4/TABLES/5

Sulaiman, S. M., Jeyanthy, P. A., & Devaraj, D. (2016, July 18). Artificial neural network based day ahead load forecasting using Smart Meter data. *2016 - Biennial International Conference on Power and Energy Systems: Towards Sustainable Energy, PESTSE 2016*. https://doi.org/10.1109/PESTSE.2016.7516422

Wood, S. N. (2011). Fast stable restricted maximum likelihood and marginal likelihood estimation of semiparametric generalized linear models. *Journal of the Royal Statistical Society: Series B (Statistical Methodology)*, *73*(1), 3–36. https://doi.org/10.1111/j.1467-9868.2010.00749.x

Lagged-kNN based data imputation approach for multi-stream building systems data

Ojas PRADHAN[1]*, David HÄLLEBERG[2], Zhelun CHEN[3], Jin WEN[4], Teresa WU[5], K. Selcuk CANDAN[6], Zheng O'NEILL[7]

[1] Drexel University, Department of Civil, Architectural and Environmental Engineering,
Philadelphia, PA, USA
omp28@drexel.edu

[2] KTH Royal Institute of Technology, Department of Civil and Architectural Engineering,
Stockholm, Sweden
halleb@kth.se

[3]Drexel University, Department of Civil, Architectural and Environmental Engineering,
Philadelphia, PA, USA
zc98@drexel.edu

[4]Drexel University, Department of Civil, Architectural and Environmental Engineering,
Philadelphia, PA, USA
jw325@drexel.edu

[5]Arizona State University, School of Computing, Informatics, Decision Systems Engineering,
Tempe, AZ, USA
teresa.wu@asu.edu

[6]Arizona State University, School of Computing, Informatics, Decision Systems Engineering,
Tempe, AZ, USA
kasim.candan@asu.edu

[7] Texas A&M University, Department of Mechanical Engineering,
College Station, TX, USA
zoneill@tamu.edu

* Corresponding Author

ABSTRACT

Increasing advancements in building digitization, smart sensing, and metering technologies have allowed large amount of data to be collected and saved for monitoring, analyzing, and controlling building systems. However, due to sensors or communications failure, the data collected are often incomplete and poor in quality. Data imputation approaches to replace the missing values, specifically based on both statistical and predictive models have been widely adopted for multivariate datasets. It is hence of interest to find an effective way to impute building system data by leveraging the mutual information from strongly correlated sensors. In this paper, we evaluate multiple data imputation approaches using data collected from a medium sized, mixed-use institution building situated in Stockholm, Sweden. Sensors with widely varying characteristics from the case study building were selected to test the imputation methods. Artificial test data with ground-truth was first created for validation by removing randomly selected portions of data. The imputation accuracy was computed for each method and the impact of the chosen method on a short-term building forecasting model was evaluated. Results demonstrate that incorporating time-lagged cross correlations within the k-nearest neighbor (kNN) model provide the most accurate imputations.

1. INTRODUCTION

Recent advances in monitoring systems, communication and information technology make it possible to collect and store large amounts of time series data. Such databases are utilized by data-driven methods and algorithms for solving complex problems in different disciplinary fields. One of the sectors that has greatly benefitted from the advances in big data is building sciences. Use of artificial intelligence, machine learning, and deep learning analysis have been proved to have remarkable impacts on various research domains such as building energy forecasting, pattern recognition, and fault detection and diagnosis (Ma et al., 2020).

An essential prerequisite for implementing these technologies is high-quality data set from the building automation system (BAS). However, the databases obtained from the BAS are usually incomplete due to equipment failures, communication/transmission issues or data corruption, thus, leading to loss of valuable information. The issue of missing data exists in almost all kinds of data sets, and the size of the missing data can significantly affect the research outcomes since most data analysis and statistical tools are not designed to handle incomplete data. Before exploring methods to reconstruct the missing values, it is important to first understand the missing data mechanism. Rubin (1976) first described and categorized the types of missing data based on the assumptions for the missing data. In the literature, missing data mechanisms are generally divided into three categories (Allison, 2003; Rubin, 1976):

- Missing Completely at Random (MCAR) – data are defined as MCAR when the probability of missingness is not related to either the specific values or the set of observed responses. If data are missing completely at random, then dropping cases with missing data does not bias the inferences.
- Missing at Random (MAR) – defined when the probability of missingness depends on the set of observed responses, but the mechanism of data missingness is traceable or predictable from other variables in the database.
- Not Missing at Random (NMAR) – defined when the probability of missingness depends on the missing variable itself. A model for the missing data must be created for the specific dataset to handle this type.

Many approaches have been developed for handling missing values that are available in the literature. The simplest approach is to simply ignore it and perform the analysis based on the available data (also known as listwise deletion). This procedure, however, can lead to loss of efficiency due to discarding the incomplete observations and may output biased and unreliable results due to the systematic differences between the available and missing data. Hence, an effective technique for estimating missing values is through data imputation which retains the original sample size and allows subsequent analysis to be carried on the entire dataset and produce much reliable results. The imputation approaches vary from simple methods such as mean imputation to some more robust methods that leverage the relationships among variables within the dataset. The following section presents some popular imputation methods for data precleaning (Allison, 2003; Rubin, 1976; Schafer & Graham, 2002):

- Mean or mode imputation: This method replaces the unknown value for a given variable by the mean (quantitative variable) or mode (qualitative variable) of all known values of that variable. A major drawback of this method is that replacing all missing records with a single value distorts the input data distribution.
- Hot deck imputation: In this method, the missing values are substituted from the variables that are closest to the variable consisting of the missing values. This method attempts to preserve the input data distribution since the substituted values are based on the similarity with the other variables in the dataset.
- Prediction models: These methods involve creating a predictive model to estimate values that will replace the missing data. The incomplete variables are used as target, and the remaining variables are used as inputs for the model. These methods leverage the relationships (correlations) among the different variables within the dataset. Those correlations can be used to create a predictive model for either classification or regression. If there are no strong correlations between the incomplete variables and the remaining variables, the imputed values will not be precise. Also, many prediction models must be designed for cases with many incomplete variables, making it computationally expensive.

This paper aims to explore potential data imputation method for building systems data and is structured as follows: Section 2 provides a review of existing literature on data imputation methods. Section 3 describes the proposed method and Section 4 present the case study building used to evaluate and compare different imputation methods. The results are discussed in Section 5 and Section 6 presents the current conclusions and future work.

2. LITERATURE REVIEW

Existing literature on data imputation methods for building data has primarily been focused on univariate time series data using statistical methods. For example, Jin et al. (2006) used a stochastic binning method to estimate the missing weather data (outdoor temperature) for the estimation of building energy. Ouyang et al. (2017) implemented linear interpolation to tackle the missing values when predicting wind power. Kasam et al. (2014) focused on a Gaussian-distributed auto-regressive model to interpolate the missing meteorological data for use in building simulations. Although statistical methods are easy to implement and perform well for low rates of missing data, their performance is limited when the datasets are more complicated since these methods are based on linear assumptions and do not account for the nonlinearities of the dataset.

In order to address this issue, nonlinear machine learning and deep learning methods are utilized to fill in the missing values. For example, Garnier et al. (2012) implemented Artificial Neural Network (ANN) to estimate the missing values for energy resources management in buildings. Ma et al. (2020) have proposed a bi-directional missing data imputation scheme based on deep learning and transfer learning for building energy data. Rahman et al. (2018) and Yang et al. (2019) have used deep Recurrent Neural Networks (RNN) models to first perform imputation on the missing data and then forecast the building energy usage.

Since the single imputation method may not work for various types of sensors in a building system, ensemble methods that include both statistical and learning methods that utilize multiple imputation methods to make an improved prediction of the missing values are used. Inman et al. (2015) utilized two data imputation methods prior to performing clustering analysis on building energy consumption data. More recently, Zhang (2020) developed a pattern recognition-based ensemble framework to first create validation data that have similar characteristics with the missing data and then, use the optimal imputation method for each sensor. Research on ensembles methods have shown good performance compared to single imputation methods for building system data.

However, there remains a few key problems that the existing literature fails to address. First, most of the existing studies only explores univariate time series data, hence, these methods do not leverage the mutual information from other variables when a multivariate dataset is available. Since many building system sensors are strongly correlated with each other, there is potential in utilizing information from other variables. Secondly, almost all the studies and developed methods only consider short time periods of missing data and the more challenging case of continuous missing data for long time periods is not explored by these methods.

Given the limitations in existing literature for handling multivariate datasets in BAS data, studies from other fields such as biomedical, traffic flow, and radio transmission are investigated. A probabilistic principal component analysis (PPCA) based method is developed for traffic flow dataset by Qu et al. (2009). In PPCA-based imputation, the PCA is used to separate the significant and dominant parts of the traffic flow, whereas the maximum likelihood estimation (MLE) is applied to estimate the missing values based on the obtained significant parts determined by the PCA. A drawback of this method is the requirement of a large amount of historical data to train the model.

Another widely adopted method was the k-Nearest Neighbor (kNN) imputation method in which k nearest neighbors for a missing instance are identified using the observed instances of other variables. The missing instance is then estimated by combining the k estimates using approaches such as the weighted average or a kernel function (Rahman et al., 2015). These kNN-based methods are most appropriate for a multivariate dataset, and when the missing variables are correlated with the other observed variables. One drawback of kNN is that, because it relies on the values of other variables, it cannot impute a value when all variables are missing in an instance and may be less accurate as more variables are absent.

To overcome this issue, Rahman et al. (2015) developed a combined method using the Fourier transforms which uses past values of each variable and an extension of kNN called the lagged-kNN which also considers the time lagged cross-correlation between each of the variables. The Fourier lagged-kNN method (FL-kNN) overcomes the limitation of the nearest neighbors methods which require observed data to be presented at each instance and improves accuracy by handling both MAR and NMAR missing data. In their research, the imputation accuracy for multiple biomedical data using this method was reported to be the highest when compared with other imputation methods for up to 50% of the dataset missing.

Motivated by the limitations in existing literature for imputing continuous missing data for long time periods and the potential of the FL-kNN method, in this paper, we applied five different imputation methods to a multi-stream building dataset. Artificial test data with ground-truth was first created for validation by removing randomly selected portions of data. The imputation accuracy was computed for each method. Considering that the sensor measurements are typically used for further data analytics, it is of interest to also examine how data repaired by different imputation schemes might affect the quality of data analytics. Hence the impact of the chosen method on a short-term building forecasting model was also evaluated. The imputation schemes chosen for the study are as follows: (i) linear interpolation (from *pandas* library), (ii) kNN imputer (from *sklearn* library), (iii) lagged-kNN, (iv) Fourier method and (v) combined Fourier lagged-kNN imputation method (FL-kNN).

3. COMBINED FOURIER LAGGED-KNN METHOD

The combined Fourier lagged-kNN method is a combination of two imputation methods presented by Rahman et al., (2015). This imputation method consists of an extension of kNN imputation with lagged correlations, and the Fourier transformation. For each missing data point, the final estimated value is calculated by averaging the estimates from the lagged-kNN imputation and the Fourier method.

3.1 Lagged-kNN Method

First, a kNN with a time lag parameter (p) is developed to incorporate the time dependent correlations that may persist between the variables. In this step, cross-correlation is used to identify which variables are correlated and at which time lags. The cross-correlation r_{xy} between variables x and y, for time delay d is defined as:

$$r_{xy}(d) = \frac{C_{xy}(d)}{\sqrt{(C_{xx}(0)C_{yy}(0)}} \tag{1}$$

$$c_{xy}(d) = \begin{cases} \dfrac{1}{T-d}\sum_{t=1}^{T-d}(x_t - \bar{x})(y_{t+d} - \bar{y}),\ if\ d \geq 0 \\ \dfrac{1}{T+d}\sum_{t=1-d}^{T-d}(x_t - \bar{x})(y_{t+d} - \bar{y}), otherwise \end{cases} \tag{2}$$

where T is the length of the series, $\bar{x}$ and $\bar{y}$ are the mean of x and y respectively, d varies from - $(D-1)$ to $(D-1)$ and D is the maximum time delay. Based on the strength of the correlation, matrices are constructed for each of the p lags, with the correlations ordered from $1\ldots p$ by decreasing strength. For each pair of variables, (L_1) contains the lag, d, with the strongest correlation $(max\ |r_{xy}|)$ and L_p the lag with the weakest. Each L is an NxN matrix, where elements represent the time lags for each correlation between the N variables.

In contrast to kNN, the multiple lags that differ across variable pairs must be accounted for while forming the training and testing vectors with candidate values. In the lagged-kNN method, for a variable x that has missing value at time t and has a time lagged relationship with variables y and z, with lags l_{xy} and l_{xz} respectively, the test vector is formed using the values of y and z at $t + l_{xy}$ and $t + l_{xz}$. Training vectors are formed in a similar way and the values of x, which are the candidate values for imputation, are stored separately. Training vectors are generated from the existing values of x and the time instances resulting after adding the lags must be within 1 to T (length of data).

A weighted modification of the Euclidean distance is used as a proximity measure while finding the nearest neighbors for each missing instance. Since the strength of the correlation between variables and across the p lags may differ substantially, a weight is incorporated into the distance measure. This ensures that neighbors based on highly correlated variables with their associated lags are given more weight than weakly correlated variables. The distance between instance x and y is calculated through:

$$d(x,y) = \frac{\sqrt{\sum_{i=1}^{N}[(x_i \wedge y_i)(x_i - y_i)^2 \times w_i]}}{\sum_{i=1}^{N}(x_i \wedge y_i)} \tag{3}$$

where N is the number of variables, and w_i is the weight, which is the normalized correlation coefficients between missing variables and i^{th} variable. The result is p sets of k nearest neighbors (one set of neighbors for each L matrix). We then average the values for the k neighbors with the lowest weighted distance (out of the set of p x k neighbors).

3.2 Fourier Method

The Fourier imputation method uses past values of each variables to impute each missing value. First, a data segment is formed with the data from the beginning of the signal up to the last non-missing data point. For example, if values v_1 through v_{p-1} are present (or imputed), and $v_p \ldots v_q$ are missing, the Fourier descriptors are obtained with:

$$F_k = \sum_{j=1}^{p-1} v_j \times e^{-(2i\pi/p-1)(j-1)(k-1)} \tag{4}$$

where F_k is the kth Fourier descriptor with $1 \leq k \leq (p-1)$ and $i = \sqrt{-1}$ Then, the imputed value for time m, where $p \leq m \leq q$, calculated from the Fourier descriptors with:

$$v_m = \frac{1}{p-1} \sum_{k=1}^{p-1} F_k \times e^{(2i\pi/p-1)(j-1)(m-1)(k-1)} \tag{5}$$

The proposed method estimates each missing value based on the observed data, hence, it is found that if the given data does not capture the high frequency components (i.e. sampling frequency is less than 2 * Nyquist frequency), the imputed value will not be accurate.

4. REAL BUILDING CASE STUDY

4.1 Multi-Stream Building Data

The study utilized data from a 13,434 sq. meter, mixed-use institution building situated in Stockholm, Sweden. It is acclimatized by an intelligent demand-control ventilation (DCV) system where the airflow can independently vary within and between different rooms, carbon- and temperature sensors controlling the air volume within in conference rooms, modern building management system (BMS) and energy meters throughout the building. Datapoints collected from the building include the heating, ventilating and air conditioning (HVAC) system, occupancy data for individual offices, and the energy-meter.

The RealEstateCore (REC) ontology is utilized for data integration in the building. RealEstateCore, which is described in more detail by Hammar et al. (2019), is a universal language that facilitates data-driven building control and the development of new services. REC implementation in the case study building enabled two-way communication which additionally allows operators to send control signals back to the BMS (called SAIA) and the DCV system (called Lindinvent) in REC format (Halleberg & Martinac, 2020). Both the BMS and the DCV system communicate by Modbus Transmission Control Protocol (TCP) and are translated using Internet-of-Things (IoT)-Edge Modules. The modules are written in C# and can be configured to poll the source system at different intervals.

A Microsoft Azure digital platform was also developed for the case study building by combining the potential of the cloud with security using an IoT-Edge solution (Halleberg & Martinac, 2020). Data from BMS and DCV system are extracted (according to the configuration specified in the two REC tag lists) using an IoT-Edge module running on an Edge Device physically located in the building. Additionally, data from the tenant's system for managing space booking, forecasted weather data from Swedish Meteorological and Hydrological Institute (SMHI) and energy management system (called Energiportalen) are imported and processed via application programming interfaces

(APIs) within an Azure Data Factory that triggers every hour. This digital platform solution enables storing, processing big data, and hosting various automated data-driven strategies for the building in the future.

To study the data quality and compare multi-stream data imputation strategies, thirteen (13) different sensors from the case study building were selected to test the imputation methods. A systematic feature selection procedure developed by Zhang & Wen (2019) was used to identify the air handling units (AHUs) and zones with the largest impact on the building electricity and heating energy. Additionally, typical sensors found in weather station, BMS and DCV system that cover different sensor types were included in study. The sensor list is presented in Table 1. Data collected from March 1st to 7th, 2021, with a 5-minute sample interval was used.

Table 1: Sensors used to evaluate data imputation accuracy

System	Sensor Name	Sensor Type	Description
BMS	LB100/GP11/MV	Pressure sensor	AHU supply air static pressure
BMS	LB101/GF11/MV	Flow sensor	AHU supply air flowrate
BMS	LB101/GT11/MV	Temperature sensor	AHU supply air temperature
BMS	LB101/TF11/R	Fan speed sensor	AHU supply fan speed
BMS	LB101/SV21/R	Valve position	AHU cooling coil valve position
BMS	VS130/GT11/MV	Temperature sensor	Radiator hot water temperature
Weather Station	R1B17/GM51/MV	Hygrometer	Outdoor humidity
Weather Station	VS110/GT31/MV	Flow sensor	Outdoor temperature
DCV	3B13/TD10:61/GF	Flow sensor	Actual supply-air flow
DCV	3B13/TD10:61/GT	Temperature sensor	Actual supply-air temperature
DCV	3B13/TD10:61/GP	Pressure sensor	Actual supply-air pressure
DCV	3B13/TD10:61/GT	Temperature sensor	Actual room temperature
DCV	3B13/TD10:56/CO	Carbon-dioxide sensor	Actual carbon dioxide concentration

4.2 Creating Validation Dataset

Artificial validation dataset was created by generating a non-uniform set of random indices through the Python 'pd.DataFrame.sample' function for each sensor. To ensure that the same index is removed for each imputation method, 'rng default' function was used to select the random seed.

'Randomized missing' validation sets for up to 40% missing ratios were created by removing the datapoints at the randomly generated indices. Similarly, 'continuous missing' validation sets were created by generating a random index, then removing the next instances of datapoints until the desired missing ratio wass met. Figure 1 presents the two categories of validation datasets created.

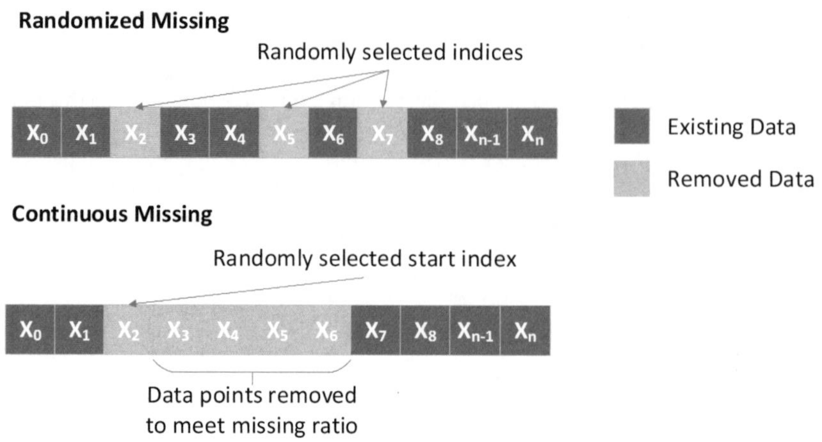

Figure 1: Categories of validation dataset for evaluating data imputation methods

4.3 Energy Forecasting Model

Following the data imputation, a building energy forecasting model was used to evaluate the impact of the imputation method on the energy forecasts. The energy modeling procedure utilizes a multi-input-single-output regression model called Multivariate Adaptive Regression Splines (MARS). MARS is a nonparametric regression which constructs underlying relationship from a set of coefficients and basis function that are determined by training data. Since MARS operates as multiple piecewise linear regression, it can handle the nonlinearities in the datasets. Mathematically, MARS is defined as:

$$f(x) = \gamma_0 + \sum_{i=1}^{m} \gamma_i h_i(x) \tag{6}$$

where, x represents the features, m is the number of features, γ is the constant coefficient of the combination whose value is jointly adjusted to give the best fit to the data and the basis function $h_i(x)$ (Gints, 2011). The basis function h_i can be represented as:

$$h_i(x) = \prod_{k=1}^{K_m} [S_{k,m} \cdot (x_{v(k,m)} - t_k, m)]_+^q \tag{7}$$

where K_m is the number of splits given to the mth basis function, $S_{k,m} = \pm 1$ indicates the right/left sense of the associated step function, $v(k,m)$ is the label of the features, and $t_{k,m}$ represents values (often called knot locations) of the corresponding variables. The superscript q and subscript $+$ indicate the truncated power functions with polynomials of lower order than q. More details of fundamentals of MARS can be found in Gints (2011).

The forecasting model was trained to predict the building electricity and heating energy using data collected from the case study building from March 1st to 7th, 2021, and tested using data collected from March 8th to 11th, 2021

4.4 Performance Metric

Normalized Root Mean Squared Error (NRMSE) was used as the performance indicator to measure both the imputation accuracy and the energy forecast accuracy. The NRMSE is defined as:

$$NRMSE(y,x) = \frac{1}{mean(x)} \sqrt{\frac{\sum_{i=1}^{n} (y_i - x_i)^2}{n}} \tag{8}$$

where, y is the imputed or predicted value, x is the actual value and n is the number of imputed data points. The overall performance metrics of the imputation method was calculated by averaging the metric values for each sensor.

5. RESULTS AND DISCUSSION

The performance of the data imputation method and the impact on the short-term energy forecast was evaluated using the NRMSE. The results are summarized in Table 2 and Table 3, respectively. The best performing imputation method under each category is highlighted in bold.

Table 2: Averaged imputation error for each validation dataset

Validation Dataset	Imputation Method				
	Linear	kNN Imputer	Lagged kNN	Fourier	Fourier-Lagged kNN
10p-Continous	0.772	0.291	**0.207**	1.229	0.664
20p-Continous	0.630	0.486	**0.273**	0.935	0.525
30p-Continous	0.874	0.436	**0.338**	1.092	0.594
40p-Continous	1.089	0.455	**0.210**	1.113	0.545
10p-Scatter	0.087	0.161	0.098	0.111	**0.081**
20p-Scatter	0.105	0.193	**0.063**	0.166	0.099
30p-Scatter	0.127	0.292	**0.086**	0.257	0.153
40p-Scatter	0.118	0.430	**0.093**	0.269	0.160

Table 3: Averaged energy forecasting error for each validation dataset

Validation Dataset	Imputation Method				
	Linear	kNN Imputer	Lagged kNN	Fourier	Fourier-Lagged kNN
10p-Continous	0.025	0.025	**0.024**	0.025	0.025
20p-Continous	0.042	0.038	**0.036**	0.038	0.039
30p-Continous	0.066	0.076	**0.063**	0.074	0.072
40p-Continous	0.082	0.091	**0.072**	0.076	0.079
10p-Scatter	**0.021**	**0.021**	**0.021**	0.022	**0.021**
20p-Scatter	**0.020**	**0.020**	**0.020**	**0.020**	**0.020**
30p-Scatter	0.024	0.024	0.024	0.024	**0.023**
40p-Scatter	0.025	0.025	**0.024**	0.025	0.025

From the results presented above, several conclusions can be drawn in terms of imputation and energy forecasting accuracies:

- The lagged-kNN method has the best performance in 7 out of the 8 validation sets evaluated for data imputation. Identifying the strongly correlated building system measurements using the kNN algorithm helps the method to utilize the shared mutual information among the different measurements, hence, resulting in more accurate estimation of missing values.
- Incorporating the time-lagged cross-correlation in the lagged-kNN method helps to significantly improve the imputation accuracy as well. Since building system measurements usually have a delayed response to changes in external weather conditions or supervisory control signals, using a set of time-lagged nearest neighbors to perform the data imputation result in better performance.

- The combined Fourier-lagged kNN method has comparable performance with the lagged-kNN method for lower missing ratios in the randomized scattered set, however, the error increases when applying the combined method for larger missing ratios. Since the Fourier imputation uses past values of each variable to impute each missing value, this method requires a large amount of historical data which is not satisfied by the 8-days of data collected from the BAS. The inaccuracies from the Fourier method, thus, leads to larger overall error for the combined method.
- While statistical methods (such as linear interpolation, backward fill, forward fill etc.) are simple and relatively easy to implement, these methods do not perform well for datasets with high missing ratios and/or when data is continuous missing. These methods also lack the capability to utilize the strong correlations that exist between various building system measurements.
- Results from the building energy forecasting model again show that the dataset imputed using the lagged-kNN has the best performance. However, the forecasting accuracy is comparable among all imputation methods. Since the forecasting model was tested using limited data under similar building operating and weather conditions, the choice of imputation method does not have a significant impact on the short-term building energy forecast.

6. CONCLUSIONS

This paper evaluates different data imputation methods using data collected from a multi-stream building dataset. Since building system sensors are often correlated with each other, there is potential in utilizing information from other variables to impute the missing values in a dataset. Recent development in machine learning models such as kNN methods allow to identify such existing correlations that improve the estimates of the missing values by leveraging the shared information.

To evaluate the effectiveness of such methods, five different imputation methods of varying sophistication were tested by creating artificial validation sets that consists of different ratios and patterns of missing data. Results demonstrate that incorporating time-lagged cross correlations within the kNN framework help to significantly improve the imputation accuracy. In terms of the impact of the chosen imputation method on building energy forecasting, results from limited testing showed comparable accuracy between the different imputation methods. Although the choice of the imputation method showed minimal impact on the performance of the forecasting model reported in the paper, additional validation is required for forecasting models that use larger datasets under diverse system operating conditions. Furthermore, the impact on other data-driven applications such as fault detection or fault diagnosis, where the strategies require complete dataset and are more sensitive to the data quality needs to be further investigated in the future.

REFERENCES

Allison, P. D. (2003). Missing Data Techniques for Structural Equation Modeling. *Journal of Abnormal Psychology*. https://doi.org/10.1037/0021-843X.112.4.545

Garnier, A., Eynard, J., Caussanel, M., & Grieu, S. (2012). Missing data estimation for energy resources management in tertiary buildings. *2nd International Conference on Communications Computing and Control Applications, CCCA 2012*. https://doi.org/10.1109/CCCA.2012.6417902

Gints, J. (2011). *ARESLab: adaptive regression splines toolbox for Matlab/Octave*. 1–19. http://www.cs.rtu.lv/jekabsons/

Halleberg, D., & Martinac, I. (2020). Indoor Climate as a Service: A digitalized approach to building performance management. *IOP Conference Series: Earth and Environmental Science*, *588*(3), 0–8. https://doi.org/10.1088/1755-1315/588/3/032013

Hammar, K., Wallin, E. O., Karlberg, P., & Hälleberg, D. (2019). The RealEstateCore Ontology. In C. Ghidini, O. Hartig, M. Maleshkova, V. Svátek, I. Cruz, A. Hogan, J. Song, M. Lefrançois, & F. Gandon (Eds.), *The Semantic Web -- ISWC 2019* (pp. 130–145). Springer International Publishing.

Inman, D., Elmore, R., & Bush, B. (2015). A case study to examine the imputation of missing data to improve clustering analysis of building electrical demand. *Building Services Engineering Research and Technology*, *36*(5), 628–637. https://doi.org/10.1177/0143624415573215

Jin, Z., Yezheng, W., & Gang, Y. (2006). A stochastic method to generate bin weather data in Nanjing, China. *Energy Conversion and Management*, *47*(13–14), 1843–1850. https://doi.org/10.1016/j.enconman.2005.10.006

Kasam, A. A., Lee, B. D., & Paredis, C. J. J. (2014). Statistical methods for interpolating missing meteorological data for use in building simulation. *Building Simulation*, *7*(5), 455–465. https://doi.org/10.1007/s12273-014-0174-7

Ma, J., Cheng, J. C. P., Jiang, F., Chen, W., Wang, M., & Zhai, C. (2020). A bi-directional missing data imputation scheme based on LSTM and transfer learning for building energy data. *Energy and Buildings*, *216*. https://doi.org/10.1016/j.enbuild.2020.109941

Ouyang, T., Zha, X., & Qin, L. (2017). A combined multivariate model for wind power prediction. *Energy Conversion and Management*. https://doi.org/10.1016/j.enconman.2017.04.077

Qu, L., Li, L., Zhang, Y., & Hu, J. (2009). PPCA-based missing data imputation for traffic flow volume: A systematical approach. *IEEE Transactions on Intelligent Transportation Systems*, *10*(3), 512–522. https://doi.org/10.1109/TITS.2009.2026312

Rahman, A., Srikumar, V., & Smith, A. D. (2018). Predicting electricity consumption for commercial and residential buildings using deep recurrent neural networks. *Applied Energy*, *212*(December 2017), 372–385. https://doi.org/10.1016/j.apenergy.2017.12.051

Rahman, S. A., Huang, Y., Claassen, J., Heintzman, N., & Kleinberg, S. (2015). Combining Fourier and lagged k-nearest neighbor imputation for biomedical time series data. *Journal of Biomedical Informatics*, *58*, 198–207. https://doi.org/10.1016/j.jbi.2015.10.004

Rubin, D. B. (1976). Inference and missing data. *Biometrika*, *63*(3), 581–592. https://doi.org/10.1093/biomet/63.3.581

Schafer, J. L., & Graham, J. W. (2002). Missing data: Our view of the state of the art. *Psychological Methods*, *7*(2), 147–177. https://doi.org/10.1037/1082-989X.7.2.147

Yang, J., Tan, K. K., Santamouris, M., & Lee, S. E. (2019). Building energy consumption raw data forecasting using data cleaning and deep recurrent neural networks. *Buildings*, *9*(9). https://doi.org/10.3390/buildings9090204

Zhang, L. (2020). A pattern-recognition-based ensemble data imputation framework for sensors from building energy systems. *Sensors*, *20*(5974), 1–16. https://doi.org/10.3390/s20205947

Zhang, L., & Wen, J. (2019). A systematic feature selection procedure for short-term data-driven building energy forecasting model development. *Energy and Buildings*, *183*, 428–442. https://doi.org/10.1016/j.enbuild.2018.11.010

ACKNOWLEDGEMENT

The work presented in the paper has been carried out with the support of National Science Foundation (NSF) under the Partnerships for Innovation – Research Project (PFI-RP): Data – Driven Services for High Performance of Sustainable Buildings (Award no. 2050509).

District-level validation of a shoeboxing algorithm for Urban Building Energy Modeling

Federico BATTINI[1]*, Giovanni PERNIGOTTO[2], Andrea GASPARELLA[3]

[1]Free University of Bozen-Bolzano, Faculty of Science and Technology,
Bolzano, Italy
Tel: +39 0471 017888, Fax: +39 0471 017009, email: federico.battini@natec.unibz.it

[2]Free University of Bozen-Bolzano, Faculty of Science and Technology,
Bolzano, Italy
Tel: +39 0471 017632, Fax: +39 0471 017009, email: giovanni.pernigotto@unibz.it

[3]Free University of Bozen-Bolzano, Faculty of Science and Technology,
Bolzano, Italy
Tel: +39 0471 017200, Fax: +39 0471 017009, email: andrea.gasparella@unibz.it

* Corresponding Author

ABSTRACT

In the framework of Urban Building Energy Modeling, several procedures have been proposed to reduce the computational burden to evaluate the thermal behavior of multiple buildings. In a previous work, we presented a novel simplification algorithm capable of abstracting a randomly-shaped building into a representative shoebox. Differently from most of the approaches available in the literature, the proposed method is supposed to go beyond the preliminary assessment of the buildings' performance in the early design stage or on an annual basis. Indeed, it makes it possible to predict accurately not only the annual or seasonal energy needs but also the hourly thermal load and indoor temperature profiles. In addition to that aspect, the proposed algorithm has been specifically developed to deal with buildings of complex shape and geometry, as they are common in the European building stock, accounting for adjacencies and reciprocal shading effects.

After a first validation step carried out at single-building level, the proposed algorithm has been tested on many fictional parametrically-generated neighborhoods and several climates. In such a way, it has been possible to assess the reliability of the algorithm at city scale, with the goal to understand whether the application at urban level, characterized by multiple mutual interactions among buildings, could undermine its accuracy. Furthermore, the role played by solar radiation modelling in the framework of the shoebox simplification and its impact on the outcome of the algorithm have been discussed, presenting two alternatives with different accuracy levels.

As a whole, the algorithm showed to be effective in estimating hourly and annual thermal loads and indoor temperatures in district-scale simulations, reducing up to 30 times the simulation time for the considered configurations.

1. INTRODUCTION

Urban Building Energy Modelling *UBEM* represents the most recent development of Building Performance Simulation *BPS*. Indeed, the main feature of *UBEM* is the analysis on a larger physical domain, encompassing groups of buildings, districts, or whole cities. Among the methodologies that can be employed to estimate the energy use of buildings at urban scale, *UBEM* is the one having the greatest level of flexibility as it can consider new technologies and different energy efficiency designs (Johari *et al.*, 2020). In this way, it is possible to assess the buildings' energy performance at urban scale to drive energy policies for reducing the greenhouse gas emissions of the building stock.

The increase of the size of the simulation domain, however, brings a series of issues which prevents the application of the very same models and methods used in *BPS* because they would require too much time and information. Furthermore, the interactions among buildings (e.g., heat transfer through adjacent components, mutual radiative exchanges, shadows cast by adjacent building, solar radiation reflections, etc.) make the characterization of the boundary conditions more complex compared to single-building analyses. Combining both aspects, it looks clear that *UBEM* is more computationally demanding compared to traditional *BPS* (Reinhart and Davila, 2015). To cope with that, several approaches have been proposed in the literature to simplify models and simulation domains (Abbasabadi & Ashayeri, 2019). The most interesting methodologies aim at converting the geometrical massing into simplified energy models, rather than the fully detailed traditional model employed in *BEM.*
The first work going in this direction has been proposed by Dogan and Reinhart (2013), who developed an automated workflow to convert architectural massing models into sets of thermal shoebox models. Such procedure is known as Shoeboxer algorithm (Dogan and Reinhart, 2017) and it is employed in the urban modeling interface *umi* (Reinhart *et al.*, 2013) to estimate the thermal performance of the buildings in the model. Starting from the tridimensional massing model, the Shoeboxer computes the incident solar radiation on the envelope of each building and performs a clustering analysis to find the regions experiencing similar solar loads. For each cluster a shoebox thermal zone is created and placed in correspondence of the point on the façade whose radiation value is closest to the cluster's mean. For each shoebox, only the visible external shading surfaces are included into the thermal model. To obtain the overall building demand, the results of every shoebox are aggregated and weighed according to the related cluster area.
Later, Zhu *et al.* (2019) developed a new method called Building Blocks Energy Estimation *BBEE* based on the typical shoebox zone concept proposed by the Shoeboxer. The *BBEE* is built on top of the hypothesis that the energy needs of a building is close to the sum of each part of it. More specifically, the *BBEE* method consists in the *BBEE* algorithm which is coupled with an energy database of typical zones. Thus, the *BBEE* method employs the *BBEE* algorithm to transform complex building forms into an assembly of typical zones, coded according to their properties. Finally, instead of performing energy simulations, the method relies on the database to retrieve the energy use of each typical zone and sums them up to get to the final building's needs.
More recently, İşeri and Dino (2020) proposed another method that reorganizes the building geometries with functional clustering and radiation analysis scaling. At first, an adapted version of the solar radiation analysis performed in the Shoeboxer is implemented to group the buildings' units according to the radiation values on the envelope. Afterwards, the clustered units are further subdivided depending on their position in the building, i.e., ground, middle and top floor, and their space conditioning.
Finally, Zhu, *et al.* (2021) proposed a hybrid metamodel-based method to be used in the early design stage to find high-performance building forms. The method subdivides the building into several zones as metamodels and estimates the total energy needs by summing up the results of all metamodels. Differently from other metamodels, it accounts for the surrounding shading and reflection effects by using the received solar radiation values as inputs. Moreover, GPU acceleration technology is used to speed up the radiation calculation, and the accuracy of the metamodels is improved by implementing a machine learning algorithm screening framework.
The aforementioned methodologies showed how it is possible to introduce simplifications in *UBEM* to reduce the computational time of the simulation. Nonetheless, there is still room for improvement, especially for what regards the spatial and temporal resolution of the outcome, e.g., properly predicting hourly temperature profiles permits to perform thermal comfort evaluations at urban scale. Indeed, given that *UBEM* is a relatively new field of research, a well-established approach to optimize its computational efficiency has still to be found (Hong *et al.*, 2019).
In a previous work, the authors presented a novel simplification algorithm capable of abstracting a randomly shaped building into a representative shoebox with a thermal zone per floor (Battini *et al.*, 2021a). Differently from most of the approaches available in the literature, the proposed method has been development not only for the assessment of the buildings' performance in the early design stages or on an annual basis. Indeed, it makes it possible to accurately predict not only the annual or seasonal energy needs but also the hourly thermal loads and indoor temperature profiles. Even though the algorithm has been already validated for stand-alone buildings, it has been designed to be employed for urban level simulations. Therefore, in this work, the approach has been tested on many neighborhoods' configurations and several climates to assess its reliability at city scale.

2. METHODOLOGY

The methodology for assessing the performance of the proposed shoeboxing algorithm at urban scale consists of three parts: (i) overview on the simplification procedure, (ii) development of a set of fictional districts for testing, and (iii) simulation and comparison.

2.1 Shoeboxing algorithm

In this work, two workflows have been followed to perform the simplification. At first, the standard procedure accounting for the self and context shadings by means of an annual radiation analysis has been followed. Then, an improved and more detailed approach capable of accounting for the obstructions on a monthly basis has been pursued. The two procedures are presented in the following.

2.1.1 Basic shoeboxing algorithm description: A detailed description of the workflow employed to develop the algorithm can be found in a previous work by the authors (Battini *et al.*, 2021a). Nonetheless, the steps to take in order to employ the approach for simplifying buildings into representative shoeboxes can be summarized as follows:

1. *Shoebox generation*: starting from the three-dimensional building massing, three geometrical indicators able to meaningfully represent the building are computed. A system of three equations in three unknowns (i.e., the three indicators and the three shoebox dimensions) is solved to obtain the simplified model's dimensions prior performing the simulation. The representative shoebox is generated according to the dimensions found and it is characterized by the same window-to-wall ratio and floor height as the starting building. Also, the thermophysical properties are assigned accordingly.
2. *Contextual- and self-shading incorporation*: to account for the reduced amount of incoming solar radiation due to the building's self-shadings and surroundings also in the simplified model, a portion of the windows is substituted by an opaque surface characterized by the same thermal transmittance of the window. To properly size the shading surfaces, a direct radiation analysis is performed on both the starting geometry and the resulting shoebox according to the all-weather Perez sky model. For sake of simplicity, only window surfaces are considered for the detailed massing while the entire geometry is considered for the shoebox. The average annual beam radiation is computed for every orientation and floor, and the equivalent obstruction ratios obtained. For every window of the simplified model, the opaque area is calculated from the corresponding obstruction ratio. The opaque surfaces are modelled as elements having the same base of the window, while the height is calculated to match the equivalent area previously found.
3. *Adjacent buildings inclusion*: building adjacencies are treated as adiabatic surfaces. The presence of adjacencies is checked for each floor and orientation, and adiabatic ratios are computed. Adiabatic ratios can range between 0 and 1, i.e., surfaces can be totally exposed to the external environment or be completely adiabatic. As for windows, the adiabatic areas of the shoebox's surfaces are computed from the adiabatic ratios. Adiabatic surfaces have the same height of the parent surface and the base length calculated to match the equivalent area obtained.

For both obstructions and adjacencies, since a shoebox can have only four vertical surfaces, the orientation of the façades of the starting building are considered with a ± 45° tolerance range in order to aggregate walls with similar orientations. Windows and their corresponding obstruction surfaces are only applied to non-adiabatic façades.
The shoeboxing algorithm has been developed employing Rhinoceros as CAD tool, its plug-in Grasshopper, and EnergyPlus as *BPS* engine. More specifically, customized Grasshopper components written in Python programming language have been utilized. RhinoCommon API (Robert McNell & Associates, 2021) has been used for the geometrical modeling and Ladybug Tools core SDK (Ladybug Tools LLC, 2021) to perform the radiation analysis and the conversion into thermal models. Figure 1 shows an example of shoeboxes with the related boundary conditions.

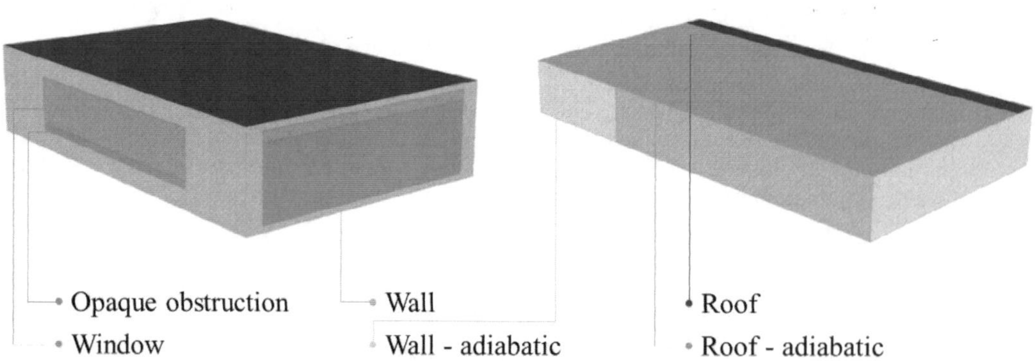

Figure 1: Example of shoeboxes

2.1.2 Improved and more detailed shoeboxing procedure: In a previous work by the authors (Battini *et al.*, 2021b), it has been proven that buildings' geometrical details and context have a greater impact on the cooling needs than on the heating ones. In order to improve the prediction accuracy of the shoebox models, especially for what regards the cooling demand, a different approach has been tested to size the opaque elements obstructing solar radiation on windows. The second step of the basic workflow has been adapted to account for the global radiation instead of the direct one; then, monthly solar radiation analyses have been employed instead of an annual one. To accomplish this step, the radiation analysis has been performed using EnergyPlus for consistently sizing the shoebox's obstructing surfaces by employing the same modeling tool as for the simulation. Differently from the basic procedure in which the radiation analysis and energy simulation have been carried out with two different engines.

To complete the process, in both workflows detailed (i.e., the starting building) and simplified (i.e., the resulting shoebox) models are automatically exported as .idf files and prepared for the simulation by means of Python scripts with the aid of the library Eppy (Philip, 2020), a scripting language for EnergyPlus input and output files. Both detailed and simplified models have one thermal zone for each floor. Nonetheless, the zoning discretization is not a built-in feature of the algorithm, thus it can be changed if needed. For what regards the second procedure, since each month has a specific set of obstruction ratios, 12 different simplified .idf files, differentiated by the size of the obstructing surfaces, are generated. This is necessary due to a limitation of EnergyPlus which does not accept to schedule variable values for this feature. Then, to properly compare the outputs of the simplified models with those of the detailed ones and save computing time, the simulations of the shoebox models are performed just for the month of interest starting one week before the first day of the month, in order to avoid the initial conditions to lead to misleading results.

2.1 Districts for testing

Testing the algorithm at the urban scale aims to explore as much as possible the space of the solutions for the shoeboxes. Therefore, it has been decided to create fictional districts to assess the capabilities of the procedure. A literature review has been performed to find workflows capable of creating from scratch neighborhoods of complex and different building shapes.

Lyu *et al.* (2019) developed nine Local Climate Zone models to find a correlation between sky view factor and microclimate. The models have the same simulation area and represent typical districts of the considered case study. Natanian *et al.* (2019) created an automated workflow for performance-driven urban design to evaluate the performance of both building and urban design parameters of residential and office buildings. Pomponi *el al.* (2021) analyzed the life cycle greenhouse gas emissions of cities decoupling density from tallness. They collected several inputs for real-world neighborhoods to create synthetic urban environments. Trepci *et al.* (2021) systematically evaluated the impact of the urban surroundings on the energy performance of buildings in a hot climate environment by applying different contexts to base case models changing the context's distance and height, as well as the model orientation. Shi *et al.* (2021), with the aim of investigating interactions between solar energy use and urban design, developed the Urban Block Generator to parametrically model typical vernacular block typologies featuring various combinations of block dimensions, building patterns, floor area ratios, and site coverage. In the mentioned studies, even though several metrics have been used to build up the districts, buildings have simple or repetitive shapes and have mainly rectangular footprints. Moreover, individual buildings or building blocks are repeated to obtain generate neighborhoods.

Since the target of this work are complex shaped buildings, a more viable solution is the Building Modular Cell *BMC* technique introduced by Javanroodi *et al.* (2018) to generate buildings based on a cuboid module according to a series of form generation rules. The *BMC* has already been used to enhance the energy efficacy of high-rise building in urban areas (Javanroodi *et al.*, 2019) and to develop the Urban Cell concept (Perera *et al.*, 2021), creating neighborhoods based on realistic morphological parameters. To create buildings, the *BMC* uses a 4x4 grid of cells along with a coverage indicator and two basic decision constrains: (a) the selected cells must be connected from their sides only (no corners), and (b) only one shape must be obtained from the grid (two or more separate footprints from one grid are not allowed).

In this study, a base district similar to the Urban Cell has been created and then modified as parametrically as possible to explore a wider space of the solutions. The district is a 6x6 grid of blocks composed by 9 buildings each, resulting in 324 buildings. Buildings' blocks are separated by fictional streets with the same width for the entire district. The buildings' footprints have been generated by adapting the *BMC* technique using a 4x4 cells grid having 4 m as side length for each cell and employing a cell coverage ratio of 75 %, in order to ensure touching buildings in each block. The variables that have been parametrically edited to create the batch of districts for the test of the algorithm are the buildings' height and scale, and the street's width.

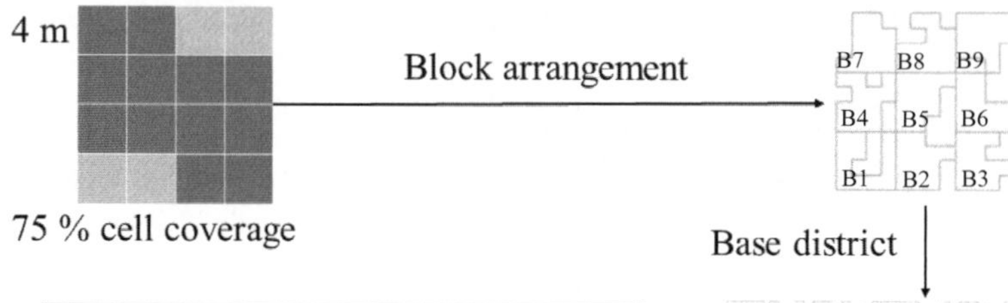

Variables	
Buildings' height	1, 2, 3 and 4 floors
Street width	10, 20, 30 and 40 meters
Buildings' scale	1, 2, 3, and 4
Climate	Bolzano, Denver, Messina
Combinations	
324 buildings per 64 district	
20736 building to simulate for each climate	

Figure 2: Base district generation process, district variables and combinations

Changing the three variables, it is possible to move from a very dense urban setting with small buildings to a less dense one with large massings, allowing to properly evaluate which are the algorithm's strength and weaknesses for different urban layouts. In each model, buildings have the same height and they can be 1, 2, 3 or 4 floors tall, given a floor height of 3 m. Moreover, buildings have been scaled by a factor equal to 1, 2, 3 or 4. Finally, the street's width have been set equal to 10, 20, 30, or 40 meters. Combing all variables leads to 64 districts to be simulated, thus 20 736 buildings in total. The district generation process, a visualization of the base district's buildings footprints and the variables' ranges coupled with the resulting combinations are reported in Figure 2.

2.1 Simulation and comparison

The 64 districts have been simulated in the three climates of Bolzano and Messina, Italy, and Denver, U.S. Bolzano and Messina have been chosen to have both heating- and cooling-dominated climates respectively, while Denver has been selected since it is usually employed for validating *BPS* tools, and it is characterized by large daily temperature variations. Detailed and simplified models according to the shoeboxing algorithm have been simulated for every district in each climate.

As regards detailed models, in order to consider only the part of the district relevant for each building, the surrounding context has been taken into account up to a distance equal to the length of the blocks' diagonal. All models have been characterized by the same energy-related properties, as in the previous work on stand-alone buildings (Battini *et al*, 2021a).

The outputs of detailed and simplified models have been compared to assess the algorithm performance. Even though both temperature and energy needs differences have been evaluated, in this paper, the comparison has been carried out only for what regards the energy demand to properly discuss the procedure's goodness and reliability at building level. Thus, the energy needs have been analyzed at annual and hourly scale. Annually, absolute and relative differences between detailed and shoebox models have been calculated, while the Root Mean Square Error *RMSE* has been employed to assess the performance at hourly scale.

3. RESULTS AND DISCUSSION

3.1 Overview

The energy needs differences between detailed and simplified models are presented as boxplots for all three climates, given the large amount of data analyzed. At first, the results from the standard shoeboxing procedure with one fixed opaque obstructing surface are presented. Then, the outcome for the improved configuration with monthly opaque shadings is reported. For both methodologies, the introduction of the shoeboxing algorithm has reduced the energy simulation time from 10 to almost 30 times depending on the district, regardless of the differences between the models. In this way, it has been possible to reduce the time required to simulate a district from several hours to tens of minutes in the worst case.

3.2 Basic simplification results

The comparison of energy needs has been performed for all 64 districts simulated in the three climates of Bolzano, Denver and Messina. First, the differences have been reported for all simulated buildings without considering individual districts, buildings or blocks. Then, the analysis has increased the level of detail to identify the conditions leading to the poorest predictions.

Figure 3 shows the relative annual differences for all buildings in all districts for the three considered climates. Overall, the heating demand is the better predicted output yielding results within ±20 % in all three climates neglecting outliers (defined as the points lying outside 1.5 times the interquartile range past the low and high quartiles), and, thus, its accuracy of this kind of simplification can be considered acceptable. Moreover, heating boxplots are often narrower, meaning that the prediction is consistent for all buildings. On the other hand, the cooling demand experiences larger deviations. For all three climates, only half of the buildings are within ±20 % deviation. Nonetheless, the remaining part of the studied buildings yields much larger differences, even ±100 %. The considerations related to the annual differences can be extended to the hourly deviations reported in Figure 4 by the *RMSEs*. Indeed, the *RMSEs* for the heating demand are much lower compared to the ones of the cooling demand.

In Figure 5 and Figure 6 the relative annual differences and *RMSEs* have been grouped by building position in the blocks for a sample district. From the obtained boxplots, it is clear that their position in a block does not influence the deviations in the three climates. The same process has been followed to check the influence of the blocks' position in the district obtaining the same outcome.

From this first comparison, the basic configuration of the shoeboxing algorithm shows to be reliable enough when employed for assessing the heating demand, regardless the considered climate. However, it is still not able to ensure high accuracy when adopted for the cooling needs estimation in all the possible cases. To understand which are the conditions in which the accuracy of the procedure drops, a more in-depth analysis has been performed to check the variables employed in generating the districts leading to the poorest predictions.

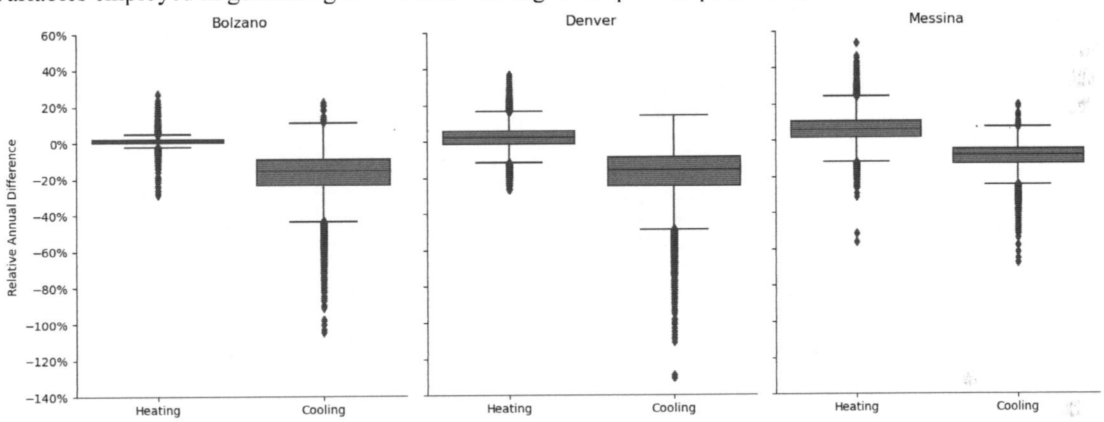

Figure 3: Boxplots reporting annual heating and cooling needs relative differences for all buildings in all districts in the three climates

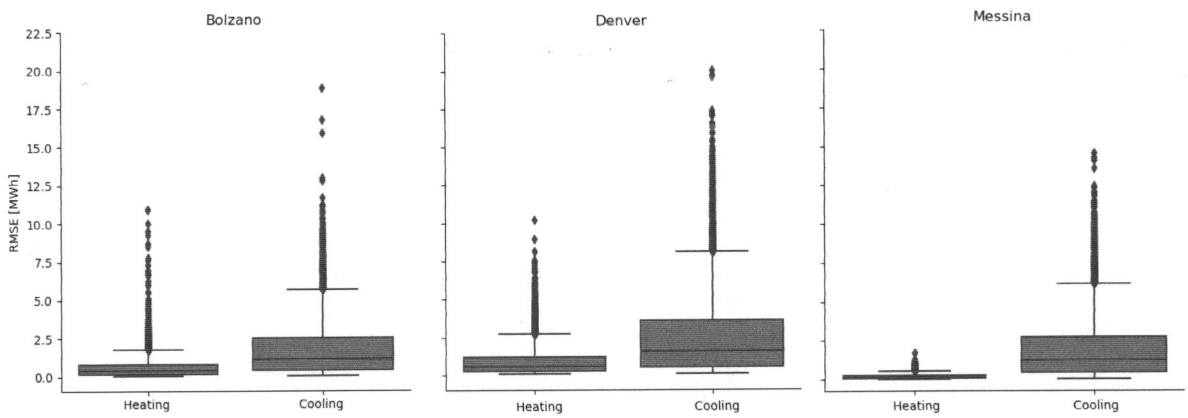

Figure 4: Boxplots reporting hourly heating and cooling *RMSEs* for all buildings in all districts in the three climates

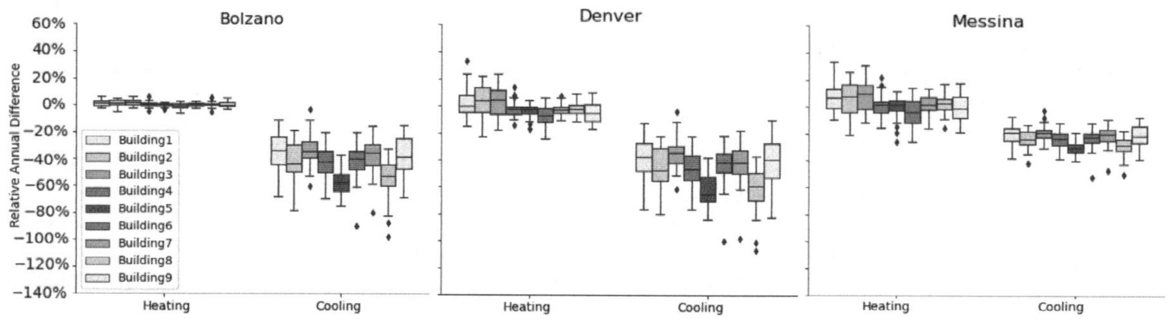

Figure 5: Relative annual differences in the three climates for the district with buildings of 4 floors, scaled by a factor of 1, and with a street width of 10 m

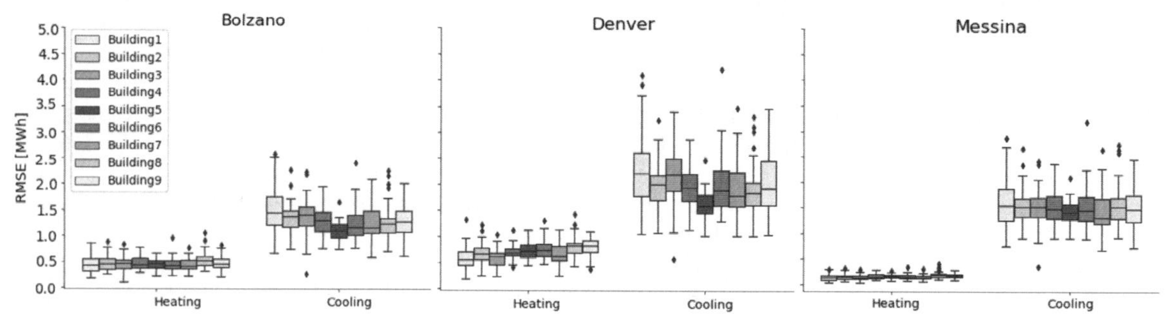

Figure 6: *RMSEs* in the three climates for the district with buildings of 4 floors, scaled by a factor of 1, and with a street width of 10 m

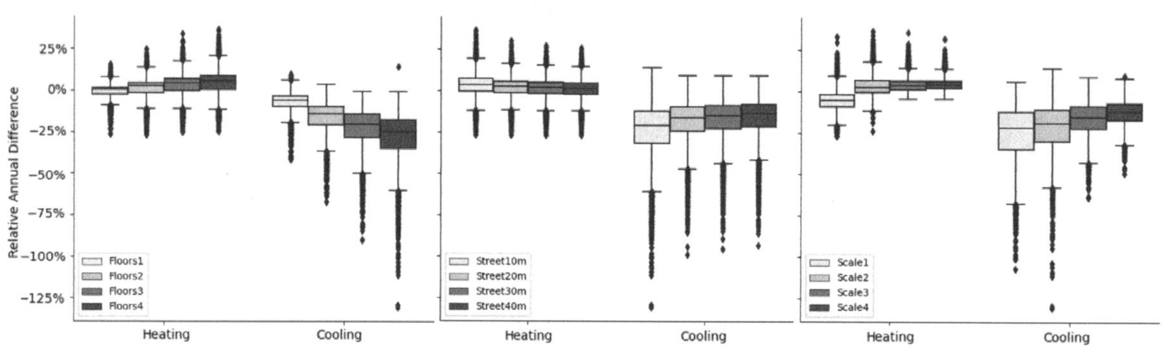

Figure 7: Relative annual differences boxplots for all districts in the climate of Denver, with buildings' deviations grouped according to the building height, street width and building scale

Specifically, the results of annual relative differences have been grouped according to the three variables utilized to create the districts: buildings' scale and height, and street width. The results are consistent in all the three climates and, as an example, the boxplots for the climate of Denver are reported in Figure 7. From Figure 7 it is possible to see that the largest differences occur for smaller and taller buildings with smaller street widths. Indeed, taller buildings obstruct more the radiation on the lower floor resulting in larger differences. In the same way, smaller street widths lead to a denser urban context shading more radiation on all buildings. Moreover, smaller buildings have poorer predictions because they have a smaller volume, in fact the impact of the solar radiation entering the thermal zone on the air-node heat balance is greater respect to larger ones. Hence, considering all the above, the worst prediction accuracy occurring for the cooling needs seems to be correlated to the modelling of the solar radiation entering into the thermal zones. For this reason, an improved solution accounting for the radiation on a monthly basis has been tested.

3.3 Improved simplification results

To assess the improvements of the shoeboxing algorithm with the new shading configuration, instead of running again all 64 districts, only the two districts yielding the worst overall prediction accuracy have been selected.

Among all districts, the two districts experiencing the poorest predictions have both buildings of 4 floors and a street width of 10 m. As mentioned before, districts with smaller buildings are those with the largest differences (Figure 5 and Figure 6), thus districts with buildings scaled by a factor of 1 and 2 have been selected. These two districts have been simulated in the three climates.

Figure 8 and Figure 9 show the comparison in accuracy prediction between the basic simplification procedure and the improved one for the two districts considered in the three climates. The results show that sizing the obstruction opaque surface monthly remarkably reduces the differences, both annually and hourly. The cooling needs estimation has even become more accurate than the heating one with annual cooling relative differences within ± 20 % for every building in both districts, independently of the considered climate. The cooling *RMSEs* are comparable to the heating ones. In general, the new approach improved the prediction accuracy of the shoeboxing algorithm for all the considered targets, as shown not only by the lower deviations on both hourly and annual basis, but also by the much narrower boxplots. In the basic procedure, the cooling demand was generally underestimated by the shoeboxes, while for the heating demand a clear trend was not detected, with heating needs either over or underestimated. Differently, in the improved approach, while the cooling demand neglecting outliers is generally overestimated, the heating demand predictions are mainly underestimated, and they also show differences larger than 20 %.

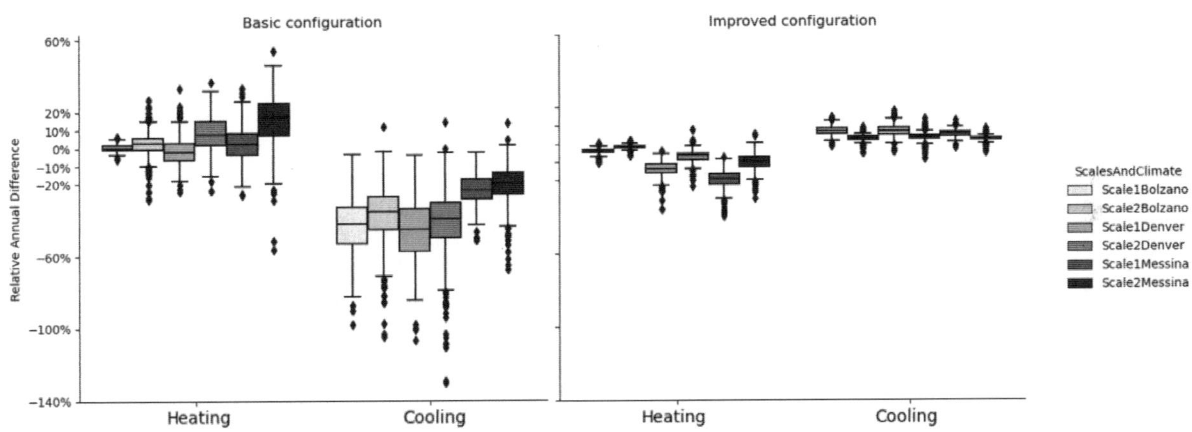

Figure 8: Relative annual differences comparison for the two districts considered in the three climates between the basic algorithm configuration and the improved one

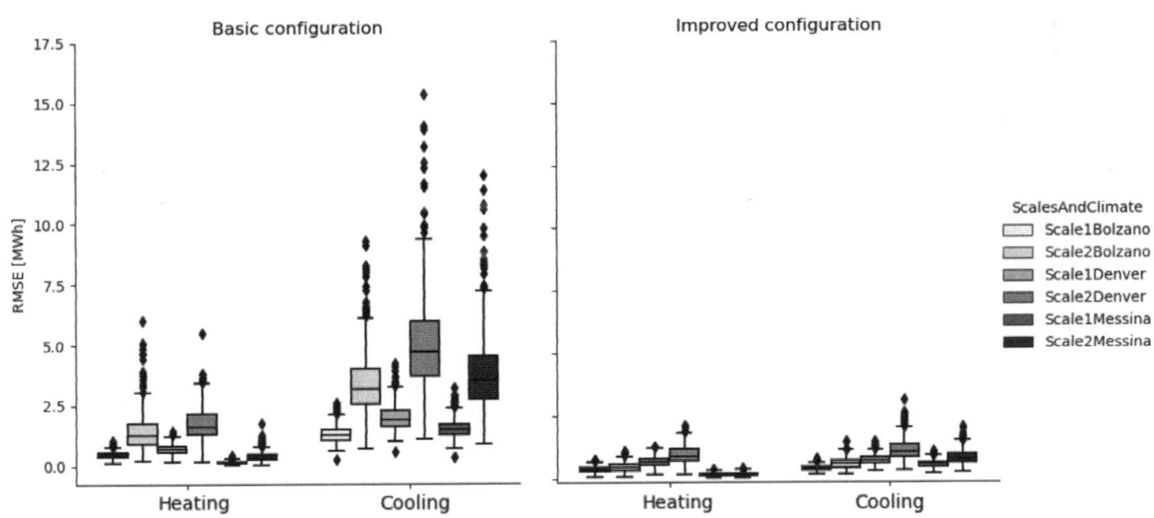

Figure 9: *RMSEs* comparison for the two districts considered in the three climates between the basic algorithm configuration and the improved one

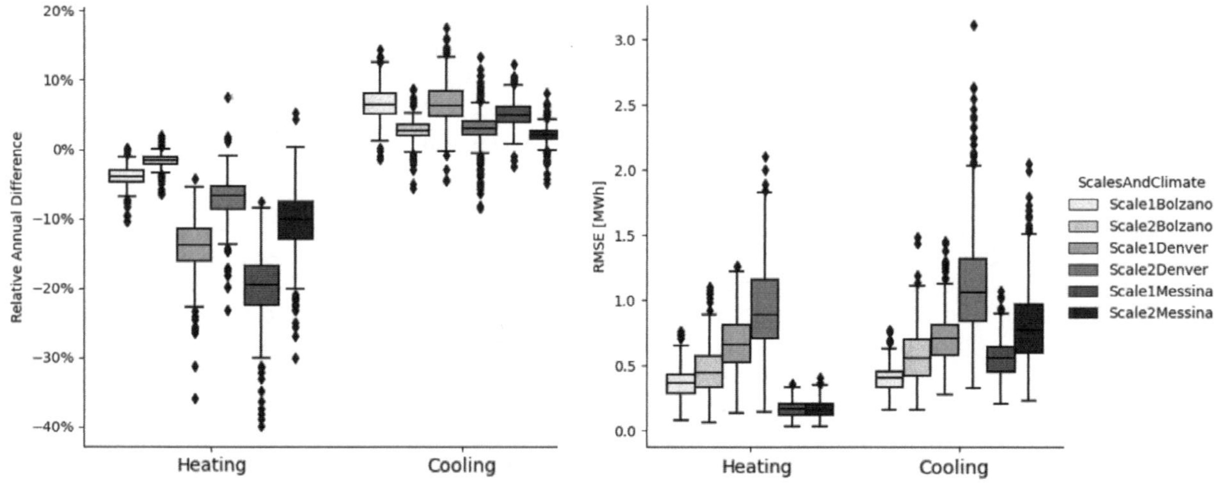

Figure 10: Detail on the differences of the improved configuration for the two districts considered in the three climates

Overall, the heating demand prediction has improved with respect to the basic procedure but not as much as the cooling one. Figure 10 reports more in detail the hourly and annual differences occurring in the six cases. The *RMSEs* boxplots highlight that the deviations between heating and cooling are comparable for all cases, excepts for Messina in which the heating demand magnitude is very low.

To conclude, the results obtained by a more detailed modelling of the incoming radiation have improved more than any other solution (e.g., changing the grid-spacing used for the radiation analysis or employing opaque elements instead of shading surfaces outside the window for casting the solar radiation) the algorithm's accuracy. More specifically, since the deviations are similar in the considered cases, the algorithm's performances are less dependent on the external boundary condition as it is for the basic configuration. It is also expected that the improvements achieved in the two analyzed districts will be confirmed in the remaining ones, already yielding better results. As for now, the improved shoeboxing algorithm is capable of properly predicting the energy needs of complex shaped buildings at the urban level, even in a wide range of conditions that could undermine its accuracy.

4. CONCLUSIONS

In this work, a novel simplification algorithm to reduce the computational cost of building energy simulations at the urban scale has been tested to evaluate its accuracy at the district level. The algorithm consistently simplifies every building in an urban model into a representative shoebox and accounts for the building's surroundings and adjacencies. Since the way in which the incoming solar radiation is modelled has a major effect on the outcome of the simplification approach's accuracy, two solutions to consider the building's self and context shadings have been investigated. The impacts of obstructions have been considered by means of opaque surfaces reducing the transparent building envelope generated according to obstruction ratios resulting from a radiation analysis. At first, such obstructions have been created from an annual direct radiation analysis. Then, an improved configuration considering monthly global radiation has been employed. The algorithm has been tested by means of a set of fictional districts, featuring complex shaped buildings arranged in blocks, that have been built to explore as much as possible the space of the solutions. To assess the algorithm's performance under different boundary conditions, three different climatic conditions have been used. For every building in each district, detailed and simplified energy models have been compared on hourly and annual basis in terms of heating and cooling needs.

Regardless of the kind of adopted radiation modelling (i.e., direct on annual basis or global on monthly basis), the algorithm has been able to cut the energy simulation time up to 30 times. The basic algorithm has proved to be effective in estimating the heating needs. However, it has not provided the same level of accuracy and reliability for what regards the cooling demand. For this reason, the improved methodology has been applied to the two districts yielding the worst results. Accounting for the solar radiation on a monthly basis significantly improved the cooling predictions, while still obtaining adequate results for the heating needs. Moreover, the results obtained were characterized by similar deviations regardless of the considered climate, making the algorithm's performance independent by external

boundary conditions.
To conclude, the shoeboxing algorithm showed to be adequate in predicting the energy needs of buildings of complex shapes at the urban level, even at fine time scale such as hourly. Further studies will be carried out in order to fully test the improved algorithm and to look for further improvements by studying outliers.

REFERENCES

Abbasabadi, N., & Ashayeri, M. (2019). Urban energy use modeling methods and tools: A review and an outlook. *Building and environment 161*. doi: https://doi.org/10.1016/j.buildenv.2019.106270

Battini, F., Pernigotto, G., & Gasparella, A. (2021a). Development of a shoeboxing approach for Urban Building Energy Modeling. *International High Performance Buildings Conference*, (p. 385). Purdue.

Battini, F., Pernigotto, G., & Gasparella, A. (2021b). A parametric analysis of the impact of thermophysical, geometry and urban features on the energy demand of a simplified building shoebox model. *Building Simulation 2021*. Bruges.

Dogan, T., & Reinhart, C. (2013). Automated conversion of architectural massing models into thermal 'shoebox' models. *Proceedings of BS2013: 13th Conference of International Building Performance Simulation Association, Chambéry, France, August 26-28*, (pp. 3745-3752). Chambéry.

Dogan, T., & Reinhart, C. (2017). Shoeboxer: An algorithm for abstracted rapid multi-zone building energy model generation and simulation. *Energy and Buildings, 140*, 140-153. doi: https://doi.org/10.1016/j.enbuild.2017.01.030

Hong, T., Chen, Y., Luo, X., Luo, N., & Lee, S. (2019). Ten questions on urban building energy modeling. *Building and Environment, 168*. doi: https://doi.org/10.1016/j.buildenv.2019.106508

İşeri, O. K., & Dino, İ. G. (2020). An Algorithm for Efficient Urban Building Energy Modeling and Simulation. *2020 Proceedings of the Symposium on Simulation for Architecture and Urban Design*, (pp. 455-462).

Javanroodi, K., Mahdavinejad, M., & Nik, V. M. (2018). Impacts of urban morphology on reducing cooling load and increasing ventilation potential in hot-arid climate. *Applied Energy*, 714-746.

Javanroodi, K., Mahdavinejad, M., & Nik, V. M. (2019). A novel design-based optimization framework for enhancing the energy efficiency of high-rise office buildings in urban areas. *Sustainable Cities and Society*.

Johari, F., Peronato, G., Sadeghian, P., Zhao, X., & Widén, J. (2020). Urban building energy modeling: State of the art and future prospects. *Renewable and Sustainable Energy Reviews*. doi: https://doi.org/10.1016/j.rser.2020.109902

Ladybug Tools LLC. (2021, July 20). Retrieved from Ladybug Tools: www.ladybug.tools

Lyu, T., Buccolieri, R., & Gao, Z. (2019). A Numerical Study on the Correlation between Sky View Factor and Summer Microclimate of Local Climate Zones. *Atmosphere*.

Natanian, J., Aleksandrowicz, O., & Auer, T. (2019). A parametric approach to optimizing urban form, energy balance and environmental quality: The case of Mediterranean districts. *Applied Energy*, 113637.

Perera, A. T., Javanroodi, K., Wang, Y., & Hong, T. (2021). Urban cells: Extending the energy hub concept to facilitate sector and spatial coupling. *Advances in Applied Energy*, 100046.

Pomponi, F., Saint, R., Arehart, J. H., Gharavi, N., & D'amico, B. (2021). Decoupling density from tallness in analysing the life cycle greenhouse gas emissions of cities. *npj Urban Sustainability*, 33.

Reinhart, C., & Davila, C. (2015, December 2). Urban building energy modeling - A review of a nascent field. *Building and Environment, 97*, 196-202. doi: https://doi.org/10.1016/j.buildenv.2015.12.001

Reinhart, C., Dogan, T., Jakubiec, J., Rahka, T., & Sang, A. (2013, August). Umi - an Urban Simulation Evironment for Building Energy Use, Daylighting and Walkability. *13th Conference of International Building Performance Simulation Association*. Chambéry, France: IBPSA.

Robert McNell & Associates. (2021, June 14). *RhinoCommon API*. Retrieved from Rhino Developer Docs: https://developer.rhino3d.com/api/RhinoCommon/html/R_Project_RhinoCommon.htm

Shi, Z., Fonseca, J. A., & Schlueter, A. (2021). A parametric method using vernacular urban block typologies for investigating interactions between solar energy use and urban design. *Renewable Energy*, 823-841.

Trepci, E., Maghelal, P., & Azar, E. (2021). Urban built context as a passive cooling strategy for buildings in hot climate. *Energy and Buildings*, 110606.

Zhu, P., Yan, D., Sun, H., An, J., & Huang, Y. (2019). Building Blocks Energy Estimation (BBEE): A method for building energy estimation on district level. *Energy and Buildings, 185*, 137-147.

Zhu, S., Ma, C., Zhang, Y., & Xiang, K. (2021). A hybrid metamodel-based method for quick energy prediction in the early design stage. *Journal of Cleaner Production, 320*.

Building Energy Model Calibration With Functional Inputs And Outputs For Performance Monitoring

Thomas CERBELAUD[1*, 2], Bruno DUPLESSIS[2], Pascal STABAT[2], Riad ZIOUR[1]

[1] Openergy R&D,
4 Rue Dolorès Ibarruri, 93100 Montreuil, France
thomas.cerbelaud@egis.fr, riad.ziour@egis.fr

[2] Mines Paris, PSL Research University, CES - Centre for Energy Efficiency of Systems,
60 bd Saint Michel, 75006 Paris, France
bruno.duplessis@minesparis.psl.eu, pascal.stabat@minesparis.psl.eu

* Corresponding Author

ABSTRACT

Building continuous performance monitoring is becoming a cornerstone in ensuring energy efficiency and sobriety of existing, retrofitted and newly built buildings. Although it may help convince investors in energy efficiency projects or bridge the gap between expected and actual performance, continuous monitoring - sometimes referred to as "Advanced M&V" or continuous commissioning – is still the exception rather than the rule. Recent efforts to continuously characterize building performance usually rely on building-level analyses: previous works include leveraging a Building Energy Model (BEM), monthly calibrated on building heating, ventilation and lighting consumption using real weather data and fine grain occupancy data, for daily monitoring. While BEM calibration against sub-daily frequency data has been increasingly studied in recent years, it is, to our knowledge, seldom used for building continuous monitoring. It is, however, particularly tailored for this task, to the extent it extracts embedded physics within the BEM into actionable insights for fault detection and diagnosis.

Fine grain calibration of BEM faces a number of challenges in the recent literature, among which are (i) accounting for time varying dependent functional inputs - e.g. electric equipment and lighting energy consumption altogether with building occupancy - but for sensor data in the calibration algorithm, and (ii) treating functional outputs as functional stochastic variables when comparing simulation outputs with real data. Our contribution is to enhance building-level performance monitoring by introducing a stochastic model inversion scheme, also referred to as stochastic calibration, to support robust preventive fault detection and diagnosis. Our approach extends the current state-of-the-art on Bayesian calibration of BEM by accounting for dependent functional inputs and outputs in both selecting the most influential parameters and calibrating the model, and deals with uncertainties in functional inputs such as daily profiles of lighting and electric equipment energy consumption. This methodology is illustrated against a medium-size real secondary school building, located in Rennes, France, and equipped with an Advanced Meter Infrastructure (AMI) with hundreds of sensors. A comparison between a classic calibration process and the described methodology is presented and the benefits of accounting for the functional nature of the inputs and outputs in both the Design of Experiment (DoE) and the calibration process are illustrated against this case study.

1. INTRODUCTION

The advent of Advanced Metering Infrastructure (AMI) and cloud computing is paving the way towards more affordable advanced building performance monitoring, notably reducing the cost of data acquisition (Granderson et al., 2016). Besides, Touzani et al. (2019) highlighted the necessity of accounting for non-routine event (NRE) in comparing metered consumption to a baseline, e.g. within an IPMVP-based process. It is only a short step from there to Automated Fault Detection and Diagnosis (AFDD) methods, where one is interested in monitoring and preventing performance drifts in an automated and continuous manner. FDD-related research has been quite active in the past years: on one hand, it mostly focuses on heavily instrumented commercial buildings and data-driven (Frank et al., 2019) approaches to precisely predict specific faults such as duct fouling - Frank et al. (2016) also tried using a hybrid or data-driven approach with a minimal set of metered data to diagnose specific faults; on the other hand, Jradi et al. (2018) used a model-based approach, where an EnergyPlus (EnergyPlus, 2019) model is dynamically calibrated by passing real time metered data to the simulation, and compares simulated and metered data to help detecting and diagnosing drifts in performance. However, while BEM calibration against sub-daily frequency data has been increasingly studied in recent years (Chong et al., 2017), it is, to our knowledge, seldom used for building continuous monitoring without heavy data metering to account for dynamic variables such as occupancy. It is, however, particularly tailored for this task, to the extent it extracts embedded physics within the BEM into actionable insights for fault detection and diagnosis.

In this paper, we propose to extend the current state-of-the-art calibration process to account for dynamic variables, so that building-level performance monitoring algorithms may be developed for building equipped with a small number of sensors not necessarily accounting for dynamic variables such as occupancy. The following sections are organized as follows. Section 2 describes the case study used to illustrates our methodology. Section 3 digs into the technical details of how the calibration process is extended to account for functional and dynamic inputs and outputs, and defines the baseline to be compared with. Eventually, section 4 compares how the exposed methodology improves the baseline.

2. CASE STUDY OVERVIEW

The case study is a medium-size real secondary school building, *Lycée Brequigny*[1], located in Rennes, France, and equipped with an Advanced Meter Infrastructure (AMI) with hundreds of sensors. For the sake of clarity and simplicity, we only describe the building main features, and tested our methodology on the heating energy consumption only. The BEM model used in this work is built using EnergyPlus (EnergyPlus, 2019).

2.1 Building's description and thermal zoning

The building is a recently retrofitted educational building, comprised mostly of classrooms and offices, and is one of three buildings that together form the *Lycée Brequigny*. The building total floor area is 9426 square meters. 3D view and thermal zoning for the ground floor are displayed in Figures 1a and 1b.

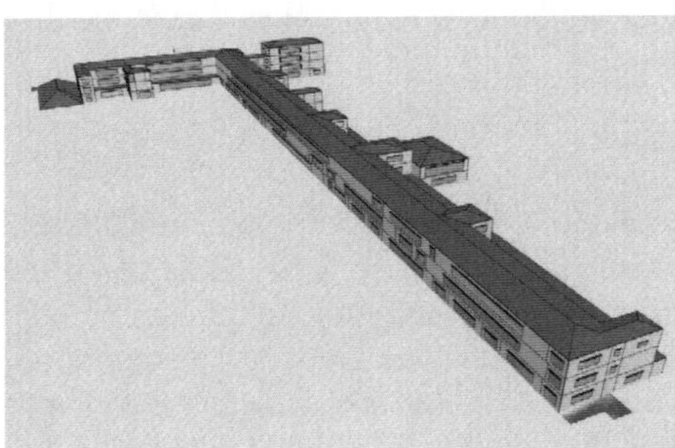

(a) 3D view of *Lycée Brequigny*

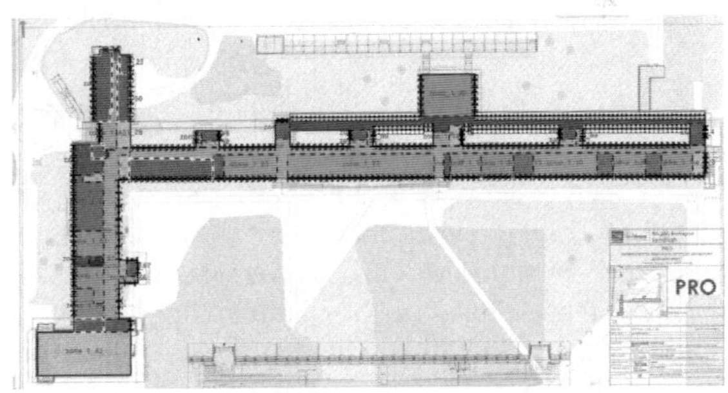

(b) Ground floor thermal zoning

Figure 1: Thermal zoning of the case study

The building was externally insulated in 2020, and is heated by hot water radiators connected to four district heating substations. The inlet temperature of the secondary loop is regulated on the outdoor dry bulb temperature.

The study is particularly focused on the parameterization of occupancy, ventilation, electric equipment, lights and indoor temperature, which impact the heating consumption. As the retrofit came with an Energy Performance Contract (EPC), a BEM was built during the design stage of the retrofit. Therefore, the BEM built in our study took over all the parameters from the previous model, and updated them according to the building as built[2]. Part of these parameters will be calibrated further down the road, except for the - measured - indoor temperature that is taken as a boundary condition used during calibration.

Heating energy is delivered to the building by a heat exchanger connected to a heating district and that feeds hot water to the radiators throughout the building. Classrooms are equipped with motion detectors and temperature sensors that control the hot water radiator terminal regulation and the mechanical ventilation flow rate based on pre-configured

[1] It was renovated in 2020 and was awarded several national and international awards.

[2] as opposed to the building as designed.

values. All other rooms are regulated using thermostatic valves and pre-configured ventilation schedules except for some teachers rooms that are ventilated according to measured CO_2 levels.

2.2 Advanced Meter Infrastructure (AMI) overview

Thousands of data points are collected through the Building Management System (BMS). The site is operational since 2019, and the sensor data from November 2020 to March 2022 are used in this study. For the purpose of the demonstration, the ten first days of December 2020 are extracted from the data set. Because of COVID-19, the site regulation and occupancy were particularly disturbed, in that the variability of the occupancy and ventilation flow rate was high - in particular the overall occupancy was exceptionally low -, and the classrooms were over-ventilated (natural ventilation) during this time period to prevent the spread of the virus. Only a small subset of the AMI data was used, and is described in Table 1.

Table 1: Data used from the AMI to illustrate the methodology

Data	# sensors	Use in the proposed methodology
Indoor temperatures	71	The 71 time series are averaged out and used as observed input data in the calibration process.
Heating energy consumption	3	The consumption of three substations out of four are aggregated altogether. The fourth substation, feeding the conference hall, is voluntarily left out.

Besides sensor data, dry bulb temperature and horizontal irradiance are also used as inputs to the model calibration procedure. Eventually, a categorical temporal variable is used to indicate whether or not, for each day, the heating system is turned on.

The high variability in building uses makes it even more critical to account for these changes during monitoring in order to properly detect and diagnose potential fault during building operation.

3. METHODOLOGY

While the methodology relies on a well-known tried and tested succession of steps[3] (Kennedy and O'Hagan, 2001; Heo et al., 2012; Chong et al., 2017; Monari, 2016), our contribution dwells in bringing to BEMs calibration new developments in stochastic model inversion so (i) functional variables are accounted for (Nanty, 2015; Nanty et al., 2016) and (ii) the process is compatible with building performance monitoring. In addition, we emphasise on the choices made for use within a building performance monitoring workflow. This section describes the methodology laid out to achieve BEM calibration with functional inputs and outputs. The methodology breaks down as follows: (i) first a Design of Experiments (DoE) is generated; (ii) then a surrogate model is trained on the simulation in order to perform a Global Sensitivity Analysis (GSA) to select a relevant subset of features; (iii) and eventually a new surrogate is trained on the reduced subset and (iv) Bayesian calibration is performed using the emulator. This section ends up with describing the benchmark procedure of our methodology.

Table 2 displays the parameters that were chosen to conduct this work. $\mathcal{N}_{[a,b]}(\mu, \sigma)$ is the truncated normal distribution on the interval $[a, b]$ with mean μ and standard-deviation σ, and $\mathcal{U}_{[a,b]}$ is the uniform distribution on the interval $[a, b]$.

3.1 Modeling input and output functional variables

To account for functional input variables, the methodology leverages Functional Data Analysis (FDA) algorithms to decompose multivariate time series data into a reduced basis of functions. When applied to a single day profile, the decomposition allows to find the functional basis $(\varphi_i)_{i=1..k}$ and mean function e such that

[3]In the literature, while the calibration processes may slightly differ from one work to another, they basically all follow these three steps: (i) reduce the number of parameters to calibrate, (ii) optionally define a surrogate model (to train before or at calibration time) and (iii) explore posterior distribution of unknown parameters

[4]Only classes are regulated with a setback and comfort setpoint

$$f_j \approx e + \sum_{i=1}^{k} a_{i,j}\varphi_i,$$

That minimizes

$$\sum_{j=1}^{n} \int_I \left(f_j(t) - e(t) - \sum_{i=1}^{k} a_{i,j}\varphi_i(t) \right)^2 dt$$

In this work the functional decomposition is extended to multiple and dependent functions, using the Simultaneous functional Principal Component Analysis (SPCA) algorithm (Ramsay et al., 2005), consequently representing all dependent functions in Table 2 with a small number of scalars $\alpha = (a_1 \dots a_k)$, that will be further used in the BEM calibration process.

Besides, for the methodology to be compatible with continuous monitoring, all functional metered data - inputs and outputs, e.g. dry bulb temperatures or hourly heating energy consumption - were segmented into hourly time series of one day.

As a consequence, a datum in the training and testing sets is no longer a simulation run period long, as it would have been following Higdon et al. (2008a), or several months long, as suggested in Monari (2016), but a day long, bringing the number of samples in the training set up to 365 times the number of simulations. Therefore, the number of training

Table 2: Model parameters description and distributions.

Symbol	Parameter description	Distribution
Scalar parameters		
υ_{occ}	Multiplicative factor of the occupancy nominal value	$\mathcal{N}_{[0.3,1.6]}(1.0, 0.5)$
υ_{vent}	Multiplicative factor of the ventilation nominal value	$\mathcal{N}_{[0.4,0.6]}(0.5, 0.2)$
$\upsilon_{natvent}$	Multiplicative factor of the natural ventilation nominal value	$\mathcal{U}_{[0.5,1.5]}$
υ_{li}	Multiplicative factor of the lights nominal value	$\mathcal{N}_{[0.6,1.4]}(1.0, 0.2)$
υ_{eq}	Multiplicative factor of the equipments nominal value	$\mathcal{N}_{[0.6,1.4]}(1.0, 0.2)$
b_{vent}	Multiplicative factor of the ventilation baseload	$\mathcal{N}_{[0.6,1.4]}(1.0, 0.2)$
b_{li}	Multiplicative factor of the lights baseload	$\mathcal{N}_{[0.6,1.4]}(1.0, 0.2)$
b_{eq}	Multiplicative factor of the equipments baseload	$\mathcal{N}_{[0.6,1.4]}(1.0, 0.2)$
λ	Multiuplcative factor for all opaque surface conductivities	$\mathcal{N}_{[0.8,1.2]}(1.0, 0.1)$
U_w	Multiplicative factor for all windows U-value	$\mathcal{N}_{[0.8,1.2]}(1.0, 0.1)$
S_w	Multiplicative factor for all windows solar heat gain	$\mathcal{N}_{[0.8,1.2]}(1.0, 0.1)$
η	Efficiency of the heat exchanger of the heating district	$\mathcal{N}_{[0.7,1.0]}(0.9, 0.1)$
q_{inf}	Infiltration flow rate under 4Pa	$\mathcal{U}_{[0.6,1.4]}$
t_c	Indoor comfort temperature offset from nominal value	$\mathcal{N}_{[-2.0,3.0]}(0.0, 0.5)$
t_s	Indoor setback temperature offset from nominal value for classrooms	$\mathcal{U}_{[-3.0,0.0]}$
s_{blinds}	Threshold of direct solar radiation for closing / opening the blinds	$\mathcal{U}_{[100.0,800.0]}$
s_{li}	Threshold of illuminance for turning on/off the lights	$\mathcal{U}_{[200.0,700.0]}$
h_s	Time offset from system nominal start time in hours	$\mathcal{U}_{[-2.0,1.0]}$
h_e	Time offset from system nominal shut down time in hours	$\mathcal{U}_{[-1.0,2.0]}$
Functional parameters		
$f_{offices}$	Normalized day profile used for offices	GMM on the coefficient of the functional decomposition of randomly generated profiles
f_{class}	Normalized day profile used for classrooms	
f_{temp}	Day profile of temperature setpoint in classes[4]	

points in the simulation data set grows larger than 10,000 data points for several hundreds of simulations, making the training of Gaussian Processes (GP) (Kennedy and O'Hagan, 2001; Rasmussen and Williams, 2005; Higdon et al., 2008a; Chong et al., 2017) unfeasible. A Sparse Variational Gaussian Process (Titsias, 2009; Hensman et al., 2013) was used instead, as it scales to potentially millions of training points, and is easily expressed as a probability distribution over the model.

Figure 2 illustrates the functional decomposition by showing the effects of positively and negatively varying the coordinates a_i on the decomposed mean function, here the building energy use day profile, for the three first eigen functions.

3.2 Accounting for functional variables in the Design of Experiment (DoE)

Let $X = (x_1, \ldots x_n)$ and $Z = (z_1(t) \ldots z_m(t))$ be respectively the unknown scalar and functional variables (in our case time series). Usually, an optimized Latin Hypercube Design (LHD) is created on a bounded domain of X using some distance metric (see Jin et al. (2005); Morris and Mitchell (1995), and is used to perform GSA (Saltelli et al., 2008; Iooss and Lemaître, 2015), which assumes the random variables are independent. Time series, besides being high dimensional, are auto-correlated, i.e. each time step depends from the previous ones, and may be dependent from each other, e.g. building lighting and equipment daily electricity consumption are likely to be high when the building is occupied and low when it is not. As such, we cannot build a LHD on the functional variables.

We relied on the work of Nanty et al. (2016) for building our full DoE. Using the SPCA algorithm, an optimized *maximin* DoE on the coefficients $a_{i,j}$ was built on a non hypercubic domain as in Auffray et al. (2012). Then the DoEs on Z and X were combined adapting an algorithm by Muehlenstaedt et al. (2016)

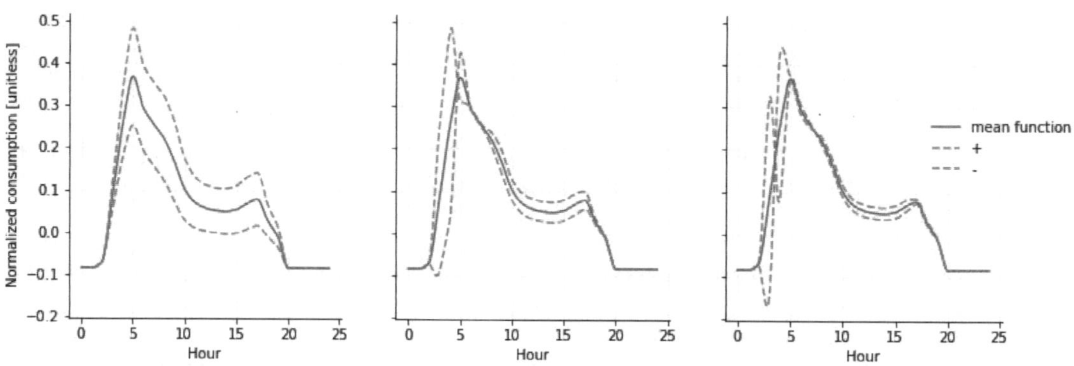

Figure 2: Illustration of the functional decomposition of the normalized heating energy use hourly day profiles. The three figures show the effect of positively (+) and negatively (-) varying the coordinates a_i on the mean decomposed function for the three first eigen functions φ_i. The first eigen function handles the overall amplitude of the day profile, and the second and third ones focus on the peak amplitude and phase variations.

3.3 Global Sensitivity Analysis (GSA)

In the standard framework originally exposed in Kennedy and O'Hagan (2001), a Global Sensitivity Analysis (GSA) (Saltelli et al., 2008) can be used to discard non influential parameters and calibrate only a handful of relevant parameters, thus helping to sanitize an otherwise under-determined problem. However, the Sobol' indices[5] (Saltelli et al., 2008; Janon et al., 2014) are, in their original formulation, only valid between independent variables and scalar outputs. Following the work of Nanty (2015), all coordinates a_i originating from the same functional decomposition were grouped, so only the influence of the group was quantified. Furthermore, the indices computation was also modified to account for the multivariate / functional outputs case as in Gamboa et al. (2014).

3.4 Bayesian BEM calibration

The input data was comprised of:

- Metered data D: these comprise daily profiles of weather - dry bulb temperature and global horizontal irradiance

[5]For model calibration, the total-order indices are used to screen out non-influential variables.

- and indoor temperature, projected onto two functional bases learnt on the simulation outputs from the DoE.

- Unknown data θ: vector of parameters to calibrate. It encompasses both scalar - i.e. we assume the value of each parameter is constant throughout the calibration period - and functional - each day has a different vector of values - variables. Assuming there are n_s scalar variables that the functional decomposition of the functional variables has n_f eigen functions, then there are $n = n_s + n_f \times L$ parameters to calibrate, L being the length, in days, of the calibration period.

The inputs were bound together by the following probabilistic model (Kennedy and O'Hagan, 2001)[6]:

$$y = \eta(D, \theta) + \delta(D) + \varepsilon \tag{1}$$

The probabilistic model from Higdon et al. (2008a) was modified to account for the modularity as in Bayarri et al. (2007, 2009) - since the surrogate model is trained before model calibration - while accounting for the emulator response and the discrepancy model respective functional decomposition. The PPL Numpyro (Bingham et al., 2018) was used in this work to factorize the likelihood on the observed days and to use enumeration of discrete latent variables[7].

3.5 Benchmark

The methodology developed in this work was benchmarked against a standard calibration setting, where neither functional nor dynamic parameters were accounted for (see Heo et al. (2012); Chong et al. (2017) for examples), that is referred to as Static Prior (SP), as opposed to the Dynamic and Functional Prior (DFP) setting. Although the model and its modifications in the SP setting - described in section 3.4 - is not quite the standard described in Higdon et al. (2008a), we felt the comparison would not have been fair otherwise, or at least it would have been blurred by the differences in the calibration procedure, besides the difference in the prior distribution.

In addition to the CVRMSE, NMBE and Q^2 indicators, that inform on the goodness of fit of the calibrated model, because our final goal is to estimate reasonable variation ranges for the observed data, and to detect and diagnose abnormal behaviors, indicators that capture the predictive capability of the model were computed, namely the empirical frequency coverage (EFC) (Kyzyurova et al., 2018) and the reliability-based metric ($\Pr(|y_i - \hat{y}_i| < \varepsilon)$) (Ling et al., 2014).

4. RESULTS

4.1 Comparison of posterior predictive distributions

The posterior predictive distributions inform on how well the observed data is explained by the EnergyPlus model - through the EnergyPlus emulator η - and what sorts of discrepancies still remain between the observed data and the EnergyPlus model - through the discrepancy model δ. While a visual analysis is well suited for exploring discrepancies, s.a. phase shifts in the energy use day profile - see Figure 3 for the posterior predictive distribution of η -, the indicators displayed in Table 3 provide a more rigorous comparison as to which model explained the data best.

Table 3: Indicators computed on the posterior predictive distributions of the emulator for both models, SP and DFP.

	Q^2	CVRMSE	NMBE	$EFC_{90\%}$	$\Pr(\lvert y_i - \hat{y}_i\rvert < \varepsilon)$
SP	0.93	0.32	0.02	**0.90**	0.77
DFP	**0.97**	**0.22**	**0.01**	0.91	**0.91**

From Table 3, it seems that the DFP setting is better suited to explain the observed data in almost every respect, and is hence likely to be of better use in an AFDD context. Indeed, indicators such as Q^2, CVRMSE and NBME appear to show better predictive power for the DFP setting. Furthermore, they abide by ASHRAE guildeline 14 about calibrated models, contrary to the SP setting. Besides, the reliability-based metric seems to go the same route and provide a better prediction interval at fixed ε. However, it should be noted that the SP setting provides slightly better uncertainty coverage as the $EFC_{90\%}$ equals its target value while the DFP setting seems to slightly overestimate the uncertainty.

[6] As in Higdon et al. (2008b), the discrepancy is built as a linear combination of 5 radial basis functions that smoothly adjust the emulator response.
[7] Which is needed when considering a GMM prior on the functional variables.

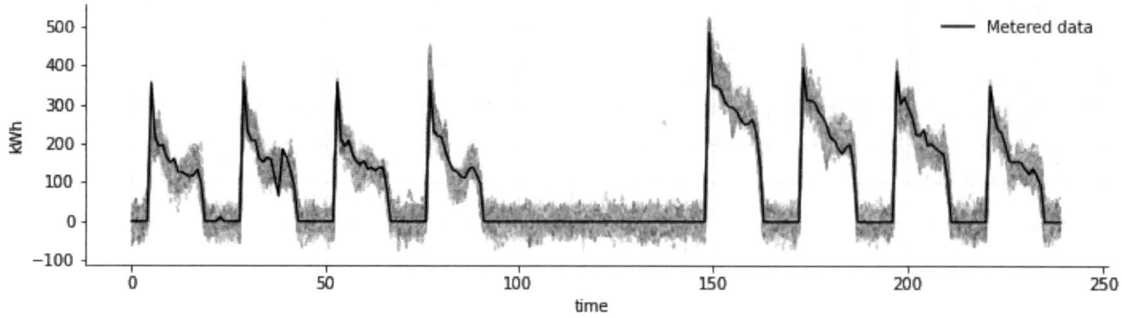

(a) $\eta + \varepsilon$, DFP

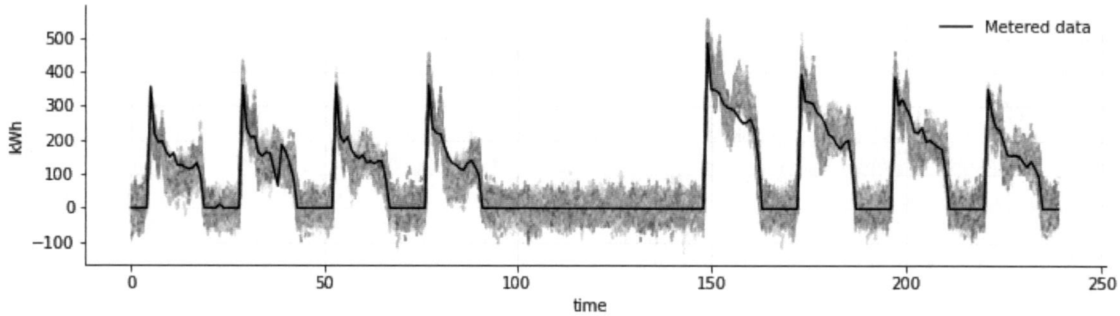

(b) $\eta + \varepsilon$, SP

Figure 3: Posterior predictive distributions of the emulator for both SP (3b) and DFP (3a) models. For the reliability-based metric, ε is arbitrarily taken equal to 50 kWh.

4.2 Comparison of posterior distributions

9 model parameters were selected using Sobol's total-order indices, and were calibrated against the building real heating energy use. Table 4 displays the posterior distributions - mean and standard deviation - for the static parameters (that are common to both settings). Mean estimates, along with calibrated high density intervals (HDI), seems robust since they are quite similar between both models.

Table 4: Posterior distributions for static variables, for both models SP and DFP

	DFP				SP			
	mean	sd	$q_{3\%}$	$q_{97\%}$	mean	sd	$q_{3\%}$	$q_{97\%}$
h_e	-0.050	0.010	-0.069	-0.031	0.255	0.344	-0.117	0.825
h_s	0.066	0.056	-0.011	0.168	0.154	0.139	-0.038	0.419
η	0.724	0.013	0.700	0.745	0.717	0.013	0.700	0.739
$\upsilon_{natvent}$	1.465	0.036	1.399	1.500	1.460	0.042	1.379	1.500
υ_{occ}	0.678	0.188	0.358	0.982	0.446	0.102	0.350	0.619
t_c	1.195	0.104	1.018	1.403	1.151	0.125	0.949	1.404
t_s	-0.054	0.055	-0.156	-0.000	-0.084	0.077	-0.228	-0.000
υ_{vent}	0.620	0.049	0.529	0.698	0.516	0.097	0.352	0.670

It is worthwhile noting that sometimes parameters HDI overlap, e.g. for parameter t_c, sometimes it seems to be shifted, e.g. for parameter h_e. When they do overlap, most of the time the DFP setting reduces the uncertainty on the said parameter, as can be expected since the functional variables add a degree of freedom per day for the emulator to adjust its response to observed data. In all cases, parameters HDI are not too different between both models.

4.3 Uncertainty reduction on functional variables

It is also worthwhile noting how the functional variables prior is updated with the metered data. Figure 4 displays the resulting posterior distribution after calibration along with the standardized profile from the simulation model, used in the SP model. While the prior distribution spans multiple kinds of profiles - half-days, full days, and holidays - the posterior distribution focuses much more on one type of profile, sometimes two, showing potential change in building occupancy from what would be expected. What is even more interesting is that the posterior seems to capture the fact that, during the calibration period, *Lycée Brequigny* was largely under-occupied due to the pandemic. However, while the school was under-occupied globally at this period, no record of individual days was kept, so it is difficult to assess whether or not the DFP accurately recovers the occupancy profiles.

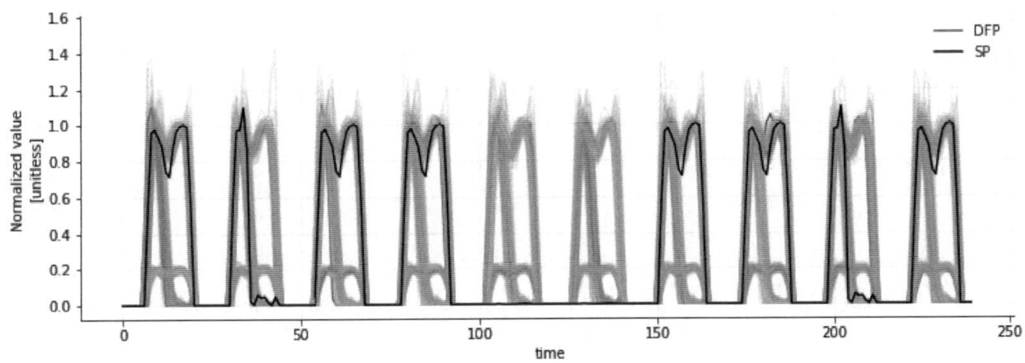

(a) Functional prior for DFP model and SP values

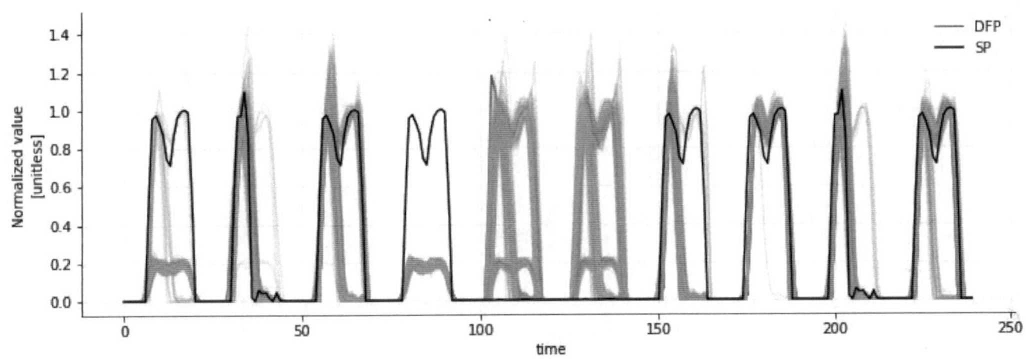

(b) Functional posterior for DFP model and SP values

Figure 4: Functional prior (4a) and posterior (4b) for standardized occupancy functional variable. In black are the base values from the EnergyPlus model, that are taken in the SP model.

5. CONCLUSIONS & PERSPECTIVES

Our study introduces an extension of the usual calibration process exposed in (Kennedy and O'Hagan, 2001; Higdon et al., 2008a), that accounts for dynamic states of the building throughout the calibration period. The methodology is competitive with the current state-of-the-art, and seems to possess superior flexibility in fitting and explaining the observed data.

However, the methodology needs to be tested against a larger set of both real life and simulated experiments to be completely validated, in terms of parameter estimation and predictability. Besides, since the calibration period exhibits only weak variability on the functional variables, little improvement is shown on the case study. We expect our methodology to be much more valuable in cases where, for instance, occupancy would change more abruptly. Eventually, testing this methodology for continuous monitoring and diagnosis in an online setting seems promising to update our knowledge of how the building is operated, and help diagnose faults at building level.

NOMENCLATURE

Built environment

AFDD	Automated Fault Detection and Diagnosis
AMI	Advanced Meter Infrastructure
BEM	Building Energy Model
BMS	Building Management System
EPC	Energy Performance Contract
HVAC	Heating, Ventilation and Air Conditioning

Computer model analysis

DoE	Design of Experiment
GSA	Global Sensitivity Analysis
LHD	Latin Hypercube Design

Math terms

DFP	Dynamic and Functional Prior calibration setting
GMM	Gaussian Mixture Model
PPL	Probabilistic Programming Languages
SP	Static Prior calibration setting
SPCA	Simultaneous functional Principal Component Analysis

Symbols

δ	Discrepancy model
η	Emulator
θ	Unknown and calibrated variables
D	Known variables, s.a. dry bulb temperature
t	Time index

REFERENCES

Auffray, Y., Barbillon, P., and Marin, J.-M. (2012). Maximin design on non hypercube domains and kernel interpolation. *Statistics and Computing*, 22(3):703–712.

Bayarri, M. J., Berger, J. O., and Liu, F. (2009). Modularization in bayesian analysis, with emphasis on analysis of computer models. *Bayesian Analysis*, 4(1).

Bayarri, M. J., Berger, J. O., Paulo, R., Sacks, J., Cafeo, J. A., Cavendish, J., Lin, C.-H., and Tu, J. (2007). A framework for validation of computer models. *Technometrics*, 49(2):138–154.

Bingham, E., Chen, J. P., Jankowiak, M., Obermeyer, F., Pradhan, N., Karaletsos, T., Singh, R., Szerlip, P., Horsfall, P., and Goodman, N. D. (2018). Pyro: Deep universal probabilistic programming.

Chong, A., Lam, K. P., Pozzi, M., and Yang, J. (2017). Bayesian calibration of building energy models with large datasets. *Energy and Buildings*, 154:343 – 355.

EnergyPlus (2019). *EnergyPlus Software, Version 9.1.0.*

Frank, S., Heaney, M., Jin, X., Robertson, J., Cheung, H., Elmore, R., and Henze, G. (2016). Hybrid model-based and data-driven fault detection and diagnostics for commercial buildings.

Frank, S., Lin, G., Jin, X., Singla, R., Farthing, A., and Granderson, J. (2019). A performance evaluation framework for building fault detection and diagnosis algorithms. *Energy and Buildings*, 192:84–92.

Gamboa, F., Janon, A., Klein, T., and Lagnoux, A. (2014). Sensitivity analysis for multidimensional and functional outputs. *Electronic Journal of Statistics*, 8(1):575–603.

Granderson, J., Touzani, S., Custodio, C., Sohn, M. D., Jump, D., and Fernandes, S. (2016). Accuracy of automated measurement and verification (m&v) techniques for energy savings in commercial buildings. *Applied Energy*, 173:296 – 308.

Hensman, J., Fusi, N., and Lawrence, N. D. (2013). Gaussian processes for big data.

Heo, Y., Choudhary, R., and Augenbroe, G. (2012). Calibration of building energy models for retrofit analysis under uncertainty. *Energy and Buildings*, 47:550 – 560.

Higdon, D., Gattiker, J., Williams, B., and Rightley, M. (2008a). Computer model calibration using high-dimensional output. *Journal of the American Statistical Association*, 103(482):570–583. cited By 356.

Higdon, D., Nakhleh, C., Gattiker, J., and Williams, B. (2008b). A bayesian calibration approach to the thermal problem. *Computer Methods in Applied Mechanics and Engineering*, 197(29-32):2431–2441.

Iooss, B. and Lemaître, P. (2015). *A Review on Global Sensitivity Analysis Methods*, pages 101–122. Springer US, Boston, MA.

Janon, A., Klein, T., Lagnoux, A., Nodet, M., and Prieur, C. (2014). Asymptotic normality and efficiency of two sobol index estimators. *ESAIM: Probability and Statistics*, 18:342–364.

Jin, R., Chen, W., and Sudjianto, A. (2005). An efficient algorithm for constructing optimal design of computer experiments. *Journal of Statistical Planning and Inference*, 134(1):268 – 287.

Jradi, M., Arendt, K., Sangogboye, F., Mattera, C., Markoska, E., Kjærgaard, M., Veje, C., and Jørgensen, B. (2018). Obepme: An online building energy performance monitoring and evaluation tool to reduce energy performance gaps. *Energy and Buildings*, 166:196 – 209.

Kennedy, M. C. and O'Hagan, A. (2001). Bayesian calibration of computer models. *Journal of the Royal Statistical Society: Series B (Statistical Methodology)*, 63(3):425–464.

Kyzyurova, K. N., Berger, J. O., and Wolpert, R. L. (2018). Coupling computer models through linking their statistical emulators. *SIAM/ASA Journal on Uncertainty Quantification*, 6(3):1151–1171.

Ling, Y., Mullins, J., and Mahadevan, S. (2014). Selection of model discrepancy priors in bayesian calibration. *Journal of Computational Physics*, 276:665 – 680.

Monari, F. (2016). *Sensitivity Analysis and Bayesian Calibration of Building Energy Models*. PhD thesis, University of Strathclyde.

Morris, M. D. and Mitchell, T. J. (1995). Exploratory designs for computational experiments. *Journal of Statistical Planning and Inference*, 43(3):381 – 402.

Muehlenstaedt, T., Fruth, J., and Roustant, O. (2016). Computer experiments with functional inputs and scalar outputs by a norm-based approach. *Statistics and Computing*, 27(4):1083–1097.

Nanty, S. (2015). *Stochastic methods for uncertainty treatment of functional variables in computer codes : application to safety studies*. Thesis, Université Grenoble Alpes.

Nanty, S., Helbert, C., Marrel, A., Pérot, N., and Prieur, C. (2016). Uncertainty quantification for functional dependent random variables. *Computational Statistics*, 32(2):559–583.

Ramsay, J., Silverman, B., Media, S. S., and Silverman, H. (2005). *Functional Data Analysis*. Springer Series in Statistics. Springer.

Rasmussen, C. E. and Williams, C. K. I. (2005). *Gaussian Processes for Machine Learning*. MIT Press Ltd.

Saltelli, A., Ratto, M., Andres, T., Campolongo, F., Cariboni, J., Gatelli, D., Saisana, M., and Tarantola, S. (2008). *Global Sensitivity Analysis. The Primer*. Johyn Wiley & Sons, Ltd.

Titsias, M. (2009). Variational learning of inducing variables in sparse gaussian processes. In van Dyk, D. and Welling, M., editors, *Proceedings of the Twelth International Conference on Artificial Intelligence and Statistics*, volume 5 of *Proceedings of Machine Learning Research*, pages 567–574, Hilton Clearwater Beach Resort, Clearwater Beach, Florida USA. PMLR.

Touzani, S., Ravache, B., Crowe, E., and Granderson, J. (2019). Statistical change detection of building energy consumption: Applications to savings estimation. *Energy and Buildings*, 185:123 – 136.

Generalized Disjunctive Programming-based, Mixed Integer Linear MPC Formulation for Optimal Operation of a District Energy System for PV Self-consumption and Grid Decarbonization: Field Implementation

Donghun Kim[1*], Tianzhen Hong[1], Mary Ann Piette[1]

[1] Building Technology & Urban Systems Division,
Lawrence Berkeley National Laboratory, Berkeley, California, United States of America
Contact Information: donghunkim@lbl.gov

* Corresponding Author

ABSTRACT

Thermal energy storage (TES) for a cooling plant serving a campus or district of buildings can shift cooling energy from peak to non-peak hours so that it is able to reduce electrical operation costs in buildings and increase stability and operational efficiency in the grid. Traditionally, heuristic control approaches, such as the storage priority control, have been widely used in practice. Those simple control approaches have shown reasonable or even near-optimal performance in the past. However, the increasing penetration of renewables changes the situation: exposing the electrical grid to the growing supply demand challenge which encourages greater consumption at times when renewable energy is plentiful and lesser when dirtier supply is used. This paper presents a model predictive control (MPC) strategy, site demonstration and evaluation results for optimal operation of a chiller plant, TES and behind-meter photovoltaics for a campus-level district cooling system. The MPC was formulated as a mixed-integer linear program using the Generalized Disjunctive Programming for better numerical and control properties. Results show that the MPC can effectively utilize PV generated electricity to drive chillers for charging the TES during the day and avoid sending excess electricity back to the grid. Compared with the storage priority control, the MPC reduces the excess PV power by around 25%, the greenhouse gas emission by 10%, and peak electricity demand by 10%.

1. INTRODUCTION

Traditional control strategies for TES cooling plants, i.e., the storage priority or chiller priority controls, fully charge TES using chillers during OFF peak price period (typically a night time period) while discharging TES to meet cooling load partially or fully during ON peak price period (typically a day time period). Although the rule-based controls have shown near-optimal behavior and provided benefits to facilities, buildings, and the grid in the past by lowering the total electrical load during peak price period and hence requiring less spinning/non-spinning reserves and other back-up power plants in the grid, the rapid penetration of renewable energy resources is changing the situation. The intermittent nature of behind-meter renewable energy sources and the grid's movement toward more dynamic price signals challenge building or facility operators to rely on the rule-based operations.

Despite many studies of optimal operation of TES with behind-meter renewable and energy storages, a majority of them are using simulations as shown in the table, and therefore real performance of MPC including self-consumption ratio are not clear. Although there are some papers with experimental assessment of MPCs coordinating TES and HVAC systems, they are limited to small-scale systems, e.g., laboratory or small-sized building (e.g., residential) levels. For large-scale systems, there are very few papers that demonstrate an MPC for renewable energy integration and present site performance. In addition, there is a lack of knowledge about applying MPC to mitigate CO2 emissions (Tarragona et al., 2021).

This paper fills the gaps by implementing an MPC for a campus-level cooling TES plant and presenting on-site performance compared with carefully selected, historical rule-based operation data for the plant. The MPC aims at promoting the self-consumption of the on-site renewable and minimizing CO2 in the grid. Because utility cost reduction is one of the key motivations for the facility, the MPC is also formulated to reduce peak demand. In this paper, we present a novel MPC formulation (Mixed Integer Linear form using the Generalized Disjunctive Programming) and the site performance evaluation.

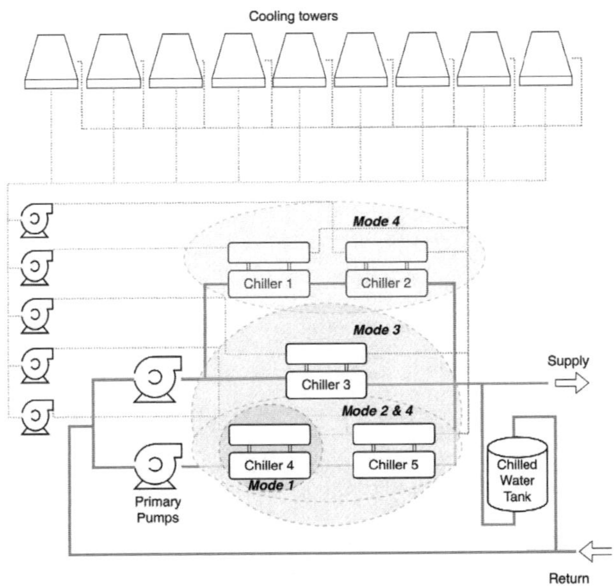

Figure 1: The equipment configuration of the central cooling plant at University of California, Merced

2. SITE DESCRIPTION AND BASELINE CONTROL STRATEGY

2.1 Site description

The central plant of the University of California, Merced is the demonstration site located in California, U.S. The cooling TES plant, shown in Fig. 1, has five chillers with a total cooling capacity of 17.6 MW, 5000 ton of refrigeration, serving around 40 buildings. It has the primary-secondary (decoupled) chilled water loop configuration, where separate variable speed pumps are dedicated to the primary and secondary water loops. The chillers in the primary water loop, where each chiller has rated cooling capacity from around 1,000 to 1,500 ton, are high-efficiency units with centrifugal compressors and internal capacity controls using variable speed compressor, inlet guide vane and diffuser modulations, and hot-gas bypass for part-load operations. The chiller configuration is a combination of series and parallel connections. Depending on a selection of plant modes, the plant could operate a single chiller, two serially connected chillers, three chillers (two series chillers and one parallel chiller), and four chillers (two parallel chiller groups of chillers where each group has two series chillers). The variable frequency drives for the secondary water pumps adjust the pump speed to regulate the differential pressure between the supply and return pipes to transport the chilled water to multiple building load terminals.

A two million gallon chilled water tank (sensible energy storage) is located at the bypass line (or decoupler) between the primary and secondary water loops. The tank is a stratified cool storage where warmer and less denser water floats on the top of colder and denser water. There are no isolation valves on the bypass line, and therefore the mismatch between primary water flow rate and secondary flow rate determines the charging or discharging flow rate. In order not to disturb the thermocline in the water tank, the chilled water supply temperature setpoint should be maintained at around 39^oF (around 4^oC). This constraint on the chiller leaving temperature together with no-isolation valves on the bypass line results in that the primary water flow rate setpoint as the only decision variable that defines charging and discharging heat rate when chillers are running.

In addition, the UC-Merced campus has more than 4 MW of on-site solar farm which covers 8.5 acres southeast of the campus and an approximately 960-space parking lot. The PV has a single-axis sun tracking system, which captures up to 30% more sunlight than conventional fixed-tilt systems by following the sun throughout the day.

It is important to mention the utility tariff structure which is one of the key driving forces for MPC. Because of the scale of energy consumption, the energy price rate is much more complex than a typical TOU (time of use) structure. It purchases energy from selected energy service providers within the energy wholesale market through a negotiation process. This program makes 1) the time variation of the energy price rate small, and 2) the demand charge is much

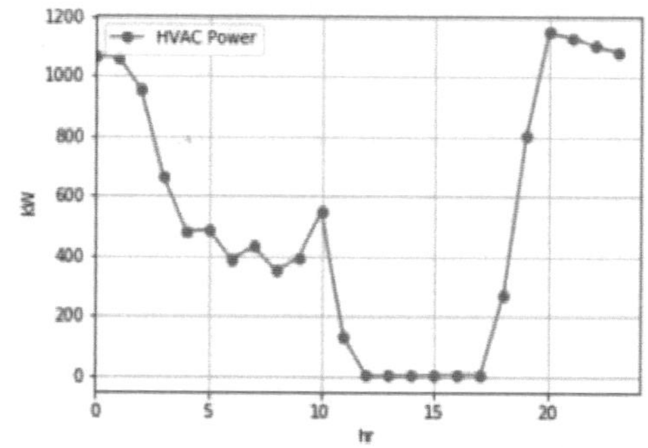

Figure 2: Daily-averaged plant power at the UC-Merced

more dominant than the energy charge at the site.

In this study, the optimization variables to the system are the primary volume flow rate setpoint (V_p) and the plant operation mode (s_j) as depicted in Fig. 1. Note that s_j does not indicate the ON/OFF status of a chiller. Instead, the model activates and deactivates some chiller(s), pumps(s), cooling towers(s) and isolation valve(s), pre-defined by the existing control sequence.

2.2 Description of baseline control

The basic operation strategy for the cooling TES plant is the storage priority control that runs chiller(s) during the nighttime to store cooling energy in the chilled water tank and uses the cooling energy during the day time to fully meet the daytime campus cooling load. Fig. 2 shows a representative plant electrical power profile of the chiller plant operation. To generate this plot, the daily profiles of the total plant electrical power consumption, which aggregates chiller compressor powers, cooling tower fan powers, pump powers for the primary and condensor water pumps, over the month of August were collected and averaged. Note that the plant power is zero from 12:00 to 17:00 hour. This means the chilled water tank completely serves the campus load, which clearly characterizes the control and system as the storage priority control and full storage system. It is important to remember that the storage priority control in general does not account for behind-meter, on-site renewable energy.

3. MIXED INTEGER LINEAR MPC FORMULATION

3.1 Objective function

The control objective for the chiller plant is to minimize utility cost and CO2 emission in the grid region (California Independent System Operator, CAISO), and self-consume the behind meter PV generation.

$$min \sum_{k=0}^{N_p-1} E[k] \times max\{P_{net}[k], 0\} + \omega_d \times d, \tag{1}$$

where E represents a time varying grid signal which could be utility price rate, a carbon emission rate, or a weighted combination of them. P_{net} is the net power consumption of the entire campus measured at the grid connection point which is the total power consumption subtracted by the on-site PV generation, i.e.,

$$P_{net} = P_{plant} + P_{nonplant} - P_{PV}. \tag{2}$$

When P_{net} is positive, the campus imports the power from the grid while the campus exports the power to the grid when negative.

The first term represents time-integrated GHG emission [MTCO2eq], energy cost [$] or a combination of them over the prediction horizon (N_p) depending on the definition of E. Note that we integrate $max\{P_{net}[k], 0\}$ rather than P_{net}. This is to promote the self-consumption of on-site renewable energy by not incentivising power export. ω_d is a weight on the max peak power (d)[1] over the prediction horizon.

[1]Strictly speaking, d is an upper bound of the max peak power over the prediction horizon.

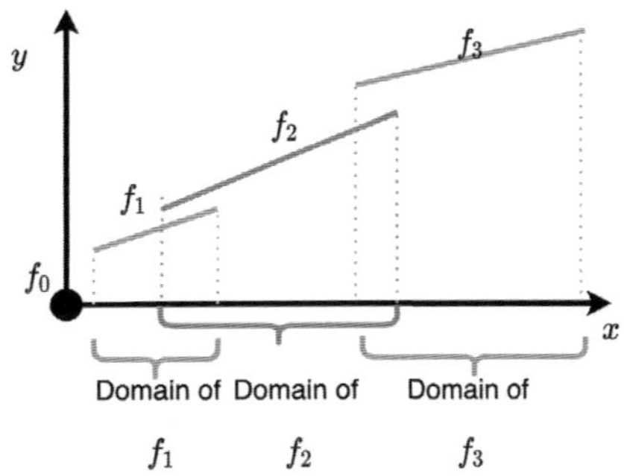

Figure 3: Conceptual plot for a switching system. A feasible set in (x, y)-space switches depending on a mode selection.

3.2 Modeling total plant power as a switching system

Because the plant mode determines a different set of chillers, cooling towers, pumps, and isolation valves, the plant power and COP behaves differently depending on plant modes. Due to this nature, one has to model the total plant power as a switching system regardless of modeling approaches (i.e., white-box, gray-box or black-box models).

We chose the following model structure for the total plant power modeling.

$$P_{plant} = \begin{cases} 0 & \text{for the plant mode 0} \\ a_1 Q_{CH} + b_1(T_{WB}) & \text{for the plant mode 1} \\ \vdots & \\ a_J Q_{CH} + b_J(T_{WB}) & \text{for the plant mode J} \end{cases} \tag{3}$$

where Q_{CH}, T_{WB} are the total chiller load (or the sum of chiller loads), and the outdoor wet-bulb temperature, respectively. a_j is a constant and b_j is a function of T_{WB}, respectively. There is no restriction on the model structure for b_j but the linear affine form, i.e., $b_j = b_{j,0} + b_{j,1} T_{WB}$, is selected in this paper. Since T_{WB} is not an optimization variable, we denote $b_j(T_{WB})$ as b_j for the simplicity of notations.

The technical challenge of a switching system for mathematical programming, in general, is that some constraints appear or disappear depending on a selection of modes because of the IF-THEN nature. For our plant power model of Eqn. (3), the feasible set of (P_{plant}, Q_{CH}) switches via a mode as depicted in Fig. 3. A straightforward approach is to model it as $P_{net} = \sum_j (a_j Q_{CH} + b_j) \times s_j$, but it results in a mixed integer nonlinear programming problem which suffers from the lack of convergence, global optimality (Wolsey, 1998; Elsido et al., 2017), computational efficiency, and MPC closed-loop stability. To resolve this issue, we utilized the Generalized Disjunctive Programming approach as described in the next section.

3.3 Treatment of Switching Plant Power Model using Generalized Disjunctive Programming Approach

3.3.1 Recap of Generalized Disjunctive Programming: The Generalized Disjunctive Programming (GDP) is an alternative mathematical programming approach to the mixed integer programming. It was first introduced by Raman and Grossmann (1994) who extended the Disjunctive Programming (Balas, 1985). Unlike typical programming approaches which only allow algebraic expressions, GDP allows formulating optimization problems with logical expressions. The logical constraints consist of disjunctions and logic propositions. Once an optimization problem is formulated with the logical expressions, GDP provides methods of transforming the disjunctions into mixed-integer (nonlinear/linear) equality and/or inequality constraints, and the logic propositions into integer-only equality and/or inequality constraints. Then, the converted mixed-integer (nonlinear/linear) program is solved by available mixed-integer programming solvers.

There are several representations of a mathematical structure of GDP in the literature, but we introduce the form introduced in (Grossmann and Ruiz, 2012).

$$\min f(x) + \sum_{k=1}^{K} c_k \quad \text{Objective Function} \tag{4}$$

$$g(x) \leq 0 \quad \text{Algebraic Constraints} \tag{5}$$

$$\bigvee_{i \in D_k} \begin{bmatrix} Y_{i,k} \\ A_{i,k} x_k \leq b_{ik} \\ c_k = \gamma_{ik} \end{bmatrix} \quad \text{Disjunctions} \tag{6}$$

$$\Omega(Y) = True \quad \text{Propositions} \tag{7}$$

$$L \leq x \leq U, x \in \mathbb{R}^n, c_k \in \mathbb{R}^1 \quad \text{Continuous Variables} \tag{8}$$

$$Y_{i,k} \in \{True, False\} \quad \text{Boolean Variables} \tag{9}$$

- $f(: \mathbb{R}^n \rightarrow \mathbb{R}^1)$, $g(: \mathbb{R}^n \rightarrow \mathbb{R}^m)$ are a function and algebraic constraints of continuous variables x
- K represents the number of disjunctions
- Y_{ik} is a Boolean variable that controls the feasible space where the global continuous variable, x, lies (represented by the inequality constraint) and the cost represented by the equality constraint in Eqn. (6). That is, when $Y_{i,k} = True$, the constraints of $A_{i,k} x_k \leq b_{ik}, c_k = \gamma_{ik}$ apply. Otherwise, they are ignored.
- D_k is a set of integers associated with the k-disjunction
- Ω is a propositional logic involving only the Boolean variables

Note Y_{ik} is controls the feasible space depending on the value of Y_{ik}. This is exactly the problem for a switching system discussed in Section 3.2.

The conversion of the above disjunctions and logic propositions to algebraic expressions can be done systematically by utilizing either the Big-M method or the Convex-Hull approach. The latter approach finds the convex hull (that is the minimal convex envelope) of the disjunctive sets, and hence provides a tighter convex envelope than that of the Big-M approach. In addition, it does not require tuning parameters as opposed to the Big-M method. The Convex-Hull approach leads to the following mixed-integer program.

$$\min f(x) + \sum_{k=1}^{K} c_k \tag{10}$$

$$g(x) \leq 0 \tag{11}$$

$$\sum_{i \in D_k} y_{i,k} = 1 \quad k \in \{1, \cdots, K\} \tag{12}$$

$$A_{i,k} v_{i,k} \leq b_{i,k} y_{i,k} \quad k \in \{1, \cdots, K\}, i \in D_k \tag{13}$$

$$x_k = \sum_{i \in D_k} v_{i,k} \quad k \in \{1, \cdots, K\} \tag{14}$$

$$c_k = \sum_{i \in D_k} \gamma_{i,k} y_{i,k} \quad k \in \{1, \cdots, K\} \tag{15}$$

$$0 \leq v_{i,k} \leq U_v y_{i,k} \quad k \in \{1, \cdots, K\}, i \in D_k \tag{16}$$

$$Ay \geq a \tag{17}$$

$$L \leq x \leq U, x \in \mathbb{R}^n, c_k \in \mathbb{R}^1 \tag{18}$$

$$y_{i,k} \in \{0, 1\} \tag{19}$$

where $y_{i,k}$ is the binary variable that corresponds to $Y_{i,k}$, $v_{i,k}$ is an axially variable and U_v is an upper bound for $v_{i,k}$. Please refer to Grossmann and Ruiz (2012) for detailed discussions.

3.3.2 Application of Generalized Disjunctive Programming: According to the GDP modeling approach, Eqn. (3) can be expressed as follows.

$$\begin{bmatrix} S_0 \\ Q_{CH}=0 \\ P_{plant}=0 \end{bmatrix} \vee \begin{bmatrix} S_1 \\ Q_{min,1} \le Q_{CH} \le Q_{max,1} \\ P_{plant} = a_1 Q_{CH} + b_1 \end{bmatrix} \vee \cdots \vee \begin{bmatrix} S_J \\ Q_{min,J} \le Q_{CH} \le Q_{max,J} \\ P_{plant} = a_J Q_{CH} + b_J \end{bmatrix}, \tag{20}$$

where S_j is the Boolean variable indicating an active plant mode. That is, when $S_j = True$, the jth constraints become active, and vice versa.

Applying the Convex-Hull approach and eliminating the redundancy associated with the 0th mode result in the following desired, mixed-integer constraints.

$$\sum_{j=1}^{J} s_j \le 1 \tag{21}$$

$$s_j Q_{min,j} \le v_j \le Q_{max,j} s_j, \quad j \in \{1, \cdots, J\} \tag{22}$$

$$\mu_j = a_j v_j + b_j s_j, \quad j \in \{1, \cdots, J\} \tag{23}$$

$$Q_{CH} = \sum_{j=1}^{J} v_j \tag{24}$$

$$P_{plant} = \sum_{j=1}^{J} \mu_j \tag{25}$$

where $v_1, \cdots, v_J$ are the auxiliary variables for Q_{CH}, and $\mu_1, \cdots, \mu_J$ are the auxiliary variables for P_{plant}.

Note that, when any k ($\in \{1, \cdots, J\}$) is selected, then $s_{j,j \neq k} = 0$ and $v_{j,j \neq k} = 0$ from Eqns. (22) and (23), and therefore $Q_{CH} = v_k$, $Q_{min,k} \le v_k \le Q_{max,k}$, and $P_{plant} = a_k Q_{CH} + b_k$ which are the desired algebraic constraints. When none of modes is selected, i.e., $s_1 = 0, \cdots, s_J = 0$, $Q_{CH} = 0, P_{plant} = 0$. Note also that all equality and inequality constraints are linear in v_j, μ_j, s_j, Q_{CH} and P_{plant}.

3.4 Start-up and shut-down constraint

It is well known that frequent mode switching causes negative impacts on the life of equipment and part-load performance. An intuitive approach is to add a penalty whenever the mode changes. However, it was not trivial to tune the weights, and it is not applicable to our case since one has to shut down the chiller whenever the storage is fully charged for example. To resolve this, we adopted the mathematical programming approach from the Unit Commitment problem (Rajan et al., 2005; Knueven et al., 2020) in the electrical power system. This method enforces minimal ON or OFF times for each unit operation: that is, once a power generation unit is activated (deactivated), it has to maintain the ON (OFF) status at least for a predefined minimal ON (OFF) time period. Application to our system is described as follows.

For the jth mode and a time-step k, let $\delta_{ON,j}[k]$ ($\in \{0,1\}$) and $\delta_{OFF,j}[k]$ ($\in \{0,1\}$) be turn-ON and turn-OFF variables, respectively. The relationship between those variables is

$$s_j[k] - s_{j-1}[k] = \delta_{ON,j}[k] - \delta_{OFF,j}[k] \ (k \in \{0, \cdots, N_p - 1\}, j \in \{1, \cdots, J\}). \tag{26}$$

In addition, we impose the following constraint to prevent the situation where turning ON and OFF occur simultaneously.

$$\delta_{ON,j}[k] + \delta_{OFF,j}[k] \le 1 \tag{27}$$

Let OT_j be the minimal run-time for the j mode. The formulations to incorporate the minimal run-time in the MPC formulation are as follows.

1) For $k \geq OT_j - 1$,

$$\sum_{l=0}^{OT_j-1} \delta_{ON,j}[k-l] \leq s_j[k] \tag{28}$$

$$\sum_{l=0}^{OT_j-1} \delta_{OFF,j}[k-l] \leq 1 - s_j[k]. \tag{29}$$

2) For $k < OT_j - 1$,

$$\sum_{l=1}^{OT_j-k-1} \delta_{ON,j}^{history}[-l] + \sum_{l=0}^{k} \delta_{ON,j}[k-l] \leq s_j[k] \tag{30}$$

$$\sum_{l=0}^{OT_j-k-1} \delta_{OFF,j}^{history}[-l] + \sum_{l=0}^{k} \delta_{OFF,j}[k-l] \leq 1 - s_j[k]. \tag{31}$$

3.5 Reformulation of the objective function

Remember that we tend not to incentivize the negative net power by using the max function in Eqn. (1) in order to promote the self-consumption of the on-site PV. Technically, the term $\sum_{k=0}^{N_p-1} E[k] \times max\{P_{net}[k], 0\}$ constitutes a mixed integer convex program. It would be greatly useful to have an equivalent mixed integer linear program, if possible, for the use of fast and reliable solvers (e.g., CPLEX, GLPK, Gurobi, LPSOLVE) and for better scalability to the increase of the number of variables.

Like the epi-graph formulation (Boyd et al., 2004), we may define an upper bound of $max\{P_{net}[k], 0\}$ for each k, put inequality constraints to replace each max term with the upper bound, and minimize the upper bounds. This approach was strictly proved (the proof is not omitted in this paper due to the page limit). The conclusive, mixed integer linear formulation for Eqn. (1) is;

$$min \quad \sum_{k=0}^{N_p-1} \left(E[k] \times P_{net}^{+}[k]\right) + \omega_d \times d \tag{32}$$

$$P_{plant}[k] + P_{nonplant}[k] - P_{PV}[k] \leq P_{net}^{+}[k] \ (k \in \{0, \cdots, N_p - 1\}) \tag{33}$$

$$0 \leq P_{net}^{+} \ (k \in \{0, \cdots, N_p - 1\}) \tag{34}$$

where $P_{net}^{+}[\cdot]$ are auxiliary variables.

4. SITE PERFORMANCE OF MPC

The presented MPC has been implemented for the campus cooling TES plant for several test periods, and sample test results for a week in May 2021 are shown and discussed in this section. We start describing our approach to evaluate performance of MPC compared with the baseline control described in Section 2.2.

4.1 Selection of baseline control period for actual performance assessment

To make fair comparisons between two controllers and evaluate actual performance for the real site, a baseline period was carefully selected considering the followings: 1) the baseline control was implemented, 2) uncontrollable disturbances including campus cooling load, MOER signal and on-site solar generation were consistent to that of the MPC period, and 3) initial and final SOC were close enough.

4.2 MPC performance assessment

4.2.1 Comparisons of closed-loop responses between baseline control and MPC: Fig. 4 compares the closed-loop responses of the baseline control and the MPC. The first sub-figure shows the campus net power consumptions, i.e., total consumption subtracted by the on-site PV generation, and the second sub-figure shows the profiles of the state of charge (measured from the relative height of the thermocline) for the two control periods. The blue and red colors are dedicated to the baseline control and MPC, respectively. The shaded areas indicate when the MOER is higher than 0.3 [mTonCO2e/MWh] that is equal to the average level of GHG emission for a natural gas power plant in the U.S. according to the EPA's Emissions and Generation Resource Integrated Database (eGRID). In other words, the shaded

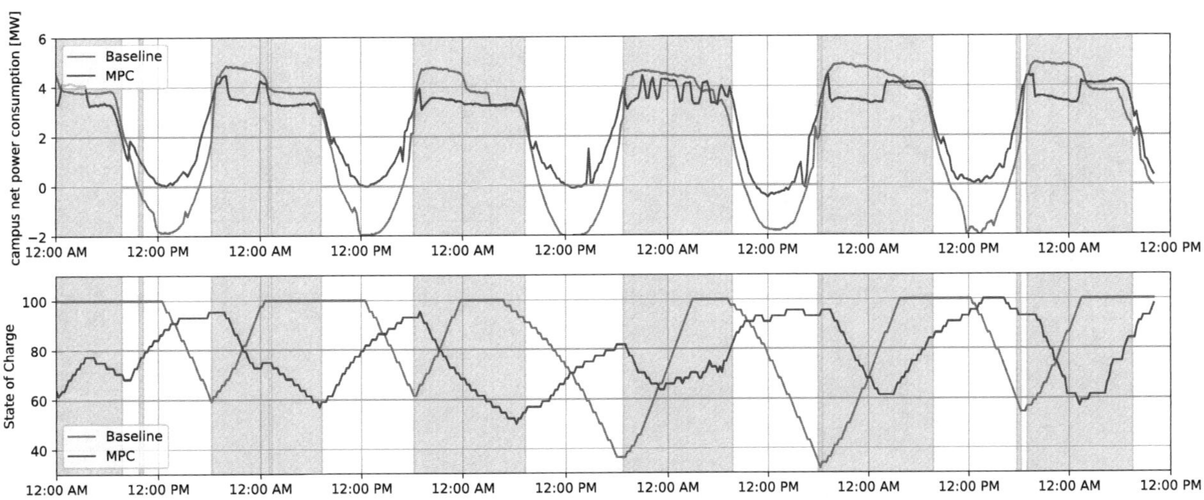

Figure 4: Comparisons of experimental net power consumption and state of charge between the baseline and MPC at UC-Merced (red: baseline, blue: MPC)

areas are when the grid is dirty, and non-shaded areas are when the grid is clean. First, look at the afternoon period for the first day where the state of the charge decreases from 100% to 60% for the baseline control. This means that the control utilizes the stored thermal energy to meet the entire campus cooling load during the daytime period. For the time being, the chillers were off, causing significant excessive power around 2 MW that was sent back to the grid as shown in the first subfigure. Compare the behavior of the state of charge with the MPC for the same period. The MPC actually increased the state of the charge which implies that the chillers were ON, met the entire campus cooling load and charged the storage. Since the MPC increased the chiller power during the period, the excessive power export was eliminated as shown in the first subfigure. Second, when the grid condition got dirtier, the baseline control increased the state of charge and this charging process continued until the TES was fully charged (see the evening time for the first day in the second subfigure). On the other hand, the MPC consumed less energy when the grid was dirty by leveraging the stored energy from the on-site renewable. This pattern repeats for the entire test period as shown in the figure.

It should be mentioned that facility operators requested to preserve 50% to 60% of the charge level during the MPC implementation period which significantly limits savings potentials. This can be confirmed in the second subfigure.

The baseline control which is independent of renewable consumed 23.7 MWh (daily average) out of 32.6 MWh (daily average) total solar generation, resulting in 72.8% self-consumption ratio. On the other hand, the MPC control consumed almost all on-site PV energy resulting in a 99.2% self-consumption ratio. In other words, the increment of the self-consumption ratio via MPC is 26.4%.

The comparison of GHG emission rates are shown in Fig. 5. One can see visible reduction in the emission rate. The daily average value of the carbon reductions using the MPC is 1 [mTonCO2e] per day. The unit metric ton CO2e is equivalent to the emission rate driving a car for around 2,500 miles or 4,023 km (roughly San Francisco, California to Washington, DC). Again, the reduction was achieved by coordinating operations of cooling TES plant with on-site PVs in response to the grid carbon emission signal. The calculated GHG savings is around 10% but it is anticipated that more than 20% reduction can be achievable for the plant once the minimum state of charge constraint (varied between 50% to 60%) can be relaxed.

The peak power reduction was around 10%. It is not that significant considering the size of the tank (2 million gallons of water). This is mainly because the facility operators requested to preserve 50% to 60% of the charge level during the MPC implementation period. This significantly limits the capability of a peak demand reduction and CO_2 reduction,

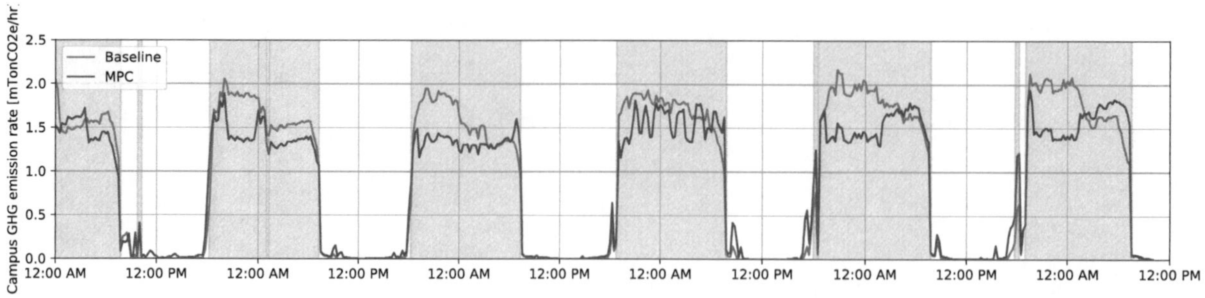

Figure 5: Comparisons of GHG emission rate between MPC and baseline storage-priority control at the UC-Merced

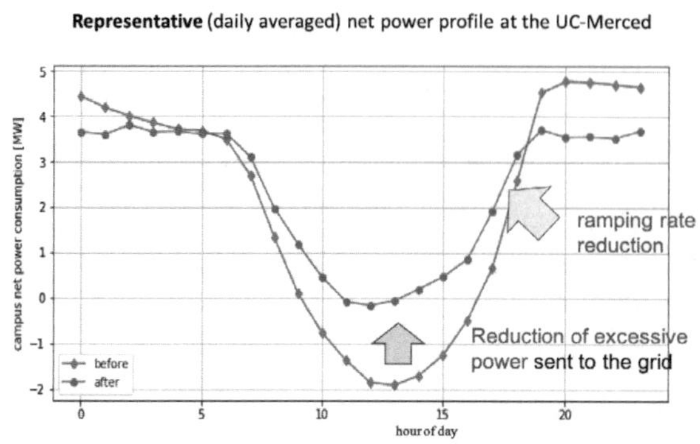

Figure 6: Comparisons of representative (daily averaged) net power profiles between the baseline control and MPC at the UC-Merced for the weeks of evaluation in May

because during the nighttime chillers have to run to maintain the minimum charge level. If the strong constraint has been relaxed, MPC would not need to operate chillers for the nighttime and would increase the savings. It should be mentioned that even with the conservative 10% peak reduction, it attracts the interest of facility operators because of the magnitude of power consumption for the campus-level cooling plant. For the UC-Merced case, the utility cost savings are around ten thousand dollars per month.

4.2.2 Benefits of MPC to the grid: Although the time series comparison provides comprehensive information, it is difficult to capture the overall behavior of the MPC. To more clearly characterize this behavior, we averaged daily power consumption profiles for the two controls and they are shown in Fig. 6. Like before, the blue and red lines are associated with the baseline control and MPC. It is interesting to see that the representative (daily averaged) profiles at the site look similar to the duck-curve which the grid is experiencing. This figure clearly shows that the MPC consumes and stores more energy during the daytime using the clean on-site solar, and consumes less during evening and night time when the grid gets dirty by leveraging the stored energy.

The smooth load curve could provide significant benefits to the grid, because 1) The higher consumption during the daytime addresses the issue of high renewable curtailment in CAISO, 2) The proper operation of the chilled water storage in response to the on-site renewable allows tolerating the increasing PV penetration to the grid, 3) The reduced ramping rate during the evening time addresses the issue of running gas turbines to follow the high ramping rate of the neck of the duck-curve as discussed in Introduction, and 4) The self-consumption of the behind-meter solar reduces the concern of voltage regulation in the distribution line and improves the grid reliability and efficiency.

5. CONCLUSIONS

Cooling TES plants are one of the oldest and reliable energy storage technologies. This paper shows that cooling plants with existing TES could be one of the most cost-effective resources to achieve state and government' carbon neutrality goals, since it does not require installing new energy storage but only requires changing the operation. We presented a novel, Generalized Disjunctive Programming-based MPC and its site performance for a campus-level cooling TES plant that coordinates operations of multiple chillers, chilled water tank and behind-meter PVs. It aims at self-consuming on-site 4 MW solar farms, lowering carbon emission at the grid and minimizing utility bills. The performance of MPC was assessed by comparing with a carefully selected baseline period where the storage priority control has been implemented. The MPC reduced the excess PV power by around 25%, the greenhouse gas emission by 10%, and peak demand by 10%. The savings estimates were conservative, since only a partial usage of the chilled water tank was allowed during the MPC implementation period. The capability of achieving utility peak demand in addition to lowering GHG emission is the uniqueness of the MPC. Future work will include a longer-term implementation of the MPC (> 1 year) to provide a more comprehensive savings assessment and reliability of the MPC technology.

REFERENCES

Balas, E. (1985). Disjunctive programming and a hierarchy of relaxations for discrete optimization problems. *SIAM Journal on Algebraic Discrete Methods*, 6(3):466–486.

Boyd, S., Boyd, S. P., and Vandenberghe, L. (2004). *Convex optimization*. Cambridge university press.

Elsido, C., Bischi, A., Silva, P., and Martelli, E. (2017). Two-stage minlp algorithm for the optimal synthesis and design of networks of chp units. *Energy*, 121:403–426.

Grossmann, I. E. and Ruiz, J. P. (2012). Generalized disjunctive programming: A framework for formulation and alternative algorithms for minlp optimization. In *Mixed Integer Nonlinear Programming*, pages 93–115. Springer.

Knueven, B., Ostrowski, J., and Watson, J.-P. (2020). A novel matching formulation for startup costs in unit commitment. *Mathematical Programming Computation*, 12(2):225–248.

Rajan, D., Takriti, S., et al. (2005). Minimum up/down polytopes of the unit commitment problem with start-up costs. *IBM Res. Rep*, 23628:1–14.

Raman, R. and Grossmann, I. E. (1994). Modelling and computational techniques for logic based integer programming. *Computers & Chemical Engineering*, 18(7):563–578.

Tarragona, J., Pisello, A. L., Fernández, C., de Gracia, A., and Cabeza, L. F. (2021). Systematic review on model predictive control strategies applied to active thermal energy storage systems. *Renewable and Sustainable Energy Reviews*, 149:111385.

Wolsey, L. A. (1998). Integer programming john wiley & sons. *New York, NY*, 4.

ACKNOWLEDGEMENT

This work was supported by the Assistant Secretary for Energy Efficiency and Renewable Energy, Building Technologies Office and the U.S. China Clean Energy Research Center, Building Energy Efficiency (CERC-BEE) program, of the U.S. Department of Energy under Contract No. DE-AC02-05CH11231. The authors would like to thank Tim Olson, Michael Pearson, Robert Stefanski, Danny Ward, Francisco Cazares-Moreno, Eric Cardoza and Victor Zaragoza from University California, Merced for their strong support for this site demonstration work. Special thanks to James Brugger and Michael Logsdon for constructive feedback and technical support, and to Erika Gupta from Building Technologies Office of U.S. Department of Energy for her support on the project.

Numerical Investigation of Solar and Geothermal Powered Desiccant Assisted Air Conditioning in Full-Year Operation

Peter NIEMANN[1*], Arne SPEERFORCK[1]

[1] Hamburg University of Technology, Institute of Engineering Thermodynamics,
Hamburg, Germany
+49 (0)40 8090365-210, peter.niemann@tuhh.de
+49 (0)40 42878-3144, arne.speerforck@tuhh.de

* Corresponding Author

ABSTRACT

In this study, a desiccant assisted air conditioning system relying on geothermal and solar thermal energy is investigated numerically with the help of dynamic annual simulations at different climates. The selected locations are characterized by different climate conditions. A system configuration with almost exclusive use of environmental energy for heating and cooling is proposed. System performance is compared with a conventional air conditioning system of the same size. The results are evaluated according to energetic and ecological criteria. Two approaches are presented which enable a combined assessment of the use of energy and the environmental impact. Operational space requirements for energy provision are taken into account. The investigated system achieves savings in primary energy of up to 71 % compared with a conventional reference process. The analysis of CO_{2e} emissions results in annual savings of up to 70 % (4.7 to). Due to the lower energy density of renewable energy sources, a higher operational land use of the investigated system has to be taken into account. The additional required land use is evaluated using characterizing numbers that are used for air conditioning systems for the first time.

1. INTRODUCTION

Increasing greenhouse gas emissions are considered to be the primary factor influencing global warming. Great efforts are therefore being made worldwide to limit the impact of carbon dioxide emissions. More than 40 % of the total primary energy demands of the western nations (EU, USA) are caused by the building sector. Around 50 % of the corresponding final energy demand is accounted for by applications for heating, ventilation and air conditioning (Pérez-Lombard *et al.*, 2008). Currently, most air conditioning systems are based on cooling outside air below its dew point temperature, using a vapor compression chiller. Due to high energy demands of such systems, alternative air conditioning concepts are investigated. The advantages of desiccant assisted systems in summer mode are well-established in literature. Only a few studies address winter and full-year operation, even though winter mode is an essential part of full year operation, especially for heating dominated regions. Desiccant assisted air conditioning is potentially beneficial against other air conditioning processes in terms of air humidification during winter. Within a desiccant assisted system, moisture recovery by means of the existing hygroscopic materials is possible and the related energy demand is limited to auxiliary energies (e.g. ventilators, pumps, wheel drives). Furthermore, the overall process can be operated efficiently relying on renewable heat sources and heat sinks.

The results achieved in dehumidification mode in our investigations are basically comparable with the results of other studies, although the underlying boundary conditions and system configuration can be very different in each case. The maximum thermal coefficient of performance of the air handling unit reaches values up to about $COP_{th} = 2.0$ and the maximum electrical COP of the ventilation system is $COP_{el} = 9.6$. The average dehumidification coefficient of performance is $DCOP = 1.2$. During the summer period, in a comparative conventional system with compression chiller, the electrical energy demand is over 84 % higher compared with the investigated desiccant assisted system with geothermal cooling and solar powered regeneration of the desiccant wheel.

In contrast, the use of desiccant wheels for enthalpy transfer has been less in focus so far. This circumstance is reflected in a small number of investigations on enthalpy transfer for air humidification in winter. Zeng *et al.* (2014) simulated the year-round operation of a desiccant evaporative cooling system in Shanghai based on a dynamic system model in

TRNSYS, which is also used in the study by La *et al.* (2011). By suitable process control, the supply air is provided by the air mass flow on the adsorption side in summer and the air mass flow on the desorption side of the desiccant wheel in winter. Due to site conditions, air humidification is significantly underrepresented compared to air dehumidification. Therefore, the authors refer only to the limitations of the studied system with single-stage sorption process in dehumidification mode. Preisler and Brychta (2012) investigated the year-round operation of a solar and desiccant assisted system with evaporative cooling in Austria. For moisture recovery in winter, the desiccant wheel is operated as an enthalpy wheel. During the period of one year, 87.5 % of time the system was operated with enthalpy recovery. The remaining time was assigned to air dehumidification. The authors give a linear annual duration curve of air humidification, from which it is deduced that in 61 % of the operation time with enthalpy recovery the humidity ratio of the process air was increased by at least 2.5 $g_w\ kg_{da}^{-1}$. Compared to a modeled reference system with air dehumidification by dew point undershoot and heating based natural gas, the primary energy demand was reduced by 73.9 % in one year. Using annual simulations, Wrobel (2014) investigated the primary energy advantage of a configuration of the system on which this study is also based compared to a conventional reference process for different locations. The reduction in primary energy demand was indicated between 24 % and 47 %, depending on the location. The results show a correlation between the energy advantage and the annual operating hours in dehumidification mode. However, heating and humidification is not considered. The author addresses an additional land use required by the solar thermal system.

The available literature shows a need to further investigate winter and full-year operation of desiccant assisted systems. Therefore, this study aims to test the applicability of the mentioned dessicant assisted system for different climates. Our investigations have shown that significant energetic advantages of the desiccant wheel are not only achieved in dehumidification operation. While the electrical energy demand of spray humidifiers is comparable, it was possible to demonstrate a significantly reduced use of electrical energy compared to steam humidifiers. The electrical energy demand for humidification is increased by an average factor of up to 13.7 compared with the enthalpy wheel. This means that humidification can account for more than 30 % of the total electrical energy demand of the associated air conditioning system. Dynamic annual simulations were used to evaluate energetic and ecological advantages and limitations of the geothermal and solar powered system compared with a conventional air conditioning process. Furthermore, the additional land use due to the integration of solar thermal and geothermal energy is analyzed. A new approach to evaluate air conditioning systems according to space requirements is presented.

2. SYSTEM LAYOUT AND CONTROL

The numerically investigated air conditioning system is derived from a test facility built on the campus at Hamburg University of Technology. The test facility is built as a set of eight 20 ft standard containers with four containers on two separated floors. Figure 1 presents an external view of the test facility and the interior layout of the air conditioned space, located in the upper four containers. Its net floor area is 54 m^2 with a clear room height of 2.64 m.

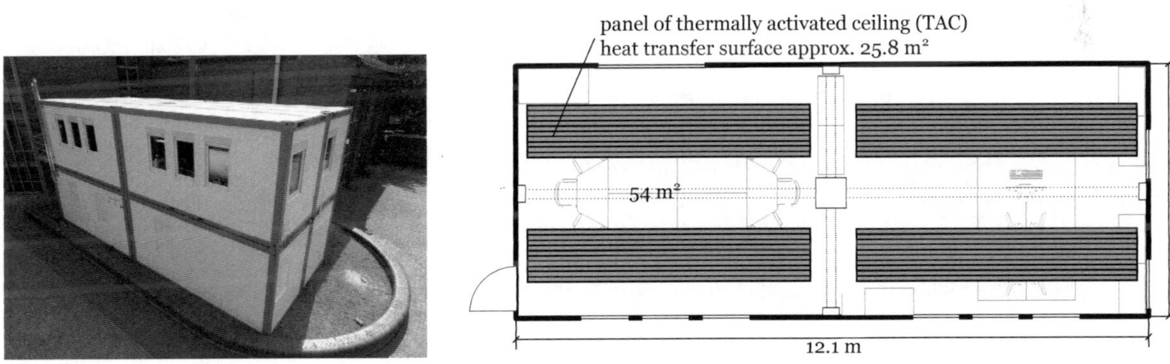

Figure 1: Exterior view (a) and interior schematic representation (b) of the air conditioned space

A system model for the numerical representation of the annual operation is created for the investigation and evaluation of the presented air conditioning concept at locations with different climate conditions. The base system model as presented in Speerforck *et al.* (2017) is focused on dehumidification mode. The energy concept of the investigated system is shown in Figure 2.

2.1 System Layout

The air handling unit is designed as a hybrid system that combines an open desiccant assisted process with closed-loop heating and circuits. The air handling unit is built as a two-wheel system relying on a rotating regenerative heat exchanger (RHE) and a desiccant wheel (DW), operated in enthalpy recovery mode during winter. The reference room is equipped with thermally activated ceiling (TAC) elements for heating and cooling puposes. The main characteristics are briefly described in the following. Further information can be found in Niemann *et al.* (2019).

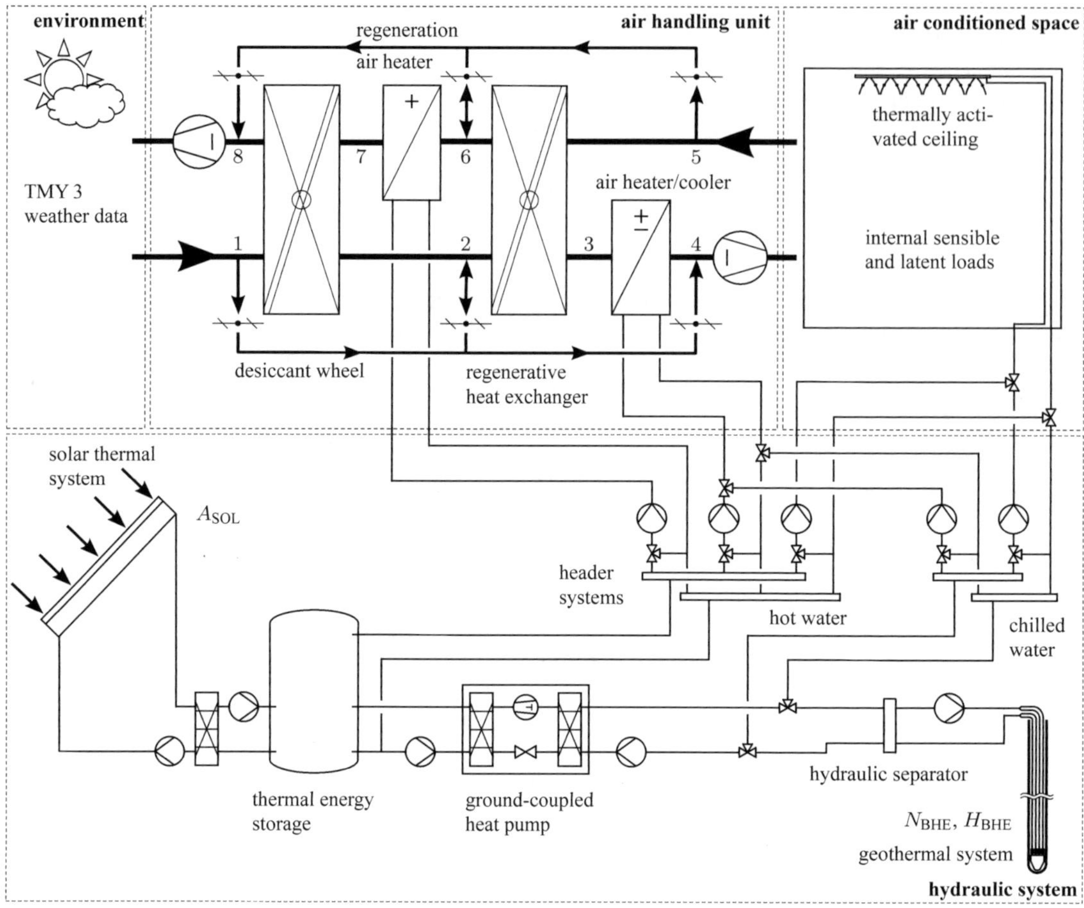

Figure 2: Layout of the of the desiccant assisted system

Outside air (oda) is dehumidified within the desiccant wheel (1–2) and precooled by the rotating heat exchanger (2–3). Water vapor is adsorbed at the hygroscopic coating of the desiccant wheel. Silica gel is used as the sorbent material. The process air is finally cooled or heated to the desired supply air (sup) temperature (3–4) within the air heater/cooler (AHC). Extract air (eta) from the room is preheated by the RHE (5–6). It is further heated to the required regeneration air temperature within regeneration air heater (6–7). Finally, the regeneration air is used to regenerate the desiccant material (7–8), before the exhaust air (exa) is emitted to the environment. For winter operation, the desiccant wheel is used in enthalpy recovery mode at higher rotational speed for coupled heat and mass transfer (1–2). This is a passive air humidification process relying on the enthalpy difference between the air mass flows connected. The regeneration air heater is not operated for enthalpy recovery. So, outside air is humidified and preheated, using extract air from the room. The air heater is used to adjust the desired temperature of the supply air mass flow (3–4). Hot and chilled water is provided by the solar thermal system (SOL) and the geothermal system (GEO) which is linked to the corresponding ground-coupled heat pump for additional heating energy supply during winter. The geothermal system is made of vertical double U-tube borehole heat exchangers (BHE). The number and length of the BHE are varied depending on the site conditions. During summer, the desiccant regeneration process is fed exclusively with solar thermal energy.

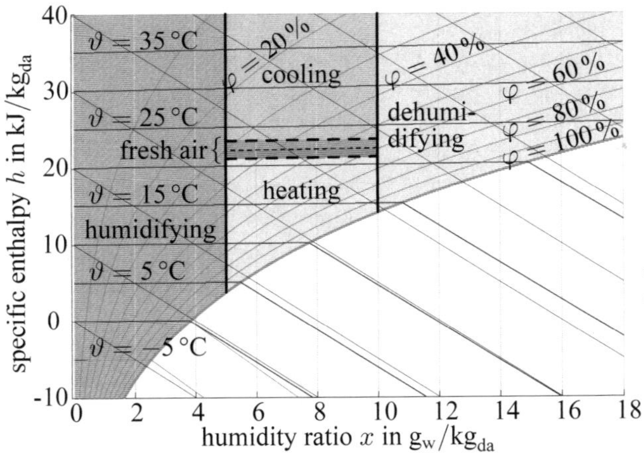

Figure 3: Air handling control strategy

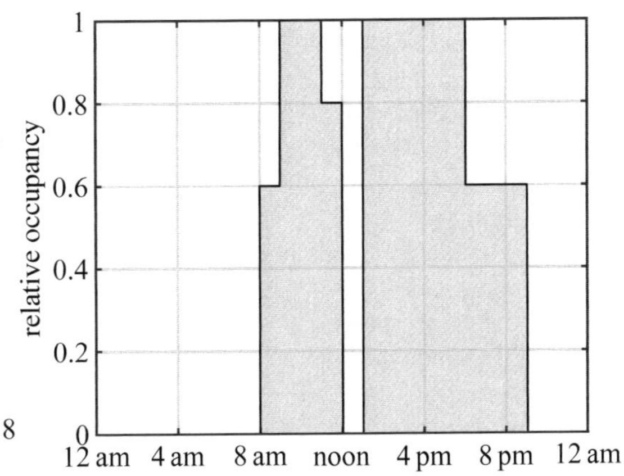

Figure 4: Occupancy profile

The required area of flat-plate collectors A_{SOL} is varied depending on the site conditions, respectively. The BHE are operated with free cooling to supply the air cooler and the TAC. Sufficient sensible heating of the room is achieved through the TAC during winter.

2.2 System Control

To achieve efficient plant operation, different components of the air handling unit can be bypassed as shown in Figure 2. The air handling control strategy is depending on the conditions of outside air as visualized in Figure 3. Five different operation modes are used. The average temperature of outside air over the past day is used to determine the operating mode. The seasonally dependent mode of operation of individual components corresponds to the concept of the test facility, which defines the operation of heat pump, air handling unit and TAC. The components are activated according to the operation modes presented in Table 1 and shown in Figure 3. The notified indoor air conditions correspond to the requirements of comfort category I ($21 - 25.5\,°\mathrm{C}$, $4 - 10\,\mathrm{g_w\,kg_{da}^{-1}}$) according to DIN EN 15251 (2012). In order to limit the system size given the very different climatic boundary conditions of the locations considered, comfortable indoor air conditions are extended up to comfort range III ($19 - 27\,°\mathrm{C}$, $3 - 12\,\mathrm{g_w\,kg_{da}^{-1}}$). The system is designed to keep the number of unmet load hours (ULH) during occupancy at ULH < 300 within one year of operation, according to ASHRAE Standard 90.1 (2016). Internal loads are determined based on the assumption that the room is used by a maximum of 15 persons. Each person is modeled with a constant heat load (sensible heat flow $\dot{Q}_{\mathrm{p}}^{\mathrm{sen}} = 65\,\mathrm{W}$, latent heat flow $\dot{Q}_{\mathrm{p}}^{\mathrm{lat}} = 55\,\mathrm{W}$). A constant value of $\dot{q}_{\mathrm{other}}^{\mathrm{sen}} = 10\,\mathrm{W\,m^{-2}}$ is assumed for other sensible room loads. The occupancy profile shown in Figure 4 is based on empirical values and information from literature.

Table 1: Boundary conditions for modeling the thermodynamic air treatment functions depending on the operating mode

operation mode		outside air conditions		active system components				
		ϑ_{oda} in °C	x_{oda} in $\mathrm{g_w\,kg_{da}^{-1}}$	DW	HRW	RHE	AHC	TAC
winter	heating	< 22	–	–	x	–	x	(x)
	humid.	–	< 5	x	–	–	x	(x)
	fresh air	≥ 22	≥ 5	–	–	–	–	(x)
summer	heating	< 22	≤ 10	–	–	–	x	(x)
	cooling	≥ 22	–	–	x	–	x	(x)
	dehumid.	–	> 10	x	x	x	x	(x)
	fresh air	≤ 22	≤ 10	–	–	–	–	(x)
TAC	heating (winter): $22\,°\mathrm{C} \leq \vartheta_{\mathrm{room}} \leq 22.8\,°\mathrm{C}$, cooling (summer): $\vartheta_{\mathrm{room}} \geq 24.5\,°\mathrm{C}$							

2.3 Site Specific System Size

Based on simulations, the smallest system size that meets the thermal comfort requirements (ULH < 300) is determined during the sizing process. Knowledge of the smallest system size for different climates can be used to estimate the design of the desiccant assisted system without prior system simulation. These studies have shown that the climatic conditions in industrialized nations offer potential for widespread use of the investigated system configuration. Suitable locations were selected based on their potential for full-year use of the examined air conditioning concept. For this purpose, preliminary investigations were carried out to determine the climate suitability of the investigated system on the basis of annual simulations in different climate regions, and a climate suitability map was created on the basis of these simulations, see Figure 5. Each studied site is assumed to be representative of the associated climate classification. In order to achieve the highest possible utilization of the system components, no investigations were carried out for climate zones in which the humidity ratio and the temperature range of the outside air are low. Based on these findings, the climatic conditions of New York City (USA), Bordeaux (France), Sapporo (Japan) and Stockholm (Sweden) are selected for the system evaluation. The climatic conditions of these locations are decribed by the outside air parameters (ϑ_{oda}, x_{oda}), as shown in Table 2. The climatic conditions are based on TMY 3 weather data. The aim of the design is to determine the smallest system size with which the requirements for thermal comfort in annual operation can be met. The characteristic variables of the solar thermal system (A_{SOL}) and the geothermal system (N_{BHE}, H_{BHE}) are therefore designed. The area of the solar thermal system A_{SOL} is determined exclusively by the thermal energy demand of the regeneration process to regenerate the desiccant in dehumidification mode. The design parameters of the geothermal system depend on the application as a heat source and sink. In the design process, the limitation of uncomfortable room air conditions due to deviating temperatures is rated higher than deviations in room air humidity. The length of a single vertical geothermal probe is limited to $H = 100\,\mathrm{m}$ and the geology of the soil is identical to the known soil properties at the installation site of the test facility. Moreover, for $N_{\mathrm{BHE}} > 1$, the horizontal distance is set sufficiently large that thermal interactions can be neglected. The volume of the thermal storage system is set constant to $V = 2\,\mathrm{m}^3$.

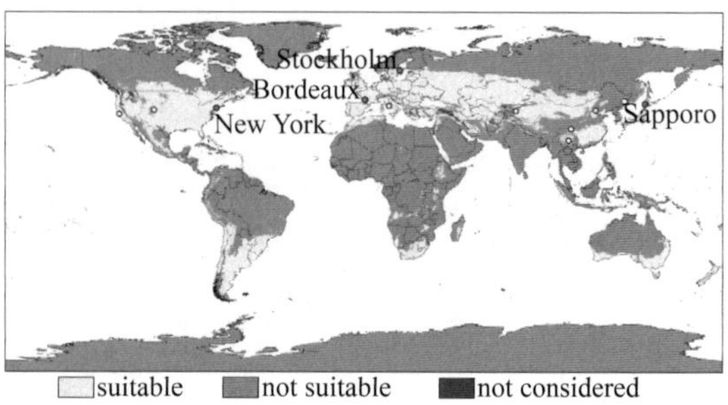

Figure 5: Climate suitability map

Table 2: Climatic conditions at selected locations

location	ϑ in °C			x in $\mathrm{g_w\,kg_{da}^{-1}}$		
	$\vartheta_{\mathrm{oda}}^{\mathrm{min}}$	$\vartheta_{\mathrm{oda}}^{\mathrm{max}}$	$\overline{\vartheta}_{\mathrm{oda}}$	$x_{\mathrm{oda}}^{\mathrm{min}}$	$x_{\mathrm{oda}}^{\mathrm{max}}$	$\overline{x}_{\mathrm{oda}}$
New York	-15.6	35.0	12.5	0.4	22.5	7.2
Bordeaux	-8.2	34.0	13.2	1.7	16.9	7.5
Sapporo	-15.2	33.5	8.8	0.6	19.2	6.0
Stockholm	-17.0	27.1	6.5	0.7	13.9	6.5

3. MODELING APPROACH

The basis of the system model for simulating annual operation is a system model for winter operation in combination with the system model designed for summer operation, as presented by Speerforck *et al.* (2017). The system model is made up of various validated component models, using the object-oriented and equation-based modeling language standard MODELICA®.

3.1 Building Model

The system model of the room includes the modeling of the building envelope, the room air volume and vapor diffusion as well as the heating and cooling ceiling follows the approach of the resistance-capacity model. The building envelope is modeled by thermal resistance and capacity, whose thermal and geometric properties correspond to the wall structure of the test facility. The indoor air model is made up of seven horizontal layers of air with different vertical dimensions. Each layer of air is considered to be a volume of still air through which energy in the form of heat is transferred solely by conduction. The calculation of the room air humidity ratio results from a water balance of the room. Ideal mixing of the room air is assumed. Measurement data from plant operation were used to validate the overall building model. During a period of one year, the mean deviation of the simulated room air temperature at a height of 0.1 m above the floor is $-0.42\,\mathrm{K}$ with a standard deviation of $\pm 1.5\,\mathrm{K}$. At a height of 2 m, the evaluation showed a deviation of $-0.29\,\mathrm{K}$ with a standard deviation of $\pm 1.24\,\mathrm{K}$.

3.2 Air Handling Unit

The model of the air handling unit consists of various component models linked by interfaces. These component models depict all relevant functional scopes based on the air conditioning system of the test facility. The core component is the desiccant wheel, whose modeling is based on the effectiveness method. This method is derived from the concept of the effectiveness of a real heat exchanger compared to the ideal reference. Due to the coupled processes of heat and mass transfer the method requires two independent effectiveness parameters given in Equation (1) to describe the global sorption process with sufficient accuracy.

$$\eta_q = \frac{q_{\text{pro,in}} - q_{\text{pro,out}}}{q_{\text{pro,in}} - q_{\text{reg,out}}} \qquad \eta_h = \frac{h_{\text{pro,in}} - h_{\text{pro,out}}}{h_{\text{pro,in}} - h_{\text{reg,in}}} \tag{1}$$

The equilibrium water load q of the sorbent is defined as the mass ratio of the absorbed water mass and the mass of the sorbent $q = m_{\text{w}}/m_{\text{sorbent}}$. Investigations into the creation of the effectiveness model for silica gel showed that the effectiveness parameters can be assumed to be constant for various input and operating parameters. The correlations for η_q and η_h are determined from empirical studies, $\eta_q = 0.86$ and $\eta_h = 0.82$ for enthalpy transfer (winter) and $\eta_q = 0.9$ and $\eta_h = 0.25$ for active dehumidification (summer). To validate the effectiveness model, 3100 different measured air states in dehumidification and enthalpy transfer operation were used. The results of the effectiveness model show good agreement with the measured temperatures and humidity ratios. The deviations are mainly within the limits of the associated measurement uncertainties. The regenerative heat exchanger is modeled as an effectiveness model according to the ϵ-NTU method, based on the equations of a counterflow heat exchanger. The AHC and RHE models are also modeled according to the ϵ-NTU method. Both are modeled as counterflow heat exchangers, but unlike the RHE model, no rotation dependency is implemented. Equation (2) is used to determine the effectivenesses.

$$\epsilon_{\text{RHE}} = \frac{\dot{C}_{\text{pro}} \cdot (\vartheta_{\text{pro,in}} - \vartheta_{\text{pro,out}})}{\dot{C}_{\text{min}} \cdot (\vartheta_{\text{pro,in}} - \vartheta_{\text{reg,in}})} \qquad \epsilon_{\text{AHC,RAH}} = \frac{\dot{C}_{\text{a}} \cdot (\vartheta_{\text{a,out}} - \vartheta_{\text{a,in}})}{\dot{C}_{\text{min}} \cdot (\vartheta_{\text{w,in}} - \vartheta_{\text{a,in}})} \tag{2}$$

The obtained values for the three heat exchangers are $\epsilon_{\text{RHE}} = 0.8$, $\epsilon_{\text{AHC}} = 0.66$ and $\epsilon_{\text{RAH}} = 0.62$. Bypass systems are modeled as logical switching elements whose state is coupled to the operating mode of the air handling.

3.3 Hydraulic System

For modeling the heat pump, a black-box approach is used in which the system behavior is described by characteristic diagrams. For this purpose, the characteristic diagrams of several heat pumps are implemented, whose characteristic curves are determined manufacturer specifications. The heat flow transferred in the condenser is determined by interpolation for a given inlet temperature at the evaporator and outlet temperature at the condenser. The thermal energy storage system is modeled as a one-dimensional system discretized into ten equivalent layers as volume segments. Complete mixing is assumed in each layer. Heat and mass transfer take place exclusively between the layers. One-dimensional heat conduction through the storage wall is assumed. Heat transfer from the solar thermal system is modeled using the ϵ-NTU method. The modeling approaches of the solar thermal system and the borehole heat exchangers can be found in Speerforck *et al.* (2017).

3.4 Electrical Power Input

To model the electrical power consumption of the heat pump, a characteristic diagram was created based on manufacturer data, as shown in Figure 6. The empirically determined electrical power input of other components are listed in Table 3 depending on the operating mode of the overall system.

3.5 Conventional Reference System

For the evaluation of the investigated system, the process of a conventional air conditioning (CAC) system is created. To cool and dehumidify outside air, the model of a compression chiller is implemented. The heat pump and the solar thermal system are replaced by a natural gas fired condensing boiler. Instead of a thermally activated ceiling, the conventional system is modeled with floor heating, which is used exclusively for sensible heating. Cooling of the room takes place exclusively through the supply air in summer, whereby the supply air temperature is reduced compared to the desiccant assisted process. The control of the reference process is tuned to achieve maximum agreement with the implemented control of the desiccant assisted system. The operation of the compression chiller, humidifier, and floor heating in the conventional system depends on the seasonal operating mode of the desiccant assisted system. The air cooler is operated in cooling mode when only sensible heat loads have to be dissipated ($x_{\text{oda}} \leq 10\,\text{g}_{\text{w}}\,\text{kg}_{\text{da}}^{-1}$) or in dehumidification mode ($\vartheta_{\text{oda}} > 22\,°\text{C}$, $x_{\text{oda}} > 10\,\text{g}_{\text{w}}\,\text{kg}_{\text{da}}^{-1}$). In the reheater, the supply air temperature is set to 16 °C

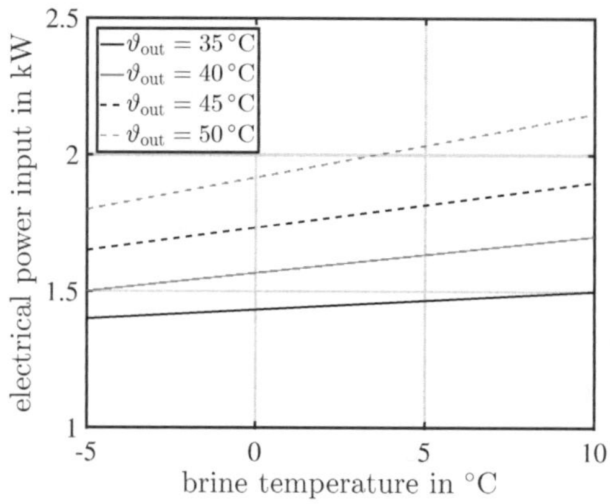

Figure 6: Electrical power of the heat pump

Table 3: Electrical power inputs of further system components

component	power input winter	power input summer
fan, supply air	160 W	300 W
fan, exhaust air	193 W	400 W
drive DW	86 W	16 W
drive RHE	93 W	93 W

Table 4: System design and sizing

location	parameter $[N \mid H]_{BHE}$	A_{SOL}
New York	$3 \mid 80$ m	27.6 m^2
Bordeaux	$1 \mid 100$ m	5 m^2
Sapporo	$3 \mid 80$ m	20.1 m^2
Stockholm	$4 \mid 90$ m	–

Table 5: Primary energy factors and specific CO_{2e} emission factors according to German standards

energy source	PE factor (non-renewable)	CO_{2e} emission factor in to CO_{2e} MWh^{-1}
gas	1.1	0.201
electricity	1.8	0.427
solar/geotherm.	0	–

when ambient air conditions require dehumidification of the process air. The electrical energy demand of the chiller is determined with a constant energy efficiency ratio of 3. The gas condensing boiler is modeled as a simple heat source that transfers a constant heat flow $\dot{Q}_{boiler} = 12.5$ kW to a closed volume. An annual efficiency of 0.87 based on the calorific value is assumed. To humidify the process air, the model of an isothermal steam humidifier is implemented. A linear relationship is assumed between the electrical power input and the introduced mass flow of water vapor. An efficiency of 0.91 is assumed. The control of the humidifier is equivalent to the operation of the enthalpy wheel in the desiccant assisted system. To ensure comparability of the systems, the humidification capacity is controlled to ensure identical humidity ratio of the supply air.

4. SIMULATION RESULTS AND DISCUSSION

Table 4 summarizes the site-specific system design results based on the dimensioning described in Section 2.3. The design process ensures that thermal comfort requirements (ULH < 300) are met. The design leads to very different units of the geothermal and solar thermal system.

4.1 Evaluation of Energy Demands and Emissions

Figure 7 shows the primary energy (PE) demands and CO_{2e} emissions, summed over one year. The underlying primary energy factors and the specific CO_{2e} emission factors are listed in Table 5.

At all locations, the solar and geothermal powered system shows savings in primary energy demand compared to the reference process. The highest relative savings potentials are achieved at New York ($PE_{DAAC}/PE_{CAC} = 0.25$) and Sapporo ($PE_{DAAC}/PE_{CAC} = 0.29$). The primary energy demand caused by the DAAC system is less site-dependent than for the conventional system. The variation is mainly caused by the gas demand. The primary energy advantage of the DAAC system increases with high utilization of the humidification and dehumidification operation. Using Stockholm as an example, it is clear that locations with cold climates have a lower relative potential for primary energy use reduction for the selected system configuration. In summary, the primary energy demand represents an indicator that correlates with the environmental impact of a system. However, this comparison does not take into account the additional land use under which the respective primary energy savings are made possible. From the combined view of Figure 7a and Table 4, it can be seen that the absolute and relative primary energy savings correlate with the land use

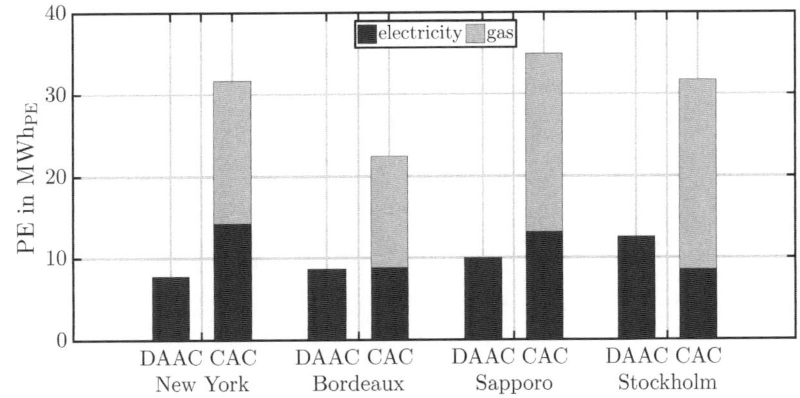

(a) primary energy demand

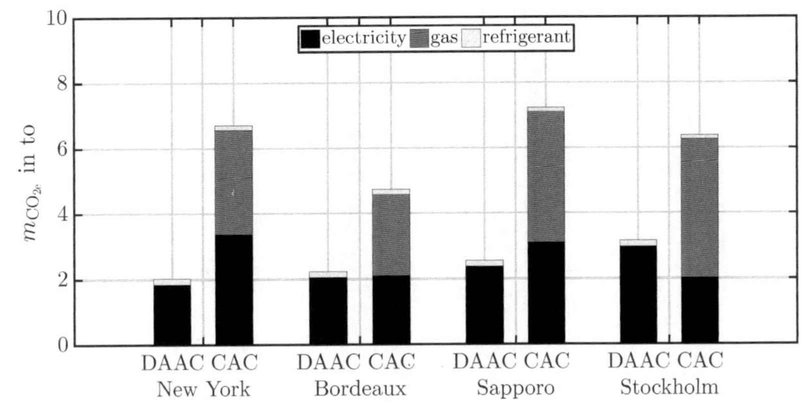

(b) CO_{2e} emissions

Figure 7: Comparison of primary energy demands and CO_{2e} emissions based on annual simulations

of the DAAC system. Similar to the primary energy analysis, the plot of CO_{2e} emissions in Figure 7b indicates a lower environmental impact of the DAAC system. At all locations, the CO_{2e} emissions of the reference process are higher. The largest emission saving is achieved in New York and corresponds to a relative saving of 70 % ($\Delta m_{CO_{2e}} = 4.7$ to). The share of CO_{2e} emissions from refrigerant loss is less than 0.2 to at all sites, corresponding to a maximum relative share of 9 % (DAAC system, heat pump) and 3.5 % (CAC system, chiller). Operational emissions of the CAC system correlate with the heating and cooling loads of the conditioned space. Accordingly, maximum emission levels are achieved in New York and Sapporo. The evaluation of CO_{2e} emissions is subject to similar limitations as the primary energy analysis. This is due to the limited significance of the emission factor for electrical energy with high shares of renewable energy sources of electricity generation. Taking into account the degree of decarbonization of the region-specific electricity generation, the DAAC system enables relative CO_{2e} emission savings of up to 90 %.

4.2 Evaluation of Land Use

To extend the system evaluation, the combined approach of a characterizing variable is applied. This quantity is described as useful energy with respect to emissions and land use and is defined according to Equation 3.

$$q^A = \frac{Q_{\text{use}}}{m_{CO_{2e}} \cdot \left(1 + \frac{A_{\text{land}}}{A_{\text{room}}}\right)} \qquad A_{\text{land}} = f_{\text{GEO}} \cdot A_{\text{GEO}} + f_{\text{SOL}} \cdot A_{\text{SOL}} \tag{3}$$

Factors f_{GEO} and f_{SOL} evaluate the availability of land and the impact of land use on a site-specific basis. The weighting of land use with $f_{\text{GEO}} = f_{\text{SOL}} = 0$ applies to sites with almost unlimited open space, while in case of a location with no available land $f_{\text{GEO}} = f_{\text{SOL}} = 1$ are set, for example in densely built-up cities. The environmental impact due to additional land use describes the site-specific and operation-related installation area of the geothermal and solar thermal system. The environmental impact due to land use during upstream manufacturing processes from raw material extraction to installation at the utilization site is not considered. The operational areas used are assumed to be sealed land, although the land can usually be put to additional use. For the geothermal system, the area orthogonal to the boreholes is taken into account including a circular barrier area with a radius of 5m to exclude thermal interactions. For the investigation of the influence and the sensitivity of q^A, three installation variants are considered:

1. installation of SOL and GEO preferably within the footprint of the air-conditioned room
2. installation of SOL within and GEO outside the footprint of the air-conditioned room
3. installation of SOL and GEO outside the footprint of the air-conditioned room

The results of the system evaluation according to q^A are shown in Figure 8. For the presentation of the results, the weighting factors of the land use are set to $f_{\mathrm{GEO}} = f_{\mathrm{SOL}} = 1$.

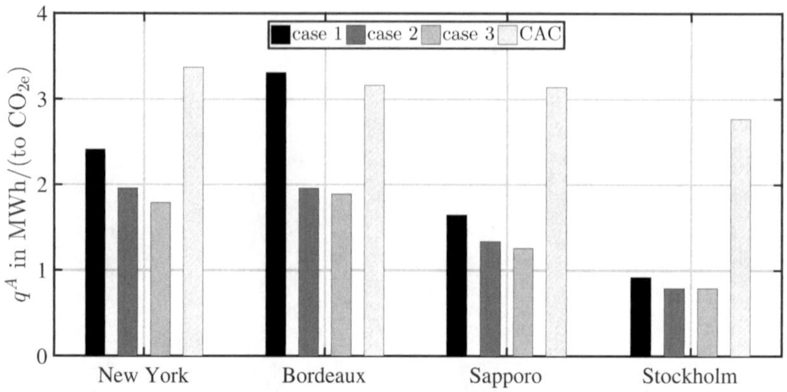

Figure 8: Area-specific useful energy for different installation variants of the DAAC system in comparison with the reference values of the CAC system.

The results show, with one exception, advantages of the conventional process, resulting in higher values for q^A_{CAC}. For the DAAC system, q^A is sensitive to the land use by the geothermal system. If no integrated installation of the geothermal system is possible, no advantages can be achieved by the DAAC system. As the number of boreholes increases, the advantage of the conventional system increases. Accordingly, the lowest values of q^A_{DAAC} are achieved in Stockholm. At the same site, q^A_{DAAC} is higher by a factor of 2.3 for integrated installation ($f_{\mathrm{GEO}} = 0.1$) and characterizes a significantly higher area-based power capacity of the DAAC system. The results for Bordeaux show that a DAAC system advantage is possible with integrated installation. This represents the benefits of small system sizes. Similarly, the benefit of the DAAC system increases with reduced weighting of land use by the geothermal system. For a factor of $f_{\mathrm{GEO}} < 0.4$, the DAAC system is beneficial at all location in terms of q^A.

5. CONCLUSIONS

The combination of desiccant assisted air conditioning with geothermal and solar thermal systems can be used under different climatic conditions and ensures a high level of comfort in annual operation with ULH < 300 in the selected comfort range. The proposed system is beneficial in terms of environmental impacts. Due to reduced electricity and gas demands, savings in primary energy and CO_{2e} emissions could be proven at all locations. The environmental benefits of the proposed system increase with the annual operating hours of humidifying and dehumidifying the process air due to the high electricity demands of the chiller and the steam humidifier in the conventional system. The evaluation of land use shows that the presented method represents a conservative approach to the evaluation of air conditioning systems considering operational land use. There is a high sensitivity to the land definition for the installation of boreholes. Under unfavorable installation conditions, which at the same time represent a conservative limit consideration, advantages of the investigated system are hardly possible.

NOMENCLATURE

A	area	(m^2)
$\dot{C}$	thermal capycity flow	($J\,s^{-1}\,K^{-1}$)
(D)COP	(dehum.) coefficient of performance	(–)
h	specific enthalpy	($kJ\,kg_{da}^{-1}$)
H	depth	(m)
m	mass	(kg)
N	quantity	(–)
P	electrical power	(W)
PE	primary energy	(kWh_{PE})
q	equilibrium water load	($kg\,kg^{-1}$)
Q	thermal energy	(kWh)
q^A	specific useful energy	($MWh\,(to\,CO_{2e})^{-1}$)
$\dot{Q}$	heat flow	(W)
$\dot{q}$	specific heat flow	($W\,m^{-2}$)
x	humidity ratio	($g_w\,kg_{da}^{-1}$)
ϵ	efficiency	(–)
η	effectiveness parameter	(–)
φ	relative humidity	(%)

Subscripts and Abbreviations

a	air	p	person
AHC	air heater/cooler	pro	process air
BHE	borehole heat exchanger	RAH	regeneration air heater
CAC	conventional air conditioning	reg	regeneration air
DAAC	desiccant assisted air conditioning	RHE	regenerative heat exchanger
DW	desiccant wheel	sen	sensible
el	electrical	SOL	solar thermal system
GEO	geothermal system	TAC	thermally activated ceiling
in	inlet	th	thermal
lat	latent	TMY	typical meteorological year
oda	outside air	ULH	unmet load hour
out	outlet	w	water

REFERENCES

ASHRAE Standard 90.1 (2016). *Energy standard for buildings except low-rise residential buildings*. Atlanta, Georgia/USA: ASHRAE.

DIN EN 15251 (2012). *Indoor environmental input for design and assessment of energy performance of buildings addressing indoor air quality, thermal environment, lightning and acoustics*. Berlin: Beuth.

La, D., Dai, Y., Li, H., Li, Y., Kiplagat, J. K., & Wang, R. (2011). Experimental investigation and theoretical analysis of solar heating and humidification system with desiccant rotor. *Energ Buildings*, *43*, 1113-1122.

Niemann, P., Richter, F., Speerforck, A., & Schmitz, G. (2019). Desiccant-assisted air conditioning system relying on solar and geothermal energy during summer and winter. *Energies*, *12*, no. 3175.

Pérez-Lombard, L., Ortiz, J., & Pout, C. (2008). A review on buildings energy consumption information. *Energ Buildings*, *40*(3), 394-398.

Preisler, A., & Brychta, M. (2012). High potential of full year operation with solar driven desiccant evaporative cooling systems. *Enrgy Proced*, *30*, 668-675.

Speerforck, A., Ling, J., Aute, V., Radermacher, R., & Schmitz, G. (2017). Modeling and simulation of a desiccant assisted solar and geothermal air conditioning system. *Energy*, *141*, 2321-2336.

Wrobel, J. (2014). *Entwicklung und Bewertung eines Konzeptes zur sorptionsgestützten Klimatisierung von Gebäuden*. Doctoral thesis. Hamburg University of Technology.

Zeng, D. Q., Li, H., Dai, Y. J., & Xie, A. X. (2014). Numerical analysis and optimization of a solar hybrid one-rotor two-stage desiccant cooling and heating system. *Appl Therm Eng*, *73*(1), 474-483.

Experimental Study of Electrospray for Exhaust Gas Treatment

Soyeon KIM[1], Minkyu Jung[2], Sanghun Jeong[3], Donik Ku[4], Soojin Bae[5], Gijeong Seo[6], Minsung Kim[*]

[1,2,3,4,5,*]Department of Intelligent Energy and Industry, Chung-Ang University, 06975, Seoul, Republic of Korea
([1]nuri36@cau.ac.kr, [2]Mamba24@cau.ac.kr, [3]suj531@cau.ac.kr, [4]ehsdlr@cau.ac.kr, [5]sooj980809@cau.ac.kr)

[6,*] School of Energy Systems Engineering, Chung-Ang University, 06975, Seoul, Republic of Korea
([6]koki8235@cau.ac.kr)

* Corresponding Author (+82-2-820-5973, [*]minsungk@cau.ac.kr)

ABSTRACT

The demand for high-efficiency dust collectors has rapidly increased to remove particulate matter (PM) from exhaust gas emission facilities, such as thermal power plants, steel mills, and industrial cogeneration plants, as the permission standards have been strengthened. Electrospray gas treatment technology is a promising method due to high-performance for PM removal and low water consumption rather than scrubbers and wet ESP. It is an atomization technique that only requires DC voltage and negligible pressure/pumping power to maintain the spray. The spray conditions of electroscrubbers have been studied are not suitable to treat the actual exhaust gas. For dust removal, the flow rate over 10 mL/min per nozzle is required, which ranges hundreds of times in typical electrospray fields. The electrospray of high flow rate has a completely different spray shape from the low flow rate condition, and it was visualized through various figures such as corona discharge photographs and shadow images. The corona discharge caused by the high-electric field strength can increase the removal efficiency of charged dust particles, and the removal efficiency was expected to be high as the corona discharge current increased with the high-applied voltage, flow rate, and electrical conductivity of the liquids. However, at voltages above a certain range, spark discharges occurred and proceed to arc discharge (short circuits). As a result of the PM removal performance evaluation based on optimal operating conditions derived from the experiments, the removal efficiency of PM10 was 99.65%, PM2.5 was 97.59%, and PM1.0 was 90.95%. Through this study, it has the potential to be practical reference for the electroscrubber in actual plants.

Keywords: Electrospray, electroscrubber, Particulate Matter, Corona discharge, Visualization

1. INTRODUCTION

For highly effective exhaust gas treatment, scrubbers have been used widely in industries. They tend to consume a lot of energies and water to remove submicron particles. Even spray type scrubber, high hydraulic pressure is required to atomize the water in the nozzle. Electroscrubbers have been proposed for many years to solve this problem of submicron particles removal. However, only a few studies reported and the main issue regarding water consumption and filtering of contaminated water remain unsolved. This content is based on configuration of arranging multi nozzles for electrospray in a cyclone scrubber. In the cyclone electrospray scrubber, large and heavy particles over 100 μm such as fly ash and coal dust can be removed by centrifugal force and gravity through the cyclone. On the other hand, the smaller and lighter ones such as PM10 which are floating inside cyclone can be treated by electrospray. They collide with charged droplets from electrospray, and flow down. The clean gas is emitted to the center pipe outlet. Electrospray can easily make droplets by electrical force without pressure and pumping power to maintain spray. Because the droplets are charged, they also collect the light particles by electrical attraction, not only inertial force. So, high efficiency for particle removal is achieved with very little water.

This content purposes to observe the shape of water electrospray for dust particles removal according to voltage, mass flowrate, and electrical conductivity of liquid (Mora and Loscertale, 1994). The flowrate set this study is very less than the water used in general spray scrubber, but it is at least tens of times larger than that in conventional water electrospray field (Lopez-Herrera *et al.*, 2004). Also, the spray shape in electrospray is easily changed by parameters such as applied voltage, flowrate, conductivity, etc., so it is essential to visualize the spray mode for use in scrubber. This study also includes discharge current experiment. In fact, there is a technical limit to measure the amount of charge of droplets in mid-air. By observing the discharge current of the dust collecting plate and comparing according to electrical conductivities of liquid, the charge of droplets was estimated. As a result, in general, the higher the electrical conductivity, the more charge the droplet has, but it is thought that there is a limit above a certain level.

Fig. 1. represents how the particles removed. When the liquid is passing through negatively charged nozzle, it is also be negatively charged. Ejected liquid is dispersed into droplets by repulsive force. They collide with charged PM particles inertially and inductively. At last, they are removed by cleansing water at the collecting electrode.

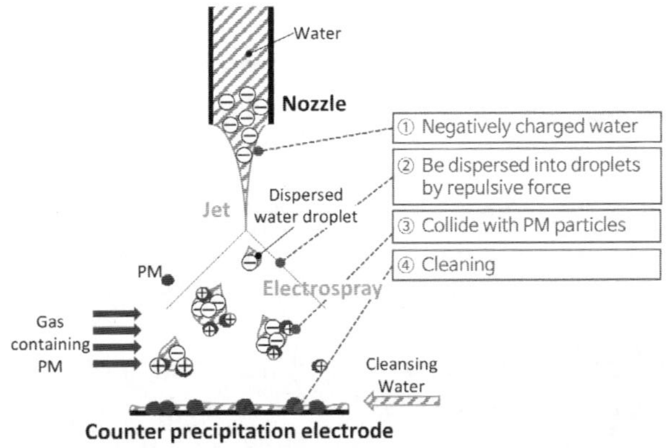

Figure 1: Mechanism of PMs removal with electrospray

2. EXPERIMENT ON WATER ELECTROSPRAY

2.1 Experimental apparatus

An experimental set-up of electrospray with single unit nozzle is illustrated in Fig. 2. Water passing through the negatively charged metal nozzle receives electric charges. That is located on the surface of the liquid and atomize to droplets due to electrical repulsive force. The droplets move toward the ground plate and the current flowing through a resistor is measured by oscilloscope. Spray shape of water is visualized using DSLR camera with strobe light source that has very short flash time. Behaviors of the droplets and jet which means water stream near the nozzle tip can be observed. Electricity shielding is the most important thing for accuracy of current measurement and safety against current leakage in designing the experiment. Because the leakage impinges on the electrical device, spraying system is surrounded by sub-conductor, insulation glass tub, and the capillary support that held the needle is composed of Teflon and acrylic.

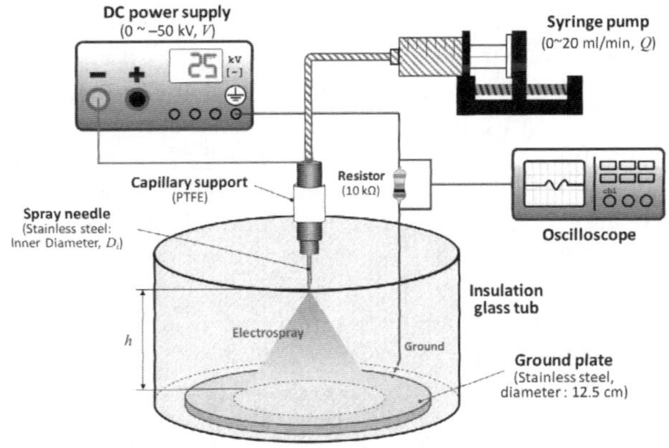

Figure 2: Schematic diagram of experiment on water electrospray

Table 1 summarizes experimental conditions. Applied voltage ranges from 0 to -50 kV with a minimum step of 1 kV. Volumetric flow rate of water is set in range of 1-20 mL/min to observe all the spray shape and current characteristics well. In this study, when voltage was just applied to the metal needle without injecting a liquid, it was referred to as a "dry condition". To presume the effect of the current generated by electrospray on corona discharge, a comparison of the discharge characteristics between in electrospraying condition and in dry condition was performed.

Table 1: Experimental conditions of spray in this study

Applied voltage [–kV]	Volumetric flow rate [mL/min]	Needle diameter [mm]	Distance from needle to plate [mm]	Diameter of plate [mm]
0–50	0 (dry) 1–20 (wet)	0.67 (inner) 1.07 (outer)	50	120

2.2 Modes of water electrospray

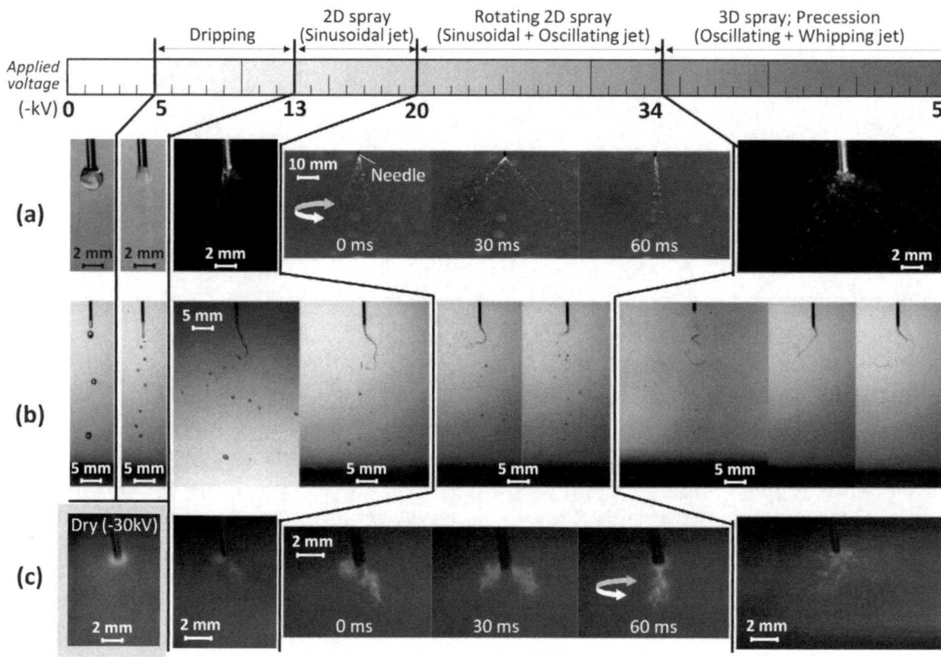

Figure 3: Electrospray mode classification (tap water, Q = 10 mL/min, V = 0 to −50 kV). (a) Trajectories of droplets; (b) Snapshots of electrospray; (c) Corona discharge

In Fig. 3, the electrospray modes are classified according to the applied voltage increasing in a single nozzle. The right side indicates the high voltage, and the flowrate of tap water is 10 mL/min to observe the variation of spray shape clearly. Fig. 3 (a) is images of trajectories of the droplets. It seems what we actually see. Fig. 3 (b) is the images with the short light exposure time to closely observe the changing droplets and jets (Kim *et al.*, 2014). Fig. 3 (c) shows the corona discharge. The occurrence of corona can be observed with eyes from the electrode or water surface, because when the secondary electrons are emitted from the electrode surface (the progress of corona discharge), and they pop out with the photoelectrons which glowing with purple. The most prominent characteristics of water electrospray under flowrate condition in this paper is appearance of two- and three-dimensional (2D, 3D) spatial distribution sprays because of its high surface tension. These features are appeared in limited high flow rate over 5 mL/min, and the glow discharge is dominant in the low flow rate of water electrospray (Borra *et al.*, 2004). When the voltage is getting higher, the droplets are accelerated, and jet which is the liquid flowed out form the tip of nozzle. It also make the droplets be smaller (Kim *et al.*, 2011). The droplets and jet are seemed like a sinuous wave and the spray looks like a two-dimensional plane as shown in -13~20 kV. The voltage is getting over -20 kV, distortion causes the jet to oscillate. It causes the jet to rotate around the axis of the needle. When over -34 kV, the jet starts to be whipping forming a helix and looks like a 3D cone shape. It is expected that the PM removal effect will be best at 3D mode which has a wide spray range, so the probability of collision between submicron particles and droplets increases.

2.3 Discharge of electrospray according to liquid electrical conductivity

The mechanism of negative corona discharge under the conventional dry needle condition and electrospray is illustrated in Fig. 4. The negative corona discharge is described by Chen and Davidson (2003) in Fig. 4 (a). The corona discharge appears when potential difference between electrodes exceeds the dielectric strength of the surrounding air, which become ionized (Jaworek and Krupa, 1997). electrons existed between the electrodes are accelerated by electric field and generate positive ions from neutral gas molecules by inelastic collision. The positive ions also become energetic, and they collide with the surface of the negatively charged needle. Simultaneously, secondary electrons are released, move to ground plate, and are measured as current at the plate. On the other hand, during electrospray, the secondary electrons are extracted from the surface of liquid not the needle. It means that the liquid surface is charged, and the purple light originates from the liquid surface like Fig. 3 (c). In this situation, droplets carry the electric charges to ground plate.

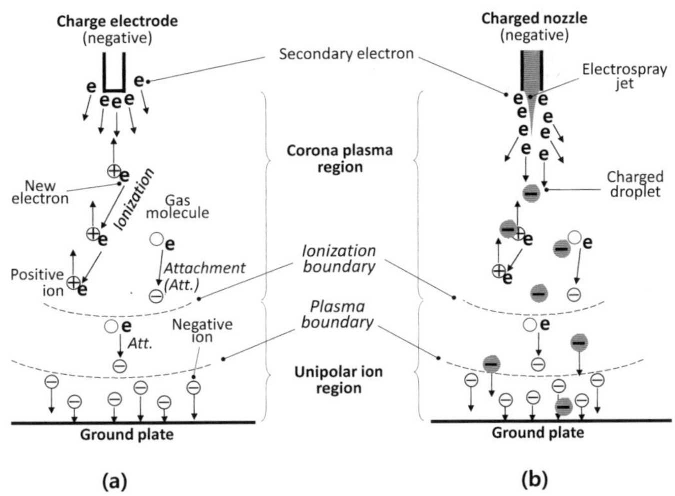

Figure 4: Mechanism of negative corona discharge. (a) dry condition (Chen and Davidson); (b) electrospray

Table 2 summarizes the properties of NaCl aqueous solutions to test the effect on the electrical conductivity of liquids. There are some studies about electrical conductivity of liquid, but they are much lower range than this study (Yazdekhsti *et al.*, 2019). The flowrate is limited until the spark discharge occurs. The power consumption can be estimated from the measured current and the average applied voltage.

Table 1: Properties of working fluids according to electrical conductivity in this study

Fluid	Temperature [°C]	Salinity [%]	Electrical conductivity [mS/cm]	Flow rate (at –34 kV) [mL/min]	Applied voltage (at 10 mL/min) [–kV]
distilled water		-	-	1–20	0–50
tap water	25	0.10	0.25 ± 0.01		
saline solution		0.85	14.05 ± 0.05	1–16	0–40

Fig. 5 shows the current-voltage (*I-V*) curves at a flowrate of 15 mL/min for three liquids. The corona current increases via a quadratic function as the applied voltage increased by the Townsend avalanche (Townsend, 1914). As the charged droplets transfer the electric charge to ground plate during the electrospray, the current is larger than dry condition (black line). The liquid which has the higher electrical conductivity transfers more charge and that of current is higher (Smith, 1986). *Spark over* is occur when the potential difference between the electrodes is over the certain range. It refers to the point at which spark discharge begins. However, since the distilled water has very low conductivity (~0.003 mS/cm), the applied voltage just atomizes liquid into droplets, and the generated droplets are hardly charged. Electron charges are released less at surface of liquid than at metal needle. That's why the current of distilled water is lower than the dry condition.

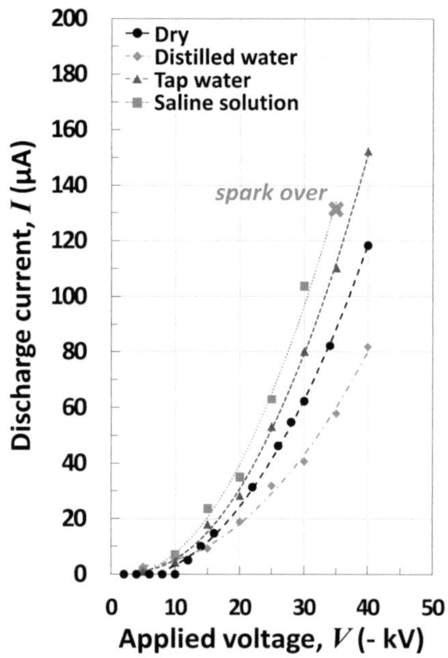

Figure 5: I–V characteristics of the electrospray with electrical conductivity (Q = 15 mL/min).

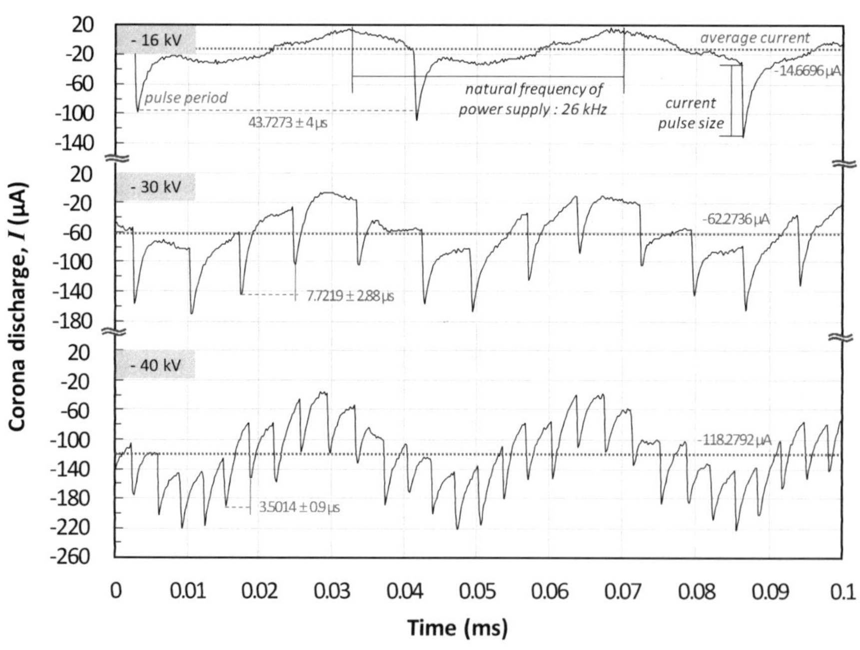

Figure 6: Waveforms of corona discharge with Trichel pulses.

Fig. 6 is the waveforms of corona discharge at -16, 30, and 40 kV in the dry condition. Corona discharge is a phenomenon in which secondary electrons drift out through the locally conductive air, as explained earlier. The emission of electrons appears as a current pulse called as Trichel pulse (Trichel, 1938), and it represents a certain frequency. The frequency of Trichel pulse is getting higher with voltage increasing, and it develops to spark or arc discharge which can make melt the needle. This is shown in Fig. 7.

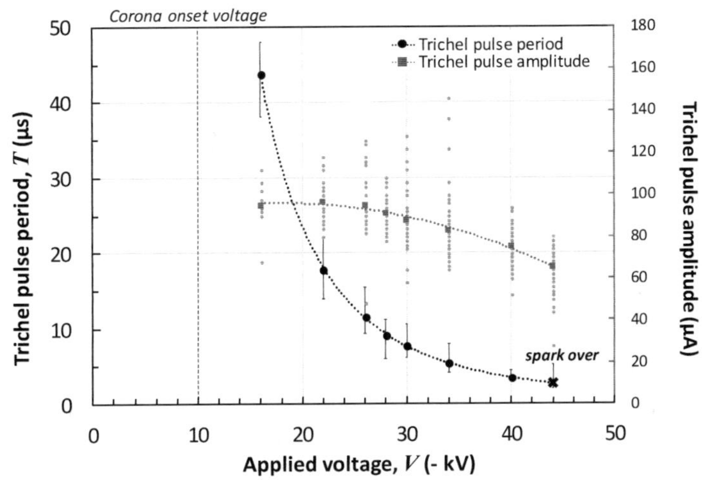

Figure 7: Trichel pulse period and amplitude with voltage increasing in dry condition

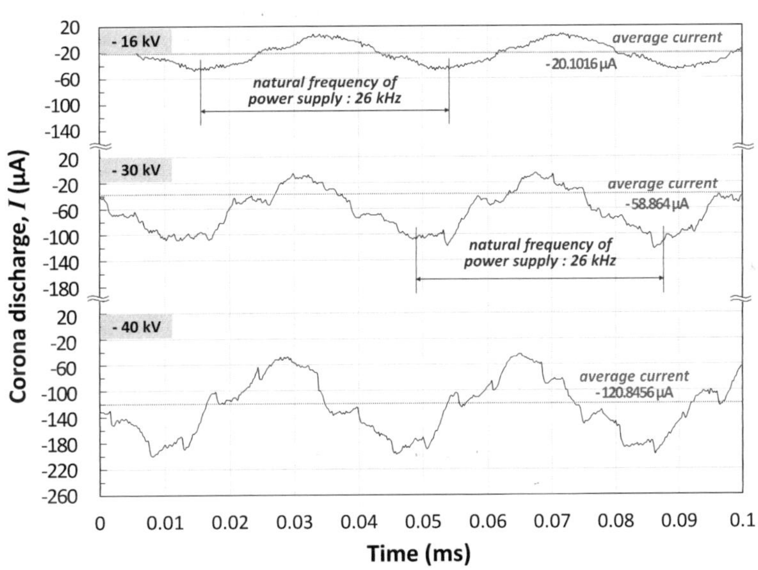

Figure 8: Waveforms of discharge current with tap water electrospray (Q = 10 mL/min).

Fig. 8 is the waveforms of tap water electrospray when voltages of -16, 30, and 40 kV are applied for 0.1ms. There is a vibration at 26 kHz in a few mV of amplitude that represents natural frequency of DC power supply. The more voltage is applied, the more average discharge current flows. Because the flow of charged droplets compensate for the potential difference between the electrodes, Trichel pulse that feature of negative corona discharge in dry condition is not appear. The Trichel pulse is that a large amount of charge is poured out instantaneously to relieve the electric field in the air as it increases. Under a fixed voltage, a current of constant magnitude appears with a constant period, and the greater the voltage, the shorter the period. However, in the case of distilled water, its droplets have few charges, and strength of electric field is continuously getting larger. So, the Trichel pulses appear which is shown in Fig. 9 and the amplitude and period of pulse at this point followed a random nature because of randomly generated droplets.

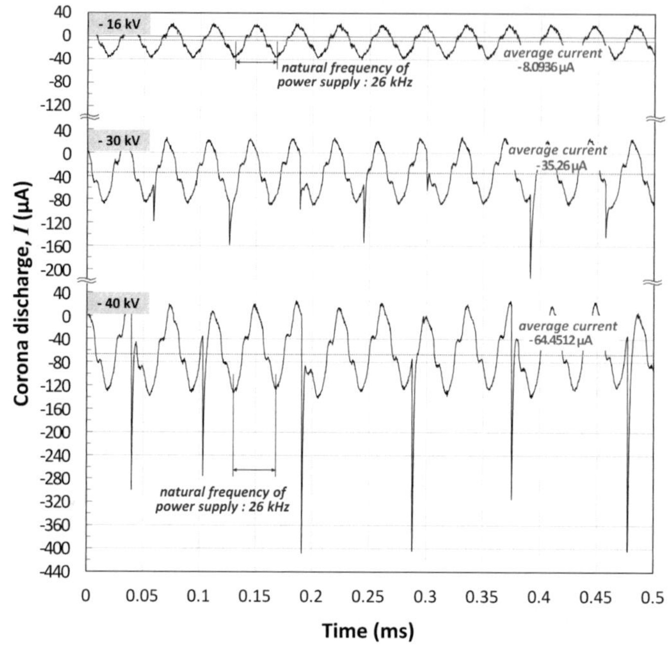

Figure 9: Waveforms of Trichel pulses of the distilled water electrospray (Q = 10 mL/min).

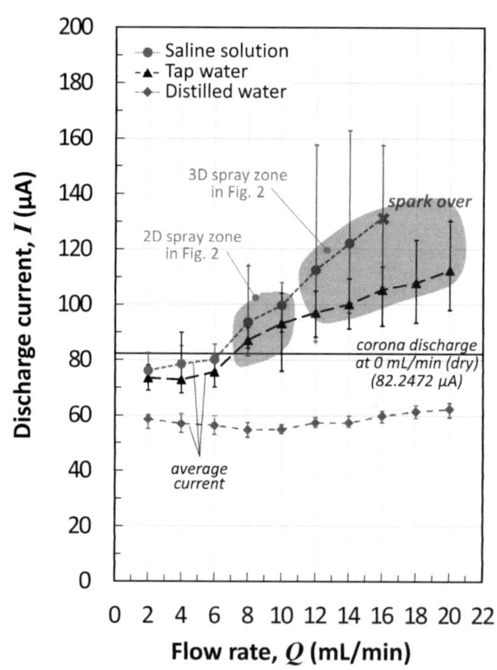

Figure 10: Discharge current according to electrical conductivity and flowrate (V = -34 kV).

Fig. 10 shows the discharge current of electrospray with flowrate increasing. The experiment is conducted at -34 kV using three different electrical conductivity of spray fluid (distilled water, tap water, and saline solution). At flow rates above 6 mL/min, the flow rate and discharge current were proportional. In other words, the contribution of current delivered by the charged droplets is dominant over corona discharge. This implies that it is easier to collect dust particles by electrical attraction as the flow rate increases. For distilled water which has very low conductivity, the current is lower than dry condition (82.2472 μA). It means that the droplets are not in charged state and cannot capture the particles dust inductively (Kim *et al.*, 2020). So, it is not suitable as working fluid for dust collection.
The power consumption of single nozzle electrospray system is analyzed based on the electrospray corona current in Fig. 11. In this graph, the higher power consumption means that the droplet can contain more electric charge when the same voltage is applied at liquids. Thus, the droplets can attract more dust particle of submicron, and it can have better performance of dust removal. The power consumption tends to increase with applied voltage and electrical conductivity of aqueous solution. However, when the electrical conductivity is higher than around that of tap water, the rate of increase in power consumption is small compared to the increase in conductivity. Even, as you can see the case of saline solution, it quickly reaches to *spark over* and it is hard to use. The tap water is suitable for application in electrospray scrubber. As a result of the PM removal performance evaluation of cyclone electroscrubber with electrospray based on optimal operating conditions derived from the lab-scale experiments, the removal efficiency of PM10 was 99.65%, PM2.5 was 97.59%, and PM1.0 was 90.95%.

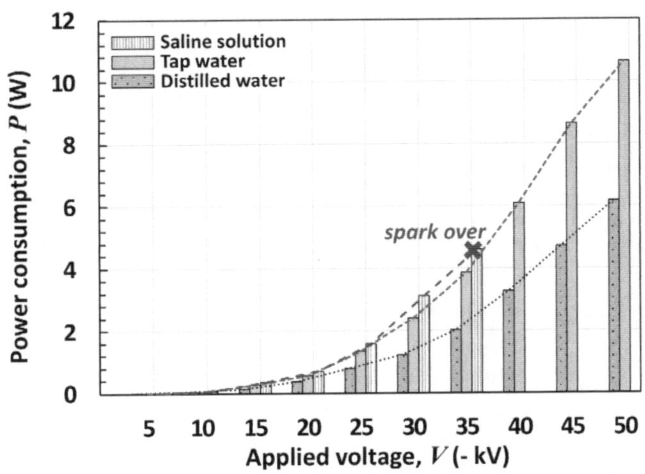

Figure 11: Relationship between power consumption and applied voltage (Q = 10 mL/min).

3. CONCLUSIONS

In this paper, electrospray of aqueous solution is visualized, and discharge characteristics are experimented for designing electrospray scrubber. The electrospray modes are classified through the visualization and the 3D spray is more suitable to dust collect because of wide spray area. During electrospray, liquid droplets carry the electron charge, and it is represented as discharge current. Liquid has higher electrical conductivity, the more the current flows. The Trichel pulse the does not appear except for distilled water which has very low electrical conductivity. Comparing flow rate and current graph, the flow rate and discharge current are proportional at flow rates above 6 mL/min. The more droplet charged means the dust removal efficiency can be higher. Through the comparison of power consumption in tap water and saline solution, there is a limit to amount of charge that can exist on a surface of droplet. As the result, tap water is suit for use not only terms of conductivity but also economic, and accessibility.

NOMENCLATURE

D	nozzle diameter	(mm)
I	discharge current	(μA)
P	power consumption	(W)
Q	flow rate	(mL/min)

V	applied voltage	(kV)

Greek symbol

τ	Trichel pulse period	(μs)

REFERENCES

Borra, J,-P., Ehouarn, P., & Boulaud, D. (2004). Electrohydrodynamic atomisation of water stabilised by glow discharge-operating range and droplet properties. *J. Aerosol. Sci.*, *35*, 1313-1332.

Chen, J., & Davidson, J.H., (2003). Model of the Negative DC Corona Plasma: Comparison to the Positive DC Corona. *Plasma Chem. Plasma Process*, *23*, 83-102.

De La Mora, J., & Loscertales, I. (1994). The current emitted by highly conducting Taylor cones. *J. Fluid Mech.*, *260*, 155-184.

Jaworek, A., & Krupa, A. (1997). Studies of the corona discharge in ehd spraying. *J. Electrostat.*, *40-41*, 173-178.

Kim, H. H., Kim, J. H., & Ogata, A. (2011). Polarity effect on the electrohydrodynamic (EHD) spray of water. *J. Aerosol. Sci.*, *42*, 249-263.

Kim, H. H., Teramoto, Y., Negishi, N., Ogata, A., Kim, J. H., Pongrac, B., Machala, Z, Alfonso, M., & Canan-Calvo (2014). Time-resolved high-speed camera observation of electrospray. *J. Aerosol. Sci.*, *42*, 249-263.

Kim, S., Jung M., Choi, S., Lee, J., Lim, J., & Kim, M., (2020). Discharge current of water electrospray with electrical conductivity under high-voltage and high-flow-rate conditions. *Exp. Therm. Fluid Sci.*, *118*, 110151.

Lopez-Herrera, J. M., Barrero, A., Boucard, A., Loscertales, I. G., & Marquez, M. (2004). An experimental study of the electrospraying of water in air at atmospheric pressure. *J. Am. Soc. Mass Spectrom*, *15*, 253-259.

Smith, D. P. H., (1986). The electohydrodynamic atomization of liquid. *IEEE Trans. Ind. Appl.*, *22*(3), 527-535.

Townsend, J. S., (1914). Electricity in gases, *Oxford University Press*, Cambridge.

Trichel, G. W., (1938). The mechanism of the negative point to plane corona near onset, *Phys. Rev.*, *54*, 1078-1084.

Yazdekhsti, A., Pishevar, A., & Valipouri, A. (2019). Inverstigating the effect of electrical conductivity on electropray modes. *Exp. Therm. Fluid Sci.*, *100*, 328-336.

ACKNOWLEDGEMENT

This research was jointly supported by National Research Foundation of Korea (NRF No. 2019R1A2C108869414) and by Korea Institute of Energy Technology Evaluation and Planning (KETEP) (20202020900290, 20214000000280, 20212050100010) funded by Ministry of Trade, Industry and Energy. This research was also supported by the Korea Environmental Industry & Technology Institute (KEITI No. 2020003060005) Authors sincerely appreciate their supports.

Developing Learning-Based Models for Occupant Centric Control

Rich Kimball[1]*, Jin Wen[1], Zheng O'Neill[2], Tao Yang[2], Yicheng Li[1]

[1]Drexel University, Department of Civil, Architectural, and Environmental Engineering, Philadelphia, Pennsylvania, United States of America
Contact Information (215.895.1918)

[2]Texas A&M University, J. Mike Walker '66 Department of Mechanical Engineering, College Station, Texas, United States of America
Contact Information (979.845.1251)

* Corresponding Author

ABSTRACT

Use of advanced building control strategies, including model predictive control, is an enabling strategy to achieve Grid-interactive Efficient Buildings (GEB). Many literature-reported control strategies are designed around an ideal building and do not account for the behavior of occupants. Yet research and field studies have shown that occupant behaviors have strong impact on building operation and energy consumption. Occupants who are uncomfortable with the control strategy will often adjust the thermostat, open/close a window, or use a personal fan/heater to better suit their comfort. How to incorporate occupant behaviors into advanced control strategies has been a focus in many of the recent occupant centric control (OCC) studies. Major challenges for OCC development include forecasting occupants' thermal comfort and behaviors and forecasting building energy with the consideration of occupant behavior. This study explores the feasibility of employing machine learning techniques, including active learning, Artificial Neural Network (ANN), and feature selection, to develop energy forecasting models that incorporate the occupant behaviors into the forecasting. To generate training and testing data needed for the control model formation, a co-simulation virtual building testbed, which utilizes a DOE Prototype residential building model developed in the EnergyPlus environment is developed. The virtual testbed also includes an Occupant Behavior Module (OBM) which is based on a previously reported agent-based-model to simulate occupants' thermal behavior in the MATLAB SIMULINK environment. Functional Mockup Units (FMU) is used to interface between the EnergyPlus environment and the MATLAB Simulink environment. The virtual testbed is used to generate both training and testing data for typical summer weather. The accuracy and scalability (under different weather and operation conditions) of the ANN-based control models are reported and compared with conventional control models. How to select and evaluate the architecture of the ANN model that is computationally efficient but also can capture the complexity of the interaction between building systems and occupants, is discussed.

1. INTRODUCTION

Advanced building control strategies have been developed in recent years with promising performance that enables Grid-interactive Efficient Buildings (GEB) to be realized. Thereunto, high-fidelity building energy forecasting modeling is critical for those control strategies and energy abnormality detection. Data-driven energy forecasting modeling approaches, especially those that use artificial intelligence methods, have demonstrated better cost-effectiveness and ease of application in the field, when compared with traditional physics-based methods (Zhang and Wen, 2021). However, these developed strategies are often developed without accounting for the behavior of its occupants. Building control strategy can be disrupted if it does not account for these stochastic occupant behaviors (Wei *et al.*, 2018). Occupant-centric Control (OCC) has therefore been a focus of many recent studies attempting to develop control strategies that incorporate occupants' needs and behaviors.

Recent OCC studies with implementation in real buildings have largely focused on individual zone level controls (e.g. thermostat, light switches) of commercial and academic buildings. These studies focused on sensing and collection of energy usage data with occupant comfort (collected by survey) considered (Park *et al.*, 2019). Other studies have used personal occupant thermal comfort sensors (e.g. wearables, smartphone apps) to collect data from occupants in real commercial buildings for use in training several types of machine learning-based models for forecasting of occupant thermal comfort (Xie *et al.*, 2020). Many studies that include machine learning-based methods, namely artificial neural networks (ANN), and support vector machines (SVM), for OCC focus on developing forecasting models for a group of occupants' thermal comfort. Studies focusing on personal comfort models (PCM) for individual occupant comfort rather than group models were the next largest category, surpassing OCC studies focused on developing models for optimization of the HVAC systems. These machine learning-based studies were found to outperform traditional predictive mean vote (PMV) occupant comfort models by significant margins (Fard *et al.*, 2022).

Based on these review papers, a gap exists in the literature, which is a lack of high fidelity and cost-effective energy forecasting model for MPC purposes in residential buildings when occupant behavior is included. Additionally, the use of machine learning methods, such as active learning, to increase the accuracy of these forecasting models, has been identified as a gap in the literature. The focus of this study has therefore been the development of a high-fidelity energy forecasting model that considers occupant behavior in residential buildings for use in a future occupant centric advanced control strategy. This study focuses heavily on the impact of occupant behavior and active learning on the accuracy of energy forecasting models. Data-driven approaches are the focus of this study based on their cost-effectiveness reported in the literature. In lieu of real building data, a virtual testbed is developed and used to generate data for this study.

The rest of this paper is organized as follows: First, the virtual testbed and its virtual measurements are described. Following is background information on the feature selection and active learning processes. Secondly, we discuss machine learning models used throughout this study, followed by an analysis on the test cases designed and the generation of data for these cases. Finally, the results of the model testing are presented.

2. VIRTUAL TESTBED

This study utilized data generated from a DOE Prototype Residential Building (DOE) virtual testbed developed in EnergyPlus. The building is a two-story detached single-family home in Houston, TX, with one conditioned living zone and an unconditioned attic zone. Within the conditioned zone, there are three occupants and typical residential appliances, while the HVAC system components are primarily housed in the attic zone. The HVAC system is a dual-setpoint controlled central air heat pump system, using schedule-based fixed setpoint control. The three occupants, including their thermal comfort and behaviors, were simulated using a literature-reported occupant behavior model (OBM) based on Bayes' theorem and adopted as an agent-based simulation, as reported by Langevin (2015). Based on the simulated indoor environmental conditions, the OBM forecasts occupants' thermal sensation. Occupants' thermal comfort was determined by comparing their thermal sensation with their thermally acceptable range based upon the Predicted Mean Vote curve detailed by De Dear (1998). If an occupants' thermal sensation was outside the acceptable range, the occupant would have a probability of taking actions to bring their sensation back into their acceptable range. Table 1 summarizes each occupant's thermally acceptable range. Figure 1 highlights the OBM occupant thermal behavior process.

Table 1: Occupant thermally acceptable ranges

Occupant	Thermally Acceptable Range	Notes
1	-1 to 1	Prefers neutral temperatures
2	-2 to 0	Prefers cooler temperatures
3	0 to 2	Prefers warmer temperature

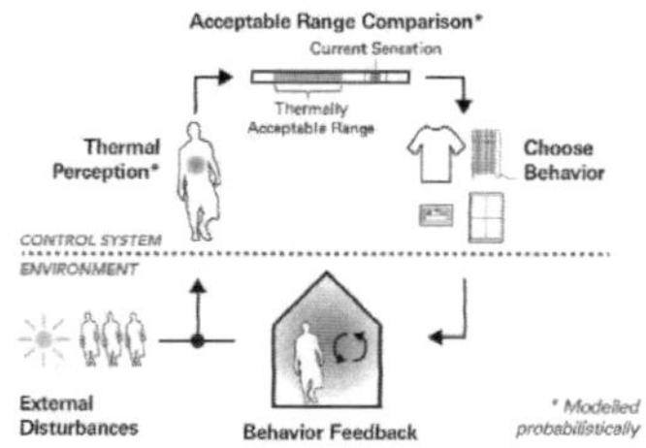

Figure 1: Occupant thermal behavior model using agent-based modeling (Langevin, 2016

The behavioral actions available to occupants' were changing their clothing level, drinking a cold/hot beverage, using a personal fan/heater, opening/closing the windows/doors/blinds, and adjusting the thermostat setpoint. These behavioral actions were assigned to both "action" and "action reversal" hierarchies that determined the order in which available behaviors were taken and/or reversed. Action reversal means adjusting the device back to its original state (e.g. occupant turned on a personal fan but now feels cold and will turn off the fan before other behaviors).

2.1 Co-Simulation Environment

A co-simulation environment was developed to incorporate the two components of the virtual building testbed, i.e., the EnergyPlus model, and the OBM in MATLAB. Moreover, two more components were connected with the virtual building testbed in this study: 1) active learning module coded in MATLAB, used for generating training data as explained later; and 2) developed energy forecasting models coded in both MATLAB and PYTHON. To allow for co-simulation of EnergyPlus and MATLAB/Python, Functional Mockup Units (FMU) were used. Figure 2 illustrates the active learning part (discussed below) of the co-simulation environment, while other parts of the co-simulation environment follow the same structure.

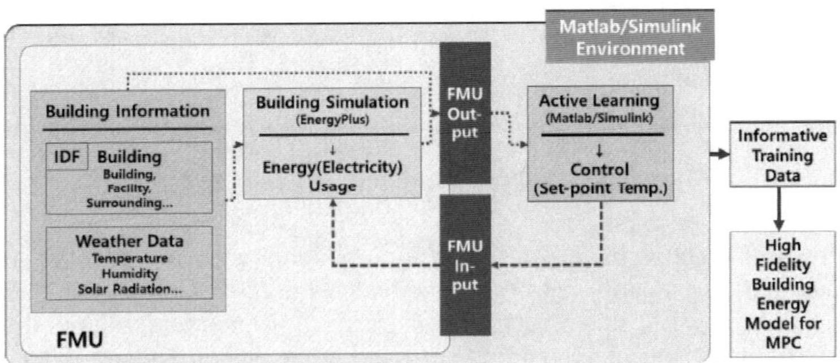

Figure 2: FMU co-simulation environment integrating EnergyPlus and MATLAB for use with active learning (Zhang and Wen, 2019).

3. VIRTUAL MEASUREMENTS AND CANDIDATE FEATURES

To develop a machine learning based model, inputs, i.e., features, are needed, which are typically selected from building measurements. The virtual testbed produces many outputs, namely virtual measurements. The candidate features used for this study were limited to features that could be measured in a real residential building. Some of the features, such as those related solar radiation, may not be easily measured for a residential building, but could be potentially obtained from a local weather station. The occupants' thermal comfort values were assumed to be obtained from market-available mobile tools, such as a smartphone app. These candidate features are summarized in Table 2. A feature selection process was used to identify the most relevant features for the energy forecasting model.

The features provided in Table 2 are shown with units in both the I-P and SI systems, denoted by [I-P, SI]. Features provided with empty [] are unitless. All features in the candidate feature set were provided at times t (current time), t-1 (previous time step), t-2 (two time steps earlier), and time t-3 (three time steps earlier). Time lag features were included because they can provide a richer dataset to capture any delayed effects on the target feature. The measurement sampling time used in this study was 1-hour for the development of an hour-ahead energy forecasting model. However, the virtual building, including its systems and occupants, were simulated with a one-minute time interval. The developed framework can be easily adapted for other energy forecasting horizons, such as 15-min ahead or 3-hour ahead. The candidate features also include control variables that are determined based on the need of the model predictive control strategy. The first three features in Table 2 represent the three setpoint control features used in this study while the fourth feature in Table 2, Whole Building Electric Energy Usage, is the target feature that the energy forecasting models predict, at time t+1 (one time step ahead).

Table 2: Candidate Feature Set

Feature Acronyms	Feature Name	Feature Acronym	Feature Name
ZTCSP	Zone Thermostat Cooling Setpoint Temperature [°F, °C]	SRDIR	Site Direct Solar Radiation Rate per Area [hp/ft^2, W/m^2]
LIGHT	Interior Lighting Level Fraction []		
DISH	Dishwasher Operational Fraction []	VMFR	Zone Ventilation Mass Flow Rate [slug/s, kg/s]
BLDGEE	Whole Building Electric Energy Usage [kWh, J]	ZAT	Zone Air Temperature [°F, °C]
OADB	Outdoor Air Drybulb Temperature [°F, °C]	ZARH	Zone Air Relative Humidity [%]
OAWB	Outdoor Air Wetbulb Temperature [°F, °C]	HPSS	Central Air Heat Pump Staged Signal []
OABP	Outdoor Air Barometric Pressure [psi, Pa]	SFMFR	Supply Fan Mass Flow Rate [slug/s, kg/s]
OAWS	Site Wind Speed [ft/s, m/s]	OCC1	Occupant 1 Thermal Comfort Value []
OAWD	Site Wind Direction [deg]	OCC2	Occupant 2 Thermal Comfort Value []
SRDIF	Site Diffuse Solar Radiation Rate per Area [hp/ft^2, W/m^2]	OCC3	Occupant 3 Thermal Comfort Value []

4. FEATURE SELECTION PROCESS

A systematic multi-step feature selection process, using historic building data (generated from the virtual testbed in this study), was performed firstly to identify the best set of features used to forecast the building energy. Figure 3 shows the steps of the process. Step 1 of this process was performed offline and results in the candidate feature set shown in Table 2. Step 2 utilized Pearson's Correlation Coefficients to remove irrelevant and redundant features while Step 3 used a machine learning based model structure, which is MARS/ANN for this study, to determine the final feature set with the best goodness of fit. The features selected from this process are specific to each test case described below.

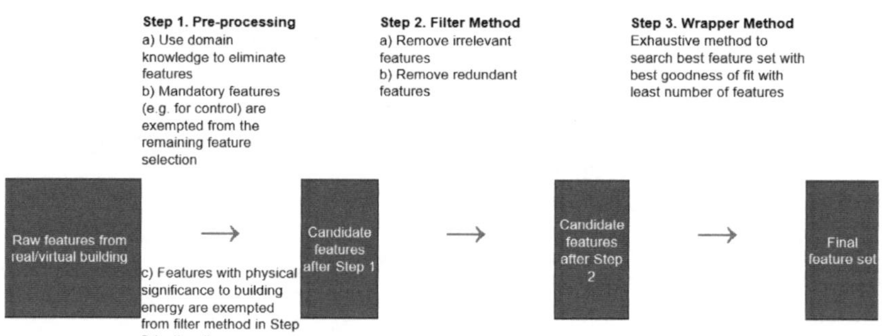

Figure 3: Systematic Multi-Step Feature Selection Process (Recreated from Zhang and Wen, 2019)

5. ACTIVE LEARNING PROCESS

A commonly reported issue with data-driven model development using historical building data is data bias. Buildings often are operated under very limited operation range. For example, zone temperature setpoint typically is not varied during normal operation. Models trained with such biased data do not have good generalization when used for MPC. To reduce data bias and increase training data quality, active learning, a machine learning technique was used to generate information-rich data for some of the forecasting models described below. A detailed description of active learning can be found from Zhang and Wen (2021). In brief, the active learning algorithm perturbs building operation in a cost-effective manner, to produce information-rich building operation data that are outside of normal operation conditions. Active learning can be used for both real buildings and virtual buildings (in this study). These conditions generated through the active learning algorithm would be similar to how the building would perform under demand response conditions. This process requires a baseline model (Model 0 below) to be used as a starting point for the forecasting model. The process is shown through pseudo-code in Figure 4 below. In a real building, active learning could be applied during the unoccupied time.

Algorithm: Expected Error Reduction in the Context of Building Energy Forecasting Model

1: Input: Normal operation data L (labeled), all possible setpoint combinations to be selected from U (unlabeled)

2: Repeat:

3: Train the building energy forecasting model $P_L(y_i|(x_i, w))$ on L. For each $x_i \in C$; x_i is controllable inputs (setpoints), w is disturbances, y_i is predicted building energy consumption.

4: for each $x_i \in U$, and its corresponding $y_i \in C$:

5: Re-train the energy forecasting model on $L' = L \cup k^*(x_i, w, y_i)$, k is the weight of added training data. The new model is marked as $P_{L'}(y_i|(x_i, w))$.

6: Calculate $V(x_i, w, y_i)$, which is the cross-validation error of $P_{L'}(y_i|(x_i, w))$.

7: end for

8: The selected (x_i, w, y_i) with the lowest $V(x_i, w, y_k)$ is marked as (x_k, w, y_k)

9: Query (or get energy consumption when applying the selected setpoint into the building) the instance (x_k, w, y_k), its energy consumption is labeled as y_k^*. The new training data is $L_i = (x_k, w, y_k^*)$. Update $L = L \cup k^*L_i$. Remove the selected combination of setpoints from the remaining setpoint combinations in the pool: $U = U \setminus x_k$.

10: Until running out of budget

Figure 4: Pseudocode of expected error reduction in the context of active learning for building energy forecasting models (from Zhang and Wen, 2019).

6. MACHINE LEARNING MODELS

Many regression or machine learning models have been reported in the literature for energy forecasting modeling (Zhang et al. 2020). For this study, two types of models are selected: Multivariate Adaptive Regression Spline (MARS) models, and Artificial Neural Network (ANN) models. MARS models had shown good performance for energy forecasting models without occupant behavior in previous studies (Cheng and Cao, 2014). ANN, however, have been reported in the literature to be able to capture stochastic occupant behavior better than other ML models (Fard *et al.*, 2022).

6.1 MARS

MARS models are non-parametric regression models that create a series of non-linear basis functions to predict target values. These basis functions are hinge functions that allow non-linearity. A final MARS model is created in two stages, a forward phase which creates a host of candidate basis functions, and a backward phase which deletes basis functions until the final selection of functions is made. They were developed by Friedman (1991). MARS does not require too much training data but can be sensitive to input data behavior.

6.2 ANN

ANN models are designed to mimic the neural networks found in the human brain and are comprised of layers of neurons that use weights and biases to learn representations of data. Each neuron in a network contains an activation function that processes the inputs and results in new outputs. A common activation function is a Sigmoid function (Seo *et al.*, 2019). ANN models can solve incredibly complex problems by creating a network with varying layers of connected neurons to map the data. The structure of an ANN is split into three categories, the input layer, the hidden layers, and the output layer. The computation is performed within the hidden layers and includes the activation functions. These ANNs require significant amounts of training data to accurately learn the problem without overfitting. ANN models are widespread in forecasting problems and have shown excellent performance (Afram *et al.*, 2017). For this study, a simple feed forward MLP ANN was used for most of the study, and a deep ANN – in the form of a recurrent neural network (RNN) - was introduced late into the study for comparison's sake.

7. TEST CASES FOR MODEL COMPARISON

In order to compare model performance under different scenarios, two test cases were designed for this study to understand how occupant behavior and active learning would affect energy forecasting accuracy. The two cases are summarized below in Table 3.

Table 3: Test Cases and Model Information for Energy Forecasting Model Comparison

	Model 0			Model 1		
Test Case	**Name**	**Model Type**	**Training Data**	**Name**	**Model Type**	**Dataset**
Case 1 (Ideal)	C1M0	MARS	Normal operation	C1M1	MARS	Normal operation + active learning operation
Case 2 (Real)	C2M0	MARS	Normal operation with occupant behavior	~~C2M1~~	~~MARS~~	~~Normal operation + active learning operation, with occupant behavior~~
	C2A0	ANN	Normal operation with occupant behavior	~~C2A1~~	~~ANN~~	~~Normal operation + active learning operation, with occupant behavior~~

Each of the two test cases were designed to provide a different combination of datasets with regard to occupant behavior. Test case 1 was the ideal building test case in which the occupants could not disrupt the building control system at any time, neither during the normal operation period nor the active learning operation period (both described below). As a result, the features related to occupant comfort are not included in Test case 1. Test case 2 represents a real building scenario where the occupants could disturb the building control system (e.g., changing their clothing level, drinking a cold beverage, operating a small personal fan, adjusting the thermostat) at any time. Both test cases are proposed to include a Model 0 and a Model 1. Model 0 is trained only on normal operation data while Model 1 is trained on an enriched dataset that includes normal operation data as well as active learning operation data.

During the study, it was found that the MARS model structure can be sensitive to input data. If the input data for a specific feature does not have enough variability, the MARS model building forward phase will fail to create any basis functions in that feature dimension and thus will not consider that feature in its forecast. While there is merit to the model not selecting a feature that is not informative to the forecast, the model failing to select a control feature (defined above) disrupts the ability for the OBM to interface with the building control scheme. To combat this sensitivity of MARS to input data, the MARS model is trained both with and without active learning data to provide a richer, more informative dataset that is less sensitive. Results are reported below for Case 1 Model 0 (C1M0) and Case 1 Model 1 (C1M1) on the performance of forecasting Whole Building Electric Energy Usage at time t+1. The ANN models trained do not have as great sensitivity to input data, rather are influenced more by the amount of training data and thus for this study, the training of Case 2 ANN Model 1 (C2A1) was not deemed necessary (given the performance of C2A0 below) and is not included but will be the topic of future studies.

8. DATA GENERATION

The generation of the data sets discussed in Table 3 is detailed in this section. All data for this study was generated using the above co-simulation environment of EnergyPlus and MATLAB/SIMULINK, designed around the summer season in Houston, TX. For Case 1 which utilizes the MARS models and does not include occupant disruption, the normal operation dataset was generated by co-simulating the virtual testbed during the first week of July (7/01 – 7/08) for summer conditions and was used to train C1M0. The active learning operation was generated during the following weekend in July (7/09 – 7/10) and utilizing C1M0 as a baseline, is used to trained C1M1. For Case 2 which utilizes an ANN model and does include occupant disruption, more historic building data was required for training due to the nature of ANN model development. Thus, the normal operation dataset was generated by co-simulating the virtual testbed for the month of June (6/01 – 6/30) to represent these conditions.

For both cases, two testing periods were used to assess the performance of the models. Each testing period lasted one week and represented different building operation conditions. The first testing period was generated during the week immediately after the active learning weekend (7/11 – 7/18) and represented normal operation conditions. The second testing period was generated during the following week (7/19 – 7/26) and represented demand response conditions. A dataset was generated for each test case for each testing period. For both cases, the three occupants are always present within the house. The occupants are sleeping between 10 PM and 6 AM. During normal operation, the thermostat controls the indoor temperature to be 72 [°F] (22.22 [°C]) when occupants are awake and 75 [°F] (23.88 [°C]) when occupants are asleep, for the summer season. The dishwasher appliance and the interior lighting system are both operated following a pre-defined schedule within EnergyPlus that represent a fraction (0 to 1) of their peak energy usage.

During the active learning operation, the algorithm perturbs the setpoints to be at varying levels between 59 [°F] (15 [°C]) and 84 [°F] (28[°C]). Given that under normal operation conditions, the dishwasher and lighting setpoints vary through most of the available range, the active learning algorithm perturbs these setpoints to simulate on/off control (e.g., 0.1 or 0.9). This perturbation occurs every 2-hours to allow the building and system to stabilize between perturbations. During the testing period that attempts to simulate a demand response operation, the three control setpoints are perturbed randomly. This random perturbation has the same perturbation frequency and setpoint range as those used in the active learning algorithm.

All other conditions are the same between test case 1 and 2, except that in test case 2, occupants can disrupt the system according to their thermal comfort. As discussed earlier, occupants are modeled using an agent-based occupant behavior module. A hierarchy exists in term of the thermal behavior that occupants can adopt. For this study, the rules ranked occupant disruption of the zone thermostat cooling setpoint as the last measure taken to address thermal discomfort. Occupants would try other behaviors such as changing their clothing level, drinking a cold beverage, and operating a small personal fan (low electricity consumption with negligible impact on whole building values) before adjusting the thermostat setpoint. Figure 5 below shows the effect of occupant disruption on the normal operation conditions for zone thermostat cooling setpoint temperature between Case 1 and Case 2. Figure 6 shows the corresponding whole building energy values for the same time period.

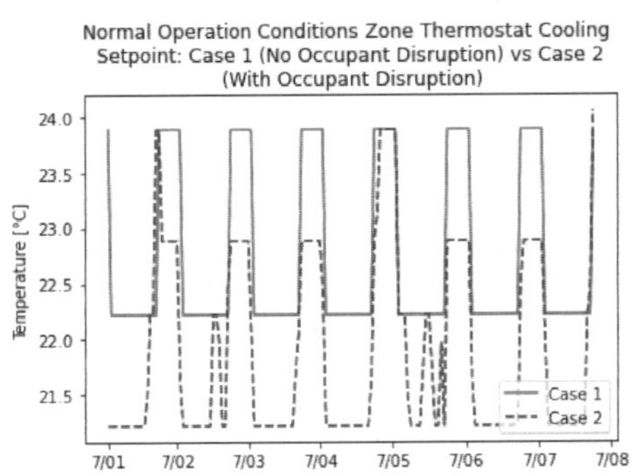

Figure 5: Normal Operation Conditions Zone Thermostat Cooling Setpoint: Case 1 (No Occupant Disruption) vs Case 2 (With Occupant Disruption)

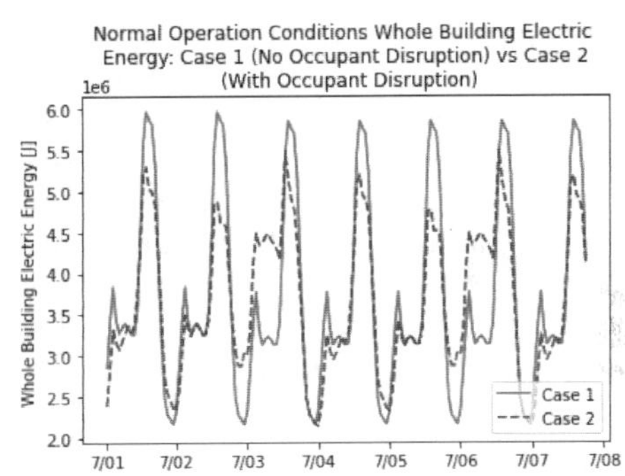

Figure 6: Normal Operation Conditions Whole Building Electric Energy: Case 1 (No Occupant Disruption) vs Case 2 (With Occupant Disruption)

Normalized Root Mean Square Error (NRMSE) of the predicted Whole Building Electric Energy at time t+1 (predicted using the models training data) vs the actual values, is used to evaluate model accuracy. Equation (1) is provided below: where $\hat{y}_i$ is the predicted value, and y_i is the actual value.

$$NRMSE = \frac{\sqrt{\sum_{i=1}^{n} \frac{(\hat{y}_i - y_i)^2}{n}}}{max_i y_i - min_i y_i} \qquad (1)$$

9. RESULTS

9.1 MARS Model Results

For Case 1 (Ideal building), two MARS models were created, C1M0, trained on the normal operation data set, and C1M1, trained on the enriched data set, which is obtained from an AL process. Both models were tested on the Case 1 with normal operation dataset as well as enriched dataset. Their performance is summarized in Table 4.

Table 4: Case 1 (Ideal) MARS Prediction NRMSE

Dataset	C1M0	C1M1	C2M0
Normal Operation Training	2%	5%	32%
Enriched Training	20%	6%	-
Normal Operation Testing	4%	6%	-
Demand Response Testing	81%	19%	-

The C1M0 model, trained on the normal operation dataset performs very well when tested on the normal operation dataset but does not reach the desired performance when tested on the demand response data set. The C1M1 model performs well on both datasets. The C2M0 model was not able to meet desired performance when occupant behavior was included. Figure 7 below shows the C2M0 model predicted Whole Building Electric Energy at Time t+1 vs. the actual values. While use of active learning to generate a more informative dataset showed good results for the C1M0 and C1M1 models, based on the magnitude of the C2M0 inaccuracy when evaluated using its own training data, it was determined that for this study, rather than study the impact of active learning for generation of a C2M1 model, the study would instead shift to focus on ANNs for Case 2. It was concluded that MARS was not comprehensive enough to capture the impact of occupant behavior on building energy consumption and thus ANNs were further studied.

9.2 ANN Model Results

For ANN models, the structure and hyperparameters of the model have a significant impact on the performance of the model, whereas for MARS models, the structure isn't as impactful as the input data is. For this study, significant testing (132 structures tested) for the best ANN structure and hyperparameters for this forecasting problem was performed based on Zhang et al., 1998) as shown in the following order: (1) testing to determine the best macro structure of the number of hidden layers (1, 2, or 3 layers) and the number of hidden neurons in each hidden layer; (2) testing to determine the most efficient optimizer for the ANN (Adam, SGD, etc.); (3) testing to determine the best activation function for the given structure and optimizer (Sigmoid, ReLU, SeLU, etc.); (4) testing to determine best learning rate; (5) retesting of the above for different training epochs if lack of convergence or model overfitting were found. The results this showed that the best structure for C2A0 was three hidden layers with 32, 4, and 1 neurons respectively. Each neuron utilized the Sigmoid activation function. The model was trained for 500 epochs with a learning rate of 0.01, used the ADAM optimizer and contained a 20% cross-validation data split. The model achieved a 3% NRMSE when evaluated on the Case 2 normal operation testing data set. Figure 8 shows the selected structure that predicted Whole Building Electric Energy at Time t+1 vs. the actual values.

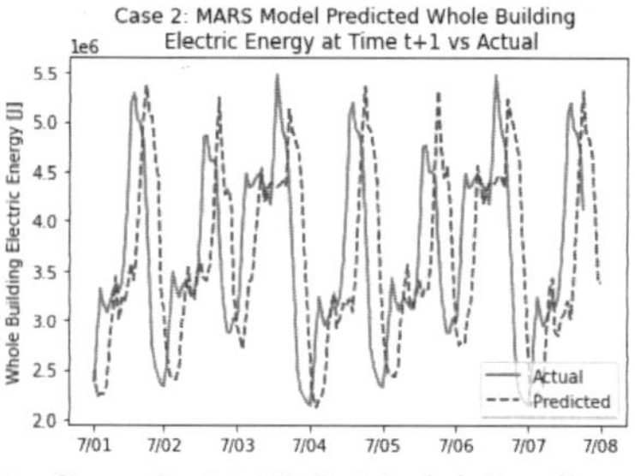

Figure 7: Case 2: MARS Model Predicted Whole Building Electric Energy at Time t+1 vs Actual

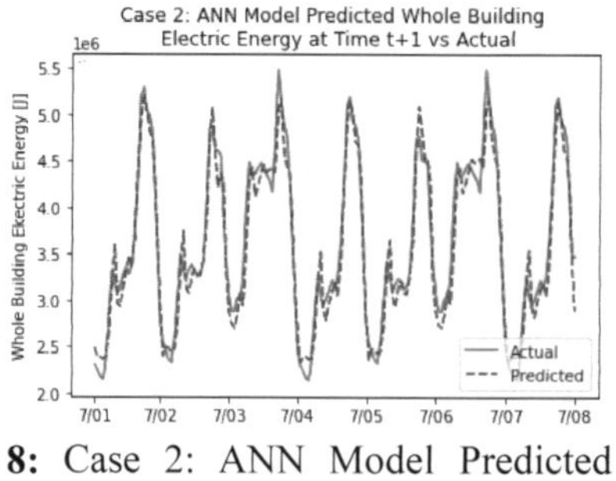

Figure 8: Case 2: ANN Model Predicted Whole Building Electric Energy at Time t+1 vs Actual

At this point it was obvious that a well-tuned simple ANN would have sufficient complexity to accurately forecast building energy when occupants' thermal behavior is included. However, for this study, it was of interest to investigate if a deep ANN would perform better and therefore literature was reviewed and a Recurrent Neural Network (RNN) was found as a suitable candidate for testing (Fan *et al.*, 2019). A model, C2R0 was trained and tested to compare. While extensive testing was required to find the best structure for a simple ANN to forecast building energy with occupant behavior, using an RNN with default parameters (per the Tensorflow Keras package in Python) achieved near equivalent performance (2.91% on Case 2 normal operation testing dataset). Only one RNN test was performed as the initial structure was sufficient.

9.3 Summary of Results

Table 5 below shows a summary of the results of the different test cases and model structures. The Case 2 models do not include enriched data sets and are not tested under demand response conditions but will be the topics of future studies.

Table 5: Summary of Model Results

	Model Type	Occupant Behavior	Data Type	Selected Features	Model NRMSE on Normal Operation Testing Data	Model NRMSE on Demand Response Testing Data
C1M0	MARS	No	Normal Operation	ZTCSP @t, LIGHT @t, DISH @t, BLDGEE @t, BLDGEE @t-1, SFMFR @t-2, SRDIF @t-4, SFMFR @t-4, ZAT @t, SFMFR @t-3, SFMFR @t-1, OADB @t-4, VMFR @t-4	4%	81%
C1M1	MARS	No	Normal Operation + Active Learning	ZTCSP @t, BLDGEE @t, BLDGEE @t-1, SFMFR @t-2. SFMFR @t-4, SFMFR @t-3, SRDIR @t-4, SRDIF @t-4, OADB @t-4	6%	19%
C2M0	MARS	Yes	Normal Operation	ZTCSP @t, BLDGEE @t, BLDGEE @t-1, SFMFR @t-1, OARH @t-4, SFMFR@t-3, ZARH@t, ZARH@t-1	38%	-
C2A0	ANN	Yes	Normal Operation	ZTCSP @t, BLDGEE @t, BLDGEE @t-1, SFMFR @t-2, OADB @t-3, SFMFR @t-1, SFMFR @t-3, OARH @t-3	3%	-

The models tested under Case 1 conditions required a significant number of features selected to achieve good performance. The enriched Case 1 model, C1M1, selected fewer features than C1M0, as that dataset was more informative resulting in less features selected to achieve good performance. Similarly, the models tested under Case 2 conditions (C2M0, C2A0, C2R0) required fewer features. The inclusion of occupant behavior in Case 2 resulted in more informative data but at the cost of increased complexity in model training requirements. When stochastic occupant behavior has a significant impact on building energy, MARS models are no longer sufficient and a more complex model, such as an ANN is required. However, ANNs require more training data and simple ANNs require significant testing to find the best tuned structure. Comparatively, a deep ANN (an RNN in this study) can achieve equivalent performance using default settings and little to no tuning. The use of these models can allow for high fidelity building energy forecasting models that can handle the inclusion of stochastic behavior and allow for advanced occupant centric building control systems.

10. CONCLUSION

This study explores the potential of using ML-based models to develop high-fidelity building energy forecasting models for a residential building, when occupant behavior is considered. The study explores several ML model structures, i.e. MARS and ANN (simple and RNN) for use in forecasting Whole Building Electric Energy at 1-hour ahead. A virtual building testbed that includes a detailed agent-based occupant behavior simulation model is used to generate both training and testing data. The impact of training data, i.e., those obtained from normal operation and from an active learning process, on model performance and model generalization is evaluated. Two testing scenarios are adopted using the developed models to forecast energy under normal operation conditions and demand response operation conditions both when occupants can and cannot impact the building control (i.e. adjust the thermostat). Comparisons are made between the different model structures and their ability to accurately forecast energy under the different testing scenarios.

The conclusions of this study are as follows. Using MARS models for energy forecasting is sufficient for forecasting building energy usage provided there is no significant stochastic occupant behavior that impacts building energy strongly. MARS model can achieve sufficient performance for demand response conditions and be able to forecast a robust dataset when active learning is used. MARS models can do so with significantly less data than required for proper training of an ANN model. However, MARS models have high sensitivity to training data and some method, such as active learning, needs to be performed to enrich the training dataset.

Use of simple ANN models will result in good performance for forecasting of building energy with or without the impact of significant stochastic occupant behavior. There is a need for significant amounts of training data as well as comprehensive tuning of model parameters to ensure that a simple ANN model is properly trained. In lieu of a simple ANN model, a deep ANN, such as an RNN, can be used with little to no parameter tuning for comparable model performance. More efforts are needed to understand why for this study, ANN models with Sigmoid activation functions perform better than those with ReLu activation functions, although literature has suggested that ReLu activation functions often perform better for nonlinear systems.

ACKNOWLEDGMENTS

This study is partially funded by the United States Department of Energy via grant EE-0008694.

REFERENCES

Afram, A., Janabi-Sharifi, F., Fung, A.S., Raahemifar, K. (2017). Artificial neural network (ANN) based model predictive control (MPC) and optimization of HVAC systems: A state of the art review and case study of a residential HVAC system. *Energy Build*, 141(4), 96-113.

Cheng, M., Cao, M-T. (2014). Accurately predicting building energy performance using evolutionary MARS. *Applied Soft Computing Journal*, 22, 178-188.

De Dear, R.J. (1998). A global database of thermal comfort field experiments. *ASHRAE Transactions* (104, 1141). Atlanta, United States of America.

DOE. Residential Prototype Building Models. https://www.energycodes.gov/prototype-building-models#Residential.

Fan, C., Wang, J., Gang, W., Li, S. (2019). Assessment of deep recurrent neural network-based strategies for short-term building energy predictions. *Applied Energy*, 236, 700-710.

Fard, Z.Q., Zomorodian, Z.S., Korsavi, S.S. (2022). Application of machine learning in thermal comfort studies: A review of methods, performance, and challenges. *Energy & Buildings*, 256(111771).

Friedman, J.H, 1991. Multivariate Adaptive Regression Splines. *The Annals of Statistics*, 19(1), 1-67.

Langevin, J., Wen, J., Gurian, P. (2016). Quantifying the human–building interaction: Considering the active, adaptive occupant in building performance simulation. *Energy and Buildings,* 117(4), 372-386.

Langevin, J., Wen, J., Gurian, P. (2015). Simulating the human-building interaction: Development and validation of an agent-based model of office occupant behaviors. *Building and Environment,* 88(6), 27-45.

Park, J.Y., Ouf, M.M., Gunay, B., Peng, Y., O'Brien, W., Kjærgaard, M.B., Nagy, Z. (2019). A critical review of field implementations of occupant-centric building controls. *Building and Environment*, 165:106351.

Seo, B., Yoon, Y.B., Song, S., Cho, S. (2019). ANN-based thermal load prediction approach for advanced controls in building energy systems. *ARCC Conference Repository*. Retrieved from https://arcc-journal.org/index.php/repository/article/view/657

Wei, S., Yong, J., Ng, B., Tindall, J., Lu, Q., Du, H. (2018). Occupant Adaptive Behavior: An Effective Method Towards Energy Efficient Buildings. *Presented at the 2018 CIBSE Technical Symposium*. London, UK: April 12-13, 2007.

Xie, J., Li, H., Li, C., Zhang. J., Luo, M. (2020). Review on occupant-centric thermal comfort sensing, predicting, and controlling. *Energy and Buildings,* 226(110392).

Zhang, G., Patuwo, B.E., Hu, M.Y. 1998. Forecasting with artificial neural networks: The state of the art. International Journal of Forecasting. 14:35-62.

Zhang, L., Wen, J. (2019). A systematic feature selection procedure for short-term data-driven building energy forecasting model development. *Energy and Buildings,* 183:428-442.

Zhang, L., Wen, J., Li, Y., Chen, J., Ye, Y., Fu., Y., Livingwood., W. (2020). A Review of Machine Learning in Building Load Prediction. *Applied Energy*, 285(116452).

Zhang, L., Wen, J. (2021). Active Learning Strategy for High Fidelity Short-Term Data-Driven Building Energy Forecasting. *Energy and Buildings*, 244(111026).

Object Tracking-based Droplet Characterization on High Flowrate Electrospray for PM Removal

Minkyu Jung[1], Soyeon Kim[2], Donik Ku[3], Sanghun Jeong[4], Soojin Bae[5], Gijeong Seo[6], Minsung Kim*

[1,2,3,4,5,*]Department of Intelligent Energy and Industry, Chung-Ang University, 06975, Seoul, Republic of Korea
([1]mamba24@cau.ac.kr, [2]nuri36@cau.ac.kr, [3]ehsdlr@cau.ac.kr, [4]suj531@cau.ac.kr, [5]sooj980809@cau.ac.kr)

[6,*]School of Energy Systems Engineering, Chung-Ang University, 06975, Seoul, Republic of Korea
([6]koki8235@cau.ac.kr)

* Corresponding Author (+82-2-820-5973, *minsungk@cau.ac.kr)

ABSTRACT

Electrospray is a phenomenon of liquid atomized into small droplets when subjected to high electrostatic potential difference. The cone-jet mode, which forms a stable cone and produces fine monodisperse droplets has been mostly focused on the previous studies. Flowrate of the cone-jet mode were limited at only few microliter/min scale to prevent corona discharge. However, in the electrospray mode for particulate matter removal, high flowrate of several milliliter/min is used to obtain a sufficient liquid-to-gas ratio. In this high flowrate condition, the applied voltage required to atomize the liquid increases, and 2D and 3D spraying mode that have not been shown in previous studies appears. In this study, the electrospray droplet diameter under high flowrate condition was measured by the shadowgraph method using a high-speed camera. Multi-layer image processing algorithm was employed in order to measure the droplet diameter from pixel occupied in the original image. Bias caused by the difference in the probability of being included in the image depending on the speed of the droplet, was corrected through object tracking. Finally, corrected droplet diameter distribution of each spraying mode under high flowrate condition was measured. Results of this study provide new guidance for calculating the droplet lifetime surviving evaporation and dust collection efficiency in an actual scrubber condition.

1. INTRODUCTION

As the social interest on particulate matter has been increased nowadays, environmental standards and regulations at emission of air pollutants are being strengthened all around the world. Power plants, emits large amount of air pollutants, use wet scrubbers as exhaust gas treatment systems. In order to overcome its large water consumption, electrospray was adopted. Electrospray is a phenomenon that the working fluid sprayed into small droplets when passing through the nozzle tip with high electrostatic charge induced. Fig. 1 shows the mechanism of PM removal with electrospray. After dispersion, charged droplets collide with positively charged PM, and removed by cleansing water at the collecting electrode. Flowrate of the working fluid and the applied voltage have the greatest influence on determining the electrospray mode. In previous electrospray research, 'Cone-jet mode', which is the most characteristic mode among the various modes has been focused (Borra, 2018). Stable cone and the jet are formulated at the end of the nozzle when the electric shear stress, surface tension and the hydraulic pressure are at the equilibrium (Jaworek and Krupa, 1999). Taylor (1966) named the stable cone as "Taylor cone" and calculated the theoretical half-angle of the cone to be 49.3°. Then the liquid elongated into thin jet by electrostatic repulsive force and atomized into fine droplets.

Cone-jet mode has been applied in various fields such as coating and EHD printing because the stability of the spray is very high, and the produced droplets are fine and uniform (Tang and Gomez, 1994). The flowrate conditions of the cone-jet mode studies were limited under few microliters per minute (Cloupeau, 1994). This is because, as the flow rate increases, the applied voltage required for atomization also increases and causes corona discharge that prevents stable cone formation (Kim *et al.*, 2011, Park et al., 2014). However, the flowrate of the electroscurbber has to be up to several milliliter per minute in order to increase the contact area with the flue gas for better collection efficiency. Moreover, required applied voltage also increased up to several tens of kV and having a different spray mode from the previous electrospray studies. Cho *et al.* (2019) classified the spray modes of high flowrate electrospray for PM removal, and Kim *et al.* (2019) visualized each mode and studied about the discharge characteristics with respect to electric conductivity of working fluids.

In the case of high flowrate spraying mode (2D 3D spraying mode) for PM collection, range of the spray is wider, and the spray density is lower than the cone-jet mode. Therefore, conventional method for measuring the size of the electrospray droplets, for example PDA method, cannot be used. This is because PDA measures the size of a particle that passing the measuring point, so in the case of sparse spray like 3D spray, the representativeness of the distribution is very low. Also, as a result of the measurement, the distribution was not well measured, so that the study on the measurement of the spray droplet size of high-flowrate electrostatic spraying for dust collection through the imaging method was conducted (Jung *et al.*, 2022).

With the development of cameras, the accuracy of the imaging method of measuring the size through the pixels occupied by the object captured in the picture has increased. And the several research has been conducted to measure the size distribution of particles through the imaging method in various particle generations. However, when the droplet size distribution is measured using the imaging method, an error in the distribution according to the particle velocity occurs. In the case of droplets with different velocities, there is a difference in the probability of being included in the image, and a biased distribution is obtained. In the present study, various spraying modes at high flowrate conditions were visualized through a high-speed camera, and the size of the spray droplets were measured through multi-stage image processing. After image processing at one frame, same process has been applied sequentially to the next frame image. Object track has been constructed through every frame with detected droplet data. Biased droplet diameter distribution and the corrected distribution were measured and compared. As a result, smaller-lighter droplets were underestimated, and the larger-heavier droplets were overcounted to the distribution. When measured without an object tracking method, a bimodal distribution appears unlike the actual spray distribution, and the actual spray distribution follows a lognormal distribution. In addition, the velocity distribution of the electrospray droplets was measured through the analysis of each frame of the high-speed camera.

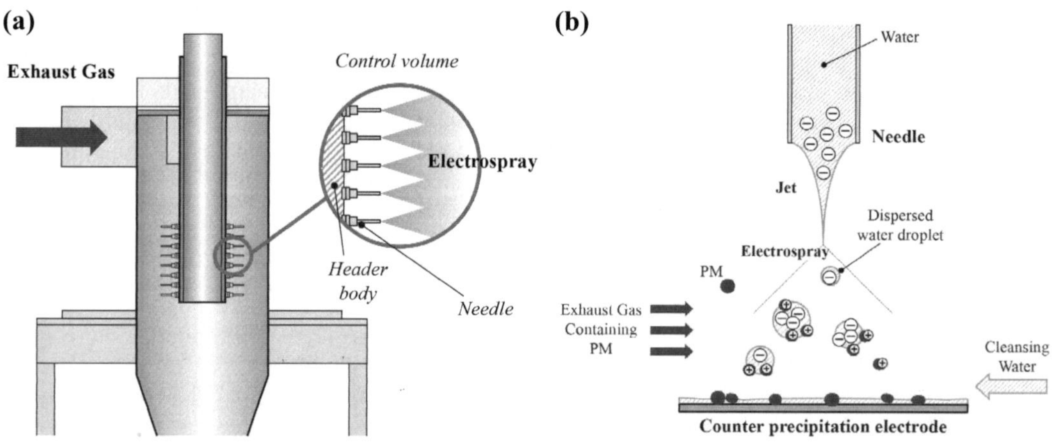

Figure 1: (a) Diagram inside the electroscrubber, (b) Mechanism of electrospray precipitation

2. EXPERIMENTAL SETUP

Fig. 2(a) shows the experimental setup of present study. Flowrate of the working fluid and applied voltage were set with a syringe pump and a high-voltage DC power supply. The high-voltage power supply is attached to a stainless-steel nozzle so that the working fluid passing through the nozzle is charged into negative voltage. To prevent the electricity leakage, the nozzle and the ground plate is located in acrylic chamber because the voltage condition of this study is high enough to exceed the electrical breakdown voltage of air. Experimental conditions and properties of the working fluid were summarized at Table 1. As mentioned above, flowrate of this study was 10 and 15 mL/min, which is tens of times larger than typical nanoscale electrospray studies. Applied voltage was adjusted in –5 kV increments from 0 kV to –30 kV, until the arc discharge initiate.

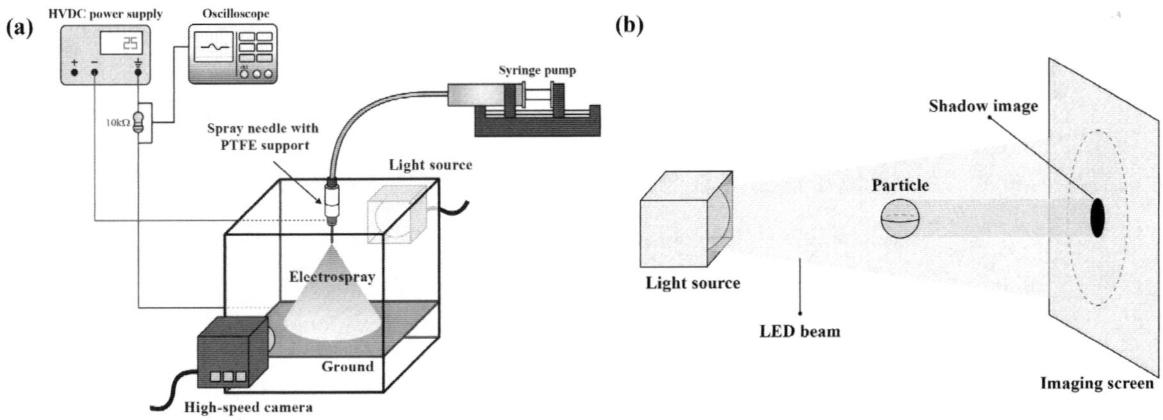

Figure 2: (a) Schematic diagram of experimental setup, (b) Working principle of shadowgraph

Table 1: Experimental conditions and parameter setup

Parameter	Specification	Range	Unit	Remarks
Electrospray	Applied voltage (V)	0 ~ 30	-kV	Negative DC
	Flow rate (Q)	5 ~ 15	mL/min	
	Nozzle to plate distance (h)	50	mm	
Nozzle	Gauge	18/21	G	Stainless steel
	Inner diameter (d_0)	0.86/0.5	mm	
	Outer diameter	1.26/0.8	mm	
Working fluid	Density (ρ)	997	kg/m^3	Distilled water
	Electric conductivity (K)	1.20×10^{-4}	S/m	
	Surface tension (γ)	0.072	N/m	
	Permittivity of vacuum (ε_0)	8.854×10^{-12}	F/m	
	Dielectric constant (ε)	80.2	-	
High-speed camera	Frame rate	4000	fps	

Trajectories of the moving droplets were visualized with high-speed camera due to the high velocity of electrospray droplets. High shutter speed not only makes object tracking easier, but also enables to measure the droplet velocity. Shadowgraph is a non-intrusive method for particle size measurement that is independent of particle shape and spray structure. A light source, particles, and the imaging device must be aligned on the same line as shown in Fig. 2(b) to obtain a shadowgraph image. A light that does not pass through the particles is reflected or refracted, creating a shadow of the particles in the image.

In high flowrate electrospray in this study, the spray range of droplets sprayed from a single nozzle is very wide, about a circle with a diameter of 30 cm. Therefore, in order to capture all the droplets in this range, the distance between the camera and the nozzle must be relatively long. At this time, in order to be able to shoot all the small droplets, the maximum shutter speed that does not reduce the image size was set to 4000. In addition, a 100mm macro lens was used for the lens used for image shooting. Moreover, in order not to blur the droplets in front and back as much as possible based on the nozzle, the aperture was tightened to secure a wide field of view.

Fig. 3 shows typical images of high flowrate electrospray with the optical configuration explained before. In Fig. 3, distilled water was supplied to the nozzle with a flowrate of 5mL/min, and the applied voltage was set as -12.8kV. Each image in Fig. 3 is an image extracted from a larger set of experiments, at the time intervals of 2.5μs to make it easy to check the movement of spray droplets.

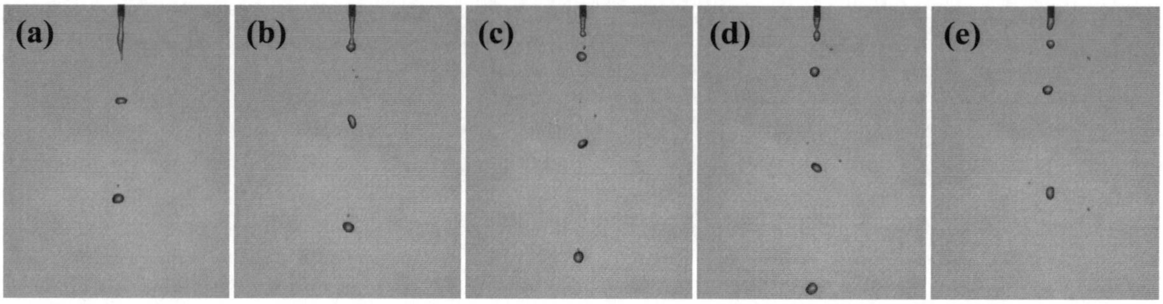

Figure 3: Examples of high-speed camera images of electrospray droplets

3. EXPERIMENTAL RESULTS

3.1 Image Processing

Sequence of image processing of high-speed camera image was shown in Figure 4. Entire image processing sequence were done with python languages and OpenCV libraries. Figure 4(a)~(f) show the original image, grayscale image, threshold image, noise removed image, edge detection image and the final labeled image, respectively.

Image processing has the greatest dominance in object tracking performance beyond entire process. At first, original RGB images were normalized in order to obtain homogenous background illumination. Normalized image was converted into 256-bit grayscale image and then binarized into B&W image with adaptive gaussian thresholding. Small salt and pepper noises included into binary image during the thresholding, were removed with median filter. Then droplets were subtracted from the background through the Suzuki's contour finding algorithm.

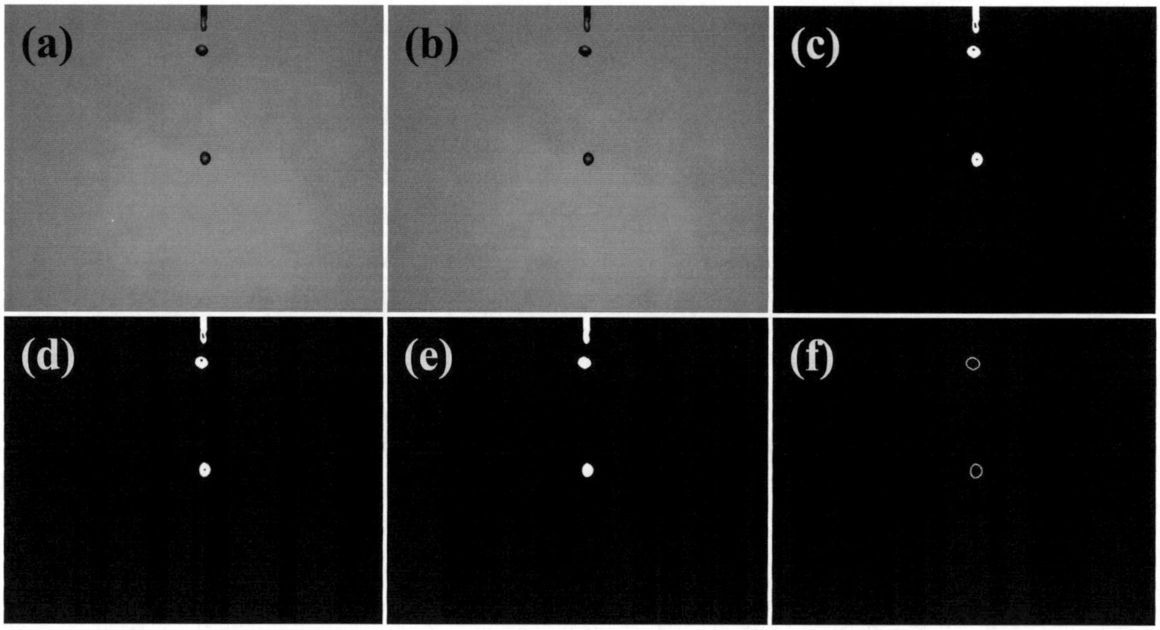

Figure 4: Sequences of image processing: (a) Original image, (b) Grayscale image, (c) Threshold image, (d) Noise removed image, (e) Edge detection image, (f) Final image

Unnecessary elements such as nozzle and the jet were filtered through the circularity. After detecting the droplets and the edges, the droplet diameter was measured by calculating the area of the separated droplet parts based on the outer diameter of the nozzle.

3.2 Object Tracking

As illustrated on the introduction, bias at the distribution due to the difference in velocities occurs. Taking a large and slow droplet as an example, in case of using high-speed camera that takes several shots in a short time, the droplet will be accumulated several times and results in overcounted in the size distribution. In addition, even if a normal DSLR shadowgraph is used or selecting the image randomly, there still is a difference in the probability that the droplet will be included in the region of interest within the instant the photo is taken, so this bias is unavoidable when using the imaging method. There are two possible solutions; first is to correct the bias with correlations between the size and velocity, and the second is to eliminate the bias by selecting the representative value among the track elements.

Object tracking from the recorded videos are comprises three stages. The image processing process described earlier is this stage. Second is to identify the detected objects of the next frame and then associate the same object. In this study, we designed an algorithm that can track the droplet through the physical characteristics of the droplet captured in each frame. In the first step, the size and centroid coordinates of the droplet were calculated through the image processing process, and the droplet with the smallest change in the size, moving distance, and moving angle of the droplet captured in each frame was judged as the same droplet. The last is to connect to the last frame of the video to form a track for each droplet.

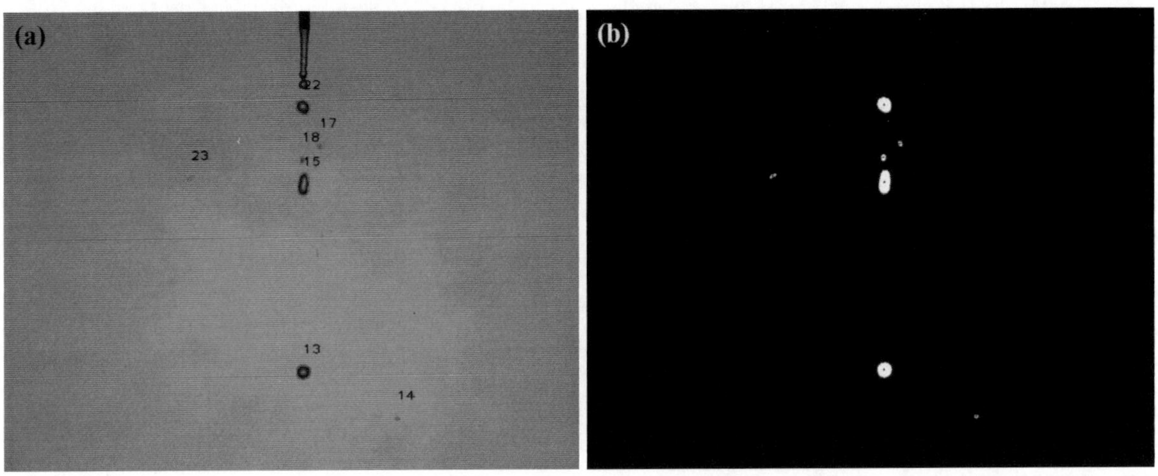

Figure 5: (a) Object tracking result, (b) Detected droplets at image processing during object tracking

Fig. 5 shows the snapshot of the tracking results of high flowrate electrospray image. 2180 frames of high-speed camera images were analyzed, and 10437 droplet data were obtained before tracking. As a result of object tracking, those 10437 raw data were identified as 417 droplets.

3.3 Droplet Characterization

Through image processing, droplet diameter distribution of distilled water electrospray (5 mL/min, -12.8 kV) was measured. The droplet size distribution obtained by analyzing the high-speed camera image in the usual way, that is, the bias distribution and the new droplet size distribution obtained using the object tracking method, that is to say the corrected distribution, were measured. Fig. 6 shows the comparison of biased distribution and the corrected distribution. According to the difference in the amount of data of the two distributions, both distributions were normalized and displayed as a histogram.

Biased distribution showed a bimodal distribution near 500μm and 1800μm. In addition, the peak probability density of small and large droplets was found to be similar. However, in the actual spray image, there are relatively more small droplets than large ones, indicating that there is an error in the distribution. On the other hand, in the case of the corrected distribution, it can be seen that small droplets of 500μm are predominantly generated compared to the large droplets. Droplets of 1000μm rarely exist, and droplets larger than 1500μm appears frequently. This is because large droplets are slow and thus overcounted in a biased distribution as they are captured much more in the high-speed camera image and are more included in the raw data. On the other hand, small droplets were underestimated at the biased distribution compared to the corrected distribution. Because small droplets are light in weight, so they are accelerated more due to the electrostatic field and highly charged liquid jet. According to the high-speed camera image, it was confirmed that small droplets disappear rapidly from the center of the nozzle to the outside. As following, if the distribution error caused by the speed difference is not corrected, a large bias in the droplet distribution occurs.

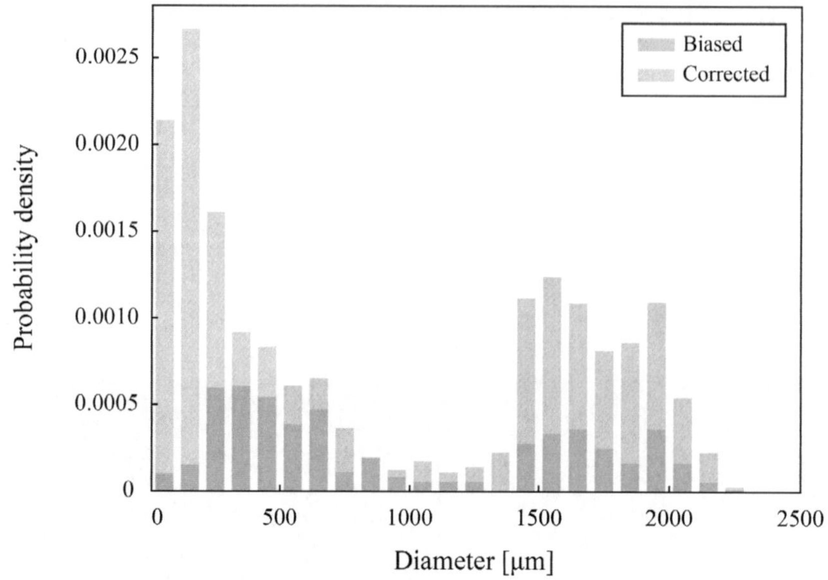

Figure 6: Comparison of biased and corrected droplet distribution

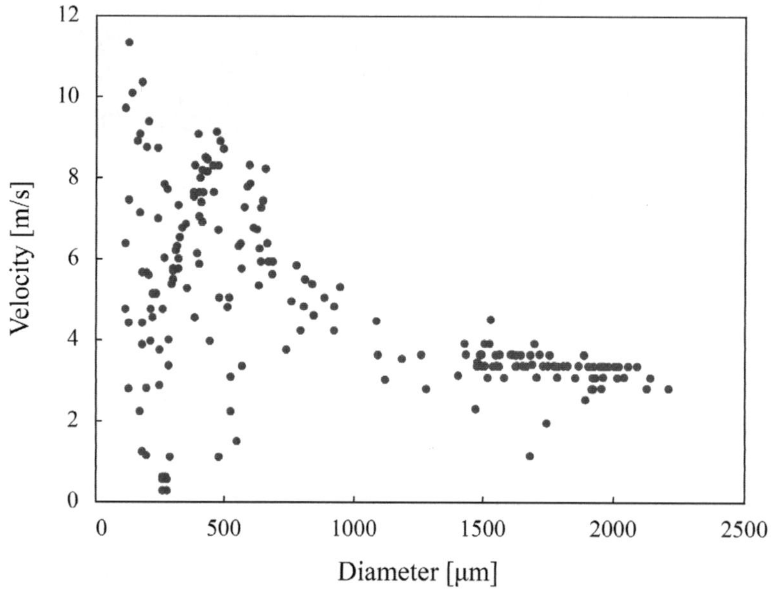

Figure 7: Droplet velocity verses droplet diameter

The velocity of each droplet was measured based on the centroid of droplets of the frames obtained through the object tracking. The average size and average velocity of the traced droplet from the start to the end of the track are shown in Fig. 7. Since the depth of field (DOF) in the image is negligible, the velocity of the droplet is calculated based on 2D trajectory. As shown in Fig. 7, the maximum velocity of the droplet tends to be inversely proportional to the size of the droplet. This explains why the distribution error appears. However, in the case of droplets smaller than 500μm, droplets with low average velocity were observed. This is a special case of electrospray, and it occurs because the droplets are negatively charged after spraying. Fig. 8 is a diagram showing an example of the behavior of droplets during the distilled water electrospray, and the velocity distribution of droplets A and B after atomization. As shown in Fig. 8(a), droplet B is accelerated downward at a very high speed immediately after being atomized. However, after the frame 21, the velocity decreases and disappears due to the electrostatic repulsive force with the charged droplet A as it approaches the large droplet A that exists below vertically.

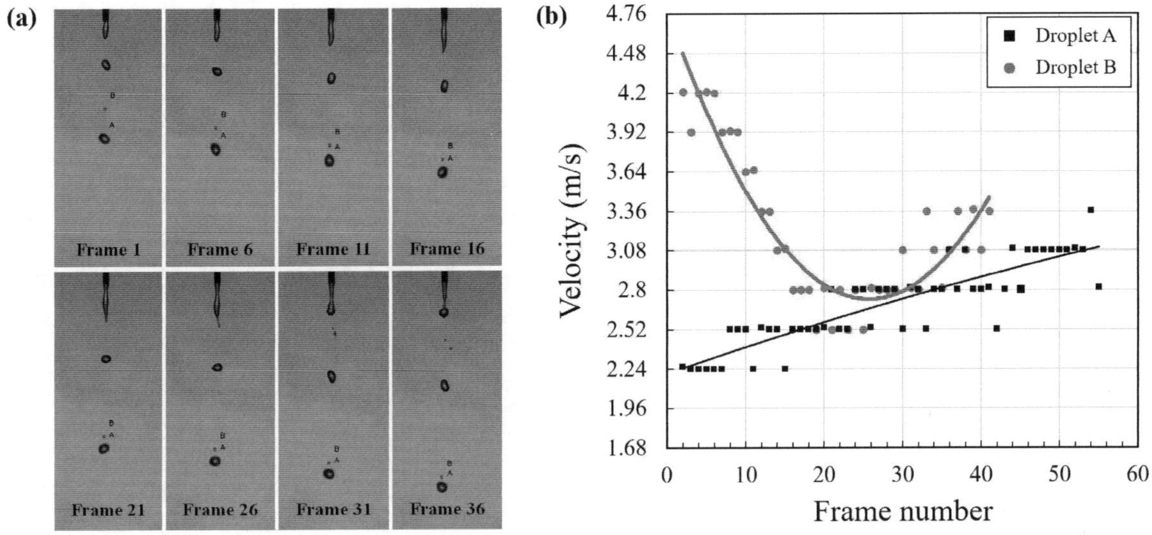

Figure 8: (a) Behavior of the electrostatic spray droplets, (b) Velocity changes of two droplets

For this reason, in the case of electrostatic spray droplets, it is difficult to clearly define the relationship between the droplets and the velocity. Therefore, when using the imaging method, the object tracking method should be used to correct the error caused by the speed difference.

4. CONCLUSIONS

In this study, a visualization-based study was conducted to measure the characteristics of droplets generated from the mode of high flowrate electrospray that can be applied to the electroscrubber precipitation system. Diameter of the spray droplet was measured by calculating based on the black pixel that spray droplet occupied in the high-speed camera image. High-speed camera images were preprocessed through multi-stage image processing algorithms. Additionally, biased of droplet diameter distribution due to the velocity difference were recognized and corrected with object tracking method.

Small, light droplets are rapidly accelerated by the electromagnetic field and disappear from the image plane, underestimated in the distribution measured in the usual way. On the other hand, since large droplets are overcounted for the same reason, the overall distribution is flattened, resulting in a bias. In case of electrospray, not only the jet, but also the sprayed droplets are charged, so the size and velocity relationship are not easily defined. In conclusion, a tracking method should be used for bias correction at droplet characterization with imaging method. As a follow-up study, electrostatic spray research on temperature and humidity in consideration of the above flue gas atmosphere characteristics is in progress.

NOMENCLATURE

V	applied voltage(negative)	(-kV)
Q	flow rate	(mL/min)
K	electric conductivity	(S/m)
h	distance of nozzle to plate	(mm)
d_0	nozzle inner diameter	(mm)

d_d	droplet diameter	(μm)
v_d	droplet velocity	(m/s)

Greek symbol

ρ	density	(kg/m^3)
γ	surface tension	(N/m)
ε_0	permittivity of vacuum	(F/m)
ε	dielectric constant	(–)

REFERENCES

Borra, J.P., 2018, Review on water electro-sprays and applications of charged drops with focus on the corona-assisted cone-jet mode for high efficiency air filtration by wet electro-scrubbing of aerosols. Journal of Aerosol Science 125, 208-236.

Cho, Y., Kim, S., Lim, H., Choi, S., Kim, M., 2019, Experimental study of electrostatic spray modes of high-flowrate water with horizontal nozzle. Journal of Mechanical Science and Technology 33(9), 4563–4572.

Cloupeau, M., 1994, Recipes for use of EHD spraying in cone-jet mode and notes on corona discharge effects, Journal of Aerosol Science 25(6), 1143-1157.

Jaworek, A., Krupa, A., 1999, Classification of the Modes of EHD Spraying. Journal of Aerosol Science 30(7), 873-893.

Jung, M., Kim, S., Lim, J., Lee, J., Jeong, S., Ku, D., Na, I., Kim, M., 2022, Droplet Characterization of High-Flowrate Water Electrospray using Shadowgraph Image Analysis. Journal of Mechanical Science and Technology 36 (4) (2022), 2139~2148.

Kim, H.H., Kim, J.H., Ogata, A., 2011, Time-resolved high-speed camera observation of electrospray. Journal of Aerosol Science 42(4), 249-263.

Kim, S., Jung, M., Choi, S., Lee, J., Lim, J., Kim, M., 2020, Discharge current of water electrospray with electrical conductivity under high-voltage and high-flow-rate conditions. Experimental Thermal and Fluid Science 118, 110151.

Park, I., Kim, S.B., Hong, W.S., Kim, S.S., 2015, Classification of electrohydrodynamic spraying modes of water in air at atmospheric pressure. Journal of Aerosol Science 89, 26–30.

Tang, K., Gomez, A., 1994, On the structure of an electrostatic spray of monodisperse droplets. Physics of Fluids, 6(7), 2317–2332.

Taylor, G.I., 1966, The force exerted by an electric field on a long cylindrical conductor, Proceedings of the Royal Society A 291(1425), 145-158.

ACKNOWLEDGEMENT

This research was jointly supported by National Research Foundation of Korea (NRF No. 2019R1A2C108869414) and by Korea Institute of Energy Technology Evaluation and Planning (KETEP) (20202020900290, 20214000000280, 20212050100010) funded by Ministry of Trade, Industry and Energy. This research was also supported by the Korea Environmental Industry & Technology Institute (KEITI No. 2020003060005) Authors sincerely appreciate their supports.

Contagion Risk Assessment For COVID-19 Variants With A Dynamic Approach For A Multizone Building Model Of University Classrooms

Riccardo ALBERTIN[1]*, Giovanni PERNIGOTTO[2], Andrea GASPARELLA[3]

[1]Free University of Bozen-Bolzano, Faculty of Science and Technology,
Bolzano, Italy
Tel: +39 0471 017886, Fax: +39 0471 017009, email: riccardo.albertin1@natec.unibz.it

[2]Free University of Bozen-Bolzano, Faculty of Science and Technology,
Bolzano, Italy
Tel: +39 0471 017632, Fax: +39 0471 017009, email: giovanni.pernigotto@unibz.it

[3]Free University of Bozen-Bolzano, Faculty of Science and Technology,
Bolzano, Italy
Tel: +39 0471 017200, Fax: +39 0471 017009, email: andrea.gasparella@unibz.it

* Corresponding Author

ABSTRACT

Due to the ongoing COVID-19 pandemic, different strategies have been employed to reduce the risk of contagion for the occupants of the built environment, in particular in educational buildings. The adoption of higher ventilation rates and personal protection devices, such as facial masks, are among the most common solutions to reduce the infection risk and allow students to keep attending classes. Since several studies, dedicated to developing strategies to limit the COVID-19 airborne transmission, have been focusing on isolated environments under steady-state conditions, the possibility for pathogens to spread to adjacent environments or the effects of dynamic occupancy schedules have been marginally investigated, albeit being of special interest for school buildings.

In a previous work, a statistical model able to evaluate the risk of infection was coupled with a TRNSYS building simulation model of three classrooms at the campus of the Free University of Bozen-Bolzano (UNIBZ), Italy. Two out of three classrooms are part of the UNIBZ Living Labs, where the indoor environmental conditions are monitored by a network of sensors. Different scenarios had been investigated, by varying ventilation strategy, number of occupants and mask utilization, and assessing for each case the airborne contagion risk. The study showed a strong correlation between a poorly ventilated environment and high risk of infection, underling also the possibility for the pathogen to spread into the adjacent environments in case of dynamic occupancy or if the classrooms doors are left open.

Drawing on the results of the previous research, this work expanded the scope of TRNSYS and TRNFLOW, extending the building model and the airflow network to the whole floor, including classrooms of different size and some offices. In such a way, the internal airflows were assessed dynamically by considering also cross-ventilation effects, improving the accuracy of the calculated infiltration rates and concentrations of COVID-19 pathogen. After a new calibration and validation against the experimental data collected in the UNIBZ Living Labs, a series of transient scenarios were simulated by accounting for the strategies listed before, as well as by considering the effects of three different vaccines – Pfizer, Moderna, AstraZeneca – with a fixed coverage percentage of full/partially/not-vaccinated occupants, new COVID-19 variants, and introducing portable air cleaners. The spread of new COVID-19 variants, which in some cases have proved to be more easily transmitted, made it necessary to re-evaluate and, possibly, adapt the strategies implemented so far, also accounting for different relative vaccination efficacies.

High risk of infection was found related to environments with poor ventilation, especially if the Omicron variant was present, even with the additional protection provided by the vaccines. Air purifiers were a viable solution to reduce the risk of infection in those cases where mechanical ventilation and/or natural ventilation was missing also for the Omicron variant, especially if combined with the effects of a mixed ventilation strategy and mask utilization.

1. INTRODUCTION

COVID-19 outbreak has been widely analyzed in the recent years to understand how to hinder the spread of the virus, especially in crowded places, as classrooms and offices, where the occupants spend long periods of time. An example is given by Di Gilio *et al.* (2021), who assessed the risk of contagion for 11 classrooms in 9 schools using the CO_2 concentration as a proxy. Similarly, Park *et al.* (2021) assessed the risk of contagion under different ventilation strategies, employing a tracer gas to evaluate the airflow rates. Finally, Buonanno *et al.* (2020a) developed the "Airborne Infection Risk Calculator" (*AIRC*), a tool able to evaluate the individual contagion risk for occupants in a room under different ventilation rates. As a whole, the researches in the literature prioritized the development of strategies aimed at lowering the probability of contagion due to the airborne transmission of COVID-19, among which there is the possibility to remove some of the airborne particles in indoor spaces with air purifiers, as investigated by Burgmann and Janoske (2021).
Despite the availability of different solutions, it is however important to remember that ventilation-based strategies may be compared in terms of airborne contagion risk reduction only if it is assumed that the airborne transmission is the main route of contagion (Buonanno *et al.*, 2020b). Usually, the tools able to evaluate the risk of contagion in a specific scenario focus on one room with a fixed occupation schedule and a given ventilation rate. Furthermore, these tools cannot account for probabilistic aspects as the presence of asymptomatic infected occupants or a vaccination coverage, as well as for dynamic aspects related to ventilation rates and occupancy schedules.
In order to overcome these limitations, in a previous work a Monte Carlo (*MC*) model was developed to evaluate different ventilation strategies in terms of contagion risk reduction for three rooms of the Free University of Bozen-Bolzano (Albertin *et al.*, 2021). The model was able to account for multiple zones, dynamic factors (such as occupancy schedules and ventilation rates), presence of asymptomatic infected occupants, variable contagious period and symptoms onset day, and the possibility for the infected to be contagious also before the symptoms onset day. The aim of the present work is to expand the capability of this model by considering also (1) the possible movements of the students from one classroom to another, (2) the vaccination coverage and the effectiveness of three popular vaccines (Pfizer, Moderna, AstraZeneca), (3) the latest COVID-19 variants (Delta and Omicron), and (4) the usage of portable air cleaners.

2. CASE STUDY

The case-study is based on three classrooms located on the fifth floor of the main building of the campus of the Free University of Bozen-Bolzano, Italy. The three rooms, namely E5.20, E5.21, E5.22, are connected by a hallway and are adjacent to each other, with E5.21 being the one in the middle. All classrooms have a floor area of approximately 56 m^2, a door with an opening area of 1.98 m^2 and share the same mechanical ventilation system, with a nominal supply of fresh air of 0.1 $m^3 s^{-1}$ (or 1.8 ACH). The mechanical ventilation system is operative only during occupancy hours, which have been scheduled from Monday to Friday, from 8 am to 12 pm and, in the afternoon, from 2 pm to 6 pm. The openable window area has been limited to two turn windows present in all classrooms, with an opening area of 1.2 m^2 each. An openable tilt window is also present in each classroom; however, it has been neglected since the occupants rarely use it for space ventilation.
The classrooms are part of the Living Labs of the University, equipped with sensors that continuously monitor indoor air variables such as temperature, humidity and CO_2 concentration. Furthermore, a weather station is installed on the roof, collecting data on the external ambient conditions (Pernigotto and Gasparella, 2021).

3. MODEL DEVELOPMENT

The original statistical model able to evaluate the risk of contagion due to airborne transmission of COVID-19 (Albertin *et al.*, 2021) has been modified in order to reduce the computational time required to evaluate each scenario, and to account for new features, such as vaccine coverage and utilization of portable air purifiers. Furthermore, the building model that was previously limited to the three classrooms and the hallway has been extended to the whole floor, in order to evaluate also the impacts of the airflows exchanges with the other rooms on the COVID-19 pathogenic concentration in the hallway.

3.1 Building Simulation Model

The building model of the previous work, developed in TRNSYS 18, has been extended to the whole floor, considering the airflow couplings with several other rooms and the hallway.

The airflow network (*AFN*), previously developed in TRNFLOW to dynamically evaluate the airflows in the building, has been expanded to the whole floor as well. It is important to specify that, while the doors of the additional rooms have been fully included in the airflow network, for those spaces not used for teaching purposes (e.g., bathrooms, storage rooms, offices) the windows have been modelled as always closed due to the lack of a representative opening schedule. For these spaces, only the infiltration rates have been simulated.
These additional rooms have not been included in the risk assessment calculations since a preliminary analysis showed that the COVID-19 concentrations in these rooms due to a contagious occupant in E5.20, E5.21 or E5.22 were negligible. However, they allowed to properly evaluate the airflows coupling of the three main classrooms with the airflow network and to ensure a realistic air change of the hallway.
The introduction of the new rooms in the model has come with the definition of new connections. The same components used for the three main classrooms have been utilized also for the new rooms, in order to simplify the model calibration carried out with the data provided by the sensors of the Living Labs. Specifically, the air temperature profiles in the three classrooms have been used to calibrate the infiltrations across the windows, the discharge coefficient of the doors and the mechanical ventilation efficacy. The results have been found similar to those of the previous calibration (Albertin et al., 2021). In detail, the crack air mass flow coefficient *Cs* has been set to $6e^{-5}$ kg s^{-1} m^{-1} at 1 Pa, the doors' discharge coefficient to 0.56 and the mechanical ventilation efficacy to 61 %.
The calibrated model has been used to create a database of airflows for each room and for each scenario, for a simulation period of one month and a time-step of 5 minutes. The database has been then utilized in the Monte Carlo model for the evaluation of COVID-19 concentrations in the three classrooms and in the hallway through an air mass balance.

3.2 Occupancy Scenarios

It is necessary to make some hypothesis before being able to run the statistical model. Also these hypotheses have been slightly changed with respect to the previous work and are briefly summarized in this paragraph. To this end, students of each classroom are from now on designed as class-group to differentiate it from the physical space of the classroom.
The number of students is 25 for each class-group and does not change in number among simulations, iterations or scenarios. Prior each iteration, a randomly generated schedule assigns each class-group to the three classrooms, simulating the movement of students between classrooms that might happen during the break. Each class-group has a designed weekly schedule that might change randomly from iteration to iteration, allowing the class-group to change classroom at break hours during a day, and from day to day during a simulation. No classroom is left empty at any time.
Additionally, 5 professors for each class-group have been set with a random schedule created at the beginning of each iteration. All 40 hours of occupancy are covered by the professors, for a total of 8 hours of lecture each. The class-group schedule affects also the professors' ones: if a class-group changes classroom within a day, if the professor associated with that class-group is the same from morning to afternoon, then also the professor changes the classroom, following the class-group.
All other hypotheses have been left unchanged. Occupancy hours are set from 8 am to 12 pm and from 2 pm to 6 pm, with break hours set from 12 pm to 2 pm, from Monday to Friday, to represent as a worst-case scenario an intensive course's schedule. It is supposed that the doors are opened only during the break hours and left closed otherwise. With respect to the previous work, only the winter schedule was kept for the natural ventilation. When active, the windows are opened with an opening fraction of 70 % of the total area. The winter schedule set the windows as open during the break hours and for 15 minutes in the morning (9:45 am – 10:00 am) and in the afternoon (3:45 pm to 4:00 pm).
Additionally, air purifiers have been introduced in a subset of simulations. The model selected for this work is suited for rooms up to 79 m^2 and has a clean air delivery rate of 333 m^3 h^{-1} when operating at nominal capacity. When present, a portable purifier is active only during occupancy hours for each classroom.

4. MONTE CARLO ANALYSIS

The Monte Carlo model was modified with respect to its previous version to account for vaccines, COVID-19 variants, air purifiers, random schedule of the class-group and to lower the computational time required to evaluate the airflows. These modifications did not affect the general structure of the algorithm, which is described in this chapter.

4.1 Scenario Initialization

Firstly, a scenario is initialized: the selected ventilation strategy (defined by the scenario) is automatically implemented in the airflow network, which evaluates the airflows of each room with a time step of 5 minutes for a period of a month. After the end of the simulation, a database is created with the airflows for each time step and for each component of the airflow network. After the creation of the airflows database, the first iteration is initialized.

4.2 Iteration Initialization

The schedules for each class-group are created, as well as the schedules for the professors. While a schedule for a class-group defines the classroom where the class-group is located for each half day of the week, the schedule for the professors defines in which half days each professors has lecture hours with a specific class-group. So, the definition of a class-group schedule determines also the location of the relative professors.

Then, the quanta emission rate (*QR*, defined as "the dose of airborne droplet nuclei required to cause infection in 63 % of susceptible persons" by Buonanno *et al.*, 2020a) is evaluated for all the occupants. The *QR* calculation is done *a priori* for all subjects (students and professors), before the start of each iteration, to save computational time. During the simulations, if a subject is not infected, or if he/she is infected but the simulation day is outside his/her contagious period, the *QR* for this subject is set temporarily equal to zero. The *QR* is randomly calculated with a lognormal distribution curve which parameters depend on the activity performed by the subject (Buonanno *et al.*, 2020b). The activities are divided into primary and secondary activities for both students and professors. The primary activity is supposed to be carried out for at least 70 % of the time. The effective time associated with the secondary activity is evaluated with a normal distribution curve, with mean value equal to zero, and three standard deviations equal to 30 % (the maximum percentage of time that can be possibly allocated to the secondary activity). At this point, the percentage of time of the primary activity is obtained by 100 % complement. Then, two *QR* are evaluated, one for the primary activity and one for the secondary with a lognormal distribution curve whose parameters depend on the selected activity (Table 1). Finally, the effective *QR* for a subject is the scalar product of the *QR* values of the two activities with the respective fractions of time.

Table 1: list of activities with relative parameters and typology.

Activity	Log mean	Log standard deviation	Subjects	Type
resting-breathing	-0.43	0.73	Students	Primary
resting-speaking	0.24	0.72	Students and Professors	Secondary
standing – loudly speaking	1.08	0.72	Professors	Primary

When a COVID-19 variant is considered in a given scenario, the mean value of the activities' distribution curves is multiplied by a factor of 2, as suggested in the AIRC tool's manual.

Then, all the occupants are categorized *a priori* as either asymptomatic or symptomatic, even if they are not infected at this point. The early categorization is done to save computational time and does not imply that a not infected subject is contagious. The asymptomatic subjects are selected with a random sorting from a uniform distribution and from a normal distribution curve with mean 40.5 % and standard deviation 3.5 % (Ma *et al.*, 2021). If the value obtained from the uniform distribution curve is lower than the value of the normal distribution curve, then the subject is considered asymptomatic.

All the occupants are then further categorized in fully (2 received doses), partially (1 received dose) or not vaccinated with respect to a vaccination coverage set accordingly to the global database of COVID-19 vaccinations (Mathieu *et al.*, 2021). The vaccination coverage selected for all scenarios is equal to 72 % fully vaccinated, 5.8 % partially vaccinated and 22.2 % unvaccinated, representative of the situation in Italy on November 1st, 2021. Furthermore, the vaccination effectiveness for the fully and partial vaccinated is selected with respect to the COVID-19 variant set by the scenario and to three different vaccines: Pfizer, Moderna and AstraZeneca (Andrews *et al.*, 2022). A random extraction from a uniform distribution is done to determine the status of each occupant. If the extracted value falls below a given threshold, the subject cannot be infected anytime during the iteration. The threshold is selected with a normal distribution curve, which mean and standard deviation depend on the vaccination typology (fully or partial), the COVID-19 variant (Delta or Omicron) and the received vaccine (Pfizer, Moderna or AstraZeneca).

Some additional hypotheses are made at the beginning of each iteration: the initial concentrations of COVID-19 quanta are zero for all internal rooms and an asymptomatic infected student is located in classroom E5.21, being part of the class-group assigned to this specific classroom on the morning of the first day. From now on, the first infected student will be referred as subject 0. It is also hypothesized that day 1 corresponds to the symptoms' onset day of subject 0, which implies that subject 0, even if asymptomatic, is contagious. The concepts of contagious period and symptoms' onset day are further detailed in the next paragraphs.
At this point, the initialization process of an iteration is complete, and the first simulation starts.

4.3 Simulation Start

Each iteration starts at day 1, set to be a Monday, and the first 12 hours of the day are simulated. For each time-step a COVID-19 quanta mass balance calculation of the three classrooms and the hallway is performed. To this mean, the database of airflows is used to evaluate the air exchange from one room to another. Within the mass balance, it is accounted also for Covid-19 quanta generation due to the presence of infected subjects within their contagious period, and only during occupancy hours. If active, the air purifiers are also included in the balance, one in each classroom acting as a COVID-19 quanta sink.
At the end of the first 12 hours, it is checked if the class-groups change classrooms. If so, class-groups previously not in contact with COVID-19 virus might be located in the afternoon in a classroom with a COVID-19 quanta concentration different from zero. Furthermore, it is also checked which professors of each class-group has lectures in the afternoon. Then, the remaining hours of the day are simulated until 12 am. At this point, the simulation ends, and before starting with a new day, the risk assessment analysis is performed.
Firstly, it is evaluated the dose received from all the subjects with Equation 1 (Buonanno *et al.*, 2020b):

$$D_q = IR \sum_t n(t) \tag{1}$$

where *IR* is the inhalation rate and *n(t)* the quanta concentration at time-step *t*. The dose received changes from class-group to class-group, and also from professor to professor within each class-group, dependently to the assigned schedule and to the COVID-19 quanta concentrations in the three classrooms. The inhalation rate was not changed with respect to the previous work and kept to 0.49 $m^3\ h^{-1}$ for students and 0.54 $m^3\ h^{-1}$ for professors. Given the dose received by the susceptible subjects (i.e., the subjects that can be infected), it is possible to evaluate the probability of infection as in Equation 2 (Buonanno *et al.*, 2020b):

$$P_i(\%) = 1 - e^{-D_q} \tag{2}$$

The probability of infection is reduced by 1/3 if all the occupants are wearing mask during the occupancy hours (AIRC User's Manual).
At this point, a uniform distribution is used to evaluate the newly infections. Only healthy subjects, who were not previously infected and have no immunity status given by the vaccination, can be infected. A random value is extracted from the uniform distribution for each subject who fulfills the aforementioned conditions, and, if its value is lower than the probability of infection, then the subject is considered to be infected. For all new infections, the symptoms' onset day and the contagious period are evaluated.
The symptoms' onset day for each newly infected subject is calculated with a gamma distribution curve, whose parameters (shape and scale) are set accordingly to Lauer *et al.* (2020). Then, the starting and ending days of the contagious period are evaluated, the former with a gamma distribution curve, whose parameters are set as defined by He *et al.* (2020). It is important to notice that an infected subject can be contagious also before the symptoms' onset day. The ending day of the contagious period is always set 9 days after the onset day (both for symptomatic and asymptomatic subjects), in order to consider the worst-case scenario.
Finally, it is checked if one of the two ending conditions is met: either a symptomatic subject reaches his symptoms' onset day, leading to the discovery of the virus in the classrooms, or all the asymptomatic subjects are no longer contagious, completely halting the spread of the virus. If an ending condition is met, the iteration comes to an end. The attack rate is then calculated as the ratio of the infected subjects with respect to the total number of susceptible subjects. Furthermore, the number of elapsed days from the start of the iteration to its end is evaluated. If no ending condition is met, another day is simulated. The *QR* for each infected subject is set equal to zero if the next day lies outside his/her contagious period, otherwise it is set to the value assigned during the initialization of the iteration.

Once that 1000 iterations have been performed for each scenario, the attack rate and number of elapsed days distributions are evaluated and compared.

5. SIMULATION PLAN

The Monte Carlo model was used to evaluate 48 scenarios, accounting for: ventilation type (none, natural, mechanical, mixed), COVID-19 variant (Delta, Omicron), vaccination (none, Pfizer, Moderna, AstraZeneca) and presence of purifiers in the classrooms (Table 2). Furthermore, all 48 scenarios were evaluated a second time accounting also for mask utilization, a third time by increasing the number of occupants by 30 % and a fourth time by decreasing the classrooms air volume by 30 %, for a total of 192 scenarios.

Table 2: main scenarios settings.

Scenario	Covid variant	Vaccine	Ventilation	Purifiers
1	Delta - Omicron	None	Off	Off
2	Delta - Omicron	None	MV	Off
3	Delta - Omicron	None	NV	Off
4	Delta - Omicron	None	Mixed	Off
5	Delta - Omicron	None	Off	On
6	Delta - Omicron	None	MV	On
7	Delta - Omicron	None	NV	On
8	Delta - Omicron	None	Mixed	On
9	Delta	Pfizer	Off	Off
10	Delta	Pfizer	MV	Off
11	Delta	Pfizer	NV	Off
12	Delta	Pfizer	Mixed	Off
13	Delta	Pfizer	Off	On
14	Delta	Pfizer	MV	On
15	Delta	Pfizer	NV	On
16	Delta	Pfizer	Mixed	On
17	Delta	Moderna	Off	Off
18	Delta	Moderna	MV	Off
19	Delta	Moderna	NV	Off
20	Delta	Moderna	Mixed	Off
21	Delta	Moderna	Off	On
22	Delta	Moderna	MV	On
23	Delta	Moderna	NV	On
24	Delta	Moderna	Mixed	On
25	Omicron	Pfizer	Off	Off
26	Omicron	Pfizer	MV	Off
27	Omicron	Pfizer	NV	Off
28	Omicron	Pfizer	Mixed	Off
29	Omicron	Pfizer	Off	On
30	Omicron	Pfizer	MV	On
31	Omicron	Pfizer	NV	On
32	Omicron	Pfizer	Mixed	On
33	Omicron	Moderna	Off	Off
34	Omicron	Moderna	MV	Off
35	Omicron	Moderna	NV	Off
36	Omicron	Moderna	Mixed	Off
37	Omicron	Moderna	Off	On
38	Omicron	Moderna	MV	On
39	Omicron	Moderna	NV	On
40	Omicron	Moderna	Mixed	On

41	Omicron	AstraZeneca	Off	Off
42	Omicron	AstraZeneca	MV	Off
43	Omicron	AstraZeneca	NV	Off
44	Omicron	AstraZeneca	Mixed	Off
45	Omicron	AstraZeneca	Off	On
46	Omicron	AstraZeneca	MV	On
47	Omicron	AstraZeneca	NV	On
48	Omicron	AstraZeneca	Mixed	On

In the first 8 scenarios the vaccination coverage is set equal to zero. Since both COVID-19 variants are accounted for by doubling the mean value of the *QR* distribution curves, which do not change from scenario to scenario, the first 8 scenarios are representative of both.
The ventilation has four possible settings: *none*, which specifies the absence of any kind of ventilation; *MV* (Mechanical Ventilation), which specifies a fixed amount of fresh air supply during occupancy hours (8 am – 12 am, 2 pm – 6 pm); and finally, *NV* (Natural Ventilation), which sets the windows status as open with a winter schedule (9:45 am – 10:00 am, 12:00 am – 2:00 pm, 3:45 pm – 4:00 pm). The last setting is *mixed*, which sets both natural and mechanical ventilation for the given scenario.
A purifier within each classroom might also be considered for a given scenario. If active, it is supposed to operate at maximum capacity, providing 333 $m^3 h^{-1}$ of clean air rate during occupancy hours.
Finally, three vaccines were implemented in the model, each with a specific effectiveness against COVID-19 variants, both for partially and fully vaccinated occupants. If a vaccine is considered for a scenario, it is supposed that only the considered vaccine was used for all the vaccinated occupants, and with a coverage that is always set as 72 % fully vaccinated, 5.8 % partially vaccinated and 22.2 % unvaccinated.
All 48 scenarios described above were evaluated three additional times. Firstly, the mask utilization was accounted for by decreasing the probability of infection of each occupant by 1/3. Secondly, the number of students for all classrooms was increased by 30 % by keeping all the other factors unchanged (for instance, no mask utilization), to simulate an inadequate usage of the rooms. Finally, the volume of the classrooms was reduced by 30 % to simulate a small and crowded environment.

6. RESULTS

In this chapter, the results of the scenarios are compared in terms of attack rate for classroom E5.21 and number of elapsed days. Given the probabilistic nature of a Monte Carlo model, the results are presented with boxplot graphs (1000 values for each metric and for each scenario). It is extremely important to highlight that the numbers of infected occupants that are presented in this chapter are due only to the airborne transmission of the virus, and only during occupancy hours. Furthermore, it was hypothesized that the students do not interact with each other. Given these considerations, it can be safely assumed that, by considering all possible pathway of contagion, the effective number of infected occupants at the end of each iteration may be substantially higher.
Due to space limitations, the graphs representing the number of elapsed days for all cases were omitted, as well as the graphs reporting the attack rate for the two set of 48 cases with increased number of occupants and reduced air volume of the classrooms.
Figure 1 shows the attack rate boxplot for the 48 scenarios of the standard case, which does not consider a reduced volume for the classrooms or an additional number of occupants. As expected, the absence of any typology of ventilation combined with the lack of a strategy aimed at reducing the risk of infection due to COVID-19 variants (as vaccines, purifiers or mask utilization) leads to high values of attack rate. It can be expected form Scenario 1 a number of infected occupants up to 10 people in a range of 2 to 18 days from the start of the iteration. It is worth mentioning that, with the same conditions but without considering a COVID-19 variant, the expected number of infected occupants is approximately half (Albertin *et al.*, 2021). Scenarios 2 to 4 show how mechanical and natural ventilation are not able alone or combined to reduce substantially the attack rate. On the other hand, Scenario 5 shows how an air purifier leads to the same attack rate distribution with respect to a mixed ventilation, highlighting its potential as a strategy to prevent the spread of the virus in locations where the mechanical ventilation is not present, and the natural ventilation is ineffective. Scenarios 9 to 24 also include the effects of vaccines while considering the Delta variant: Pfizer in Scenarios 9 to 16 and Moderna in Scenarios 17 to 24. As it shown in Figure 1, the vaccines are highly effective in terms of risk reduction, especially if combined with an air purifier and any type of ventilation strategy. The number of elapsed days for these cases drops quickly to a range of 9 – 12 days, or

even collapse to a single value equal to 10 days. Lowering the number of days of an iteration means that not only the occupants infected through an airborne transmission of virus are less, but also the occupants infected by any other virus pathway.
Scenarios 25 to 48 show the attack rate in case of Omicron variant, for three vaccines: Pfizer (25 - 32), Moderna (33 - 40), AstraZeneca (41 - 48). The overall effectiveness of vaccines coverage is lower with respect to the Delta variant. If no ventilation system is present, nor the air purifier, the attack rate is just slightly lower with the respect to the case with no vaccines. The only effective way to lower the risk of contagion is to adopt a combined action of vaccines, mixed ventilation and air purifiers.
Figure 2 show as mask utilization greatly reduce the risk of infection for the occupants. However, Scenario 1 shows how their effect is still limited in cases without any kind of ventilation. If adopted in combination with a ventilation strategy and an air purifier, the results on the reduction of the attack rate are similar to the effects of Pfizer or Moderna vaccines (Scenarios 6 - 8). Furthermore, even if an air purifier is not present but a ventilation strategy is, the number of elapsed days is greatly reduced, lowering also the risk of infections through a different pathway of contagion. The combined adoption of mask utilization, mixed ventilation strategy and a vaccine coverage leads to a risk value which is almost zero (Scenarios 16 and 24). Since the effect on the contagion risk of mask utilization is independent with respect to the considered COVID-19 variant, from Scenarios 25 to 48 it is possible to observe how mask utilization is extremely useful to lower the risk of contagion in case of Omicron variant. Not only the scenarios with a ventilation strategy and the utilization of an air purifier are comparable to the respective cases with Delta variant, but it is now sufficient to have at least one ventilation strategy to greatly reduce the likelihood of an unacceptable outcome, especially if air purifiers are used (Scenarios 25 to 27, 33 to 35, 41 to 43).
The results of other two sets of scenarios (a case with an increased number of occupants and another with an air volume of the classrooms reduced by 30 %, which could not be reported due to space limitations) highlight the importance of the combined action of a mixed ventilation strategy, air purifier and vaccines, especially with the Omicron variant. An increase in the number of the occupants leads to an overall higher mean value of attack rate with respect to the standard case, while the spread of the distribution curves is higher in the case of a reduced air volume. It is important to mention that a given value of attack rate represents a different number of infected occupants in the two cases, since the total number of susceptible subjects is different (30 occupants in the former, 38 in the latter).

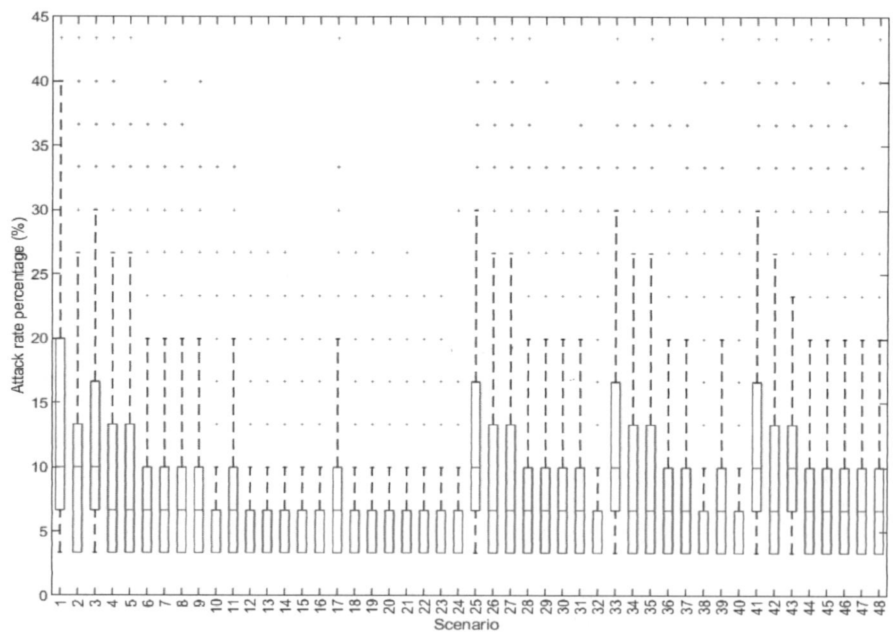

Figure 1: attack rate of each scenario for the standard case.

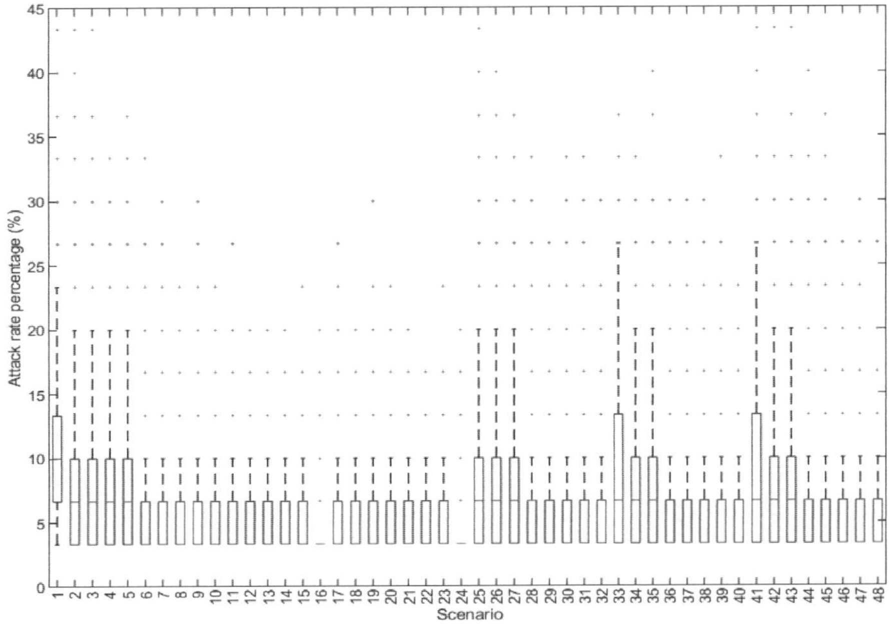

Figure 2: attack rate of each scenario for the case which consider mask utilization.

7. CONCLUSION

In this work, a previously developed Monte Carlo model able to assess the risk of contagion due to the airborne transmission of COVID-19, was further expanded to consider also vaccine coverage, utilization of portable air purifiers, COVID-19 variants and the possible movement of students between classrooms. The case-study are three classrooms of the Free University of Bozen-Bolzano campus' main building, as for the previous work. However, the building model used to assess the airflows needed to evaluate COVID-19 quanta concentrations was extended to the whole floor, as well as the airflow network required for the airflow's dynamic evaluation. The additional complexity introduced with the new rooms and their connections in the airflow network led to the necessity to perform a new calibration, which was carried out against measured data. The calibrated building model was successively used to create a database of airflows for each room, required by the Monte Carlo model to evaluate with a mass balance the concentrations of COVID-19 quanta for the three classrooms and the hallway that connects them. The airflows database allowed to evaluate a total of 48 scenarios according to a probabilistic model based on 1000 iterations for each scenario. We got as outputs the attack rate (i.e., the number of final infected occupants with respect to the total number of susceptible subjects) and number of elapsed days from the start of each iteration. Furthermore, all 48 scenarios were considered three additional times by also considering mask utilization, increasing the number of students in all classrooms by 30 %, and, finally, by reducing the air volume of the classrooms by 30 %, for a total of 192 scenarios. The results showed the danger posed by the COVID-19 variants, especially in environments without any kind of ventilation. The portable air purifiers showed some potential to effectively reduce the risk of contagion, whenever mechanical ventilation is not present and/or natural ventilation is not efficient. Vaccines were observed to be able to greatly reduce the risk of contagion, especially if used in combination with air purifiers, but only in case of the Delta variant, due to a low effectiveness of the vaccines relative to the Omicron variant. If the Omicron variant is considered, the only possible solution to keep the attack rate to low values is by combining the effects of vaccines, air purifiers, as well as a mixed mechanical ventilation. Mask utilization was identified as being able to greatly reduce the risk of contagion also due to Omicron variant, and, so, masks should be employed when possible to keep the risk of contagion contained, also in those buildings without purifiers or an adequate ventilation strategy. Finally, two additional set of the main 48 scenarios, which consider an increased number of occupants or a reduced air volume for the classrooms, show that, in both cases, the combined action of mixed ventilation strategy, air purifiers, and vaccines coverage, is not able anymore to keep the risk of contagion at an acceptable value, discouraging an improper use of the classrooms and limiting the occupancy densities.

REFERENCES

Airborne Infection Risk Calculator (AIRC) User's Manual, v.3.0 Beta, April 2021.

Albertin, R., Pernigotto, G., & Gasparella, A. (2021). Assessment of the Covid 19 contagion risk in university classrooms with TRNSYS and TRNFLOW simulations. *Proceedings of IAQ 2020: indoor environmental quality performance approaches*, May 4-6, 2022, Athens, Greece.

Andrews, N., Stowe, J., Kirsebom, F., Toffa, S., Rickeard, T., Gallagher, E., Gower, C., Kall, M., Groves, N., O'Connell, A. M., Simons, D., Blomquist, P. B., Zaidi, A., Nash, S., Iwani Binti Abdul Aziz, N., Thelwall, S., Dabrera, G., Myers, R., Amirthalingam, G., Gharbia, S., … Lopez Bernal, J. (2022). Covid-19 Vaccine Effectiveness against the Omicron (B.1.1.529) Variant. *The New England journal of medicine*, NEJMoa2119451. Advance online publication. https://doi.org/10.1056/NEJMoa2119451

Buonanno, G., Morawska, L., & Stabile, L. (2020b). Quantitative assessment of the risk of airborne transmission of SARS-CoV-2 infection: Prospective and retrospective applications. *Environment international,* 145, 106112. https://doi.org/10.1016/j.envint.2020.106112

Buonanno, G., Stabile, L., & Morawska, L. (2020a). Estimation of airborne viral emission: Quanta emission rate of SARS-CoV-2 for infection risk assessment. *Environment international*, 141, 105794. https://doi.org/10.1016/j.envint.2020.105794

Burgmann, S., & Janoske, U. (2021). Transmission and reduction of aerosols in classrooms using air purifier systems. *Physics of fluids* (Woodbury, N.Y.: 1994), 33(3), 033321. https://doi.org/10.1063/5.0044046

Di Gilio, A., Palmisani, J., Pulimeno, M., Cerino, F., Cacace, M., Miani, A., & de Gennaro, G. (2021). CO2 concentration monitoring inside educational buildings as a strategic tool to reduce the risk of Sars-CoV-2 airborne transmission. *Environmental research*, 202, 111560. https://doi.org/10.1016/j.envres.2021.111560

He, X., Lau, E., Wu, P., Deng, X., Wang, J., Hao, X., Lau, Y. C., Wong, J. Y., Guan, Y., Tan, X., Mo, X., Chen, Y., Liao, B., Chen, W., Hu, F., Zhang, Q., Zhong, M., Wu, Y., Zhao, L., Zhang, F., … Leung, G. M. (2020). Temporal dynamics in viral shedding and transmissibility of COVID-19. *Nature medicine*, 26(5), 672–675. https://doi.org/10.1038/s41591-020-0869-5

Lauer, S. A., Grantz, K. H., Bi, Q., Jones, F. K., Zheng, Q., Meredith, H. R., Azman, A. S., Reich, N. G., & Lessler, J. (2020). The Incubation Period of Coronavirus Disease 2019 (COVID-19) From Publicly Reported Confirmed Cases: Estimation and Application. *Annals of internal medicine*, 172(9), 577–582. https://doi.org/10.7326/M20-0504

Ma, Q., Liu, J., Liu, Q., Kang, L., Liu, R., Jing, W., Wu, Y., & Liu, M. (2021). Global Percentage of Asymptomatic SARS-CoV-2 Infections Among the Tested Population and Individuals with Confirmed COVID-19 Diagnosis: A Systematic Review and Meta-analysis. *JAMA network open*, 4(12), e2137257. https://doi.org/10.1001/jamanetworkopen.2021.37257

Mathieu, E., Ritchie, H., Ortiz-Ospina, E., Roser, M., Hasell, J., Appel, C., Giattino, C., & Rodés-Guirao, L. (2021). A global database of COVID-19 vaccinations. *Nature human behaviour*, 5(7), 947–953. https://doi.org/10.1038/s41562-021-01122-8

Park, S., Choi, Y., Song, D., & Kim, E. K. (2021). Natural ventilation strategy and related issues to prevent coronavirus disease 2019 (COVID-19) airborne transmission in a school building. *The Science of the total environment*, 789, 147764. https://doi.org/10.1016/j.scitotenv.2021.147764

Pernigotto, G., & Gasparella, A. (2021). Analysis of the potential of smart ventilation controls: application to a university classroom in Bolzano. *Proceedings of ROOMVENT 2020: energy efficient ventilation for healthy future buildings*, February 15-17, 2021, Turin, Italy.

Preliminary Analysis on The Perception Of The Four Comfort Domains Of People With Autism Spectrum Condition

Arianna Marzi[1]*, Luca Zaniboni[1], Marco Caniato[1], Andrea Gasparella[1]

[1]Free University of Bozen, Faculty of Science and Technology,
Bozen, Italy

*Contact author e-mail: arianna.marzi@natec.unibz.it

ABSTRACT

Autism Spectrum Condition (ASC) is a neurodevelopmental disorder affecting one child over 160 worldwide. People in the spectrum may show different sensitivity in the perception of the different stimuli provided by the surrounding environment. On the one hand, this can affect the perceived level of comfort and increase the environmental requisites, also in relation to the dangerous effects of stressful conditions. On the other hand, in the case of reduced sensitivity, this can harm the subject unknowingly exposed to inappropriate or unhealthy conditions. Furthermore, if individual comfort has been typically investigated by means of subjective approaches, such as questionnaires, administered to individuals in laboratory conditions, individuals on the spectrum can face some difficulty with communication and, even worst, can be endangered by harmful reactions to discomfortable or stressful environments. This makes it more difficult to collect direct feedbacks on the perception of their own well-being and prevents from conducting experiments involving the most fragile individuals for whom the research would be particularly beneficial. In this framework, an analysis of the Indoor Environmental Quality (IEQ) is here provided with attention to acoustic, thermo-hygrometric, visual environments and indoor air quality and their relative importance in determining comfort or discomfort to individuals with ASC. In order to investigate the subjective sensitivity of individuals with severe comorbidities or communication problems, this study interviewed people who most directly are taking care of them and follow their activities daily. In more details, a controlled group of 19 young adults with a diagnosis of autism were analyzed by collecting the opinions provided by both family members (parents) and professional caregivers, by means of questionnaires evaluating the sensitivity to the four comfort domains. Biases in answers due to the different proxy respondents involved were analyzed.

1. INTRODUCTION

Design of buildings requires particular attention not only to ensure structural stability or energy efficiency, but also to guarantee the occupants the indoor comfort necessary to carry out the activities (Lai & Yik, 2009) (Frontczak, Schiavon, et al., 2012). The indoor comfort is therefore of paramount importance as it influences the quality of life (Pinto et al., 2017) health (Kim et al., 2015), productivity (Wang & Hong, 2020) and performance (Catalina & Banu, 2014) in residences, offices, educational and therapeutic environments affecting both psychological and physiological factors. To ensure this, there are a number of measures that international standards tend to monitor in the design phase, providing suggestions, suitable ranges of indoor conditions and minimum criteria to be met (Comite'Europe'en de Normalisation, 2007; European Committee for Standardization, 2011; European Committee for Standardization, 2019).

When managing indoor comfort perception and evaluation, it is a common procedure to collect physical measurements by means of monitoring campaigns in order to associate the objective data measured by the instrumentation to the subjective data from interviewing the individuals (Fransson et al., 2007; Castaldo et al., 2017). In this way, parameters such as temperature, relative humidity, mean radiant temperature, illuminance, sound pressure levels, clothing insulation or metabolic rate can be collected and associated with the data on the perception of the occupants (American Society of Heating, Refrigerating and Air Conditioning Engineers, 2010; Castaldo et al., 2018). The use questionnaires is recommended to both collect data on clothing to be linked to the metabolic rate and link the physical data observed by the instrumentation to the stimuli (temperature, humidity, intensity of natural or artificial light, air quality, acoustic environment quality) perceived by occupant (Frontczak, Andersen, et al., 2012).

A suitable indoor comfort is even more important when approaching either therapeutic or healthcare environments (Pinter & Kiss, 2019; Wong et al., 2014; Leung et al., 2019) or when considering people with special needs (Vilcekova et al., 2017; Van Hoof et al., 2008). Thus, understanding the environmental stimuli perception of people with special needs is an increasingly important requirement in the development of user-centered design, user well-being and perceived comfort.

Particularly, Autism Spectrum Condition (ASC) is a neurodevelopmental disorder increasingly growing among world population (Johnson et al., 2018; Maenner et al., 2020). Thus, understating the perception process of this special users will help building an inclusive approach to design and optimize the occupants' comfort of those who are not always able to manifest their perceptions, because of difficulties in communication and expression. For the above-mentioned reasons, people with ASC are not only acknowledged as neurodiverse, but also identified by the Declaration of World Medical Association as fragile individuals. Therefore, an indirect methodology of investigation is the most suitable to evaluate their perception and wellbeing. Besides increased sensitivity to specific stressors, people on the spectrum may feature hyposensitivity to some comfort domains (American Psychiatric Association, 2013).

Particularly, William et al (Williams et al., 2019) tested the hypothesis that individuals with ASD exhibit significantly elevated warm and cool detection thresholds. The study focused on thermal perception (heat and cold) of two sample groups: individuals with ASC and Typically Developed (TD) groups of occupants. The participants completed a self-reported questionnaire measuring autistic traits and sensory features and then carried out a thermal detection task conducted according to the Declaration of Helsinki. Results showed that no difference was found in thermal detection thresholds between ASC and TD groups despite large differences between groups in sensory reactivity questionnaire. Conversely, Duerden et al. (Duerden et al., 2015) focused on the pain hypo-reactivity that in some cases could distinguish ASC individuals. They studied a group of adolescents: 20 with a diagnosis of ASC and 56 typical development (TD) adolescents composing the control group. The probands to the study were subjected to 4 separated thermal trials (warm, cool, heat and cold) using the method of limits protocol of thermal detection and pain thresholds. From the analysis of the results, it was highlighted that adolescents with ASC were less sensitive to warm and cold stimuli compared to the TD group.

Kanakri et al. (Kanakri et al., 2017) investigated the impact of noise on ASC learning process in three schools. A survey on impact noise in learning environment was distributed to teachers and the results showed that sound insulated walls and the presence of carpets in the classroom were the most important solution for children with ASC. The study indeed showed how, when modifying the built environment in acoustically friendly ways, the treatment and education of children with autism may be greatly enhanced. Although this study considers hypo-sensitive hearing behavior, in which phenomena such as tapping or making noise (that may highlight that the individual is enacting to increase sensory stimulation) are analyzed, a dedicated hyposensitivity scale or analysis was not adopted.

Another study focusing on the impact of noise on ASC people was performed by Keith et al. (Keith et al., 2019) who examined the relationship between sensory processing, arousal, and cognitive performance in adolescents with ASD and TD controls matched on age, gender, and Full-Scale IQ. The results were obtained by performing two cognitive tasks of different difficulties and varying the noise level in background. The results obtained suggested that individuals with ASC could experience an energizing effect of auditory stimulation during manageable tasks. Furthermore, an hyperresponsivity to noise was observed in ASC probands.

Then, approaching to the four comfort domains, Noble et al. (Noble et al., 2018) explored the perception of environmental stimuli with a comprehensive survey. By means of the collected results, it was highlighted that participants on the autism spectrum reported that people, people noise and indoor/other noise were the greatest causes of discomfort and avoidance, followed by glare and electric light. From the results, it was also showed that the thermal comfort and IAQ may not be the most important factors for people on the autism spectrum. In this case mention is made of a possible hyper response of individuals on the spectrum, but an analysis of hyposensitivity has not been addressed. A former study was already conducted by the authors (Caniato et al., 2022, p. 1; Caniato et al., 2022, p. 2). This explored the perception of people on the spectrum among the 4 domains of comfort, highlighting how the acoustic environment is the most critical for this type of users both in care facilities and in households. Furthermore, the analysis performed showed how gender, severity of autism, age and presence of other disorders among co-morbidities, strongly influenced the perception of the indoor comfort for people with ASC.

In the present study, a smaller sample is analysed (19 individuals). Due to the frequent communication disorder that may occur, the questionnaire was developed to be completed by proxy respondents, namely parents and professional caregivers (named as "caregivers" from now on). The results collected were then analysed to understand if the reported perception could vary based on the proxy respondents. Furthermore, an analysis was conducted to understand only whether the average grades provided by the two types of respondents were similar (with a Mann-Whitney U-test), but also to assess whether response trends can be assimilated to an increasing or decreasing monotone function (through Spearman's test). In addition, a further analysis was conducted on understanding the influence of personal data of individuals with ASC on reported perceptions. This survey was conducted in accordance with the Declaration of Helsinki and the Ethical Committee of the Free University of Bozen.

2. MATERIALS AND METHODS

In order to investigate the ASC users' indoor comfort, questionnaires on the environment perception are a common tool among experts and researchers (Katafygiotou & Serghides, 2014; Dykes & Baird, 2013; Fransson et al., 2007). Unfortunately, questionnaires may not be suitable to investigate the indoor comfort of people with communication difficulties, who would need support in their daily activities and life. For this reason, also thanks to the contribution of a psychologist expert in this field, it was possible to develop a questionnaire asking parents and caregivers the perception of the ASC individuals to the different indoor comfort domains. Thus, it was asked to report if the cared individuals perceived any disturbances related to thermo-hygrometric, acoustic, visual environments and indoor air quality. Moreover, it was also possible to collect personal data such as age, information on the severity of autism and presence of other co-morbidities. The questionnaire was structured in order to provide answers both on persistent aspects of perception and on hyposensitivity and sporadic conditions. Figure 1 summarizes the *ad hoc* perception scale which was built for the survey.

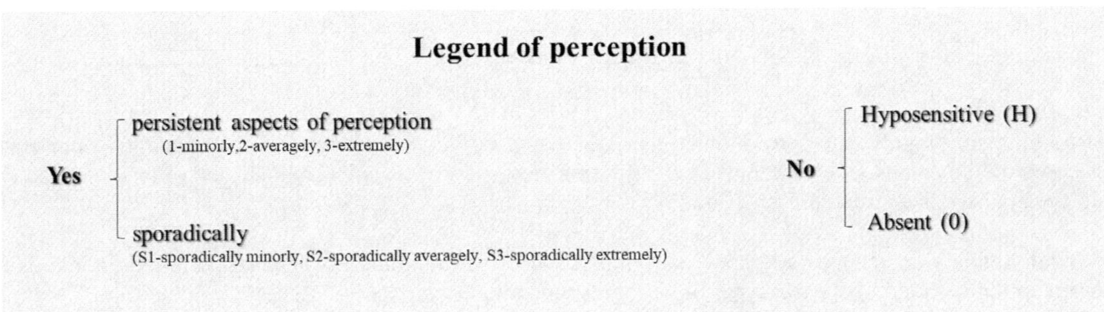

Figure 1: Perception scale designed for the questionnaire

2.1 Data collection

Parents provided answers related with the experiences of individuals in their own households, while caregivers focused on the same sample of persons in the care facility. This coupled analysis was useful to explore whether any variation in assessment due to personal biases or environmental features was present. The care facility involved in this study is an environment dedicated to young adults in which experienced psychologists and caregivers take care of them in order to increase their independence and autonomy in communication, activities and daily life.

Figure 2: Pictures of the care facility of the Association involved in this study.

A total of 19 users were involved, therefore 38 evaluations were provided by the proxy respondents (19 by parents and 19 by caregivers). Moreover, since the users involved in this study were all in the pre-adolescent or adolescent age, in which the need for autonomy and the difficulties linked to the autistic spectrum most influence behavior and perception of the environment, two different groups were considered: under 15 and over 15 years old. Indeed, Hume et al. (Hume et al., 2018) explained how, when 15 years old, an individual with ASC starts not only to process stimuli (learning, perceptual and emotional stimuli) but also to manifest them through behavioral or physical manifestations (motor stereotypies). Furthermore, the influence of the spectrum severity and the presence of co-morbidities were here considered. A summary is shown in Table 1.

Table 1: Personal data of the individuals in the spectrum reported by the two types of proxy respondents

		Parents [n°]	**Caregivers [n°]**
Age threshold	Under 15 y.o.	9	10
	Over 15 y.o.	10	9
Severity of autism	A	8	6
	B	7	9
	C	4	4
Co-morbidities	OCD	0	1
	ID	9	6
	AD	3	0
	Other	1	0

2.2 Data analysis

In order to compare properly the perception of the ASC users in the two environments, a qualitative preliminary analysis based on the comparison of frequency distributions followed by a statistical analysis, were conducted . The methodological analysis is structured as follows:

1) A preliminary percentage analysis was performed in order to highlight whether a difference could exist in the perception detected by the two types of proxy respondents.

2) A Mann-Whitney test (McKnight & Najab, 2010) was performed to explore the possibility of obtaining different reports of perceptions from the two proxy respondents. In this case the null hypothesis in the Mann-Whitney test was that the two perceptions were drawn from the same ASC population and therefore that their probability distributions were equal.

3) A Spearman's rank correlation test (Artusi et al., 2002) was performed in order to measure the degree of correlation between the perception detected by the two types of proxy respondents on each comfort domain. The Spearman's coefficient does not measure a linear relationship but permits to determine how well a relationship between two variables can be described using a monotonic function. In this case it is a useful test to see if, even if in different perceptual degrees, there is a similar response trend between the two perceptions measured by the proxy respondents. Thus, Spearman's test provides information not only whether the mean value assigned for that domain by the proxy respondents is similar, as in the case of the Mann-Whitney test, but describes how similar are the two response trends reported by the two proxy respondents.

4) Finally, a further analysis with a Spearman's correlation test was performed in order to evaluate if there were correlations between the perceptions and personal data of the ASC individuals (age, presence of comorbidities and severity of autism). In particular, the correlations between the answers of parents and caregivers were studied restricting the sample to individuals with and without co-morbidities, under- and over-15, with low severity and high severity autism.

3. RESULTS AND DISCUSSION

Figure 3 shows that the perception reported by parents tended to be always higher than the one reported by caregivers in the care facility. Particularly, caregivers reported higher response rates "0 not at all" for almost all 4 comfort domains and lower presence of sporadic perception of discomfort. As regards hyposensitivity, it can be seen that while parents report an equal share of hyposensitive individuals for thermal, visual and indoor air quality, caregivers report it only for indoor air quality and thermal environment (a higher percentage with respect to parents' in the latter case). In both cases, the acoustics seems to be the comfort domain with the highest percentage of "2- averagely" and "3- extremely", highlighting that it is the comfort field to which users with ASC seem to require more precautionary measures.

These results can indeed be justified precisely by the fact that the care facility was designed *ad hoc* to reduce possible discomfort for ASC individuals, whereas the same adjustments could be more difficult to be present in private households.

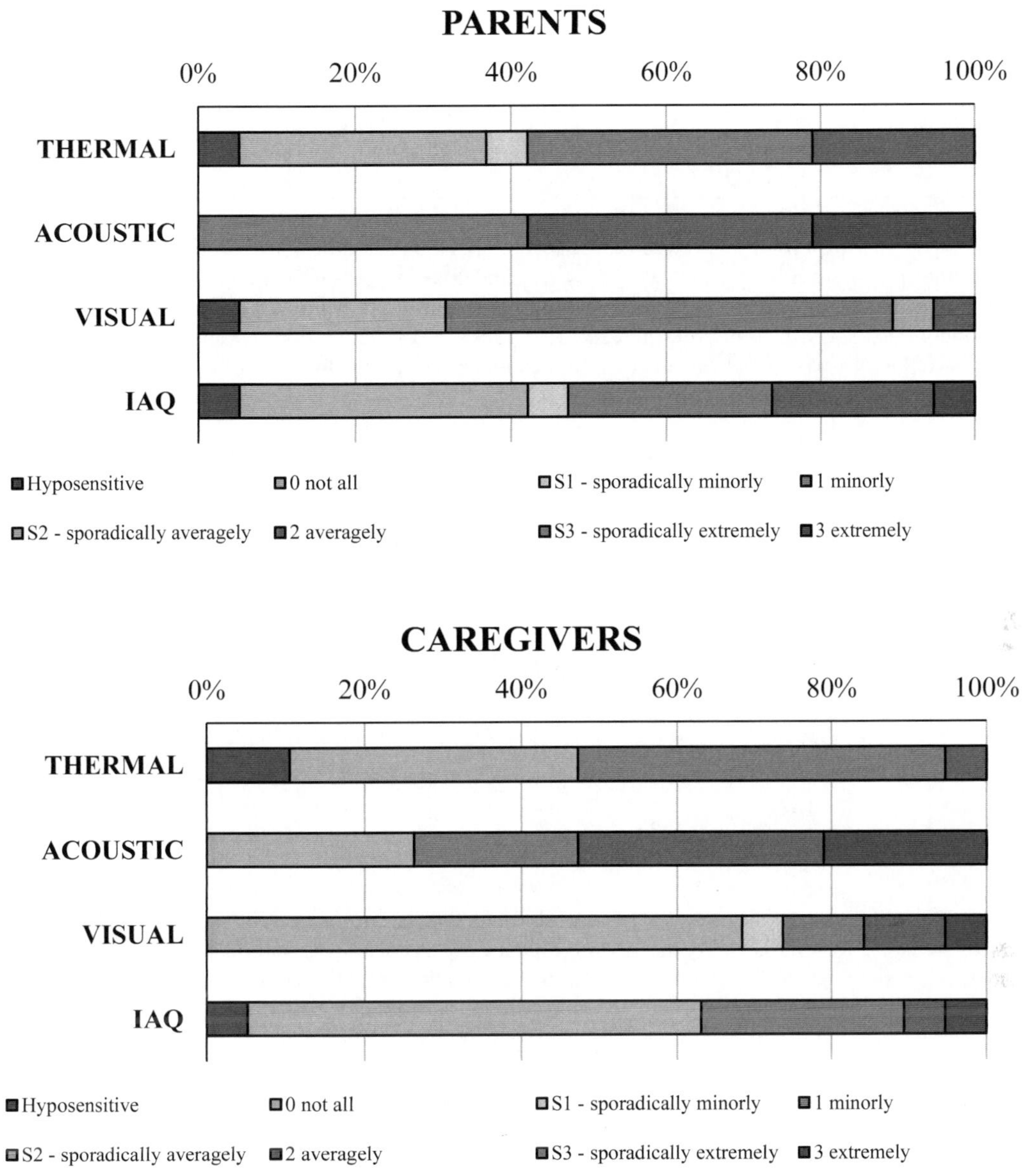

Figure 3: Answers provided by the proxy respondents- comparison between Parents and Caregivers

Table 3 depict the results of the Mann-Whitney test. This statistical analysis shows that, with the exception of visual environment, the differences between answers in the everyday life (provided by parents) and care environments (provided by caregivers) were not significant. In terms of the visual indoor environment, the results depicted in Figure 3 show a major perception of visual environment provided by parents, in the range of "minor" perception (60 % by parents vs. 10% by caregivers). This difference can be correlated to the design of the care unit, specifically designed to be utilized by ASC people, such as the presence of proper illumination, lamps and light colors, or proper daylight exploitation (e.g., rooms with large fenestration areas and a proper view to the outside). This could be in contrast with normal conditions in domestic environments.

Table 3: Mann-Whitney test between the answers provided by proxy respondents

	Control sample (N_P =19; N_C =19)
Thermal	0.405
Acoustic	0.435
Visual	**0.098***
IAQ	0.203

Finally, Table 4 explores the correlation between the answers provided separately by parents and caregivers to check if a similarity in the trends provided by the two proxy respondents was present. In this case, a good correlation on indoor air quality was detected. The correlation coefficient was positive, meaning that when one perception detected by a proxy respondent increases, the other one increases too. This last result on indoor air quality shows that, regardless of the facility where the user with ASC was located, the response was quite similar. In fact, the IAQ domain seemed to be the one causing the least discomfort in both cases.

Table 4: Spearman's correlation test between the answers provided by parents and caregivers on the four comfort domains

Cross Correlation between Parents and Caregivers				
		Caregivers		
	Domain	Thermal	Acoustic	IAQ
Parents	Thermal	corr. coeff=0.186 p=0.445		
	Acoustic		corr.coeff=0.023 p=0.925	
	IAQ			**corr. coeff=0.527 p=0.020**

Table 5 explores the influence of age, presence of comorbidities and spectrum severity with a dedicated Spearman's rank correlation analysis. Results showed that in three cases the reported perception by the two different proxy respondents were similar. In particular, acoustic and indoor air quality perceptions of ASC individuals over 15 years old were depicted as quite similar. Furthermore, acoustic perception was also identified by Spearman's statistical test to be perceived similarly by individuals with a low severity ASC by both parents and caregivers. In all these cases, the cross-correlation coefficient was positive, meaning that, as one perception detected by a proxy respondent increases, the other one increases too.

No correlations related with the presence or absence of comorbidities were identified.

Table 5: Spearman's correlation test between the answers provided by parents and caregivers on the four comfort domains, restricting the sample according to different personal data of individuals

Cross Correlation between Parents and Caregivers							
		Age		**Comorbidities**		**Severity**	
Parents	**Caregivers**	Under 15 y.o.	Over 15 y.o.	With comorbidities	Without comorbidities	Low Severity	High Severity
Thermal	Thermal	cross.coeff= -0.035 p=0.929	cross.coeff= 0.500 p=0.170	cross.coeff= -0.542 p=0.063	cross.coeff= 0.484 p=0.206	cross.coeff= -0.530 p=0.358	cross.coeff= -0.272 p=0.418
Acoustic	Acoustic	cross.coeff=	**cross.coeff=**	cross.coeff=	cross.coeff=0	**cross.coeff=**	cross.coeff=

		0.052 p=0.895	**0.797 p=0.010**	-0.024 p=0.754	p=0.99	**0.968 p=0.007**	0.225 p=0.505
Visual	Visual	cross.coeff= -0.082 p=0.834	cross.coeff= 0.456 p=0.217	cross.coeff= -0.310 p=0.398	cross.coeff=0 p=0.99	cross.coeff= 0.408 p=0.495	cross.coeff= -0.420 p=0.198
IAQ	IAQ	corr.coeff =0.635 p=0.066	**corr.coeff =0.673 p=0.047**	corr.coeff= 0.561 p= 0.116	corr.coeff= 0.074 p= 0.838	corr.coeff<0. 001 p= 0.99	corr.coeff= 0.11 p= 0.974

4. CONCLUSIONS

Results show that, with exception made for the visual environment, differences between the perception reported by the two proxy respondents are not significant. The difference detected by the Mann-Whitney test in the perception of the visual environment can be directly linked to the attention paid in the care facility design of lights and daylight exploitation.

Moreover, analyzing the correlation between the responses provided by both proxy respondents as regards the perception of the four comfort domains, it was found that when the IAQ perception determined in the households increases, it also increases in care facilities.

Finally, in order to understand if the results were connected to the personal data of the ASC individuals, a further Spearman's correlation analysis was performed. The results of this test showed that, for individuals over 15 years old, the trend of responses described by the two proxy respondents is similar for both the IAQ domain and the perception of the acoustic environment. Moreover, for individuals featuring low autism severity, a cross-correlation between the answers provided by parents and caregiver was found for the acoustic domain, showing that the answers provided for these individuals had similar trends. No relationship as regards co-morbidities was found.

Since the sample was rather restricted, the future perspective of the presented research is to enlarge the control sample in order to evaluate if, on a larger population, there are further correlations between the perceptions detected by parents and caregivers in diverse environments (household and care facility).

REFERENCES

American Psychiatric Association. (2013). Diagnostic and statistical manual of mental disorders—Edition Fifth. *American Psychiatric Association*, *21*, 591–643.

American Society of Heating, Refrigerating and Air Conditioning Engineers. (2010). *ASHRAE Standard 55-2010: Thermal Environmental Conditions for Human Occupancy*.

Artusi, R., Verderio, P., & Marubini, Ejti. (2002). Bravais-Pearson and Spearman correlation coefficients: Meaning, test of hypothesis and confidence interval. *The International Journal of Biological Markers*, *17*(2), 148–151.

Caniato, M., Zaniboni, L., Marzi, A., & Gasparella, A. (2022). Evaluation of the main sensitivity drivers in relation to indoor comfort for individuals with autism spectrum disorder. Part 2: Influence of age, co-morbidities, gender and type of respondent on the stress caused by specific environmental stimuli. *Energy Reports*, *8*, 2989–3001. https://doi.org/10.1016/j.egyr.2022.01.011

Castaldo, V. L., Pigliautile, I., Rosso, F., Cotana, F., De Giorgio, F., & Pisello, A. L. (2018). How subjective and non-physical parameters affect occupants' environmental comfort perception. *Energy and Buildings*, *178*, 107–129.

Castaldo, V. L., Pigliautile, I., Rosso, F., Pisello, A. L., & Cotana, F. (2017). Investigation of the impact of subjective and physical parameters on the indoor comfort of occupants: A case study in central Italy. *Energy Procedia*, *126*, 131–138.

Catalina, T., & Banu, T. (2014). Impact of indoor environmental conditions on students intellectual performance. *Buletinul Institutului Politehnic Din Lasi. Sectia Constructii, Arhitectura*, *60*(3), 23.

Comite'Europe'en de Normalisation, C. E. N. (2007). Indoor environmental input parameters for design and assessment of energy performance of buildings addressing indoor air quality, thermal environment, lighting and acoustics. *EN 15251*.

Duerden, E. G., Taylor, M. J., Lee, M., McGrath, P. A., Davis, K. D., & Roberts, S. W. (2015). Decreased sensitivity to thermal stimuli in adolescents with autism spectrum disorder: Relation to symptomatology and cognitive ability. *The Journal of Pain*, *16*(5), 463–471.

Dykes, C., & Baird, G. (2013). A review of questionnaire-based methods used for assessing and benchmarking indoor environmental quality. *Intelligent Buildings International*, *5*(3), 135–149.

European Committee for Standardization. (2011). *EN 12464-1. Light and lighting—Lighting of work places—Part 1: Indoor work places.*
European Committee for Standardization. (2019). *EN 16798-1. Energy performance of buildings—Ventilation for buildings—Part 1: Indoor environmental input parameters for design and assessment of energy performance of buildings addressing indoor air quality, thermal environment, lighting and acoustics.*
Fransson, N., Västfjäll, D., & Skoog, J. (2007). In search of the comfortable indoor environment: A comparison of the utility of objective and subjective indicators of indoor comfort. *Building and Environment*, *42*(5), 1886–1890.
Frontczak, M., Andersen, R. V., & Wargocki, P. (2012). Questionnaire survey on factors influencing comfort with indoor environmental quality in Danish housing. *Building and Environment*, *50*, 56–64.
Frontczak, M., Schiavon, S., Goins, J., Arens, E., Zhang, H., & Wargocki, P. (2012). Quantitative relationships between occupant satisfaction and satisfaction aspects of indoor environmental quality and building design. *Indoor Air*, *22*(2), 119–131.
Hume, K., Dykstra Steinbrenner, J., Sideris, J., Smith, L., Kucharczyk, S., & Szidon, K. (2018). Multi-informant assessment of transition-related skills and skill importance in adolescents with autism spectrum disorder. *Autism*, *22*(1), 40–50.
Johnson, J., Atkinson, M., Rosenberg, R., Schoon, I., Joshi, H., & Smith, K. (2018). *Global Research on Developmental Disabilities Collaborators. Developmental disabilities among children younger than 5 years in 195 countries and territories, 1990–2016: A systematic analysis for the Global Burden of Disease Study 2016. Lancet Glob Health 2018; 6: E1100-21.[PubMed: 30172774][Cross Ref].*
Kanakri, S. M., Shepley, M., Varni, J. W., & Tassinary, L. G. (2017). Noise and autism spectrum disorder in children: An exploratory survey. *Research in Developmental Disabilities*, *63*, 85–94.
Katafygiotou, M. C., & Serghides, D. K. (2014). Indoor comfort and energy performance of buildings in relation to occupants' satisfaction: Investigation in secondary schools of Cyprus. *Advances in Building Energy Research*, *8*(2), 216–240.
Keith, J. M., Jamieson, J. P., & Bennetto, L. (2019). The influence of noise on autonomic arousal and cognitive performance in adolescents with autism spectrum disorder. *Journal of Autism and Developmental Disorders*, *49*(1), 113–126.
Kim, S.-K., Hwang, Y., Lee, Y. S., & Corser, W. (2015). Occupant comfort and satisfaction in green healthcare environments: A survey study focusing on healthcare staff. *Journal of Sustainable Development*, *8*(1), 156.
Lai, J. H., & Yik, F. W. (2009). Perception of importance and performance of the indoor environmental quality of high-rise residential buildings. *Building and Environment*, *44*(2), 352–360.
Leung, M., Wang, C., & Chan, I. Y. (2019). A qualitative and quantitative investigation of effects of indoor built environment for people with dementia in care and attention homes. *Building and Environment*, *157*, 89–100.
Maenner, M. J., Shaw, K. A., & Baio, J. (2020). Prevalence of autism spectrum disorder among children aged 8 years—Autism and developmental disabilities monitoring network, 11 sites, United States, 2016. *MMWR Surveillance Summaries*, *69*(4), 1.
McKnight, P. E., & Najab, J. (2010). Mann-Whitney U Test. *The Corsini Encyclopedia of Psychology*, 1–1.
Noble, B., Isaacs, N., & Lamb, S. (2018). The impact of IEQ factors on people on the autism spectrum. *International Conference of the Architectural Science Association*, 27–33.
Pinter, J. M., & Kiss, M. L. (2019). Determination and measurement of parameters affecting indoor comfort. *2019 20th International Carpathian Control Conference (ICCC)*, 1–6.
Pinto, S., Fumincelli, L., Mazzo, A., Caldeira, S., & Martins, J. C. (2017). Comfort, well-being and quality of life: Discussion of the differences and similarities among the concepts. *Porto Biomedical Journal*, *2*(1), 6–12.
Van Hoof, J., Kort, H. S., Duijnstee, M. S., Schoutens, A. M., Hensen, J. L., & Begemann, S. H. (2008). The indoor environment in relation to people with dementia. *Indoor Air*, 17–22.
Vilcekova, S., Meciarova, L., Burdova, E. K., Katunska, J., Kosicanova, D., & Doroudiani, S. (2017). Indoor environmental quality of classrooms and occupants' comfort in a special education school in Slovak Republic. *Building and Environment*, *120*, 29–40.
Wang, Z., & Hong, T. (2020). Learning occupants' indoor comfort temperature through a Bayesian inference approach for office buildings in United States. *Renewable and Sustainable Energy Reviews*, *119*, 109593.
Williams, Z. J., Failla, M. D., Davis, S. L., Heflin, B. H., Okitondo, C. D., Moore, D. J., & Cascio, C. J. (2019). Thermal perceptual thresholds are typical in autism spectrum disorder but strongly related to intra-individual response variability. *Scientific Reports*, *9*(1), 1–14.
Wong, J. K.-W., Skitmore, M., Buys, L., & Wang, K. (2014). The effects of the indoor environment of residential care homes on dementia suffers in Hong Kong: A critical incident technique approach. *Building and Environment*, *73*, 32–39.

ACKNOWLEDGEMENT

This work was partially financed by the European Interreg SENSHome project, ITAT 1088 CUP: I54I18000310006. The authors want to thank Dr. Michele Borghetto for his precious help in defining the correct questions form and all the partners of the European Project SENSHome, who helped us with their contributions. In particular, the authors thank the section lead by Eng. Daniela Krainer of Carinthia University of Applied Sciences and by Prof. Giuseppina Scavuzzo of University of Trieste for the fundamental contributions in the redaction of the questionnaires. The authors would like to thank Fondazione Progetto Autismo and Elena Bulfone for their precious help. This paper was written according to the language acceptability indications of Autism-Europe (Autism-Europe, 2021). The study was conducted after the approval of the Ethics Committee of the Free University of Bozen-Bolzano. The procedure was implemented in order to comply with the Declaration of Helsinki (World Medical Association, 1991).

Assessment Of The Accuracy Of Solar Irradiance Models In Mountain Locations: The Case Of Bolzano, Italy

Giovanni PERNIGOTTO[1]*, Alessandro PRADA[2], Andrea GASPARELLA[3]

[1]Free University of Bozen-Bolzano, Faculty of Science and Technology,
Bolzano, Italy
Tel: +39 0471 017632, Fax: +39 0471 017009, email: giovanni.pernigotto@unibz.it

[2]University of Trento, Department of Civil Environmental and Mechanical Engineering,
Trento, Italy
Tel: +39 0461 282516, Fax: +39 0461 282672, email: alessandro.prada@unitn.it

[3]Free University of Bozen-Bolzano, Faculty of Science and Technology,
Bolzano, Italy
Tel: +39 0471 017200, Fax: +39 0471 017009, email: andrea.gasparella@unibz.it

* Corresponding Author

ABSTRACT

Two main groups of solar irradiance models can be identified in the literature: the "*horizontal diffuse irradiance models*", whose goal is to process the global horizontal irradiance to calculate the beam and the diffuse components, and the "*irradiance models for tilted surfaces*", which are used to determine the solar radiation incident on generic surfaces starting from the horizontal measured data, if beam and diffuse components are available, or from the results of a model of the first group. According to the state-of-the-art, there is no pair of "*horizontal diffuse irradiance model*" and "*irradiance model for tilted surfaces*" which can give the same level of accuracy worldwide, since they were developed starting from solar irradiance data of specific meteorological stations, with typical weather and sky cover conditions. In addition, this becomes particularly critical for locations in mountain regions, characterized by many natural obstacles altering the view of the sky, its radiance distributions, as well as the ground reflections in a very dynamic and localized way, with an impact on the solar diffuse and beam radiation available.
In this research, the performance of 22 "*horizontal diffuse irradiance models*" and 12 "*irradiance models for tilted surfaces*" were discussed, using as benchmark the global solar irradiance data collected for both horizontal and vertical surfaces in a multi-year solar monitoring campaign performed in Bolzano, a city in the Italian Alps. After identifying the most and the least accurate models, they were adopted to simulate the energy needs for space heating and cooling for a dataset of 48 simplified building configurations, in order to discuss the impact of the calculated incident solar irradiance on the simulated building energy performance.

1. INTRODUCTION

In order to minimize its energy demand, a high performance building has to carefully manage the solar radiation received on its external envelope, especially on the transparent elements. Indeed, solar gains are crucial to reduce the space heating load during the cold season, even if proper controls are required to avoid overheating and cooling peak loads during the hot one. As a whole, an optimal control of solar gains can ensure the achievement of higher energy performance, preserving or increasing thermal and visual comfort. Nevertheless, extensive application of advanced solar controls can be technically challenging or too expensive to adopt, making still very common solutions based on the assessment of vertical solar irradiance on the main façade or of horizontal irradiance on the roof. In both cases, the measured irradiance is the global one, without distinguishing between beam and diffuse components, albeit their different impact on the performance and on specific control algorithms. In these contexts, solar irradiance models can be used to increase the information detail.

Several empirical models were proposed in the literature, to distinguish the horizontal solar irradiation into beam and diffuse components (*horizontal diffuse irradiance models*) and to calculate the irradiation on tilted surfaces (*irradiance models for tilted surfaces*). Nevertheless, their empirical nature means that the model accuracy is closely tied to site characteristics, since every model was developed minimizing the deviation to experimental data collected in specific locations. For this reason, constant and accurate reliability cannot be taken for granted by changing location and climate.

In (Dervishi and Mahdavi, 2012), the authors assessed the model accuracy by comparing the models' predictions against experimental data, drawing some conclusions about the best model for a given locality. Other researches (Prada *et al.*, 2014a, 2014b, 2015; Pernigotto *et al.*, 2015 and 2016), instead, quantified the effect of the different models on the Building Energy Simulation *BES* prediction by highlighting the building characteristics increasing the sensitivity of *BES* to the selection of solar irradiance models. This is, for instance, a problematic issue when irradiance models, mostly derived in flat regions, are used to evaluate the micro-climatic boundary conditions in mountain contexts, where the complex irradiation patterns depend on the orography and the multiple terrain reflections.

Further expanding previous analyses, in this work we investigated the effect of the choice of solar irradiation models on the uncertainty of the predicted energy performance in the Alpine city of Bolzano, Italy. 22 *horizontal diffuse irradiance models* were coupled with 12 *irradiance models for tilted surfaces* and tested against the experimental data collected in Bolzano. A subset of models with extreme behavior were selected from the statistical comparison between model predictions and measurements on vertical surfaces oriented towards the main cardinal points. The different solar radiation patterns were then used as input in TRNSYS 18 for the simulation of the energy performances of a dataset of 48 simplified residential buildings, built varying parametrically insulation and thermal inertia of opaque components, windows surface and orientation, and solar heat gain coefficient of glazing.

2. METHODS

This section describes the methods adopted for the development of this research. After the description of the case-study mountain location of Bolzano, Italy, as well as the features of the considered weather station, the list of the selected solar irradiance models is presented. Finally, the dataset of building models simulated with TRNSYS 18 to assess the accuracy of a subset of solar irradiance models is illustrated.

2.1 Case-study location and weather station

Bolzano is an Italian municipality in the middle of the Alps (46.500° N, 11.350° E), specifically in a basin where three valleys - Sarntal, Eisacktal, and the Adige Valley, and their rivers, respectively, Talfer-Talvera, Eisack-Isarco, and Adige, meet. The settlement area occupies a surface of about 30 km^2, where almost 110'000 people live. The municipality spreads from 232 m to more than 1600 m above the sea level; however, the city center is located at an altitude of 262 m.

The weather station considered in this study was designed by the Research Group in Building Physics of the Free University of Bozen-Bolzano and installed in 2017 on the flat green roof of one of the buildings of the university campus in the city center, specifically the Building E (46.498° N, 11.349° E), characterized by a height of about 25 m. Figure 1 shows a map of the basin of Bolzano with the university campus in the city center (red dot in the figure).

The urban layout of the city center of Bolzano is pretty homogeneous, with buildings surrounding the university campus with similar or lower heights compared to the Building E. Consequently, the main obstacles reducing the solar irradiance received on the roof where the weather station is installed are the natural ones, i.e., the mountains around the basin of Bolzano (Figure 2). In detail, it can be observed that the obstacles south of the weather station are small (with an altitude generally lower than 5°) while those found on the north-east and north-west are more significant. In particular, the obstacles' maximum altitude (about 30°) can be located on the north-west direction.

As regards the weather station, it is equipped with 5 EKO MS-802 pyranometers, respectively one installed horizontally for the measurement of the global horizontal irradiance and four installed vertically towards the main cardinal directions for the collection of the global irradiance incident on those orientations. Furthermore, the weather station includes a EKO MS-56 pyrheliometer and a EKO MS-802 pyranometer, fixed on a STR-22G EKO sun-tracker station, for the measurement of the direct normal irradiance *DNI* and the diffuse component of the horizontal irradiance. Despite the availability of *DNI* and diffuse horizontal irradiance, the study presented in this paper considered only the global irradiance values recorded by the MS-802 pyranometers, representing data more commonly found in other weather stations in this mountain region, facilitating comparisons in the next developments of this research.

As mentioned above, the weather station was installed in 2017 and has collected solar data with 1-minute time-discretization since then. In the current research, data referred to years 2018, 2019 and 2021 were included. The year 2020 was not considered since the pandemic lockdowns prevented periodical maintenance of pyranometers.

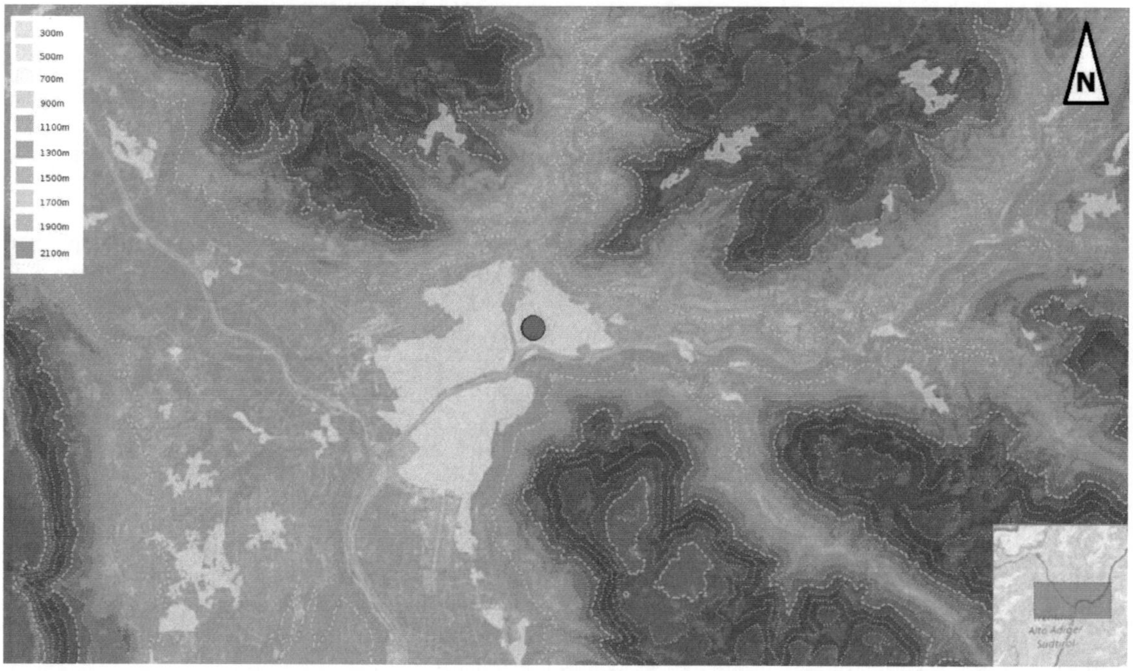

Figure 1: Basin of Bolzano: the different colors (light green to brown) identify the altitude while the yellow the urban areas; the red dot indicates the position of the university campus. Map developed from the Geobrowser Maps by the Autonomous Province of Bolzano

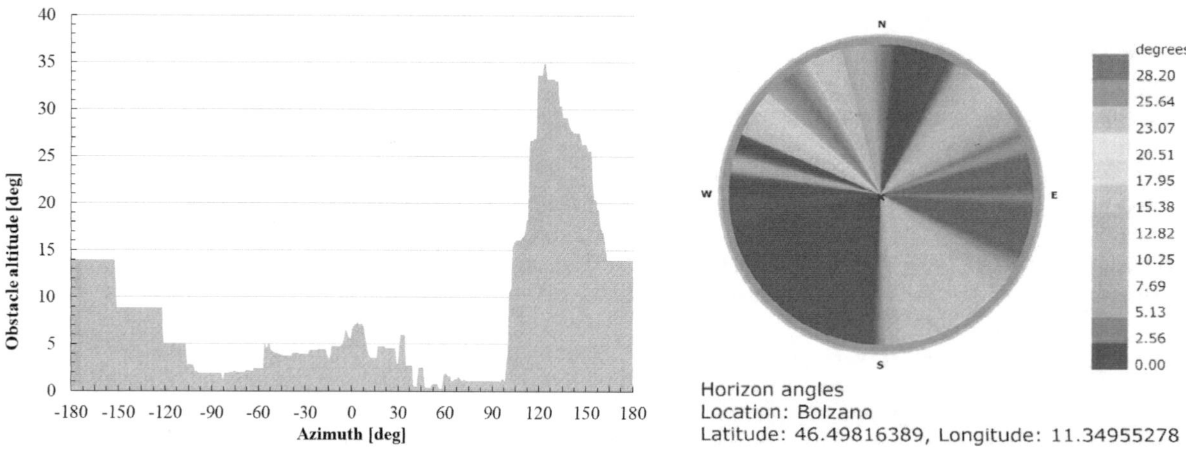

Figure 2: Altitude and position of the natural obstacles as determined with Grass GIS considering the Digital Surface Model *DSM* 2.5 m and the Digital Terrain Model *DTM* 2.5 m by the Autonomous Province of Bolzano

2.2 Solar irradiance models

As a follow-up to previous researches on this topic (Prada *et al.*, 2014a and 2014b; Pernigotto *et al.*, 2015 and 2016), we considered the same 22 *horizontal diffuse irradiance models* and 12 *irradiance models for tilted surfaces* analyzed before and reported in Table 1. In agreement with the approach in (Pernigotto *et al.*, 2016), the two groups of irradiance models were combined, for a total of 264 alternatives.

Table 1: Solar irradiance models

ID	*Horizontal diffuse irradiance models*	ID	*Irradiance models for tilted surfaces*
1	Erbs *et al.* (1982)	A	Liu & Jordan (1960)
2	Orgill & Hollands (1977)	B	Burgler (1977)
3	Reindl *et al.* (1990a) – Model 1	C	Temps & Coulson (1977)
4	Reindl *et al.* (1990a) – Model 2	D	Klucher (1978)
5	Reindl *et al.* (1990a) – Model 3	E	Hay & Davies (1980)
6	Lam & Li (1996)	F	Ma & Iqbal (1983)
7	Boland *et al.* (2008)	G	Skartveit & Olseth (1986)
8	Hawlader (1984)	H	Gueymard (1986)
9	De Miguel *et al.* (2001)	I	Reindl *et al.* (1990b)
10	Karatasou *et al.* (2003)	J	Perez *et al.* (1990)
11	Chandrasekaran & Kumar (1994)	K	Muneer (2006) – Model 1
12	Oliveira *et al.* (2002)	L	Muneer (2006) – Model 2
13	Soares *et al.* (2004)		
14	Muneer *et al.* (1984)		
15	Spencer (1982)		
16	Chendo & Maduekwe (1994) – Model 1		
17	Chendo & Maduekwe (1994) – Model 2		
18	Skartveit & Olseth (1987)		
19	Maxwell (1987)		
20	Perez *et al.* (1992) – Model 1		
21	Perez *et al.* (1992) – Model 2		
22	Perez *et al.* (1992) – Model 3		

2.3 Dataset of building configurations

A dataset of 48 simplified buildings was employed to assess the impact of the solar irradiance models on the simulated energy needs for space heating and cooling. All 48 building configurations have the same geometry, i.e., a thermal zone with a square floor area of 100 m^2, an internal height of 3 m, and the façades oriented towards the main cardinal directions. For each case, all windows are positioned on the same façade. Both sides of the vertical walls and the internal side of the roof have a solar absorptance of 0.3 while the external side of the roof and the internal side of the floor have 0.6.

All opaque components have the same composition, i.e., a two-layer structure with insulation on the external side and a massive layer whose thermal resistance is around 0.8 $m^2\ K\ W^{-1}$. The insulating layer is polystyrene (thermal conductivity: 0.04 $W\ m^{-1}\ K^{-1}$; density: 40 $kg\ m^{-3}$; specific heat capacity: 1470 $J\ kg^{-1}\ K^{-1}$) while the massive layer can be either timber (thickness: 0.10 m; thermal conductivity: 0.13 $W\ m^{-1}\ K^{-1}$; density: 399 $kg\ m^{-3}$; specific heat capacity: 1880 $J\ kg^{-1}\ K^{-1}$) or concrete (thickness: 0.30 m; thermal conductivity: 0.37 $W\ m^{-1}\ K^{-1}$; density: 1190 $kg\ m^{-3}$; specific heat capacity: 840 $J\ kg^{-1}\ K^{-1}$). Windows are composed by a double-pane glazing with a *U*-value of 1.1 $W\ m^{-2}\ K^{-1}$ and a timber frame (20 % of the window's area) with a *U*-value of 1.2 $W\ m^{-2}\ K^{-1}$.

Internal gains and ventilation rate are assumed constant and equal, respectively, to 4 $W\ m^{-2}$, half radiative and half convective, and to 0.3 air changes per hour, as indicated by the Italian technical specification UNI/TS 11300-1:2014 (UNI, 2014) for residential buildings. An ideal system provides all the power needed to maintain the zone internal air temperature between the heating and the cooling setpoints of 20 °C and 26 °C. Although for the climate of Bolzano the heating season starts on October 15th and finishes on April 15th, we assumed space heating and cooling available all year, whenever needed. Further details about the dataset of buildings can be found in (Pernigotto *et al.*, 2021). Table 2 reports an overview of the variables differentiating the buildings in the dataset.

Table 2: Variables differentiating the buildings in the dataset

Insulation thickness and U-value	Materials and thermal inertia c	Windows' $SHGC$	Windows' size and window-to-wall ratio WWR	Windows orientation
5 cm ($U = 0.45$ W m^{-2} K^{-1})	Timber ($c = 75$ kJ m^{-2} K^{-1})	$SHGC = 0.35$	14.56 m^2 ($WWR = 48.5$ %)	East
15 cm ($U = 0.21$ W m^{-2} K^{-1})	Concrete ($c = 300$ kJ m^{-2} K^{-1})	$SHGC = 0.61$	29.12 m^2 ($WWR = 97.1$ %)	South
-	-	-	-	West

2.4 Analysis methodology

The first step of the analysis regarded the data collected by the pyranometers of the weather station on top of the Building E of the university campus. A quality check was performed, in order to identify outliers and missing entries. For instance, values exceeding the solar constant and positive values before dawn and after dusk were removed. Input data, with 1-minute time discretization, were processed in order to get hourly profiles of solar irradiation expressed in watthours per square meter, as they are conventionally found in weather data for building performance simulation (see, for instance, the EnergyPlus .epw weather files). Minor missing entries (i.e., one or few hours of missing solar irradiation data) were fixed by linear or cyclic interpolation, depending on the case. On the contrary, larger missing data entries, longer than a day, were not fixed and simply not involved in the analysis.
The second step was dedicated to the assessment of the irradiance models against the measured solar data. The hourly profiles of global solar horizontal irradiation of 2018, 2019 and 2021 were used as inputs to calculate the global solar irradiation on four vertical surfaces oriented towards the main cardinal directions, using each one of the 264 combinations of pairs of *horizontal diffuse irradiance models* and *irradiance models for tilted surfaces*. The determined hourly profiles of global solar irradiation on vertical surfaces were compared to the measured ones, calculating for each orientation the Mean Absolute Error MAE. This index was exploited to identify the best- and the worst-performing pairs of irradiance models, as well as those *horizontal diffuse irradiance models* and *irradiance models for tilted surfaces* most frequently found among the best- and the worst-performing ones (i.e., those with the minimum and the maximum values of MAE).
The last step considered those *horizontal diffuse irradiance models* and *irradiance models for tilted surfaces* most frequently found among the best- and the worst-performing ones. Those models were used to determine the solar irradiance incident on the external envelope of the dataset of 48 buildings, respectively for 2018, 2019 and 2021, and simulate with TRNSYS 18 their energy needs for space heating and cooling. The goal of this final assessment was to discuss the variability which can be found in building performance simulation because of the selection of solar irradiance models when applied in mountain environments, as well as those building configurations more affected by the choice of solar irradiance models for the specific location of Bolzano.

3. RESULTS AND DISCUSSION

3.1 Dataset of solar irradiation measurements

The quality check for the three available years allowed us to identify those series of hourly data of solar irradiation with missing or wrong entries, as reported in Table 3. As it can be noticed, although the year 2021 is characterized by a larger share of missing data, its numerosity can be still considered adequate for a robust comparison. As specified in Section 2.4, these series were discarded in the rest of the analysis and, as regards the last step, the corresponding months (February 2018, August 2019, April and May 2021) not even simulated.

3.2 Comparison between simulated and measured solar irradiation values

The Mean Absolute Error was used as index to identify, for each year and for the whole series, which pairs of solar irradiance models gave the closest estimates to the actual recordings for the four vertical cardinal directions.

Table 3: Quality check on the solar irradiation measurements

Year	Sunshine hours for vertical surfaces [h]				Missing data [h]	Period missing
	South	East	North	West		
2018	4641	4652	4654	4657	83	8 days in February
2019	4661	4661	4660	4660	81	5 days in August and 1 day in September
2021	4217	4212	4216	4218	522	34 days in April and May

Table 4: Best- and worst-performing pairs of solar irradiance models and respective *MAE* (Wh m^{-2}) for the three considered years and the four cardinal orientations. In grey those pairs confirmed in multiple years. See Table 1 for the model identification

	Best-performing pairs of irradiance models				Worst-performing pairs of irradiance models			
	South	**East**	**North**	**West**	**South**	**East**	**North**	**West**
2018	A5 41.7	C20 78.6	J18 22.7	B20 163.5	C20 59.6	F15 134.5	D20 76.9	F5 335.0
2019	H18 44.1	C18 70.7	J13 31.7	B20 160.7	F15 75.3	I15 111.6	J20 129.4	J20 469.2
2021	H18 45.7	C20 81.5	J18 22.9	B20 172.5	F18 69.1	F15 147.6	D20 77.8	F18 377.5
Similarity among the years	**67 %**	**83 %**	**83 %**	**100 %**	**33 %**	**83 %**	**83 %**	**33 %**
ALL	H18 43.8	C20 79.3	J18 26.1	B20 165.3	I15 67.4	F15 130.6	D20 79.9	F18 384.1

As it can be noticed in Table 4, the *MAEs* assessed over the three years combined vary significantly changing orientation and comparing best- and worst-performing pairs of solar irradiance models. The south orientation is well-predicted, with an average *MAE* around 40-45 Wh m^{-2} for the best-performing pairs and around 60-75 Wh m^{-2} for the worst-performing ones. Although the same index for the north-orientation is smaller, its relative incidence is high, considering that this orientation receives predominantly diffuse irradiance. Moreover, the worst-performing pairs of models show *MAEs* which are more than three times those of the best-performing models. East and west-orientation are characterized by larger errors in the irradiation estimation. Specifically, the *MAEs* for east-oriented walls range from about 80 Wh m^{-2} (best-performing pairs of models) to 110-150 Wh m^{-2} (worst-performing ones). Finally, the west-orientation is the most critical one, with Mean Absolute Errors larger than 160 Wh m^{-2} for the best-performing models and in the range 330-470 Wh m^{-2} for the worst-performing ones. This is somehow expected: indeed, as observed in Section 2.1, north-east and north-west show the presence of the most important obstacles
The best- and the worst-performing models are found different for each orientation and, considering that the different years have various missing entries, not always constant along the time-series. Nevertheless, some trends can be identified. Gueymard + Skartveit & Olseth (**H18**), Temps & Coulson + Perez *et al.* Model 1 (**C20**), Perez *et al.* + Skartveit & Olseth (**J18**) and Burgler + Perez *et al.* Model 1 (**B20**) are the best solutions, respectively for the south-, east-, north- and west-orientation. On the contrary, Ma & Iqbal + Spencer (**F15**) and Klucher + Perez *et al.* Model 1 (**D20**) show the worst performance for east- and north-orientations.

Table 5 and Table 6 show, respectively, the *horizontal diffuse irradiance models* and *irradiance models for tilted surfaces* most frequently found among the best- and the worst-performing ones.
For east- and west-oriented vertical components, the Perez *et al.* Model 1 (**20**) is often the most accurate *horizontal diffuse irradiance model* while the very same model happens to be the worst-performing one, when it comes to the

north-oriented vertical elements. As a whole, the Spencer model (**15**) is generally the worst-performing *horizontal diffuse irradiance model* for the remaining orientations.
The best-performing *irradiance model for tilted surfaces* depends on the considered vertical orientation: the Burgler model (**B**) is the best-performing one for south- and west-oriented elements, the model by Temps & Coulson (**C**) is more accurate for the east-facing ones, while the Perez model (**J**) is superior for the north orientation. As regards the worst-performing *irradiance model for tilted surfaces*, the model by Ma & Iqbal (**F**) is the worst-performing one for all orientations but the north, for which the model by Klucher (**D**) is more frequently the one leading to the largest errors.

In consideration of the results found in this section, the following list of irradiance models was considered for the last step of the analysis involving building performance simulation:

- *horizontal diffuse irradiance models*: the Spencer model (**15**) and the Perez *et al.* Model 1 (**20**);
- *irradiance model for tilted surfaces*: the Burgler model (**B**), the model by Temps & Coulson (**C**), the model by Klucher (**D**), the model by Ma & Iqbal (**F**), and the Perez model (**J**).

Table 5: Best- and worst-performing *horizontal diffuse irradiance models* for the three considered years and the four cardinal orientations. In grey those models confirmed in multiple years. See Table 1 for the model identification

	Best-performing horizontal diffuse irradiance models				**Worst-performing horizontal diffuse irradiance models**			
	South	**East**	**North**	**West**	**South**	**East**	**North**	**West**
2018	18	20	13	20	15	15	20	15
2019	5	20	15	20	15	15	20	5
2021	18	20	18	20	15	15	20	15
Similarity among the years	**67 %**	**100 %**	**0 %**	**100 %**	**100 %**	**100 %**	**100 %**	**67 %**
ALL	5	20	6 / 13	20	15	15	20	5 / 15

Table 6: Best- and worst-performing *irradiance models for tilted surfaces* for the three considered years and the four cardinal orientations. In grey those models confirmed in multiple years. See Table 1 for the model identification

	Best-performing irradiance models for tilted surfaces				**Worst-performing irradiance models for tilted surfaces**			
	South	**East**	**North**	**West**	**South**	**East**	**North**	**West**
2018	B	C	J	B	F	F	D	F
2019	K	C	J	B	F	F	D	F
2021	B	C	J	B	F	F	D	F
Similarity among the years	**67 %**	**100 %**	**100 %**	**100 %**	**100 %**	**100 %**	**100 %**	**100 %**
ALL	B	C	J	B	F	F	D	F

3.3 Analysis of building energy needs

TRNSYS simulations were run for each of the 48 building configurations and the three considered years. As mentioned in Section 2.4, the months with long series of missing data reported in Table 3 were discarded: simulation start was fixed at the next month after those series (i.e., March, September and June, respectively for 2018, 2019 and 2021) and simulation stop before them (i.e., January, July and March, respectively for 2018, 2019 and 2021). Energy needs for space heating and cooling were analyzed at both monthly and annual basis. Table 7 reports the minimum and the maximum deviations found simulating the dataset of buildings with the 10 different pairs of solar irradiance models selected in Section 3.2.

Table 7: Minimum and maximum monthly and annual deviations of heating and cooling needs for the simulated dataset of 48 buildings. The grey cells indicate the non-simulated months while the crossed ones represent those months without energy demand

	Heating need deviations [kWh m^{-2}]						**Cooling need deviations [kWh m^{-2}]**					
	2018		**2019**		**2021**		**2018**		**2019**		**2021**	
	min	**max**	**min**	**max**	**min**	**max**	**min**	**max**	**min**	**max**	**min**	**max**
Jan	0.4	4.1	0.5	5.2	0.5	5.3	0.0	1.2	0.0	1.7		
Feb			0.9	4.0	0.8	3.6			0.0	4.6	0.0	3.6
Mar	0.7	2.4	0.0	2.8	0.0	3.8	0.0	3.9	0.0	9.6	0.0	9.6
Apr	0.0	1.1	0.0	1.4			0.0	7.0	0.0	5.6		
May			0.0	1.5			0.0	6.7	0.0	10.2		
Jun							1.5	7.2	2.4	44.9	1.4	6.4
Jul							1.6	7.7	2.1	33.7	1.4	7.7
Aug							1.1	9.6			0.9	9.4
Sep							0.3	8.7	0.0	6.2	0.3	9.2
Oct	0.0	1.0	0.0	1.4	0.0	2.2	0.0	5.9	0.0	4.5	0.0	5.3
Nov	0.4	3.3	0.3	2.6	0.6	4.4	0.0	1.1			0.0	2.4
Dec	0.4	4.9	0.3	4.2	0.4	4.9						
Period	**2.8**	**14.3**	**4.7**	**16.2**	**4.4**	**19.6**	**5.0**	**51.9**	**4.7**	**100.7**	**4.2**	**47.9**

As far as the heating needs are concerned, the largest deviations are usually found in the coldest months of the year and are around 5 kWh m^{-2} m^{-1}. Considering the whole simulated period, the largest deviations can range from 14 to 20 kWh m^{-2} a^{-1}. However, they are not related to the configurations with the largest energy needs for space heating but to the one with only 5-cm insulation on a concrete structure and large south-oriented windows with high *SHGC*. Analyzing the configurations with annual heating deviations larger than 10 kWh m^{-2} a^{-1}, it can be noticed that buildings with large south-oriented windows with high *SHGC* and poorly insulated concrete structures are more sensitive to the choice of solar irradiance models in the mountain location of Bolzano. If just the best-performing solar irradiance models are considered (i.e., **B20**, **C20**, **J20**), the largest annual deviations are reduced to the range from 4 to 8 kWh m^{-2} a^{-1}.

The cooling needs are characterized by larger deviations, usually occurring during the summer months (May to September) and equal to the range 6-10 kWh m^{-2} m^{-1}. June and July 2019 represent an exception, with extremely large cooling deviations, respectively equal to almost 45 kWh m^{-2} m^{-1} and almost 34 kWh m^{-2} m^{-1}. As regards the whole simulated period, the largest deviations are around 50 kWh m^{-2} a^{-1}, with the exception of the year 2019, which can reach 100 kWh m^{-2} a^{-1}. This time the largest deviation occurs with the configuration with the largest cooling energy need, i.e., the well-insulated one with a timber structure and large west-oriented windows with high *SHGC*. Analyzing the configurations with annual cooling deviations larger than 20 kWh m^{-2} a^{-1}, it can be noticed that buildings with large west-oriented windows with high *SHGC* and well-insulated envelopes are more sensitive to the choice of solar irradiance models in the considered location. If just the best-performing solar irradiance models are considered (i.e., **B20**, **C20**, **J20**), the magnitude of the largest annual deviations is reduced for 2018 and 2021, within the range from 20 to 22 kWh m^{-2} a^{-1}. On the contrary, no improvement is registered for 2019. Indeed, for this year larger *MAEs* were observed in Section 3.2 compared to the other two years, especially for north- and west-orientations, whose worst-performing pair of solar irradiance models was **J20**. Further analyses will be performed for this year in the next development of this research, studying the time series of solar irradiance values in order to identify the issues leading to these very large deviations.

The next figure shows, as an example, the cumulative distribution functions of the monthly energy needs for space heating and space cooling simulated for the 48 buildings, respectively for December 2018 and August 2018. As it can be seen, the selection of the *horizontal diffuse irradiance model* can largely affect the simulated energy needs. For instance, when model **15** (the Spencer model) is chosen, the differences among the considered *irradiance models for tilted surfaces* are more limited, for both heating and cooling needs. If model **20** (the Perez *et al.* Model 1) is picked, a larger sensitivity can be detected. As regards the heating needs, for instance, model **F** (the Ma & Iqbal model) gives different results from the rest of the group for most of building configurations. The same is true also in the case of cooling needs and a larger variability can be observed for all configurations characterized by high cooling energy demands.

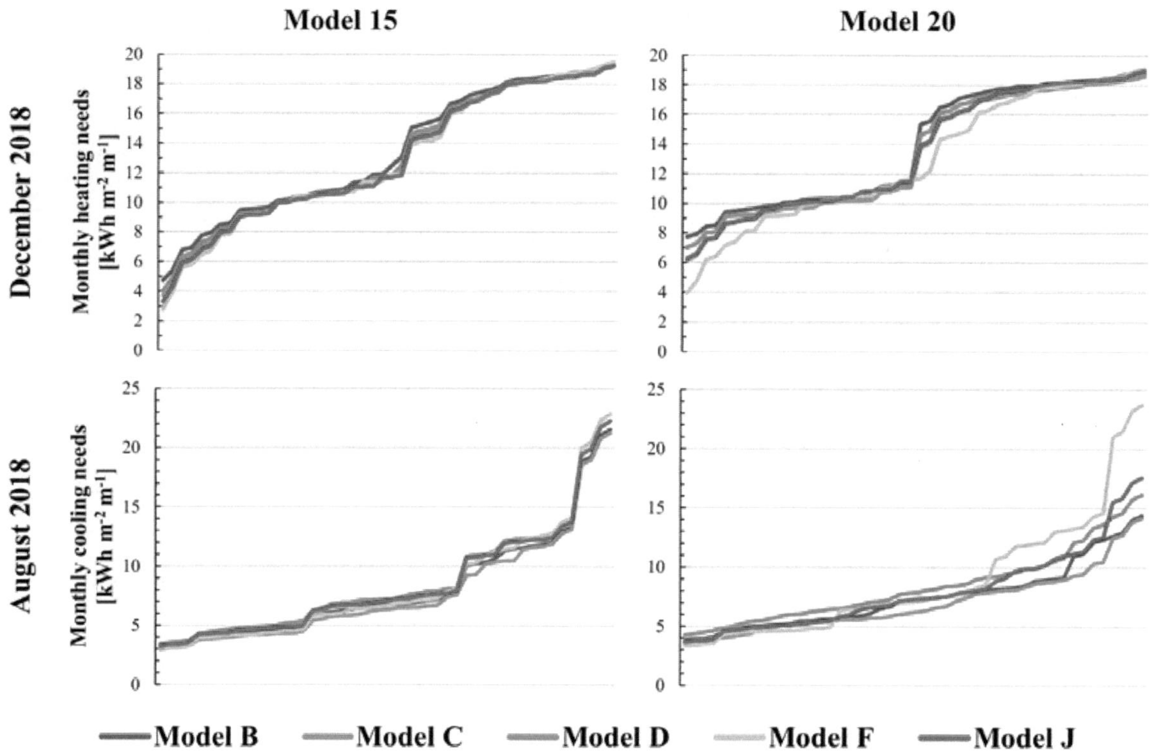

Figure 3: Cumulative distribution functions of the monthly energy needs for space heating and cooling simulated for the 48 buildings, respectively for December 2018 (top) and August 2018 (bottom). The results are distinguished according to the selected irradiance model for tilted surfaces (**B**, **C**, **D**, **F**, or **J**) and by the horizontal diffuse irradiance model (**15** on the left and **20** on the right)

4. CONCLUSIONS

In this research we analyzed the capabilities of 22 *horizontal diffuse irradiance models* and 12 *irradiance models for tilted surfaces* commonly found in the literature and in many building performance simulation codes for the calculation of the solar irradiance incident on the building envelope. The specific focus of the research was put on the simulation of solar irradiance in mountain environments, characterized by complex irradiation patterns depending on the orography and the multiple terrain reflections. In order to discuss the models' capabilities, we compared the solar irradiance calculated by all combinations of *horizontal diffuse irradiance models* and *irradiance models for tilted surfaces* with the global irradiance measured in the Alpine location of Bolzano, Italy, in the years 2018, 2019 and 2021, on four vertical surfaces oriented towards the main cardinal directions. Thanks to the analysis

of the hourly Mean Absolute Errors, the best- and the worst-performing pairs of models were identified for each year and orientations. Then, this subset of irradiance models was employed to simulate with TRNSYS 18 the energy needs for space heating and cooling for a dataset of 48 simplified buildings.
We found that:

- The performances of the pairs of irradiance models can be very different, depending on the considered orientation. The west-oriented surfaces showed the worst overall performance, probably due to the presence of natural obstacles in that direction.
- None of the models in the literature was found able to ensure the same level of accuracy for all the four vertical cardinal orientations at the same time. In some cases, a model able to minimize the errors with respect to the collected solar data in a given orientation was identified as the worst-performing one for another orientation.
- The impact of the selection of solar irradiance models on the simulated energy performance is clearly affected by the building's features. As regards the heating performance, poorly insulated buildings with massive walls and large south-oriented windows with high *SHGC* were identified as more sensitive to how the solar irradiance is modelled in the considered case-study mountain location. On the contrary, buildings with large west-oriented windows with high *SHGC* and well-insulated envelope proved to be more sensitive when it comes to the cooling performance.

Considering all the above, the next developments of this research will integrate also other weather stations and other measurements from the weather station considered in this work (e.g., beam and diffuse components of solar irradiance, besides the global values already discussed), in order to better identify the weaknesses of the existing irradiance models and suggest potential modifications to increase their capabilities first for the considered case-study location of Bolzano, Italy, and then for other Alpine environments.

REFERENCES

Boland, J., Ridley, B., & Brown, B. (2008). Models of diffuse solar radiation. *Renewable Energy*, *33*, 575–584.

Bugler, J.W. (1977). The determination of hourly insolation on an inclined plane using a diffuse irradiance modelbased on hourly measured global horizontal insolation. *Solar Energy*, *19*(5), 477–491.

Chandrasekaran, J., & Kumar, S. (1994). Hourly diffuse fraction correlation at a tropical location. *Solar Energy*,*53*(6), 505-510.

Chendo, M. A. C., & Maduekwe, A. L. (1994). Hourly global and diffuse radiation of Lagos, Nigeria - correlationwith some atmospheric parameters. *Solar Energy*, *52*(3): 247–251.

de Miguel, A., Bilbao, J., Aguiar, R., Kambezidis, H.D., & Negro, E. (2001). Diffuse solar irradiation modelevaluation in the North Mediterranean Belt area. *Solar Energy*, *70*(2), 143–153.

Dervishi, S., & Mahdavi, A. (2012). Computing diffuse fraction of global horizontal solar radiation: A modelcomparison. *Solar Energy*, *86*, 1796–1802.

Ente Nazionale Italiano di Normazione (UNI). (2014). *UNI/TS 11300-1:2014 – Energy performance of buildingsPart 1: Evaluation of energy need for space heating and cooling*, Milan, Italy: UNI.

Erbs, D. G., Klein, S. A., & Duffie, J. A. (1982). Estimation of the diffuse radiation fraction for hourly, daily andmonthly-average global radiation. *Solar Energy*, *28*(4), 293–302.

Gueymard, C.A. (1986). An anisotropic solar irradiance model for tilted surfaces and its comparison with selectedengineering algorithms. *Solar Energy*, *38*(5), 367 – 386.

Hay, J.E., & Davies, J.A. (1980). Calculation of the solar radiation incident on an inclined surface. *Proc. FirstCanadian Solar Radiation Data Workshop*, 59 – 65.

Hawlader, M.N.A. (1984). Solar diffuse, global and extraterrestrial solar radiation for Singapore. *InternationalJournal of Ambient Energy*, *5*(1), 31 – 37.

Karatasou, S., Santamouris, M., & Geros, V. (2003). Analysis of experimental data on diffuse solar radiation inAthens, Greece, for building applications. *International Journal of Sustainable Energy*, *23*(1), 37–41.

Klucher, T. M. (1979). Evaluation of models to predict insolation on tilted surfaces. *Solar Energy*, *23*(2), 111–114.

Lam, J.C., & Li, D.H.W. (1996). Correlation between global solar radiation and its direct and diffuse components. *Building and Environment*, 31(6), 527–535.

Liu, B.Y.H., & Jordan, R.C. (1960). The interrelationship and characteristic distribution of direct, diffuse and total solar radiation. *Solar Energy*, 4(3), 1 – 19.

Ma, C.C.Y., & Iqbal, M. (1983). Statistical comparison of models for estimating solar radiation on inclined surfaces. *Solar Energy*, 31(3): 313–317.
Maxwell, E.L. (1987). A quasi-physical model for converting hourly global horizontal to direct normal insolation. *Technical Report of Solar Energy Research Institute, SERI/TR-215-3087.*
Muneer, T., Hawas, M., & Sahili, K. (1984). Correlation between hourly diffuse and global radiation for New Delhi. *Energy Conversion and Management*, 24(4), 265–267.
Muneer, T., & Younes, S. (2006). The all-sky meteorological radiation model: proposed improvements. *AppliedEnergy*, *83*(5), 436–450.
Oliveira, A., Escobedo, J.F., Machado, J.A., & Soares, J. (2002). Correlation models of diffuse solar-radiation applied to the city of São Paulo, Brazil. *Applied Energy*, *71*(1), 59–73.
Orgill, J.F., & Hollands, K.G.T. (1977). Correlation equation for hourly diffuse radiation on a horizontal surface. *Solar Energy*, 19(4), 357 – 359.
Perez, R.R., Ineichen, P., & Maxwell, E.L. (1992). Dynamic global-to-direct irradiance conversion models. *ASHRAE Transactions*, 98(1), 354-369.
Perez, R.R., Ineichen, P., & Seals, R. (1990). Modeling daylight availability and irradiance components from direct and global irradiance. *Solar Energy*, 44(5), 271–289.
Pernigotto, G., Prada, A., Baggio, P., Gasparella, A., & Mahdavi, A. (2015). Impact of solar irradiation models on simulated hourly energy performance of buildings, *BS 2015 - 14th International Conference of the International Building Performance Simulation Association*, Hyderabad, India.
Pernigotto, G., Prada, A., Baggio, P., Gasparella, A., & Mahdavi, A. (2016). Solar irradiance modelling and uncertainty on building hourly profiles of heating and cooling energy needs, *IV International High Performance Buildings Conference at Purdue*, West Lafayette, Indiana, U.S.
Pernigotto, G., Gasparella, A., & Hensen, J.L.M. (2021). Assessment of a weather-based climate classification with building energy simulation, *BS 2021 - 17th International Conference of the International Building Performance Simulation Association*, Bruges, Belgium.
Prada, A., Pernigotto, G., Gasparella, A., & Mahdavi, A. (2014a). Combined effects of diffuse fraction and tilted surface radiation models, *ECPPM 2014 - 10th European Conference on Product & Process Modelling*, Vienna, Austria.
Prada, A., Pernigotto, G., Baggio, P., Gasparella, A., & Mahdavi, A. (2014b). Effect of Solar Radiation Model on the Predicted Energy Performance of Buildings, *III International High Performance Buildings Conference at Purdue*, West Lafayette, Indiana, U.S.
Prada, A., Pernigotto, G., Cappelletti, F., & Gasparella, A. (2015). Impact of solar irradiation models on building refurbishment measures from multi-objective optimization, *BS 2015 - 14th International Conference of the International Building Performance Simulation Association*, Hyderabad, India.
Reindl, D., Beckman, A., & Duffie, J. A. (1990a). Diffuse fraction correlations. *Solar Energy*, 45(1), 1–7.
Reindl, D. Beckman, A. & Duffie, J. A. (1990b). Evaluation of hourly tilted surface radiation models. *Solar Energy*, 45(1), 9–17.
Skartveit, A., & Olseth, J.A. (1986). Modelling slope irradiance at high latitudes. Solar Energy, 36(4), 333 – 344.
Skartveit, A., & Olseth, J.A. (1987). A model for the diffuse fraction of hourly global radiation. *Solar Energy*, 38(4), 271–274.
Soares, J., Oliveira, A., Boznar, M., Mlakar, P., Escobedo, J.F., & Machado, J.A. (2004). Modeling hourly diffuse solar radiation in the city of São Paulo using a neural-network technique. *Applied Energy*, 79(2), 201–214.
Spencer, J. (1982). A comparison of methods for estimating hourly diffuse solar radiation from global solar radiation. *Solar Energy*, 29(1), 19–32.
Temps, R.C., & Coulson, K.L. (1977). Solar radiation incident upon slopes of different orientations. *Solar Energy*, 19(2), 179 – 184.

ACKNOWLEDGEMENT

This research was funded by the internal project of the Free University of Bozen-Bolzano "SOMNE - Bolzano Solar Irradiance Monitoring Network" (CUP: I56C18000930005; CRC Call 2018).

Energy Model of an Air Source Heat Pump to Explore Performance Improvements under Cold Conditions: a Python Framework

Conrado ERMEL[1,2*], Marcus V.A. BIANCHI[1], Paulo S.SCHNEIDER[2]

[1] National Renewable Energy Laboratory - NREL,
Golden, Colorado, USA
conradoermel@gmail.com; Marcus.Bianchi@nrel.gov

[2] Federal University of Rio Grande do Sul, Department of Mechanical Engineering,
Porto Alegre, Rio Grande do Sul, Brazil
pss@mecanica.ufrgs.br

* Corresponding Author

ABSTRACT

Replacing combustion appliances with heat pumps can be part of a strategy to electrify buildings. When their electricity is supplied by renewables, the adoption of heat pumps for space heating may help curb the CO_2 emissions inherent to traditional heating systems that burn natural gas or coal. When operating in very cold climates, however, heat pump performance deteriorates, sometimes requiring backup heating supplied by electric resistances or other sources to continue to function. To avoid the use of electric resistances, complementary technologies, like thermal energy storage, solar collectors, or alternative refrigerants, have been explored in recent years. In this study, we present an open-source Python-based numerical model developed to evaluate the annual hourly performance of an air source heat pump (ASHP) operating in the USA. Simulation results include coefficient of performance and heating capacity. The model, validated with experimental results, will be employed in the future to investigate the feasibility of integrating thermal energy storage and other technologies into ASHPs operating in cold climates.

1. INTRODUCTION

Heat pumps can support the transition from fossil fuel appliances to electrified heating systems, which may foster the decarbonization of the building sector (Le et al., 2019). Heat pumps can supply heating demands up to five times more efficiently than electric resistances by running a vapor compression cycle (Dincer et al., 2017). The higher efficiency aligned with the fast growth of renewables' share in the world energy mix suggest an increase in the deployment of heat pumps in the near future (Bianco et al., 2017). There have been several recent scientific publications exploring the adoption of heat pumps to replace combustion appliances. Most of those studies rely on simulation models, a less expensive option than experiments, and an increasing number of simulation platforms for heat pumps have been documented.

Heat pump models can be structured as unified solutions (one single expression correlating the performance with independent parameters). Unified solutions consist of using regressions and generic correlations with experimental data to obtain the heat pump coefficient of performance (COP). These black-box models lead to very simple expressions, which can be solved with low computational effort. However, because they are curve fittings for specific equipment, the results are limited because no extrapolation can be made (Lyden, 2020).

Alternatively one can create separate models for each component because it facilitates exchanging the components (Huang et al., 2020). The strategy relies on using thermodynamic sub-models for each device of the heat pump, explicitly considering their construction and operational parameters being as complex as required. Such simulation can capture transient behaviors of the cycle (Han et al., 2008) but, to simplify the simulation, a quasi steady-state approach has been commonly adopted to build scenarios or to assess equipment performance (Hirmiz et al., 2020).

Due to their user-friendly interfaces and powerful solvers, commercial software like TRNSYS®, MATLAB®, and Dymola® have been extensively used for simulating the operation of heat pumps (Palomba et al., 2019). Despite their wide adoption, commercial platforms can be expensive, and sometimes integration with other systems can be limited, therefore instigating the development of open-source packages to simulate thermal systems. Open-source codes were employed in the study of heat pump parameters (Moreno et al., 2014; Shen et al., 2018; Cimmino and Wetter, 2017), or to build forecast scenarios for different systems integrating heat pumps (Carroll et al., 2020; Hilpert, 2020).

Among the available programming languages, Python is a high-level, general-purpose language that serves both linear and object-oriented programming (Lyden, 2020). Python is included in the Open Energy Modelling Initiative (Open-mod, 2022), and some packages have been developed to model heat pumps. PyLESA, for instance, is a local energy system platform for Python that can be used for the simulation of heat pumps integrated to thermal and electric storage (Lyden et al., 2021). This open-source modeling package offers a Microsoft Excel interface that facilitates its use. Users may select among three approaches for COP calculation: (i) a Generic Regression Model, (ii) the Lorentz Model, and (iii) a Standard Test Regression Model. All of these are gray- or black-box models require manufacturer data or experimental results to build the regressions to predict the heat pump COP and capacity. TESPy—Thermal Engineering Systems in Python (Witte and Tuschy, 2020)—is another available collaborative package that embeds several thermal systems, is object-oriented, and includes heat pump components. Although the thermodynamic cycle is solved at each time-step, in the current available library the condenser and the evaporator are modeled as having only one phase in the refrigerant side. Moreover, the user interface is complex, and the learning curve may be steep. Other libraries are available in the official Python repository (Foundation, 2022), however, most of the options lack documentation.

In this work we present a quasi-steady heat pump model developed in Python 3 to simulate the operation of air-air and air-water heat pumps. The model includes construction parameters of the heat exchangers like tube length, fin ratio, and fin efficiency. Compressor rotation, displacement, and isentropic and volumetric efficiencies are considered to solve the thermodynamic cycle in each time step. Model validation and an application example are also provided in the paper. The proposed structure is flexible, allowing for simulating ASHP-TES with different components and operation conditions. The heat exchanger model is more detailed when compared with the existent packages, however, computational time is not compromised.

2. MODELING

Heat pumps operate a vapor compression cycle to transport energy from a relatively colder outside environment to the indoors through a vapor compression cycle. Although equipment presents several additional peripheral components, the basic cycle includes of a compressor, a condenser, an expansion device, and an evaporator. Aiming flexibility and future extrapolation the model was conceived in an object-oriented philosophy: each component (compressor, condenser, evaporator, and expansion vale) is an object, and they are combined in the greater object to represent the heat pump operation. The whole system is then composed of a heat pump, the residential thermal load, and a thermal storage device.

2.1 Vapor Compression Cycle

Figure 1a shows the schematic of a basic air-to-water heat pump integrated with sensible TES (a water tank). The heat pump is used to charge the TES, and water is recirculated from the TES tank to radiators to distribute heat in the home. All the equipment are modeled separately, hence, they can be replaced by regressions, gray-box, finite-volume (FV), or any other type of model.

Figure 1b shows the temperature-entropy diagram of the vapor compression cycle. The refrigerant mass flowrate in $\mathrm{kg\,s^{-1}}$ is calculated by

$$\dot{m}_r = \rho_1 V_{comp} \,(n/60)\, \eta_{vol}, \tag{1}$$

where V_{comp} is the displacement [m^3], n the compressor rotation speed [rpm], η_{vol} the volumetric efficiency. The energy balance on the condenser can be expressed by

$$\dot{Q}_c = \dot{m}_r(h_2 - h_3) \quad = \quad \dot{m}_r U_c A_c (T_{m_c} - T_{air_{indoor}}), \tag{2}$$

with h_2 and h_3 the specific enthalpies [kg/kJ] in the condenser inlet and outlet, respectively, U_c the overall heat transfer coefficient [$\mathrm{W\,m^{-2}\,K^{-1}}$], and A_c the condenser area [$\mathrm{m^2}$]. The average temperature of the refrigerant in the condenser is denoted by T_{m_c}. The expansion device is considered isenthalpic, therefore $h_4 = h_3$. In the evaporator, the energy

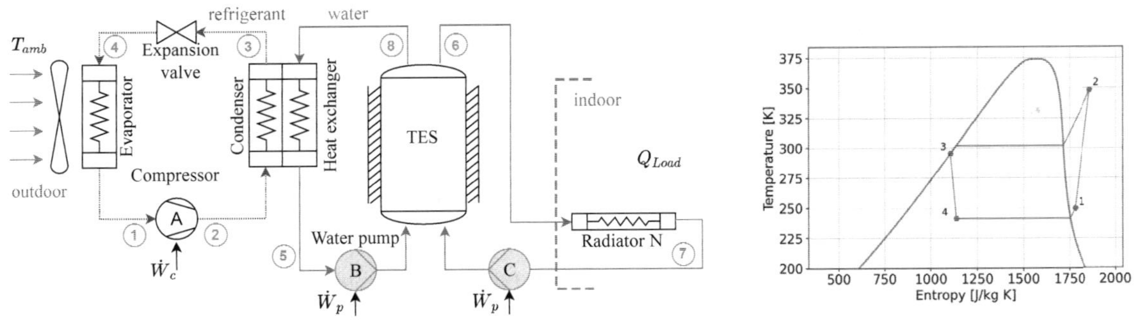

(a) System schematic

(b) T-S diagram

Figure 1: Air-water air source heat pump integrated with sensible thermal energy storage.

balance is similar to the condenser so that

$$\dot{Q}_e = \dot{m}_r(h_1 - h_4) \quad = \quad \dot{m}_r U_e A_e (T_{m_e} - T_{air_{outdoor}}), \tag{3}$$

where U_e and A_e are the evaporator overall heat transfer coefficient and external surface area, respectively. T_{m_e} is the average temperature of the refrigerant in the evaporator. The heat exchange model for both condenser and evaporator is presented in the next section.

2.2 Heat Exchangers

A simple way of modeling heat exchangers is to assume they have a fixed effectiveness (Witte and Tuschy, 2020). Li et al. (2020), for instance, investigated the performance of a ASHP when charging a PCM storage tank, with a MATLAB® and TRNSYS® model with constant effectiveness in the condenser and evaporator. More detailed models can be employed to improve the accuracy of the results through finite volumes or finite elements, but they require a larger simulation time, which may limit their usage (Bonamente et al., 2016; Huang et al., 2020). In the current study for refrigerant-air heat exchangers (evaporator and condenser) we adopted the Kriging-assisted three-zone model proposed by Huang et al. (2020). This model can be up to 600 times faster than a finite volume approach, yet presents good accuracy (≤0.9 %). This model cannot be used for a refrigerant-water heat exchanger because the water temperature cannot be assumed as uniform. Therefore, for the refrigerant-water condenser we considered a constant effectiveness model.

The Kriging-assisted three-zone model considers the presence of three different regions in the heat exchanger with their respective phases, as presented in Figure 2. The refrigerant in the condenser is expected to have a super-heated (SH) refrigerant in the inlet, a 2-phase state in the middle region, and a sub-cooled (SC) state in the outlet. The length of each region will depend on the state of the refrigerant at the HX inlet and the external air temperature.

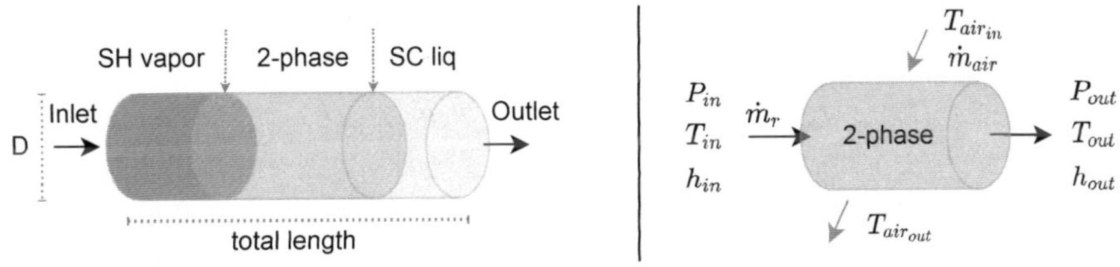

Figure 2: Kriging-assisted three-zone model (Huang et al., 2020) applied to the condenser.

The overall heat transfer coefficient, given in W m^{-2} K^{-1}, is calculated for each zone as (Li et al., 2020)

$$U_{zone} = \left(\frac{1}{h_z} + \frac{1}{h_{air}\eta_{fin}R_{fin}} \right)^{-1}, \tag{4}$$

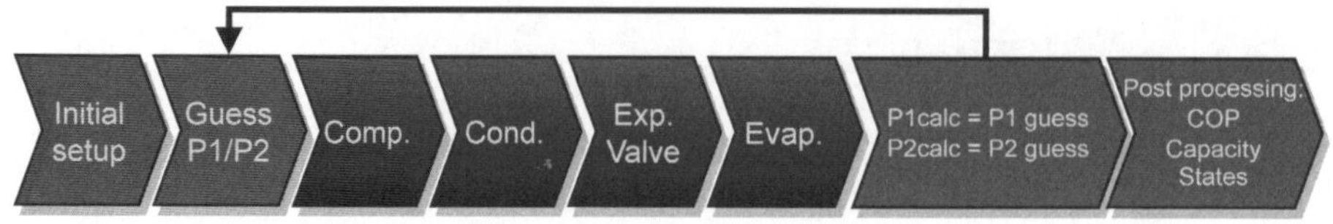

Figure 3: Vapor compression solution scheme.

where h_z and h_{air} are the refrigerant-side and air-side heat transfer coefficients, respectively. η_{fin} stands for the fin's efficiency, and R_{fin} is the fin ratio. For the vapor and liquid phases, we assumed a representative heat transfer coefficient of $h_{vap} = h_{liq} = 600\,\mathrm{W\,m^{-2}\,K^{-1}}$, while for the 2-phase region it is $h_{2ph} = 3000\,\mathrm{W\,m^{-2}\,K^{-1}}$. h_{air} is assumed to be $100\,\mathrm{W\,m^{-2}\,K^{-1}}$. The adopted values are representative for the studied case, although they can vary depending on the fluids and the flow conditions. By performing an energy balance on each zone one can calculate the zone length (Eq. 5), outlet enthalpy (Eq. 6), or the outlet temperature (Eq. 7) (Huang et al., 2020).

$$L_{zone} = \frac{\dot{m}_r dh}{dT U_{zone} \pi D} \tag{5}$$

$$h_{zone} = h_{in} - U_{zone} A_{zone} \frac{dT}{\dot{m}_r} \tag{6}$$

$$T_{zone} = \frac{U_{zone} A_{zone} (T_{air_{in}} - 0.5 T_{in}) + \dot{m}_r * C_{p_r} * T_{in}}{(\dot{m}_r C_{p_r} + 0.5 U_{zone} A_{zone})} \tag{7}$$

The zone area is $A_{zone} = L_{zone} \pi D$ for a tube with diameter D and zone length L_{zone}, both in m. dh is the enthalpy variation along the zone, and dT is the difference between the zone mean temperature and $T_{air_{in}}$. The refrigerant-specific heat C_{p_r} ($\mathrm{J\,kg^{-1}\,K^{-1}}$) is evaluated at the zone inlet.

2.3 Thermal Load

The residential load factor method from standard 169-2021 (ANSI/ASHRAE, 2021) was used to simulate the heating load considering a detached single-family $120\,\mathrm{m^2}$ dwelling with the following features: walls ($A_w = 132\,\mathrm{m^2}$ and $U_w = 0.51\,\mathrm{W\,m^{-2}\,K^{-1}}$) and roof ($A_r = 120\,\mathrm{m^2}$ and $U_r = 0.18\,\mathrm{W\,m^{-2}\,K^{-1}}$). The thermal load is then calculated by

$$\dot{Q}_{load} = A_w U_w (T_{amb} - T_{sp}) + A_r U_r (T_{amb} - T_{sp}) \tag{8}$$

The indoor temperature setpoint T_{sp} is considered constant in the home.

2.4 Solution Scheme

Figure 3 displays the solution scheme, which was based on the enthalpy marching solution procedure (Winkler et al., 2008). After setting the system parameters, the pressures at points 1 and 2 are initially guessed. The energy balance is then applied for each equipment in the cycle, with a sub-cooling condition imposed to the condenser outlet (3) as well as a super-heating level imposed to the compressor inlet (1).

The SciPy optimization package (Virtanen et al., 2020) was employed to numerically solve the cycle, and the calculation is considered complete when the convergence criteria (residual $< 1 \times 10^{-6}$) is satisfied. Usually, the COP and the system capacity are relevant results when assessing heat pumps (Bianco et al., 2017; Li et al., 2020). The former expresses the device efficiency when transporting thermal energy from the air outside to the indoor environment, and it is necessary to evaluate the compressor energy required to meet the heating loads.

3. MODEL ASSESSMENT

3.1 Validation with Experimental Data

The heat pump model was validated with experimental data reported by Shoukas et al. (2022), where a 35,000 Btu/h (10 263 W) commercial heat pump unit was evaluated in a controlled environment at the National Renewable Energy Laboratory (NREL). The model was fed with the parameters of the heat pump, so the predictions of the COP and the system capacity could be compared. Model and experimental results are presented in Figure 4 with the measurement

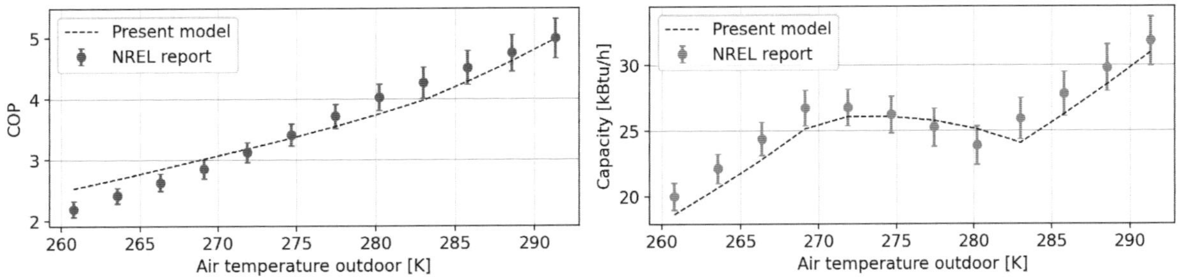

Figure 4: Validation with experimental data (Shoukas et al., 2022).

uncertainties. A maximum 10 % relative deviation was found for COP results, while the mean deviation for the entire range was 1.43 %. Regarding the system capacity, the model deviates 1.18 % from the experimental results on average, and the maximum deviation was 6.58 %. Model results show good agreement with the experimental data for the entire operation range, including the low temperature region, with the most significant deviations in the range of 275 K to 285 K.

3.2 Comparison with Heat Pump Simulator—Oak Ridge National Laboratory (ORNL)

We also compared the model with simulation results from the Heat Pump Design Model from Oak Ridge National Laboratory (2019) to increase our confidence on its accuracy. The parameters in Table 1 were provided to both models to compare COP and heating capacity, displayed in Figure 5. Over the simulated range of outdoor temperatures 264 K

Table 1: Parameters of the heat pump simulated in the ORNL model (Oak Ridge National Laboratory, 2019).

Parameter	Value	Unit	Parameter	Value	Unit
Compressor speed	3500	rpm	Condenser air flow	0.47	$m^3\,s^{-1}$
Isentropic efficiency	0.63	-	Evaporator air flow	1.18	$m^3\,s^{-1}$
Volumetric efficiency	0.95	-	Condenser area	20.72	m^2
Displacement	5.37×10^{-5}	m^3	Evaporator area	41.87	m^2
Refrigerant	R134a	-	Indoor temperature	294	K

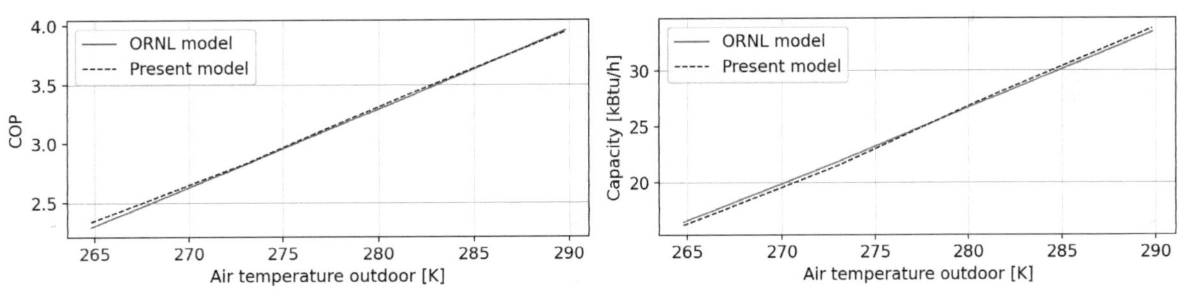

Figure 5: Comparison between ORNL heat pump simulator and the present model: COP and system capacity (Oak Ridge National Laboratory, 2019).

to 289 K, the maximum and average deviations for COP results were 1.98 % and 0.63 %, respectively. For the system capacity, the deviations were 1.79 % (max) and 0.4 % (average).

3.3 Sensitivity Analysis

A sensitivity analysis was performed with the independent input variables $T_{air,outdoor}$ varying from 250 K to 290 K and the condenser inlet temperature T_8 in the range of 295 K to 330 K. A Sobol's sampling with n^{12} resulted in 24,576 simulated points (Saltelli et al., 2008). Polynomial regressions were built from 70 % of the data set, while the remaining

30 % was used to verify the regression quality, which presented a 0.019 root mean square error. Results indicate that the COP has a similar correlation to $T_{air,outdoor}$ and T_8 (approx. 0.51), as shown in Figure 6a. Heat capacity is strongly correlated to $T_{air,outdoor}$ and weakly correlated to T_8, Figure 6b.

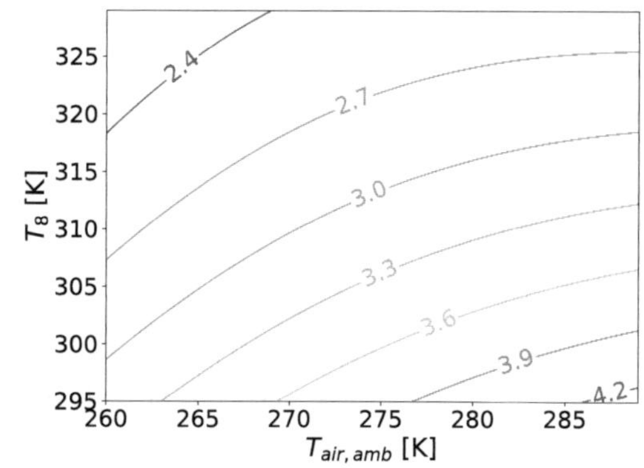

(a) COP

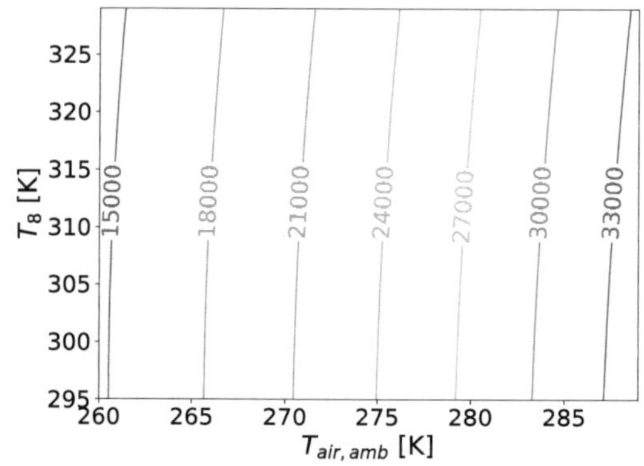

(b) Heating capacity [BTU/h]

Figure 6: Contour plots for COP and heating capacity as a function of the ambient air temperature $T_{air,outdoor}$, and the condenser temperature T_8.

4. ASHP-TES INTEGRATION

To illustrate the model capability of representing the integration between a heat pump and a thermal storage device, an application example is presented. Table 2 shows the parameters adopted for the system (Figure 1). An artificial sinusoidal temperature profile varying from 265 K to 285 K was imposed to T_{amb} to assess the model response. Moreover, an initial state of charge (SOC) = 15 % was considered for the TES.

Table 2: System parameters adopted in the example.

Device	Parameter	Value
Compressor	n	3600
	$Disp$	5.37×10^{-5}
	η_{iso}	0.63
	η_{vol}	0.95
Condenser	Effectiveness	0.7
Water tank	Area	$6.28\ m^2$
	Volume	$1.17\ m^3$
	Overall heat transfer coeff. U	$1\ W\,m^{-2}\,K^{-1}$
Pump B	Mass flow rate	$0.1\ kg\,s^{-1}$

The example considered the system operating under four different modes, as observed in Table 3. If $T_{outdoor} > 270$ and $Q_{load} > 0$ mode 1 is adopted. Mode 2 is adopted if the ambient temperature is lower than 270 K and the SOC of TES is > 10 %. Whenever $Q_{load} = 0$, no heat transfer to the indoors takes place (mode 3). In this situation if the SOC is less than 95 % the heat pump is used to charge the TES. Mode 4 is assumed when $\dot{Q}_{load} = 0$ and SOC > 95 %. The logic of Table 3 is a basic approach to operate an ASHP-TES system. More elaborated strategies can be explored and depends on the system goals. For example, if the electricity rate differs depending on the hour of the day, one may want to shift the heating load by using the system of Figure 3. Another example is the integration with renewables like solar and wind. In this case the system can introduce flexibility by supplying the heating loads in times when renewables are not available.

Figure 7a displays the state of charge of the TES, the operation mode, and the outdoor temperature over 24 h of oper-

Table 3: ASHP-TES operation logic adopted for this example.

Mode	Heat pump	Thermal storage	Temp. outdoor [K]	Thermal load [W]
1	ON	Charging/Discharging	> 270	> 0
2	OFF	Discharging	< 270	> 0
3	ON	Charging	> 270	= 0
4	OFF	OFF	> 270	= 0

ation. Due to the imposed rules, the system only assumed operation modes 3, 1, and 2 while mode 4 was never in use. During the first 10 hours the system is charging the TES and supplying $\dot{Q}_{load}$, modes 3 and 1. In the following hours

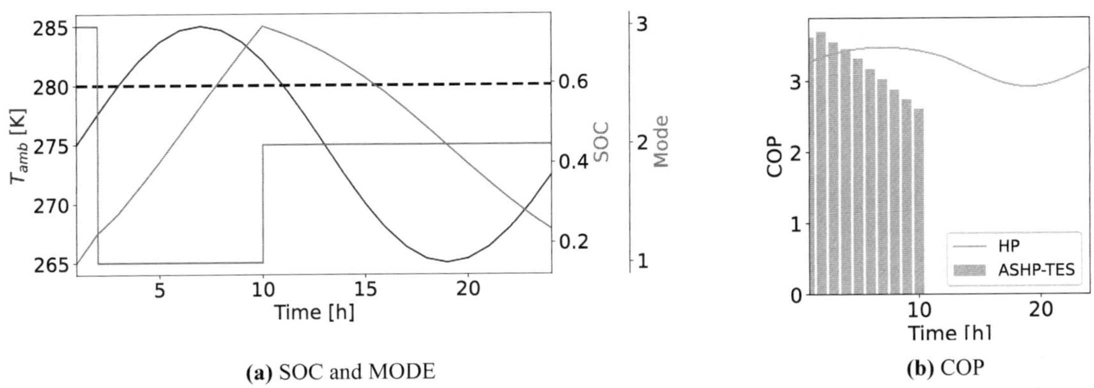

Figure 7: System operation mode and COP results over time.

mode 2 is activated and TES is discharged from SOC = 80 % to 25 %. Figure 7b displays the system COP for the same time range. It is interesting to note that although T_{amb} is increasing in time 0 h to 10 h the system COP is decreasing. This occurs because T_8 is also increasing with the TES SOC. Additionally, from time 10 h to 24 h the system COP is zero because the TES is in discharge mode and the heat pump is off.

Figure 8 shows the temperature of the relevant points of the vapor compression cycle and the water circuit for the same time window. T_2 is the temperature of the refrigerant entering the condenser, hence it is only presented when the heat pump is on. T_5, the water leaving the condenser, follows T_2 trend, while T_6 and T_8 represent the water temperature in the tank. Regarding computational time, the model took 8 s to complete the 24 h simulation in a windows-based computer (CPU Intel® Core i7-9750H CPU @2.6 GHz, 16 Gb RAM DDR4), which indicates computational time of 600 s for annual simulations. Several arrangements and operation strategies may be envisioned for ASHP-TES systems. It may be used for peak-shaving and load shifting, specially observing different electricity rates. The use of TES may also prevent the heat pump from operating under extremely low temperatures, where the heat pump performance is lower.

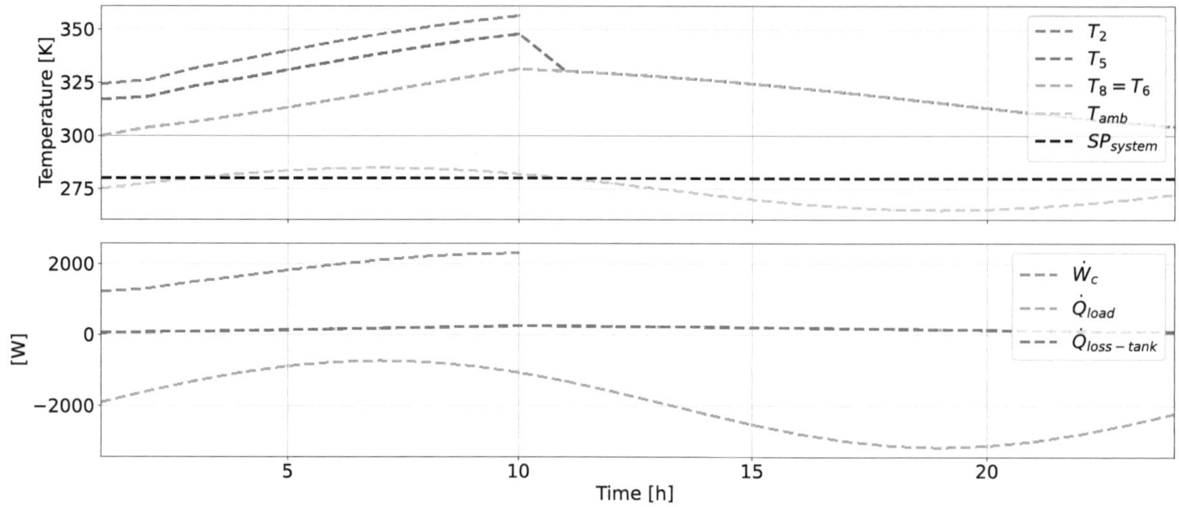

Figure 8: Temperatures in relevant points of the system and the heat exchange levels.

5. CONCLUSIONS

In this study we developed a Python package to simulate air source heat pumps integrated with thermal energy storage. The model solves the thermodynamic states of each component by following an enthalpy marching solver solution procedure. The framework was built in a modular philosophy to facilitate further investigation with different equipment and sub models. A Kriging-assisted three-zone model was adopted to solve the refrigerant-air condenser and evaporator. For the refrigerant-water condenser, the effectiveness was set constant. Validation with experimental data obtained from a real heat pump is provided, with maximum and average deviations of 10 % and 1.43 % for COP results and 6.58 % and 1.18 % for the system capacity. An example of how the model can be used to assess AHSP-TES integration was presented, with the system operating over 24 h. Results demonstrate the model capability in representing the system operation and allows for simulations in different time-steps. Given the low computational effort further analyses in a annual base can now be developed for different locations in a reasonable time (approx. 600 s).

NOMENCLATURE

Latin Symbols

A	Area	(m^2)
C_p	Specific heat at constant pressure	($J\,kg^{-1}\,K^{-1}$)
h	Specific enthalpy	($J\,kg^{-1}$)
h	Convective heat transfer coefficient	($W\,m^{-2}\,K^{-1}$)
ID	Inner diameter	(m)
L	Length	(m)
$\dot{m}$	Mass flow rate	($kg\,s^{-1}$)
n	Angular speed	(rpm)
$\dot{Q}$	Heat transfer rate	(W)
R_{fin}	Fin thermal resistance	($W\,m^{-2}\,K^{-1}$)
T	Temperature	(K)
U	Overall heat transfer coefficient	($W\,m^{-2}\,K^{-1}$)
V	Displacement	(m^3)

Greek Symbols

η	Efficiency	(-)
ρ	Specific mass	($kg\,m^{-3}$)

Subscripts

amb	Ambient
comp	Compressor
c	Condenser
e	Evaporator
m	Average
p	Pressure
r	Refrigerant
vol	Volumetric
z	Zone

REFERENCES

ANSI/ASHRAE (2021). ANSI/ASHRAE Standard 169-2021 - Climatic Data for Building Design Standards. Technical report, ANSI/ASHRAE, Peachtree Corners.

Bianco, V., Scarpa, F., and Tagliafico, L. A. (2017). Estimation of primary energy savings by using heat pumps for heating purposes in the residential sector. *Applied Thermal Engineering*, 114:938–947.

Bonamente, E., Aquino, A., and Cotana, F. (2016). A PCM Thermal Storage for Ground-source Heat Pumps: Simulating the System Performance via CFD Approach. *Energy Procedia*, 101(September):1079–1086.

Carroll, P., Chesser, M., and Lyons, P. (2020). Air Source Heat Pumps field studies: A systematic literature review. *Renewable and Sustainable Energy Reviews*, 134.

Cimmino, M. and Wetter, M. (2017). Modelling of Heat Pumps with Calibrated Parameters Based on Manufacturer Data. *Proceedings of the 12th International Modelica Conference, Prague, Czech Republic, May 15-17, 2017*, 132(2002):219–226.

Dincer, I., Rosen, M. A., and Ahmadi, P. (2017). *Optimization of Energy Systems*. John Wiley and Sons Ltd, 1 edition.

Foundation, P. S. (2022). Python Repository.

Han, Z., Zheng, M., Kong, F., Wang, F., Li, Z., and Bai, T. (2008). Numerical simulation of solar assisted ground-source heat pump heating system with latent heat energy storage in severely cold area. *Applied Thermal Engineering*, 28(11-12):1427–1436.

Hilpert, S. (2020). Effects of decentral heat pump operation on electricity storage requirements in Germany. *Energies*, 13(11).

Hirmiz, R., Teamah, H. M., Lightstone, M. F., and Cotton, J. S. (2020). Analytical and numerical sizing of phase change material thickness for rectangular encapsulations in hybrid thermal storage tanks for residential heat pump systems. *Applied Thermal Engineering*, 170.

Huang, R., Ling, J., and Aute, V. (2020). Comparison of approximation-assisted heat exchanger models for steady-state simulation of vapor compression system. *Applied Thermal Engineering*, 166(November 2019):114691.

Le, K. X., Huang, M. J., Shah, N., Wilson, C., Artain, P. M., Byrne, R., and Hewitt, N. J. (2019). High temperature air source heat pump coupled with thermal energy storage: Comparative performances and retrofit analysis. *Energy Procedia*, 158:3878–3885.

Li, Y., Zhang, N., and Ding, Z. (2020). Investigation on the energy performance of using air-source heat pump to charge PCM storage tank. *Journal of Energy Storage*, 28.

Lyden, A. (2020). *Modelling and Design of Local Energy Systems Incorporating Heat Pumps, Thermal Storage, Future Tariffs, and Model Predictive Control*. PhD thesis, University of Strathclyde.

Lyden, A., Flett, G., and Tuohy, P. G. (2021). PyLESA: A Python modelling tool for planning-level Local, integrated, and smart Energy Systems Analysis. *SoftwareX*, 14:100699.

Moreno, P., Sole, C., Castell, A., and Cabeza, L. F. (2014). The use of phase change materials in domestic heat pump and air-conditioning systems for short term storage: A review. *Renewable and Sustainable Energy Reviews*, 39:1–13.

Oak Ridge National Laboratory (2019). DOE/ORNL Heat Pump Design Model.

Openmod (2022). Open Energy Modelling (openmod) Initiative.

Palomba, V., Varvagiannis, E., Karellas, S., and Frazzica, A. (2019). Hybrid adsorption-compression systems for air conditioning in efficient buildings: Design through validated dynamic models. *Energies*, 12(6).

Saltelli, A., Ratto, M., Andres, T., Campolongo, F., Cariboni, J., Gatelli, D., Saisana, M., and Tarantola, S. (2008). *Global Sensitivity Analysis. The Primer.* John Wiley & Sons Inc, West Sussex, England, 1 edition.

Shen, B., Nawaz, K., Baxter, V., and Elatar, A. (2018). Development and validation of quasi-steady-state heat pump water heater model having stratified water tank and wrapped-tank condenser. *International Journal of Refrigeration*, 87:78–90.

Shoukas, G., Bonnema, E., Paranjothi, G., Faramarzi, R., Klun, L., Shoukas, G., Bonnema, E., Paranjothi, G., Faramarzi, R., and Klun, L. (2022). Performance Assessment of High Efficiency Variable Speed Air-Source Heat Pump in Cold Climate Applications. Technical Report February, National Reneable Energy Laboratory (NREL), Golden, CO.

Virtanen, P., Gommers, R., Oliphant, T. E., Haberland, M., Reddy, T., Cournapeau, D., Burovski, E., Peterson, P., Weckesser, W., Bright, J., van der Walt, S. J., Brett, M., Wilson, J., Millman, K. J., Mayorov, N., Nelson, A. R. J., Jones, E., Kern, R., Larson, E., Carey, C. J., Polat, I., Feng, Y., Moore, E. W., VanderPlas, J., Laxalde, D., Perktold, J., Cimrman, R., Henriksen, I., Quintero, E. A., Harris, C. R., Archibald, A. M., Ribeiro, A. H., Pedregosa, F., van Mulbregt, P., and SciPy 1.0 Contributors (2020). SciPy 1.0: Fundamental Algorithms for Scientific Computing in Python. *Nature Methods*, 17:261–272.

Winkler, J., Aute, V., and Radermacher, R. (2008). Comprehensive investigation of numerical methods in simulating a steady-state vapor compression system. *International Journal of Refrigeration*, 31(5):930–942.

Witte, F. and Tuschy, I. (2020). TESPy: Thermal Engineering Systems in Python. *Journal of Open Source Software*, 5(49):2178.

ACKNOWLEDGEMENT

This study was financed in part by the Coordenação de Aperfeiçoamento de Pessoal de Nível Superior - Brasil (CAPES) - Finance Code 001. Conrado Ermel acknowledges Conselho Nacional de Desenvolvimento Científico e Tecnológico (CNPq) for his Ph.D. grant. Paulo S. Schneider acknowledges CNPq for his research grant (PQ305357/2013-1). This work was authored in part by the National Renewable Energy Laboratory, operated by Alliance for Sustainable Energy, LLC, for the U.S. Department of Energy (DOE) under Contract No. DE-AC36-08GO28308. Funding provided by the U.S. Department of Energy Office of Energy Efficiency and Renewable Energy Building Technologies Office. The views expressed in the article do not necessarily represent the views of the DOE or the U.S. Government. The U.S. Government retains and the publisher, by accepting the article for publication, acknowledges that the U.S. Government retains a nonexclusive, paid-up, irrevocable, worldwide license to publish or reproduce the published form of this work, or allow others to do so, for U.S. Government purposes.

Field-test performance of Solid Oxide Fuel Cells (SOFC) for residential cogeneration applications

Nicolas PAULUS[1]*, Vincent LEMORT[1]

[1]University of Liège, Thermodynamics Laboratory,
Liège, Belgium
nicolas.paulus@uliege.be,

* Corresponding Author : nicolas.paulus@uliege.be

ABSTRACT

Much needed energy transition brings special focus on fuel cell micro-combined Heat and Power (mCHP) systems for residential uses, one of which is a SOFC, fed by natural gas, designed to provide continuously 1.5 kW_{el} (up to about 13 000 kWh yearly). However, this constant power output can be modulated as desired down to 500 W_{el}. With this machine, heat can also be recovered to partially contribute to the heat demand of the household. One main advantage of this appliance is that the heat recovery system is designed to be purely optional and it can be shut down, removed or added as desired, even if the machine is running. The machine is driven and completely monitored by the manufacturer remotely. This study is monitoring two of those installations in residential houses in Belgium for the whole year 2021. It focuses on the comparison of the actual field test performance with the targets expected for this technology, electrical efficiency of 60% at nominal power output being the main selling argument that has been verified in the real onsite applications studied in this work.

Since the financial incentive represents a major factor in the investor's decision towards such a technological investment, focus is brought on an economic indicator based on an average Belgian household energy bill, with promising results of about 1,3-1,4 k€ of utilization cost savings per year. At last, ecological indicators are established in an attempt to foresee the place this mCHP technology in the energy transition challenge of today towards the carbon neutral future everyone should aim for. The ecological balances can either be considered as positive, if the system is supposed to replace Combined-Cycle Gas Turbine (CCGT) power plant electrical production, or negative, compared to the actual Belgian electrical mix.

1. INTRODUCTION

Although its high operating temperature could be seen as a drawback (Ellamla, et al., 2015), SOFC technology has nevertheless a lot of advantages in comparison to other fuel cells, as summarized in the following lines :

- Higher electrical efficiency (Ellamla, et al., 2015).
- Fuel flexibility (Hauch, et al., 2014) because practically all hydrocarbons can be used with SOFC thanks to internal reforming allowed by the tremendously high temperatures.
- Low sensitivity to impurities. Carbon deposition is indeed less an issue in this system thanks to internal high reforming temperature whereas it is one of the main contaminants of Proton-exchange membrane fuel cells (Paulus & Lemort, 2022a). However, sulphur compounds are still absolute poison for SOFC's and fuel desulphurization is often required when the system is not fed with pure hydrogen (Fernandes, et al., 2018).
- No necessary use of platinum, rhodium and other precious metals as catalysts (Peng, et al., 2021).

The studied system is a commercialized SOFC for residential applications. Its expected specifications are presented in Table 1. It is not supposed to be turned OFF and ON and it is supposed be ran at a constant power output that can be chosen between 0.5 and 1.5 kW_{el} (nominal output power). The machine is remotely controlled by the manufacturer over an Ethernet port and, to adapt its system power output, the owner must send an email or give a phone call.

Internal schematics and working principle are unknown for intellectual property protection. However, base principle is quite common. Such as any fuel cell, this system provides electricity through redox reactions between a hydrocarbon fuel (here, natural gas) and an oxidizing agent (here, as often, oxygen). Since those redox reactions have combustion-like resulting equations (Shao, et al., 2005), there are thus exothermic by nature (Schmidt-Rohr, 2015) and release heat in addition of the produced electricity. Cogeneration is thus possible.

Specific literature review on mCHP SOFC's combined with little reverse engineering performed by observing the system with its protective cover unmounted has nevertheless allowed for assuming the most probable internal scheme of the system, which is presented in Figure 1. This scheme's lay-out is strongly inspired from literature (Braun, et al., 2006) cross-check against a specific study conducted along with this particular SOFC's manufacturer (Wagner, 2019). Adjustments to the literature scheme have been performed for relevance with the physical observations of the system without its covers (indeed, for example, the system involves two water outlets periodically actioned).

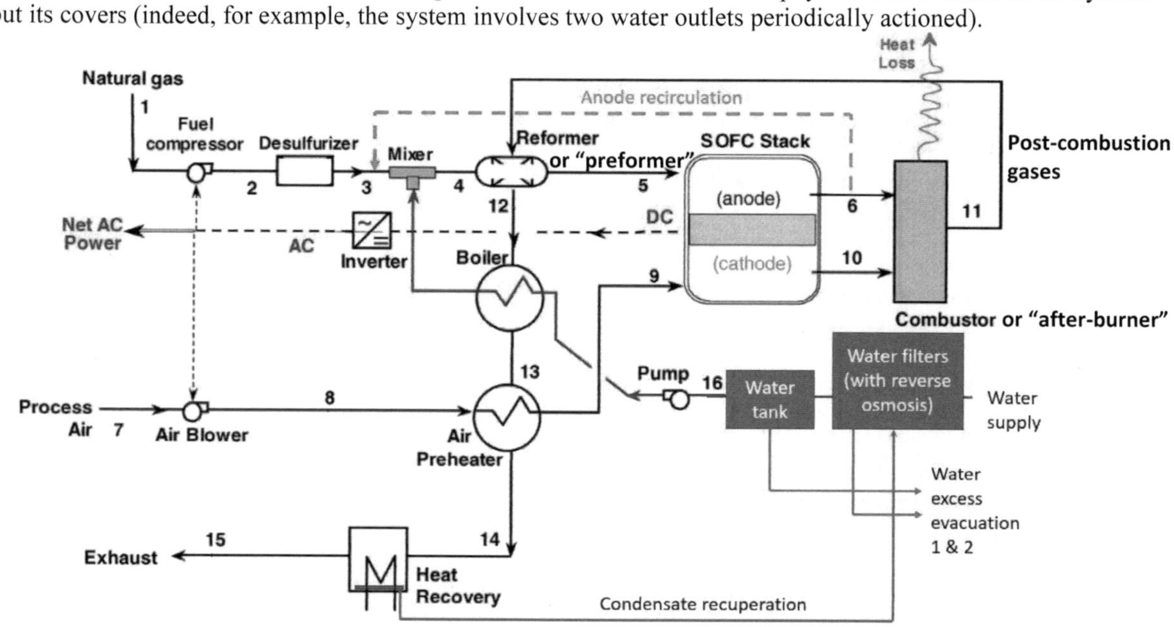

Figure 1: Most probable internal scheme. Lay-out strongly inspired from literature (Braun, et al., 2006)

Important elements are presented in the assumed configuration of Figure 1:

- Natural gas fuel supply is desulfurized as it is an absolute poison for SOFC's (Fernandes, et al., 2018).
- Natural gas is mixed with water vapor in order for steam reforming to take place (a reformer is implemented before the stack, also called "preformer" for this reason). Therefore, the reforming process in not only internal (directly onto the stack).
- Some of the fuel (mainly methane) is not directly consumed by the stack. It is rather consumed in an "after-burner" in order to generate the heat required for steam generation (for the reformer).
- The post-combustion gases are then used to provide heat to the preformer, to the boiler (as explained) and at last, before the exhaust, to the external heat recovery auxiliary circuit (offering the CHP capability of the system).
- Figure 1 does not show that the exhaust gases are providing their remaining heat to the inlet air thanks to a "double walled chimney" (Lichtenegger, et al., 2015), in addition to the probable "air preheater".
- Anode recirculation is most likely implemented to increase fuel utilization (Chen, et al., 2021). Indeed, the gases at the outlet of the anode still contain H_2, CO and CH_4 that can still react and provide power, but they also contain H_2O usable for the required steam reforming upstream of the stack. Maybe cathode recirculation has also been implemented but this is not the strongest assumption one could make because it complexifies

the system. It is unknown to the authors if cathode recirculation already exists in commercialized units whereas it is known that this SOFC involves anode recirculation (Wagner, 2019).

- The system requires a connection to water mains for steam reforming purposes. The system periodically draws water from the main, processes it and stores it in a tank. However, if heat recovery is implemented (which is actually optional with this system, that can function as an electrical generator only), water can be recuperated from the condensation of the flue gases. This reduces the grid water consumption. Potential excess of water in the tank can be evacuated through one specific water outlet.
- Water impurities are processed through several filters involving reverse osmosis (Fritzmann, et al., 2007). It is believed that, at a certain point, the water impurities concentration in the periodical inlet water volume becomes too high compared to the pressure used for the reverse osmosis. Thus, the remaining inlet water (with high levels of impurities) has to be thrown away for another periodical grid fresh water inlet to take place (and to be submitted to the reverse osmosis filter).

Table 1: Studied SOFC expected targets (data provided by manufacturer)

Type	Technical specifications
Operation mode	Power-led, continuous (approx. 8,700 h per year)
Fuel type	Natural gas, bio-methane
Fuel consumption[1]	2.51 kW
Electrical efficiency[1,2] (output)	Up to 60 % (1.5 kW)
Thermal efficiency[1,2] (output)	Up to 25 % (0.6 kW)
Electrical and thermal energy generated per year	~ 13,000 kWh_{el} | ~ 5,220 kWh_{th}
Weight, Dimensions (H x W x D)	195 kg, 1010 x 600 x 660 mm
Service interval[3]	12 months

[1] - Low Heating Value (LHV) based figures
[2] - At maximum electrical efficiency, nominal output of 1.5 kW
[3] - Replacement of filters depending on local water, air and gas quality

As explained, two systems have been monitored in Belgian households for the whole year 2021 (one in Riemst, East of Belgium and the other one in Duffel, North of Belgium). The monitoring architecture is presented in Figure 2. Contrary to other commercialized fuel cells in Belgium (Davila, et al., 2022), there is no internal DHW tank in the system so the standby losses, which can be significant with such systems (Paulus & Lemort, 2022b), do not affect the thermal output measure. Sensor reference, precision and resolution of the acquired data are presented in Table 2.

Table 2: Reference of the monitoring sensors. The data logger (cloud connection) is a Viltrus MX-9

Sensors	Reference	Resolution	Accuracy
Indoor & Outdoor temperature and humidity	Weptech Munia	0,1 K 0,1 %	± 0,3 K ± 2 %
Heat recovery counter	Qalcosonic E1 Qn2,5 qi=0.025 m^3/h L=130mm	1 kWh 1 L | 0,1 K	Accuracy Class 2 (OIML R 75-, 2002)
Machine 2-ways electrical energy counter	Iskraemeco MT174-D2A42-V12G22-M3K0	10 Wh	Accuracy Class 1 (IEC 62053-21, 2003)
House 2-ways electrical energy counter	Iskraemeco MT174-D2A42-V12G22-M3K0	10 Wh	Accuracy Class 1 (IEC 62053-21, 2003)
Gas volume counter	BK-G4T DN25 Qmax 6 m^3/h	10 L	<0.5%

Last very important parameter not shown in Table 2 is the sampling rate, the frequency of the acquisition, of the measurements. It could not be smaller than 5 minutes to prevent the battery inside the sensors to empty. Except for temperatures and humidity, all of those meters are computing energy index values (always increasing). The heat counter is preferably placed on the return line (Paulus, et al., 2022a).

It is worth mentioning that the equivalent energy of the consumed gas is established thanks to hourly heating values given by the gas provider. The method used for this energy computation has been described in a parallel study (Paulus & Lemort, 2022c).

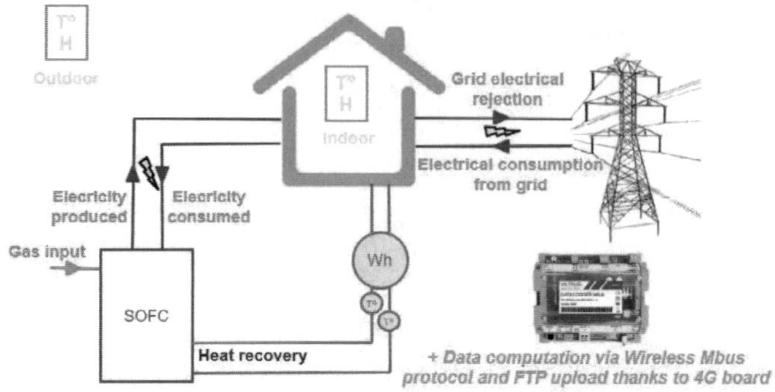

Figure 2: Monitoring architecture of both field-test households

2. METHODOLOGY

2.1 Energetical performance

Electrical efficiencies have been computed without considering the monitored electrical consumption of the system. Only the equivalent energy of the consumed gas has been considered at the denominator of the efficiency calculations. It is worth mentioning that a consumption signal at the system's output can only be seen if the machine is not producing electricity. Indeed, the system provides electricity to its own auxiliaries in running mode and only the net electrical production is measured. Since the system is supposed to be ran constantly, its remaining measured electrical consumption is not relevant in the efficiency calculations.

2.2 Economical and ecological performance

The chosen methodology for establishing performance indicators is similar to what the European Parliament recommends for calculation of primary energy savings of CHP (European Parliament, 2012). Their directive recommends to study the performance of a cogeneration by comparison to state-of-the-art separate heat and electrical energy productions. Based on this European directive, the Walloon energy regulator (in Belgium) has stated that the reference state-of-the-art system for heat production is a gas condensing boiler of constant 90% LHV efficiency (CWaPE, 2005).

Reference electrical state-of-the-art production system will consider the actual electrical Belgian mix in most cases or, as considered by the Walloon energy regulator (CWaPE, 2005), following European directives (European Parliament, 2012), a CCGT plant of constant 55% LHV efficiency (CWaPE, 2005). This assumption is only considered once, for one single ecological indicator. The CCGT plants considered in this work as reference electrical production systems are never assumed to be used as CHP, as it is still not that common.

Each indicator of this paper has been established by considering the actual pricing or ecological performance in the field-test (computed with given assumptions) subtracted to the assumed performance that would be achieved by reference machines based on the same field-test heat demands (and local electrical production). Positive indicators imply better performance than reference machines.

Economical indicator subsequently consists in utilization cost savings (€ saved/year) compared to the reference machine, which is, as stated, a classical gas condensing boiler. In this work, one has neither taken into account the investment costs of the machine nor the one of its installation (or its removal). The pricing assumptions for this economical indicator have been provided by the Belgian energy regulator, named CREG (*Commission de Régulation de l'Électricité et du Gaz*), that oversees the whole energy market and parties. The federal regulator is providing those energy pricing based on the average Belgian household energy bills (Paulus, et al., 2022b). For 2021, only the second semester average prices have been considered and applied to the whole monitoring data of the year, in order to partially consider the impact of the energy crisis (Statistisches Bundesamt - Destatis, 2021) on the economical performance of the machine. The energy prices, considered constant for the whole year, are thus 0.333€/kWh_{el} for electrical energy and 0.093€/kWh_{HHV} for natural gas (CREG, 2021). It must be pointed out that European natural gas prices even rose by almost 70% after Russia invaded Ukraine in February 2022 (Emerald Expert Briefings, 2022) but that is not yet considered in this work.

Other important pricing assumption is that the electrical energy transport and distribution costs are rounded to 0.15€/kWh$_{el}$ (0.1€/kWh$_{el}$ for distributions and 0.05€/kWh$_{el}$ for transport) as considered by the Walloon regulator (named CWaPE) in its tariffication plans (CWaPE, 2020). This means that rejected energy on the grid will not be bought back to the residential customer at the same kWh price mentioned earlier. This is because the customer uses the grid to sell its extra energy and it must pay for it. The selling price of the rejected electrical energy is thus equal to the electrical price mentioned earlier minus the 0.15€/kWh$_{el}$ to account for transport and distribution costs.

As stated, the environmental indicators are based on the same method of comparison with reference systems. The pricing assumptions are simply replaced by emission factors set through Life Cycle Assessment studies (Paulus, et al., 2022b). Two sets of assumptions are considered in this work. The first one is used by the Walloon energy regulator (CWaPE, 2005) to promote CHP. It overestimates on purpose the reference electrical emission factor by considering that the local electrical production allowed by the system replaces CCGT power plants (of 55% LHV efficiency) whereas actual Belgian mix is far more decarbonated, as it involves renewables and significant part of nuclear energy considered as "low carbon" (Sovacool, 2008). Indeed, Belgian CHP investors can beneficiate from public grants depending on those CO_2 savings. The second ecological indicator considers the hourly emission factor for the Belgian consumption electrical mix (data provided for the whole year 2021 by www.Electricitymap.org for academic purposes). Therefore, the first indicator implies an electrical emission factor of 456 gCO_{2eq}/kWh$_{el}$ whereas the second evolves hourly. For information, the statistical electrical emission factor for the whole 8760 hours of 2021 is 167 gCO_{2eq}/kWh$_{el}$ (Paulus, et al., 2022b), well below the assumption of the first indicator.

Regarding the emission factor for natural gas consumption/combustion, it is quite close between the two sets of assumptions : 251 gCO_{2eq}/kWh for the first indicator as used again by the Walloon regulator (CWaPE, 2005) and, for the second one, relevant with the www.Electricitymap.org data, 254 gCO_{2eq}/kWh, (Paulus, et al., 2022b).

Further explanations on the method used to establish economical and ecological indicators have been presented in a parallel study for a different system (Paulus, et al., 2022b). That study also involves further explanations on emission factors. It is worth mentioning that the ecological indicator used in this work accounts for all greenhouse gases (expressed in CO_{2eq} and not in CO_2 only). Also, even if the first dataset has been established in 2005, it is still valid and used as there has neither been any game changer regarding the efficiency of the reference systems for energy production (nor regarding natural gas production and importation). Actually, the only emission factor that is likely to become obsolete in the few years to come is the electrical one considered in the second set of assumption (from the www.Electricitymap.org data). This is because electrical mix evolves constantly, especially with increased penetration of renewables.

At last, to the understanding of the authors, grid transportation and distribution electrical losses, which can reach about 6-7% in EU (Psomopoulos, et al., 2010), have not been considered in any of the electrical emission factors. However, this is assumed to be compensated by the fact that it could also be considered that the extra gas consumption for the decentralized SOFC electrical production is subjected to fugitive methane emissions, which implies a high 100-year GWP of about 28 (Paulus, 2022), on longer gas network distances. For information, fugitive losses can be estimated to 5.4×10−6kg for the transport of 1kg of natural gas for a distance of 1km (NETL, 2010).

3. RESULTS

Performance results are presented in Table 3. Overall, performance of both field-test machines are quite similar.

The first observation is that the Duffel system is really close to the expected electrical efficiency whereas the Riemst system is a little below. The main reason is that the Riemst system's power output has been modulated down for some significant amounts of time through the year (for the vacation of the occupants) whereas the system in Duffel was running at nominal power the whole year. This can be perceived by the lower yearly electrical production (Table 3). Indeed, partial electrical load induces a reduction in electrical efficiency, as it can be seen in Figure 3. Second possible explanation is the probable intrinsic statistical efficiency difference between produced units. Indeed, the difference between percentile 0.9 and percentile 0.1 can account for a 2.5 percentage points efficiency difference for this system at nominal output power (Föger, 2011).

Second observation is that thermal efficiency (and heat recovered) is higher for Riemst. This is mainly again because of the fact that the SOFC electrical output power in Riemst was modulated down to the minimal value (500 W_{el})

during the vacation time of the owner. Indeed, the lower the electrical output, the higher the thermal recuperation, as it can be seen in Figure 3. In both cases, by looking only at the absolute yearly amount of heat recovered (and not the daily profiles nor the working temperature), the heat recovered with this system is greater than the amount required by the "M" DHW profile from European Standard EN 16147:2011 (Fuentes, et al., 2018). It is also greater than the average household's DHW consumption for a lot countries, such as the USA, the UK, Finland or Spain (Fuentes, et al., 2018). The fact that this system might be able to provide all the DHW demand of the household might allow significant capital savings as it simplifies the DHW production system. For example, there would be no need for a dedicated heat pump water heater or extra investments for the gas condensing boiler or the heat pump to be able to provide DHW in addition to its space heating function. It seems however preferable to at least include a backup heating resistance in the upper part of the DHW tank in order to ensure minimum DHW delivery temperature. Especially if the household DHW demand occasionally peaks, the system's DHW capability might not suffice.

Table 3: 2021 monitoring performance of the field-tested SOFC's. Utilization savings indicators consider a gas condensing boiler of 90% constant LHV efficiency as reference for heat production

Monitoring performance	Riemst	Duffel	Monitoring performance	Riemst	Duffel
HHV energy consumed (kWh)	25031	24273	LHV Electrical efficiency (%)	52,4	59,0
Electrical production (kWh)	11843	12922	LHV Thermal efficiency (%)	15,8	11,6
Electrical consumption (kWh)	11	2	LHV Total efficiency (%)	68,2	70,6
Heat recovered (kWh)	3569	2549	Utilization CO_{2eq} savings – 1st dataset ($kgCO_{2eq}$)[2]	723	1107
Utilization cost savings (€)[1]	1429	1308	Utilization CO_{2eq} savings – 2nd dataset ($kgCO_{2eq}$)[3]	-3013	-2969

[1] - Electrical price: 0.333€/kWh_{el}. Gas price 0.093€/kWh_{HHV}.

[2] - 1st dataset emission factors : 251 gCO_{2eq}/kWh for gas and 456 gCO_{2eq}/kWh_{el} for electricity.

[3] - 2nd dataset emission factors : 254 gCO_{2eq}/kWh for gas and hourly data from Belgian electrical consumption mix from www.Electricitymap.org for electricity.

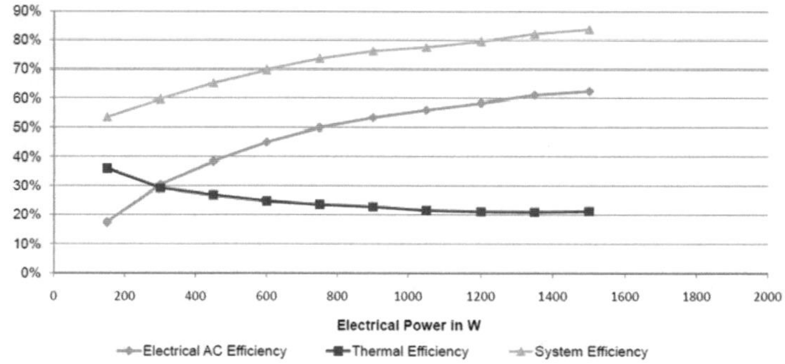

Figure 3: LHV efficiencies of the BlueGen (for 30°C return temperature) reproduced from literature (Föger, 2011)

It is worth mentioning that utilization cost savings depend greatly on the ability of the household to consume the electricity the system is producing while it is producing (as it is financially not that interesting to reject electricity on the grid with such fixed electrical tariffication assumptions). Therefore, cost savings of Table 3 are strongly case-dependent: both households involve an electric car that truly helps increasing the "self-consumption" of the electricity produced, also called "supply cover factor" (Paulus, et al., 2022a), to about 60%. This improves the economical indicator. In addition to local spark spread (Näsäkkälä & Fleten, 2005) and household supply cover factor, utilization cost savings should be looked at considering the capital costs of the system, as they will define the Return On Investment (ROI). For instance, with those results, if an investor wanted to ensure reaching ROI before 10 years, the maximum capital costs of the system should not reach more than about 13-14 k€ (this is considering that the owner already has a gas condensing boiler as main space heating appliance and this is not considering the additional capital savings of the already stated simplified DHW production). This example is also valid for households that do not need space heating at all (passive houses). If one particular household needed to invest in a space heating appliance such as a gas condensing boiler and was also considering this SOFC, that example shall not directly be considered as valid. Indeed, that household's choice would therefore lie between a single gas condensing boiler (also able to provide DHW) against the combination of the SOFC system and a gas condensing boiler (or another space heating appliance). For the 10-years ROI to be reached, the addition of both appliances shall not exceed the stated 13-14 k€, which is more restrictive than the previous cases. Potential maintenance costs have not been considered in those economical analyses. Although this SOFC's capital costs are unknown to the authors, it is interesting to point out that back in 2015, residential SOFC technology cost (with auxiliaries) was estimated to about 10 k€/kW (Napoli, et al., 2015). This would mean that this 1,5 kW_{el} system would cost about 15 k€, slightly above the 10 years utilization savings stated earlier. It is possible that current and future SOFC technology costs have been reduced and will continue to reduce (with higher technology penetration and prices learning curve).

At last, from an environmental point of view, the emission factors have significant influences. The ecological balances can either be considered as positive or negative. On one hand, the fact that the first dataset gives positive CO_{2eq} savings means that these decentralized electrical production systems are more environmentally friendly than centralized CCGT plants. On the other hand, fact that the second dataset gives negative CO_{2eq} savings means these systems are ecologically worse than the average local electrical mix (with its renewables and its nuclear energy). As these systems are currently used with constant output power, the second emission factor dataset with the resulting negative CO_{2eq} balances is more relevant. However, since remote control is already possible, these SOFC systems have the ability to only be powered up to replace CCGT electrical production, which mainly occurs for grid peak load (thanks to the high flexibility allowed by CCGT). In addition, with potential dynamic electrical production contracts, for example based on day-ahead hourly prices (Huisman, et al., 2019), this SOFC potential flexibility service could lead to higher rejection prices for the household and subsequent quicker ROI.

4. UNCERTAINTY ANALYSES

The method used for uncertainty propagations in this report is the one advised by the National Institute of Standards and Technology (Taylor & Kuyatt, 1927) and to do so, the EES (Equation Engineering Solver) software has been used.

In order to compute the propagation of uncertainties, one must establish it for every variable. For each monitored sensor, it comes directly from Table 2 and its referenced standards. However, the uncertainty has not been established for the hourly HHV figures given by the gas provider for each field-test site. Therefore, it is assumed that it is directly linked to the variance of all the given HHV figures: the uncertainty has been set to thrice the standard deviation of the given HHV figures for the whole monitored year (actually, the standard deviation has been calculated from the population of the hourly HHV figures provided for 2020 and 2021). According to statistics theory, fictively assuming that the actual HHV is constant all year long and that the values given by the gas provider follow a normal distribution around it, this would mean that the obtained uncertainty window accounts for 99.7% of the given HHV figures, which is, as often considered in statistics, sufficient. In reality, the actual HHV is not constant and slightly varies all year long but one can assume that the uncertainty window established here remains valid around the evolving actual value.

This gives an uncertainty on HHV figures of approximately ±234 Wh/m^3 for Riemst and ±105 Wh/m^3 for Duffel. There is a need for separate uncertainty analyses between the sites because the gas type provided to the machine by the grid is different: according to the gas provider, Duffel has still been provided with lean gas (called "type L" in Belgium) whereas Riemst already has a rich gas supply (called "type H" in Belgium). There is about a 10% difference in the given HHV figures between the two types of gas. Such a difference in the chemistry of the gas could indeed have a significant impact on the accuracy of the method used by the gas provider to establish HHV figures.

Thanks to the uncertainty on the HVV figures and the accuracy of the gas meter (about ±0.5% according to Table 2), one can propagate the uncertainties to the daily HHV energy consumed and then to the total yearly HHV energy consumed presented in Table 3 (which corresponds to 8760 observations of hourly HHV equivalent energy consumption). Thanks to the accuracy of the energy meters deduced from Table 2 (±5% for heat meters and ±1% for electrical energy meters), one obtains, for the site of Riemst, uncertainties of about ±0.5 percentage points for the electrical efficiency, ±0.8 percentage points for the thermal efficiency and ±1 percentage points for the total efficiency. For the site of Duffel, one obtains respectively ±0.6, ±0.6 and ±0.8 percentage points for the electrical, thermal and total efficiency.

One limitation of those uncertainty calculations is that they have not considered the accuracy of the correction factor that must be applied to the metered volume to convert it in the reference conditions used by the gas provider in its

HHV definition (Paulus & Lemort, 2022c). According to this study, the metered volume has to be multiplied by a correction factor established from the real gases equation of state. This correction factor, close to one, involves thus pressure ratio, temperature ratio and compression ratio between standard conditions and delivery conditions. Only the delivery conditions account for uncertainties. Delivery temperature uncertainty can be neglected as the gas meter used in the field-test (BK-G4T, as stated in Table 2) implements temperature compensation. According to the values given in ISO 13443:1996, it is unlikely that the compression factor ratio would have an uncertainty greater than ±0.0002 (absolute). At last, delivery pressure depends on the setting of the pressure regulator upstream of the system and on the atmospheric pressure at the field-test site. The pressure regulator setting is either 21 or 25 mbar depending on the gas type in Belgium (Paulus & Lemort, 2022c), so a relevant maximum uncertainty can realistically be assumed to about ±2 mbar. Regarding the atmospheric pressure, it comes from hourly measurement from Royal Meteorological Institute (in Belgium). The real uncertainty of those pressure measurements is unknown, especially since the pressure measurement are not performed at the field-test sites but at a meteorologic station nearby. Therefore, it has been assumed that the uncertainty of each pressure measurement is again equal to thrice the standard deviation of all the hourly pressure measurements that have been provided (again for two years: 2020 and 2021). This leads to an uncertainty of about ±3000 Pa. The propagated uncertainty on the correction factor is thus equal to ±0.028. This is actually too small to have an influence on the propagated uncertainties on the efficiencies stated before.

The emission factor and pricing assumptions are not assumed to be subjected to any uncertainty (uncertainty only propagated on monitored measures and HVV figures). Since the economical and ecological balances are resulting mainly from the efficiencies of the system, it can be assumed that the maximum relative efficiency stated earlier can be considered for all the economical and ecological balances of this work. Actually, the relative uncertainty on thermal efficiencies is of about ±5% for both field-test system.

In conclusion, it is safe to consider that the propagated uncertainties for all the results of this work are at about ±5%. Of course, this is only based on the measured values and the intrinsic accuracy of the sensors. There is no estimation of the uncertainty induced by potential unoptimized installations of the sensors or unoptimized fluidic conditions, especially for the temperature probes of the heat flow meter. Indeed, literature usually finds it is better to mount the probe within the pipe in an elbow to allow it to be parallel to the incoming flow (ASME, 2010). However, that is not the case for the monitored sites, where probes are mounted perpendicularly to the pipe-wall (easier installation as they are commercialized monitoring sensors and not quite laboratory equipment). This probe configuration is still accepted in literature. Indeed, those two possibilities of mounting temperature probes on a hydraulic pipe (along with a third angular one) are shown in Figure 4 (a) (Klason, et al., 2014). Also, as the temperature probes are singularly placed in the pipe, the assumption is made that the single in-pipe temperature point accounts for all the fluid on the measured section of the pipe. As shown in Figure 4 (b) (Klason, et al., 2014), this might not be the case as wall effects as well as temperature gradients may happen. Therefore, there might be a temperature distribution along the section of the pipe that one single temperature measurement might not reliably represent. This is another example of a source of uncertainty due to the quality of the installation and of the monitoring equipment that is neglected in this report uncertainty studies. At last, literature has established the need for the depth of the probe in the pipe to be set precisely, down to 45 to 50% of the pipe diameter (Kolpatzik, et al., 1998). This time, this is likely to be ensured for the monitored sites, as the probes are sold with the corresponding portion of pipe that they are supposed to be screwed in. A mechanical stop is also involved within this probe-pipe assembly, for which a picture is shown on Figure 4 (c). So, an unoptimized depth could only come from the unlikely unproper mounting of the technician that may not have screwed the probe all the way to its mechanical stop. Unfortunately, this could still happen and there is no way of knowing for sure remotely without visiting the installations. Still, as best practices and sensor mounting manufacturer recommendations have been followed, these kinds of eventual misplacements can be neglected.

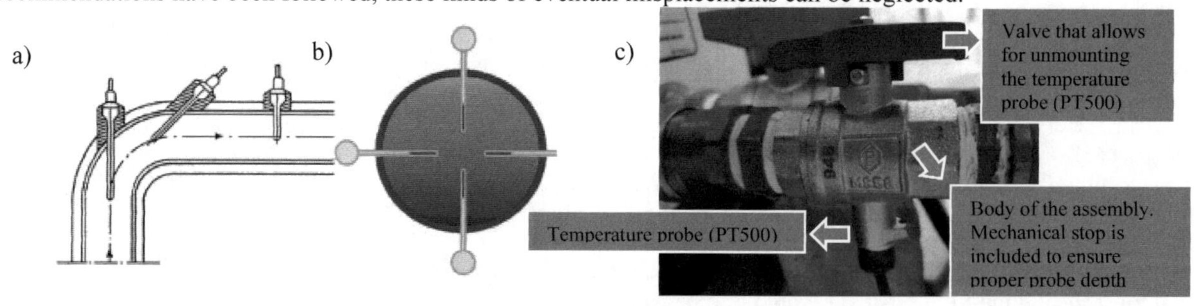

Figure 4: a) The three recommended options to mount a temperature sensor in a pipe (Klason, et al., 2014) b) Example of the distribution of the temperature of the fluid within a pipe along with an example of multiple probes installation in order to study it (Klason, et al., 2014) c) Example of in-pipe temperature probe assembly used in this study

5. LIMITATIONS AND CONCLUSIONS

The strongest limitation of this work is that the resulting economical indicators are strongly case dependent and that considering only two systems does not allow for drawing statistical conclusions on the system's performance (especially the economical ones). The results developed here shall thus be considered as case studies. However, the ecological balances are unlikely to change much from one household to the other as it does not depend on supply cover factors and as the system is mainly used with a constant and nominal power output.

This 2021 field-test monitoring study has demonstrated in real applications the 60% LHV electrical efficiency of this residential SOFC mCHP system (at least for one of the two SOFC's, the other being close to that figure probably because of intrinsic variance between manufactured products, because it has been functioning at partial load on the studied timeframe and because of aging, as it was commissioned one year before the other unit). Demonstrated LHV thermal efficiency is not higher than 12-15%, mainly because the heat recovery is mostly used for DHW production (at high delivery temperature). However, the amount of heat still recovered has been proven to be quite significant and sufficient, for example, to cover DHW demand of the average USA household (backup heating appliance is still advised to ensure sufficient temperature delivery at all times). Resulting yearly utilization costs savings (not considering the impact of the conflict in Ukraine on the energy prices) are significant and at about 1.3 - 1.4 k€/year (for 2021 Belgian energy prices assumptions). This figure has been achieved thanks to a quite high supply cover factor of about 60% for each household (mainly allowed thanks to one electric car). Potential dynamic day-ahead tariffication could even improve those utilization cost savings. From an environmental point of view, two considerations could be made: compared to the actual Belgian consumption mix, the CO_{2eq} balance are way negative because of the intrinsic use of natural gas (that remains a carbonated fossil fuel). However, the system has better efficiencies than CCGT power plants and its power output can be modulated from distance. Therefore, it can partially be considered that this system allows for decreasing the electrical demand on centralized CCGT power plants. This consideration changes totally the CO_{2eq} balances that become quite positive.

NOMENCLATURE

CCGT	Combined Cycle Gas Turbine
CREG	Commission de Régulation de l'Électricité et du Gaz
DHW	Domestic Hot Water
HHV	High Heating Value, [J/kg]
LHV	Low Heating Value, [J/kg]
mCHP	micro-combined Heat and Power
ROI	Return On Investment, [years]
SOFC	Solid Oxide Fuel Cells

REFERENCES

ASME, 2010. *Thermowells - Performance test codes.* s.l.:American Society of Mechanical Engineers.

Braun, R., Klein, S. & Reindl, D., 2006. Evaluation of system configurations for solid oxide fuel cell-based micro-combined heat and power generators in residential applications. *Journal of Power Sources,* 158(2).

Chen, K. K. et al., 2021. Membrane processes for CO2 removal and fuel utilization enhancement for solid oxide fuel cells. *Journal of Membrane Science,* Volume 620.

CREG, 2021. *Analyse semestrielle de l'évolution des prix de l'énergie – 2e semestre 2021,* s.l.: CREG.

CWaPE, 2005. *Décision CD-5j18-CWaPE relative à "la définition des rendements annuels d'exploitation des installations modernes de référence, ...",* s.l.: CWaPE.

CWaPE, 2020. *Communication - FAQ – Tarif Prosumer,* s.l.: CWaPE.

Davila, C., Paulus, N. & Lemort, V., 2022. *Experimental Investigation of a Micro-CHP Unit Driven by Natural Gas for Residential Buildings.* Purdue, s.n.

Ellamla, H. R. et al., 2015. Current status of fuel cell based combined heat and power systems for residential sector. *Journal of Power Sources,* Volume 293.

Emerald Expert Briefings, 2022. *Ukraine crisis will bring heavy costs for Europe,* s.l.: Oxford Analytica.

European Parliament, 2012. *DIRECTIVE 2012/27/EU on on energy efficiency,* s.l.: s.n.

Fernandes, M. et al., 2018. SOFC-APU systems for aircraft: A review. *International Journal of Hydrogen Energy,* 43(33).
Föger, K., 2011. *Ceramic Fuel Cells Residential Generator BlueGen - Ultra-efficient distributed power generation in Smart Grid Environment.* Seoul, s.n.
Fritzmann, C., Löwenberg, J., Wintgens, T. & Melin, T., 2007. State-of-the-art of reverse osmosis desalination. *Desalination,* 216(1-3).
Fuentes, E., Arce, L. & Salom, J., 2018. A review of domestic hot water consumption profiles for application in systems and buildings energy performance analysis. *Renewable and Sustainable Energy Reviews,* 81(1).
Hauch, A., Hagen, A., Hjelm, J. & Ramos, T., 2014. Sulfur Poisoning of SOFC Anodes: Effect of Overpotential on Long-Term Degradation. *Journal of The Electrochemical Society,* 161(6).
Huisman, R., Huurman, C. & Mahieu, R., 2019. Hourly electricity prices in day-ahead markets. *Energy Economics,* 29(2).
IEC 62053-21, 2003. *Electricity metering equipment (a.c.) – Static meters for active energy (classes 1 and 2).*
Klason, P. et al., 2014. Measuring temperature in pipe flow with non-homogeneous temperature distribution. *International Journal of Thermophysics,* Volume 35.
Kolpatzik, S., Hilgenstock, A., Dietrich, H. & B., N., 1998. The location of temperature sensors in pipe flows for determining the mean gas temperature in flow metering applications. *Flow Measurement and Instrumentation,* 9(1).
Lichtenegger, K. et al., 2015. The role of leak air in a double-wall chimney. *Heat Mass Transfer,* 51(6).
Napoli, R., Gandiglio, M. L. A. & Santarelli, M., 2015. Techno-economic analysis of PEMFC and SOFC micro-CHP fuel cell systems for the residential sector. *Energy and Buildings,* Volume 103.
Näsäkkälä, E. & Fleten, S.-E., 2005. Flexibility and technology choice in gas fired power plant investments. *Review of Financial Economics,* 14(3-4).
NETL, 2010. *Life Cycle Analysis: Natural Gas Combined Cycle (NGCC) Power Plant,* s.l.: US Dpt. of Energy.
OIML R 75-, 2002. *Heat meters. Part 1: General requirements.* s.l.:International Organization of Legal Metrology.
Paulus, N., 2022. Will Greenhouse Gas emission commitments in France and Wallonia respect IPCC's carbon budget ?. *Energy Policy,* Under review.
Paulus, N., Davila, C. & Lemort, V., 2022a. *Grid-impact factors of field-tested residential Proton Exchange Membrane Fuel Cell systems.* Rotterdam, doi: 10.34641/clima.2022.176
Paulus, N., Davila, C. & Lemort, V., 2022b. *Field-test economic and ecological performance of Proton Exchange Membrane Fuel Cells (PEMFC) used in residential micro-combined heat and power applications (micro-CHP).* Copenhagen, s.n.
Paulus, N. & Lemort, V., 2022a. Review of probable degradation mechanisms and recovery procedures for reversible performance losses in a residential cogeneration polymer electrolyte membrane fuel cell. *To be submitted.*
Paulus, N. & Lemort, V., 2022b. *Correlation between field-test and laboratory results for a Proton Exchange Membrane Fuel Cell (PEMFC) used as a residential cogeneration system.* Valenciennes, doi: 10.25855/SFT2022-119
Paulus, N. & Lemort, V., 2022c. Field-test performance models of a residential mi-cro-cogeneration system based on a proton exchange membrane fuel cell and a gas condensing boiler. *To be submitted.*
Peng, J. et al., 2021. Solid oxide fuel cell (SOFC) performance evaluation, fault diagnosis and health control: A review. *Journal of Power Sources,* Volume 505.
Psomopoulos, C. et al., 2010. Electricity savings and CO2 emissions reduction in buildings sector: How important the network losses are in the calculation?. *Energy,* 35(1).
Schmidt-Rohr, K., 2015. Why Combustions Are Always Exothermic, Yielding About 418 kJ per Mole of O2. *Journal of Chemical Education,* 92(12).
Shao, Z. et al., 2005. A thermally self-sustained micro solid-oxide fuel-cell stack with high power density. *Nature,* 435(7043).
Sovacool, B.-K., 2008. Valuing the greenhouse gas emissions from nuclear power: A critical survey. *Energy Policy,* 36(8).
Statistisches Bundesamt - Destatis, 2021. *Prices - Data on energy price trends,* s.l.: s.n.
Taylor, B.-N. & Kuyatt, C.-E., 1927. *Guidelines for Evaluating and Expressing the Uncertainty of NIST Measurement Results,* s.l.: s.n.
Wagner, P. H., 2019. *Integrated Design, Optimization, and Experimental Realization of a Steam-Driven Micro Recirculation Fan for Solid Oxide Fuel Cell Systems (PhD Thesis),* s.l.: Ecole Polytechnique de Lausanne.

Cooling Concepts for Residential Buildings: A Comparison Under Climate Change Scenarios

Essam ELNAGAR[1,*], Vincent LEMORT[2]

[1]Thermodynamics Laboratory, Aerospace and Mechanical Engineering Department, Faculty of Applied Sciences, University of Liège, Liège, Belgium
essam.elnagar@uliege.be

[2]Thermodynamics Laboratory, Aerospace and Mechanical Engineering Department, Faculty of Applied Sciences, University of Liège, Liège, Belgium
vincent.lemort@uliege.be

* Corresponding Author

ABSTRACT

Buildings in the EU account for 40% of our energy consumption and 36% of our greenhouse gas emissions, with HVAC systems being the biggest contributors. With the predicted increase in global air temperature up to 4.8 K by the end of this century, cooling will be the most rapidly increasing energy-consuming technology in buildings. Heat pumps are one of the most energy-efficient heating and cooling systems. In this paper, the energy performance of three different cooling concepts including an air-to-air heat pump in a residential building is assessed with the impact of climate change on the heating and cooling energy demands under future climatic scenarios in Belgium. The paper presents the results obtained by simulating heating and cooling systems based on using DesignBuilder while taking into consideration the influence of climate change on the performance of the split AC system. This study uses several weather data sets one forced by a reanalysis model on the past period (1980-2020) and three forced by Earth System Models (ESM) on past (1980-2014) and future periods (2015-2100). The paper also presents a performance comparison between the passive cooling technologies such as natural ventilation and active cooling system such as split AC systems in Belgium. The Indoor Overheating Degree (IOhD) indicator is used to assess the thermal comfort and overheating discomfort in the building. The obtained results showed that the climate plays an important role in the final energy end-use for heating and cooling, the final heating energy end-use decreases by 40% while the final cooling energy end-use increases by 187% by the end of the century. The results also showed that the IOhD using passive cooling scenarios such as natural ventilation decreased by 77% in 2090s compared to the base case where there is only a mechanical ventilation system, while using an active cooling system such as a split AC system could maintain the thermal comfort almost all the time in the future weather scenarios. This paper is part of an ongoing study, the objective of the ongoing study, of which some results are presented in this paper, is to upscale the impact of climate change on the Belgian residential building stock and to evaluate its influence on the future heating and cooling energy demands.

1. INTRODUCTION

The recent decades have seen a major concern about global warming and climate change. According to the Intergovernmental Panel on Climate Change (IPCC) sixth assessment report, global warming of 1.5 K and 2 K will be exceeded during the 21st century relative to 1850-1900 and the average global surface air temperature in (2081-2100) is predicted to increase within a range of 1 to 5.7 K in relation to (1850-1900) period under various CO_2 emission scenarios (*Climate Change 2021*, 2021; Eyring et al., 2016). This increase in temperature leads to a significant increase in energy demand for HVAC systems. Due to the increased CO_2 emissions, humidity, wind, and solar radiation are also expected to change over the next years (Karl et al., 2009; Ruosteenoja & Räisänen, 2013; Wyard et al., 2018).

To assess the possible changes in climate in the future, there are different emissions scenarios for the future climate (Nakicenovic et al., 2000; Riahi et al., 2017). These emissions scenarios are calculated using various models for society development which include technological and economic factors. There are five narrative scenarios and a set of radiative forcing. Each scenario consists of different groups characterizing alternative energy technology developments and thus carbon emissions (Riahi et al., 2017).

Buildings in the EU account for 40% of our energy consumption and 36% of our greenhouse gas emissions. Several studies showed that climate change has a major impact on heating and cooling demands. Few studies assessed the impact of climate change on the CO_2 emissions of heating and cooling systems (Radhi, 2009). In most European countries, the amount of energy required for heating is greater by far than the energy used for space cooling (Ürge-Vorsatz et al., 2015). In a temperate climate, one of the most predicted effects of climate change is an increase in overheating events and the related health risks for building occupants. Therefore, it is necessary to assess the future thermal performance of buildings and implement robust cooling measures to ensure occupants' comfort.

Several studies showed that, active cooling systems will be necessary for buildings in the future to maintain thermal comfort conditions. (Wang & Chen, 2014) investigated the impact of climate change on heating and cooling consumptions in different residential and commercial buildings in all seven climate zones in the united states, the study showed that by 2080 the heating demand will decrease by 30-65% and the cooling demand will increase by 50-150%, the study also showed that by 2080s, the impact of natural ventilation will not be significant in some cities. (Berardi & Jafarpur, 2020) studied the potential impact of climate change on 16 ASHRAE prototype buildings in Toronto. The results showed that, there is a drop in the annual heating energy use intensity and an increase in the cooling energy use intensity for the different building types, by 2070 the heating demand will decrease by 18%–33% and the cooling demand will increase by 15-126%. (Invidiata & Ghisi, 2016) evaluated the future energy consumption of Brazilian houses in three cities in 2020, 2050 and 2080. The study showed that there will be an increase in the annual energy demand by 112-185% in 2080 in the three cities, and the annual heating demand will decrease by 94% in the coldest city, the study also investigated the possibility of passive strategies to reduce the future annual heating and cooling demand up to 50% compared to the building without strategies. (Elnagar et al., 2022) studied the potential of natural ventilation in Belgium to reduce the internal cooling loads during the summer period. The results show that without natural ventilation, thermal comfort could be reached on average 53.7% of the time, while it could reach 66.2% of the time using single-sided ventilation and 78.8% with cross ventilation.

Despite the numerous studies on assessing the climate change impact on the energy demand in buildings, there have only been a few studies assessing the thermal comfort. As part of the International Energy Agency (IEA) EBC Annex 80 – "Resilient cooling of buildings", this paper assesses the performance of different cooling concepts; passive concepts such as ventilative cooling and active concepts such as air source heat pumps under different climate change scenarios in Belgium. This study uses several weather data sets: one forced by a reanalysis model on the past period (2000-2020) and three forced by Earth System Models (ESM) on past (1980-2014) and future periods (2015-2100). The results presented in this work respond to a growing need for researchers and decision-makers to investigate the relationship between changing future weather conditions and their impact on the energy demand, especially for cooling demand and thermal comfort in buildings.

2. METHODOLOGY

The conceptual framework of the study is presented in Figure 1. The first part of the framework introduces the literature review that has been conducted to identify the cooling concepts in this study, three cooling technologies are considered in this work; natural ventilation, mechanical ventilation and air source heat pumps. The second part presents the case study as well as the weather data used in the framework. The building and HVAC models are created with DesignBuilder software, which is a complete and Graphical User Interface (GUI) for the EnergyPlus simulation engine.

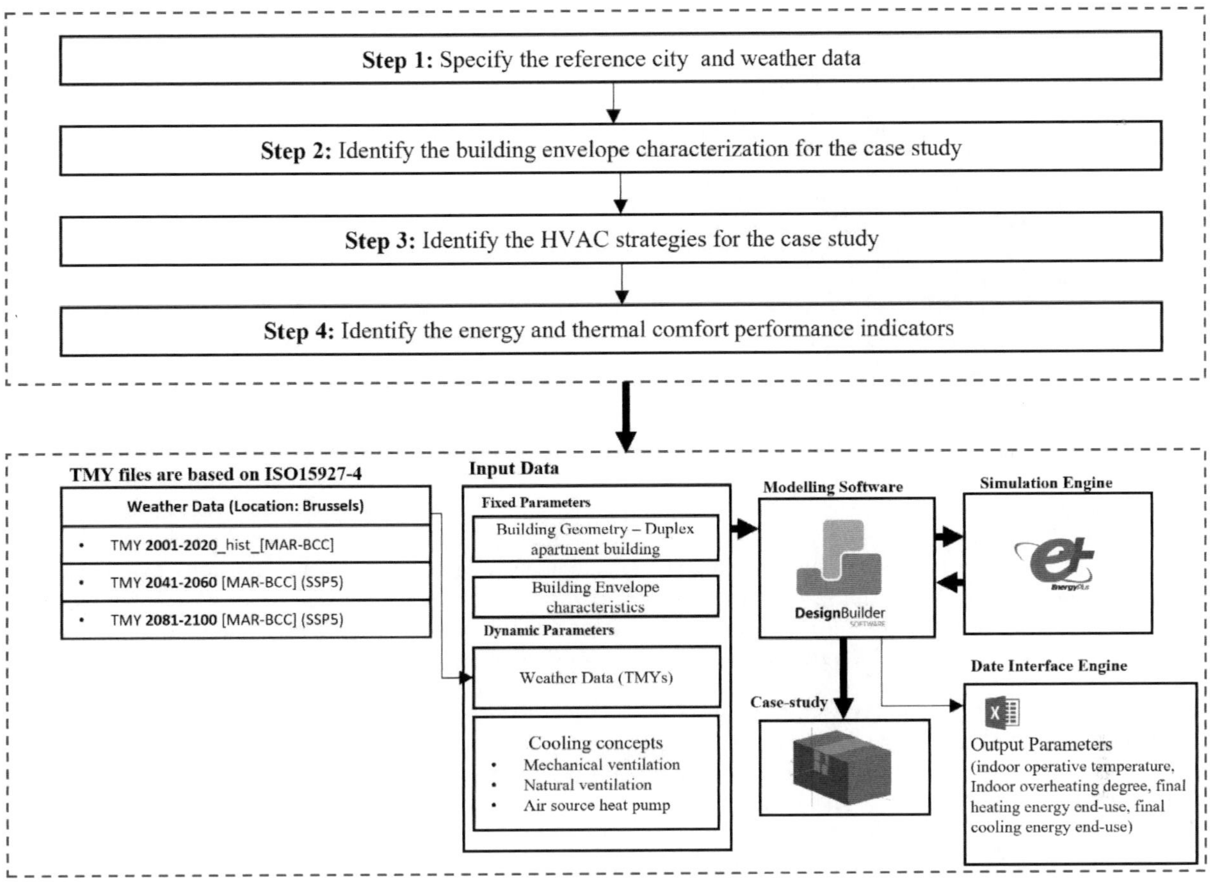

Figure 1: Conceptual framework of the study

2.1 Building model

This section gives the most important information about the geometry, envelope characteristics and operation features of the studied dwelling. Figure 2 (b) is an upper view of the studied duplex. The building has a North-South orientation and a large window-to-wall ratio (0.6) on the South façade. The dwelling has been divided into seven thermal zones, which have been defined depending on the zone usage, temperature setpoint and orientation. (Elnagar et al., 2022) presented the general characteristics of the building envelope.

The EPBD imposes energy performance standards for buildings. The building walls are compliant with the EPBD 2020 (*Exigences PEB du 1er juillet 2019 au 31 décembre 2020*, 2019). The glazing has been defined in the EPB file of the building as triple-glazing. The frame of the window, made of wood and aluminum, has a U-value of 0.9 $W/m^2 \cdot K$, resulting in a net U-value of 0.6 $W/m^2 \cdot K$, in compliance with the EPBD. The duplex is equipped with a balanced mechanical ventilation system with a heat recovery module of 85% efficiency and a by-pass system. The heating system is traditional with water radiators fed by a gas boiler. There is no mechanical cooling system.

Regarding the operation of the building, it is assumed that the occupants are four adults each emitting 100 W. For lighting, it is good practice to consider the consumption of 6 W/m^2, according to norm NBN EN 15193 (The Bureau for Standardisation in Belgium, 2021). The energy consumption linked to electric appliances is assumed to be 3 W/m^2. Some operation/occupancy schedules can be applied to that energy consumption. They can be found in standard ISO 17772 (International Organization for Standardization, 2017)

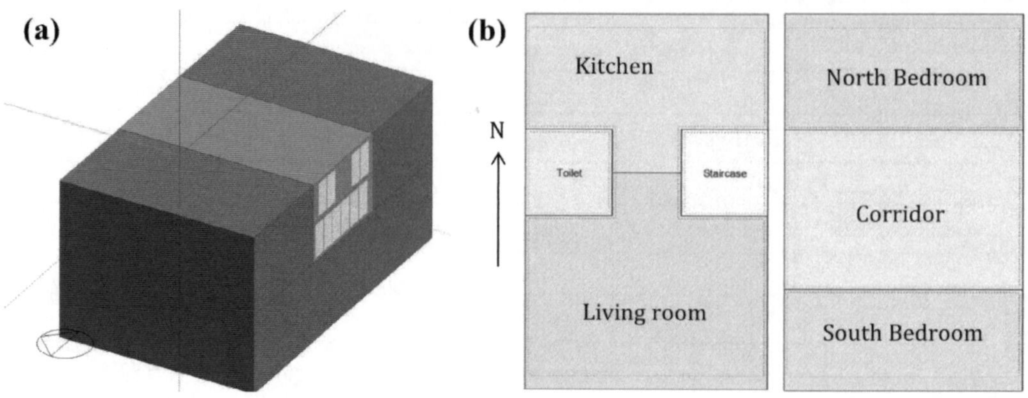

Figure 2: The representative case study (a) 3D model of the building (b) Upper view of the building.

2.2 Climate data

This study uses "Modèle Atmosphérique Régional" model (hereafter called "MAR") in its version 3.11.4 (Doutreloup et al., 2022). MAR is a three-dimensional atmospheric model coupled to a one-dimensional transfer scheme between the surface, vegetation, and atmosphere (Ridder & Gallée, 1998). This model is validated over the Belgium territory by several studies (Doutreloup et al., 2019; Wyard et al., 2021). For this study, the spatial resolution of MAR is 5 km over an integration domain (120 x 90 grid cells) centered over Belgium as shown in Figure 3.

This MAR model, like any regional model, must be forced at its boundaries by a global model, either a reanalysis model or an ESM. Firstly, the MAR model is forced by ERA5 (called hereafter MAR-ERA5 [30]) in order to have past simulation (1980-2020), and then by 3 Earth System Models (ESMs) from CMIP6 database namely BCC-CSM2-MR (MAR-BCC), MPIESM.1.2 (MAR-MPI), and MIROC6 (MAR-MIR) (Doutreloup et al., 2022)

These ESMs forced MAR simulations can be used to anticipate future periods. MAR simulations can represent the current climate and its interannual variability with success, except MAR-MIR which significantly overestimates temperature and solar radiation in summer. MAR-MPI can be considered as the coldest MAR simulation and MAR-BCC is the ensemble mean of all the MAR simulations (Doutreloup et al., 2022).

Typical Meteorological Years (TMY) are used for the 3 MAR-Simulations for the Shared Socioeconomic Pathway (SSP5.85) (Doutreloup & Fettweis, 2021). There are 3 main scenarios, SSP2-4.5, SSP3-7.0, and SSP5-8.5, which are respectively increasingly warming for 2100. This study used SSP5-85 and MAR-BCC. TMY for the period (2001-2020) based on MAR-BCC simulation is used as the reference scenario.

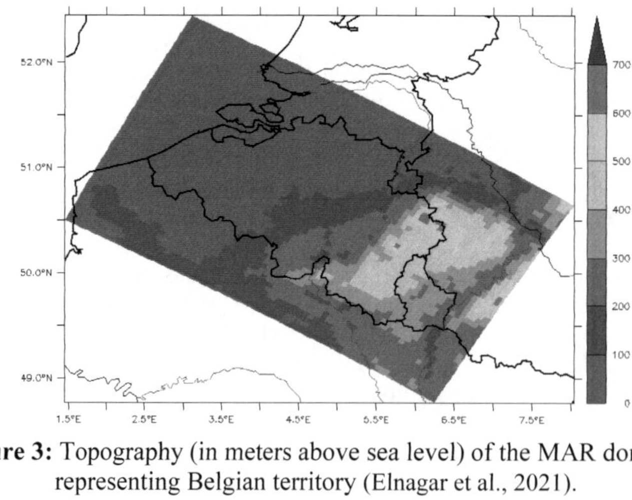

Figure 3: Topography (in meters above sea level) of the MAR domain representing Belgian territory (Elnagar et al., 2021).

2.3 Thermal comfort

In this work, two comfort models are used to determine the overheating risk in the dwelling and the efficiency of passive cooling. The first comfort model is based on the PMV (according to norm EN USI 7730). The PMV is an index that predicts the mean value of the votes of a large group of persons on the 7-point thermal sensation scale (from -3 to 3), based on the heat balance in the steady-state of the human body. A Steady-state is obtained when the internal heat production in the body is equal to the loss of heat to the environment. The PMV can be used to check whether a given thermal environment complies with comfort criteria, and to establish requirements for different levels of acceptability. Thermal comfort is ensured if the PMV is within the range [-0.5;0.5], which gives a rate of dissatisfied people of 10%.

Contrary to the PMV which is a "static" measurement of thermal comfort, the adaptive comfort theory states that the comfort temperature range can be broadened because the occupants can adapt their behavior depending on the outdoor conditions. The adaptive comfort model is described in norm EN 15251. In a naturally ventilated building without any air-conditioning, the occupants feel more connected to the outdoor environment and can bear higher temperatures in summer. The maximum adaptive comfort temperature can be expressed by a relationship depending on the mean outdoor temperature:

$$T_{comfort,max} = 0.33\, T_{rm} + 21.8 \tag{1}$$

where T_{rm} is a weigh average of the mean outdoor temperature during the previous seven days.

Indoor Overheating Degree (IOhD) indicator is selected to assess the thermal comfort and overheating discomfort in the building as suggested by (Hamdy et al., 2017).

$$IOhD = \frac{\sum_{z=1}^{Z} \sum_{i=1}^{Nocc(z)} [(T_{in,z,i} - TL_{comf,z,i})^{+} \times t_{i,z}]}{\sum_{z=1}^{Z} \sum_{i=1}^{Nocc(z)} t_{i,z}} \tag{2}$$

where z is the building zone counter, Z is the total number of conditioned zones in a building, i is the occupied hour counter, $N_{occ(z)}$ is the total occupied hours in zone z, $T_{in,z,i}$ is the indoor operative temperature in zone z at hour i, $TL_{comf,z,i}$ is the comfort temperature limits in the zone z at the hour i, t is the time step (typically it is 1 h). Only positive differences of ($T_{in,z,i}$ - $TL_{comf,z,i}$) are taken into the summation

2.3 Cooling concepts

Three different cooling concepts are applied in the building. The first concept evaluated the cooling potential of the mechanical ventilation system with heat recovery (basic system in the building) which provides a minimum fresh air flow rate of 25 m3/h according to NBN D50-001 with a minimum temperature difference between inside and outside of 2 K.

The second cooling concept is natural ventilation. The natural ventilation maximum air change per hour (ACH) is explicitly defined for all the naturally ventilated zones. The ACH is modified by controlling the minimum indoor temperature, the temperature difference between the indoor and outdoor temperature (ΔT) in the different scenarios and the operation schedule. The Air change per hour (ACH) was set equal to 5 ACH.

In the third cooling concept, a split Air conditioner (AC) is added to the base case to provide cooling without changing the existing gas boiler system in the building. The conventional split AC consists of an indoor and an outdoor unit. The outdoor unit includes an air-cooled condenser, a compressor and a condenser fan. The indoor unit includes an evaporator, a distribution fan an expansion valve. The split AC in DesignBuilder is implemented by using a unitary air-to-air heat pump module. The unitary air-to-air heat pump model simulates a constant volume of direct-expansion cooling via the DX cooling coil, with rated SHR, rated total cooling capacity, rated air volume flow rate, and rated EER as inputs. The HVAC model input parameters are shown in Table 1. The rated total cooling capacity, rated air volume flow rate and rated SHR are auto-sized.

Table 1: HVAC Model input

	Specifications
Zones (heating)	Kitchen, living, north bedroom, south bedroom
Zones (cooling)	Living, north bedroom, south bedroom
Zones (ventilation)	Kitchen, living, north bedroom, south bedroom
Mechanical ventilation rates	25 m^3/h
Natural ventilation rates	5 ACH
Setpoint temperatures	[Min: 20°C, Max: 22°C] for heating (living, kitchen) [Min:19°C, Max:21°C] for heating (north bedroom, south bedroom) [Min: 22°C, Max: 25°C] for cooling (living, north bedroom, south bedroom)
Fuel type	Natural gas (heating) and electricity (cooling)
Rated EER	5.8
Boiler Nominal thermal efficiency	0.88

3. RESULTS

3.1 Evolution of climate

Figure 4 shows the evolution of the daily outdoor temperature representing the distribution of the outdoor temperature in the reference period (2001-2020), the period (2041-2060) and the period 2081-2100. It can be seen that, there is a significant increase in the annual air temperature by 1.6 K in (2041-2060) period and 3.2 K in (2081-2100) period. Following the adaptive comfort model in the ASHRAE 55-2017 Standard (ASHRAE, 2017), the figure also shows the identified range of temperatures (90% and 80%) acceptability limits. The acceptable limits of 80% are for common applications. When a higher level of thermal comfort is desired, the 90% acceptability limits can be used.

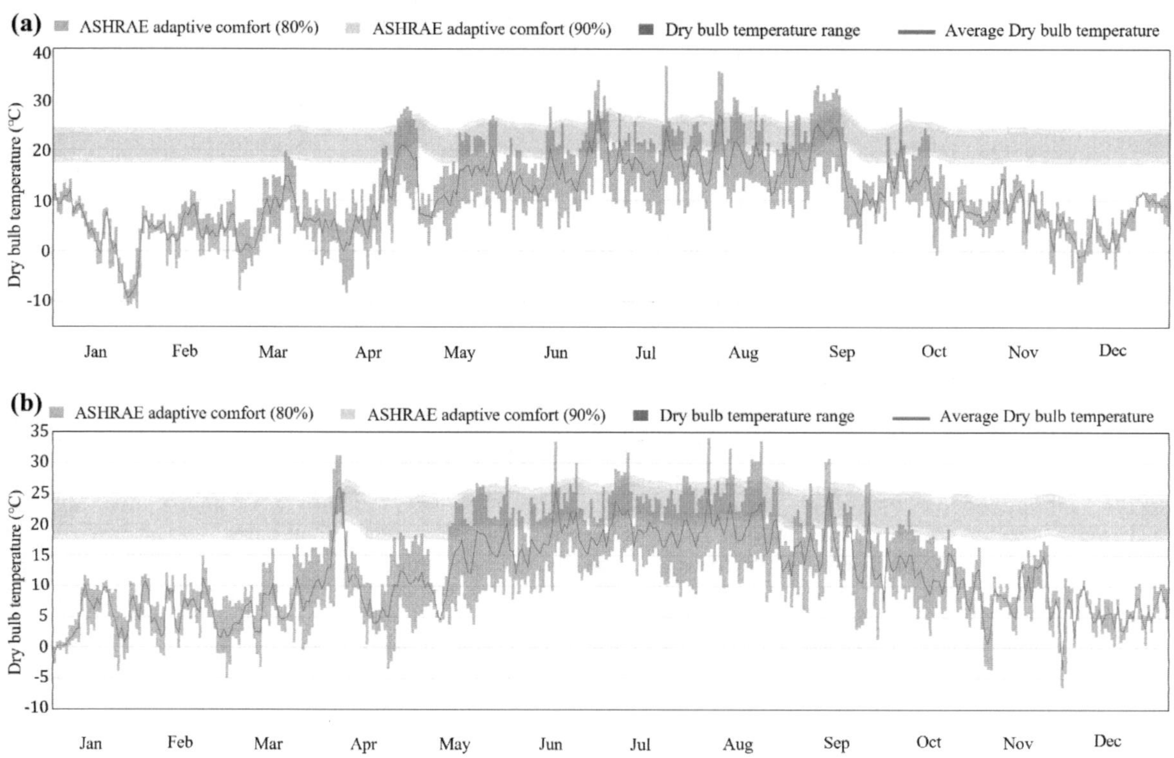

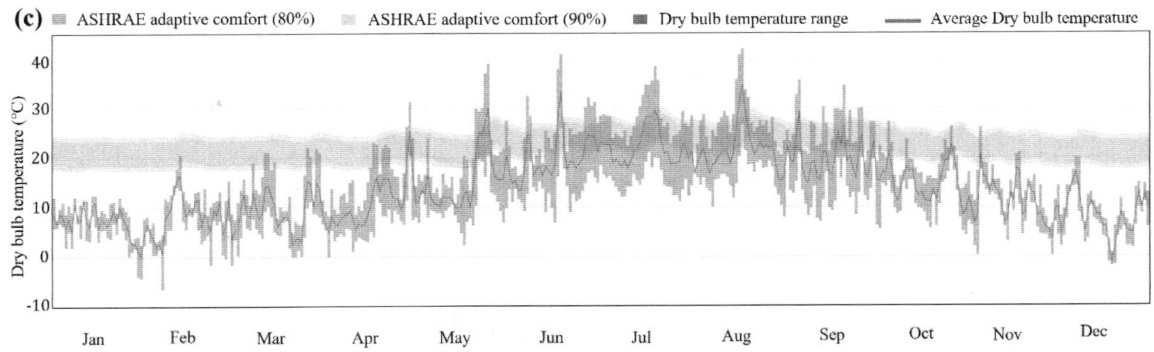

Figure 4: Evolution of daily outdoor temperature with the adaptive comfort model for three periods (a) TMY 2001-2020, (b) TMY 2041-2060 and (c) TMY 2081-2100

3.2 Thermal comfort assessment

In this section, the result of the thermal comfort assessment for the case study by analyzing the Indoor Overheating Degree (IOhD). Figure 5 presents the IOhD for the different cooling concepts and different weather scenarios. The static comfort model is used for the bedrooms and the adaptive comfort model is used for other zones based on ISO 17772-1. It can be seen that in the base case with only a mechanical ventilation system installed in the building, the IOhD increases from 0.13°C to 0.8 °C. In the second scenario using natural ventilation, the IOhD increased from 0.03 °C to 0.17 °C while in the third scenario by applying an active cooling system using split AC, the IOhD reached only 0.01 by (2081-2100).

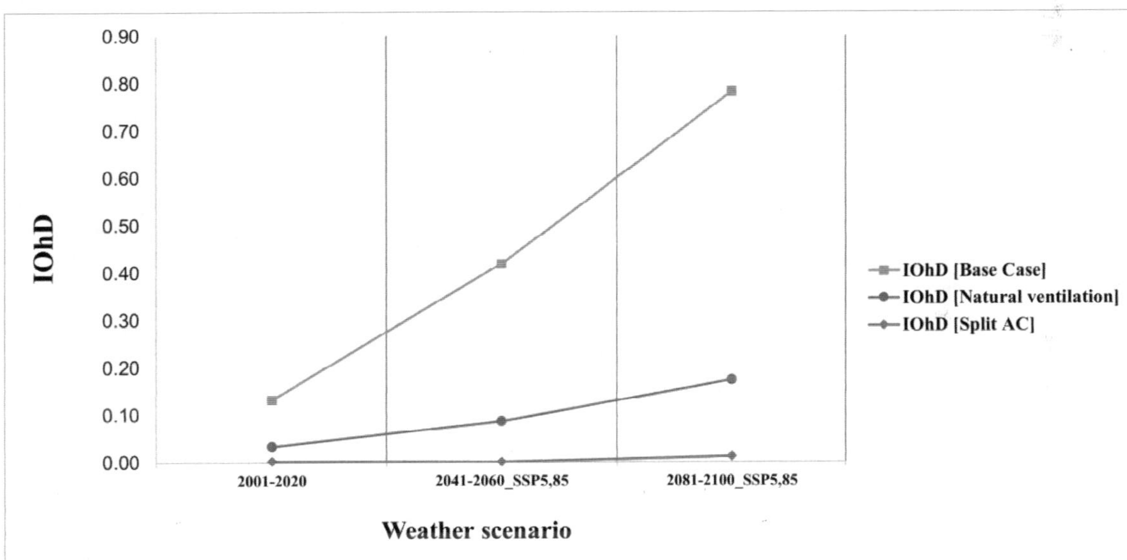

Figure 5: IOhD for the different cooling concepts and weather scenarios

3.2 Future heating and cooling energy end-use

The final heating and cooling energy end-uses are presented in this section for the third cooling concept where a split AC system is used in the building. Figure 6 shows that, the final heating energy end-use decreases by 26% by (2041-2060) and 40% by (2081-2100) compared to (2001-2020), while the final cooling energy end-use increases significantly by 82% in (2041-2060) and 187% in (2081-2100) compared to (2001-2020). It can also be seen that the final energy end-use decreases in the future scenarios by 32% in (2081-2100) compared to (2001-2020).

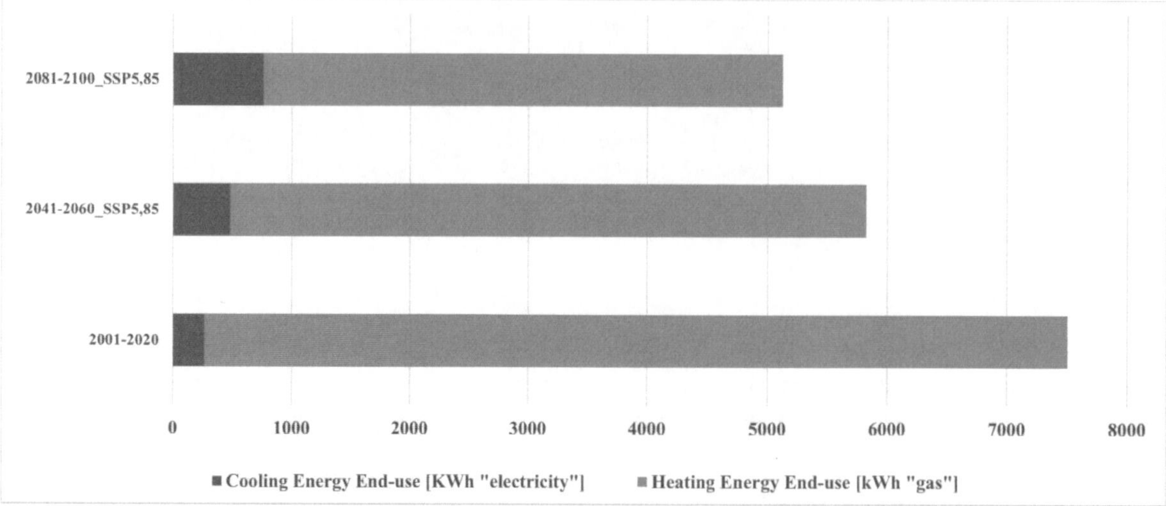

Figure 6: Final heating and cooling energy end-use for different weather scenarios for Split AC cooling concept

4. DISCUSSION

4.1 Findings and recommendations

According to the Intergovernmental Panel on Climate Change (IPCC), global warming of 1.5 K and 2 K will be exceeded during the 21st century. In this paper, using the future weather scenarios in Brussels city for the MAR model, the mean annual air temperature increases by 3.2 K in TMY (2081-2100) compared to TMY (2001-2020). With this significant increase in the temperature and the more frequent heatwaves, it became very necessary to study the future cooling energy demand and the related thermal comfort in buildings.

In this paper, IOhD indicator is used to assess the impact of climate change on thermal comfort using a multi-zonal approach in the case study. Two comfort models are used, the static comfort model for the bedrooms and the adaptive comfort model for other zones. The results showed that the IOhD using passive cooling scenarios such as natural ventilation decreases by 77% in 2090s compared to the base case where there is only a mechanical ventilation system, while using an active cooling system such as a split AC system could maintain the thermal comfort in the future weather scenarios as shown in Figure 5.

Figure 6 showed the final heating and cooling energy end-use. There is a significant increase in the cooling energy end-use and a reduction in the heating energy end-use. As mentioned earlier that buildings in the EU have a major share in the total energy consumption.

To summarize the key findings and provides some recommendations to the building designers and owners regarding the choice and installation of cooling systems:

- It's necessary to consider highly efficient active cooling systems in combination with passive cooling techniques during the construction of new buildings and the renovation of existing ones.
- Cooling systems coupled with air handling units allow to improve the indoor air quality and are generally suited for residential and commercial buildings.
- IOhD indicator is recommended to be used for thermal comfort assessment.

4.2 Strengths and limitations

- The study's first strength is the validity of the weather data used for the simulations using the MAR model.
- This study also compares different passive and active cooling concepts using a multizonal approach.
- The limitation of this study relies on neglecting the degradation of the HVAC systems in the future.

5. CONCLUSION

This paper studied the potential of three cooling concepts (mechanical ventilation, natural ventilation and Split AC) to cover the internal cooling loads during the summer period through different weather scenarios using the TMYs for 2001-2020 (2010s), 2041-2060 (2050s), and 2081-2100 (2090s). The paper showed the necessity to install active cooling systems along with passive measures to maintain a comfortable environment for the occupants. By comparing the three cooling concepts, the Split AC system showed the lowest overheating discomfort estimated by an IOhD of 0.01°C in (2081-2100). The paper also showed that, the final heating energy end-use decreases by 40% while the final cooling energy end-use increases by 187%.

NOMENCLATURE

T	Temperature	(°C)
IOhD	Indoor Overheating Degree	(°C)
TL	Temperature limit	(°C)
EER	Energy Efficiency Ratio	(–)
SHR	Sensible Heat Ratio	(–)

Subscript

rm	weigh average of the mean outdoor temperature
z	buildings zones counter
i	occupied hours counter
in	indoor conditions
t	time step

REFERENCES

ASHRAE. (2017). *ANSI/ASHRAE Standard 55-2017—Thermal Environmental Conditions for Human Occupancy.* https://www.ashrae.org/technical-resources/bookstore/standard-55-thermal-environmental-conditions-for-human-occupancy

Berardi, U., & Jafarpur, P. (2020). Assessing the impact of climate change on building heating and cooling energy demand in Canada. *Renewable and Sustainable Energy Reviews*, *121*, 109681. https://doi.org/10.1016/j.rser.2019.109681

Climate Change 2021: The Physical Science Basis. Contribution of Working Group I to the Sixth Assessment Report of the Intergovernmental Panel on Climate Change. (2021). Cambridge University Press.

Doutreloup, S., Fettweis, X., Rahif, R., Elnagar, E., Pourkiaei, M. S., Amaripadath, D., & Attia, S. (2022). *Historical and Future Weather Data for Dynamic Building Simulations in Belgium using the MAR model: Typical & Extreme Meteorological Year and Heatwaves* [Preprint]. Atmosphere – Meteorology. https://doi.org/10.5194/essd-2021-401

Doutreloup, S., Wyard, C., Amory, C., Kittel, C., Erpicum, M., & Fettweis, X. (2019). Sensitivity to Convective Schemes on Precipitation Simulated by the Regional Climate Model MAR over Belgium (1987–2017). *Atmosphere*, *10*(1), 34. https://doi.org/10.3390/atmos10010034

Doutreloup, Sébastien, & Fettweis, X. (2021). *Typical & Extreme Meteorological Year and Heatwaves for Dynamic Building Simulations in Belgium based on MAR model Simulations* (version 1.0.0) [Data set]. Zenodo. https://doi.org/10.5281/ZENODO.5606983

Elnagar, E., Doutreloup, S., & Lemort, V. (2021, September 1). Modeling the Impact of Climate Change on the Future Heating Demand in Different Types of Buildings in the Belgian Residential Building Stock. *Proceeding of the International Building Simulation Conference*. Building Simulation 2021 Conference. https://orbi.uliege.be/handle/2268/263135

Elnagar, E., Zeoli, A., & Lemort, V. (2022). Performance Evaluation of Passive Cooling in a Multi- Zone Apartment Building Based on Natural Ventilation. *CLIMA 2022 Conference*. https://doi.org/10.34641/clima.2022.263

Exigences PEB du 1er juillet 2019 au 31 décembre 2020. (2019). https://energie.wallonie.be/fr/exigences-peb-du-1er-juillet-2019-au-31-decembre-2020.html?IDD=136456&IDC=7224

Eyring, V., Bony, S., Meehl, G. A., Senior, C. A., Stevens, B., Stouffer, R. J., & Taylor, K. E. (2016). Overview of the Coupled Model Intercomparison Project Phase 6 (CMIP6) experimental design and organization. *Geoscientific Model Development*, *9*(5), 1937–1958. https://doi.org/10.5194/gmd-9-1937-2016

FETTWEIS, X., WYARD, C., DOUTRELOUP, S., & BELLEFLAMME, A. (n.d.). *Noël 2010 en Belgique: Neige en Flandre et pluie en Haute-Ardenne* [Text]. https://popups.uliege.be/0770-7576. Retrieved January 20, 2022, from https://popups.uliege.be/0770-7576/index.php?id=4568&format=print

Hamdy, M., Carlucci, S., Hoes, P.-J., & Hensen, J. L. M. (2017). The impact of climate change on the overheating risk in dwellings—A Dutch case study. *Building and Environment*, *122*, 307–323. https://doi.org/10.1016/j.buildenv.2017.06.031

International Organization for Standardization. (2017). *ISO 17772-1:2017—Energy performance of buildings—Indoor environmental quality*. ISO. https://www.iso.org/cms/render/live/en/sites/isoorg/contents/data/standard/06/04/60498.html

Invidiata, A., & Ghisi, E. (2016). Impact of climate change on heating and cooling energy demand in houses in Brazil. *Energy and Buildings*, *130*, 20–32. https://doi.org/10.1016/j.enbuild.2016.07.067

Karl, T. R., Melillo, J. M., & Peterson, T. C. (2009). *Global climate change impacts in the United States*. Cambridge University Press. https://www.cabdirect.org/cabdirect/abstract/20103085717

Nakicenovic, N., Alcamo, J., Davis, G., Vries, B. de, Fenhann, J., Gaffin, S., Gregory, K., Grubler, A., Jung, T. Y., Kram, T., Rovere, E. L. L., Michaelis, L., Mori, S., Morita, T., Pepper, W., Pitcher, H., Price, L., Riahi, K., Roehrl, A., … Zhou, D. (2000). *Special Report on Emissions Scenarios. A SpecialReport of Working Group III of the Intergovernmental Panel on ClimateChange*. Cambridge University Press, Cambridge, UK and New York, NY, USA,. https://escholarship.org/uc/item/9sz5p22f

Radhi, H. (2009). Evaluating the potential impact of global warming on the UAE residential buildings – A contribution to reduce the CO2 emissions. *Building and Environment*, *44*(12), 2451–2462. https://doi.org/10.1016/j.buildenv.2009.04.006

Riahi, K., van Vuuren, D. P., Kriegler, E., Edmonds, J., O'Neill, B. C., Fujimori, S., Bauer, N., Calvin, K., Dellink, R., Fricko, O., Lutz, W., Popp, A., Cuaresma, J. C., Kc, S., Leimbach, M., Jiang, L., Kram, T., Rao, S., Emmerling, J., … Tavoni, M. (2017). The Shared Socioeconomic Pathways and their energy, land use, and greenhouse gas emissions implications: An overview. *Global Environmental Change*, *42*, 153–168. https://doi.org/10.1016/j.gloenvcha.2016.05.009

Ridder, K. D., & Gallée, H. (1998). Land Surface–Induced Regional Climate Change in Southern Israel. *Journal of Applied Meteorology and Climatology*, *37*(11), 1470–1485. https://doi.org/10.1175/1520-0450(1998)037<1470:LSIRCC>2.0.CO;2

Ruosteenoja, K., & Räisänen, P. (2013). Seasonal Changes in Solar Radiation and Relative Humidity in Europe in Response to Global Warming. *Journal of Climate*, *26*(8), 2467–2481. https://doi.org/10.1175/JCLI-D-12-00007.1

The Bureau for Standardisation in Belgium. (2021, June). *Standard NBN EN 15193-1:2017—Energy performance of buildings—Energy requirements for lighting*. https://www.nbn.be/shop/en/standard/nbn-en-15193-1-2017-a1-2021_104645/

Ürge-Vorsatz, D., Cabeza, L. F., Serrano, S., Barreneche, C., & Petrichenko, K. (2015). Heating and cooling energy trends and drivers in buildings. *Renewable and Sustainable Energy Reviews*, *41*, 85–98. https://doi.org/10.1016/j.rser.2014.08.039

Wang, H., & Chen, Q. (2014). Impact of climate change heating and cooling energy use in buildings in the United States. *Energy and Buildings*, *82*, 428–436. https://doi.org/10.1016/j.enbuild.2014.07.034

Wyard, C., Doutreloup, S., Belleflamme, A., Wild, M., & Fettweis, X. (2018). Global Radiative Flux and Cloudiness Variability for the Period 1959–2010 in Belgium: A Comparison between Reanalyses and the Regional Climate Model MAR. *Atmosphere*, *9*(7), 262. https://doi.org/10.3390/atmos9070262

Wyard, C., Scholzen, C., Doutreloup, S., Hallot, É., & Fettweis, X. (2021). Future evolution of the hydroclimatic conditions favouring floods in the south-east of Belgium by 2100 using a regional climate model. *International Journal of Climatology*, *41*(1), 647–662. https://doi.org/10.1002/joc.6642

ACKNOWLEDGEMENT

This research was partially funded by the Walloon Region under the call 'Actions de Recherche Concertées 2019 (ARC)' and the project OCCuPANt, on the Impacts Of Climate Change on the indoor environmental and energy PerformAnce of buildiNgs in Belgium during summer. The authors would like to gratefully acknowledge the Walloon Region and Liege University for funding. This study is a part of the International Energy Agency Annex 80 project activities to define resilient cooling in residential buildings.

Comparison study of high-performance rule-based HVAC control with deep reinforcement learning-based control in a multi-zone VAV system

Xing Lu[1], Yangyang Fu[1], Shichao Xu[2], Qi Zhu[2], Zheng O'Neill[1]*, Zhiyao Yang[1]

[1]Texas A&M University,
College Station, Texas, USA

[2]Northwestern University,
Evanston, USA
* Corresponding Author

ABSTRACT

The design, commissioning, and retrofit of heating, ventilation, and air-conditioning (HVAC) control systems are crucially important for energy efficiency but often neglected. Generally, designers and control contractors adopt ad-hoc control sequences based on diffused and fragmented information and therefore the majority of the existing control sequences are diverse and sub-optimal. ASHRAE Guideline 36 (GDL36), High-performance Sequences of Operation for HVAC Systems, is thus developed to provide standardized and high-performance rule-based HVAC control sequences with the main focus on maximizing energy efficiency. However, these high-performance rules-based control sequences are still under-development, and only a few studies verify their overall effectiveness. In addition, the performance evaluations in most existing studies only focus on the energy-saving potentials compared with the conventional rule-based control strategies. In this study, the high-performance rule-based control sequences from GDL36 was compared to the state-of-the-art deep reinforcement (DRL) control in terms of the energy efficiency in a multi-zone VAV system. The system-level supervisory controls (i.e., supply air temperature and supply differential pressure setpoint) in ASHRAE GDL36 were replaced by the counterpart in the DRL control, of which action space is a bi-dimensional continuous space. A five-zone medium office building model in Modelica was utilized as a virtual testbed. Particularly, the plant side power consumption uses a regression model to reflect the real condition of the plant loop operation, Proximal policy optimization (PPO) was selected as the DRL algorithm due to its stable performance for the continuous space and easiness of the hyper-parameter tuning. The DRL algorithm was implemented using the Tianshou library in Python. A containerized OpenAI gym environment was leveraged to enable the connection between the Modelica building model and the DRL algorithm. Typical load conditions in Chicago, 5A (high and mild load weeklong simulation) were considered. The simulation results show that control sequences from GDL36 perform comparable performance in terms of energy efficiency and thermal comfort as the DRL controls.

1. INTRODUCTION

Innovations in building controls at the supervisory level have great potential to achieve the whole-building level energy savings on the order of 30% and higher (Pritoni et al., 2020). Despite the significant role of the HVAC control systems in energy efficiency, its design, commissioning, and retrofit have long been an intricate and complicated issue, considering that only diffuse and fragmented information on system operation is available for decision making in most of the scenarios. Due to this limitation, designers and control contractors can only rely on ad-hoc control sequences for system operation in practice, which is one of the major reasons why buildings are operated sub-optimally. To provide standardized and high-performance HVAC control sequences, the American Society of Heating, Refrigerating and Air-Conditioning Engineers (ASHRAE) has developed the Guideline 36 (GDL36) High Performance Sequences of Operation (SOO) for HVAC Systems to maximize energy efficiency.

Control sequences collected in GDL36 belong to prescriptive and feedback-based reactive control. In this type of control, different setpoints or schedules are determined at the supervisory level based on the heuristic rules and then the Proportional-Integral-Derivative (PID) local controls are used to track the setpoints. The energy efficiency performance of GDL36 has been demonstrated by limited lab/field tests and simulation-based studies (Paliaga et al., 2020; Pritoni et al., 2020; Wetter et al., 2018), as tabulated in Table 1. Rodriguez (Rodriguez, 2019) conducted an experiment in a commercial building air handling unit (AHU) and showed that more outside air was used and building cooling energy use was reduced after GDL36 was implemented. The control retrofit using the GDL36 airside SOO in 555 County Center (Talyor Engineering, 2020), a five-story office building in California, indicated that 15% of whole building electricity use and 56% of natural gas use were saved in the first year of the testing. A medium building in Vallejo, California, was retrofitted by implementing GDL36 SOO (Kiriu & Stein, 2021). The control sequence retrofits reduced annual energy bills by over $200,000 and heating energy use by more than 55%. These two field demonstrations showed a simple payback year of 6.7 and 8.9 years, respectively.

Wetter et al. (Wetter et al., 2018) implemented the airside GDL36 SOO in a single-floor medium office building model and reported a 30% annual site energy usage saving for the HVAC system with acceptable thermal comfort compared to the old SOO published in 2006 by ASHRAE (ASHRAE, 2005). Zhang et al. (Zhang et al., 2020) implemented and verified both airside and waterside control sequences in a Modelica-based simulation environment for a single-zone VAV system. Their simulation results showed that the GDL36 yielded 17.3 % of annual HVAC energy saving compared to the conventional baseline control strategy. In a follow-up study (Zhang et al., 2022), they estimated the energy saving of the control retrofit for multi-zone variable air volume systems using Spawn of EnergyPlus. The results showed the GDL36 SOO could provide a wide range of HVAC energy savings with an average savings of 31% in different climates, internal loads, and HVAC system operation periods. Overall, the energy-saving potential from retrofitting existing controls to the GDL36 SOO has yet to be shown for a wide range of cases (Zhang et al., 2022).

Table 1: Summary of existing evaluation studies of GDL36 SOO on energy efficiency

Study	Approach	Test Conditions	Baseline	Results
(Wetter et al., 2018)	Simulation	Modelica-based single-floor five-zone medium office VAV system (airside)	Typical rule-based	- 30% annual site energy use of the HVAC system with acceptable thermal comfort.
(Zhang et al., 2020)	Simulation	Modelica-based single-zone system (plant & airside)	Typical rule-based	- 17.3% of annual HVAC energy saving with acceptable thermal comfort.
(Zhang et al., 2022)	Simulation	21-zone VAV system (airside)	Typical rule-based	- an average of 31% HVAC energy saving
(Rodriguez, 2019)	Field test	AHU in a commercial building	Not mentioned	- More outside air was used and building cooling energy use was reduced.
(Talyor Engineering , 2020)	Field test	555 County Center, five-story office building, CA (Airside)	Not mentioned	- In the first year, 15% of whole building electricity use and 56% of natural gas use. - Estimated payback period was 6.7 years.
(Kiriu & Stein, 2021)	Field test	A medium hospital building in Vallejo, CA (Airside & Waterside)	Not mentioned	- Reduced annual energy bills by over $200,000 and heating energy use by more than 55%. - Estimated payback period was 8.9 years.

The reviewed literature indicates the design and commissioning of the HVAC system control sequences could have a substantial impact on building energy efficiency. The airside control retrofits following high-performance SOO could achieve energy savings up to 30% and more compared to the conventional rule-based control. However, achievable savings from the high-performance SOO in other HVAC system parts are still unknown (e.g., airside controls together with waterside controls). In addition, the energy saving potential of high-performance SOO compared to the state-of-the-art intelligent controls as the benchmark is also unclear. Therefore, there is a practical need to benchmark the SOO in GDL36 with other intelligent controls such as deep reinforcement learning (DRL)-based control.

Reinforcement learning (RL) is a category of machine learning algorithms that aims to learn an optimal control policy from the direct interaction between the agent and the environment. The agent will perform empirical learning and decide on the action to drive the environment towards a favorable trajectory according to a predefined reward function. Different researchers have investigated the application of RL controllers in building systems, including single air-conditioning units (Costanzo et al., 2016; Leurs et al., 2016), VAV systems (Azuatalam et al., 2020; Jia et al., 2019; Wang et al., 2017; Yu et al., 2020; Yuan et al., 2021), radiant heating systems (Chen et al., 2019; Zhang et al., 2019), building envelope(Chen et al., 2018), and whole HVAC systems (Ahn & Park, 2020; Liu & Henze, 2006, 2007; Yang et al., 2015). Despite the reported benefits after the successful controller tuning, the model-free RL controllers are subject to the issues such as the long training period and stability issues.

There exist several studies on developing RL strategies for commercial building VAV systems. (Yuan et al., 2021) applied Q-learning-based RL in both single-zone and multi-zone VAV systems to optimize the supply airflow rate. Despite achieved energy saving over the baseline control, details on the baseline control are unclear. In addition, it is not clear whether the control design of AHU-level control loops is considered or not. Wei et al. (Wei et al., 2017) used deep Q network algorithms for optimal airflow control of multi-zone VAV systems. The simulation experiments demonstrated the DRL-based algorithm was more effective in an energy cost reduction compared with the rule-based controllers. Hanumaiah et al. (Hanumaiah & Genc, 2021) proposed a distributed multi-agent DRL framework for the optimal control of multi-zone systems. The impact of the reward ratio and the weather on different model-free RL performances was discussed. Ding et al. (Ding et al., 2020) developed a model-based RL with a model predictive path integral control method to the multi-zone VAV. The results showed that the proposed controller could achieve 10.65% more energy savings compared to the rule-based benchmark while maintaining similar thermal comfort. Furthermore, the training time was reduced significantly compared to the model-free RL benchmark.

Although aforementioned literature review demonstrates the energy saving potential of DRL, there exist few studies that compared the performance of DRL-based controllers with high-performance rule-based control sequences of operation, i.e., ASHRAE Guideline 36. To be specific, the benchmark control strategies in most existing evaluation studies are the PID and on-off controllers. For example, the benchmark rule-based controller for the zone air temperature (ZAT) controller is the on-off control (Wei et al., 2017). Essentially speaking, the GLD36 SOO is much simpler compared to the DRL controllers. The heuristic rules in GLD36 could improve the energy efficiency; however, the improvement may be constrained by the incapability of predictive and adaptive learning. OBC and RL controllers have their own obvious challenges and limitations, which prevent their wide applications in the field at the current stage. In this context, this study presents an energy performance comparison of GDL36 SOO and DRL-based control within a medium office building virtual testbed in Modelica. Section 3 describes the case study description and the simulation testbed. Section 4 details the formulation and implementation of the DRL-based controller (DRLC). Section 5 discusses the energy efficiency comparison results of the DRLC and GDL36.

2. DESCRIPTION OF CASE STUDY

This case study aims to compare energy efficiency and thermal comfort performance of DRLCs with ASHRAE Guideline 36 for a multi-zone VAV cooling system, which is a typical HVAC system configuration in commercial buildings. The energy efficiency is reflected by the cooling energy use for the whole HVAC system and total ZAT violation during the system operation hours (i.e., 7 am – 7 pm) is calculated as a thermal comfort metrics.

$$E_{Coo,tot} = \sum_{t_0}^{t_N} E_{HVAC}(t_i) = \sum_{t_0}^{t_N} E_{Fan}(t_i) + \sum_{t_0}^{t_N} E_{Plant}(t_i) \tag{1}$$

$$dt_{tot}(t_0, t_N) = \sum_{z \in Z} \sum_{t_0}^{t_N} |s_z(t_i)| \tag{2}$$

where N is the sampling number for each operation time step point t. E_{Fan} and E_{Plant} are the energy use for the AHU fan and the plant system. z is the zone index for the set of zones, and s_z is the deviation from the lower and upper setpoint temperatures. The zone air cooling temperature setpoint is 24 °C, and the allowable deviation in this study is ± 0.5 °C from the setpoint.

Since the published version of GDL36-2018 only contains the control sequences on the airside systems, this study focuses on the comparison of rule-based airside high-performance sequences with the intelligent controllers. Another assumption is that the comparison is conducted at the supervisory level, which is the overall control of the local subsystems (Wang & Ma, 2008). For the airside control SOO of multi-zone VAV systems in GDL36, there exist several critical supervisory level controls, e.g., AHU supply air temperature (SAT) reset and static differential pressure (DP) setpoint reset, and economizer damper controls. In this comparison study, the first two controls are replaced by the counterpart in two intelligent controllers. For the AHU SAT reset, the SAT is reset based on the outdoor air temperature (OAT) and the setpoint request from the zone terminal units to find a balance between the fan energy and cooling energy. To be specific, the setpoint shall be reset from minimum cooling SAT when the outdoor air temperature is maximum OAT and above, proportionally up to maximum SAT when the outdoor air temperature is minimum OAT and below. The maximum OAT is reset using T&R logic based on the zone-level reset requests between minimum and maximum cooling SAT. The static DP reset is enabled by the Trim & Respond control. Under that control logic, the system will tend towards minimum static pressure but respond to the increasing demand from the zone terminal units.

Compared to the high-performance SOO in GDL36, the SAT and the static DP are determined by control policy in DRLC. To ensure an apples-to-apples comparison, the local controls (e.g., zone-level PID controls) remain the same for the three controllers. Table 2 lists the differences of SOO between two controller types.

Table 2: Difference between two controller types

Supervisory control loop name	**GDL36**	**DRLC**
SAT setpoint reset	Rule-based, i.e., determined by the OAT and zone requests	Determined by the optimal control policy after the training
Static DP reset	Rule-based, i.e., determined from zone requests	

The simulation experiment was for Chicago, IL, USA (ASHRAE climate zone 5A) in two typical weeks of different cooling loads, i.e., a cooling week (07/24-07/31) and a shoulder week (06/09-06/16). The cooling week has a high average outdoor air temperature, and the shoulder week has a mild cooling load which enables a long operation period of the airside economizer. The simulated medium office building was the single-floor five-zone VAV system as described in the reference (Lu, Fu, et al., 2021) under both airside and waterside control sequences of GDL36. The original model was developed from the GDL36 model in Modelica Buildings Library 7.0.0 (Wetter et al., 2014). To ease the computational cost, the detailed waterside model was replaced by the data-driven regression model for both the cooling season and the shoulder season, respectively, as shown in Eq. (3) and (4).

$$P_{pla} = 11188 + 0.18 \cdot \mathrm{Q_{coo}} + 24.24 \cdot \mathrm{T_{db}} - 44.44 \cdot \mathrm{T_{wb}} - 0.05, \tag{3}$$

$$P_{pla} = 726656 + 4.12 \cdot Q_{coo} - 2816.1 \cdot T_{db} - 2638.7 \cdot T_{wb} + 26.2 \cdot H_{gh} - 0.0037 Q_{coo} \cdot T_{db} - 0.0097 Q_{coo} \cdot T_{wb} + 2.01 \cdot 10^{-5} \cdot Q_{coo} \cdot H_{gh} + 10.2 \cdot T_{db}\, T_{wb} + 0.16 \cdot T_{db} \cdot H_{gh} - 0.25 \cdot T_{wb} \cdot H_{gh}, \tag{4}$$

where Q_{coo} is the cooling load at the cooling coil, T_{db} is the dry bulb outdoor air temperature, $\mathrm{T_{wb}}$ is the wet bulb outdoor air temperature and H_{gh} is the global horizontal solar radiation. Table 3 shows the statistical metrics of the regression model. The coefficient of determination (R2), root mean square error (CV-RMSE), and normalized mean bias error (NMBE) both indicate the high accuracy of the regression models. As mentioned earlier, the system is sized under the ASHRAE climate zone 5A Chicago, IL (Lu, Adetola, et al., 2021; The U.S. Department of Energy (DOE), 2020).

Table 3: Regression model accuracy statistical results

Regression model	**Model type**	**R2**	**CV-RMSE**	**NMBE**
Cooling season model	Linear	0.99	2.1%	1.3%
Shoulder season model	Interactions linear	0.99	3.2%	2.3%

3. CONTROLLER FORMULATION AND IMPLEMENTATION

The DRLC was formulated to minimize the HVAC total energy consumption while mitigating the ZAT violation by adjusting the AHU SAT setpoint and the AHU static pressure setpoint, were bounded within [12 °C,18 °C] and [25 Pa,410 Pa]. Figure 1 depicts the DRLC formulation. The reward R for the DRL is shown in Eq. (5).

$$R_i = E_{HVAC,i} + \alpha \cdot dT_{vio,i}, \tag{5}$$

where $E_{HVAC,i}$ are the HVAC energy consumption and $dT_{vio,i}$ are and ZAT violation at the ith control interval (i.e., 15 minute each). α is the penalty coefficient that balances the energy consumption and thermal comfort. Generally, a small α corresponds to a smaller HVAC energy consumption but a larger temperature violation and vice versa. An appropriate α needs to be tuned to keep a similar level of ZAT violation with the GDL36 controller.

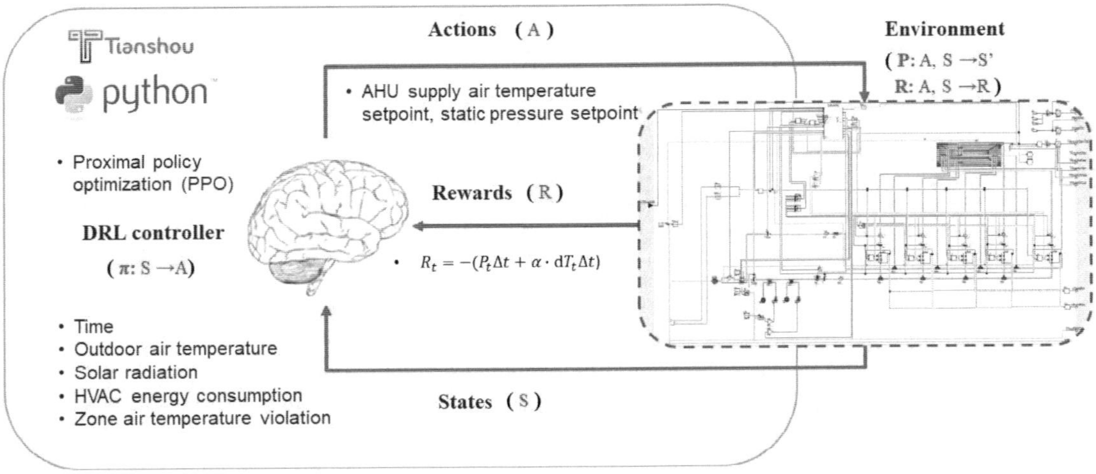

Figure 1: Schematics of DRLC formulation and implementation

The control actions are bidimensional and the action space is a continuous space to avoid the curse of the dimension in the discretized action space. The states are determined based on HVAC engineering knowledge. We consider three different combinations of states, as shown in Table 4. The commonly used states are time, outdoor air temperature, solar radiation, HVAC energy consumption, and ZAT violation. The simulation environment is the virtual medium office in Modelica which provides the reward value (R) and next observations (S') during the interaction with the DRLC.

Table 4: Summary of different state design

State Space	Number of States	State Variables
S1	7	time, outdoor air temperature, solar radiation, and HVAC energy consumption, ZAT violation, fan speed, maximum/minimum zone terminal damper position
S2	6	time, outdoor air temperature, solar radiation, HVAC energy consumption, ZAT violation, fan speed
S3	5	time, outdoor air temperature, solar radiation, HVAC energy consumption, and ZAT violation

For the implementation, a flexible containerized framework (Fu et al., 2021) was leveraged where the building model was interfaced with a state-of-the-art DRL library Tianshou (Weng et al., 2021) through the functional mockup unit (FMU). Tianshou is a highly modularized DRL library in Python based on pure PyTorch (Paszke et al., 2019) and has supported more than 20 classic algorithms. Tianshou's performances are reported to be comparable or better than the best reported results for most algorithms in the open literature (Weng et al., 2021). In this study, the Proximal Policy Optimization (PPO) (Schulman et al., 2017) is selected as the DRL algorithm because it suits the continuous bi-dimensional action space in our case. Still, several critical hyperparameters in PPO need to be tuned (Raval, 2021). For example, Step Per Collect (also called Time Horizon), i.e., how many steps to collect before adding it to the experience buffer; Batch size (also called Minibatch), i.e., how many experiences are used for each gradient descent update; Entropy coefficient i.e., a regularizer that helps the exploration; and Updated time, i.e., how many times the data collected are reused, etc. are hyperparameters that could have significant impacts on the DRLC performance.

To fine-tune the proposed DRLC, various other factors were considered, including the penalty coefficient, the state design, the neural network layer number, as suggested in this reference (Andrychowicz et al., 2020). The penalty coefficient α was first swept to determine the appropriate value that keeps a similar level of ZAT violation with the GDL36 controller. Then different common values are grid-searched in other aspects to find the best hyperparameters. Table 5 lists the sweeping parameters for tuning the DRLC. The DRL policy is trained and tested for 800 epochs in each scenario (i.e., different combinations of parameters). One epoch length is one week. The computation time for one epoch training takes around 10 minutes on a Windows 10 machine with Intel® Core™ i5-9500 @3.00 GHz CPU and 16 GB RAM. That being said, the training time for 800 epochs in single scenario would take about 5.5 days. Due to the large computation cost for the hyperparameter tuning of the DRLC, different scenarios are assigned to different cores in the high-performance computing clusters in Texas A&M University.

Table 5: Sweeping parameters for DRLC tuning considering various factors

Aspects	**State design**	**Entropy coefficient**	**Step per collect**	**Batch size**	**Repeat per collect**	**State normalization**	**Advantage normalization, Value Clip**	**Number of Neural Network Layers**
Value	S1, S2, S3	0, 0.01	384, 512	64, 128	5, 10	True, False	True, False	3, 4, 5
Number	3	2	2	2	2	2	2	3

4. RESULTS AND DISCUSSION

In this section, the experimental results for DRLC after the hyperparameter tuning are reported analyzed. Recall that for each scenario (i.e., the combination of the hyperparameters), the DRL policy has been trained and tested for 800 epochs (each epoch denotes one week).

4.1. Cooling Season Results

Figure 2 illustrates the rewards under different scenarios using the parallel coordinate plot. Each line denotes one epoch under different combinations of hyperparameters. The line color represents the reward value. The redder represents a higher reward value, while the bluer represents the lower reward value. The reward value of the GDL36 SOO is annotated in the color bar on the right side. It can be seen that the DRLC under the cooling season needs to be trained for at least 300 epochs. The larger number of Step Per Collect and Repeat Per Collect is generally beneficial to the final rewards but increases the DRLC training time.

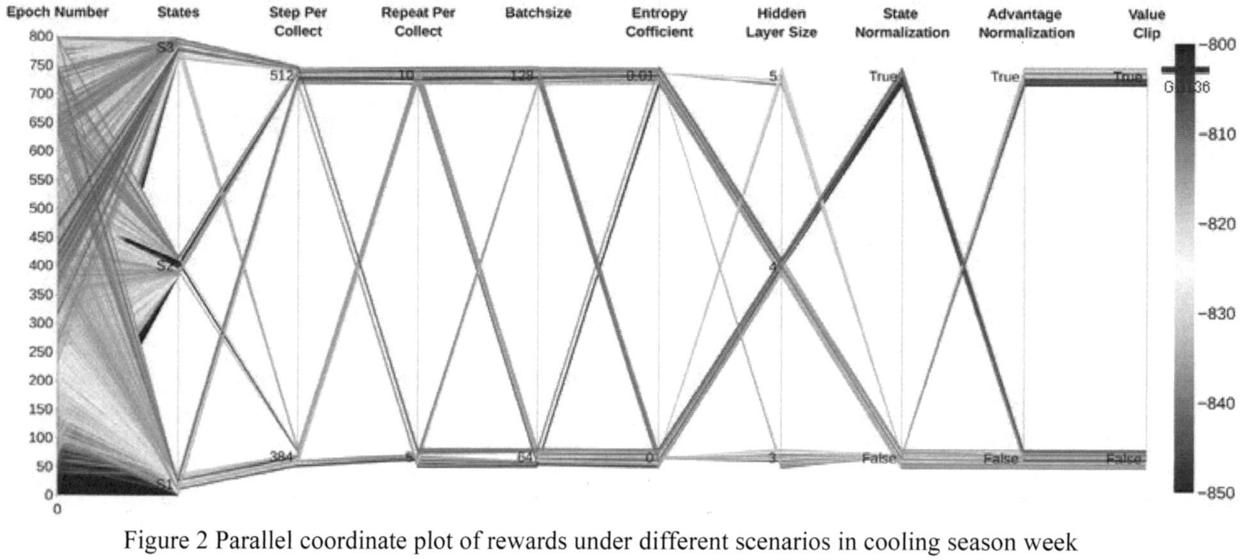

Figure 2 Parallel coordinate plot of rewards under different scenarios in cooling season week

Figure 3 shows the reward evolution throughout the epochs for the best scenario in the cooling season week. The hyperparameter setting for the best scenario is also provided. The blue line represents the entire reward for the GDL36 SOO. It can be seen that after 800 epochs of training, the DRLC performance in the cooling season week could nearly chase up with the GDL36 SOO in terms of the reward. The HVAC energy consumption for the best scenario of DRLC increases 2.2% compared to the GDL36 SOO while decreasing 0.83 K·h temperature violation in the cooling season week. This indicates the energy efficiency performance of GDL36 SOO is comparable to DRLC in the cooling week for this specific study. In addition, the DRLC is still subject to the curse of the high training time to achieve comparable performance.

State design	Entropy coeff.	Step per collect	Batch size	Repeat per collect	State normal.	Adv. normal.	Value Clip	DNN architecture
S1	0	512	128	10	False	False	False	[128]^4

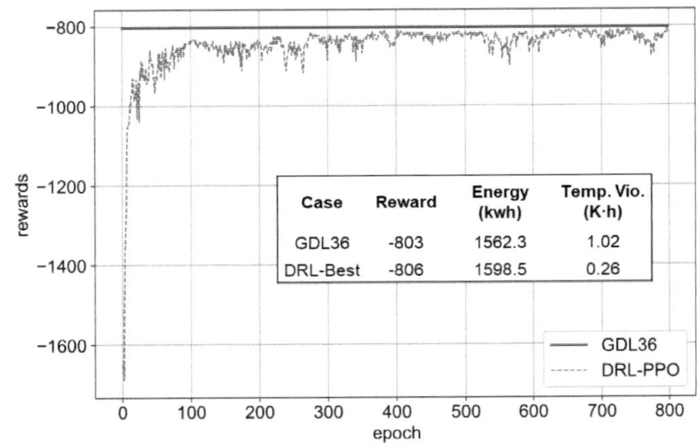

Figure 3: Reward per epoch for the best scenario in the cooling season week

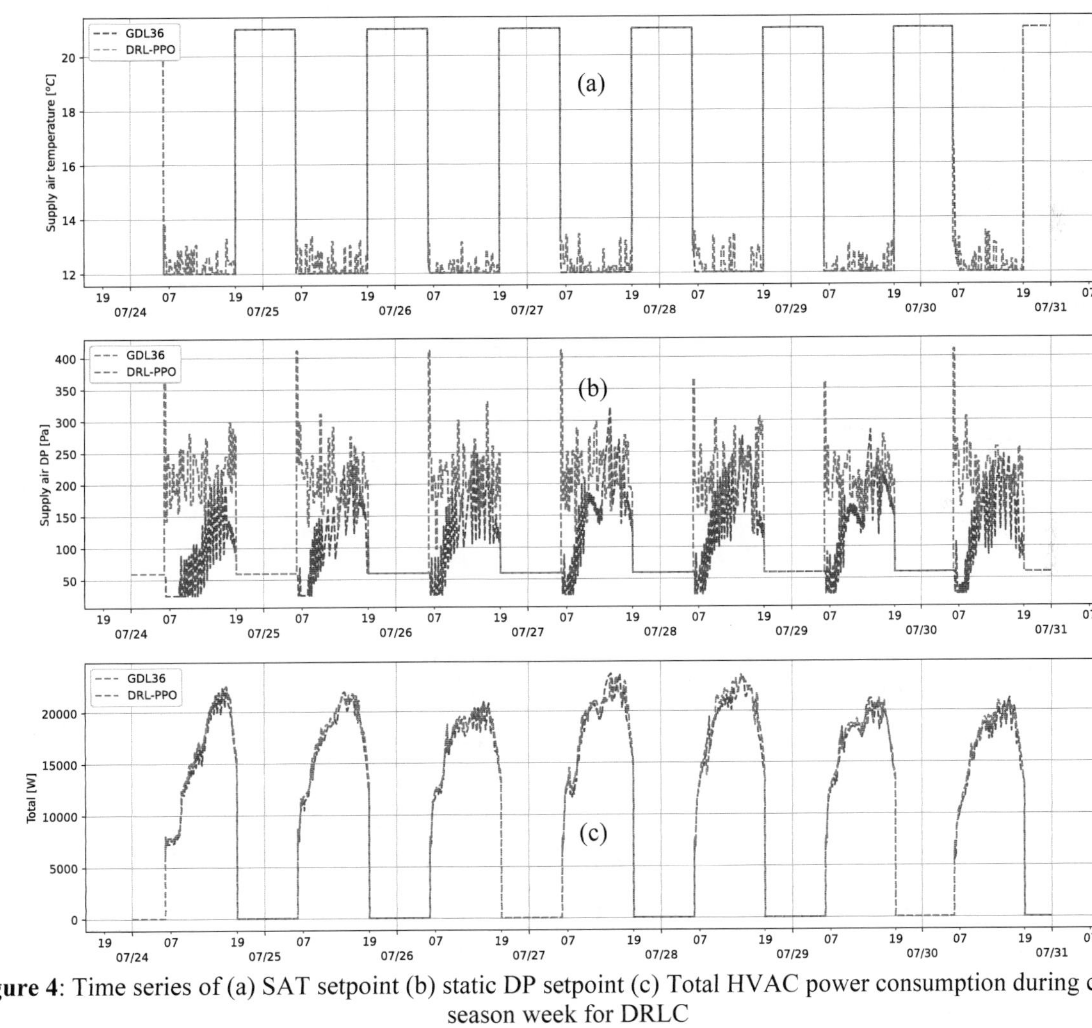

Figure 4: Time series of (a) SAT setpoint (b) static DP setpoint (c) Total HVAC power consumption during cooling season week for DRLC

Figure 4 depicts the detailed energy results for the DRLC in the cooling season week. Compared to the SAT setpoints in GDL36 staying at the lowest value (i.e., 12 °C), the SAT setpoints have a frequent variation between 12-14 °C in the case of DRLC. For the static DP setpoints, DRLC generally has a higher value throughout the operation hours. Figure 4(c) shows that the line of HVAC power consumption for DRLC overlap for most of the days with GDL36 SOO, while DRLC expending slightly less power consumption at some periods.

4.2. Shoulder Season Results

Similarly, the rewards under different scenarios are illustrated in Figure 5. The reward value of the GDL36 SOO is annotated in the color bar on the right side. Roughly after 150 epochs' training, DRLC under the shoulder season week could achieve an equivalent level performance with GDL36 SOO. Similarly, Figure 6 depicts the details regarding the best scenario in the shoulder season week. The best reward for the shoulder season week is -596, which is 4.64% higher than the baseline GDL36.

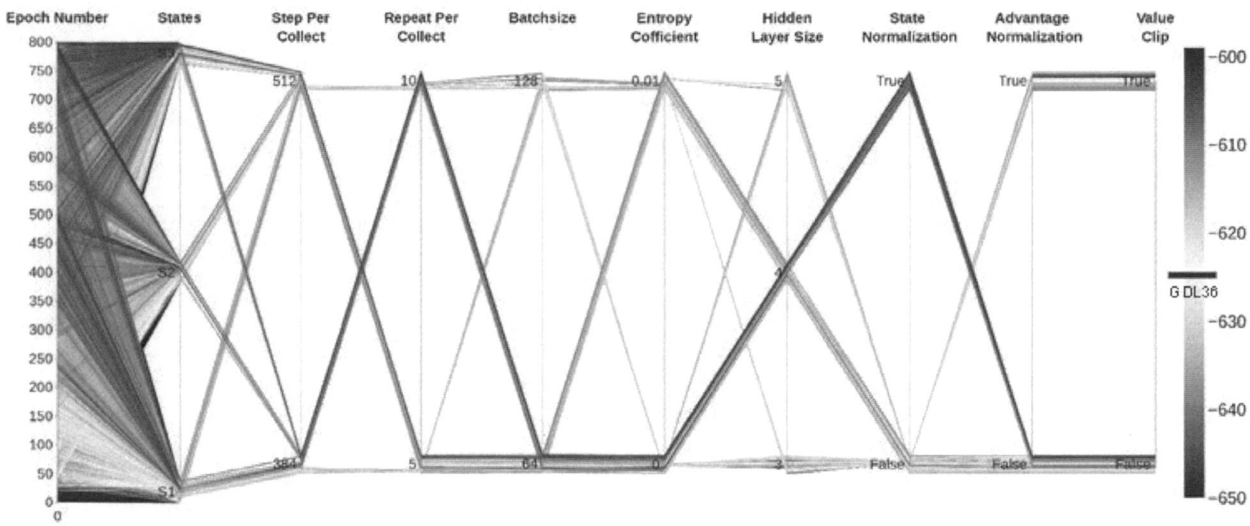

Figure 5: Parallel coordinate plot of rewards under different scenarios in shoulder season week

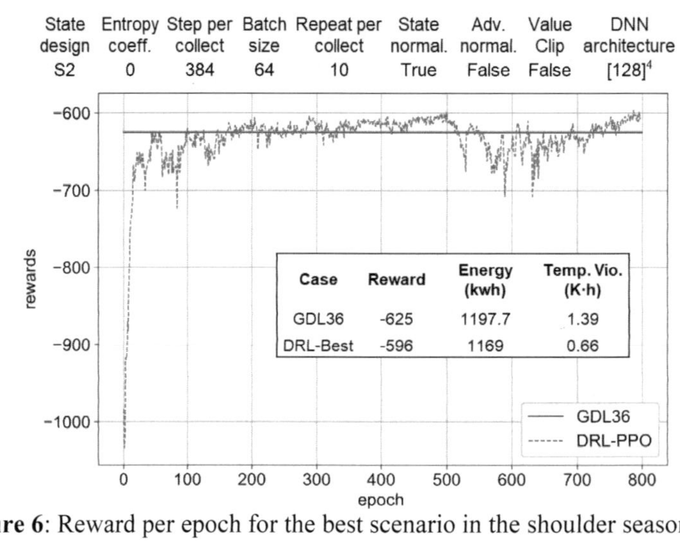

Case	Reward	Energy (kwh)	Temp. Vio. (K·h)
GDL36	-625	1197.7	1.39
DRL-Best	-596	1169	0.66

Figure 6: Reward per epoch for the best scenario in the shoulder season week

4.3. Summary

Based on the above results, following findings could be summarized. For the HVAC energy consumption in the cooling season week, the DRL consumes 2.3% more compared to GDL36 SOO. In the shoulder season week, the DRL save 2.4%. It is noted that this amount of saving is less than the penalties of uncertainties from typical sensor measurements in HVAC systems. For the thermal comfort metric, two controllers can all maintain the ZAT within the predefined comfort bounds with minor temperature violation. The DRLC have slightly less temperature violation than the GDL36 SOO.

5. CONCLUSIONS

ASHRAE GDL36 has demonstrated energy efficiency benefits over the conventional rule-based controls. In this study, the energy and thermal comfort performance of GDL36 are compared with DRLC. This study is conducted with a five-zone VAV cooling system virtual testbed in Chicago, IL. The baseline control system is implemented with the high-performance airside and waterside GDL36 SOO. The DRLC replace the airside supervisory level control loops. In other words, the optimal SAT and static DP are determined by trained control policy in DRLC. The DRLC was formulated to minimize the HVAC energy consumption and zone air temperature violations. The DRLCs with different hyperparameters in the PPO algorithm were studied and fine-tuned. The results showed that the GDL36 SOO has a comparable energy performance (within a 3% deviation) with DRLC in both high and mild cooling loads. For the thermal comfort metric, the GDL36 has slightly more ZAT violation in both typical weeks compared to DRLC.

For this case study, the GDL36 has demonstrated its comparable performance in terms of energy efficiency and thermal comfort with the two intelligent controllers. The GDL36 is good enough considering the complexity, long training time, and tuning efforts of the DRLC. However, there are several limitations of this study to be noted. First, the DRLC are formulated ideally only for theoretical comparison studies. They are not deployable for real applications. The control policies are trained and tested for the same week, which is also not realistic in practice. Second, the performance might be further improved by considering more complex aspects. For DRLCs, only the PPO algorithm is explored and other DRL algorithms are not studied. Third, the simulation-based study is only experimented in a five-zone medium office building, one single climate zone, and only cooling season. The effect of the climate, building type, internal loads, and operation time on the final results are not investigated. Therefore, the future work includes the expansion of the evaluation studies to other building types with different HVAC systems and climate zones; and the comparison studies for more complicated intelligent controllers.

REFERENCES

Ahn, K. U., & Park, C. S. (2020). Application of deep Q-networks for model-free optimal control balancing between different HVAC systems. *Science Technology for the Built Environment*, *26*(1), 61-74.

Andrychowicz, M., Raichuk, A., Stańczyk, P., Orsini, M., Girgin, S., Marinier, R., . . . Michalski, M. (2020). What matters in on-policy reinforcement learning? a large-scale empirical study. *arXiv preprint arXiv:2006.05990*.

ASHRAE. (2005). *Sequences of Operation for Common HVAC Systems*. American Society of Heating Refrigerating and Air-Conditioning Engineers.

Azuatalam, D., Lee, W.-L., de Nijs, F., & Liebman, A. (2020). Reinforcement learning for whole-building HVAC control and demand response. *Energy and AI*, *2*, 100020.

Chen, B., Cai, Z., & Bergés, M. (2019). Gnu-rl: A precocial reinforcement learning solution for building hvac control using a differentiable mpc policy. Proceedings of the 6th ACM international conference on systems for energy-efficient buildings, cities, and transportation,

Chen, Y., Norford, L. K., Samuelson, H. W., & Malkawi, A. (2018). Optimal control of HVAC and window systems for natural ventilation through reinforcement learning. *Energy and Buildings*, *169*, 195-205.

Costanzo, G. T., Iacovella, S., Ruelens, F., Leurs, T., & Claessens, B. (2016). Experimental analysis of data-driven control for a building heating system. *Sustainable Energy, Grids Networks*, *6*, 81-90.

Ding, X., Du, W., & Cerpa, A. E. (2020). Mb2c: Model-based deep reinforcement learning for multi-zone building control. Proceedings of the 7th ACM international conference on systems for energy-efficient buildings, cities, and transportation,

Fu, Y., Xu, S., Zhu, Q., & O'Neill, Z. (2021). Containerized framework for building control performance comparisons: model predictive control vs deep reinforcement learning control. Proceedings of the 8th ACM International Conference on Systems for Energy-Efficient Buildings, Cities, and Transportation,

Hanumaiah, V., & Genc, S. (2021). Distributed Multi-Agent Deep Reinforcement Learning Framework for Whole-building HVAC Control. *arXiv preprint arXiv:.13450*.

Jia, R., Jin, M., Sun, K., Hong, T., & Spanos, C. (2019). Advanced building control via deep reinforcement learning. *Energy Procedia*, *158*, 6158-6163.

Kiriu, R., & Stein, J. (2021). Medical Office Building Thrives With Advanced Control Sequences. *ASHRAE Journal*, *63*, 62-67.

Leurs, T., Claessens, B. J., Ruelens, F., Weckx, S., & Deconinck, G. (2016). Beyond theory: Experimental results of a self-learning air conditioning unit. 2016 IEEE International Energy Conference (ENERGYCON),

Liu, S., & Henze, G. P. (2006). Experimental analysis of simulated reinforcement learning control for active and passive building thermal storage inventory: Part 2: Results and analysis. *Energy and Buildings*, *38*(2), 148-161.

Liu, S., & Henze, G. P. (2007). Evaluation of reinforcement learning for optimal control of building active and passive thermal storage inventory.

Lu, X., Adetola, V., & O'Neill, Z. (2021). What are the Impacts on the HVAC System when it Provides Frequency Regulation?–A Comprehensive Case Study with a Medium Office Building. *Energy and Buildings*, 110995.

Lu, X., Fu, Y., O'Neill, Z., & Wen, J. (2021). A holistic fault impact analysis of the high-performance sequences of operation for HVAC systems: Modelica-based case study in a medium-office building. *Energy and Buildings*, *252*, 111448.

Paliaga, G., Singla, R., Snaith, C., Lipp, S., Mangalekar, D., Cheng, H., & Pritoni, M. (2020). Re-Envisioning RCx: Achieving Max Potential HVAC Controls Retrofits through Modernized BAS Hardware and Software.

Paszke, A., Gross, S., Massa, F., Lerer, A., Bradbury, J., Chanan, G., . . . Antiga, L. (2019). Pytorch: An imperative style, high-performance deep learning library. *Advances in neural information processing systems*, *32*.

Pritoni, M., Prakash, A., Blum, D., Zhang, K., Tang, R., Granderson, J., . . . Paliaga, G. (2020). Advanced control sequences and FDD technology. Just shiny objects, or ready for scale?

Raval, S. (2021). *Best Practices when training with PPO.* https://github.com/llSourcell/Unity_ML_Agents/blob/master/docs/best-practices-ppo.md

Rodriguez, A. (2019). *Utilizing ASHRAE G36 for Free Outside Air Cooling (Economizer)* The George Washington University].

Schulman, J., Wolski, F., Dhariwal, P., Radford, A., & Klimov, O. (2017). Proximal policy optimization algorithms. *arXiv preprint arXiv:.06347*.

Talyor Engineering. (2020). *Case Study: Advanced HVAC Controls*. https://taylorengineers.com/wp-content /uploads/2020/04/2018-09-18-Advanced-HVAC-Controls-Case-Study-555-County-Center.pdf

The U.S. DepartmentofEnergy(DOE). (2020). *DOE Commercial Prototype Building Models* https://www. energycodes.gov/development/commercial/prototype_models

Wang, S., & Ma, Z. (2008). Supervisory and optimal control of building HVAC systems: A review. *HVAC & R Research*, *14*(1), 3-32.

Wang, Y., Velswamy, K., & Huang, B. (2017). A long-short term memory recurrent neural network based reinforcement learning controller for office heating ventilation and air conditioning systems. *Processes*, *5*(3), 46.

Wei, T., Wang, Y., & Zhu, Q. (2017). Deep reinforcement learning for building HVAC control. Proceedings of the 54th annual design automation conference 2017,

Weng, J., Chen, H., Yan, D., You, K., Duburcq, A., Zhang, M., . . . Zhu, J. (2021). Tianshou: a Highly Modularized Deep Reinforcement Learning Library. *arXiv preprint arXiv:.14171*.

Wetter, M., Hu, J., Grahovac, M., Eubanks, B., & Haves, P. (2018). OpenBuildingControl: Modeling feedback control as a step towards formal design, specification, deployment and verification of building control sequences. Building Performance Modeling Conference and SimBuild,

Wetter, M., Zuo, W., Nouidui, T. S., & Pang, X. (2014). Modelica buildings library. *Journal of Building Performance Simulation*, *7*(4), 253-270.

Yang, L., Nagy, Z., Goffin, P., & Schlueter, A. (2015). Reinforcement learning for optimal control of low exergy buildings. *Applied Energy*, *156*, 577-586.

Yu, L., Sun, Y., Xu, Z., Shen, C., Yue, D., Jiang, T., & Guan, X. (2020). Multi-agent deep reinforcement learning for HVAC control in commercial buildings. *IEEE Transactions on Smart Grid*, *12*(1), 407-419.

Yuan, X., Pan, Y., Yang, J., Wang, W., & Huang, Z. (2021). Study on the application of reinforcement learning in the operation optimization of HVAC system. Building Simulation,

Zhang, K., Blum, D., Cheng, H., Paliaga, G., Wetter, M., & Granderson, J. (2022). Estimating ASHRAE Guideline 36 energy savings for multi-zone variable air volume systems using Spawn of EnergyPlus. *Journal of Building Performance Simulation*, *15*(2), 215-236.

Zhang, K., Blum, D. H., Grahovac, M., Hu, J., Granderson, J., & Wetter, M. (2020). *Development and Verification of Control Sequences for Single-Zone Variable Air Volume System Based on ASHRAE Guideline 36.*

Zhang, Z., Chong, A., Pan, Y., Zhang, C., & Lam, K. P. (2019). Whole building energy model for HVAC optimal control: A practical framework based on deep reinforcement learning. *Energy and Buildings*, *199*, 472-490.

Smart Low-cost Thermal Imaging Acquisition Towards Personal Comfort Prediction

Ati Soleimanijavid[1]*, Iason Konstantzos[1]

[1]Durham School of Architectural Engineering and Construction, University of Nebraska - Lincoln, Omaha, NE, USA
asoleimanijavid2@huskers.unl.edu – iason.konstantzos@unl.edu

* Corresponding Author

ABSTRACT

Ensuring occupants' thermal comfort is rapidly becoming an essential objective in building design and operation, as it plays a crucial role in well-being and productivity. Conventional HVAC system controllers operate based on predefined setpoints and schedules, or individuals' selections. However, as thermal sensation may vary, not only from person to person but also differ over time, the automated operation needs to be informed by real-time sensing. In recent years, advancements in deep learning models have provided an opportunity to deploy vision-based sensors in buildings toward occupant-centric controls (OCC). Vision-based systems are popular devices used to monitor individual thermal sensation and satisfaction due to their capacity to non-intrusively measure skin temperature, a physiological variable that is related to thermal comfort prediction. However, this advantage of remote sensing also leads to reduced accuracy compared to conventional temperature sensors. One of the critical variables responsible for the reduced accuracy is the camera's distance from the subject, also known as 'the working distance'. As the requirement of thermal cameras in front of the targets within a fixed distance is not applicable to real operational conditions, this study proposes an experimental framework to perform real-time correction of the thermal camera's temperature for targets at the distance longer than the thermal camera's calibration distance. The prediction framework uses a low-cost thermal camera and an RGB-D module to extract the target's surface temperature and their distance to the camera, and its output can be used to assess an individual's thermal comfort based on skin temperature variation. This approach can be combined with computer vision approaches to allow the continuous detection of the occupants' faces or motion patterns, providing a holistic, multi-modal sensing solution towards occupant-centric controls.

1. INTRODUCTION

A thermal camera is a non-intrusive tool used to measure surface temperatures quantitatively and/or qualitatively by displaying warm objects against a cooler background. In thermography, thermal cameras detect infrared radiation and produce images that show temperature variations across an object. In order to measure the temperature of any target, radiometric thermal cameras measure the radiant energy and create a thermal image, in which the temperature of each pixel is determined based on the radiant energy. Microbolometers are commonly used as detectors in thermal cameras due to their lower cost. They measure radiation in the long-wavelength infrared band. The use of Non-Contact Infrared Thermometers (NCIT) and IR thermal cameras has been extended for detecting higher body temperatures in public settings during infectious disease pandemics including the recent COVID-19 pandemic. After the SARS pandemic in 2003, International Organization for Standardization (ISO), The International Electrotechnical Commission (IEC), and other governing agencies like U.S. Food and Drug Administration (FDA) developed standards to ensure the success of relevant screening programs (IEC, 2017; FDA, 2019). Medical purposes require high-accuracy thermal cameras, however, there have been attempts to use low-cost thermal cameras to capture the variation of skin temperature parameters to address occupants' comfort status, paving the way to extend the use of such technologies for human comfort studies (Li *et al.*, 2018). However, the use of thermal cameras also includes uncertainties, such as the emissivity of the target, intrinsic errors, and systematic errors in the measurement procedure. To obtain the actual temperature from the apparent temperature measured by thermal cameras, it is required to compensate for the emissivity of the measured object, while there has been research proving that changing the working distance and angle of view of a thermal camera compared to the factory-calibrated values has a negative impact on accuracy. More specifically, (Cheung et al., 2012) experimental results showed that increasing the working distance between the target and the camera decreases the reading temperature by 0.26° C per meter, while (Litwa, 2010) showed that, as long as the angle of view remains lower than 50 degrees, the temperature measurement results can be reliable. In that direction, correction models were proposed to compensate for the effect of distance and angle of view. Zhang et al. (2016) proposed a theoretical formula to investigate the effect of the angle of view and working distance on the accuracy of thermal cameras. The proposed theory was validated by fitting experimental data to blackbody temperature. The proposed distance correction theory applied two working distance ranges. For the working distance of less than 7 meters, a linear relation was considered for the distance correction factor, however, a quadratic relation was proposed for the correction factor for distances larger than 7 meters. Gutierrez (2020) study was the first attempt to use low-cost handheld thermal cameras for distance correction experiments. Although the low-cost thermal cameras included higher noise and less accuracy, their compensation formula based on working distance and angle of view was able to improve thermal camera accuracy from a range of (-2.56°C to +2.31°C) at the baseline to a range of (-1.44°C to +1.6°C) for the narrow working distances considered in the study (10-50 cm). Also, Li et al. (2019) proposed a linear model for distance compensation of low-cost thermal cameras within 0.8 and 2 meters working distance while using humans as the target.

As demonstrated in a previous study, high-cost instruments such as an expensive thermal camera and/or a black body target were used for distance calibration. Human targets are also used for camera calibration purposes when using a low-cost thermal camera that cannot provide a uniform and constant temperature target.

This proven dependency of accuracy on working distance pointed out the fact that automatic distance detection is the next milestone required for the integration of thermal cameras as automated smart sensing systems. RGB-depth sensors are primarily used for distance detection, but they are also useful in human-centered applications, aiding in human detection and tracking. To be utilized in a system with a thermal camera, a registration process is required to locate the corresponding points for the two sensors that have different optical characteristics and are not in the exact same position. The computation time, complexity, and required robustness of the registration method used can all differ. The research on multi-modal registration approaches has advanced rapidly, with studies on the fusion of thermal and RGB cameras ranging from medical applications (Gonzalez-prez et al.,2021; Muller et al., 2018; Ma e al.,2019) to low-cost 3D thermal model reconstruction for monitoring and inspection purposes (Yang et al.,2018; Aryal et al., 2019). A comparison of four different feature and intensity-based image registration algorithms for diabetic foot monitoring (González-Pérez et al., 2021) found that using feature-based algorithms that rely on feature point detection can lead to higher errors, especially for low-cost thermal cameras. In contrast, intensity-based registration performed better than feature-based however they add the requirement of image segmentation.

Multi-modal stereo cameras have also been utilized in the field of occupant-centric building control applications (Li et al., 2019; Aryal et al., 2019) to measure human face skin temperature and comfort or towards glare assessment and controls (Kim et al., 2019). Off-the-shelf low-cost thermal and RGB cameras with calibration-based registration were used to extract temperature information from an object detected in an RGB image (Li et al.,2019), while the 2D transformation matrix was obtained by selecting control points for thermal and RGB cameras registration (Aryal et al.,2019).

Despite the previous research on multi-modal camera integration and correction methods for low-cost thermal cameras, the validity and applicability of the experimental methods should be investigated further. In this paper, a dual-camera registration and distance correction experiment for low-cost off-the-shelf thermal cameras is designed for various measuring distances. The stereo-calibration process for image registration and the influence of the working distance on thermal camera measurement accuracy is obtained and analyzed using the experimental results. For thermal camera calibration, we proposed a low-cost framework that did not require the use of a black-body or a human target. Based on our results, the proposed distance correction formula helps in reducing the thermal camera's temperature measurement error, while the fusion of the depth sensor allows for this correction to be seamless and automatic in real-world implementations. The distance correction model and image registration of thermal and RGB cameras will be valuable for vision-based human thermal comfort assessment, while the accuracy of low-cost thermal cameras remains low.

2. METHODOLOGY

This study employs two techniques to establish a framework that will allow assessing comfort using a low-cost-thermal camera and a depth sensor: (i) Computer vision to register two cameras so that the corresponding pixels from two different cameras can be found; and (ii) Distance Calibration of a thermal camera to compensate for the effect of distance on thermal camera accuracy through a correction function. Below, the main components are summarized:

a. Low-cost thermal camera: A low-cost off-the-shelf thermal camera with an uncooled Vox microbolometer detector was chosen for this study. This long-wave infrared camera (LWIR) can detect electromagnetic wavelengths in the range of 8 to 14 μm and produces images with a resolution of 160 by 120 pixels, the thermal camera's radiometric feature can measure the temperature of each pixel based on the intensity of the infrared signal received by the camera (Table 1).

Table 1: Thermal camera specifications

Features	Descriptions
Dimensions	11.8x12.7x7.2mm
Array Format	160x120 pixels
FOV Horizontal	56
FOV Diagonal	71
Radiometric accuracy	±5∘C or ± 5% of reading in the working distance

The camera (Figure 1 left) was hosted in a custom 3D-printed case and connected to a chipset-on-board computer through a breakout board providing 3.3 input voltage for the camera, and the camera's output displays each pixel's temperature in Kelvin.

Figure 1: Thermal capture prototype (left) – Dual camera setup (right)

b. RGB-D Sensor: To extract depth and RGB images, a stereo-vision depth camera was utilized (Table 2). After capturing images from the left and right IR sensors, the sensor undergoes disparity and triangulation to calculate depth data.

Table 2: RGB-D sensor specifications

Features	Descriptions
Operating Range	0.3-3m
Depth FOV	87x58
RGB FOV	69x42
Depth resolution	Up to 1280x720
RGB frame resolution	Up to 1920x1080
Depth accuracy	<2% at 2m

The proposed dual-camera system consists of a thermal camera and RGB-D sensor rigidly placed on a mounting plate (Figure 1 right). As the resolution and Field of View (FOV) for these two image sensors differ, image registration was a necessary step to identify the corresponding pixels in their resulting images, as described below.

2.1 Sensor Fusion

For a multi-modal system, two different image registration approaches were used: feature-based and calibration-based. Key points were detected in feature-based, and then feature mapping and image transformation were discovered. The information from the two cameras, including their intrinsic and extrinsic matrices, is required for calibration-based image registrations. The Zhang (2000) method was used to map 3D coordinates into 2D image planes, with the camera modeled as a pinhole camera, in order to determine the intrinsic and extrinsic matrices.

The intrinsic matrix expresses the relation between 3D camera coordinates and the 2D image coordinates, while the extrinsic matrix shows the relation between the 3D world coordinates and the 3D coordinates of the camera. The extrinsic matrix includes rotation and translation matrices.

Figure 2 shows a schematic diagram for a stereo camera, where P is a part of a scene in 3D world coordinates and captured with a thermal camera on an image plane IR at point P_{IR} and a RGB-D camera on an image plane RGB at point P_{RGB}.

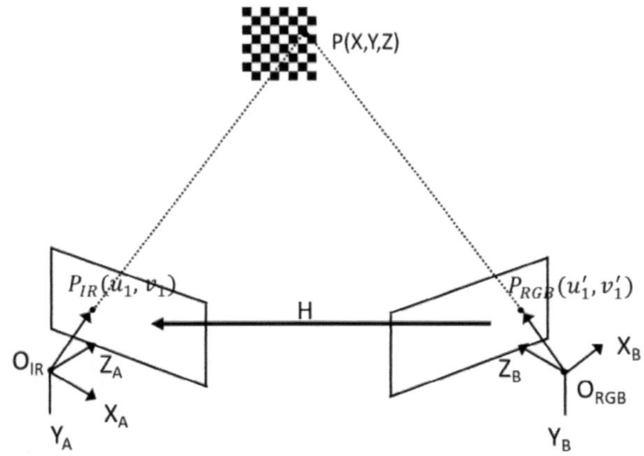

Figure 2: Stereo multi modal camera geometry

The coordinates of the corresponding point in image plane of IR as P_{IR} (u_1, v_1) and RGB as P_{RGB} (u_1', v_1') are obtained from the pinhole camera model (Zhang, 2000) in a stereo calibration process, Z_1 is the scaling factor. The Homography matrix shown in Equation (1), describes the transformation of points from one image to another image when the points lie in a plane. To find the Homography matrix, at least four pairs of corresponding points from thermal and RGB cameras are required (Swamidoss et al.,2021):

$$Z_1 \begin{bmatrix} u_1 \\ v_1 \\ 1 \end{bmatrix} = \begin{bmatrix} H_{11} & H_{12} & H_{13} \\ H_{21} & H_{22} & H_{23} \\ H_{31} & H_{32} & H_{33} \end{bmatrix} \begin{bmatrix} u_1' \\ v_1' \\ 1 \end{bmatrix}, \ H_{33} = 1 \tag{1}$$

$$u_1 = \frac{H_{11}u_1' + H_{12}v_1' + H_{13}}{H_{31}u_1' + H_{32}v_1' + 1} \tag{2}$$

$$v_1 = \frac{H_{21}u_1' + H_{22}v_1' + H_{23}}{H_{31}u_1' + H_{32}v_1' + 1} \tag{3}$$

If we have "i" corresponding points from stereo-calibration, we can change the form of Equations (2) and (3) to Equation (4), which incorporates all corresponding points:

$$
\begin{bmatrix} u_i' & v_i' & 1 & 0 & 0 & 0 & -u_i u_i' & -u_i v_i' \\ 0 & 0 & 0 & u_i' & v_i' & 1 & -v_i u_i' & -v_i v_i' \end{bmatrix}_{2iX8} \begin{bmatrix} H_{11} \\ H_{12} \\ H_{13} \\ H_{21} \\ H_{22} \\ H_{23} \\ H_{31} \\ H_{32} \end{bmatrix}_{8x1} = \begin{bmatrix} u_i \\ v_i \end{bmatrix}_{2iX1} \tag{4}
$$

In order to find the Homography matrix, the Least Square method is used to solve Equation (4).
For the stereo calibration experiment, an aluminum plate with 50cm x 40cm dimension with a chessboard pattern (8x7) on it was used. The square pattern on the chessboard had a 5cm dimension. Ten image pairs were taken from an aluminum chessboard plate by RGB-D and thermal camera in different orientations and distances from the camera. The results of the calibration are presented in Section 3.

2.2 Distance Calibration

To perform calibration for different working distances, the first step required setting up an experiment where temperature can be measured and controlled. To that end, a reptile pad was utilized, enabling to maintain the temperature at the desired temperature for an hour. As our end objective for this framework is the use for human skin temperature screening, this reptile pad was considered to be a good fit able to provide a temperature of 30-40 °C. The reptile pad was attached to an aluminum sheet painted with a flat black color spray to avoid reflection. The process started once the aluminum temperature reached 36°C, with the plate temperature being monitored through T-type thermocouples. The aluminum plate maintained its temperature within a margin of approximately 36°C, with the minor fluctuations being a result of the fluctuations of the reptile pad. The camera was fixed in place and the working distance with the target was varied from 50 cm to 2 meters in 10 cm increments. As a result, 160 thermal images were captured and processed to obtain a correction function as presented in Section3.

3. RESULTS AND DISCUSSION

Figure 3 shows the detected points of the reference chessboard by the thermal camera (camera 1) and RGB camera (camera 2) that were used for the calibration process. Despite the low resolution of the low-cost thermal camera, the mean re-projection error was kept within 0.1 pixels (Figure 4). This shows the calibration was able to detect the corners of the chessboard for both the thermal and RGB images with high accuracy.

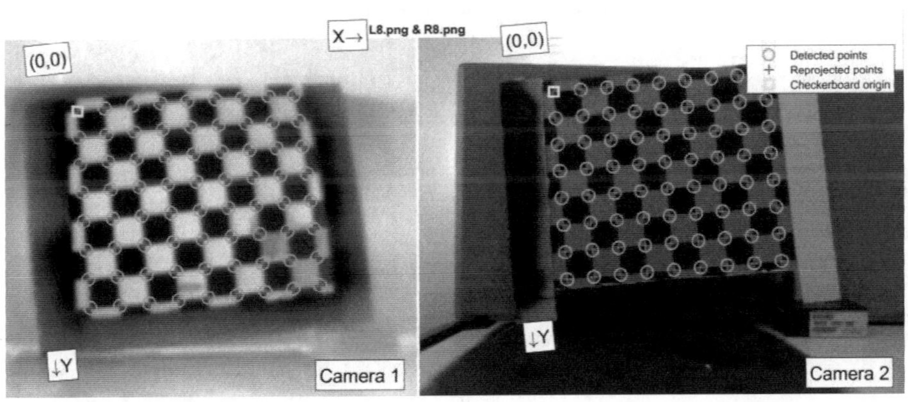

Figure 3: Detected and re-projected points after stereo calibration

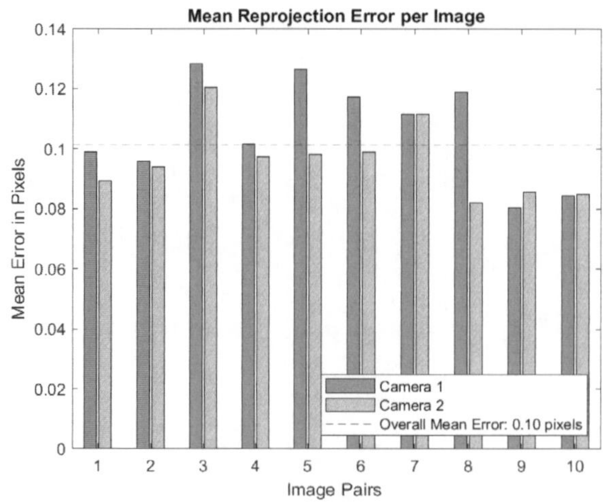

Figure 4: Mean re-projected error in pixels

The relative position and rotation of these two cameras with respect to the target orientation are shown in Figure 5, while Figure 6 reflects through rectified images how the corresponding points of thermal and RGB images are in the same row coordinates.

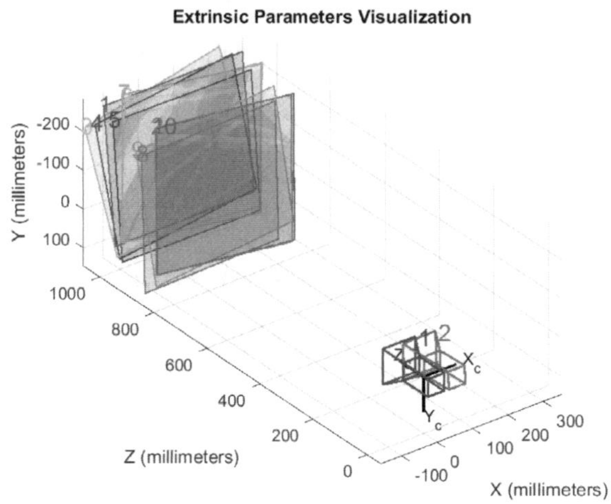

Figure 5: Extrinsic parameter visualization

Figure 6: Epipolar geometry of rectified images

The Homography matrix is calculated using the corresponding points, so by having Homography matrix, the associate points in RGB image in thermal image will be found.

$$H = \begin{pmatrix} 1.0407 & 0.0252 & 6.42 \\ -0.0016 & 1.0460 & 8.8858 \\ 0 & 0 & 1 \end{pmatrix}$$

After the successful fusion, the distance information obtained from the RGB-D sensor will be utilized to correct the resulting temperature of the thermal imaging base on the measured distance from the camera. To that end, Figure 7 illustrates an example of the image shooting process for distance correction.

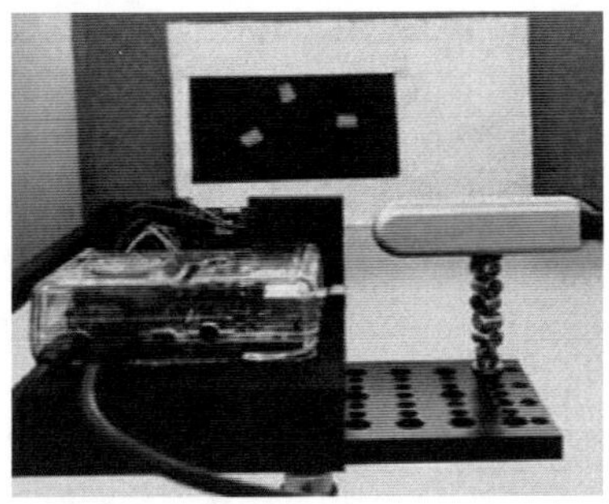

Figure 7: Experiment measurement

A noteworthy issue with low-cost thermal cameras is the occurrence of temperature peaks in standard frequencies. Figure 8 illustrates the issue, with the experience of peaks for the plate temperature reading from the thermal camera at 1m. These peaks are an inherent characteristic resulting from the Flat Filed Correction of low-cost thermal cameras and are necessary for temperature readings to cancel the impact of increasing temperature of the internal camera. After coordination with the manufacturers, these standard peaks were filtered out of the data, as they were out of the scope of our study.

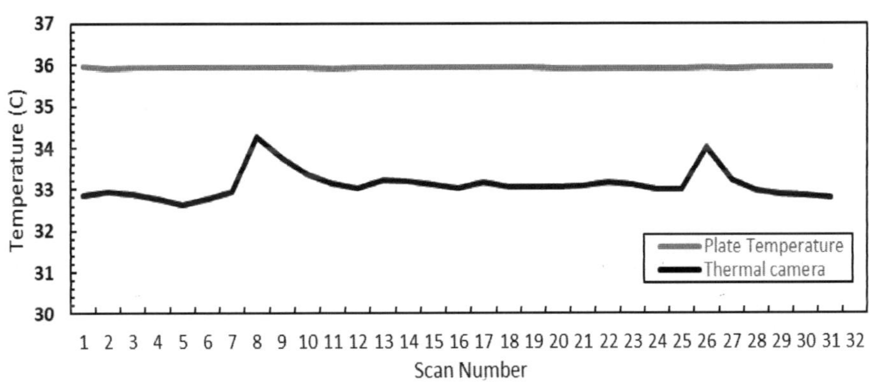

Figure 8: Thermal camera reading and Plate temperature

The plate temperature was measured by calculating the average value of the region of interest (ROI) of 10x10 pixels (T_{cam}), as shown in Figure 9. The temperature reading by the thermocouples is referred to as reference temperature (T_{REF}), while the relative error is defined as the difference between the reference temperature (T_{REF}) and thermal camera reading (T_{cam}) divided by T_{REF}.

Figure 9: ROI of 10x10 pixels in the thermal image

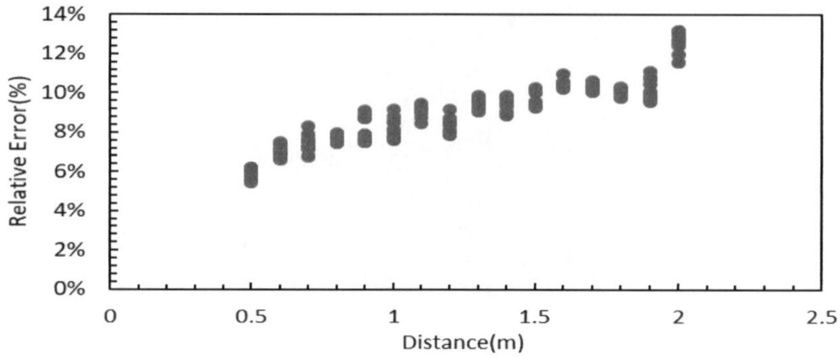

Figure 10: Thermal camera relative error in different working distances

As it can be observed in Figure 10, there is an increasing trend in the impact of working distance on the relative temperature error for the thermal camera. The behavior of this trend can be approximated through a linear relationship, which in turn advocates towards a correction function for the measured temperature by the thermal camera.

For our study, a regression model was used to correct the estimated temperature using thermal camera readings. The model used T_{REF} as one of the inputs (predictor variables) as well as working distance. The camera measurement was used as the response variable. After inferring the model, the inverse methodology can be used to obtain a corrected temperature based on the thermal camera reading and the working distance measurement. The result was assessed through 10-fold cross-validation that showed a RMSE of 0.28 with a standard deviation of 0.039.

The 80% of the collected data was chosen as the training dataset for linear regression model, and 20% of the data was used to validate the developed model. The structure of the developed model is shown in Equation (6), where D is the working distance and T_{REF} is the temperature measured by the thermocouple. The model results are shown in Table 3.

$$T_{cam} = Const + (\alpha_1)*(D) + (\alpha_2)*(T_{REF}) \quad (6)$$

Table 3: Model Parameters

Predictors	**Values**	**P-Value**
Const	12.083	0.002
Distance Coefficient (α_1)	-1.168	0.000
Reference Coefficient (α_2)	0.616	0.000
R2	0.776	
Adjusted R2	0.772	

$T_{Corrected}$ (or T_{REF} in Equation (6)) is the thermal camera's distance-corrected temperature derived from Equation (6). As a result, the corrected camera reading is estimated using Equation (7), which includes the distance and thermal camera reading.

$$T_{corrected} = \frac{1}{(\alpha_2)} [-Const + T_{cam} - (\alpha_1)*(D)] \tag{7}$$

Figure 11 shows that using the distance corrected formula on the validation dataset can reduce relative error from 9% to 0.9%.

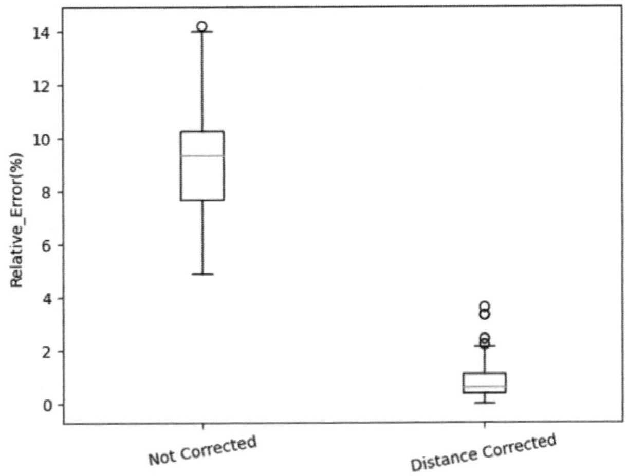

Figure 11: Relative error of thermal camera reading

4. CONCLUSION

This paper presented the formulation of a thermal imaging system consisting of a thermal camera and a depth sensor, aiming to address the deficiencies of thermal imaging with respect to errors due to varying target working distances. The system, aided by a newly developed distance correction formula, was proven to reduce the thermal camera's measurement error from 9% to 0.9%. The overarching goal and the future work of the authors is for this vision-based system to be able to provide non-intrusive information about occupant thermal comfort. Flexible temperature sensors of adequate accuracy as the proposed one can be used in that context to capture temperatures of the occupants' skin or even surrounding surfaces, which can serve as feedback in future occupant-centric building controls. Currently, the authors are also working on adding face or video patterns detection features to this framework, further expanding its usability as an integrated human comfort feedback device.

ACKNOWLEDGEMENT

The project that led to this publication was made possible by Nebraska Tobacco Settlement Biomedical Research Development Funds, as well as funding from the University of Nebraska Collaboration Initiative.

REFERENCES

Aryal, A., & Becerik-Gerber, B. (2019, November). Skin temperature extraction using facial landmark detection and thermal imaging for comfort assessment. In Proceedings of the 6th ACM International Conference on Systems for Energy-Efficient Buildings, Cities, and Transportation (pp. 71-80).

Cheung, B. M. Y., Chan, L. S., Lauder, I. J., & Kumana, C. R. (2012). Detection of body temperature with infrared thermography. Hong Kong Medical Journal, 18(4 Suppl 3), 31-34.

Enforcement policy for telethermographic systems ... - FDA. (n.d.). Retrieved April 15, 2022, from https://www.fda.gov/media/137079/download.
González-Pérez, S., Perea Ström, D., Arteaga-Marrero, N., Luque, C., Sidrach-Cardona, I., Villa, E., & Ruiz-Alzola, J. (2021). Assessment of registration methods for thermal infrared and visible images for diabetic foot monitoring. Sensors, 21(7), 2264.
Gutierrez, E., Castañeda, B., & Treuillet, S. (2020, February). Correction of temperature estimated from a low-cost handheld infrared camera for clinical monitoring. In International Conference on Advanced Concepts for Intelligent Vision Systems (pp. 108-116). Springer, Cham.
IEC80601-2-59:2017.ISO.(2022,March14).Retrieved April 15, 2022, from https://www.iso.org/standard/69346.html
Kim, M., Konstantzos, I., & Tzempelikos, A. (2019). A low-cost stereo-fisheye camera sensor for daylighting and glare control. Journal of Physics: Conference Series. Vol. 1343, No. 1, p. 012157). IOP Publishing.
Li, D., Menassa, C. C., & Kamat, V. R. (2018). Non-intrusive interpretation of human thermal comfort through analysis of facial infrared thermography. Energy and Buildings, 176, 246-261.
Li, D., Menassa, C. C., & Kamat, V. R. (2019). Robust non-intrusive interpretation of occupant thermal comfort in built environments with low-cost networked thermal cameras. Applied energy, 251, 113336.
Litwa, M. (2010). Influence of angle of view on temperature measurements using thermovision camera. IEEE Sensors Journal, 10(10), 1552-1554.
Ma, J., Ma, Y., & Li, C. (2019). Infrared and visible image fusion methods and applications: A survey. Information Fusion, 45, 153-178.
Müller, J., Müller, J., Chen, F., Tetzlaff, R., Müller, J., Böhl, E., ... & Schnabel, C. (2018). Registration and fusion of thermographic and visual-light images in neurosurgery. IEEE Transactions on Biomedical Circuits and Systems, 12(6), 1313-1321.
Nakagawa, W., Matsumoto, K., Sorbier, F. D., Sugimoto, M., Saito, H., Senda, S., ... & Iketani, A. (2014, September). Visualization of temperature change using RGB-D camera and thermal camera. In European Conference on Computer Vision (pp. 386-400). Springer, Cham.
Swamidoss, I. N., Amro, A. B., & Sayadi, S. (2021). Systematic approach for thermal imaging camera calibration for machine vision applications. Optik, 247, 168039.
Yang, M. D., Su, T. C., & Lin, H. Y. (2018). Fusion of infrared thermal image and visible image for 3D thermal model reconstruction using smartphone sensors. Sensors, 18(7), 2003.
Zhang, Y. C., Chen, Y. M., Fu, X. B., & Luo, C. (2016). A method for reducing the influence of measuring distance on infrared thermal imager temperature measurement accuracy. Applied Thermal Engineering, 100, 1095-1101.
Zhang, Z. (2000). A flexible new technique for camera calibration. IEEE Transactions on pattern analysis and machine intelligence, 22(11), 1330-1334.

Development of a Camera-based Tool to Monitor Non-binary Occupants' Interaction with Windows and Shadings

Julian Donges[1]*, Federica Morandi[1], Alessandro Prada[2], Francesca Cappelletti[3], Andrea Gasparella[1]

[1] Free University of Bozen-Bolzano, Faculty of Science and Technology,
Bolzano, Italy
Tel.: +39 0471017890, Email: julian.donges@natec.unibz.it

[2] University of Trento, Dept. of Civil Environmental and Mechanical Engineering,
Trento, Italy

[3] University IUAV of Venice, Dept. of Architecture and Arts,
Venice, Italy

* Corresponding Author

ABSTRACT

In building energy calculation, several energy fluxes can be associated to occupants interacting with building services such as light or HVAC system switching on/off, the thermostat settings as well as window and shading operation. These interactions - always caused by discomfort - are undertaken to restore comfort but can have consequences on the final energy demand. Thus, an accurate building energy simulation for both new and retrofit building design needs to correctly represent people's behavior.

In this paper, a new measurement approach has been developed to collect information about the window's opening angle and shading position, both measured continuously along time. To achieve this, a camera setup connected to a Raspberry Pi microprocessor is designed to track window movements through a color recognition algorithm coded in Matlab. The examined windows and shadings are color coded on the moving parts with rectangular targets. The camera is equipped with wide angle lenses and a physical barrier to be able to capture all present windows and shading simultaneously while ensuring the privacy of occupants.

Preliminary results show that the proposed monitoring system can monitor window opening angles and shading levels, dealing simultaneously with multiple windows. The validation of the method will allow to develop accurate behavioral models based on the obtained data to analyze and suggest improvements for the indoor air quality of classrooms. Moreover, the output data from this equipment can be used to model the window opening free area and perform calculation on the ventilation rate according to EN 16798-7.

1. INTRODUCTION

The detailed understanding of occupant behavior plays a key role towards the improvement of performance estimation and comfort in energy efficient buildings (O'Brien et al., 2020). Especially in environments gathering many people and being solely naturally ventilated by windows and doors, the profound understanding and development of effective ventilation strategies is of major importance. The opening of windows significantly influences the thermal exchange between in- and outside environment. The European standard EN 16798-7:2017 (EN, 2018) deals with the calculation methods for the ventilation of buildings. The amount of airflow through a window depends on the window open area, next to wind speed and direction, turbulence conditions, and the internal-external temperature difference (Coley, 2008). The window open area is given by the product of widow size and opening fraction which can be calculated using the opening angle. As shown in Table 1, opening angles around 20° already represent an opening fraction of around 50%. Therefore, for detailed analysis and in depth understanding the aperture angle α plays an important role as it affects the flow rate depending on the window opening area. According to EN 16798-7:2017 (EN, 2018), the air flow entering through window opening is directly proportional to this area.

Table 1: Opening angle α, opening distance d, window open area A & open fraction.

α [°]	d [m]	A [m²]	Open fraction [%]
5	0.087	0.262	13
10	0.174	0.522	26
15	0.261	0.781	39
20	0.347	1.037	52

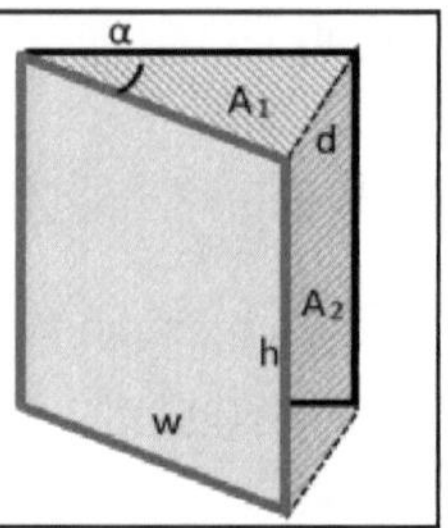

* e.g., for a window of h = 2 m and w = 1 m, calculated according to (Coley, 2008)

Furthermore, the area of the totally opened window is affected by the use of shadings as they cover a big part of the window. For accurate calculation of the natural ventilation provided by the opening of windows it is therefore essential to have information about the windows opening angle and shading position. Consequently, a detailed monitoring approach of window opening angles, shading positions and lights is proposed in the here presented research proceeding.

1.1 Comparison of already existing technologies to monitor window opening

Regarding the prediction of windows and shading operation, in literature a few models have been developed to predict the window opening angle and the shading factor according to indoor or outdoor conditions. These models are usually calibrated and validated using experimental data collected by means of accelerometers, contact sensors, Bluetooth devices, ultrasound, or laser sensors, which are used to calculate the changes of the magnitude of the net exchange surface area with increasing window opening angle and obstacles such as shadings covering a portion of the opening. Almost all these devices have two drawbacks: i. they return binary outputs or ii. they have possible information losses related to a limited working distance, viewing angle or spatial resolution.

Therefore, most research proceedings due to simplicity only consider the binary status of building services such as lights, windows, and shadings. Only a few studies (Fritsch et al., 1990; Haldi & Robinson, 2009) within the here analyzed literature consider the measurement of window opening angles. The main reasons for the underestimation of the value of angle measurements are either that the research interest is purely phenomenological or the avoidance of the experimental effort to measure the angles. The most common monitoring approaches are:

1.1.1 Contact Sensors: Sensors which close an electric circuit on contact are commonly used in most monitoring campaigns (e.g., Park & Choi, 2019). They are a cheap solution, but some disadvantages include their quite complicated wiring for each window and the binary output. Therefore, windows are assumed to be fully opened even if they are just slightly opened at only 10°.

1.1.2 Air speed measurements: Stazi et al. performed in 2017 a campaign to improve indoor air quality and comfort in classrooms using automated systems such as Pitot tubes to measure the air stream entering a window. This allows for a direct monitoring of the ventilation rate as it includes already the windows aperture and external parameters as in- and outdoor temperature difference and wind speed. However, the devices reading the pressure difference of a Pitot tube are quite expensive and it has the further disadvantage of just being able to monitor a single window.

1.1.3 Distance sensors using Bluetooth: Bluetooth sensors have been used to track the position of a certain target in a room (Clark et al., 2018). These sensors have the advantage of being quite cheap and already included in many devices such as cellphones. However, the accuracy is limited to 30 cm which excludes the use for window opening monitoring. A more suitable application would be presence monitoring.

1.1.4 Distance sensor using ultrasounds: Distance measurement using ultrasounds is a very accurate and at the same time cheap solution to measure a distance. To reflect sound correctly, the target must be perpendicular to the emitting source. This cannot be guaranteed on a rotating surface like a window that swivels around its axis.

1.1.5 Distance sensor using laser: The requirement that the surfaces must be at right angles to each other is not given here, making laser measurements an interesting technology to measure the opening of a window. However, the wiring is very annoying because il would be necessary to lay a separate cable to each window.

1.2 Possible advantages of using a camera

To minimize the effort to monitor all present windows at once and simultaneously maximize accuracy, a camera setup is being developed within the here presented work. The main feature is the possibility to recognize predefined colored targets mounted on the windows moving parts. Therefore, it is a valid alternative to the mentioned conventional methods of measuring window opening behavior which, to the authors knowledge, has not been fully explored in literature. Especially, the simultaneous non-intrusive monitoring of multiple windows, shadings, and lights at once without being much effort to mount, makes the presented device a promising alternative to conventional sensors. Combining the camera with a Raspberry Pi allows for data storage and post processing with the advantage to attach additional sensors and instrumentation if needed. A Wi-Fi port simplifies data extraction and small physical dimensions enhance the ease of installation. The development of this camera-based tool contributes to a large-scale monitoring campaign in Morlupo (Rome), aiming to fully understand the phenomena of IEQ in schools, the interactions of teachers and students with building services, and the effects that safety measures regarding viral loads have on performance, comfort, and behavior. The here presented monitoring device increases the level of detail of the monitored behavior using simple and affordable technology. The obtained data can be cross validated with alternative forms of data acquisition such as questionnaires and event checklists filled out by the students.

The general research aim therefore is to develop a versatile, not complex instrumentation to monitor in detail the main building services operated by occupants, namely windows, shadings, and lights in a continuous, nonbinary way. Further it has been a target to validate the device and explore possible additional applications of a camera-based tool, such as presence monitoring or desk illuminance. Further applications of a camera-based tool for indoor monitoring can be a occupancy count to improve management and control of ambient light, air flow and temperature as described in Alishahi et al. (2022). Luminance measurements (Kim & Tzempelikos, 2022) and forehead temperature measurements to estimate people's comfort (Speak et al., 2021) are of further interesting applications for indoor monitoring using a camera. A dynamic response to each condition could be provided by a optimal real-time monitoring control using a camera for the aforementioned applications.

2. METHODOLOGY

The study is based on three steps: the choice of suitable hardware components for the proposed monitoring device, the development of a reliable post-processing algorithm to convert the acquired images into processable data and finally the validation of the device using a sample dataset.

In the first development stage, the device will require some simple calibration to the properties of the monitored room. First, sample images showing each window fully open and closed need to be taken to determine the coordinates of the targets on the sensor's pixel pane when closed and fully open. In a next step five sample images of each possible light condition are required to define the according RGB-ranges of the red targets (windows), yellow targets (shades) and lights (white). Finally, an image cut-out area for the given room properties needs to be defined ensuring full visibility of the target's trajectory.

2.1 Technical Hardware

Regarding the hardware setup, the main components are a Raspberry Pi 4B with 8 GB Ram used as the data acquisition and handling unit as well as a high-definition camera and lens.

The newest model version of the Raspberry Pi is chosen because of its powerful working memory and processor unit. Further, it comes with Wi-Fi for remote data download. It has its own camera port and additionally a standard 40 pin general-purpose input/output (GPIO) board which allows for a simultaneous use of multiple sensors. For the present study a camera and a magnetic switch is connected to the Raspberry. The key components of the setup are the image sensor and camera lens. To choose adequately the lens aperture angle, room dimensions and sensor size must be considered carefully as the ability to successfully postprocess the images strongly depend on this parameter. Two different setups are considered for operation: (i) the standard Raspberry Pi camera model with an electrical board featuring a unit with image sensor and lens in one component, and (ii) the in Pi high quality image sensor recently released in 2020 which has a C/CS mount for interchangeable camera lens and tripod mount.

Given the larger horizontal field of view and the larger target resolution, setup 2 has the clear advantages of being able to monitor more windows simultaneously; it is more suitable for the classroom size in Morlupo and features a better target resolution. Many reviewers consider the 12.3 megapixel sensor as a noticeable upgrade providing sharper images and capturing more color (Medellin, 2020).

Table 2: Comparison of chosen Raspberry Pi Setups

Specification	**Setup 1: Raspberry Pi standard camera with integrated lens**	**Setup 2: Raspberry Pi HQ camera with 6mm wide angle lens**
Lens	Built-in	Raspberry 6 mm wide angle
Sensor	OmniVision OV5647	Sony IMX477
Sensor resolution [pixel]	2592 x 1944	4056 x 3040
Sensor image area [mm]	3.76 x 2.74	6.287 x 4.712
Horiz. field of view [°]	53.50 +/- 0.13	65°
Horiz. field of view* [m]	3.98 m	5,09 m
Target resolution* [pixel]	~ 15x42	~ 18x56
Cost [€]	118.65	202.16

* Assuming a camera to wall distance of 4 m and target size of 30 mm x 70 mm

The target size is chosen to be 30 mm x 70 mm for two main reasons: on the one hand, because the area is ensuring good visibility even if geometrically deformed if window swivels towards larger opening angles. On the other hand, the pixel resolution is sufficient to recognize target ratio and size accurately. The physical accuracy limits for angle measurements are given by the horizontal distance travelled depending on the opening angle and pixel resolution. Every error added is caused by inaccuracies during post-processing. Summing up, results need to be evaluated separately for *theoretical accuracy* connected to hardware components like maximum sensor resolution, target size and room geometry. And second, the limitations connected to the accuracy of the algorithm resulting in the *measured accuracy*.

Figure 1: Camera perspective on classroom (left); Camera & Raspberry Pi setup (right)

The tripod mount is used to attach the camera including the raspberry case to a DIN-rail mounted on the classrooms wall. A cable is run to the magnetic switch mounted on the classrooms door and connected to the Pi's GPIO board. The camera data and power cable are plugged into the raspberry camera port. Finally, the physical spacer between camera lens and sensor needs to be removed to ensure the focal length is suitable for indoor applications and shorter distances. The raspberry is run using the most recent release of the operating system Debian Linux 11 – codenamed Bullseye. Several internal files need to be adapted such as the file containing the network interface information to ensure functionality also in headless mode. Special attention needs to be given to protected institutional Wi-Fi ports referred to as EAP-Typ (Extensible Authentication Protocol). To access the Pi remotely, a VNC server needs to be installed and activated. For automated image acquisition and door sensor readings, two python scripts - *DoorLog.py & TakeImage.py* - are coded. They are embedded in Shell scripts automatically executed by a raspberry built-in functionality called *crontab* which schedules command execution. The main feature of the shell script is that it immediately names the images according to date and time. Finally, the graphical output of the raspberry is set to "experimental direct capture mode" to be able to view the camera preview image over the VNC connection. Otherwise not all video layers of the Pi's graphical interface are transferred.

2.2 Postprocessing

Image processing is a powerful tool to read the information contained in an image and automate the process of data extraction. Images can be abstracted converting them to greyscale, smoothening them or coloring areas according to their size. The main idea of the postprocessing algorithm is to monitor predefined targets on building services. In the development process of the device, the first idea to track window opening has been to follow a geometrical approach using round objects attached to the window frame (see Figure 2). The intended post-processing steps were: the definition of a Region Of Interest (ROI), the conversion into greyscale or b/w, the smoothening through the deletion of objects smaller than 20 pixels. The key component is the object identification; for round objects, this is done through the calculation of the ratio of perimeter to surface area. However, round objects are often found throughout the image due to blur or light rays resulting in circular reflections. For these reasons, this option is discarded in favor of color recognition, detailed in the following.

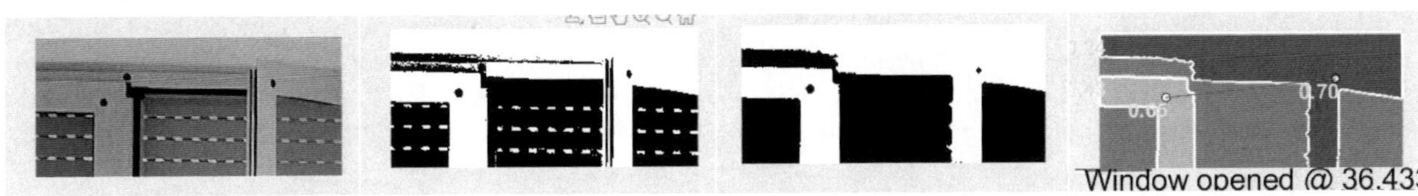

Figure 2: Abstracting an image into black/white, smoothening, and round object identification

2.3 The color recognition algorithm

Colors are more unique than the form of objects. In preliminary test series it has been shown that the color recognition is more reliable than the identification of round objects. It is to mention that the choice of a unique color - not being present in the surrounding environment included in the image - is essential.

The colored area recognition algorithm is coded in Matlab using tools of the image post-processing toolbox. To ensure privacy using a camera device, images are cut to a small region of interest, used for internal post-processing, and immediately deleted.

To find only the color-coded targets, an algorithm is run over the image filtering all pixels for certain RGB values. It must be considered that ambient conditions influence on the RGB values measured by the camera's sensor and therefore all possible light conditions as shown in Figure 3 should be evaluated individually.

Figure 3: Camera perspective on classroom (left); Camera & Raspberry setup (right)

The pixels identified to be within the specified RGB range are saved into a matrix of the same dimensions as the image, referred to as *mask*. In a next step, noise (i.e., areas measuring less than 200 pixels) is removed from the mask.

To verify that the mask corresponds to the target on the window, a further check for geometry, size, and color ratio of the mask is made. The red targets on the windows are 3 cm wide and between 5 to 7 cm high. Therefore, the aspect ratio ranges between 0.4 to 0.6. Considering the camera sensor, lens and distance to wall, the size of the targets is around 40x18 pixels as shown in Table 1. As color codes, red is used for the windows, yellow for the shades and white for the lights given the overexposure if switched on. Each target is composed of shares of red, green, and blue which vary according to the chosen color, light conditions, camera sensor, and lens properties.

Table 2: RGB margins for different light conditions and targets

	Theoretical RGB value	Direct light	Diffuse Light	Artificial Light	Dark Light
Windows (red)	R: 255 G: 0 B: 0	R: 158-195 G: 56-97 B: 72-147	R: 113-202 G: 29-70 B: 35-73	R: 98-142 G: 17-37 B: 22-41	R: 31-97 G: 8-35 B: 11-40
Shadings (yellow)	R: 255 G: 255 B: 0	R: 204-255 G: 228-250 B: 104-184	R: 95-197 G: 96-194 B: 32-99	/	R: 31-95 G: 33-97 B: 9-79
Lights (white)	R: 255 G: 255 B: 255	/	/	Mean (RGB) > 210	Mean (RGB) < 210

Given that the ambient light conditions majorly impact on the colors seen, four different RGB ranges are tested as pre-sets: direct light, diffuse light, artificial light, and dark light. While programming the post-processing algorithm, they are individually assessed through 10 sample images for each condition and summarized in Table 2. In the current version, the algorithm tests each light condition separately and uses the most suitable one in terms of target size and ratio. Future developments foresee the implementation of the white balance of the camera sensor to retrieve the necessary information on the light situation.

If the mask passes the verification of RGB values, aspect ratio, surface area and color ratio, it is recognized successfully as "target" within the image. Otherwise, the post-processing algorithm asks for manual input. Subsequently, the coordinates of the target center are received and compared to the coordinates of the window or the shading if closed. To convert this pixel value into a percentage, angle, or metric value, the following Equations are used for the windows angle (1) and the shading position (2):

$$\alpha = acos(1 - d/d_{90}) \tag{1}$$

with: $d = x_{measured} - x_{closed}$; d_{90} = distance in pixels if window fully opened; $x_{measured}$ = current coordinates of target; x_{closed} = coordinates of target if closed

$$S_x = (x_{measured} - x_{closed}) * H \tag{2.1}$$

$$S_{\%} = S_x \,/\, H \tag{2.2}$$

with: $x_{measured}$ = current coordinates of target; x_{closed} = coordinates of target if shading is closed; H = height of window

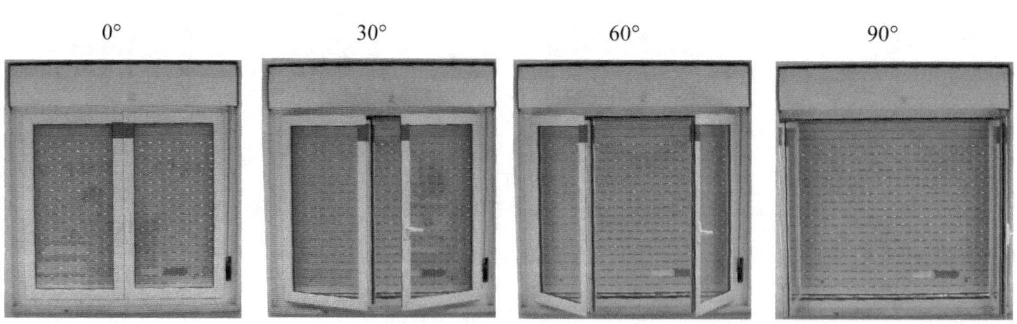

Figure 4: Images used for calibration of windows, taken in different positions at specified angles

The received information is associated to the date and timestamp in the image's filename as described in section 2.1 and finally written into a csv-file for further post-processing and studies of occupant behavior. The *main* code calls subfunctions for windows, shading and lights. The postprocessing steps done to find the targets on windows are shown exemplified in Figure 5.

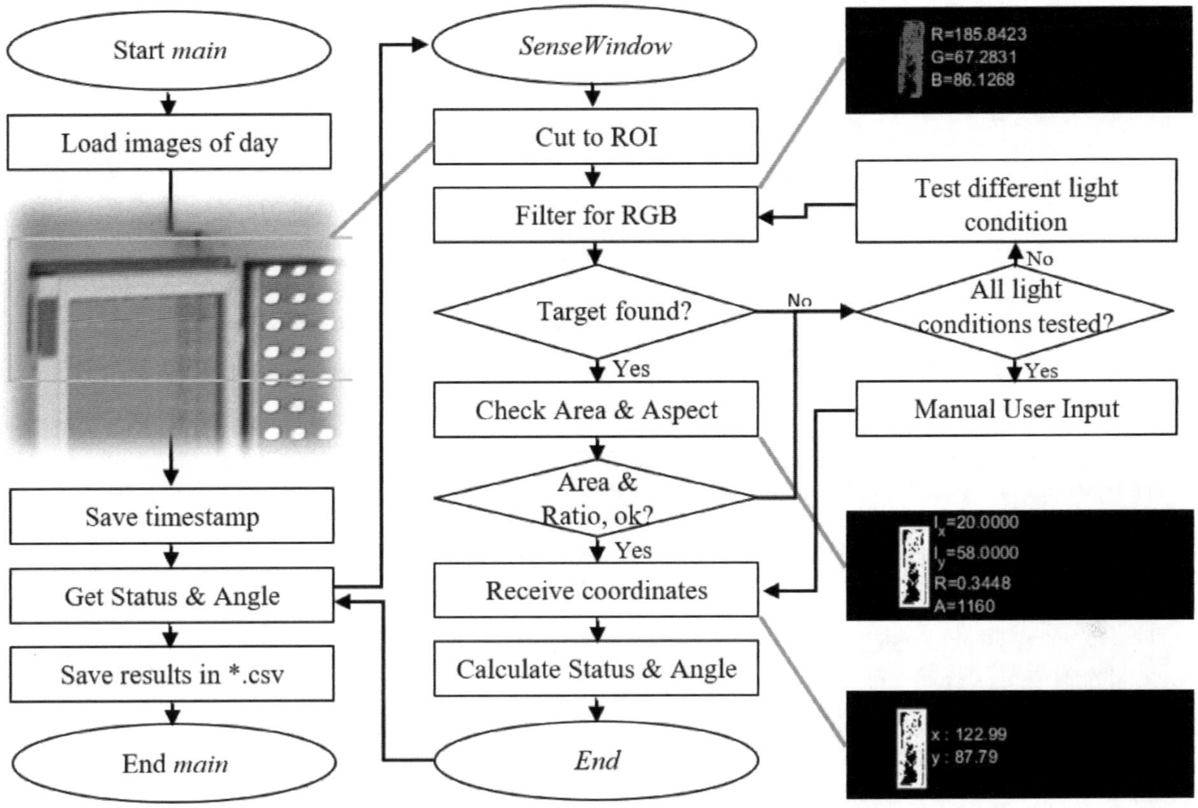

Figure 5: Flow chart showing the image post processing algorithm

It further needs to be mentioned, that for the given application the ROI's of shadings and windows might overlap. Therefore, classic edge detection algorithms would not work reliably. Colors are unique even if ROI's overlap. Additionally, monitoring lights is only possible using RGB, therefore this uniform postprocessing approach using RGB is chosen. However, we consider edge detection as a future development to increase reliability of the algorithm.

3. RESULTS

To test the device's accuracy and reliability, 35 representative sample images have been selected. Each image has been taken at different times of the day and of the year, in varying ambient and light conditions. The images were taken from the first three-month data set of the monitoring campaign. The verification dataset therefore provides 210 independent window states and angles, 105 shading state and position and 70 light state measurements. Every image has been validated manually to retrieve the exact status and position information of the monitored devices.

For windows and shades, it has been distinguished between status errors and angle or position inaccuracies. Status errors indicate that an open window has been detected as closed and vice versa. They are treated as the most relevant errors, whereas simple inaccuracies in opening angles are marked as major if greater than 20° or more than 5cm in the case of shadings. Else they are marked as minor errors.

To visualize status errors and the overall performance during verification, confusion matrices are shown in Figure 6. Each matrix shows the true positives in the upper left square, next to false negatives. In the lower squares the false positives are show on the left next to the true negatives on the right. The dataset has been balanced between opened (41.9 %) and closed windows (58.1 %) as well as turned-on (47.1 %) and turned-off lights (52.9 %). Just the shades are mostly in use (90.5 %) which is caused by the park position in which shades remained lowered at around 10 cm. If a shade lowered less than 20 cm would be considered as *in park position*, the set of shading sample images would be quite balanced, showing 42 % of shades in park position and 58 % of shades in use.

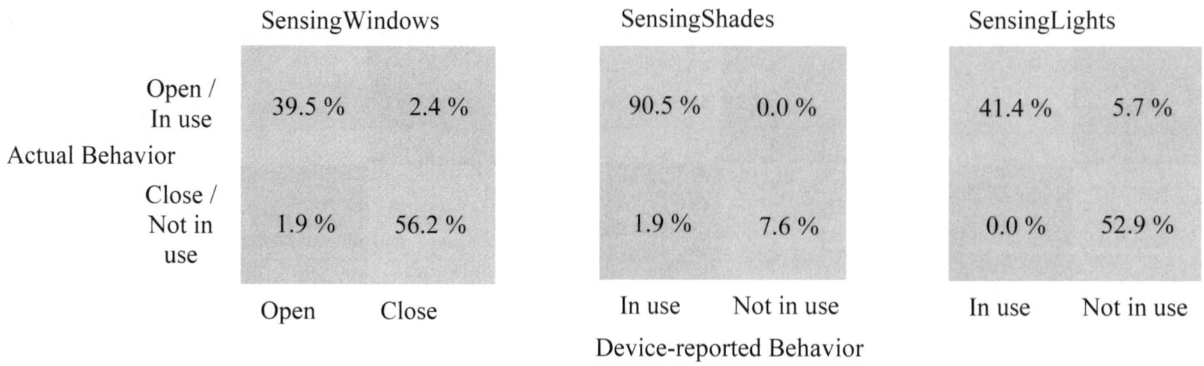

Figure 6: Confusion matrices showing status errors for windows, shades, and lights.

3.1 Windows

The status prediction of windows achieved an accuracy of 95.7 %. Evaluating further the false measurements, it should be noted that from a total of 9 status errors 8 occurred at opening angles lower than 20°. These accumulated inaccuracies at low aperture angles are due to the windows swiveling trajectory. Low angles result in a lower pixel resolution due to reduced target movement per angle of aperture, calculated in Equation 1. Therefore, slightly opened windows (< 20°) are considered as critical events.

If we add minor and major angle measurement errors to the evaluation, the accuracy of entirely correct measurement drops to 80 %. Minor errors in the angle measurement occur in 15 % of all measurements, whereas major errors in angle measurement only occur in 5 % of all cases. The summary of these results is shown as stacked column charts in Figure 7 where windows, shades and lights are numbered from left to right considering the field of view as shown in Figure 1. It is noted that the camera tends to slightly underestimate the opening angle. Further, there is evidence that major angle measurement errors are caused solely by window 2 and 5. Here the main cause of this increased errors on window 2 and 5 is that the camera position and the window's swiveling trajectory is rather unfavorable given that the target is swiveling towards the camera, moving only very little horizontally.

3.2 Shades

The status prediction of shades achieved an accuracy of 98.1 % from which 89 % have been entirely correct. 9 % of all measurements show minor position error of less than 5 cm, whereas only 2 % have been more than 5 cm beside the mark. It can be noted that no measurement errored showed more than 6 cm deviation. Furthermore, 42 % of all sample images showed shades lowered less than 20 cm, which leads to the assumption that the park position of a shade not in use is between 0 to 20 cm. Shading measurements showed the lowest fault rate with only 1.9 % of false measurements and an overall standard deviation of only 0.91 cm. It has been possible to achieve such high accuracy due to the simple geometric conception of a purely vertical moving target in contrast to a swiveling target such as the ones fixed on the windows. Figure 7 shows that the algorithm slightly tends to underestimate the shadow position which is rather not relevant given the low percentage and amount of deviation.

3.3 Lights

The status prediction of the lights achieved an accuracy of 94.3 % where the errors occurred entirely on light 1. Further investigations showed that 3 out 4 images with light error have a generally high illumination of the room with no shades in use and direct sun light resulting in a false negative measurement. Whereas 1 out 4 images showed a reflection on the frame of the light resulting in a false positive measurement.

3.4 Solutions to Common Errors

The preliminary validation of the device has been made on a three-month measurement campaign in a primary high school. Several potential improvements have been found.

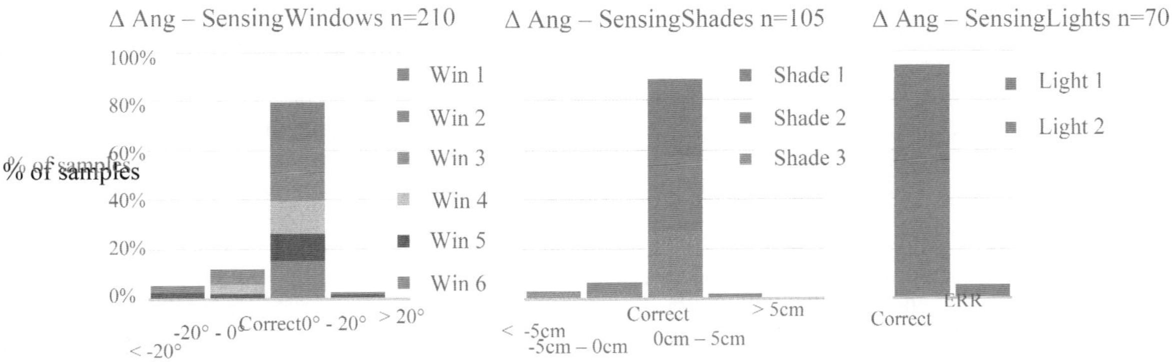

Figure 7: Stacked column charts showing measurement error categories

The accuracy of light measurements can be improved through the definition of multiple ROI's which would allow to avoid capturing the light frame avoiding reflections and double checking for illumination conformity throughout all captured ROI's. Else a manual operator interaction is requested (often at bright illuminance levels, shades completely open, etc.).

Shades could be bundled to be assumed to be *in park position* in case they are lowered less than 15 cm. However, given that the room illumination and window area in contact with the outdoor environment is affected even for small lowering distances, for detailed calculations of air flow they are significant.

In preliminary data processing attempts, it could be shown that measurement accuracy of windows can be increased through a simple check: if opening angles < 15° are persistent for multiple timesteps in a row, it is considered as truly opened at low angle. If an opening event of < 15° occurs only at a single timestep it is considered as measurement error and not being considered for postprocessing. Applying this additional rule, accuracy can be increased, given that most status errors occur at these low angles a window is considered as open only if the opening angle of < 15° remains for at least two timesteps in a row. However, these improvements will be future developments of the here presented measurement device. Finally, we will be working on accurate accuracy limitations due to resolution and geometric deformation of the image.

4. CONCLUSION

The present study reports the development of a camera-based device that is capable to monitor multiple windows, shadings, and lights operation simultaneously. The main feature is the possibility to measure opening angles and shading positions for a detailed analysis of the ventilation characteristics in a room. Furthermore, the device requires little wiring and effort to be set up. The core of the device is an image processing algorithm able to recognize color coded targets. It has been tested and evaluated in a High School in Morlupo, Italy. Considerations concerning the implementation of the device and of the algorithms are presented together with a preliminary validation based on a dataset of 35 independent images. Results show that the proposed monitoring system can monitor window opening angles and shading levels, dealing simultaneously with multiple windows, shadings, and lights. It therefore is a capable tool to monitor occupant behavior in different environments. The detailed validation showed an overall accuracy of 95.7 % for window openings, 98.1 % for shading operation and 94.3 % for the use of lights. Major attention has been paid to privacy-related issues, which are ensured through a strict image reduction to the necessary ROI and elimination of the image after having it post-processed automatically.

The analysis showed some limitations of the approach to be further addressed. The main limitations are the geometrical properties of the room, especially the window opening trajectory to the camera position within the room. A target swiveling towards the camera has a very low horizontal movement, resulting in higher inaccuracies during measurement. This is even amplified if windows are only slightly opened. Therefore, opening angles < 15° are considered as critical events and need to be handled with care. As a future development, the algorithm could be further stabilized through a double check if a critical angle < 15° occurs as a single event or for multiple timesteps in a row.

Future developments will focus on improving the reliability of the device through automized recognition of ambient conditions enabling the algorithm to automatically set the range of acceptable target ratio, color range and target area through a calibration sample dataset. Taking the post-processing algorithm one step further could involve machine learning techniques which would enable the algorithm to measure the use of building services even without mounted targets. Calibration could be done virtually using photorealistic renderings increasing automatization and requiring fewer manual interventions. A further application of the proposed device is thought to be from the outside of a building. Capturing an entire building façade would allow to monitor multiple windows simultaneously while ensuring higher levels of privacy.

The here presented monitoring device has been proven to be a capable and powerful tool to monitor behavioral patterns. Based on the obtained data accurate behavioral models can be developed to analyze and suggest improvements for the indoor air quality of classrooms. Moreover, the output data from this equipment can be used to obtain the window opening free area and perform calculation on the ventilation rate according to EN 16798-7.

REFERENCES

Alishahi, N., Ouf, M. M., & Nik-Bakht, M. (2022). Using WiFi connection counts and camera-based occupancy counts to estimate and predict building occupancy. *Energy and Buildings*, *257*, 111759. https://doi.org/10.1016/j.enbuild.2021.111759

Clark, B. K., Winkler, E. A., Brakenridge, C. L., Trost, S. G., & Healy, G. N. (2018). *Using Bluetooth proximity sensing to determine where office workers spend time at work*. https://doi.org/10.1371/journal.pone.0193971

Coley, D. A. (2008). Representing top-hung windows in thermal models. *International Journal of Ventilation*, *7*(2), 151–158. https://doi.org/10.1080/14733315.2008.11683807

EN. (2018). *UNI EN 16798-7 Energy Performance of buildings - Ventilation for buildings - Part 7: Calculation methods for the determination of air flow rates in buildings including infiltration*.

Fritsch, R., Kohler, A., Nygård-Ferguson, M., & Scartezzini, J. L. (1990). A stochastic model of user behaviour regarding ventilation. *Building and Environment*, *25*(2), 173–181. https://doi.org/10.1016/0360-1323(90)90030-U

Haldi, F., & Robinson, D. (2009). Interactions with window openings by office occupants. *Building and Environment*, *44*(12), 2378–2395. https://doi.org/10.1016/j.buildenv.2009.03.025

Kim, M., & Tzempelikos, A. (2022). Performance evaluation of non-intrusive luminance mapping towards human-centered daylighting control. *Building and Environment*, *213*(January), 108857. https://doi.org/10.1016/j.buildenv.2022.108857

Medellin, A. (2020). *The New Raspberry Pi Camera Is a Worthwhile Upgrade*. The New Raspberry Pi Camera Is a Worthwhile Upgrade. https://www.premiumbeat.com/blog/new-raspberry-camera-upgrade/

O'Brien, W., Wagner, A., Schweiker, M., Mahdavi, A., Day, J., Kjærgaard, M. B., Carlucci, S., Dong, B., Tahmasebi, F., Yan, D., Hong, T., Gunay, H. B., Nagy, Z., Miller, C., & Berger, C. (2020). Introducing IEA EBC annex 79: Key challenges and opportunities in the field of occupant-centric building design and operation. *Building and Environment*, *178*(February). https://doi.org/10.1016/j.buildenv.2020.106738

Park, J., & Choi, C. S. (2019). Modeling occupant behavior of the manual control of windows in residential buildings. *Indoor Air*, *29*(2), 242–251. https://doi.org/10.1111/ina.12522

Speak, A., Montagnani, L., Wellstein, C., & Zerbe, S. (2021). Forehead temperatures as an indicator of outdoor thermal comfort and the influence of tree shade. *Urban Climate*, *39*(August), 100965. https://doi.org/10.1016/j.uclim.2021.100965

Stazi, F., Naspi, F., Ulpiani, G., & Di Perna, C. (2017). Indoor air quality and thermal comfort optimization in classrooms developing an automatic system for windows opening and closing. *Energy and Buildings*, *139*, 732–746. https://doi.org/10.1016/J.ENBUILD.2017.01.017

Operating Cost Assessment of Space Conditioning and Water Heating Technologies in a Residential Building Across the United States

Zechao LU[1*], James E. BRAUN[1], Davide ZIVIANI[1]

[1] Ray W. Herrick Laboratories, School of Mechanical Engineering, Purdue University,
West Lafayette, Indiana, 47906, USA
*lu500@purdue.edu; jbraun@purdue.edu; dziviani@purdue.edu

* Corresponding Author

ABSTRACT

To address environmental concerns and the need for more environmentally friendly cooling and heating solutions, the U.S. Department of Energy (DOE) is supporting the development of smarter, more efficient, and affordable heat pumping systems operating with low- or no-GWP refrigerants through the Energy, Emissions, and Equity (E3) Initiative. In the literature, limited studies were found that investigated different combinations of conventional and emerging space conditioning and water heating technologies while accounting for real building loads, different climate zones, utility structures, and state-of-the-art equipment.

This paper presents operating cost analyses of eight configurations of space heating, cooling, and water heating technologies for a realistic residential building scenario under typical year-round weather conditions. Different combinations of heat pump water heaters, ground source heat pumps, cold climate heat pumps, dual fuel heat pumps, and conventional furnaces were considered in the study. A building model was developed in EnergyPlus and validated with historical data from the DC Nanogrid House at Purdue. A total of eleven climate zones were considered and both local weather conditions and utility pricing were implemented in the simulations. Climate-based equipment selection guidelines were developed and the paper includes future SEER2/HSPF2 ratings, E3 Initiative targets, and also quantitative discussions on emerging technologies such as membrane heat pumps.

1. INTRODUCTION

Residential buildings consumed 20.8 quadrillions Btu of primary energy in 2020, of which approximately 56% was associated with space conditioning and water heating loads (EIA, 2021a). Most space conditioning and water heating devices rely on vapor compression technology and heat-of-combustion due to their low initial cost and high reliability. Non-vapor compression space conditioning technologies are gradually gaining attention, but are still under scrutiny for performance, cost, and maturity. The U.S. Department of Energy (DOE) EERE Building Technology Office (BTO) recently emphasized the importance of developing highly efficient non-vapor compression HVAC&R technologies. In 2014, two reports were released by BTO to compare twenty-two non-vapor-compression HVAC technologies and outline the research opportunities in this area (Goetzler, Zogg, Young, & Johnson, 2014; Goetzler, Guernsey, & Young, 2014). In addition, the Intergovernmental Panel on Climate Change (IPCC) released the sixth assessment report (Masson-Delmotte et al., 2021), which provided updated projections on the Global Warming Potential (GWP) of refrigerants. For instance, propane (R290) now has a GWP_{100yr} of 0.02, butane (R600) has a GWP_{100yr} of 0.006, and the GWP_{100yr} value of R32 was increased from 675 to 771. Moreover, the AIM Act was enacted by U.S. Congress in 2020, which directs the EPA to reduce the usage of hydrofluorocarbons (HFCs) (EPA, 2021). The HFC production allowance will be allocated by the EPA each year, and the total production and consumption of HFCs will be phased down by 85% in 2036. The U.S. DOE also launched a national initiative, namely the Energy, Emissions, and Equity (E3) Initiative, to promote clean heating and cooling systems for buildings (DOE, 2021). This goal will be achieved by advocating advanced solutions to transform the heating and cooling market to be cleaner, more efficient, and more affordable. The effort that DOE is going to make includes promoting advanced Heat Pump Water Heaters (HPWHs), research on low- to non-GWP refrigerants, advancing cold climate heat pumps, and implementing buildings with low carbon footprints. Among those efforts, heat pump water heaters, low to near-zero GWP refrigerants, and cold climate heat pumps will be explored and discussed in this research.

(a) **(b)**

Figure 1: (a) Photo of the DC Nanogrid House; (b) 3D view of a three-story residential house EnergyPlus model for DC Nanogrid House

Qian et al. (2016) reviewed not-in-kind cooling technologies and proposed two performance indices to compare solid-state technologies (e.g., elastocaloric, magnetocaloric, thermoelectric, among others) with each other and with conventional vapor compression systems. The two performance indices that the authors proposed are 1) material COP defined as the theoretical cooling COP for the material in the specified cycle, and 2) non-dimensional latent heat defined as the ratio of the maximum useful latent heat to the heat loss. However, the analysis did not account for real system losses, such as compressor and expander efficiency, regenerator loss, external heat transfer irreversibility, pressure drop, and parasitic power (pump, fan, etc.) and provided only thermodynamic assessments. More recently, El Fil, Boman, Tambasco, and Garimella (2021) conducted a quantitative comparison of 14 air-source heat pump technologies for residential or commercial space conditioning. Thermodynamic models were implemented for each technology and three figures of merit, i.e., primary-energy-based COP, exergetic efficiency, and power density, were calculated for heating and cooling mode operation. Lu, Kim, Braun, and Ziviani (2021) also published a paper that reviewed non-vapor compression technologies quantitively. Lu et al. (2021) adopted the same strategy, which provided a quantitative comparison between the emerging technologies and state-of-the-art vapor compression technologies. In addition, new alternative technologies, such as electrochemical heat pumps were included (James, Braun, & Groll, 2019).

Based on the available literature, it is clear that there are several research shortcomings: 1) Most of the quantitative comparisons between alternative technologies and vapor compression heat pumps focused on thermodynamic performance at fixed boundary conditions (indoor/outdoor conditions). There are no studies that include the effect of building loads, geographical weather data and utilities. 2) Most of the studies were incomplete with respect to emerging technologies and a direct comparison with current state-of-the-art vapor compression heat pumps was not performed. Combinations of emerging technologies are also absent. 3) In most of the case studies, there is no reference to current and future performance rating standards such as SEER2/HSPF2 or load-based testing (Cheng, Braun, & Horton, 2021).

In this study, a residential building model was constructed in EnergyPlus and validated with measurement data for HVAC energy consumption. After that, the HVAC and water heating configurations were replaced with eight test cases, including a two-stage compression R290 heat pump designed for cold climates, ground source heat pumps (GSHP), and membrane heat pumps. An annual operating cost analysis was conducted in eleven locations across the US with TMY3 weather data. In the end, recommendations were given for each location.

2. BUILDING MODEL AND VALIDATION

2.1 EnergyPlus Building Model

In order to have a clear understanding of the advantage of each HVAC technology, it is important to obtain the performance of each technology when coupled to a building and affected by actual weather conditions. Whole building energy modeling (BEM) is often useful to compare the annual energy consumption and cost of emerging technologies. Compared with field measurements, building energy simulations are more cost-effective and time-efficient. EnergyPlus is one of the most widely used building energy modeling (BEM) tools (Crawley et al., 2001). It is maintained by the U.S. Department of Energy and certified by ASHRAE Standard 140, which ensures its accuracy. Its open-source nature and various ways for user input make it suitable for this study.

The EnergyPlus building model used in this study was modified from a PNNL typical residential building model (Mendon & Taylor, 2014). Several modifications were made to mimic a real detached single-family residential house at Purdue Campus, namely the DC Nanogrid House. According to U.S. EIA, 63.5% of the residential buildings in the U.S. are detached single-family buildings (Shi, Wang, Schwartz, & Hofbauer, 2020). The DC Nanogrid House is a historical building located in West Lafayette, which is a representative residential building in the midwest. The building was renovated in 2019 and houses three graduate students living and working inside the building (Ore, Ziviani, & Groll, 2021). The building has a total of 208 m^2 of floor space. The floor plan is 9.14m×10.67m (30ft×35ft). It has two floors, a 2.64m (104 in) tall basement, and an attic. Figure 1a is a photo of the DC Nanogrid House. Hence, the PNNL building model was modified to represent this building. The foundation of the building model was resized to a 2.64 m tall basement. The floor plan and the height for each floor were modified to mimic the real building. The 3D view of the building model is shown in Figure 1b. The whole building was divided into three thermal zones, attic, living zone, and basement, while only the living zone and the basement were conditioned. The U-values for the building insulation materials were collected and input into the EnergyPlus model. The infiltration of the DC Nanogrid House was measured using a blower door test.

A representative temperature setpoint schedule based on actual data was identified for the heating and cooling seasons throughout the year. The operation of the appliances was also modeled in the EnergyPlus building model. Appliance utility usage data was obtained from measured electricity assumption in the building for each appliance during 2019. Though the average consumption for each appliance was set to match the data, a fixed day-by-day schedule of the appliance was assumed and deviates from reality. It is assumed that only the total consumption will affect the internal gain in the building. Moreover, the mechanical ventilation of the living zone was 0.207 ACH and the infiltration rate was 0.474 ACH. These values are based on an effective leakage area of 0.0323 m^2, which was measured using a blower door test for the DC Nanogrid House. The monthly domestic hot water usage was 1959 gal/month, which is an average value from the monthly utility bills.

The vapor compression heat pump used at the DC Nanogrid House is a Trane XV18 model with a SEER of 18 and an HSPF of 10. Regression curves were generated based on manufacture's data. The HPWH is located in the heated basement and uses the air in the basement as the heat source. The HPWH is a Rheem PROPH50 T2 RH350 model, which has a UEF rating of 3.55. Based on its spec sheets, the device has a COP of 3.2 and a capacity of 4 kW at the rated conditions, where the entering air dry-bulb temperature is 29.4 °C and the entering condenser water temperature is 56.7 °C.

Several assumptions needed to be made during the modeling process. Firstly, the building model was resized from the PNNL single-family residential house. As a result, the shape of the roof and the location of the windows will be different. However, it was assumed that the roof shape and the window location would have a minor influence on energy consumption. The reasons are as follows. Changing the rooftop shape should not have a significant impact on the surface area and the heating and cooling load of the building because the projected area for the solar radiation is the same and the attic is unconditioned, resulting in little difference in building load. This minor difference would not change the conclusion drawn from the comparison. In addition, since the window-to-wall ratio is the same as the actual building, the window location on a specific wall does not have a significant influence on energy consumption. Secondly, the appliance usage data and the human occupancy data were set manually. In the reality, the appliance usage may vary on a daily basis, but the simulation schedule is the same every day. It is assumed that as long as the total appliance consumption matches the real-world data, the deviation in the daily schedule will not influence the final results.

2.2 Model Validation for DC Nanogrid House

The EnergyPlus building model was validated by comparing the simulated energy consumption with the measured data. Ore et al. (2021) measured the actual HVAC energy consumption for the DC Nanogrid House from April 2018 to 2021. The time from Jan 1st to Dec 31st, 2019, was selected for validation. The weather data measured at the location weather station was converted to an EnergyPlus weather file to make sure the building model was under the same climate condition as the actual building. EnergyPlus was able to determine the COP of the unit at different conditions using COP mapping and weather data given. Then the real measured data was used to validate the building and the heat pump performance map.

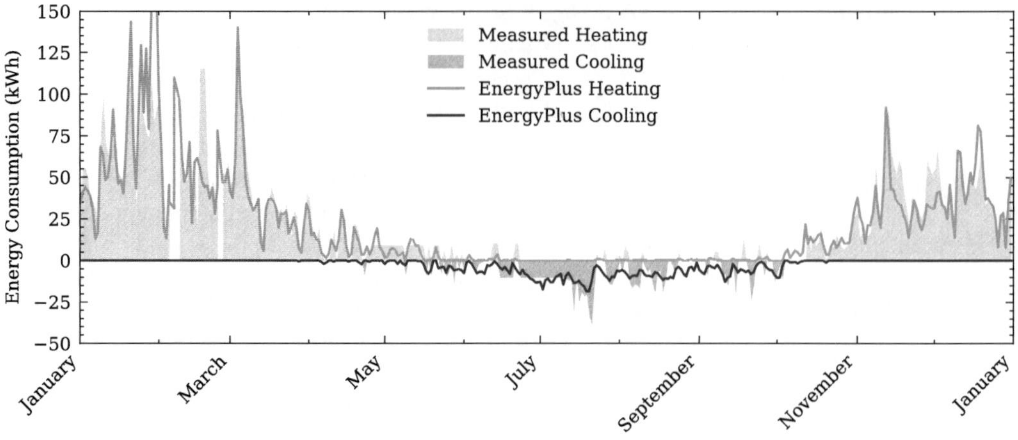

Figure 2: Comparison of EnergyPlus simulation and measurements for space heating and cooling energy consumption at DC Nanogrid House in 2019

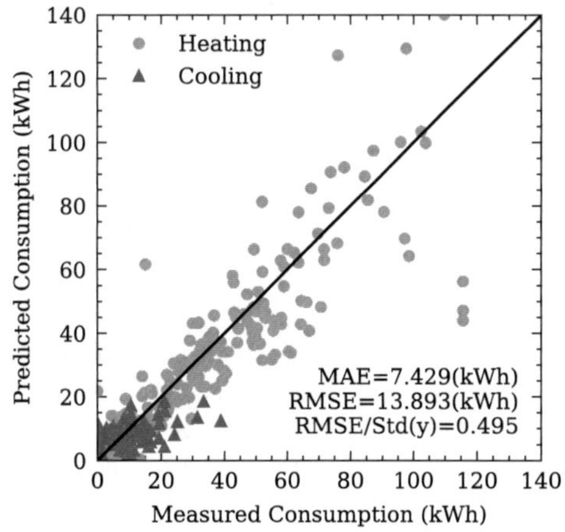

Figure 3: Parity plot for DC Nanogrid House daily heat pump electricity consumption predicted by EnergyPlus

Comparisons between the simulated and actual energy consumption are shown in Figure 2 and Table 1. EnergyPlus simulation results matched reasonably well with actual consumption qualitatively. The heating power peaks were captured for the most part. Figure 3 is a parity plot to compare the EnergyPlus predictions with the measured results. The R^2 for the correlation is 0.758. The errors are reasonably low. Possible reasons for the errors are as follows. Firstly, human interactions with the building, such as leaving windows and doors open, are not taken into account. These interactions will change the leakage area, leading to larger peaks in energy consumption. Secondly, the residents in the building may change the temperature setpoint temporarily, causing differences in instantaneous peaks or valleys in the energy consumption. Thirdly, the appliance usage and the hot water consumption may have deviations from the schedules in the model.

3. CASE STUDIES

A total of eight case studies were identified to evaluate the state-of-the-art vapor compression technologies and other feasible scenarios for emerging technologies. Table 2 shows a general overview of the case setups. Case 3 had the same HVAC characteristics as the DC Nanogrid House and was designed to validate the EnergyPlus simulated energy consumption with the measured data on-site. Case 1 served as a baseline case. Natural gas was used as the heat source for space heating and water heating in Case 1. The gas furnace and the gas boiler used in Case 1 had an AFUE of 0.8 and efficiency of 60%, respectively. A central air conditioner with SEER 13 was used for space cooling. Case

Table 1: Space heating and cooling energy consumption comparison in 2019

	Heating (kWh/year)	Cooling (kWh/year)
EnergyPlus	7,895	1,041
Measured	7,967	1,007
Deviation	-72 (0.9%)	+34 (3.4%)

Table 2: HVAC Configurations for EnergyPlus Case Studies

No.	Space Heating	Space Cooling	Water Heating
1	Gas Furnace	Vapor Compression AC	Gas Boiler
2	Vapor Compression HP	Vapor Compression HP	Resistance Heater
3	Vapor Compression HP	Vapor Compression HP	HPWH
4	Cold Climate HP	Cold Climate HP	HPWH
5	Dual Fuel HP	Vapor Compression HP	HPWH
6	Ground Source VC HP	Ground Source VC HP	HPWH
7	Gas Furnace	Membrane HP	Gas Boiler
8	Vapor Compression HP	Membrane HP	HPWH

2 and Case 3 were used to address the influence of a heat pump water heater (HPWH). The space conditioning heat pump used in Case 2 and Case 3 had a SEER of 18 and an HSPF of 10. The HPWH for Case 3 used the air in the basement as the heat source. The device delivered a COP of 3.2 and a capacity of 4 kW at the rated conditions. In Case 4, a new cold climate heat pump was used for heating and cooling. The heat pump was an R290 heat pump that can switch between single- and two-stage operation in heating mode. The unit is currently under development at the Herrick Labs. The simulated performance of the heat pump had a SEER rating of 14.1 and an HSPF rating of 10.1. The performance map and the optimization process are discussed in Section 3.1. The dual-fuel heat pump in Case 5 used the same vapor compression heat pump as Case 2 and 3. The only difference is that the supplementary heater was changed to a gas furnace with 80% efficiency. Case 6 explored the possibility of using a ground source heat pump for heating and cooling. It used a water loop to connect the water-to-air heat pump with a vertical borehole heat exchanger. The water-to-air heat pump was integrated into EnergyPlus by the equation-fit method. The gross rated COPs were 4.4 in heating mode and 6.15 in the cooling mode under ISO rated conditions. The heat pumps discussed above were auto-sized by EnergyPlus based on the heating load on Jan. 21st or cooling load on July 21st, whichever was larger. If the capacity of the heat pump did not meet the demand, a supplementary resistance heater with 100% efficiency was used.

We simulated eleven ASHRAE climate zones for space heating, space cooling, and water heating, which represent a large portion of the population across the United States, as shown in Table 3. The typical meteorological year (TMY3) weather data (Wilcox & Marion, 2008) was used for the EnergyPlus simulation, listed in Table 3 for each corresponding location.

Another important factor involved in this process is the utility rates. In general, electricity rates are higher in densely populated areas due to high demand and large demand fluctuations. Natural gas is cheaper where there are local extraction sites or major pipelines. Seasonal fluctuations also exist due to the fluctuations in demand and supply. To capture the variations, utility rates from the EIA database considering time-of-use (TOU) were used in economic analysis. Natural gas prices were obtained from U.S. Energy Information Administration (EIA) website (EIA, 2021b) to make sure the prices were up to date and accounted for the seasonal changes. The electricity price varies depending on the date and time in some places to encourage people to use electricity in off-peak hours. Table 4 presents the TOU electricity price in the eleven locations across the U.S. The start and end time and date of the peak and the number of the peak levels are dependent on the location. Detailed electricity price schedules were obtained from a database (OpenEI.org) maintained by National Renewable Energy Laboratory (NREL, 2021). In some places, demand cost was added to the utility bill to penalize high electricity demand behavior. The demand cost can also be flat or TOU. The demand charge was calculated based on the peak electricity demand of the whole building during a month. Then the contributions to the peak charges for heating, cooling, and water heating were calculated separately at the peak hour. After that, the cost of each category is added to the annual electricity bill.

Table 3: Locations for Energy Consumption Study

Climate Zone	City, State	Weather Data	TOU
1A	Miami, FL	Miami Intl AP	Time-of-use energy only
2A	New Orleans, LA	New Orleans Intl AP	Flat rate (seasonal)
2B	Tucson, AZ	Davis-Monthan AFB	Time-of use energy with time-of-use demand
3A	Atlanta, GA	Atlanta-Hartsfield Jackson Intl AP	Time of use energy with flat demand
3C	San Diego, CA	Chula Vista-Brown Field Muni AP	Time-of-use energy only
4A	New York, NY	John F Kennedy Intl AP	Time-of-use energy only
4C	Seattle, WA	Seattle Tacoma Intl AP	Flat energy
5A	Indianapolis, IN	Indianapolis Intl AP	Flat energy
5B	Denver, CO	Denver-Aurora-Buckley AFB	Time-of-use energy with time-of-use demand
6A	Minneapolis, MN	St Paul Intl AP	Time-of-use energy only
6B	Great Falls, MT	Great Falls Intl AP	Flat energy

Table 4: Example of time-of-use electricity price (The number of levels and start time of the peaks depend on the location)

City, State	Peak Cost (\$/kWh)	Off-Peak (\$/kWh)	Demand (\$/kW)	Gas (\$/kWh)
Miami, FL	0.193	0.056	0	0.073
Tucson, AZ	0.101	0.06	12.85 (Peak) 8.85 (Off-peak)	0.051
New Orleans, LA	-	0.095	0	0.047
Atlanta, GA	0.147	0.0278	7.96	0.065
San Diego, CA	0.395	0.198	0	0.048
New York, NY	0.456	0.222	0	0.048
Denver, CO	0.0417	0.0257	10.67 (Summer) 13.65 (Winter)	0.029
Indianapolis, IN	-	0.126	0	0.040
Minneapolis, MN	0.205	0.0417	0	0.030
Seattle, WA	-	0.098	0	0.040
Great Falls, MT	-	0.115	0	0.027

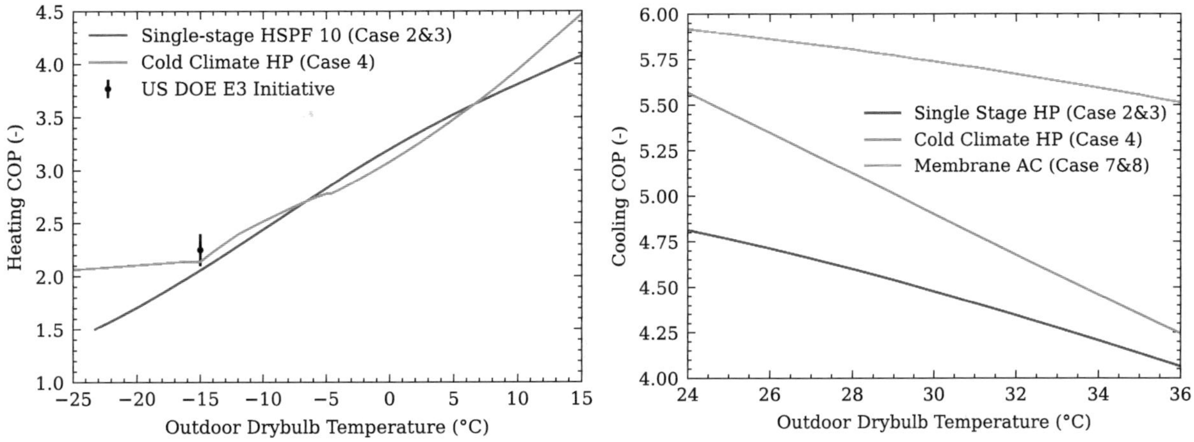

(a) **(b)**

Figure 4: (a) Heating COP comparison between the cold climate heat pump and a regular single-stage heat pump. E3 CCHP Initiative requires COP 2.1-2.4 at -15°C (5°F); (b) Cooling COP comparison between the cold climate heat pump and a regular single-stage heat pump

3.1 Air Source Vapor Compression Cold Climate Heat Pumps

The air-source vapor compression heat pump used in Case 4 from Table 2 was designed specifically for cold climates. In heating mode, the heat pump could switch from a single-stage to a two-stage cycle if the temperature is below a certain point. The two-stage cycle of the system uses an internal heat exchanger as the economizer of the system. If the temperature is higher than that point, the single-stage heating cycle which employs only the high-stage compressor only is used. In cooling mode, the heat pump only uses a single-stage compression cycle.

The performance map of the heat pump was integrated into EnergyPlus by the curve-fitting method. The capacity and power consumption of the heat pump under various outdoor dry-bulb temperature, indoor dry- and wet-bulb temperature, and indoor coil airflow rate conditions in the cooling and heating mode were filled into a spreadsheet provided by EnergyPlus to generate the coefficients of the curves which describe the performance of the heat pump. To obtain the performance map, quasi-steady state models using Engineering Equation Solver (EES) (Klein, 2018) were created to simulate the heat pump under various conditions.

After the transition point was determined, the heating and cooling performance of the cold climate heat pump was integrated into EnergyPlus through the curve fitting method. The COP curve for heating was split into three pieces through the EnergyPlus Energy Management System (EMS) due to the discontinuities. The original curve was overridden using the EnergyPlus Runtime Language (ERL) if the outdoor temperature is between −15°C and −7.5°C or below −15°C. The other curves were obtained directly from curve fitting. The heating and cooling performance of the heat pump at different indoor and outdoor conditions was determined from the coefficients of the curves. Figure 4a shows heating performance comparisons between the cold climate heat pump and a single-stage heat pump used in Case 2 and Case 3, which has an HSPF of 10. The heating performance of the single-stage HSPF 10 heat pump was also obtained from the performance data in the spec sheet provided by the manufacturer. The outdoor temperature working range of this heat pump in heating mode is from -23.3°C (-10°F) to 18.3°C (65°F). The cold climate heat pump which employs the two-stage cycle at low temperature exhibits better heating performance when the temperature is below −15°C. It indicates the advantages of the cold climate heat pump in terms of energy savings in cold climate locations. Similarly, cooling performance comparisons of the two heat pumps are shown in Figure 4b. Both of the heat pumps employ single stage compression in cooling mode.

3.2 Ground Source Vapor Compression Heat Pumps

The ground source heat pump uses a vertical borehole heat exchanger as the heat source. The vapor compression heat pump that integrates the ground-loop heat exchanger and the DC Nanogrid House is a water-to-air heat pump. The water-to-air heat pump can operate in heating mode and cooling mode based on the outdoor temperature. It was sized based on the heating or cooling load on Jan. 21st and July 21st. A supplementary resistance heater with 100%

efficiency could provide additional heating capacity if the demand is not met.

The borehole and the soil characteristics are important aspects of ground source heat pumps. Calculations for the vertical borehole length were demonstrated in Philippe, Bernier, and Marchio (2010). First, a baseline EnergyPlus model was used to obtain the heating and cooling load based on the building and weather data. Next, local soil characteristics such as porosity, classification type, and diffusivity were gathered. Then the building load data and the soil characteristics could be used to derive the vertical borehole length to accommodate the heating and cooling load. In this study, some assumptions were made to simplify the process. The soil in the discussed locations across United State was assumed to have the same properties. The number of vertical boreholes and the borehole length were assumed to be the same in all the locations, and large enough to accommodate the heating load in the coldest location. The undisturbed ground temperature for the selected locations was obtained from Neves, Cho, and Zhang (2021). The source consulted a geothermal heat pump design guideline published by Virginia Polytechnic Institute and Virginia State University.

In EnergyPlus, the ground source heat pump was handled through two parts, a condenser loop from the condenser to the ground-coupled heat exchanger, and a water-to-air heat pump. EnergyPlus was able to determine the water temperature in the loop by the water flow rate, ground temperature, and soil/borehole characteristics. The water-to-air heat pump used in this study was Bosch Greensource SM Series Residential Geothermal Heat Pump SM036. The heat pump had a heating COP of 4.4 and a cooling COP of 6.15 at the ISO 13256-2 rated conditions (Neves et al., 2021; Corp., 2021). The rated conditions were indoor DBT 20°C and inlet water temperature 20°C in heating mode, and indoor DBT 27°C and inlet water temperature 30°C in cooling mode. There are two ways to model water-to-air heat pumps in EnergyPlus, Curve Fit Method and Parameter Estimation Method. In this study, the Curve Fit Method was selected thanks to its simplicity and accuracy. Using this method, EnergyPlus was able to generate regression curves for capacities and COPs under different water and air inlet conditions based on the product spec sheet (Tang, 2005). The energy consumption of the water-to-air heat pump under various indoor and outdoor conditions could be determined from the regression curve.

3.3 Membrane Heat Pumps

The membrane heat pumps used in Case 7 and 8 from Table 2 came from Bukshaisha and Fronk (2019). The heat pump system was a recirculation system that used the indoor air as the supply air and was based totally on membranes. A membrane dehumidifier was used to dehumidify the supply air and the supply air was cooled by an air-water heat exchanger. The cooled water was supplied by a membrane evaporative cooler. The water vapor from the membrane dehumidifier and membrane evaporative cooler was extracted by a vacuum pump and delivered to a membrane outdoor unit to be released to the outdoor environment. Bukshaisha and Fronk (2019) used EES to build thermodynamic models for the membrane system. Parameters such as dehumidifier effectiveness were assumed to infer the component sizings. After the component sizings were decided, the power consumption and cooling capacity were determined for different indoor and outdoor temperature and humidity conditions. The performance of the membrane system was also compared with conventional vapor compression air conditioners which have SEER ratings of 14 and 16. To analyze the membrane system performance, the authors did building energy simulations for a 185.8 m^2 (2000 ft^2) building using eQuest. The required cooling hours for different locations were obtained from eQuest and fed into the EES model to get the annual electricity consumption. The cooling hours were divided only by three scenarios, outdoor DBT greater than 25°C (77°F), greater than 21.7°C (71°F), and great than 19.4°C (67°F). However, the indoor conditions were not involved in the electricity consumption calculation process. These limitations made the electricity consumption calculation and comparison less reliable.

In this study, the thermodynamic models for the membrane heat pump were reproduced with the same parameters and component sizings, based on the same set of equations of states as Bukshaisha and Fronk (2019). EES was used in reproducing the thermodynamic models. The cooling performance of the models was compared with the original model outputs in Bukshaisha and Fronk (2019) to make sure the two models match. The cooling COP at different outdoor dry bulb temperatures is shown in Figure 4b. This performance curve was calculated for an indoor DBT of 20°C (68°F) and an indoor air flow rate of 0.52 m^3/s (1100 CFM). To conduct an annual operating cost analysis, the performance map of the membrane heat pump was integrated into EnergyPlus through capacity and power consumption curves, with respect to the variations in indoor air flow rate, indoor WBT, and outdoor DBT. This procedure was the same as integrating the performance map of conventional vapor compression heat pumps, instead of the rough estimation based on the cooling hours as presented in Bukshaisha and Fronk (2019). Moreover, the water consumption from the evaporative cooler of the membrane heat pump in various locations was also considered in calculating the cooling cost.

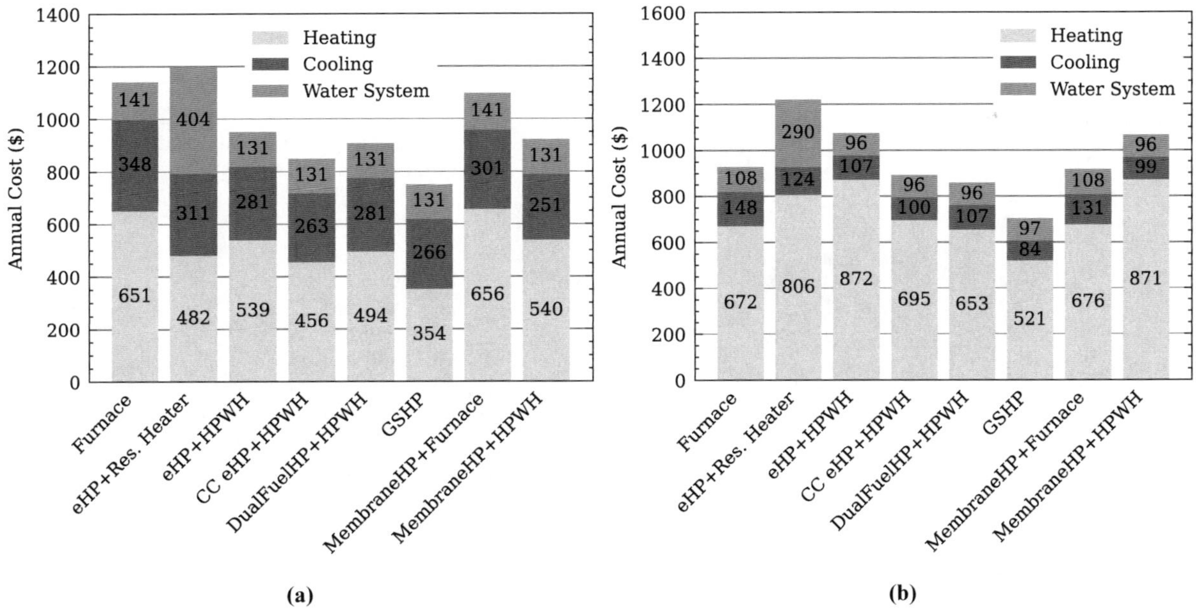

Figure 5: (a) Annual HVAC energy cost in New York, NY under TMY3 weather condition; (b) Annual HVAC energy cost in Minneapolis, MN under TMY3 weather condition

The water rates in the discussed locations were obtained from Black and Veatch (2019).

4. RESULTS

The annual energy costs for space conditioning and water heating were analyzed for eleven locations across the U.S. as listed in Table 3. Due to the page limit, the detailed cost analysis for only three locations are shown in this section. The three locations are New York, NY (Figure 5a), Minneapolis, MN (Figure 5b), and Atlanta, GA (Figure 6a)

Annual HVAC energy cost comparisons for Zone 4A (New York, NY) are shown in Figure 5a. Since the natural gas price in New York is higher than that in Indiana and California, the cases with vapor compression HP cost less for space heating. New York State Electric and Gas Corporation (NYSEG) also provide a $1000 one-time rebate when purchasing an electric HP for space heating and cooling. The cost-saving differences between the cold climate HP and the GSHP was only $100 annually, but the initial cost of the GSHP is significantly higher than the cold climate HP, which indicates the potential of the cold climate HP. The membrane HP could also save $30-$40 compared to vapor compression HPs.

Annual HVAC energy cost comparisons for Zone 6A (Minneapolis, MN) are shown in Figure 5b. Since the electricity price is high in Minneapolis, especially at peak hours, the electric vapor compression HP and HPWH cost $150 more than the furnace annually. The cold climate HP cost $36 less than the furnace. The GSHP can save $226 compared to Case 1 (gas furnace). The dual fuel HP can save $72 per year. The results indicate that using a gas furnace as the supplementary heater in extremely cold weather is a viable option, if the electricity price is not significantly higher than the gas price. The cooling load in Minneapolis, MN was quite low. Hence the membrane HP only saved $10 per year.

In Atlanta, GA, thanks to its low electricity price, the vapor compression HP was the best performer there. Since the temperature was not very low, the cold climate HP and the GSHP performed worse than the basic HP. The HPWH could also reduced the cost significantly. The combination of the vapor compression HP with SEER 18 and HSPF 10 and HPWH (Case 3) reduce the total cost by $370 when compared with Case 1 which uses a gas furnace for heating. In Denver, CO, Case 1, 3, 4, and 5 had similar energy costs annually. The cold climate HP and the dual fuel HP had small advantages. The ground source HP could save $154 each year. The membrane HP didn't provide cost savings, because of the low cooling load in Denver, CO.

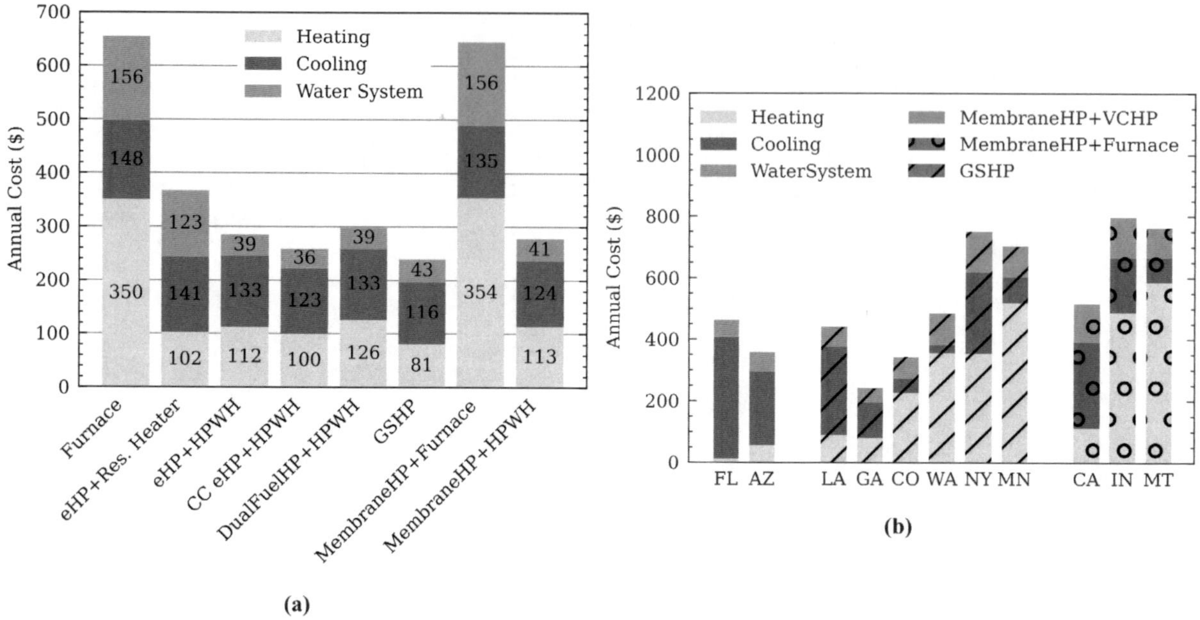

Figure 6: (a) Annual HVAC energy cost in Atlanta, GA under TMY3 weather condition; (b) Minimum cost and the recommended HVAC configuration for each location

Overall results for the eleven climate zones are provided in Figure 6b, which represent the best combination of the HVAC configurations for each climate zone. The initial simulation results provided climate-based equipment selection guidelines for the eleven locations. For instance, CCHPs with two-stage compression in heating mode save 10%-20% in annual heating cost compared with single-stage VCHPs in Climate Zone 4A, 4C, 5A, 5B, 6A, and 6B. Membrane evaporative air-conditioners could provide cooling cost savings in places where thereis a significant cooling load, such as Zone 1A, 2A, 2B, 3A, 3C, 4A, 5A, and 6A. The cost savings for membrane air-conditioners also depends on the water prices. Gas furnaces should only be used in cold places where the electricity price per kWh to gas price ratio is higher than 3. GSHP has the lowest HVAC annual energy cost in six out of eleven climate zones in the U.S. Dual fuel heat pumps are not always the most economical option but yield better average cost savings among the eleven locations. HPWHs should be recommended in areas where the electricity price to gas price ratio is below 3.

5. CONCLUSION

In this paper, annual HVAC operating cost analyses were conducted for eight HVAC and water heating combinations in a two-story single-family building in eleven locations across the U.S. The analyses were done in EnergyPlus and the building model was validated with measured HVAC energy consumption at the Purdue campus under the same weather conditions. The validated building model was expanded to cover eight HVAC and water heating combinations and eleven locations across the U.S. The HVAC options included state-of-the-art vapor compression heat pumps, such as cold climate two-stage HPs, GSHPs, and dual fuel HPs, and alternative heat pumps, such as membrane HPs. The water heating options included gas boilers, resistance heaters, and HPWHs. The eleven locations covered climate zones from 1A to 6B. Typical Meteorological Year (TMY3) weather data of the eleven locations and the actual gas price and TOU electricity price in 2021 were used in evaluating the annual energy cost. The initial simulation results provided climate-based equipment selection guidelines and quantitative operating cost assessments. This study also had two limitations. The selected building can only represent the building style in the midwest of the U.S. Certain modifications should be made when moving it to other places across the U.S. The other limitation is that payback period analysis was not performed in this study, because the initial cost of HPs using emerging technologies was hard to evaluate.

NOMENCLATURE

HVAC	Heating, Ventilation, Air Conditioning
GWP	Global Warming Potential
HPWH	Heat Pump Water Heater
COP	Coefficient of Performance
SEER	Seasonal Energy Efficiency Rating
HSPF	Heating Seasonal Performance Factor
TMY3	Typical Meteorological Year
GSHP	Ground Source Heat Pump
ASHP	Air Source Heat Pump
VRF	Variable Refrigerant Flow
DX	Direct Expansion
BEM	Building Energy Simulation
ACH	Air Changes Per Hour
HP	Heat Pump
WBT	Wet Bulb Temperature
DBT	Dry Bulb Temperature
EMS	EnergyPlus Management System
ERL	EnergyPlus Runtime Language
EES	Engineering Equation Solver
NTU	Number of heat transfer units
TOU	Time-Of-Use
VCHP	Vapor Compression Heat Pump

REFERENCES

Black and Veatch, M. C. (2019). *2018 - 2019 50 Largest Cities Water & Wastewater Rate Survey.*

Bukshaisha, A., & Fronk, B. (2019). Simulation of membrane heat pump system performance for space cooling. *International Journal of Refrigeration*, *99*, 371–381.

Cheng, L., Braun, J. E., & Horton, W. T. (2021, June). Load-based testing using a thermostat environment emulator. *International Journal of Refrigeration*, *126*, 109–122. doi: 10.1016/j.ijrefrig.2021.02.010

Corp., B. T. (2021, September). *Bosch Greensource SM Series Residential Geothermal Heat Pump.* Bosch-Climate.

Crawley, D. B., Lawrie, L. K., Winkelmann, F. C., Buhl, W. F., Huang, Y. J., Pedersen, C. O., ... Glazer, J. (2001, April). EnergyPlus: creating a new-generation building energy simulation program. *Energy and Buildings*, *33*(4), 319–331. doi: 10.1016/S0378-7788(00)00114-6

DOE. (2021, May). *Energy, Emissions and Equity (E3) Initiative.*

EIA. (2021a). *Annual Energy Outlook 2021 (AEO2021).* US Energy Information Administration. https://www. eia. gov/outlooks/aeo/pdf ....

EIA. (2021b). *Natural Gas Prices.* U.S. Energy Information Administration.

El Fil, B., Boman, D. B., Tambasco, M. J., & Garimella, S. (2021, January). A comparative assessment of space-conditioning technologies. *Applied Thermal Engineering*, *182*, 116105. doi: 10.1016/j.applthermaleng.2020.116105

EPA. (2021, October). *Phasedown of Hydrofluorocarbons: Establishing the Allowance Allocation and Trading Program Under the American Innovation and Manufacturing Act.*

Goetzler, W., Guernsey, M., & Young, J. (2014). Research & development roadmap for emerging HVAC technologies. *U.S. Department of Energy, Building Technologies Office.*

Goetzler, W., Zogg, R., Young, J., & Johnson, C. (2014). Energy savings potential and RD&D opportunities for non-vapor-compression HVAC technologies. *Navigant Consulting Inc., prepared for U.S. Department of Energy.*

James, N. A., Braun, J. E., & Groll, E. A. (2019, February). The chemical looping heat pump: Thermodynamic modeling. *International Journal of Refrigeration*, *98*, 302–310. doi: 10.1016/j.ijrefrig.2018.11.005

Klein, S. (2018). *Software F-Chart, EES-Engineering Equation Solver.*

Lu, Z., Kim, J., Braun, J. E., & Ziviani, D. (2021, June). A quantitative review of the state-of-the-art of conventional and alternative HVAC&R technologies. In *ASHRAE Transaction* (pp. VC–21A–C063). ASHRAE Transaction.

Masson-Delmotte, V., et al. (Eds.). (2021). *IPCC Climate Change 2021: The Physical Science Basis. Contribution of Working Group I to the Sixth Assessment Report of the Intergovernmental Panel on Climate Change.* Cambridge University Press.

Mendon, V. V., & Taylor, Z. T. (2014). *Development of Residential Prototype Building Models and Analysis System for Large-Scale Energy Efficiency Studies Using EnergyPlus* (Tech. Rep.). Pacific Northwest National Lab.(PNNL), Richland, WA (United States).

Neves, R., Cho, H., & Zhang, J. (2021, March). State of the nation: Customizing energy and finances for geothermal technology in the United States residential sector. *Renewable and Sustainable Energy Reviews*, *137*, 110463. doi: 10.1016/j.rser.2020.110463

NREL. (2021). *U.S. Utility Rate Database.*

Ore, J. P., Ziviani, D., & Groll, E. A. (2021). Evaluation of Energy Consumption and Efficiency in a Historical Residential Home Before and After Building Renovations. In *ASHRAE Transactions* (Vol. 127, p. 110).

Philippe, M., Bernier, M., & Marchio, D. (2010). Sizing calculation spreadsheet: Vertical geothermal borefields (zip file). *ASHRAE Journal*, *52*(7), 20–28.

Qian, S., Nasuta, D., Rhoads, A., Wang, Y., Geng, Y., Hwang, Y., … Takeuchi, I. (2016, February). Not-in-kind cooling technologies: A quantitative comparison of refrigerants and system performance. *International Journal of Refrigeration*, *62*, 177–192. doi: 10.1016/j.ijrefrig.2015.10.019

Shi, P., Wang, L.-S., Schwartz, P., & Hofbauer, P. (2020, November). State-wide comparative analysis of the cost saving potential of Vuilleumier heat pumps in residential houses. *Applied Energy*, *277*, 115547. doi: 10.1016/j.apenergy.2020.115547

Tang, C. C. (2005). *Modeling packaged heat pumps in a quasi-steady state energy simulation program* (PhD Thesis). Oklahoma State University.

Wilcox, S., & Marion, W. (2008). *Users manual for TMY3 data sets.* National Renewable Energy Laboratory Technical Report. (Publisher: Citeseer)

A Heuristic Model Predictive Control Method to Activate the Energy Flexibility of Electrically-Heated School Buildings

Navid Morovat[1]*, Andreas K. Athienitis[1], José Agustín Candanedo[1,2]

[1]Concordia University, Centre for Zero Energy Building Studies,
Montréal, Québec, Canada
n_morova@encs.concordia.ca, andreask.athienitis@concordia.ca

[2]CanmetENERGY,
Varennes, Québec, Canada
jose.candanedoibarra@canada.ca

* n_morova@encs.concordia.ca

ABSTRACT

This paper presents a methodology to model and activate the energy flexibility of electrically-heated school buildings using a heuristic model predictive control approach. The heuristic model predictive control method is developed based on data-driven grey-box models for archetype thermal zones in school buildings. The archetype zones include the zones with convective systems (e.g., classrooms, library, and kindergarten). A third-order RC network for classrooms is developed and calibrated using measured data from an archetype school building. The winter peak load in Québec (Canada) significantly strains the electrical grid. Therefore, very cold and cold winter days are clustered into three categories depending on the one-day-ahead weather prediction: sunny, semi-cloudy/intermediate, and cloudy days. Heuristic setpoint profiles are selected to achieve the optimal zone temperature profile based on forecasted weather scenarios while considering the thermal comfort of the occupants. Key performance indicators are applied to quantify the energy flexibility at the school building. The case study school (located in Québec, Canada) is an electrically-heated building with geothermal heat pumps, radiant floors, and energy storage. Preliminary results show that with an appropriate heuristic model predictive control strategy, the zones with the convective system can provide energy flexibility of 47% during on-peak hours relative to a reference as-usual profile.

1. INTRODUCTION

The share of renewable energy sources (RES) increases in parallel with extensive electrification of the energy demand. The large-scale deployment of intermittent RES could adversely affect the operation and stability of the grid. In Québec, winter's morning and evening peak load significantly strain the electrical grid, while photovoltaic (PV) electricity production peaks around noon. Therefore, controlling energy use is essential to reduce the mismatch between supply and demand (Afroz *et al.*, 2018; Klein *et al.*, 2017; Morovat *et al.*, 2019). Recently, Québec awarded a 25-year contract to power New York State (HydroQuebec, 2021), and there has been another contract with Massachusetts since 2019 (HydroQuebec, 2019). These contracts mean the province needs to emphasize load management in buildings more. In this context, energy flexibility is critical to addressing the grid's challenges of (a) balancing supply and demand and (b) incorporating renewable energy capacity. Energy flexibility is defined as "the ability to manage demand and generation according to local climate conditions, user needs, and grid requirements."
The role of buildings as flexible loads is increasingly recognized as a key component of electricity grids; they can act as energy generators, energy storage devices, and controllers of demand. A study by Choi *et al.* (2012) revealed that the best level of energy savings cannot always be achieved just by implementing new energy-saving technologies. They found that energy management strategies could be more effective than energy-saving technologies.

Commercial buildings are typically the focus of studies evaluating buildings' energy performance (Azar *et al.*, 2012; Mulville *et al.*, 2014), but few studies have been focused on school buildings in Canada. There are over 15,500 schools in Canada, with more than 5 million students and nearly 700,000 teachers and other employees (Statistic

Canada, 2017). Ouf *et al.* (2017) found that school buildings' median total energy consumption is higher than other similar Canadian benchmarks. Thus, quantification of energy flexibility in school buildings has a significant role in providing a safe and efficient operation of the future resilient grid.

Therefore, we need to develop models for school buildings that provide reliable predictions and can be generalized for widespread deployment in schools. Classical control techniques such as thermostat control (On/Off control with predefined setpoints and PID control) are popular in Building Energy Management Systems (BEMS). These controllers react to changes in weather and occupancy conditions without making any predictions. These approaches are suitable for fast responding local loops. However, they fail to efficiently control slow responding dynamic processes (e.g., radiant floor systems used for energy storage) (Afram *et al.*, 2014). Medium and large commercial buildings usually have significant thermal mass in exposed concrete or tiled concrete floors. Thus, anticipatory controls can be beneficial since they address the delay between the supplied heating/cooling and its effect on the indoor temperature (Ruusu *et al.*, 2019). MPC can enable programming the building operation based on future weather and occupant behavior. The proactive "look ahead" approach of MPC makes it possible to optimize the building operation, resulting in significant improvements in energy flexibility, IEQ, load management, and building-grid interaction.

Predictive control strategies can be classified into two types: 1) rule-based model predictive control (RBMPC), which is a Near-optimal approach; 2) model-based predictive control (MPC), which is a "proper" MPC, based on a formal mathematical optimization (O'Brien *et al.*, 2015). The first approach is typically used for actual implementations since formulating an optimization problem in real-time is challenging. Load shifting with fixed scheduling is the most common form of rule-based control to maximize energy flexibility (Carvalho *et al.*, 2015; Lee *et al.*, 2015).

This paper presents a heuristic model predictive control approach to model and activate the energy flexibility of electrically-heated school buildings in cold climate regions. The study consists of four components:

- Developing grey-box models and calibration of these models with real data,
- Clustering forecasted weather data based on the outdoor temperature and solar irradiance,
- Testing of different control scenarios to assess the impact on electricity demand and energy performance,
- Energy flexibility quantification to enable interaction between buildings, aggregators, and utilities.

2. Case study: Electrically-heated school building

An electrically-heated school building (Horizon-du-Lac) located near Montréal, Canada (45°31'N 73°56'W), was considered as a case study (Figure 1). It is a two-floor school building with twenty classrooms, five offices, one gym, one kindergarten, one kitchen, and one meeting room as shown in Figures 2 and 3.

Figure 1: Horizon-du-Lac school building in a winter cold cloudy day

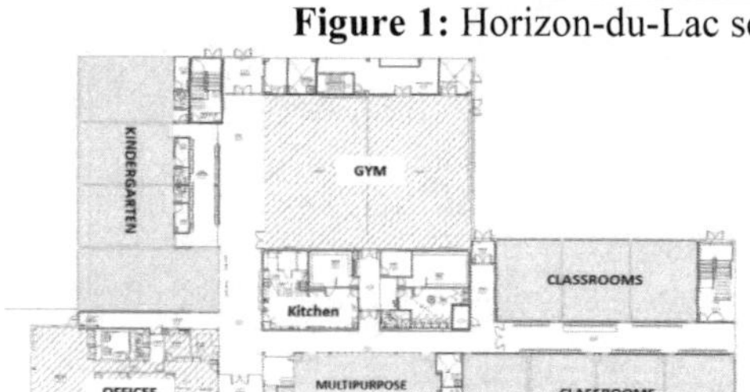

Figure 2: Plan view of the first floor (zones with convective system are highlighted)

Figre 3: Plan view of the second floor (zones with convective system are highlighted)

This study was conducted in a school with geothermal heat pumps, radiant floors, and energy storage systems. A geothermal water-water HP can provide 24 kW of heating, a high-temperature thermal storage device can store 80 kW of heating, and 36 water-air HPs can generate 182 kW of heating. Proportional-Integral control (PI) is used in the local-loop control of room air temperature. All heating systems are electrical devices and hence provide a link with the electrical grid. A predictive controller can exploit this link to help balance electricity production and demand, among other potential uses. Figure 4 presents the schematic of the building heating systems.

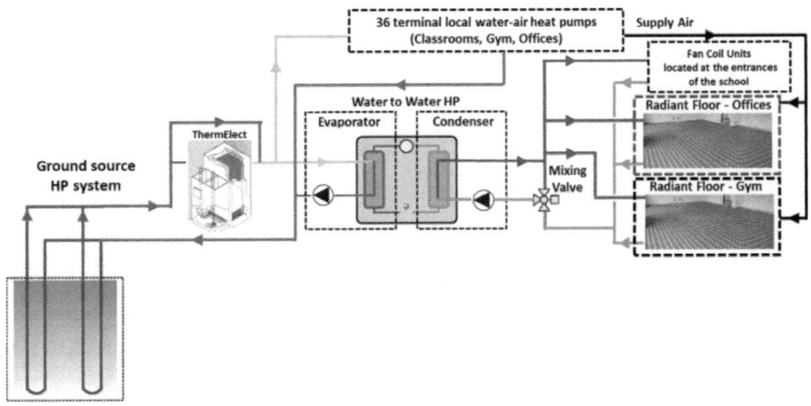

Figure 4: Process flow diagram (PFD) of the building heating systems.

3. METHODOLOGY

The heuristic MPC routine proposed in this paper includes collecting input data, developing an archetype data-driven grey-box model, clustering the weather forecasts with a prediction horizon of 24 h, developing heuristic model predictive control strategies, and quantifying energy flexibility and energy efficiency (Figure 5).

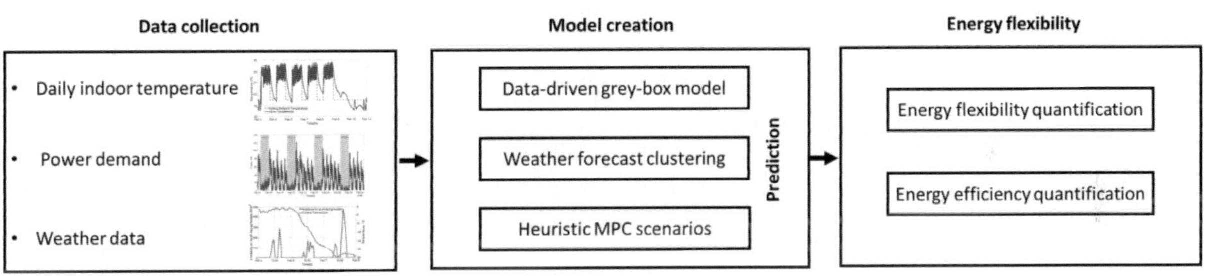

Figure 5: Model structure – Data collection, predicting, and energy flexibility quantification

2.1 Data-Driven Grey-Box Modelling

Proper identification of building thermal models with adequate resolution, robust, and acceptable computation time is fundamental for implementing MPC or other advanced control strategies in building automation systems. Reducing features to those that are most relevant can deliver the following gains:

- Improved performance.
- Reduced complexity.
- Improved interpretability of the developed models.

The grey box model structures are derived from resistance-capacitance (RC) networks analogy to electric circuits to describe the dynamics of the systems. Grey-box models rely on physical knowledge about the system dynamics to define the model structure (i.e., the layout of RC parameters). Optimization methods are then used to estimate the unknown parameters. These models are an integral part of the heuristic MPC method. By performing a heat balance on the control volume, the differential equation can be written as equation 1:

$$C_i \frac{dT_i}{dt} = Q_i + \sum_{i,j}^{n} \frac{(T_j - T_i)}{R_{i,j}} \tag{1}$$

Each node solves its energy balance using a fully explicit finite difference approach, in which the conditions determine the current temperature at the previous time step. The time derivative term is discretized as follows:

$$C_i \frac{dT_i}{dt} \approx C_i \frac{\Delta T_i}{\Delta t} = C_i \frac{(T_i^{t+1} - T_i^t)}{\Delta t} \tag{2}$$

Temporal discretization of Equation (1) can be rearranged in the following explicit manner:

$$T_i^{t+1} = \frac{\Delta t}{C_i} \left[\sum_j U_{ij}^t (T_j^t - T_i^t) + \sum_k U_{ik}^t (T_k^t - T_i^t) + \frac{C_i}{\Delta t}(T_i^t) + \dot{Q}_i^t \right] \tag{3}$$

Where $R_{i,j}$ is the thermal resistance between nodes, Q_i is the heat source generation at a node, T is the temperature of a node, and C is the thermal capacitance of a node. Using statistical indices, such as the Coefficient of Variance of Root Mean Squared Error (CV-RMSE), model predictions are compared to measurements, and RC parameters are identified. The details of the governing equations are presented in (O'Brien & Athienitis, 2015).

2.2 Weather data clustering

This paper implements a centroid-based approach (*K*-means clustering) in Python to classify weather prediction into sunny, semi-cloudy, and cloudy days. The forecasts were obtained using CanMETEO® a free program developed by Natural Resources Canada (NRCan, 2021). An illustration of the *K*-means clustering algorithm is shown in Figure 6.

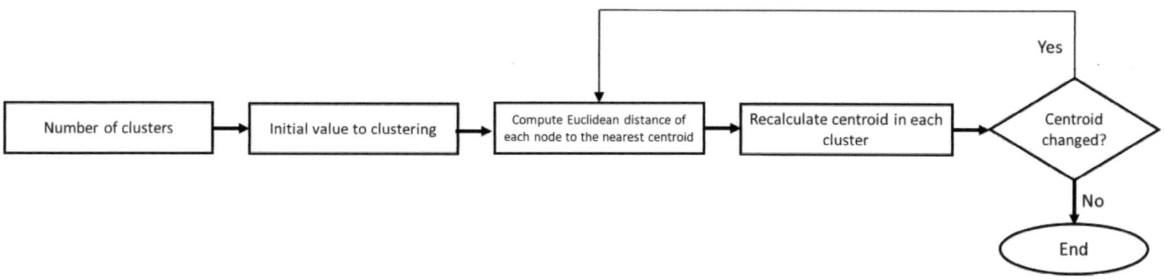

Figure 6: *K*-means clustering algorithm flowchart

The first step in *K*-means clustering is selecting the number of clusters (*K*). Then, *K* time-series data points are randomly selected from the data set to serve as the initial centroids for clusters. Each data point is assigned to a cluster based on its closest centroid. In the third step, a new centroid for each cluster is calculated. Once the centroids are recalculated, each data point is examined for closer proximity to a different centroid (a new cluster assignment). Cluster assignment and centroid updating are repeated until the cluster assignments stop changing.

2.3 Rule-based model predictive control

MPC is a control method that includes a dynamic model of the system to be controlled, forecasts of future disturbances (such as weather, occupancy), and a cost function that is minimized over a prediction horizon. Figure 7 provides a representation of the rule-based MPC approach which is used in this study.

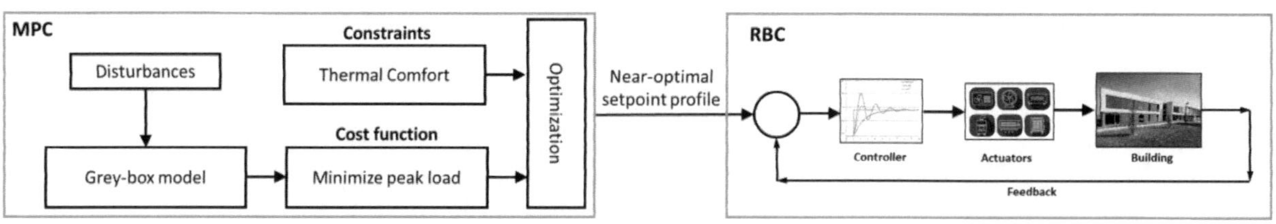

Figure 7: Conceptual representation of the rule-based model predictive Control (RBMPC)

The optimal control problem in MPC is solved by looking ahead at forecast disturbances (e.g., weather) over the prediction horizon. The prediction horizon is a period that has reasonably reliable information about the future, ranging from a few hours to a few days. An optimization routine is solved using data from the prediction horizon period, and an optimal sequence of control moves is identified through MPC implementation. A "control horizon" is applied to the building to determine schedules and controls. The objective function of the controller is to minimize a cost that may include energy and power while maintaining the thermal comfort of occupants. Equation 4 describes the general MPC framework; some describe the system's dynamics for control, while others define limits and boundary conditions.

$$
\begin{aligned}
& J = \min_{u^0,\dots,u^{N-1}} \sum_{t=0}^{N-1} l(x^t, u^t, w^t) \\
& \text{subject to } h_i(x^t, u^t, w^t) = 0, i \in \{1,\dots,m\}, \text{equality constraints} \\
& g_j(x^t, u^t, w^t) \geq 0, j \in \{1,\dots,n\}, \text{inequality constraints} \\
& x^0 = x, \text{ current state} \\
& x^{t+1} = f(x^t, u^t, w^t) = Ax^t + Bu^t + Ew^t, \text{ system dynamics}
\end{aligned} \tag{4}
$$

Where,

- x^t system variables track the system dynamics,
- u^t control variables which can be manipulated in order to improve the building performance,
- w^t exogenous inputs that can be observed but cannot be controlled, such as weather,
- $l\,(x, u, w)$ cost function, which could be to minimize the utility cost or the grid interaction.
- $h\,(x, u, w) = 0$ equality constraints; the system dynamics of the system are given by the trained model,
- $g\,(x, u, w) \geq 0$ inequality constraints; here, the inequality constraints are the boundaries of the problem.

2.1 Energy Flexibility Indicators

A building Energy Flexibility Index would help define the amount of power variation available from a building (Finck *et al.*, 2020). In this paper, energy flexibility has been calculated based on Equation (5). This Equation calculates the average BEFI under implementing the flexibility strategy and the reference as-usual profile. We presented the details of this index in (Athienitis *et al.*, 2020).

$$\overline{BEFI}(t, Dt) = \frac{\int_t^{t+Dt} P_{ref}\,dt - \int_t^{t+Dt} P_{flex}\,dt}{Dt} \tag{5}$$

The calculation of the BEFI as a percentage compares the peak power under the flexible case and the reference as usual profile (Equation 6).

$$BEFI\% = \frac{P_{ref} - P_{Flex}}{P_{ref}} \tag{6}$$

4. RESULTS AND DISCUSSION

4.1 Model development

The measured data (room temperatures, historical weather data, and power demand) were used to calibrate an archetype grey-box model of the zones with convective systems. The third-order model, shown in Figure 8, is a resistance-capacitance (RC) thermal network with three capacitances. In this model, C_1 represents the envelope capacitance, C_2 the air capacitance, C_4 the floor capacitance, $R_{1,ext}$ the thermal resistance between the envelope and the exterior air, $R_{2,ext}$ the thermal resistance between the interior and exterior air, $R_{1,2}$ the thermal resistance between the envelope and the interior air, and $R_{2,3}$ the thermal resistance between the interior air and the floor.

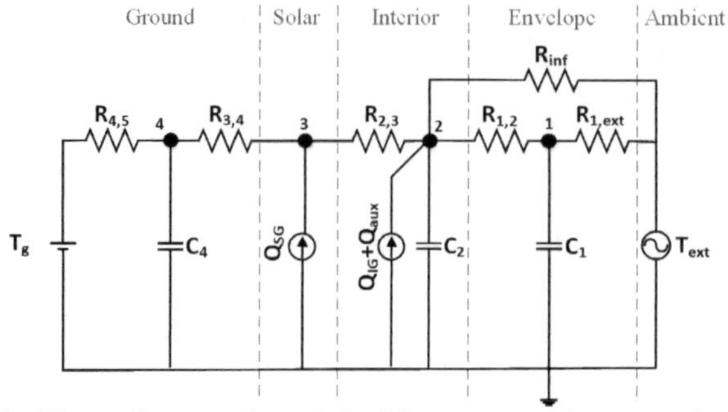

Figure 8: Thermal network model of the zones with convective system

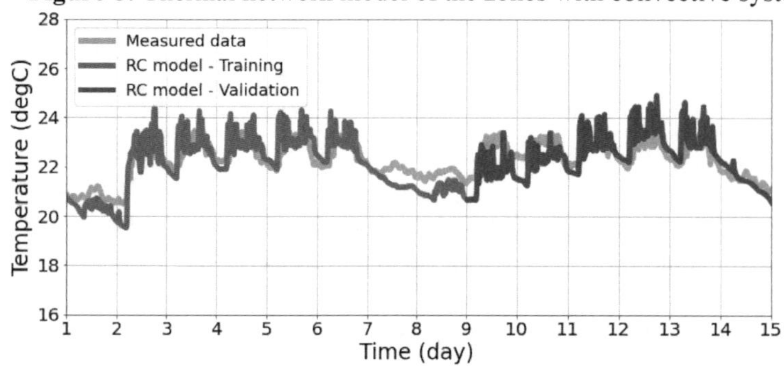

Figure 9: Calibration of thermal network model with measured data

The measured data from February 1st to February 15th were used for model learning (Figure 9). The dataset was divided into two datasets: 60% of the dataset was used for training, while 40% was kept for validation purposes. The performance of models was assessed by comparing model predictions with BAS measurements. The accuracy was evaluated in terms of coefficient of variance of the root mean square error (CV-RMSE) as a fit metric. In accordance with ASHRAE Guideline 14, the model should not exceed a CV-RMSE of 30% relative to hourly measured data (ASHRAE Guideline 14, 2002). By minimizing CV-RMSE, the optimization algorithm determines the equivalent parameters for RC circuits.

4.2 Weather forecast

Weather data for Montreal's coldest months (January and February) are selected because peak energy demand occurs under these conditions. The historical weather data is obtained from the hourly Montreal weather file [32]. *K*-means clustering is performed for the solar irradiance (SI) and outdoor temperature data, and the Silhouette index is used to determine the clustering quality. With this indicator (ranging between +1 and -1), the clustering quality is assessed according to how well data points fit within the cluster. Clusters with an average silhouette index close to 1 are dense and well-separated; negative values indicate that the dataset may have been grouped incorrectly. The number of clusters (within a specific range) that maximizes the average Silhouette is considered the optimal value. As shown in Figure 10 and 12, the average Silhouette index for the solar radiation shows three and for outdoor temperature shows two clusters as the optimal number of clusters.

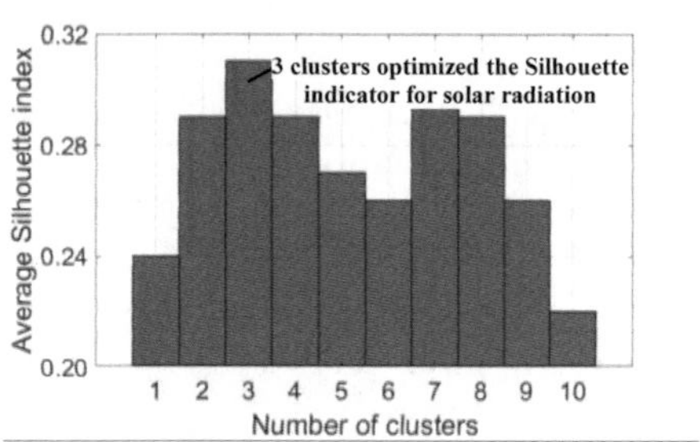

Figure 10: Average Silhouette index for the solar radiation

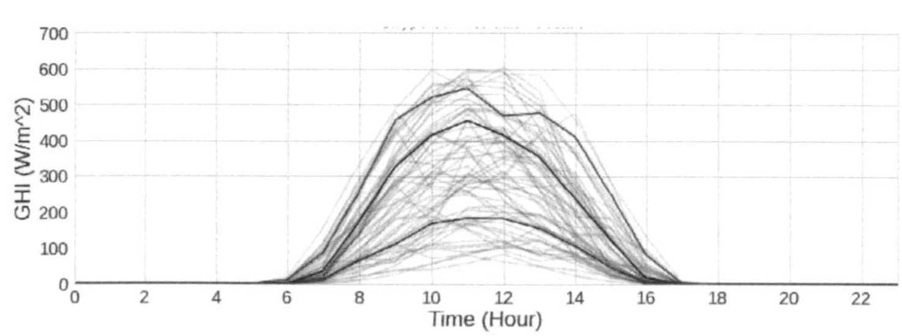

Figure 11: Three clusters and the corresponding solar radiation

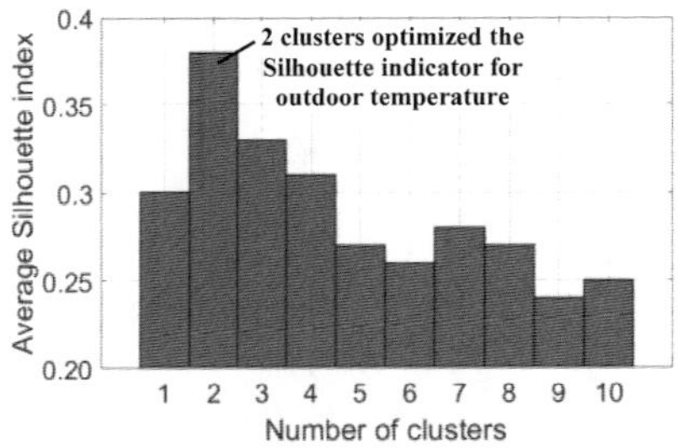

Figure 12: Average Silhouette index for the outdoor temperature

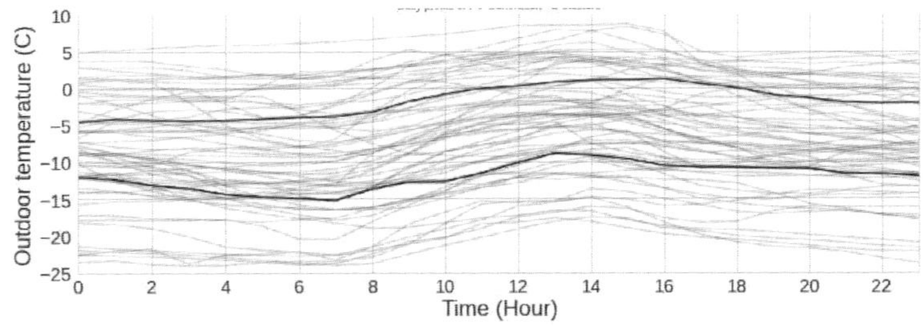

Figure 13: Two clusters and the corresponding outdoor temperature

Thus, the possible expected weather conditions are clustered into six possible categories. A predictive setpoint profile is developed with a target for each scenario to maximize energy flexibility. Table 1 presents weather forecast scenarios clustered based on the outdoor temperature and solar irradiance.

Table 1: Weather prediction classification

	Cloudiness		
Ambient Temperature	**Sunny** (SI >500 W/m^2)	**Semi-cloudy** (250 < SI < 500 W/m^2)	**Cloudy** (SI < 250 W/m^2)
Very cold day (T_{avg} = -12.5 °C)	Scenario 1	Scenario 2	Scenario 3
Cold day (T_{avg} = -2.5 °C)	Scenario 4	Scenario 5	Scenario 6

The developed model runs a simulation using forecast weather data for each scenario to estimate the associated electricity consumption over the next 24 h. The setpoint profile yielding the lowest peak load during on-peak hours is selected as the optimal control strategy for energy flexibility.

4.3 Heuristic MPC routine

Once the building model is appropriately developed, a set of setpoint profiles can be defined with an objective function. The objective of the proposed MPC is the peak load reduction in electrically-heated school buildings. Overall, this approach aims to provide a general methodology for load management in commercial and institutional buildings, thus facilitating replicability in other buildings.

The MPC routine consists of the following steps:

1. The control-oriented model estimates the building heating demand for all the considered pre-defined setpoint profiles (Figure 14).
2. Weather forecasts with a prediction horizon of 24 h, derived from CanMETEO® (developed by Natural Resources Canada).
3. Weather forecasts are clustered and used along with the control-oriented model to estimate the building heating demand for all the considered predefined set-point profiles.
4. The setpoint profile achieving the lowest peak load is identified as a "near-optimal control strategy" for energy flexibility.
5. The next day, the building is operated under the identified optimal control set-point.

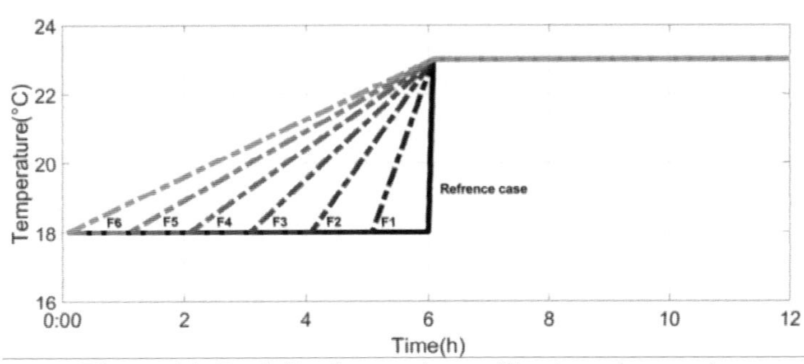

Figure 14: Predefined set-point profiles in early morning

The considered pre-defined setpoint profiles could be applied to most commercial and institutional buildings, as these buildings are usually occupied during the same period of the day, from 8:00 to 17:00, while considering thermal comfort.

4.4 Energy Flexibility

Quebec's utility rates include fees for power consumption and demand. Because of this, customers are seeking to reduce energy consumption and adequately manage their power use. Customers can enroll in a Demand Response (DR) Program during peak winter times to get financial assistance to reduce their building's demand. In order to assess the peak load reduction from the MPC implementation, benchmarking models of the building loads under business-as-usual control were developed (Figures 15a and 16a). These benchmarking models predict the daily power demand using the outdoor air temperature and solar radiation as inputs. The observed trend was studied by running the control-oriented models with weather data corresponding to two different days of the MPC implementation period: a) a very cold day (-12.5 °C average daily OAT) and b) a cold day (-2.5 °C average daily OAT). Six predefined profiles were tested on both day clusters. For example, Figures 15b and 16b present power demand for very cold days and cold days under a set-point profile with a four-hour transition ramp.

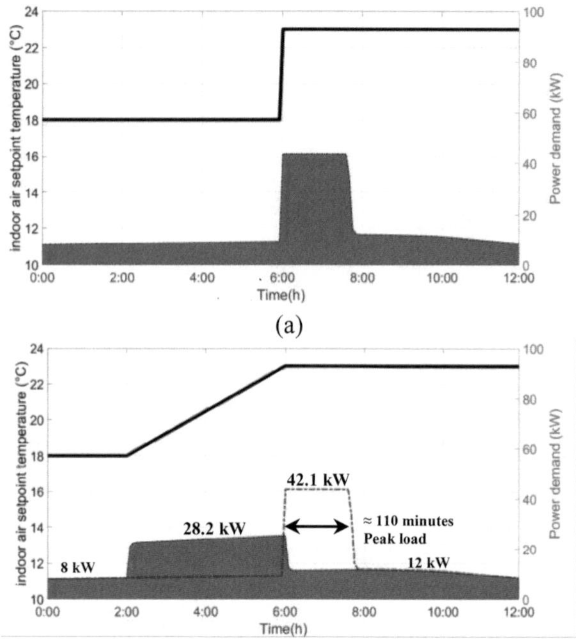

(b)

Figure 15: Power demand during very cold sunny days (Daily avergae OAT = -12 °C) with a) Step-change, b) 4-hour ramp in setpoint temperature

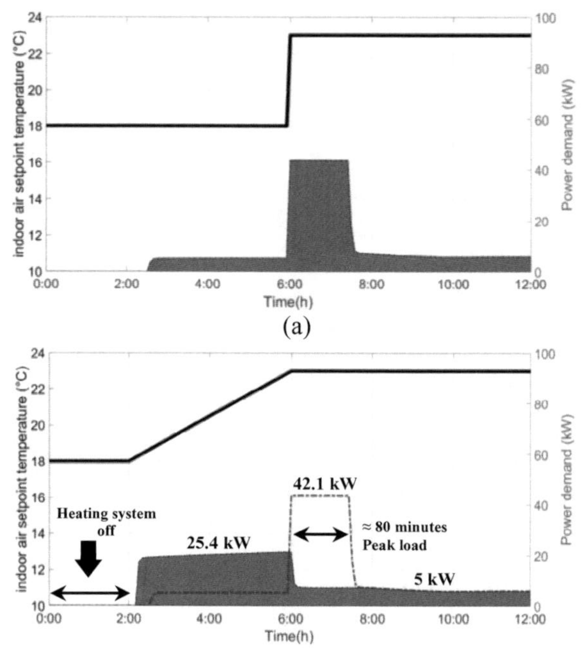

(b)

Figure 16: Power demand during cold cloudy days (OAT = -2 °C) with a) Step-change, b) 4-hour ramp in setpoint temperature

In the case of the very cold and cold days, simulations showed that when the outdoor air temperature is very low, it is beneficial to maintain a lower night set-back value to avoid peak load during on-peak hours. An almost flat set-point allows shifting the building heating load from the on-peak hour to the off-peak hours (nighttime). Furthermore, MPC improved occupants' thermal comfort by gradually increasing the indoor temperature during the night, rather than applying abrupt set-point variations just before occupancy starts.

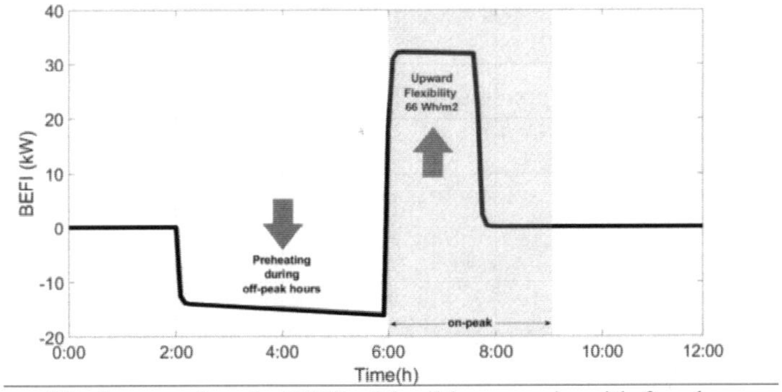

Figure 17: Average hourly BEFI – Flexible scenario with four hours ramp

As can be observed in Figure 17, by applying four hours ramp, the available hourly BEFI that can be provided to the grid during peak hours is positive, which indicates the value of power reduction available compared to the reference case. During off-peak hours (nighttime), the BEFI is negative, showing a higher power demand for preheating the zones. According to Figure 19, energy flexibility of around 66 W/m^2 in the during morning on-peak hours can be provided to the grid.

Table 2: BEFI and energy efficiency for proposed control strategies

Control Scenario	Peak load (kW)	BEFI (%)	Energy consumption (kWh)	Energy efficiency (%)
Reference case	42.1 kW	(-)	173.9	-
1-hour ramp (F1)	42.1 kW	(0 %)	174.5	- 0.34
2-hour ramp (F2)	39.2 kW	(6 %)	175.8	- 1.09
3-hour ramp (F3)	31.5 kW	(25 %)	176.4	- 1.43
4-hour ramp (F4)	25.4 kW	(39%)	177.9	- 2.30
5-hour ramp (F5)	24.1 kW	(42 %)	178.7	- 2.81
6-hour ramp (F6)	22.2 kW	(47 %)	179.9	- 3.45

According to table 2, with notification from the utility to the customer given at midnight (6 hours ahead of an event at 6 AM), a BEFI of 47% can be achieved. Depending on the utility rate structure, the peak demand during critical event hours can be reduced by 20 kW for 3 hours. A relatively short notification time of 3 hours also allows MPC to reduce peak demand during the critical hours by 25 percent and achieve a BEFI of 10.4 kW for 3 hours. In general, this approach provides a method for load management in commercial and institutional buildings, which can be replicated in other buildings. In the future work, the MPC strategy developed and presented in this paper will be improved by integrating constraints on building power demand into the objective function of the MPC routine and by considering additional indoor setpoint profiles when aiming for optimal control.

5. CONCLUSIONS

The development of control-oriented rule-based models for MPC applications requires less time, information, and technical expertise than an approach based on proper MPC. This paper presented the heuristic model-based prediction control strategies applied to an archetype electrically heated school building. Results show that rule-based MPC can be a successful alternative to the fully optimized MPC to significantly improve energy flexibility and enhance load management while eliminating the cost of modeling. Compared to simple rule-based controls, it comprises the building model, thus considering the thermal behavior under the anticipated weather conditions and taking near-optimal actions to benefit from that information maximally. This approach provides a general methodology for load management in commercial and institutional buildings, enabling replicability in other buildings.

NOMENCLATURE

BEFI	Building Energy Flexibility Index (W)	*R*	Thermal resistance (K/W)
C	Thermal capacitance (J/K)	RES	Renewable energy sources
Dt	Time (seconds/hours)	*T*	Temperature (°C)
J	Objective function	T_{SP}	Setpoint temperature (°C)
P	Electric power (W)	*x*	state

REFERENCES

Afram, A., & Janabi-Sharifi, F. (2014). Theory and applications of HVAC control systems–A review of model predictive control (MPC). *Building and Environment, 72*, 343-355.

Afroz, Z., Shafiullah, G., Urmee, T., & Higgins, G. (2018). Modeling techniques used in building HVAC control systems: A review. *Renewable and Sustainable Energy Reviews, 83*, 64-84.

ASHRAE Guideline 14. (2002). Guideline 14-2002, Measurement of energy and demand savings. *American Society of Heating, Ventilating, and Air Conditioning Engineers, Atlanta, Georgia.*

Athienitis, A. K., Dumont, E., Morovat, N., Lavigne, K., & Date, J. (2020). *Development of a dynamic energy flexibility index for buildings and their interaction with smart grids.* Paper presented at the 2020 Summer Study on Energy Efficiency in Buildings, California, USA.

Azar, E., & Menassa, C. C. (2012). A comprehensive analysis of the impact of occupancy parameters in energy simulation of office buildings. *Energy and Buildings, 55*, 841-853.

Carvalho, A. D., Moura, P., Vaz, G. C., & de Almeida, A. T. (2015). Ground source heat pumps as high efficient solutions for building space conditioning and for integration in smart grids. *Energy Conversion and Management, 103*, 991-1007.

Choi, I. Y., Cho, S. H., & Kim, J. T. (2012). Energy consumption characteristics of high-rise apartment buildings according to building shape and mixed-use development. *Energy and Buildings, 46*, 123-131.

Finck, C., Li, R., & Zeiler, W. (2020). Optimal control of demand flexibility under real-time pricing for heating systems in buildings: A real-life demonstration. *Applied energy, 263*, 114671.

HydroQuebec. (2019). Retrieved from https://www.hydroquebec.com/clean-energy-provider/markets/new-england.html#:~:text=Further%20to%20a%20clean%20energy,state's%20regulatory%20authorities%20in%202019.

HydroQuebec. (2021). Retrieved from https://news.hydroquebec.com/en/press-releases/1781/champlain-hudson-power-express-finalizes-contract-to-deliver-clean-energy-to-new-york-city/

Klein, K., Herkel, S., Henning, H.-M., & Felsmann, C. (2017). Load shifting using the heating and cooling system of an office building: Quantitative potential evaluation for different flexibility and storage options. *Applied energy, 203*, 917-937.

Lee, K.-H., Joo, M.-C., & Baek, N.-C. (2015). Experimental evaluation of simple thermal storage control strategies in low-energy solar houses to reduce electricity consumption during grid on-peak periods. *Energies, 8*(9), 9344-9364.

Morovat, N., Athienitis, A. K., Candanedo, J. A., & Dermardiros, V. (2019). Simulation and performance analysis of an active PCM-heat exchanger intended for building operation optimization. *Energy and Buildings, 199*, 47-61.

Mulville, M., Jones, K., & Huebner, G. (2014). The potential for energy reduction in UK commercial offices through effective management and behaviour change. *Architectural Engineering and Design Management, 10*(1-2), 79-90.

NRCan. (2021). CanMETEO. Retrieved from https://www.nrcan.gc.ca/energy/canmeteo/19908

O'Brien, W., & Athienitis, A. (2015). *Modeling, design, and optimization of net-zero energy buildings*: John Wiley & Sons.

Ouf, M. M., & Issa, M. H. (2017). Energy consumption analysis of school buildings in Manitoba, Canada. *International Journal of Sustainable Built Environment, 6*(2), 359-371.

Ruusu, R., Cao, S., Delgado, B. M., & Hasan, A. (2019). Direct quantification of multiple-source energy flexibility in a residential building using a new model predictive high-level controller. *Energy Conversion and Management, 180*, 1109-1128.

Statistic Canada. (2017). Elementary–Secondary Education Survey for Canada, the provinces and territories. . Retrieved from https://www150.statcan.gc.ca/n1/daily quotidien/171103/dq171103c-eng.htm.

ACKNOWLEDGEMENT

Technical support from Hydro-Québec Laboratoire des Technologies de l'énergie Shawinigan research center under NSERC/Hydro-Quebec Industrial Research Chair is greatly acknowledged. Also, technical support provided by Commission scolaire de la Seigneurie-des-Mille-Îles (CSSMI) is acknowledged with thanks. The authors would like to thank NSERC for the funding of this project through Dr. Candanedo's Discovery Grant, and the NSERC/Hydro-Québec Industrial Research Chair ("Optimized operation and energy efficiency: towards high performance buildings") held by Dr. Athienitis. The comments of our colleagues at CanmetENERGY are gratefully appreciated.

Investigating the quality of the correlation between indoor environmental factors and human perception

Ilaria PITTANA[1], Federica MORANDI[2], Francesca CAPPELLETTI[1]*, Andrea GASPARELLA[2]

[1]Department of Architecture and Arts, Iuav University of Venice, Italy
ipittana@iuav.it, francesca.cappelletti@iuav.it

[2]Faculty of Science and Technology, Free University of Bozen-Bolzano, Italy
federica.morandi@unibz.it, andrea.gasparella@unibz.it

* Corresponding Author

ABSTRACT

In building design, commissioning and operation, the prediction of people comfort level is fundamental. However, the overall effect of the indoor environment on human perception and performance depends on the combination of many environmental factors, such as acoustic, thermal, visual, and air quality stimuli. Among all different building types, in the educational ones the indoor factors can critically affect students' concentration and ability to learn, so that schools are especially worth investigating. In this framework this paper aims to investigate the correlation between the measured environmental characteristics and the subjective response obtained by means of a questionnaire, designed for the purpose of capturing students' sensation, preference, and comfort in all the four environmental domains that contribute to people indoor wellbeing in educational buildings. Fifty-five panels, i.e., groups of students interviewed in the same moment and in the same classroom, have been considered, collecting 949 questionnaires. To increase the representativeness of the responses, panels have been grouped according to similar environmental conditions and subjective responses. In each group the mean sensation and comfort votes have been compared with the mean value of each environmental parameter, namely air temperature, CO_2 concentration, illuminance level, and sound pressure level characterizing the corresponding domain. For each regression the coefficients of determination and p-values have been calculated. In order to evaluate the confidence on these parameters, and consequently to characterize the quality of the correlations, a permutation test has been applied. Some quality correlations are found between indoor environmental factors and human perception either on the same or on different environmental domains. In particular, results show that all sensation votes are correlated with the corresponding environmental parameters with a high confidence level. Concerning the cross-domain effects, air temperature and illuminance are the parameters correlated with the highest confidence to any of the other domains' sensations, while CO_2 concentration is correlated with high confidence only to thermal sensation.

1. INTRODUCTION

Inhabitants of industrialized areas spend most of their time (85–96%) inside buildings (Schweiker et a., 2020, Frontczak et al., 2011). Among them, students use to stay more than 30% of the day in classrooms (Giuli et al., 2012). In the spotlight due to the resumption of learning activities after the health emergency related to the COVID pandemic, the Indoor Environmental Quality (IEQ) of school buildings is increasingly considered. IEQ is the result of the interaction of multiple quantities related to indoor air quality, thermal, visual and acoustic environment, affecting occupants' sensation, preference, and comfort (Frontczak et al., 2011). While IEQ can be assessed through in-field measurements, the evaluation of people' perception is conducted through subjective surveys. In the last decades, several studies combined objective and subjective evaluations to characterize the comfort requisites and to find

possible correlations between physical parameters and subjective response in terms of discomfort, loss of productivity or adaptation activity. Most of the research took into consideration a single comfort aspect at a time, with a large share investigating thermal comfort only (Corgnati et al., 2007, Kim et al., 2018). More recently, researchers tried more comprehensive approaches to consider together with thermal comfort a second domain, such as acoustic comfort (De Giuli et al., 2012, Astolfi et al., 2008), Indoor Air Quality (Korsavi et al., 2020, Vilčeková et al., 2017) and visual comfort (Dorizas et al., 2015, Fassio et al. 2014). Some studies explore the cross-effects of different environmental characteristics on students' comfort and satisfaction in order to find a comprehensive comfort or satisfaction index or to understand the interference that different environmental parameters have to the subjective satisfaction (Ricciardi and Buratti, 2018, Buratti et al., 2018, Tang et al., 2020). Yet, findings on multi-domain effects of environmental parameters on sensation, comfort, or satisfaction, seem not to show a general agreement.
In this framework, this paper aims to investigate the influence of the environmental physical characteristics on students' sensation, considering all the four environmental aspects that contribute to indoor wellbeing in educational buildings. Measurements of physical quantities have been collected together with the subjective responses of the occupants in a range of different conditions. To this aim, a dedicated questionnaire was developed in order to make the answers in the different domains comparable. The dataset was processed through an original method for aggregating the responses based on the measured conditions, in order to reach a representative number of responses for each of the analyzed conditions. The mean sensation votes for each of the environmental domain have been compared with the value of each peculiar characteristic (i.e., air temperature, CO_2 concentration, illuminance, sound pressure level) evaluating direct (in-domain) and indirect (cross-domain) dependence. In order to judge the goodness of the obtained regressions, a permutation test has been applied.

2. METHODOLOGY

Eight field campaigns are included in the analysis, carried out in three different Italian educational buildings located in Bolzano, Rome, and Venice, from winter 2019 to spring 2021. The indoor conditions of some selected classrooms were evaluated by means short term measurements of the main physical parameters related to Indoor Air Quality (IAQ), thermal, visual, and acoustic environment, during transverse surveys administered to the students. A total number of 55 different panels, i.e., groups of students exposed to the same environmental conditions at the same time, were interviewed, collecting 949 individual questionnaires in total.

2.1 Data collection

The environmental parameters related to IEQ, i.e. globe temperature, indoor air temperature (T_{air}), relative humidity (RH), air velocity, Total Volatile Organic Compounds (TVOC), CO_2, CO, horizontal illuminance level and A-weighted equivalent sound pressure level ($L_{A,eq}$), were recorded during regular classes lasting approximately 1 h, starting from the beginning of the lesson until the students had completed the questionnaires in the last part of the observation period. Hygro-thermal parameters were recorded at 1-minute intervals by means of the Microclimatic Station DeltaOhm HD32.1 located in the center of the classrooms, away from heat sources and sun patches at a height of 1.1 m, as recommended by the Standard EN ISO 7726 (CEN 2001). TVOC, CO_2, CO, were measured at 1-minute intervals with the IAQ Meter Graywolf Wolfpack (IQ-610) located on the teacher's desk. The horizontal illuminance level was measured at all the students' desks with the illuminance meter Konika Minolta T-10A and the A-weighted equivalent sound pressure level ($L_{A,eq}$) was measured with B&K 2270 sound level meter, located in the center of the classrooms near the Microclimatic Station. Further specifications of the sensors are reported in Pittana et al. (2021).
The subjective survey consisted of a questionnaire, filled during regular classes while collecting the measurements of the physical parameters. The questionnaire was administered after measurement had been collected for at least half an hour, in order to check the achievement of steady conditions. The standard questionnaire considering all the comfort domains, i.e., thermal, visual, acoustic, and indoor air quality, along with the global comfort and satisfaction, was validated and presented in Morandi et al. (2021). It included sensation, preference, comfort, and satisfaction votes. The votes expressed by students in each panel have been aggregated in groups based on environmental conditions and then analyzed. In the present work only the sensation votes are considered for the analysis. The aggregation method and the analysis of the dataset is presented further on.
The collected responses underwent a quality verification by identifying and removing irrational, inconsistent answers or questionnaires filled by people with some learning, language, or acoustic impairment. Questionnaires with missing answers were analysed limited to the answered parts, both in case of single missing answer and uncompleted domain section.

2.2 Panels aggregation method and data analysis

In order to increase statistical significance and analyze a representative number of responses, the panels (i.e. groups of subjects exposed to the same indoor conditions – in the same room - at the same time) were grouped according to a methodology which takes into account similar environmental conditions and similar distribution of the subjective responses. According to this method, the mean values of indoor air temperature, CO_2 concentration, illuminance level and pressure level recorded during the corresponding observation period were calculated and assigned to the corresponding panel. The illuminance level was averaged over multiple measurements taken on each desk inside the classroom. The grouping and analysis method resumed in Figure 1 was previously tested on a smaller dataset and presented in Pittana et. al (2021). It consists of three main steps. First, the indoor conditions were binned considering specific ranges (bins) of the measured physical parameters. In details, for the indoor air temperature the intervals of 0.6 °C have been considered (CEN 2001); for the CO_2 average levels have been binned according to the categories defined by EN 16798-3 (CEN 2017), i.e. 500 ppm, 800 ppm, 1100 ppm and 1400 ppm; regarding the visual environment the data have been binned according to the recommended steps of illuminance defined by EN ISO 12464-1 (CEN 2011), i.e. 20, 30, 50, 100, 200, 300, 500, 1000 lx; finally for the acoustic data the step of 1 dB has been used. In doing so, the distribution of sensation votes of each panel in the same bin were pairwise compared by analyzing by means of Mann-Whitney U-test, in order to decide whether the distributions belong to the same population and could be merged in the same group. That was repeated independently for each of the environmental dimensions (i.e., indoor air temperature, CO_2 concentration, illuminance level and sound pressure level). Outlier panels, i.e. panels whose distributions of sensation votes were different, were identified and discharged through the same approach. In the second step, the same statistical test on students' sensation votes was used to decide whether to join groups from adjacent bins. As a third step, a correlation analysis was conducted to identify the relationship between measured environmental conditions and the subjective responses given by students. In particular the mean values of the physical parameters, i.e. air temperature CO_2 concentration, illuminance level and $L_{A,eq}$, were pairwise correlated to each of the mean thermal, IAQ, visual and acoustic sensation votes, to highlight direct and crossed effects between comfort domains. The goodness of the regressions was assessed using the coefficient of determination R^2 and p-value for the t-statistic of the hypothesis for each variable of the assumed regressions (i.e. x in case of linear regression; x and x^2 for the quadratic one). To further test the quality of the correlations, a permutation analysis of the observed points (physical parameters vs votes) was implemented. Regression of all permutations of the couples x (i.e., physical parameter) and y (i.e., subjective vote) coordinates allowed to assign a probability of randomly finding a better correlation than the observed one.

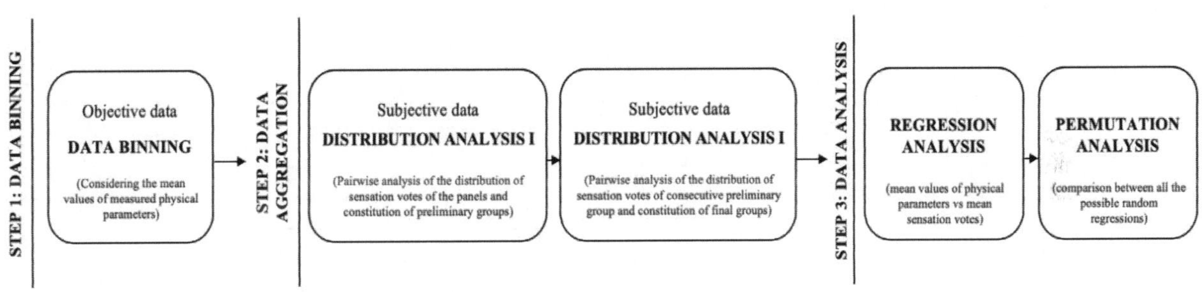

Figure 1 The three steps of the method for analyzing the dataset collected for the four environmental (Step 1: Data Binning; Step2: Data Aggregation; Step 3: Data Analysis (regression and permutation analysis of the group).

3. RESULTS

3.1 Correlation analysis between physical parameters and subjective responses

The observed sets of points for the grouped panels are shown in Table 3. In details, 5 groups (776 questionnaires) were formed with respect to the indoor air temperature, 4 (592 questionnaires) to the CO_2 concentration, 6 (773 questionnaires) based on the illuminance level and 5 (692 questionnaires) according to the A-weighted equivalent sound pressure level. The influence of the above-mentioned physical parameters on subjective responses, i.e., sensation votes, was analyzed through regression analysis (Figures 2 to 5) and evaluated by means of a permutation test. The regressions related to the effects of air temperatures, CO_2 concentrations, illuminance levels and sound pressure levels on thermal, IAQ, visual and acoustic sensations are reported in Figures 2a-d, 3a-d, 4a-d and 5a-d, respectively. In these figures the boxplots of the distribution of the measured physical parameters are also reported. In the boxplots, the red central mark is the median, the box limits denote 25th and 75th percentiles, and the upper and lower hyphens are the maximum and minimum observed values. The outcomes of the permutation tests are presented in Table 4, summarizing the values of the coefficients of determination R^2, and of the p-values for the regressions of the observed points, and in boxplot diagrams reporting the details of the distribution of R^2 and p-values of all permutations compared with those of the observed data points (Figure 3). In particular, the probabilities reported in Table 4 represent the percentage of permutations producing (i.e., the probability of) correlations with better (higher) R^2 and (lower) p-value than the observed dataset. The lower the probability, the higher the confidence that the observed R^2 and p-values are statistically significant, i.e., not originated by a random association of sensation votes with environmental conditions. In the following, probabilities lower than or equal to 10 % are associated with high confidence level (red bold font in Table 4), between 11 and 25 % with average confidence level (black bold font), between 26 % and 50 % with low confidence level (black normal font), and above 51 % with negligible confidence level (grey font). When an acceptable confidence level is found, the slope of the observed regression will give indication on the sensitivity of the sensation vote on the environmental parameter. Negligible slopes are then interpreted as independence between votes and conditions. In the same table, asterisks highlight quadratic regressions.

3.1.1 **In-domain effects**

From Table 4 it can be seen that all sensation votes are correlated with the corresponding environmental parameters, with high confidence level for what concerns both R^2 and p-values. All regressions are characterized by a non-negligible slope, meaning that the vote is sensitive to the change in the corresponding environmental parameter.

Table 3: Environmental conditions groups and subjective mean sensation votes in the same domain (in-domain, in shaded background) and in different domains (cross-domain).

Domain	**Group**	**Subjects**	**Tair [°C]**	**TSV_{mean}**	**$IAQSV_{mean}$**	**VSV_{mean}**	**ASV_{mean}**
Thermal	T1	248	20.4	-0.5	0.8	1.0	1.0
	T2	237	22.3	-0.2	0.9	0.3	1.0
	T3	183	22.9	0.2	0.9	0.6	0.9
	T4	57	24.1	0.4	1.0	0.5	0.8
	T5	53	24.7	0.8	1.3	0.2	0.7
Domain	Group	Subjects	CO2 [ppm]	TSV_{mean}	$IAQSV_{mean}$	VSV_{mean}	ASV_{mean}
Air quality	IAQ1	142	515	-0.2	0.6	0.9	0.8
	IAQ2	269	781	-0.1	0.9	0.7	1.0
	IAQ3	109	1035	-0.1	0.9	0.5	0.7
	IAQ4	72	1360	0.1	1.1	1.2	1.0
Domain	Group	Subjects	E [lx]	TSV_{mean}	$IAQSV_{mean}$	VSV_{mean}	ASV_{mean}
Visual	V1	12	19	0.1	1.1	-1.7	0.4
	V2	157	157	0.0	1.1	0.1	1.2
	V3	125	327	-0.3	0.9	0.6	0.7
	V4	351	491	-0.2	0.8	0.9	0.9
	V5	71	733	-0.2	0.8	0.9	0.9
	V6	17	983	0.7	1.8	1.4	1.5
Domain	Group	Subjects	SPL [dB]	TSV_{mean}	$IAQSV_{mean}$	VSV_{mean}	ASV_{mean}
Acoustic	A1	273	61	0.0	0.8	0.4	0.5
	A2	190	66	-0.2	0.8	0.8	0.9
	A3	37	68	0.0	0.8	0.2	0.7
	A4	122	70	0.0	1.0	1.1	1.2
	A5	70	75	-0.4	0.8	0.9	1.7

Table 4: Correlation analysis and probability from permutation test for in-domain and cross-domain correlations: the columns T_{air}, CO_2, E and LA_{eq} report the coefficients of determination and the p-values of the regressions between the sensation votes and the environmental parameter; the columns denoted with Probability report the percentage of the permuted correlations with higher R^2 and lower p-value than the observed dataset

Sensation Vote ↓	Environmental parameter →	Tair	Probability %	CO_2	Probability %	E	Probability %	LA_{eq}	Probability %
TSV	R^2	0.93	1	0.88	0	0.91*	4	0.43	**20**
	p-value	0.008	1	0.058	0	0.03*	4	0.234	**20**
IAQSV	R^2	0.68	10	0.89	8	0.83*	**11**	0.05	40
	p-value	0.084	10	0.059	8	0.075*	6	0.726	60
VSV	R^2	0.63	10	0.13	63	0.88*	6	0.29	35
	p-value	0.108	10	0.645	63	0.051*	5	0.350	35
ASV	R^2	0.79	3	0.09	58	0.46	**13**	0.87	2
	p-value	0.044	3	0.705	58	0.136	**13**	0.021	2

*quadratic regression

3.1.2 Cross-domain effects of indoor air temperature

Looking at the permutation analysis it can be seen that the air temperature is correlated with high confidence to all sensation votes in the non-thermal domains, with somehow worse performance in the visual and IAQ domains. As concerns air quality and visual sensation votes, the correlation is mutual, meaning also CO_2 concentration and illuminance appear to be affecting thermal sensation votes. Higher temperatures tend to make the occupants perceive the air quality as worse (Fig. 2b). In contrast, looking at the two other regressions (Figure 2c and d), it can be seen that both of them have slightly negative slopes, suggesting that an increased temperature is associated with a small reduction of the visual and acoustic sensation votes. In particular, votes tend to move towards neutrality (from brighter, and from noisier to darker and quieter sensation, respectively). It is likely that some of the light observed effect, is due to some of the adaptation actions and or conditions that can be associated with higher temperatures. For instance, higher temperatures may be observed when windows and shades are closed and so daylight is reduced. Similarly, the classroom might have been perceived as quieter when windows and doors were closed and so also temperatures were higher.

3.1.3 Cross-domain effects of CO_2 concentration

From the permutation analysis it can be seen that the CO_2 concentration is significantly correlated to thermal sensation votes, while correlations with other domains are poorly significant. As observed, the correlation between the two domains (i.e., thermal and IAQ) appears to be mutual, meaning that also higher temperatures appear to be affecting indoor air quality sensation votes. Higher CO_2 concentrations tend to make the occupants perceive the environment as warmer (Fig. 3a), even if the slope is not so large. Looking at the two other regressions (Figure 3c and d), it can be seen that the regression curves are not effective in representing any meaningful sensitivity of visual and acoustic sensation votes on the CO_2 changes. In contrast, it seems the illuminance level has a significant effect on the IAQ sensation vote, as discussed in the following paragraph.

3.1.4 Cross-domain effects of the illuminance level

The permutation analysis suggests that the illuminance level is significantly correlated to sensation votes in the indoor air quality domain, which does not appear to be mutual. Correlations with thermal sensation votes appears to be significant when considering a quadratic regression, while it is average with acoustic sensation votes. As with temperature effects on visual sensation votes, this could be explained in the hypothesis of an indirect effect, with higher illuminance levels associated to lower temperatures, due to windows and shades open.

3.1.5 Cross-domain effects of sound pressure level

As from the outcomes of the permutation test, there seems to be only an average confidence of correlation with the thermal sensation votes, even if with a very mild negative slope. A slightly more significant correlation was observed between temperatures and acoustic sensation votes. Cooler sensations associated with higher sound pressure levels might be explained by higher noise levels and cooler temperatures occurring when doors and windows are open.

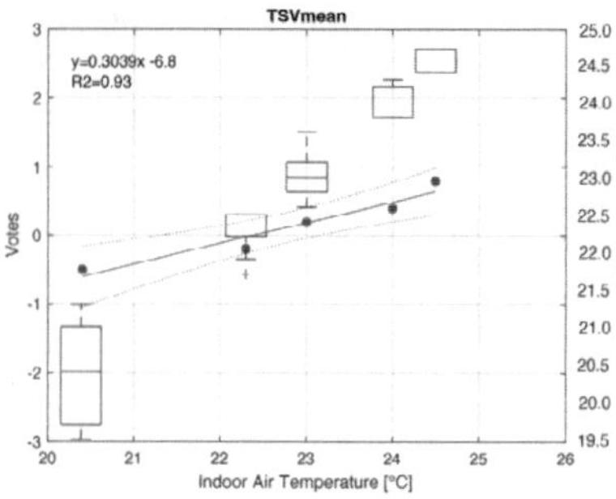

(-3) = cold; (-2) = cool; (-1) = slightly cool; 0 = nor cold/warm; 1 = slightly warm; 2 = warm; 3 = hot

a.

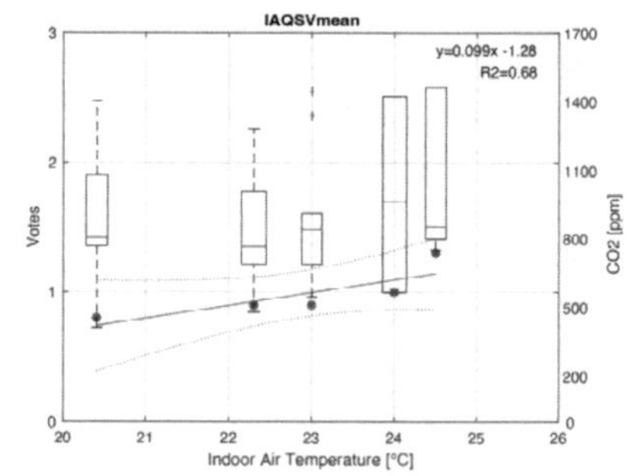

0 = fresh; 1 = slightly stuffy; 2= stuffy; 3= very stuffy

b.

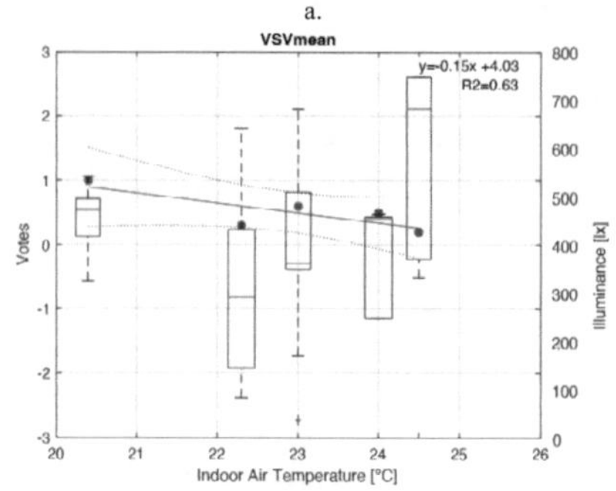

(-3) = very dark; (-2) = dark; (-1) = slightly dark; 0 = nor dark/bright; 1 = slightly bright; 2 = bright; 3 = very bright

c.

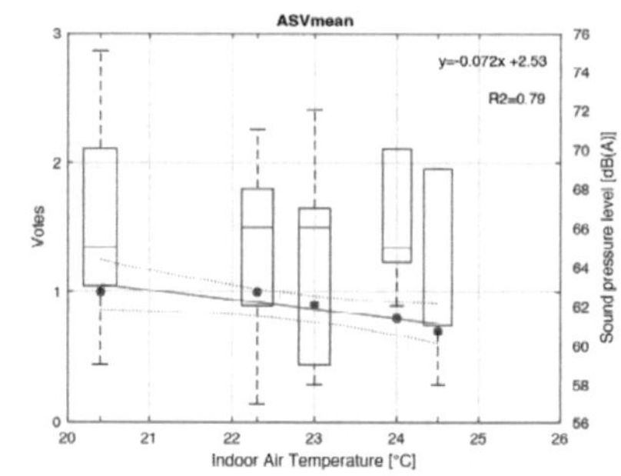

0 = quiet; 1 = slightly noisy; 2= noisy; 3= very noisy

d.

Figure 2: Indoor air temperature vs subjective votes.

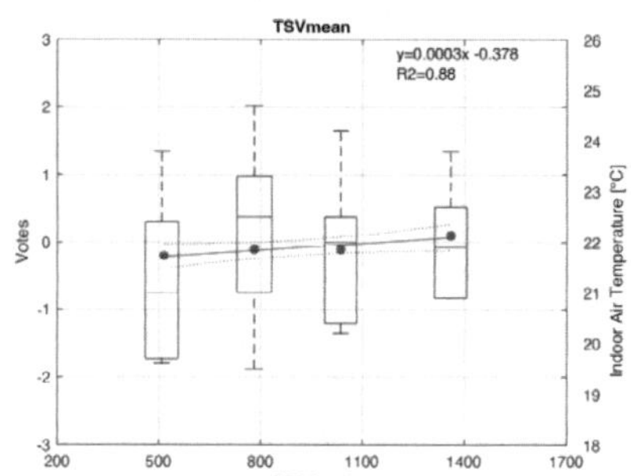

(-3) = cold; (-2) = cool; (-1) = slightly cool; 0 = nor cold/warm; 1 = slightly warm; 2 = warm; 3 = hot

a.

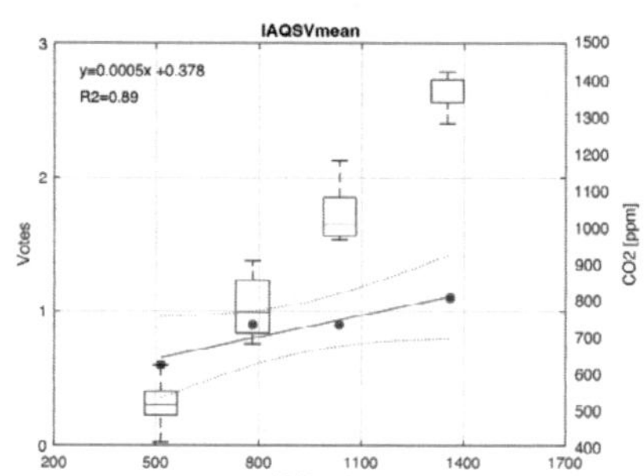

0 = fresh; 1 = slightly stuffy; 2= stuffy; 3= very stuffy

b.

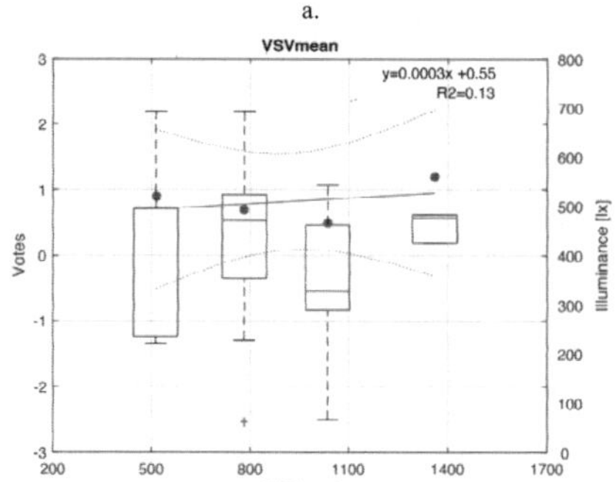

(-3) = very dark; (-2) = dark; (-1) = slightly dark; 0 = nor dark/bright; 1 = slightly bright; 2 = bright; 3 = very bright

c.

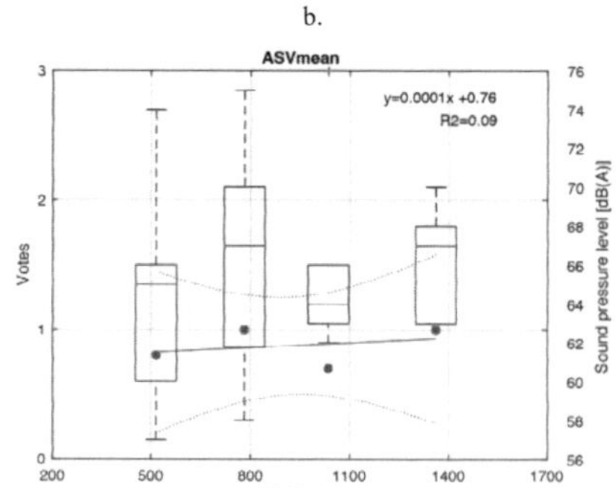

0 = quiet; 1 = slightly noisy; 2= noisy; 3= very noisy

d.

Figure 3: CO_2 level and subjective votes.

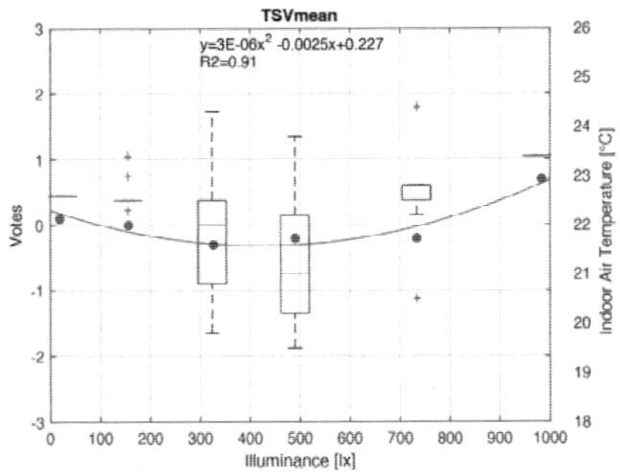

(-3) = cold; (-2) = cool; (-1) = slightly cool; 0 = nor cold/warm; 1 = slightly warm; 2 = warm; 3 = hot

a.

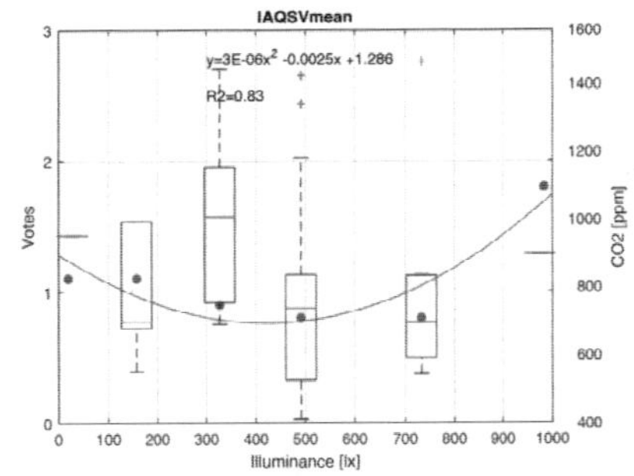

0 = fresh; 1 = slightly stuffy; 2= stuffy; 3= very stuffy

b.

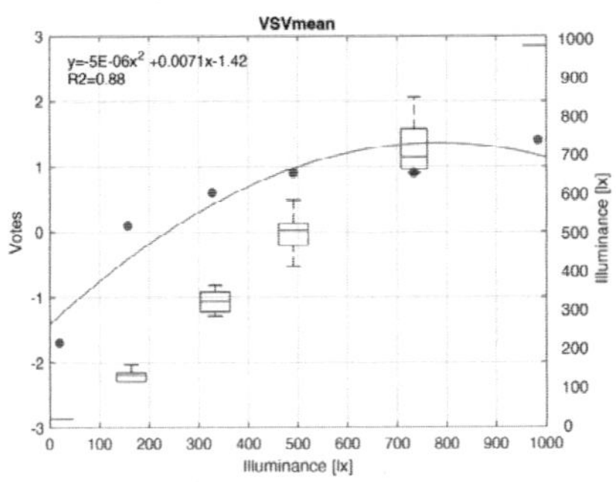

(-3) = very dark; (-2) = dark; (-1) = slightly dark; 0 = nor dark/bright; 1 = slightly bright; 2 = bright; 3 = very bright

c.

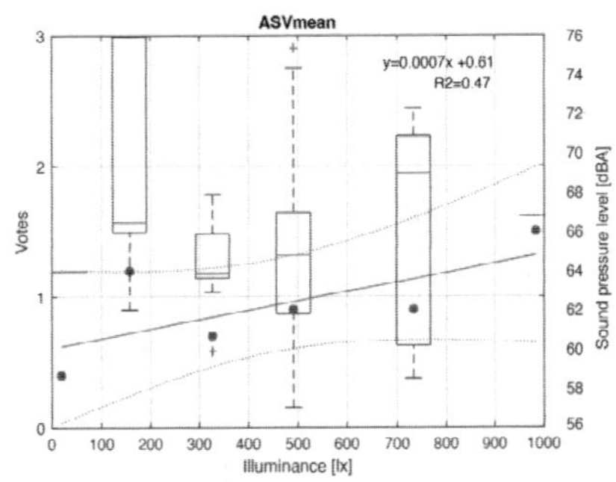

0 = quiet; 1 = slightly noisy; 2= noisy; 3= very noisy

d.

Figure 4: Illuminance level and subjective votes.

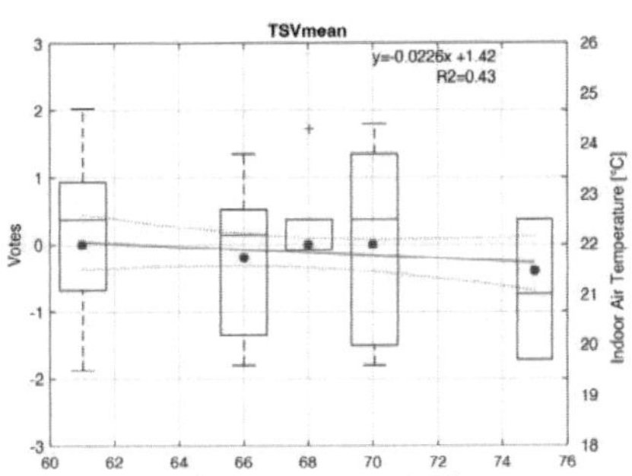

(-3) = cold; (-2) = cool; (-1) = slightly cool; 0 = nor cold/warm; 1 = slightly warm; 2 = warm; 3 = hot

a.

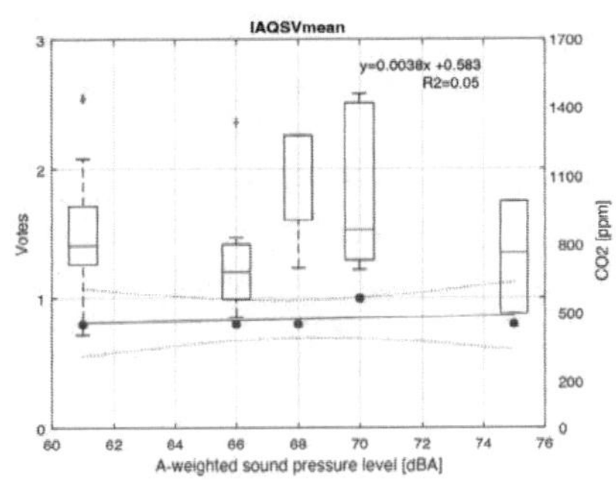

0 = fresh; 1 = slightly stuffy; 2= stuffy; 3= very stuffy

b.

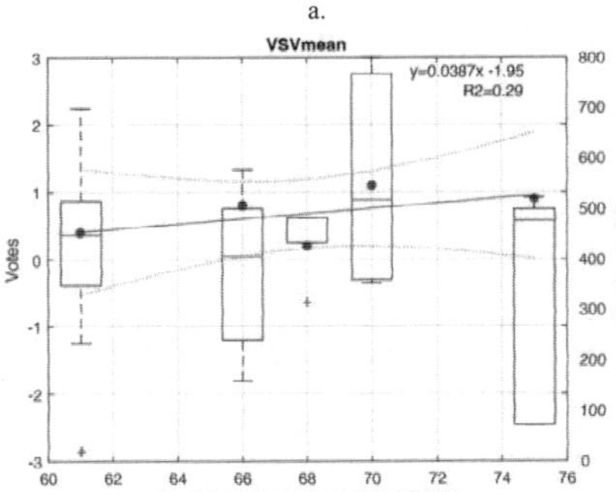

(-3) = very dark; (-2) = dark; (-1) = slightly dark; 0 = nor dark/bright; 1 = slightly bright; 2 = bright; 3 = very bright

c.

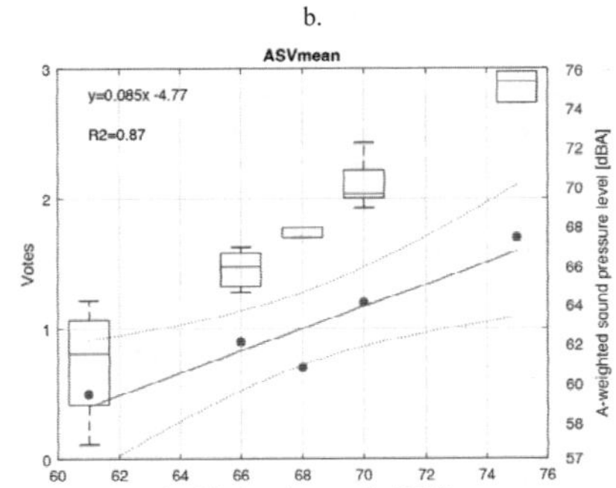

0 = quiet; 1 = slightly noisy; 2= noisy; 3= very noisy

d.

Figure 5: A-weighted L_{eq} sound level and subjective votes.

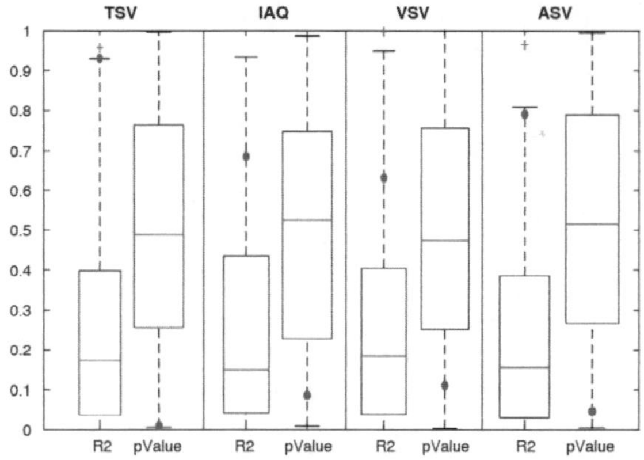

Thermal Environment
a.

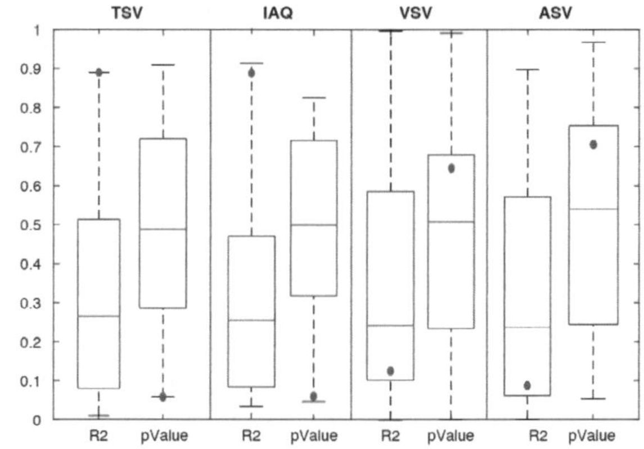

Indoor Air Quality
b.

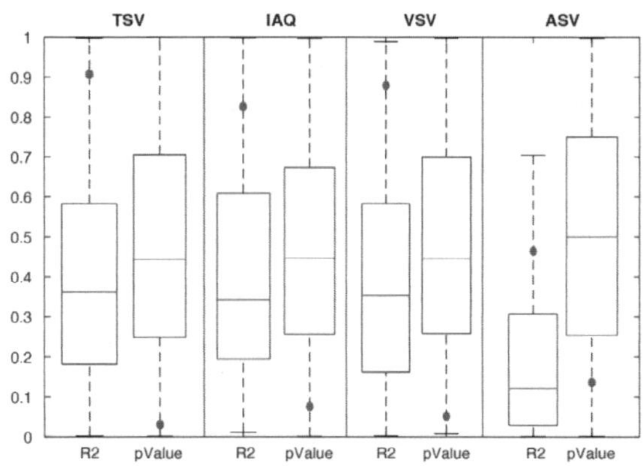

Visual Environment
c.

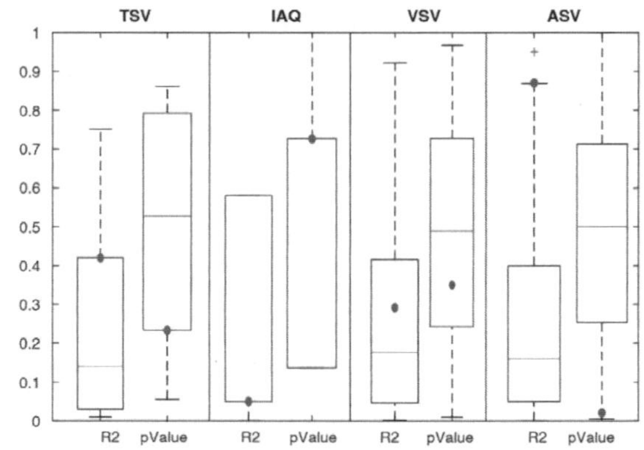

Acoustic Environment
d.

Figure 6. Distribution of the coefficients of determination and p-values referring to the permutation analysis. In the boxplots, in each box, the red central mark is the median, the box limits denote 25th and 75th percentiles, and the upper and lower hyphens are the maximum and minimum sensation votes. Blue dots represent the coefficient of determination and p-value of the regressions of the observed points.

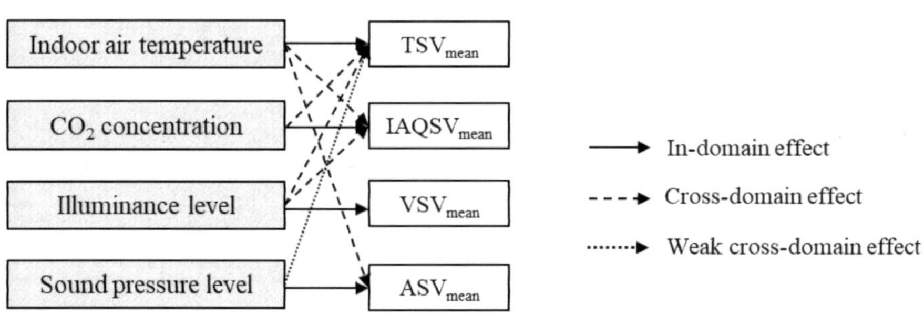

Figure 7. In-domain and cross-domain effects of environmental conditions on sensation votes.

4. CONCLUSIONS

This study investigates in-domain and cross-domain effects of indoor air temperature, CO_2 concentration, illuminance level and A-weighted sound pressure level on students' sensation. The influence of physical parameters on the sensation votes on the four comfort domains was investigated through a monitoring survey, involving objective

measurements and subjective perception questionnaires. A total number of 55 different panels, i.e., groups of students exposed to the same environmental conditions at the same time, were interviewed, collecting 949 individual questionnaires.
The relation between environmental conditions and subjective perception was inspected through a regression analysis, considering the sets of observed combinations mean parameter-mean vote. Individual votes distributions were not the focus, since, by analogy to thermal predicted mean vote, environmental comfort is assumed to be described through the average response of a group of people. In order to increase the statistical significance, a pairwise comparisons of votes distributions allowed the aggregation of panels belonging to similar mean environmental conditions (bins), repeating the process in each domain. The grouping process included the identification and discharge of outlier panels (i.e., panels showing different sensation votes distributions) and further grouping between adjacent bins. This allowed an increased sensitivity of the regressions more on the binning parameter, either in- and cross-domain. This procedure is also consistent with the definition of the Just Noticeable Differences (JNDs), i.e. ranges of environmental conditions which cannot be perceived differently, enabling subjective responses from different bins to be evaluated as a single group.
The physical parameter considered to form the groups was assumed as the independent variable and each of the group mean sensation votes for the four comfort domains were the dependent variables of a specific regression analysis. In order to evaluate the goodness of the regressions, a permutation test was implemented and the coefficient of determination and the p-value of the variable(s) of the regression were used as Key Performance Indicators of the analysis.
The analysis highlighted a dependence between students' sensation and some of the measured physical parameters, but the type of dependency and the confidence level of the regression calculated are different from one comfort domain to another, as summarized in Figure 7, in which solid black lines represent strong in-domain effects and dashed and dotted lines represent stronger or weaker cross-effects, respectively. Missing connections are due to lack of confidence in the quality of the analyzed regressions.
Regarding in-domain effects, all sensation votes appeared to be correlated with the corresponding environmental parameters with a high confidence level.
Concerning the cross-domain effects, we found high confidence in the regressions between air temperature and IAQ sensation and, vice versa, between CO_2 concentration and thermal sensation.
The change in air temperature has a slight effect on visual and acoustic sensation with a negative small slope of the regression line which was found to be statistically significant.
There is no statistical significance for the impact of changes in CO_2 concentration on visual and acoustic sensation.
The illuminance level is significantly correlated to the indoor air quality sensation and thermal sensation, while no confidence could be found for the correlation with the acoustic sensation.
Finally, the sound pressure level seems to have some effect on the thermal sensation but with an average confidence on the correlation, while it has negligible effect on IAQ and visual sensation.

5. FURTHER DEVELOPMENTS

Developments of this work will include the enlargement of the collected database through new field campaigns in other educational buildings and periods of the year in order to test the above-mentioned method under different environmental conditions and thus enlarging the ranges of measured physical parameters. Indeed, parameter ranges might have a great impact in detecting the possible effects of physical parameters on subjective responses (Tang et al. 2020). Furthermore, multi-variate correlation analysis might be necessary, in order to consider all the IEQ factors simultaneously recognizing their relative weight, understanding and characterizing students' global comfort.

REFERENCES

Astolfi, A. and Pellerey, F. 2008. Subjective and objective assessment of acoustical and overall environmental quality in secondary school classrooms. The Journal of the Acoustical Society of America 123:163-173. 10.1121/1.2816563.

Buratti, C., Belloni, E., Merli, F., Ricciardi, P. A new index combining thermal, acoustic, and visual comfort of moderate environments in temperate climates, 2018. Building and Environment 139:27–37.

CEN. 2001. EN ISO 7726:2001, Ergonomics of the thermal environment - Instruments for measuring physical quantities.

CEN. 2011. EN ISO 12464-1:2011, Light and lightling – Lighting of work places – Part 1: Indoor work places.

CEN. 2019. EN 16798-1:2019, Energy performance of buildings - Ventilation for buildings - Part 1: Indoor environmental input parameters for design and assessment of energy performance of buildings addressing indoor air quality, thermal environment, lighting and acoustics - Module M1-6.

Corgnati, S. P., Filippi, M. and Viazzo, S. 2005. Perception of the thermal environment in high school and university classrooms: Subjective preferences and thermal comfort. *Building and Environment* 42:951–959.

De Giuli, V., Da Pos, O. and De Carli, M. 2012. Indoor environmental quality and pupil perception in Italian primary schools. *Building and Environment* 56:335–345.

Dorizas, P.V., Assimakopoulos, M. and Santamouris, M. 2015. A holistic approach for the assessment of the indoor environmental quality, student productivity, and energy consumption in primary schools. *Environmental Monitoring and Assessment* 187(5).

Fassio, F., Fanchiotti, A., de Lieto Vollaro, R., Linear, Non-Linear and Alternative Algorithms in the Correlation of IEQ Factors with Global Comfort: A Case Study. Sustainability 2014, 6, 8113-8127.

Frontczak M., Wargocki P., Literature survey on how different factors influence human comfort in indoor environments. Building and Environment 2011; 46: 922-937. Building and Environment 46 (2011) 922-937.

Kim, J. and de Dear, R. 2018. Thermal comfort expectations and adaptive behavioural characteristics of primary and secondary school students. *Building and Environment* 127:13–22.

Korsavi, S. S., Zhang, D., Montazami, A. and Mumovic, D. 2020. Perceived indoor air quality in naturally ventilated primary schools in the UK: Impact of environmental variables and thermal sensation. *Indoor Air* 00:1–22.

Morandi F., Pittana I., Cappelletti F., Gasparella A., Tzempelikos A., Assessing the Overall Indoor Environmental Comfort and Satisfaction: Evaluation of a Questionnaire Proposal By Means of Statistical Analysis of Responses, IAQ Conference 2020: Indoor Environmental Quality Performance Approaches.

Pittana I., Morandi F., Cappelletti F., Gasparella A., Tzempelikos A., Understanding the Effects of Environmental Factors on Human Perception by Means of Surveys and in Field Measurements, IAQ Conference 2020: Indoor Environmental Quality Performance Approaches.

Ricciardi, P. and Buratti, C. 2018. Environmental quality of university classrooms: Subjective and objective evaluation of the thermal, acoustic, and lighting comfort conditions. *Building and Environment* 127:23–36.

Schweiker M., Ampatzi E., Andargie M S., Korsholm Andersen R., Azar E., Barthelmes V. M., Berger C., Bourikas L., Carlucci S., Chinazzo G., Prabha Edappilly L., Favero M., Gauthier S., Jamrozik A., Kane M., Mahdavi A., Piselli C., Pisello A. L., Roetzel A., Rysanek A., Sharma K, Zhang S. Review on multi-domain approaches to indoor environmental perception and behaviour. Building and Environment 2020; 176: 106804.

Tang H., Ding Y., Singer B., Interactions and comprehensive effect of indoor environmental quality factors on occupant satisfaction. Building and Environment 2020; 167: 106462.

Vilčeková, S., Kapalo, P., Mečiarová, L., Burdová, E., Imreczeová, V., The real and subjective indoor environmental quality in schools. Procedia Engineering 190 (2017) 496 – 503.

Yang W., Moon H.J., Cross-modal effects of illuminance and room temperature on indoor environmental perception, Build. Environ. 146 (2018) 280–288

Recursive Bayesian Calibration of Data-Driven Archetype Building Energy Models for Residential Sector: Application to a Research House

Seyed Matin ABTAHI[1]*, Andreas K. ATHIENITIS[1], Benoit DELCROIX[2]

[1]Centre for Zero Energy Building Studies, Department of Building, Civil, and Environmental Engineering, Concordia University,
Montréal, Québec, Canada
matinabtahi@gmail.com, andreask.athienitis@concordia.ca

[2] Hydro-Québec Research Centre, Laboratoire des Technologies de l'Energie (LTE), Shawinigan, Québec, Canada
delcroix.benoit@hydroquebec.com

* Corresponding Author

ABSTRACT

This paper studies recursive Bayesian calibration of archetype building energy models developed for optimal operation, energy flexibility and resilience in the residential sector. Real-time building performance monitoring has been facilitated recently with low cost faster instrumentations. With an online stream of measured data, recursive model calibration can be employed, where the model is initialized with prior knowledge and is updated with new measurements available to the building automation system. A Markov chain Monte Carlo approach is applied to calibrate two multi-zone archetype models with data from a research house in Québec. Most likely parameter values are estimated with an iterative Metropolis-Hastings algorithm over a 5-day train period and are validated by comparing the measured indoor air temperature in each zone with the model output over another 5-day test period. The time series of the posterior probability distribution shows that heat source activation influences the maximum likelihood values of most parameters and remarkably narrows down their associated credible intervals. Analyzing probability distribution time series helps understand the progression of acquired knowledge with new data and facilitates the comparison of various models.

1. INTRODUCTION

Archetype building/zone energy models are often used to evaluate the potential savings under near-optimal control strategies and system configurations; hence, recursive calibration is critical to ensure expected/promised savings. Calibration of archetype energy models based on sensor information is typically a non-convex offline (batch) optimization problem in which the parameters are estimated by minimizing the model error over a specific period. Alternatively, an online approach proposes initializing the model with prior knowledge and updating it with new measurements every time a sufficient number of data points are available to the building automation system (BAS). This approach enables monitoring the effect of outside ambient conditions and occupants' behaviour on uncertainty and identifiability of the parameters and establishes a direct link between new observations and information gained by the model.

Optimization techniques commonly assume the model parameters have fixed values and attempt to estimate them so that the model output fits the measurements as closely as possible. However, a Bayesian approach assumes the parameters have an underlying probability distribution and attempts to infer it by sampling from the solution space. Such an approach provides valuable insights regarding the inherent uncertainty in the model and its structural limitations. In this paper, we apply a Markov chain Monte Carlo approach – as a class of recursive Bayesian inference methods to update two multi-zone archetype energy models for (1) understanding the progression of information gain with new measurements and (2) detecting the structural shortcomings in the models.

2. METHODOLOGY

2.1 Case Study

The case study is an unoccupied research house – Experimental House for Building Energetics (EHBE) of Hydro-Québec, located in Shawinigan, Québec, Canada. This test bench is a two-storey detached house with an excavated basement. The house has outer dimensions of 7.6 m × 7.9 m and 60 m^2 footprint. It has three bedrooms and a bathroom on the second floor; the kitchen, the living room, the dining room and a small washroom are on the first floor. The wall assemblies of the building represent the typical lightweight wood-framed house in Québec. The total fenestration area is 20 m^2, consisting of vinyl framed double-glazing windows with an air gap. There are more than 150 sensors in the building and the soil around it. The available data consist of 15-minute average records of room-level electrical heating loads and 15-minute instant records of solar irradiance, outside ambient air temperature, soil temperature, and room-level indoor air temperature from 2019-01-01 to 2019-03-31.

2.2 Multi-Zone Archetype Energy Models

There are three main approaches to building energy modelling (Foucquier et al., 2013): 1) physics-based (white-box) approach, established on the principles of energy/mass conservation and heat transfer with comprehensive descriptions of building geometry, systems and material properties, 2) data-driven statistical (black-box) approach, based solely on mathematical models and measurements, with no assumption regarding the building physics and systems, and 3) hybrid (gray-box) approach, based on simplified building physics and systems performance, with parameters identified from measurements where simplified building geometry and information on the building energy systems reinforce the parameterization process. Gray-box models require much less data for calibration (Arendt et al., 2018) and are more likely to stay reliable outside the calibration range (Afroz et al., 2018) compared to black-box models. Compared to the white-box models, gray-box models are much faster to set up and calibrate. A gray-box model is not an oversimplification of the system topology but a selection of relevant information which requires 1) deep knowledge of the application and its physics and 2) coherent choice of significant states, inputs and disturbances (Abtahi et al., 2021). Essentially, an ideal gray-box model is the simplest one that describes all the dominant patterns and information embedded in the measurements (Bacher & Madsen, 2011). The computational simplicity and reasonable precision make gray-box models suitable for control applications.

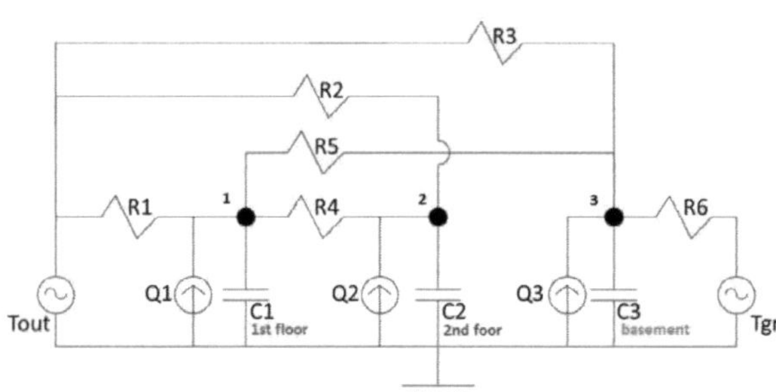

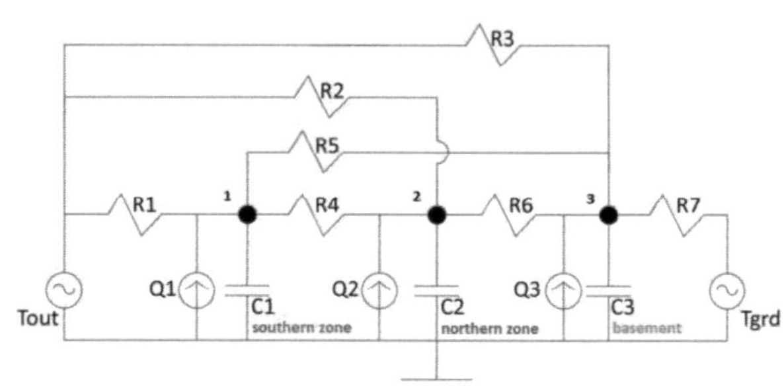

Figure 1. Left: 3C6R network model analogous to the zoning-by-floor approach – Right: 3C7R network model analogous to the zoning-by-orientation approach (Abtahi et al., 2021)

Authors have previously shown that a 3rd-order RC network model is sufficient to capture the fundamental thermal dynamics of typical single-family houses in Québec by developing two 3rd-order archetype models for the case study, with different approaches to discretize the building's indoor space (Abtahi et al., 2021): (1) Zoning-by-floor, i.e., to assume each floor of the building is a separate thermal zone. The assumption here is that the air inside each floor is thermally uniform, and the airflow between different floors is not significant; (2) Zoning-by-orientation, i.e., to assume the southern and northern zones of the above-grade space are separate thermal zones since the latter gains more solar radiation throughout the clear days, which effectively reduces its heating demand. Fig. 1-left shows the 3rd-order model analogous to the zoning-by-floor approach, where C_1, C_2 and C_3 represent the effective thermal capacity of the first floor, second floor and basement, respectively. Likewise, the model in Fig. 1-right shows the zoning-by-orientation approach where C_1, C_2 and C_3 represent the effective thermal capacity of the southern zone of the above-grade space, northern zone of the above-grade space and basement, respectively. T_{out} and T_{grd} are the outside ambient air temperature and average soil temperature, respectively.

These archetype models are adopted for this study, and the modelling steps are explained in the following. Eq. 1 is the differential equation governing the heat balance at any node (zone) *i* in the thermal RC networks, where C_i is the effective thermal capacity of zone *i*, Q_i is the aggregated heat flow to/from the air in zone *i*; *j* is all the zones connected to zone *i* by air, and R_{ij} is the thermal resistance between zone *i* and zone *j*. The weighted-average air temperature in any zone *i* is T_i. Eq. 2 is an explicit temporal discretization of Eq. 1, where *k* and *k*+1 indicate the current and the next time-step, respectively, and δt is the simulation time-interval.

$$C_i \frac{dT_i}{dt} = Q_i + \sum_j \frac{T_j - T_i}{R_{ij}} \quad (1)$$

$$T_i^{k+1} = \frac{\delta t}{C_i}\left(Q_i^{\ k} + \sum_j \frac{T_j^{\ k} - T_i^{\ k}}{R_{ij}}\right) + T_i^{\ k};\ \delta t \leq min\,(\frac{C_i}{\sum_j \frac{1}{R_{ij}}}) \quad (2)$$

2.3 Recursive Bayesian Calibration

Uncertainty is any deviation from the unachievable ideal based on deterministic knowledge about the system (Walker et al., 2003). Uncertainty sources are either aleatoric or epistemic (Iaccarino 2008). Aleatoric uncertainty is irreducible and refers to the inherently random nature of involved processes in the system. An example of an aleatoric source of uncertainty is the wind speed and direction patterns in a specific region. Epistemic uncertainty refers to the lack of knowledge in practice and can be reduced by conducting (additional accurate) measurements. For example, buildings' U-value is an epistemic source of uncertainty (Rastogi, 2016). In data-driven modelling, aleatoric and epistemic sources result in (1) experimental, (2) structural, and (3) identification uncertainty.

Experimental (observation) uncertainty is inevitable in the instrumentation and data acquisition process. Uncalibrated sensors, improper installment and communication failure in the acquisition system are among the more common sources of experimental uncertainty. Analyzing experimental uncertainty requires initializing the test/data collection many times under identical conditions.

Structural (model-inherent) uncertainty rises with insufficient knowledge of involved physical processes (e.g., unmeasured heating/cooling sources) and modelling simplifications. Common simplifications in gray-box models to facilitate the representation of buildings' thermal dynamics are linearization of heat transfer coefficients, spatial dimensionality reduction and temporal discretization (Athienitis & O'Brien, 2015). There is no generic methodology to deal with structural uncertainty due to its model-specific nature (Rajabally et al. 2002).

Identification (parameter) uncertainty refers to uncertainty in the model parameters given the available data. Analyzing identification uncertainty in building energy models is an active research topic with both frequentist and Bayesian approaches to probability and optimization. Bayesian inference methods suit control applications in buildings due to the capacity to monitor input signals' effect on parameter uncertainty in real-time. Table 1 summarizes the key differences between these two approaches:

Table 1: Key differences between frequentist and Bayesian approaches to probability and optimization (in the context of data-driven building energy modelling)

Frequentist Approach	Bayesian Approach
Suggests that probability can only be assigned to repeatable events, where the long-term frequency is measurable. Therefore, probability in the frequentist approach is related to the frequencies of repeated events.	Suggests that probability not only can be assigned to repeatable events but also helps express our belief about a particular subject given limited (sparse) observations. Therefore, probability in the Bayesian approach is related to our certainty/belief about the event/subject.
Assumes the model parameters have fixed values and attempts to estimate them so that the model output fits the observations as closely as possible.	Assumes the parameters have an underlying probability distribution and attempts to infer posterior distributions by sampling from the solution space. This approach does not perform a fit; instead, it explores the parameters space to determine their distributions without an explicit objective of refining the solution.

Calculates the variation of data (measurements) and the derived quantities given the fixed model. $P(data \mid model)$	Calculates the variation of certainty/beliefs about the model given the fixed data. $P(model \mid data) = P(data \mid model) \times \frac{P(model)}{P(data)}$
Makes a probabilistic statement about the intervals containing the unknown fixed parameter values; therefore, **Confidence** Intervals capture the uncertainty about the calculated intervals. _Example:_ If the 95% Confidence Interval for the overall U-value of a house is 400 W/°C ± 10%, it means given the model, if the data is collected 100 times under near-identical conditions and the interval that contains the parameter value is computed, 95 of the 100 intervals fit in the range of 360 W/°C to 440 W/°C.	Makes a probabilistic statement about the model parameters within the desired solution region; therefore, **Credible** Intervals capture the current uncertainty in the parameter values. _Example:_ If the 95% Credible Interval for the overall U-value of a house is 400 W/°C ± 10%, it means given the data, 95% of values that maximize the likelihood of observations and fit in the prior probability distribution are between 360 W/°C and 440 W/°C.

It is obvious from the example in Table 1 that the Bayesian approach is more intuitive for dealing with identification uncertainty in building energy modelling, as Credible Intervals address the uncertainty in parameter values and not the intervals containing them; also, collecting data under identical conditions is almost impossible. Bayesian inference methods are not adequate for fitting; however, they make it possible to thoroughly explore the parameter space around the solution after a fit has been done and help gain an improved understanding of the probability distribution. Bayesian methods refine the estimation of the most likely values for a set of parameters but do not iteratively find a better solution to minimize the model error.

For calibration, first, a uniform prior probability distribution of parameters (θ_i^0) is estimated from simplified geometry and material properties under certain assumptions about infiltration. Next, the prior probability distributions are updated by minimizing the model error over a training period with length N_{train}. Available data are transformed into the zone-level aggregated heat flow and weighted-average air temperature. HVAC output and solar gains are networked as aggregated heat flow ($\hat{Q}_i$) to the zone air node as in Eq. 3, where $\hat{Q}_{aux,i}$ is the measured heat delivered to the zone air, $\hat{G}_{vert}$ is the measured global vertical irradiance on the southern façade, and α_i is the solar gain normalizer. Eq. 4 calculates the measured zone air temperature ($\hat{T}_i$) by averaging the records of the sensors in the zone ($\hat{T}_{s,i}$) weighted by the floor area covered by each sensor (A_s), where *s* is all the sensors in zone *i*. The result of the least-square minimization in Eq. 5 is the input prior probability distribution to the Metropolis-Hastings algorithm ($N_i = 3$).

$$\hat{Q}_i = \hat{Q}_{aux,i} + \alpha_i \hat{G}_{vert} \tag{3}$$

$$\hat{T}_i = \textstyle\sum_s A_s \hat{T}_{s,i} / \sum_s A_s \tag{4}$$

$$\begin{gathered} \min_{\theta_i} \sum_{i=1}^{N_i} \sqrt{\sum_{k=1}^{N_{train}} \left(T_i^{\,k} - \hat{T}_i^{\,k}\right)^2 / N_{train}} \;;\theta_i = [R_{i,j}, C_i, \alpha_i] \\ \textit{subject to:} \\ \underline{\theta_i} < \theta_i < \overline{\theta_i} \\ T_i^{\,k+1} = T_i^{\,k} + \left(\frac{\delta t}{C_i}\right)\left(\hat{Q}_i^{\,k} + \textstyle\sum_j \frac{T_j - T_i}{R_{ij}}\right) \; k \in N_0^{\,N_{train}-1} \end{gathered} \tag{5}$$

The Metropolis-Hastings algorithm takes random samples from the parameter space regarding this prior probability distribution and refines the prior knowledge about the model by calculating a Credible Interval for the values that maximize the likelihood of given data. First, an ensemble of Markov chains is formed starting from relatively distant arbitrary points. Each chain is a stochastic process of one walker that moves around randomly in the parameter space using a proposal distribution for new steps and a policy for rejecting a fraction of the proposed moves. Therefore, the proposal distribution for each walker's new steps depends on the other walkers' relative situation. The ensemble evolves through the parameter space and shapes a posterior probability distribution. The more steps are included in the ensemble evolution, the more closely the posterior distribution matches the actual (real-life) distribution. However, as the ensemble's first steps (burn-in phase) may not reflect the posterior distribution, they must be discarded.

The choice of the proposal distribution and a proper initialization are critical for the algorithm's performance. Determining the number of steps required to converge to the stationary distribution within an acceptable uncertainty is a challenging task. Given an improper rejection policy, the ensemble convergence can take a long time as the walkers may double back and explore the space already covered.

3. EVALUATION OF METHODOLOGY

Electric space heating is widely adopted in the Québec residential sector and strongly influences the region's demand profile during cold winters. The grid's peak demand periods typically occur on extremely-cold weekday mornings from 6:00 to 9:00 (when people wake up, get ready and leave home) and evenings from 16:00 to 20:00 (when people return home, cook and rest), during which space heating stresses the grid the most. A newly introduced dynamic tariff (rate flex-D) gives the residential customers in Québec the opportunity to actively reduce the pressure on the grid and save money during the peak periods by shifting the load to the off-peak periods when electricity is approximately 12 times cheaper. Archetype energy models help residential customers take cost-optimal actions; hence, recursive calibration is critical to ensure expected savings.

The training data needs to comprise enough information regarding the building thermal dynamics within the comfort range. Therefore, a 5-day period (N_{train} = 120h) from March 2019 is selected, during which the indoor air temperature in each zone varies due to variable levels of auxiliary heat and solar gains. There is also a period of free-floating indoor air temperature, which helps understand the effect of heat source activation on the maximum likelihood values and credible intervals. The trained model is then tested over another 5 days (N_{test} = 120h) of the same month to validate its ability to predict the indoor air temperature in each zone. Figure 1 presents the training data.

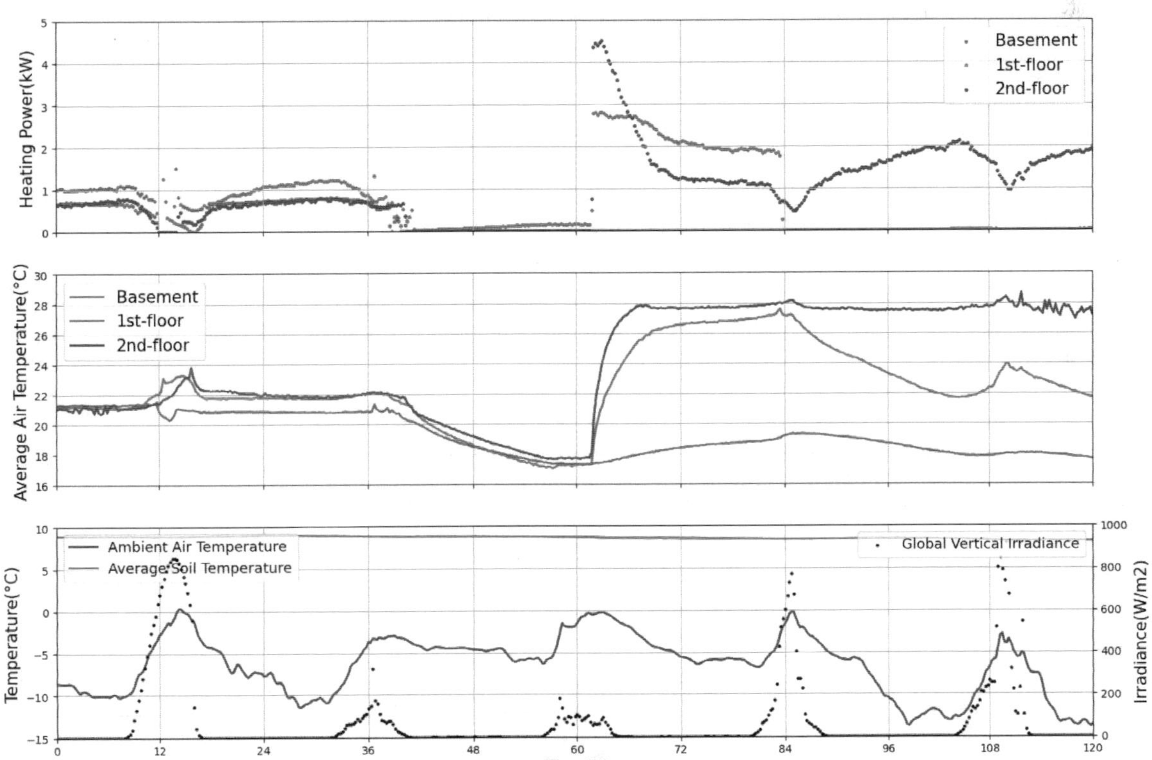

Figure 2. Training data

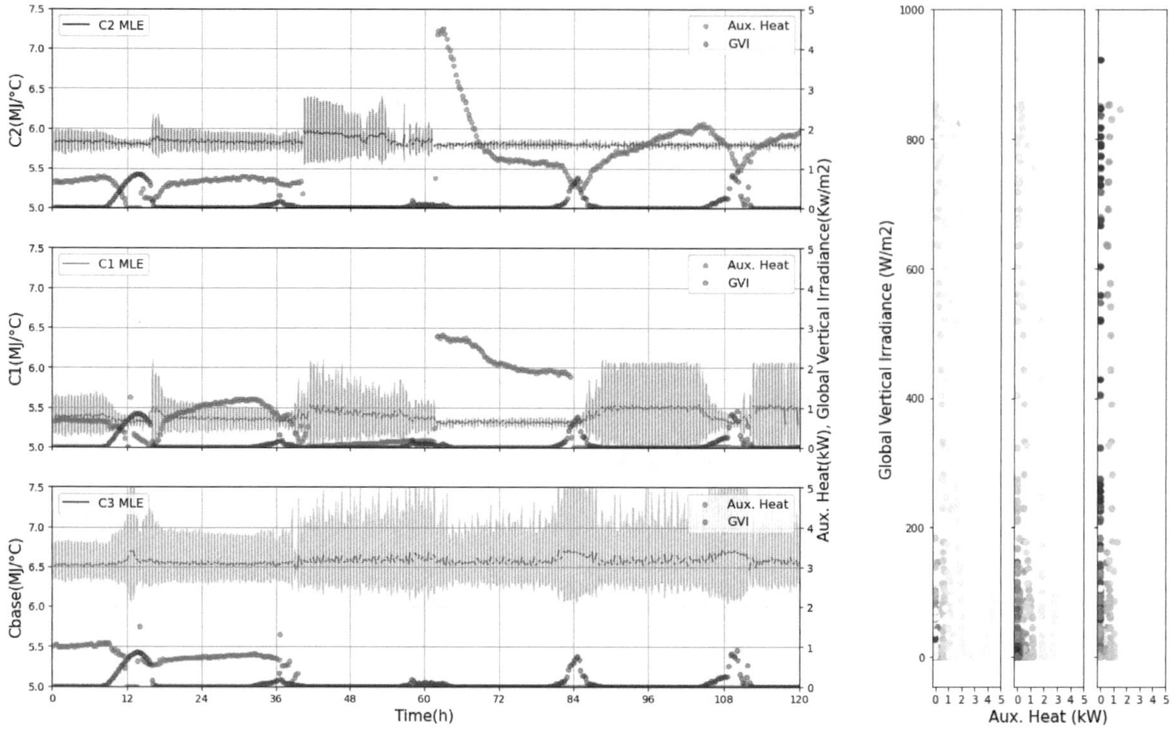

Figure 3. Time series of 95% Credible Interval and Maximum Likelihood Estimation (MLE) for effective thermal capacity of different zones: 3C6R model (left) – Corresponding normalized uncertainty (right)

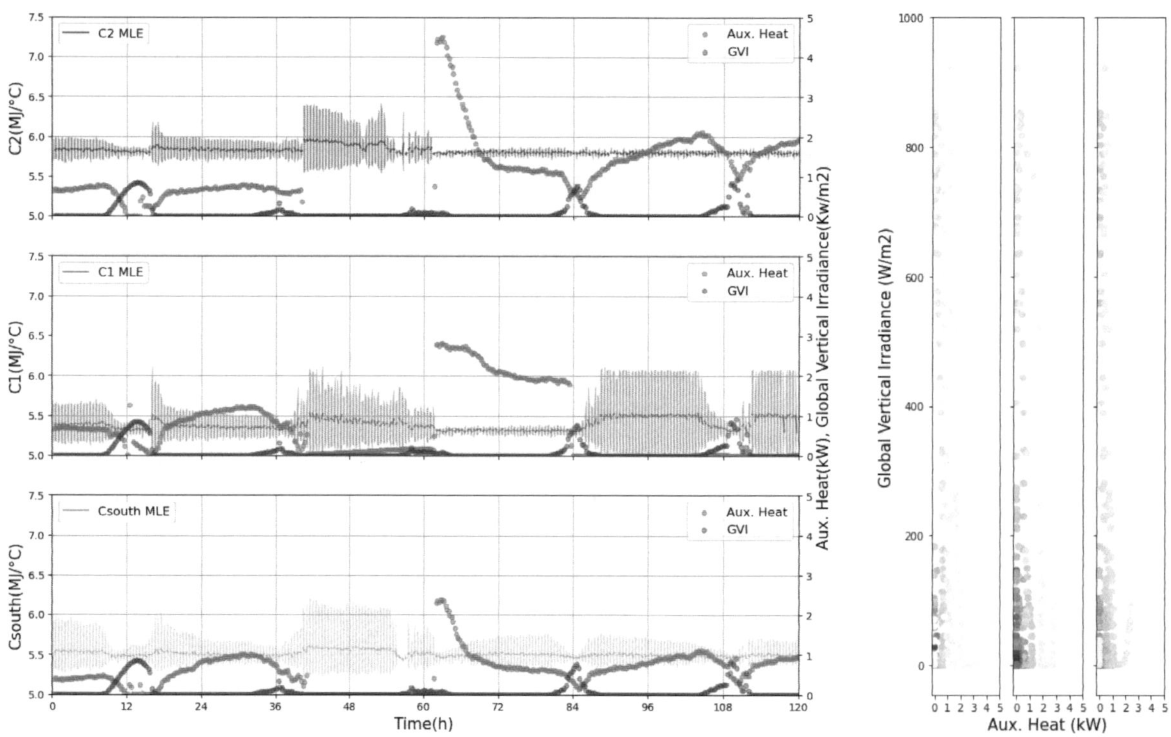

Figure 4. Time series of 95% Credible Interval and Maximum Likelihood Estimation (MLE) for effective thermal capacity of different zones: 3C6R model vs. 3C7R model (left) – Corresponding normalized uncertainty (right)

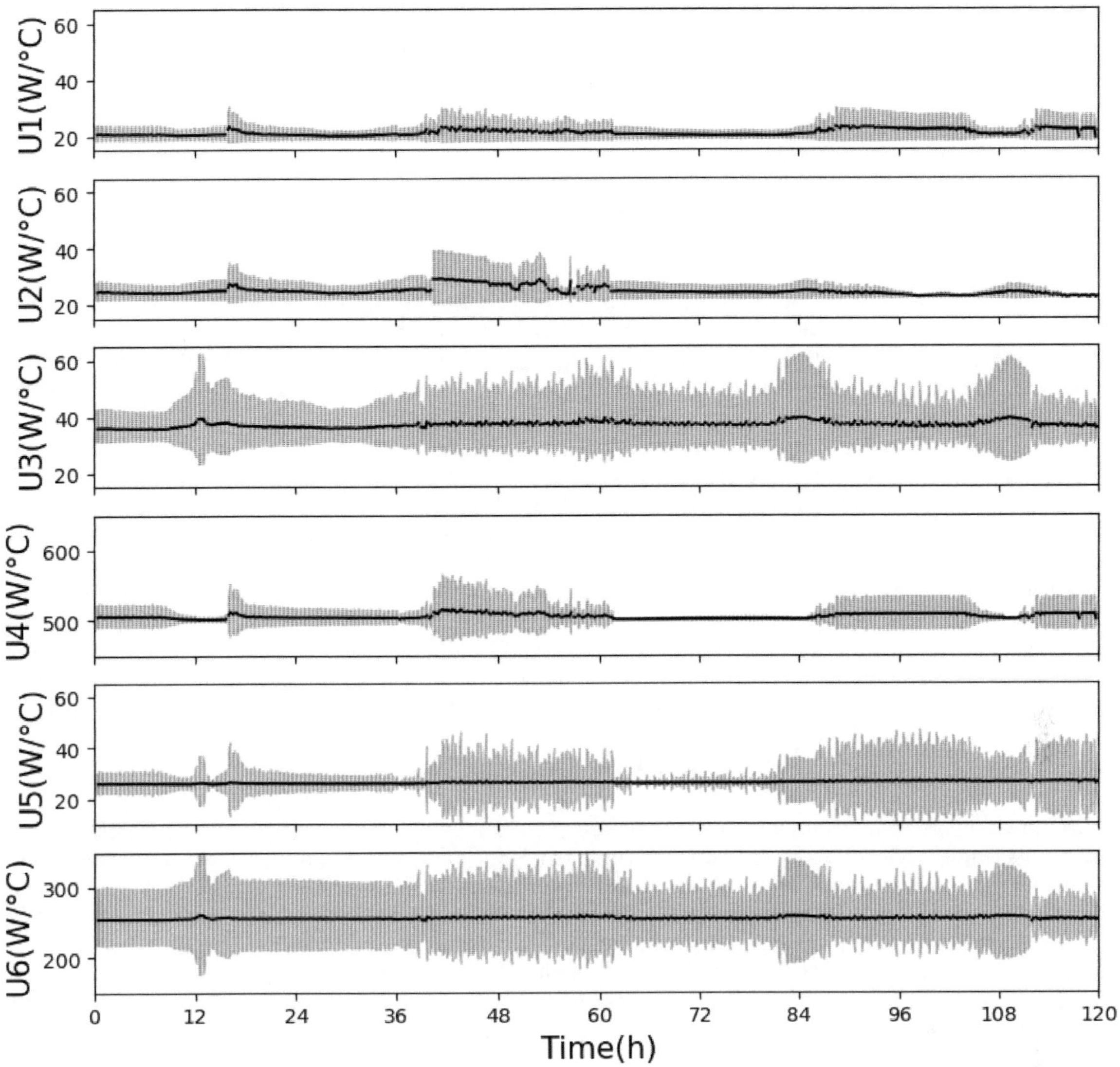

Figure 5. Time series of 95% Credible Interval and Maximum Likelihood Estimation (MLE) for thermal conductance between zones and surroundings: 3C6R model

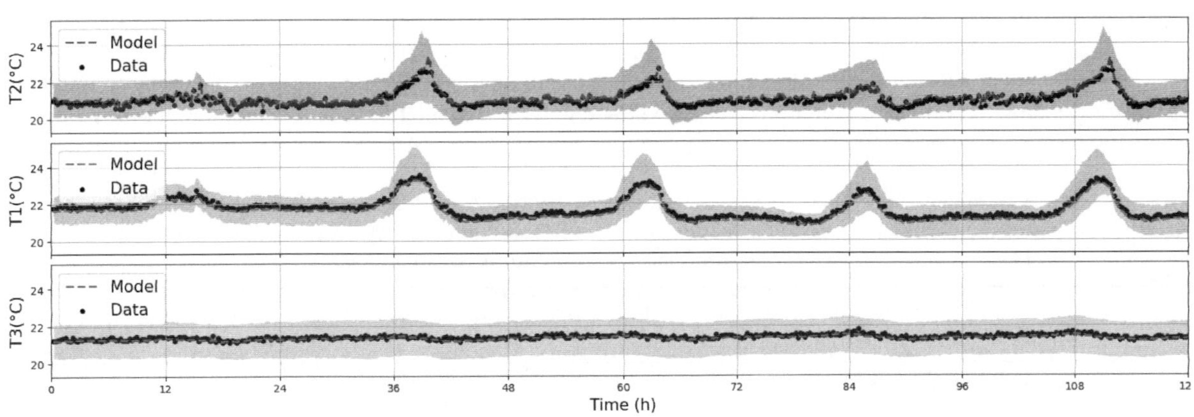

Figure 6. Predicted vs. measured indoor air temperature: 3C6R model (from top to bottom: second floor, first floor, basement. The highlighted band shows the uncertainty due to propagation of the probability distributions)

Figure 3-left presents the posterior probability distribution time series for the effective thermal capacity of the basement, first floor and second floor. Progression of 95% Credible Intervals shows an inverse correlation between activation of heat sources and parameters uncertainty. As the auxiliary heat turns off, the only intermittent heat source is the solar gains which are negligible unless it is a clear day. Therefore, the energy flow through RC networks in free-float conditions is much less than what it is in the presence of auxiliary heat. This provides less information to capture per unit of time and as a result, credible intervals start wide-ranging, which interprets to less certainty in the parameter values and models' output. Figure 3-right colour maps the normalized uncertainty of C_1, C_2 and C_3 (3C6R model) with regard to heat sources, where darker points represent less certain (more uncertain) values and vice-versa. This figure emphasizes the mentioned inverse correlation.

Figure 4 shares the concept with figure 3, however, it compares the thermal capacity of the air in the first and second floors from the 3C6R model (C_1 and C_2) with the thermal capacity of the air in the southern zone from the 3C7R model (C_{south}). Figure 4-right implies that the uncertainty of C_{south} is more dependent on the global vertical irradiance on the southern façade compared to C_1 and C_2, which verifies the idea of two different approaches to discretize the building's indoor space.

Figure 5 presents the time series of 95% Credible Interval and Maximum Likelihood Estimation (MLE) for thermal conductance between zones and surroundings (3C6R model), where $U_{ij} = 1/ R_{ij}$. This figure shows the probability distribution of U_{grd} ($1/R_{grd}$) does not change much, which means that given the training data, this parameter is non-identifiable. Generally, analyzing posterior distribution time series helps understand how models capture information with new data, what factors critically influence the uncertainty and what structural shortcomings exist in the model.

Finally, to validate the accuracy and interpretability of the methodology, the measured indoor air temperature in each floor is compared to the corresponding output of the 3C6R model over a 5-day test period. Figure 6 shows the model's outputs given the MLE and its uncertainty due to propagation of the posterior probability distribution.

4. CONCLUSIONS

This paper presented a methodology for recursive Bayesian calibration of simple archetype house energy models in the Québec residential sector. Two 3rd-order RC network models were adopted to capture the thermal dynamics of the case study, which is an unoccupied two-story research house. One model (3C6R network) assumes that each floor is a separate thermal zone, and the other model (3C7R network) assumes that the south-facing zone and the north-facing zone of the above-grade space are separate thermal zones. A Markov chain Monte Carlo approach is applied to recursively calibrate these archetype models over a 5-day train period. Minimizing the models' error in the prediction of the indoor air temperature over the training period provides a prior probability distribution, which is later refined by random sampling from the parameter space with an iterative Metropolis-Hastings algorithm.

Progression of the posterior probability distributions in time shows that activation of auxiliary heat and solar gains influences the maximum likelihood values of most parameters and effectively reduces their associated uncertainty. The training data comprised a period of free-floating indoor air temperature to investigate the effect of heat source deactivation on the credible intervals. As the auxiliary heat turns off, the only intermittent heat source is the solar gains which are negligible unless it is a clear day. Therefore, the energy flow through RC networks in free-float conditions is much less than what it is in the presence of auxiliary heat. This provides less information to capture per unit of time and as a result, credible intervals start wide-ranging, which interprets to less certainty in the parameter values and models' output.

Comparing the two models shows that zoning the above-grade space by orientation and dividing it into the southern and northern zones increases the dependence of thermal capacitances uncertainty on the global vertical irradiance on the southern façade, meaning that the 3C7R model uncertainty is more sensitive to the vertical irradiance patterns than the 3C6R model uncertainty. Generally, analyzing posterior distribution time series helps understand how models capture information with new data, what factors critically influence the uncertainty and what structural shortcomings exist in the model. In the end, to validate the accuracy of the methodology, the measured indoor air temperature in each zone is compared to the corresponding model output over a different 5-day test period.

NOMENCLATURE

$\widehat{G}_{vert}$	Measured global vertical irradiance on the southern façade (W/m^2)
$\widehat{Q}_{aux,i}$	Measured heat delivered to the air in zone *i* (W)
$\widehat{Q}_i$	Measured aggregated heat flow to the air in zone *i* (W)
$\widehat{T}_i$	Measured weighted average air temperature of zone *i* (°C)
$\widehat{T}_{s,i}$	Measured air temperature by sensor *s* in zone *i* (°C)
A_s	Floor area covered by sensor *s* (m^2)
C_i	Thermal capacity of the air in zone *i* (J/°C)
N_i	Number of the zones in the RC network (-)
N_{test}	Testing period length (sec)
N_{train}	Training period length (sec)
R_{ij}	Thermal resistance between zone *i* and zone *j* (°C/W)
T_{grd}	Average soil temperature (°C)
T_i	Predicted weighted average air temperature of zone *i* (°C)
T_{out}	Outside ambient air temperature (°C)
α_i	Solar gain normalizer of zone *i* (-)
δt	Simulation time-interval (sec)
θ_i	RC network parameters related to zone *i* (-)

REFERENCES

Abtahi M, Athienitis A, Delcroix. Control-oriented thermal network models for predictive load management in Canadian houses with on-Site solar electricity generation: application to a research house. Journal of Building Performance Simulation 2021. doi:10.1080/19401493.2021.1998223.

Afroz Z, Shafiullah G. M, Urmee T, Higgins G. Modeling techniques used in building HVAC control systems: A review. Renewable and Sustainable Energy Reviews 2018; 83:64–84.

Arendt K, Jradi M, Shaker H. R, Veje C. Comparative analysis of white, gray and black-box models for thermal simulation of indoor environment: Teaching building case study. Proceedings of the 2018 building performance modelling conference and simbuild co-organized by ASHRAE and IBPSA-USA; 173–180.

Athienitis A. K, O'Brien W. (2015). *Modelling, Design, and Optimization of Net-Zero Energy Buildings*. Wiley.

Bacher P, Madsen H. Identifying suitable models for the heat dynamics of buildings. Energy and Buildings 2011; 43:1511–22.

Foucquier A, Robert S, Suard F, Stephan L, Jay A. State of the art in building modelling and energy performance prediction: A review. Renewable and Sustainable Energy Reviews 2013; 23:272–88.

Iaccarino G. Quantification of Uncertainty in Flow Simulations Using Probabilistic Methods. VKI Lecture Series - Sept. 8-12, 2008.

Rajabally, E., P. Sen, and S. Whittle (2002). "A methodology for model dependability assessment." In: Proceedings of the Computer-Based Design: Engineering Design Conference 2002, pp. 291–300.

Rastogi P. On the sensitivity of buildings to climate: the interaction of weather and building envelopes in determining future building energy consumption. PhD thesis. EPFL, Lausanne, Switzerland 2016.

Walker, W. E., P. Harremoes, J. Rotmans, J. P.v. d. Sluijs, M. B.A. v. Asselt, P. Janssen, and M. P.K. v. Krauss (2003). "Defining Uncertainty: A Conceptual Basis for Uncertainty Management in Model-Based Decision Support." In: Integrated Assessment 4, pp. 5–17. doi:10.1076/iaij.4.1.5.16466.

ACKNOWLEDGEMENT

The financial support through an NSERC/Hydro-Québec Industrial Research Chair is gratefully acknowledged. The authors would also like to thank the research team at Hydro-Québec Research Centre Laboratoire des Technologies de l'Energie for providing the data and valuable suggestions.

A Novel Hybrid Modeling Method for Predicting Energy Use of Hydronic Radiant Slab Systems

Lichen WU[1], Liping WANG[1]*, James BRAUN[2]

[1]Department of Civil and Architectural Engineering, University of Wyoming,
Laramie, WY, 82071, United States
lwang12@uwyo.edu

[2]School of Mechanical Engineering, Purdue University,
West Lafayette, IN 47907-2088, United States

* Corresponding Author

ABSTRACT

Accurately predicting the performance of radiant slab systems can be challenging due to the large thermal capacitance of the radiant slab and room temperature stratification. Current methods for predicting heating and cooling energy consumption of hydronic radiant slabs include detailed first-principles (e.g., finite difference) and reduced-order (e.g., thermal Resistor-Capacitor (RC) network) models. Creating and calibrating detailed first-principles models, as well as detailed RC network models for predicting the performance of radiant slabs require substantial effort. To develop improved control, monitoring, and diagnostic methods, there is a need for simpler models that can be readily trained using in-situ measurements.

In this study, we explored a novel hybrid modeling method that integrates a simple RC network model with an evolving learning-based algorithm termed the Growing Gaussian Mixture Regression (GGMR) modeling approach to predict the heating and cooling rates of a radiant slab system for a Living Laboratory office space. The RC network model predicts heating or cooling load of the radiant slab system that is provided as an input to the GGMR model. Three modeling approaches were considered in this study: 1) an RC network model; 2) a GGMR model, and 3) the proposed hybrid modeling between RC and GGMR. The three modeling methods have been compared for predicting the energy use of a radiant slab system of a Living Laboratory office space using measurement data from January 15th to March 7th, 2022. The first two weeks of data were used for training, while the remaining data was used for testing of all three modeling methods. The hybrid approach had a Normalized Root Mean Square Error (NRMSE) of 15.46 percent (8.62 percent less than the RC-Model 3 alone and 19.36 percent less than the GGMR alone), a Coefficient of Variation of RMSE (CVRMSE) of 6.43 percent (3.59 percent less than the RC-Model 3 and 8.05 percent less than the GGMR), a Mean Absolute Error (MAE) of 3.61 kW (2.13 kW and 3.87 kW less than the RC-Model 3 and GGMR, respectively), and a Mean Absolute Percentage Error (MAPE) of 5.28 percent (3.85 percent and 3.92 percent lower than the RC-Model 3 and GGMR, respectively).

1. INTRODUCTION

Hydronic radiant slab systems (HRSS) have significant benefits for thermal management of conditioned spaces, including increased thermal comfort and energy savings (Joe and Karava 2019; Rhee and Kim 2015). Apart from these benefits, the large thermal storage capacity of an HRSS has a few disadvantages. One disadvantage of the large thermal time constant is that it causes cooling output to be delayed when supply water flow rates and temperature are adjusted (Liu et al. 2011). Additionally, conventional control based on room temperature feedback may consume more primary energy than a conventional air system (Sourbron et al. 2009). Moreover, an HRSS frequently experiences concurrent thermal disturbances caused by solar radiation, internal heat, and air systems (Koschenz and Dorer 1999) that when combined with conventional control approaches can lead to overcooling or overheating issues. To address these issues, an HRSS should incorporate Model Predictive Control (MPC) with accurate load prediction (Joe and Karava 2019). In general, load prediction methods for buildings fall into three categories: first-principles models, reduced-order thermal Resistor-Capacitor (RC) network models, and data-driven models, as summarized in ASHRAE (Handbook 2001) and (Dong et al. 2016). The following subsections will review those models followed by a brief statement of research objectives.

1.1 First-Principles Models

In this application, first-principles models refer to models that use Computational Fluid Dynamics (CFD) (Zhang et al. 2013) or building energy simulation software such as EnergyPlus (Crawley et al. 2001), and ESP-r (Clarke 2001). The computational cost of CFD makes them incompatible with large-scale simulation programs (Neumann, Gamisch, and Gschwander 2021; Rodríguez Jara et al. 2016). Most current building energy software requires a detailed physical and operational description of the building, as well as a well-mixed zone air assumption, to predict the performance of a building and its heating, ventilation, and air conditioning (HVAC) system.

1.2 Thermal RC Network Models

An inverse grey-box RC model strikes a balance between a physically-based model and a data-driven model (Braun and Chaturvedi 2002). An RC network model is considered as a collection of linear ordinary differential equations (ODEs). RC models are typically in the form of 2R1C, 3R2C, or lumped RC parameter models with associated self-adjusting methods (Rodríguez Jara et al. 2016). According to (O'Dwyer et al. 2016), when the resistance and capacitance values are positive, there is theoretically a guaranteed thermal passivity solution for RC models. As for the training of RC models, there is considerable research devoted to optimizing the trade-off between model accuracy and complexity (Ahn and Song 2010; Goyal, Liao, and Barooah 2011; Koschenz and Dorer 1999; Liu et al. 2011).

Nevertheless, there are some limitations in terms of the application of the RC model. The accuracy of lumped parameter methods is highly dependent on the estimation and calibration of their characteristic parameters (Rodríguez Jara et al. 2016), which requires substantial effort. Moreover, the accuracy of the RC model applied to a radiant slab degrades when the slab is subjected to rapid thermal disturbances (Neumann, Gamisch, and Gschwander 2021; Rhee and Kim 2015).

1.3 Data-driven Models

Many data-driven/machine learning algorithms have been evaluated for building energy models such as Partial Least Squares (PLS), Principal Component Analysis (PCA), Gaussian Process Regression (GPR) and Gaussian Mixture Model (GMM). For example, GPR has been used to capture the complex and highly subjective relationships between room temperature and subjective thermal perception (Guenther and Sawodny 2019). Also, GMM is widely recognized for its ability to model multimode characteristics and deal with process uncertainty (Billard et al. 2008; Li and Song 2020).

Considerable efforts have been made in the field of incremental learning Gaussian Mixture Regression (GMR), or Growing GMR (GGMR), to develop a mechanism for GMR adaptation (Bouchachia and Vanaret 2011; Cederborg et al. 2010; Karami and Wang 2018; Li and Song 2020; Wang, Kubichek, and Zhou 2018).

In this study, we propose a hybrid approach, in which the output from a simpler RC model is used as one of the inputs to a GGMR model. The proposed hybrid model combines the benefits of both the GGMR and RC models. The methodology and performance metrics for model evaluation are detailed in Section 2. Section 3 presents model development and a case study for an existing office at Purdue University followed by conclusions in Section 4.

2. METHODOLOGY

This section discusses the development of RC network models, the GGMR approach, and the hybrid modeling approach along with describing the model prediction performance criteria metrics.

2.1 RC Network Model

An RC network model is based on heat balance equations applied to temperature state variables (Braun and Chaturvedi 2002; Joe and Karava 2017). A general state-space model for estimating a radiant slab systems load is of the form

$$\frac{x}{dt} = Ax + \mathrm{Bu} \tag{1}$$

$$y = c^T x + d^T u$$

where estimated resistances, capacitances and heat flux coefficients (such as the ratio of solar radiation to the envelope state) form matrices *A, B* and vectors *c, d*. The variables *x, u, y, t* represent vectors of state, input, output variable and sampling time, respectively. For an HRSS, a single output variable is the cooling and heating load. The state vector contains all the temperature nodes. The input vector contains all the driving conditions, such as the heating or chilled water temperature, exterior air temperature, solar radiation, lighting, and occupancy schedule. The discrete version of the above state-space model can be written in terms of a recursive formula as

$$x_{(k+1)T} = A_d x_{kT} + B_d u_{kT} \tag{2}$$

$$y_{kT} = {c_d}^T x_{kT} + {d_d}^T u_{kT} \tag{3}$$

where the subscript *d* indicates these variables are the discretized forms of *A, B, c, d* in equation (1). A typical objective function for training an RC network model is to minimize the Root-Mean-Square Error (RMSE) for the training duration, denoted as the following, where *N* stands for the number of samples.

$$minimize \sqrt{\frac{\sum_{k=1}^{N_{train}} \left(y_{measured,k} - y_{predicted,k}\right)^2}{N_{train} - 1}} \tag{4}$$

2.2 GGMR Method

GMR(Sung 2004) is a regression approach that models probability distributions rather than functions. Assume the data follow the joint density

$$f_{X,Y}(x,y) = \sum_{j=1}^{K} \pi_j \, \phi(x,y;\mu_j,\Sigma_j) \tag{5}$$

where K is the number of Gaussian mixtures, π_j is the weight coefficient mean $\mu_j = \begin{bmatrix} \mu_{jX} \\ \mu_{jY} \end{bmatrix}$, covariance $\Sigma_j = \begin{bmatrix} \Sigma_{jX} & \Sigma_{jY} \\ \Sigma_{YX} & \Sigma_{YY} \end{bmatrix}$. The Gaussian mixture probability function shown in equation (5) can be presented as

$$f_{X,Y}(x,y) = \sum_{j=1}^{K} \pi_j \, \phi(y|x;m_j(x),\sigma_j^2)\phi(x;\mu_{jX},\Sigma_{jX}) \tag{6}$$

where

$$m_j(x) = \mu_{jX} + \Sigma_{jYX}\Sigma_{jX}^{-1}(x - \mu_{jX}) \tag{7}$$

$$\sigma_j^2 = \Sigma_{jYY} - \Sigma_{jYX}\Sigma_{jx}^{-1}\Sigma_{jXY} \tag{8}$$

From equation (6), the marginal density of X is

$$f_X(x) = \sum_{j=1}^{K} \pi_j \, \phi(x;\mu_{jX},\Sigma_{jX}) \tag{9}$$

The conditional probability density function of $Y|X$ is

$$f_{Y|X}(y|x) = \sum_{j=1}^{K} w_j(x) \, \phi(y;m_j(x),\sigma_j^2) \tag{10}$$

with the mixing weight

$$w_j(x) = \frac{\pi_j \phi\left(x;\mu_{jX,\Sigma_{jX}}\right)}{\sum_{j=1}^{K} \pi_j \, \phi(x;\mu_{jX},\Sigma_{jX})} \tag{11}$$

In the current study, we are interested in the expectation of *y* among all gaussian components:

$$y_{predicted|x} = m(x) = E[Y|X=x] = \sum_{k=1}^{K} w_j(x)m_j(x) \tag{12}$$

To accommodate new data in an online setting, control model complexity and allow modeling of time-varying processes, GGMR has been proposed by (Bouchachia and Vanaret 2011) with growing and shrinking mechanisms. We utilized its updating Gaussians algorithm in the present paper. More details can be seen in (Bouchachia and Vanaret 2011). The best match Gaussian is updated using the following formulas:

$$q_j = \frac{p_j}{\sum_{k=1...K} p_k} \tag{13}$$

$$c_j(t) = c_j(t-1) + q_j \tag{14}$$

$$\tau_j(t) = (1-\alpha)\tau_j(t-1) + \alpha q_j \tag{15}$$

$$\eta_j = q_j\left(\frac{1-\alpha}{c_j} + \alpha\right) \tag{16}$$

$$\mu_j(t) = (1-\eta_j)\mu_j(t-1) + \eta_j x_i \tag{17}$$

$$\Sigma_j(t) = (1-\eta_j)\Sigma_j(t-1) + \eta_j\left(x_i - \mu_j(t-1)\right)^2 \tag{18}$$

in which p_j is the match probability calculated with new input x_i and best match Gaussian $\phi_j(\mu_j, \Sigma_j)$, q_j is the expected posterior, c_j is the sum of the expected posterior for the best match Gaussian, τ_j are the weights of the best match Gaussian, η_j is the on-going learning rate for the j-th Gaussian, α is the converging learning rate.

2.3 Hybrid Approach

In the present study, we have designed the hybrid modeling approach schema as shown in Figure **1**, which illustrates the underlying structure of the hybrid approach. For each timestamp, the real-time predicted load from the RC model is used as one of inputs for the GGMR model (equations (13) – (18)). GGMR predicts system performance by updating key GMM parameters such as weighting factors, mean vectors, and covariance matrices. Sec. 3 described the detailed hybrid model inputs. In practice, the trained RC model should be started with a reasonable estimate of the initial states. As a result of non-optimal state initialization, it is expected that the first few predictions from the RC model will be chaotic. The current study proposed a warming-up period to minimize the errors caused by imperfect states initialization, and subsection 3.3 detailed how the warming-up steps were determined.

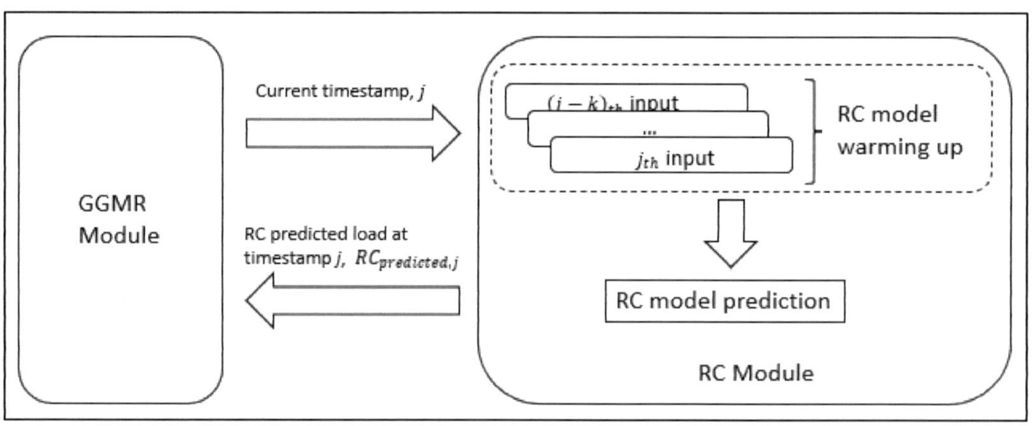

Figure 1 Underlying communication for hybrid approach.

2.4 Model Performance Evaluation Criteria

Four indices, Normalized Root Mean Square Error (NRMSE), Coefficient of Variation of RMSE (CVRMSE), and Mean Absolute Error (MAE), and Mean Absolute Percentage Error (MAPE), are used for model performance evaluation.

$$RMSE = \sqrt{\frac{\sum_{k=1}^{n}\left(y_{measured,k} - y_{predicted,k}\right)^2}{n-1}} \tag{19}$$

$$\text{NRMSE} = \frac{RMSE}{s\big(abs(y_{measured})\big)} \quad (20)$$

$$\text{CVRMSE} = \frac{RMSE}{mean\big(abs(y_{measured})\big)} \quad (21)$$

$$MAE = \frac{\sum abs\big(y_{measured,k} - y_{predicted,k}\big)}{n} \quad (22)$$

$$MAPE = \frac{1}{n}\sum abs\left(\frac{y_{measured,k} - y_{predicted,k}}{abs(y_{measured,k})}\right) \quad (23)$$

where n is the number of observations, $s\big(abs(y_{measured})\big)$ is the standard deviation of absolute measured values, and $mean\big(abs(y_{measured})\big)$ is the average of absolute measured values. Both $y_{measured,k}$ and $y_{predicted,k}$ are signed energy uses, with positive representing heating loads and negative representing cooling loads. It should be noted that the absolute prediction difference is determined using signed energy use values, whereas the prediction difference is normalized using absolute energy uses.

3. CASE STUDY

This section presents a case study that investigates the performance of the RC, GGMR and hybrid modeling approaches.

3.1 Testbed

The dataset included in-situ measurements for a living laboratory office space from January 15th to March 7th, 2022, with a 5-minute sampling rate. The first two weeks of data were used for training and the rest of the data was used for testing. The dataset was divided into two categories, onsite sensor data and estimated data. Onsite sensor data includes the following: outdoor air temperature denoted by T_{out}, façade cavity space temperature denoted by T_{cav}, slab concrete temperature denoted by T_{slab}, flowing water temperature within slab pipe denoted by T_{source}, solar radiation retrieved from a weather station denoted by $\dot{Q}_{solar}$ (AmbientWeatherNetwork2022), and Air Handling Unit (AHU) consumed heating power $\dot{Q}_{AHU}$. The estimated input values were determined using hourly schedules in accordance with (ANSI/ASHRAE/IES 90.1-2010 2010, 1), such as internal heating radiation denoted by $\dot{Q}_{int}$ based on occupancy schedule, and lighting radiation $\dot{Q}_{light}$ based on lighting schedule.

3.2 RC Network Model Development

This subsection describes the design logic for the RC model, followed by a description of the target room's physical structures and, finally, our consideration of various RC model designs and their associated performance. Ultimately, the chosen design will be detailed.

The major thermal components of the living laboratory office space include external walls, roof/ceiling, internal wall, south-facing double façade system, conditioned air from AHU system, and hydronic radiant floor system. Generally, adding complexity to an RC model can improve accuracy over a wide range of operating conditions, but at the expense of requiring additional input variables and more training data. In the present study, we considered three RC network designs by considering model robustness and various levels of complexity or model orders. As illustrated in Figure **2**, the three RC network models considered are a four-state Model 1, a five-state Model 2 and a six-state Model 3. In these network diagrams, T, C, Q, α represent temperatures, capacitances, resistances, heat fluxes and corresponding coefficients, whereas the subscripts, $out, cav, slab, source, sink, env, room, intwall, sol, int, light, AHU, rad$, represent outdoor air, façade cavity, slab concrete, hot water or chilled water within tubes, insulation below tubes, envelope, room air, internal wall, solar radiation, internal heat, lighting, air handling unit, and the amount of energy added to or extracted from the water within radiant slab pipes to maintain the space thermal comfort, respectively.

Each of the three models is composed of two components: room and concrete slab. We chose the same RC network model for the room portion of the model to effectively capture its thermal properties: a two-node envelope, one-node internal wall, one node cavity for the double façade system, and a room air node to capture disturbances due to heating or cooling from the AHU system. It is worth noting that we used the envelope node to represent the external wall and roof/ceiling to keep the model simple. For the concrete slab portion of the model, we experimented with various model orders. The detailed thermal structure of the radiant floor was omitted from Model 1 and we considered the entire slab to be a single node. In comparison to Model 1, Model 2 included an additional source node to represent the flow of water through slab pipes. Furthermore, Model 3 had one additional sink node compared to Model 2 to represent heat transfer between the source node and another space.

Figure **3** depicts predicted and actual results obtained for the testing period (10656 sampling points for 37 days). Model 1 has significantly higher errors than Models 2 and 3, which can be attributed to the oversimplified concrete slab representation. Model 3 has a lower CVRMSE than Model 2, which is consistent with the addition of a sink node. Table **2** contains a more detailed comparison of performance.

The Model 2 can be represented by a state-space model with the following state, input, and output variable definitions:

$$\boldsymbol{x}^T = [T_{env1}, T_{env2}, T_{room}, T_{intwall}, T_{slab2}, T_{sink}] \tag{24}$$

$$\boldsymbol{u}^T = \left[T_{out}, T_{slab1}, T_{cav}, T_{source}, \dot{Q}_{sol}, \dot{Q}_{int}, \dot{Q}_{light}, \dot{Q}_{AHU}, \frac{dT_{source}}{dt}\right] \tag{25}$$

$$y = \dot{Q}_{rad} \tag{26}$$

As stated in equation (4), the RC network model training is essentially an optimization problem to determine those unknown resistances, capacitances, and heat flux coefficients. The final converged values for those learning parameters have been listed in Table **1**. In the present paper, Particle Swarm Optimization (PSO) from python package (James V. Miranda 2018) was used to solve the above optimization problem.

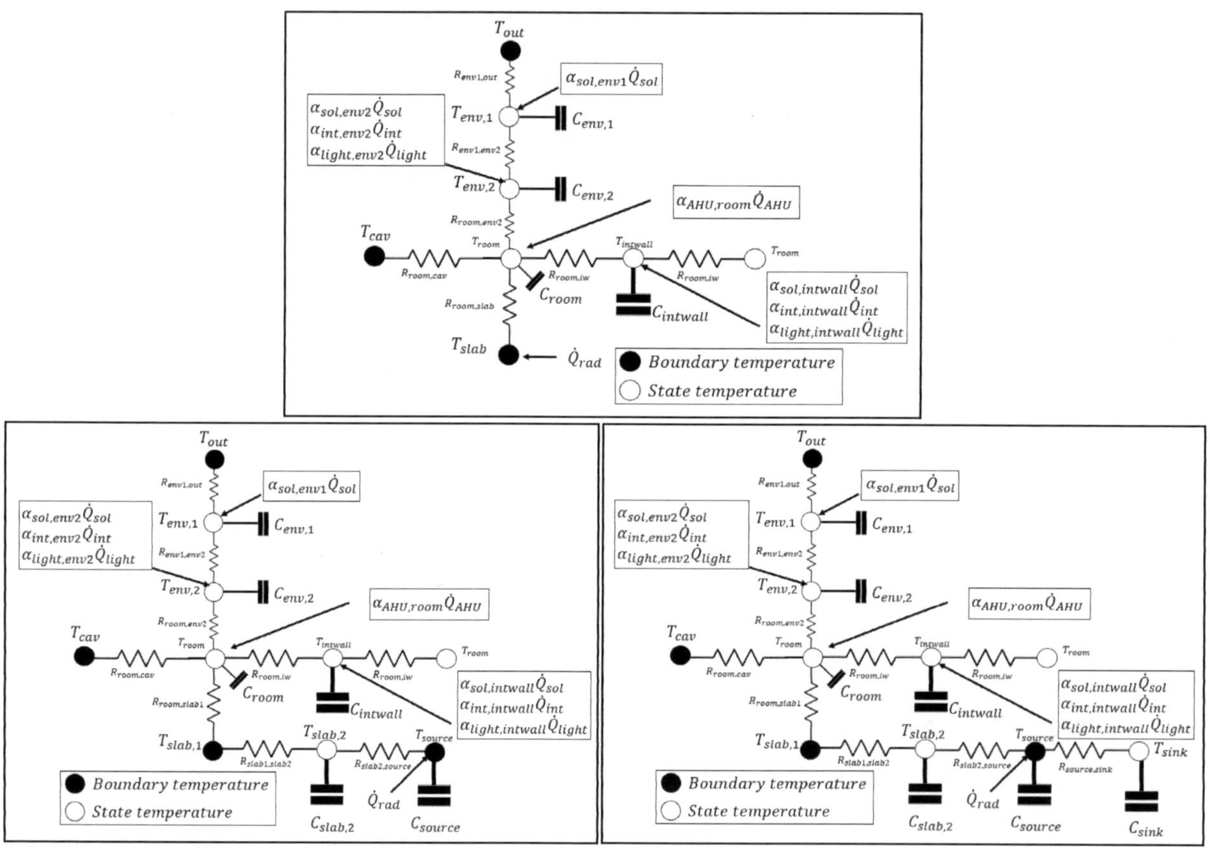

Figure 2 Structure of RC network. Top: four states Model 1; Left: five states Model 2; Right: six states Model 3.

Table 1 Final converged values of resistances (K/W), capacitances (J/K) and heat flux coefficients for Model 3.

$R_{out,env1}$ = *3.11E-1*	$R_{env2,env1}$ = *5.45E-1*	$R_{env2,room}$ = *9.94E-1*	$R_{intwall,room}$ =*-1.46E-2*
$R_{slab1,room}$ = *7.21E-1*	$R_{cav,room}$ = *4.69E-1*	$R_{slab1,slab2}$ = *5.64E-4*	$R_{source,slab2}$ = *6.44E-4*
$R_{source,sink}$ =*9.07E-4*	$C_{env1} = 2.6E6$	$C_{env2} = 1.3E6$	$C_{room} = 1E8$
$C_{iw} = 1.2E6$	$C_{slab2} = 6E6$	$C_{sink} = 2E4$	$C_{source} = 2.75E5$
$\alpha_{sol,env1} = 1E2$	$\alpha_{sol,env2}$ = *1.87E-1*	$\alpha_{int,env2} = 1.52$	$\alpha_{light,env2} = 2.46$
$\alpha_{AHU,room}$ =*1.41*	$\alpha_{sol,intwall}$ = *4.66E-1*	$\alpha_{int,intwall} = 1.01$	$\alpha_{light,intwall} = 1.78$

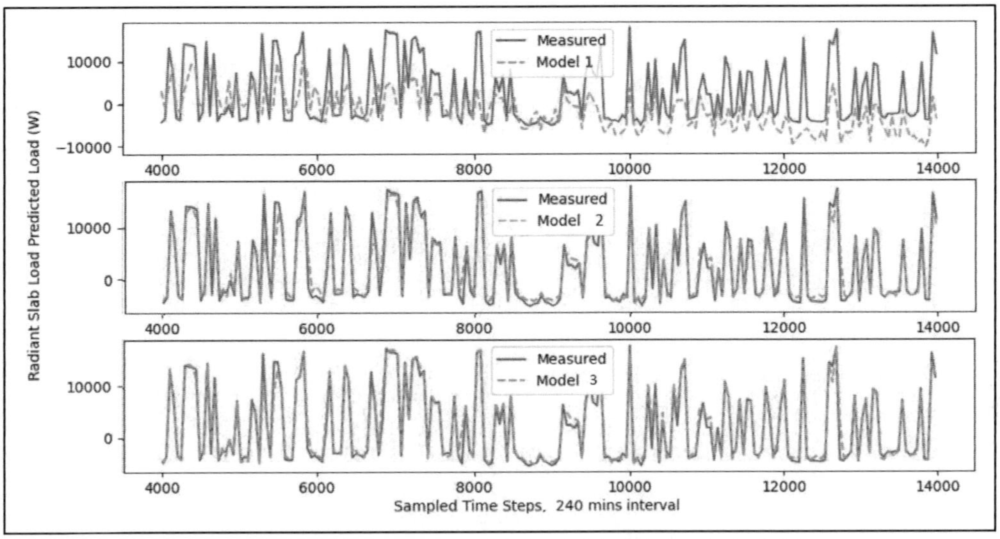

Figure 3 Testing results for Model 1, Model 2 and Model 3.

Table 2 Hourly prediction performance statistical comparison for proposed RC models.

Models	NRMSE (%)	CVRMSE (%)	MAE (kW)	MAPE (%)
Model 1	169.09	70.34	42.24	52.33
Model 2	27.38	11.39	6.90	10.68
Model 3	**24.08**	**10.02**	**5.74**	**9.13**

3.2 GGMR Model Development

This subsection primarily discusses the determination of the input variables for the GGMR model. Correlation coefficients R were used as an initial guess input variables. We experimented with various input combinations for the GGMR model. Table **3** shows example correlation coefficients for one of the cases considered, whereas the performance of the GGMR models with 3 different sets of inputs is shown in Table **4**. It is worth noting that larger correlation coefficients do not necessarily mean better prediction. For instance, the correlation coefficient of Q_{solar} is more significant than T_{out}, while the inputs including Q_{solar} did not provide additional prediction performance as shown in cases 1 and 2 of Table **4**. Moreover, it was found that better prediction performance is achieved if we provide flow rate information as an additional input. In comparison to case 1, case 3 had a 2% lower CVRMSE after adding $Flow_{Predicted,GGMR}$ from another GGMR prediction. In the end, the case 3 inputs were selected for the GGMR model.

Table 3 Correlation coefficients between radiant slab system load and input variables.

T_{out}	T_{slabs}	T_{cav}	$Valve_{cl}$	$Valve_{ht}$	Q_{solar}	$RadiantSlab_{load}$
-0.06	-0.08	-0.16	-0.89	0.35	-0.16	1

Table 4 Hourly prediction performance comparison for different GGMR inputs.

Case #	Inputs	CVRMSE (%)
1	$T_{out}, T_{slabs}, T_{cav}, Valve_{cl}, Valve_{ht}$	16.48
2	$T_{out}, T_{slabs}, T_{cav}, Valve_{cl}, Valve_{ht}, Q_{solar}$	16.49
3	$T_{out}, T_{slabs}, T_{cav}, Valve_{cl}, Valve_{ht}, Flow_{Predicted,GGMR}$	**14.48**

3.3 Hybrid Model Development

As mentioned in Sec. 2.3, the development of the hybrid approach is primarily concerned with the determination of warming-up step for the RC model. As illustrated in Figure **4**, the warming up step was statistically chosen as 15 to minimize the prediction error. Consistent with GGMR model, different input combinations for the hybrid model were also investigated as presented in Table **5** where $RC_{predicted,RealTime}$ represents the predicted load from RC model. Compared with case 1, case 2 had 0.30% lower CVRMSE, which was consistent with the results shown in Table **4.** In the end, the case 2 inputs were selected for the hybrid model.

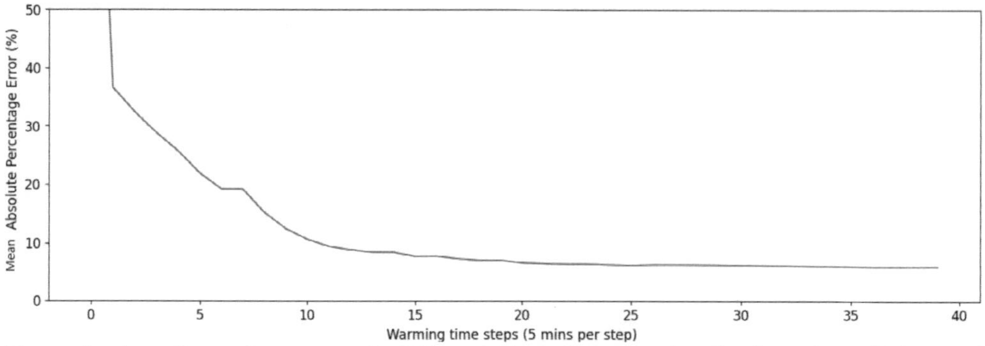

Figure 4 Determination of warming up step for RC model based on stepwise (5 minutes) prediction performance.

Table 5 Hourly prediction performance comparison for different hybrid model inputs.

Case #	Inputs	CVRMSE (%)
1	$T_{out}, T_{slabs}, T_{cav}, Valve_{cl}, Valve_{ht}, RC_{predicted,RealTime}$	6.73
2	$T_{out}, T_{slabs}, T_{cav}, Valve_{cl}, Valve_{ht}, Flow_{Predicted,GGMR}, RC_{predicted,RealTime}$	**6.43**

3.3 Performance Comparison for Proposed Models

Based on the statistical results presented in Table **5**, all three proposed models comply with ASHRAE Guideline 14 (ASHRAE 2014). Moreover, this table demonstrates that the hybrid model is the most accurate model for predicting the energy consumption of radiant slab systems, as it incorporates information from both the RC and GGMR models. Specifically, the hybrid approach has an NRMSE of 15.46 percent (8.62 percent less than the RC-Model 3 alone and 19.36 percent less than the GGMR alone), a CVRMSE of 6.43 percent (3.59 percent less than the RC-Model 3 and 8.05 percent less than the GGMR), a MAE of 3.61 kW (2.13 kW and 3.87 kW less than the RC-Model 3 and GGMR, respectively), and a MAPE of 5.28 percent (3.85 percent and 3.92 percent lower than the RC-Model 3 and GGMR, respectively).

Table 5 Hourly prediction performance comparison of proposed models.

Models	NRMSE (%)	CVRMSE (%)	MAE (kW)	MAPE (%)
RC-Model 3	24.08	10.02	5.74	9.13
GGMR	34.82	14.48	7.48	9.20
Hybrid	**15.46**	**6.43**	**3.61**	**5.28**

4. CONCLUSIONS

In this paper, a novel hybrid modeling approach has been proposed to predict the energy consumption of a hydronic radiant slab system that incorporates the advantages of both RC and GGMR models. The hybrid approach involves using an output from one simplified RC model as an input to the GGMR. The proposed method was validated using measurements from a radiant slab system operating at Purdue University. According to the case study, the hybrid model significantly outperformed the RC and GGMR models in terms of prediction performance. The proposed hybrid model had a CVRMSE of 6.43 percent (3.59 percent less than the RC-Model 3 and 8.05 percent less than the GGMR), which clearly meets the criteria for ASHRAE Guideline 14. In addition, it's worth noting that the case study makes use of a single onsite dataset source. In the future, we need to conduct additional case studies using a variety of data sources.

ACKNOWLEDGEMENT

This study was supported by the National Science Foundation EPSCoR Research Infrastructure program under Grant No. 1929209. Any opinions, findings, and conclusions, or recommendations expressed in this material are those of the authors and do not necessarily reflect the views of the National Science Foundation.

REFERENCES

Ahn, Byung-Cheon, and Jae-Yeob Song. 2010. "Control Characteristics and Heating Performance Analysis of Automatic Thermostatic Valves for Radiant Slab Heating System in Residential Apartments." *Energy* 35(4): 1615–24.

Ambient Weather Network. 2022. *Ambient Weather Network*. https://ambientweather.net/ (April 11, 2022).

ANSI/ASHRAE/IES 90.1-2010. 2010. *Energy Standard for Buildings Except Low-Rise Residential Buildings*. American Society of Heating, Refrigerating and Air-Conditioning Engineers.

ASHRAE. 2014. "ASHRAE Guideline 14: Measurement of Energy, Demand and Water Savings." : 150.

Billard, Aude, Sylvain Calinon, Rüdiger Dillmann, and Stefan Schaal. 2008. "Robot Programming by Demonstration." In *Springer Handbook of Robotics*, eds. Bruno Siciliano and Oussama Khatib. Berlin, Heidelberg: Springer, 1371–94. https://doi.org/10.1007/978-3-540-30301-5_60 (April 12, 2022).

Bouchachia, Hamid, and Charlie Vanaret. 2011. "Incremental Learning Based on Growing Gaussian Mixture Models."

Braun, James E., and Nitin Chaturvedi. 2002. "An Inverse Gray-Box Model for Transient Building Load Prediction." *HVAC&R Research* 8(1): 73–99.

Cederborg, Thomas, Ming Li, Adrien Baranes, and Pierre-Yves Oudeyer. 2010. "Incremental Local Online Gaussian Mixture Regression for Imitation Learning of Multiple Tasks." In *2010 IEEE/RSJ International Conference on Intelligent Robots and Systems*, , 267–74.

Clarke, Joseph. 2001. *Energy Simulation in Building Design*. 2nd ed. London: Routledge.

Crawley, Drury B. et al. 2001. "EnergyPlus: Creating a New-Generation Building Energy Simulation Program." *Energy and Buildings* 33(4): 319–31.

Dong, Bing, Zhaoxuan Li, S. M. Mahbobur Rahman, and Rolando Vega. 2016. "A Hybrid Model Approach for Forecasting Future Residential Electricity Consumption." *Energy and Buildings* 117: 341–51.

Goyal, Siddharth, Chenda Liao, and Prabir Barooah. 2011. "Identification of Multi-Zone Building Thermal Interaction Model from Data." In *2011 50th IEEE Conference on Decision and Control and European Control Conference*, , 181–86.

Guenther, Janine, and Oliver Sawodny. 2019. "Feature Selection and Gaussian Process Regression for Personalized Thermal Comfort Prediction." *Building and Environment* 148: 448–58.

Handbook, ASHRAE. 2001. "Fundamentals SI Edition." *American Society of Heating, Refrigerating and Air-Conditioning Engineers, Inc., Atlanta, GA*.

James V. Miranda, Lester. 2018. "PySwarms: A Research Toolkit for Particle Swarm Optimization in Python." *The Journal of Open Source Software* 3(21): 433.

Joe, Jaewan, and Panagiota Karava. 2017. "Agent-Based System Identification for Control-Oriented Building Models." *Journal of Building Performance Simulation* 10(2): 183–204.

Joe, Jaewan, and Panagiota Karava. 2019. "A Model Predictive Control Strategy to Optimize the Performance of Radiant Floor Heating and Cooling Systems in Office Buildings." *Applied Energy* 245: 65–77.

Karami, Majid, and Liping Wang. 2018. "Fault Detection and Diagnosis for Nonlinear Systems: A New Adaptive Gaussian Mixture Modeling Approach." *Energy and Buildings* 166: 477–88.

Koschenz, Markus, and Viktor Dorer. 1999. "Interaction of an Air System with Concrete Core Conditioning." *Energy and Buildings* 30(2): 139–45.

Li, Deyang, and Zhihuan Song. 2020. "A Novel Incremental Gaussian Mixture Regression and Its Application for Time-Varying Multimodal Process Quality Prediction." In *2020 IEEE 9th Data Driven Control and Learning Systems Conference (DDCLS)*, , 645–50.

Liu, Kuixing et al. 2011. "Establishment and Validation of Modified Star-Type RC-Network Model for Concrete Core Cooling Slab." *Energy and Buildings* 43(9): 2378–84.

Neumann, Hannah, Sebastian Gamisch, and Stefan Gschwander. 2021. "Comparison of RC-Model and FEM-Model for a PCM-Plate Storage Including Free Convection." *Applied Thermal Engineering* 196: 117232.

O'Dwyer, Edward et al. 2016. "Modelling and Disturbance Estimation for Model Predictive Control in Building Heating Systems." *Energy and Buildings* 130: 532–45.

Rhee, Kyu-Nam, and Kwang Woo Kim. 2015. "A 50 Year Review of Basic and Applied Research in Radiant Heating and Cooling Systems for the Built Environment." *Building and Environment* 91: 166–90.

Rodríguez Jara, Enrique Á. et al. 2016. "A New Analytical Approach for Simplified Thermal Modelling of Buildings: Self-Adjusting RC-Network Model." *Energy and Buildings* 130: 85–97.

Sourbron, M. et al. 2009. "Efficiently Produced Heat and Cold Is Squandered by Inappropriate Control Strategies: A Case Study." *Energy and Buildings* 41(10): 1091–98.

Sung, Hsi Guang. 2004. "Gaussian Mixture Regression and Classification." Ph.D. Rice University. https://www.proquest.com/docview/305155652/abstract/8C63788CCF824897PQ/1 (April 12, 2022).

Wang, Liping, Robert Kubichek, and Xiaohui Zhou. 2018. "Adaptive Learning Based Data-Driven Models for Predicting Hourly Building Energy Use." *Energy and Buildings* 159: 454–61.

Zhang, Rui, Khee Poh Lam, Shi-chune Yao, and Yongjie Zhang. 2013. "Coupled EnergyPlus and Computational Fluid Dynamics Simulation for Natural Ventilation." *Building and Environment* 68: 100–113.

Model Predictive Control for a Grid-interactive Efficient Thermal Storage-integrated Heat Pump System

Liang SHI[1], Ming QU[1]*, Xiaobing LIU[2], Jin DONG[2], Borui CUI[2], Lingshi WANG[2]

[1]Purdue University, Lyles School of Civil Engineering,
West Lafayette, IN, USA
shi448@purdue.edu
mqu@purdue.edu

[2]Oak Ridge National Laboratory,
Oak Ridge, TN, USA
liux2@ornl.gov
dongj@ornl.gov
cuib@ornl.gov
wangl2@ornl.gov

* Corresponding Author

ABSTRACT

Building heating and cooling systems can be used to overcome the mismatch between the intermittent supply of renewable power and the fluctuating demand for electricity. A novel underground thermal energy storage integrated with a dual-source heat pump has been proposed to mitigate the mismatch while meeting the thermal demand of buildings efficiently. Conventional thermostat control with heuristic rules cannot provide intelligent decisions to maximize the thermal efficiency and flexibility of the proposed system. Advanced control strategies like model predictive control (MPC) have provided a new paradigm for grid-interactive efficient building operation with the advancement of computation and sensing. This study developed an MPC for the proposed system to provide grid service for Demand Side Management and minimize the operating cost of building owners. A control-oriented dynamic model of the proposed system has been developed. Given an objective function and proper constraints, an optimization problem is formulated to determine the optimal control strategy of the system. Dynamic Programming is adopted to solve the optimization problem. A rule-based control (RBC) is also developed to achieve similar goals. Short-term simulations are conducted to compare the system performance resulting from the two controls. The simulation results indicate that the MPC performs more intelligently than the RBC in charging thermal energy storage and selecting heat pump sources by taking advantage of the predicted cooling demands of the building and the performance of the integrated system. As a result, the MPC could save energy and reduce operating costs compared with the RBC. A case study shows that, for a 3-day operation, the MPC saves 36.9% energy and reduces 38.5% operating cost compared with the RBC.

1. INTRODUCTION

The increasing penetration of intermittent renewable power raises challenges to the existing electric grids due to a mismatch between the supply and demand sides. Buildings consume 74% of all U.S. electricity, and space heating/cooling accounts for 34% and 27% of the electricity consumed in residential and commercial buildings (Schwartz et al., 2017). In addition, space heating/cooling contributes to 46% peak demand of the electric grid (Neukomm et al., 2019). Therefore, large-scale implementation of highly efficient building thermal systems with greater flexibility is a solution to mitigate the mismatch between the intermittent renewable power supply and the fluctuating electric demand in buildings. While thermal energy systems (TES) have been successfully implemented in large commercial buildings for shifting the electricity demand from the on-peak hours to the off-peak hours of the grid, few technologies have been developed for residential buildings to effectively shift electric demand without compromising the thermal comfort of buildings' occupants.

A novel thermal energy storage (TES) integrated heat pump system for residential buildings was proposed to meet the building's thermal demand while shifting the electric demand. The system can provide distributed grid service for demand-side management (DSM) by (1) providing long-term high efficiency with a newly designed heat pump; and (2) shifting electric demand diurnally using an underground TES.

To realize DSM strategies with the new system, a well-fitted control strategy should be properly designed. Conventional thermostat control with heuristic rules (rule-based control, RBC) can make control decisions depending on the current measurement and pre-defined rules. The implementation for RBC is relatively simple. However, it is difficult for the RBC to integrate all useful information (e.g., forecasted load and weather) to fully explore the system's potential on approaching high thermal efficiency and flexibility. With decreasing costs in computation and sensing, advanced control strategies like model predictive control (MPC) have become popular research topics in building operations. Improving sensing and forecasting techniques on weather, occupancy, etc., enables accurate prediction of building thermodynamics using simplified mathematical models (Drgoňa et al., 2020). Given the fact that the thermal response time of the building is moderate, MPC is effective and suitable for building thermal system control on achieving optimal operation results. Specifically, with predictable grid conditions, the building thermal system with MPC can respond intelligently to realize DSM goals while satisfying the room thermal comfort requirement.

Unlike commercial buildings, MPC for residential thermal systems conducting various DSM is less studied. The most widely used indicator for DSM is operating cost reduction. Baniasadi et al. (2019) applied a time-of-use (TOU) tariff to a smart residential building in Australia that incorporated a heat pump, TES water tank, rooftop P.V., and battery to achieve minimum cost. The results indicated that a reduced daily energy cost could be as high as 44% in summer and 18% in winter. Golmohamadi et al. (2021) applied day-ahead pricing to minimize the operating cost in a residential building in Denmark by taking advantage of the building's thermal inertia and an active TES water tank. The results showed a weekly energy cost reduction of 37% in winter. Other indicators like the primary energy factor (a reflection of renewable power penetration) can also be used to evaluate DSM performance. Wolisz et al. (2020) developed an MPC to reduce the primary energy demand by activating the building thermal mass of a residential building in Germany. The result shows that the optimization control strategy successfully activated the building thermal mass, and the primary energy demand can be reduced by 3-7% annually. D'Ettorre et al. (2019) numerically studied the impact of integrating a stratified water tank with an air-source heat pump (ASHP) for space heating in Italy. With a proposed MPC for cost minimization, the results show that a water tank with 500 L volume would lead to an energy cost saving of up to 8% and a reduction of primary energy consumption by up to 13%, which is much higher than that in Wolisz et al. (2020). It indicates that TES tanks can provide more DSM flexibility than using the thermal mass of residential buildings.

Existing literature shows that MPC can exploit the TES integrated residential building thermal systems for DSM. However, most of the studies focus on heating scenarios. In addition, the stratified water tank is the dominant TES system used in the previous studies. It has low storage capacity and requires considerable installment space, demanding for residential buildings. Moreover, the MPC developed in many previous studies decoupled the building envelope model from the mechanical system and over-simplified the component models to neglect their nonlinearity for easier optimization problem-solving.

This paper presents an MPC for the novel TES integrated heat pump system for residential buildings to provide grid service for DSM in the cooling season. The MPC uses a simple dynamic model that couples the building envelope and the mechanical system. The modeled TES tank uses phase change material (PCM) to increase its energy storage density. A customized optimization solver is developed to account for the system's inherent nonlinearity. A case study is conducted to compare the performance of the system resulting from using the MPC and a typical RBC.

2. SYSTEM CONFIGURATION

The proposed system is targeted at single-family houses. The mechanical system mainly consists of a dual-source heat pump (DSHP) and a dual-purpose underground thermal battery (DPUTB), as depicted in Figure 1. The DSHP is an electricity-driven vapor compression heat pump that utilizes either the ambient air or the ground as its heat source/sink. It runs as an air-source heat pump (ASHP) when the ambient temperature is favorable and switches to a ground-source heat pump (GSHP) when the ambient air temperature is too high or too low. The source side of the GSHP connects to the helical heat exchanger submerged in the outer tank of the DPUTB. The heating/cooling load from the helical heat exchanger will finally be rejected to the surrounding soil. At the room side, a fan coil works as the terminal that heats up or cools down the indoor air through forced convection. DPUTB is an innovative design that integrates TES into a shallow-buried ground heat exchanger (a detailed configuration of the DPUTB can be found in Shi et al., 2021). The tank is installed in the subsurface of the ground so that no TES installment space is needed in the room. The inner tank of the DPUTB works as a TES tank. Macro-encapsulated PCM is applied to enhance its storage capacity. PCM cans are installed layer by layer with limited space in between. The water flows through the gaps among PCM cans to freeze or melt PCMs. Proper flow direction is selected to maintain thermal stratification within the tank. The outer tank of the DPUTB works as a ground heat exchanger. A helical coil heat exchanger is installed in the outer tank that connects the DPUTB to the source side of the ground-source heat pump. The strong natural convection induced by the helical heat exchanger will keep the outer tank water well mixed. Therefore, the temperature of the helical coil leaving the water, in response to the heating/cooling input, is buffered by the entire water body in the outer tank.

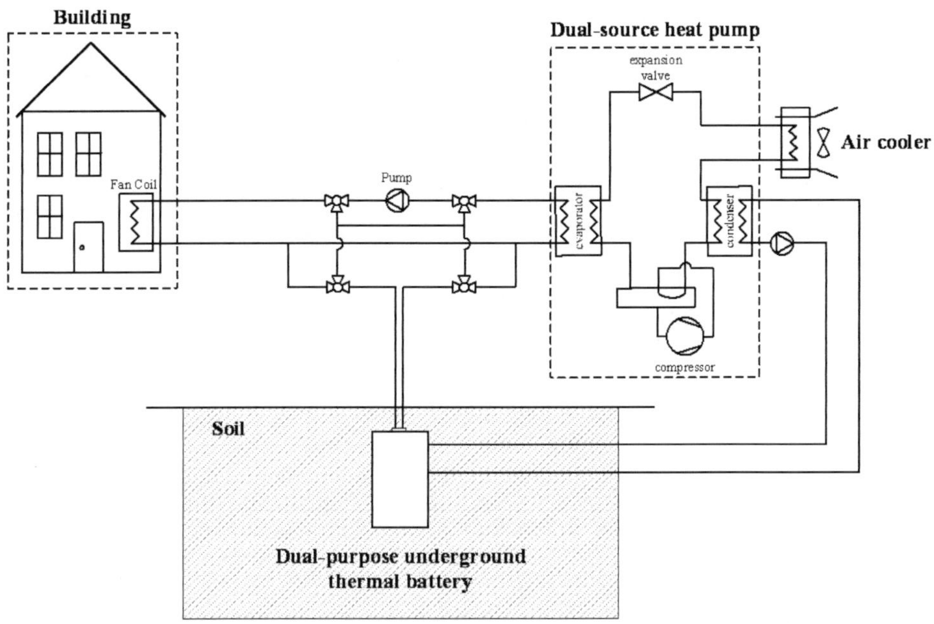

Figure 1: Schematic of a DPUTB integrated DSHP system for a single-family house

3. METHODOLOGY

3.1 Overview

MPC discretizes the prediction horizon in small sample times. Given future boundary conditions, it is capable of predicting building thermal response and guiding on control sequence so that an objective within the prediction horizon can be achieved. To implement MPC in an integrated building energy system, three key components are required: disturbance, model predictive controller, and a testbed. They are shown in Figure 2.

The disturbance is the non-controllable input of the MPC. Typically, the disturbances include weather conditions (e.g., ambient temperature, solar radiation, etc.), internal heat gains (e.g., occupancy, equipment, etc.), and grid information (e.g., electricity price). Model predictive controller development is the key component for the entire MPC scheme. A dynamic system model was developed to predict the thermal response of the building and the mechanical system. The model was simple enough so that it could be efficiently solved in the optimization problem. The cost function described the objective of the control: for the current study, minimizing the operating cost according to a TOU tariff. The output of the optimization problem was a sequence of 'future' control actions. Only the first control signal was sent to the mechanical system. With proceeding time, the prediction horizon kept being shift forward, and the optimization problem was always updated with new disturbances and states. In this way, the MPC was in a closed-loop, and this principle is called 'receding horizon control.' The testbed of the system can be either a real building energy system with measurement or a detailed building simulation model that incorporates many trustworthy sub-models. For the current study, the testbed is a simple simulation model similar to the dynamic model in the model predictive controller. Typically, an estimator (e.g., Kalman Filter) is implemented to estimate those state variables that cannot be measured directly.

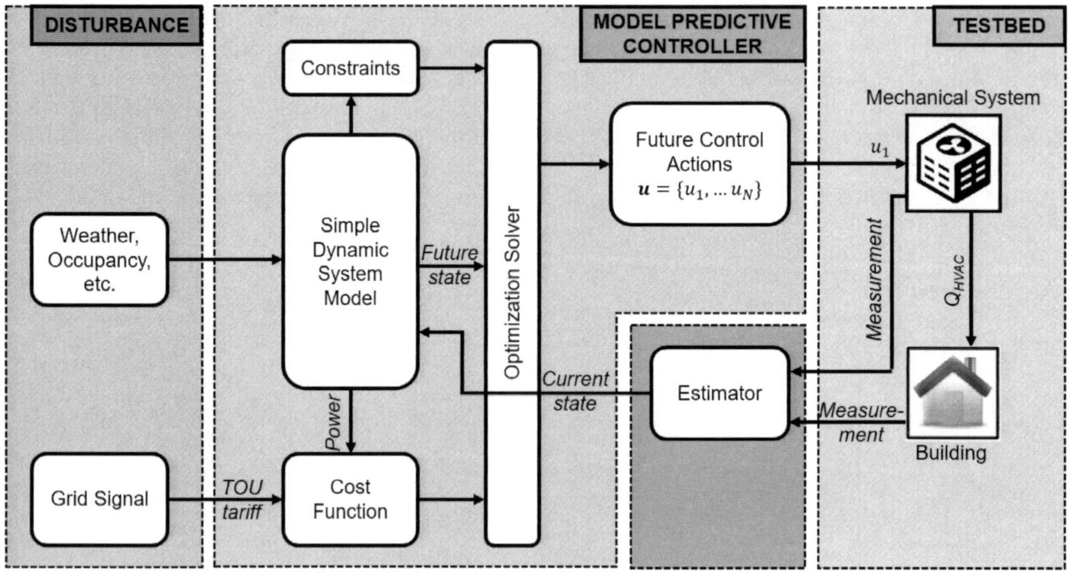

Figure 2: MPC scheme for the current study

3.2 Dynamic Model for MPC

As shown in Figure 1, to formulate our MPC controller, we need to develop dynamic models for building envelope, DSHP, and DPUTB, respectively.

3.2.1 Building envelope model. Gray-box models are widely used in model predictive control for building modeling due to their good balance of accuracy and computation speed. A typical representation of the gray-box model is the reduced-order Resistance-Capacitance (R.C.) model. Its accuracy is guaranteed by system identification from measured data or detailed building model simulation, and the computation speed can be enhanced since the number and complexity of differential equations are greatly reduced. The current study focuses on the model predictive control development of the mechanical system, but not the system identification of the R.C. model. Thus, the building model is further simplified as 0R1C with a given pre-calculated load. The model needs 1C to represent the lumped thermal capacitance of the building. The simplified building model can be expressed as:

$$C_{room}\frac{\mathrm{d}T_{room}}{\mathrm{d}t} = q_{load} + u_{FC}\cdot\left[u_{TES}\cdot mFR_{pump}\cdot c_p\cdot\left(T_{w1,in} - T_{w1,out}\right) - u_{GSHP}\cdot q_{GSHP} - u_{ASHP}\cdot q_{ASHP}\right] \quad (1)$$

3.2.2 DSHP model. For computational efficiency, a curve-fit method was adopted to simulate the DSHP. Catalog data from the heat pump manufacturers were used to generate the performance curves for the power and capacity prediction of the heat pump. Notice that the DSHP model is steady-state, and it is expressed in the algebraic equation instead of a first-order differential equation since it is fluid machinery with negligible stationary thermal mass. For the current study, a single-stage heat pump with constant flow rates on both load and source sides is applied. The heat pump performance is affected by the entering fluid temperatures at both the source and load sides. The performance can be expressed as below:

$$\dot{Q} = \left(\alpha_1 + \alpha_2\cdot\frac{T_{loa,ent}}{T_{loa,ref}} + \alpha_3\cdot\frac{T_{sou,ent}}{T_{sou,ref}}\right)\cdot\dot{Q}_{ref}\cdot s \quad (2)$$

$$P = \left(\beta_1 + \beta_2\cdot\frac{T_{loa,ent}}{T_{loa,ref}} + \beta_3\cdot\frac{T_{sou,ent}}{T_{sou,ref}}\right)\cdot P_{ref}\cdot s \quad (3)$$

3.2.3 DPUTB model. We have developed a detailed DPUTB model in a previous study (Shi et al., 2020), which can predict the thermal response of the DPUTB with good accuracy. However, it is too complex to be used in MPC. Instead, we use simple first-order differential equations to determine the required states.

The first important state is the in/outlet temperatures of the TES (inner tank of the DPUTB). Even though the inner tank is simulated as a stratified tank in the detailed model, to avoid listing numerous equations reflecting the temperature glide along the vertical direction, the inner tank water is simplified as a lumped control volume, and an average temperature of it is proposed to represent the inner tank water status. For the in/outlet temperatures, a constant temperature difference is assumed between the average temperature and the top/bottom parts of the inner tank. The first-order differential equation for the inner tank water can be expressed as:

$$m_{w1} \cdot c_p \frac{\mathrm{d}T_{w1,avg}}{\mathrm{d}t} = -u_{TES} \cdot mFR_{pump} \cdot c_p \cdot \left(T_{w1,in} - T_{w1,out}\right) - UA_{w1-PCM} \cdot \left(T_{w1,avg} - T_{PCM,avg}\right) - UA_{w1-w2} \cdot \left(T_{w1,avg} - T_{w2,avg}\right) \tag{4}$$

The second state that needs to be expressed is the status of the stored energy in the TES. Since most of the useful stored energy is in the form of latent heat in the PCM, the total PCM solid fraction is chosen as the indicating state. Currently, for simplification, the PCM is assumed to have a fixed melting point. All PCMs in the inner tank are assumed to be lumped and the dynamic change of the total average PCM solid fraction can be expressed as:

$$\frac{\mathrm{d}\lambda_{PCM,avg}}{\mathrm{d}t} = -\frac{1}{CAP} \cdot UA_{w1-PCM} \cdot \left(T_{w1,avg} - T_{PCM,avg}\right) \tag{5}$$

The last important state for the DPUTB is the returning temperature of the ground heat exchanger, which is the helical coil submerged in the outer tank water. Roughly, this temperature is only influenced by the outer tank water temperature, given the assumption that the helical coil heat exchanger has steady performance. For the current study, it is assumed that the helical coil has perfect heat transfer. That is, the outlet temperature of the helical coil is identical to the average outer tank water temperature. Thus, a first-order differential equation representing the outer tank water average temperature is adequate, which can be expressed as:

$$m_{w2} \cdot c_p \frac{\mathrm{d}T_{w2,avg}}{dt} = UA_{w1-w2} \cdot \left(T_{w1,avg} - T_{w2,avg}\right) - UA_{w2-soil} \cdot \left(T_{w2,avg} - T_{soil}\right) + u_{GSHP} \cdot q_{coil-w2} \tag{6}$$

3.3 Optimization Problem Formulation for MPC

Based on the models developed above, we can formulate our optimization problem framework, which generally includes cost function, constraints, and system dynamics. The objective of the current study is to take advantage of the proposed system and the TOU tariff to minimize the operating cost over the prediction horizon while keeping the indoor air temperature within the pre-described range. The optimal control problem can be expressed as:

$$\min_{u_1,u_2,\ldots,u_{N-1}} \sum_{k=1}^{N-1} P_{DSHP,k} \cdot \Delta t \cdot Price_k \tag{7a}$$

subject to

$$x_{k+1} = f(x_k, u_k, d_k) \tag{7b}$$
$$x_{lb,k} \leq x_k \leq x_{ub,k} \tag{7c}$$

Equation 7b defines the simplified dynamic equations for the entire thermal system. As illustrated in section 3.2, the proposed system is small-scale (4 first-order differential equations) but includes nonlinearity. Its nonlinearity is two-fold: 1. The control actions of the components are discontinuous. Most of them are on/off or switch control instead of continuous modulating control; 2. There are non-linear terms in dynamic equations that mix control variables and state variables. These characteristics of the current optimization result in a problem similar to shortest path searching, which can be solved using the dynamic programming method. Compared with using a complex non-linear mixed-integer programming (NLMIP) solver, dynamic programming can easily handle the nonlinearity of the system with acceptable computation speed due to a small-scale system.

3.4 Rule-based Control Development

To testify that the proposed system with MPC has better performance, as a reference, an RBC is designed. This control strategy is developed based on sensing the following states before each sample time: 1. The period of pricing (i.e., on-peak or off-peak), which depends on the TOU tariff structure; 2. The status of TES (i.e., fully discharged during the on-peak period or fully charged during the off-peak period) depends on the working fluid temperature at the top/bottom of the DPUTB inner tank; 3. Thermal load requirement from fan coil (i.e., 0 or 1), which depends on the room temperature or the set-point; 4. The selection of heat pump source (i.e., air-source or ground-source) depends on the ambient temperature.

4. CASE STUDY

A simulation-based case study was implemented to investigate the performance of the DPUTB integrated DSHP system under MPC and RBC strategies. A three-day simulation from July 25-27 was conducted for a single-family house located in Atlanta, GA. The prediction horizon and sample time choosing play a significant role in the performance of MPC. It is a trade-off between accuracy and computational speed. For the current study, the prediction horizon was chosen as 24 hours which always contains enough information for the next day's electricity price. The sample time was set as 15 minutes, considering the thermal response time of the building thermal mass and TES. For one-step optimization, it took the MATLAB-based dynamic programming solver 90 seconds to give optimal control outputs on a P.C. with Core i7-6700 CPU 3.41 GHz.

4.1 Testbed

The testbed of the control is either a real building with sensor measurement or a detailed simulation model that can predict the thermal response of the system. Currently, for simplicity, a simple model similar to the dynamic model in MPC was used. Equations 1-5 were applied for the simulation testbed to predict the dynamic change of room temperature, DPUTB inner tank water average temperature, PCM average charging status, and thermal output of the heat pump. For the borehole wall soil temperature, the g-function method was applied to predict its thermal response. Thus, the DPUTB outer tank water temperature was predicted by an energy balance, assuming it is in steady state at the end of each sample time.

The g-function is a non-dimensional temperature response factor proposed by Eskilson and Claesson (1988). Its value is determined by regression of the thermal response of detailed numerical or analytical borehole simulation. Given undisturbed soil temperature and the thermal input at each time step in history, the borehole wall soil temperature can be predicted using the following equation:

$$T_{soil}(t_n) = T_{ug} + \sum_{i=1}^{n} \frac{(Q_i - Q_{i-1})}{2\pi k} g(t_n - t_{i-1}) \tag{8}$$

The short-time g-function curve used in this study is a modified one from a UC Davis report (2021).

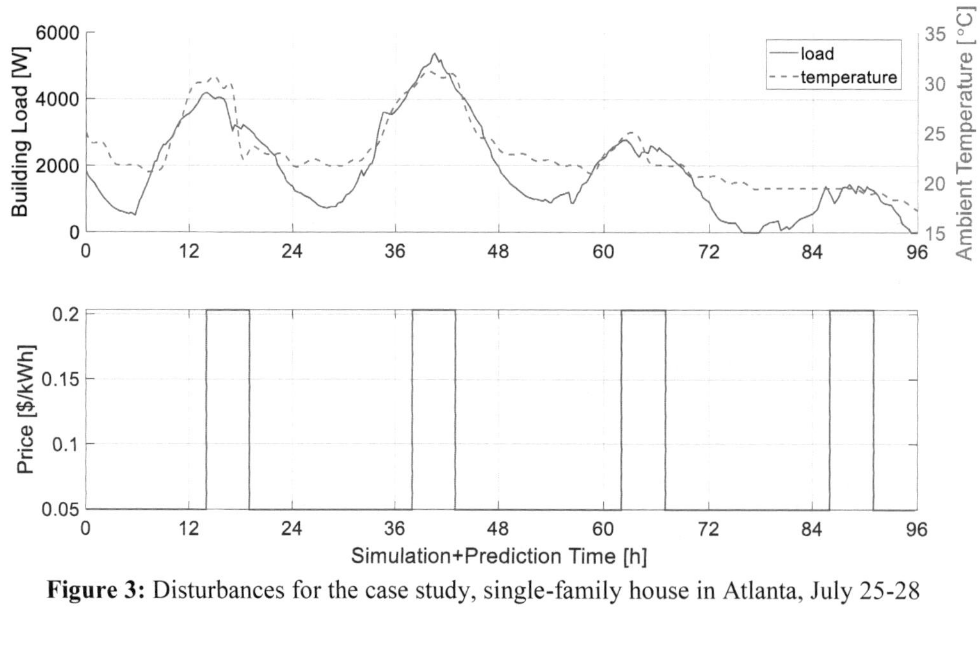

Figure 3: Disturbances for the case study, single-family house in Atlanta, July 25-28

4.2 Disturbances and Other Inputs

Hourly TMY3 data for Atlanta, GA, was used as the weather data. Given the weather condition, the hourly thermal load of a typical single-family house was determined with the EnergyPlus program using a prototype building model (DOE, 2021) in Atlanta, GA. The building load profile and ambient temperature profile from July 25-28 are shown in Figure 3.

A full-size DPUTB alone can provide the heat sink for a GSHP with 0.5-1 ton capacity. As a TES, its capacity is a combination of the PCM latent heat and the water sensible heat. The total volume of the PCM is 0.168 m^3, and it occupies 18% of the inner tank volume. The volumetric energy density of the PCM used for simulation is 300 MJ/m^3, and the total latent heat is 50.4 MJ, which is equivalent to 1-ton cooling for 4 hours. The melting point of the PCM is set as 9.5 °C. Because the maximum building thermal load is around 2 tons (from the annual building load profile), 2 full-size DPUTB would be needed. The thermal load of the building was divided by two to simplify the system simulation, and it is used to size the system. Thus only 1 DPUTB is included in the simulation. For half of the maximum building load, a heat pump with a 1-ton capacity at rated condition is large enough.

4.3 Results and Discussions

Thermal outputs of the proposed system operated with RBC and MPC, respectively, are plotted in Fig. 4 for a 3-day period. One clarification is that positive values for TES output indicate that TES is being discharged (i.e., releasing stored cooling energy), while negative values indicate it is being charged (i.e., storing cooling energy).

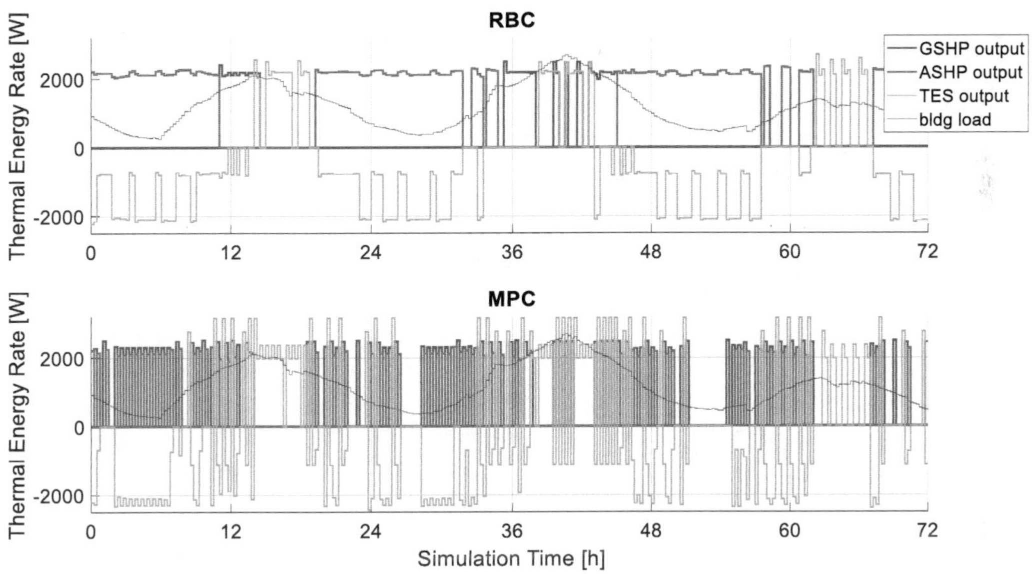

Figure 4: Thermal output profile

For RBC, the system always followed the pre-defined rule—during the off-peak period (0:00 to 14:00 and 19:00 to 24:00), the cooling demand of the building is always met soly by the heat pump, and the extra cooling output of the heat pump is used to charge the TES if it has not been fully charged; during the on-peak period (14:00 to 19:00), discharging TES takes high priority to satisfy the cooling demand unless it is fully discharged or the TES output rate is not enough to meet the cooling demand and maintain the room temperature within setpoints (21-23°C). The heat sink of the heat pump during cooling operation is selected based on a pre-defined ambient air temperature threshold, which is set to 26 °C. It can be observed that when the ambient air temperature is high (mid of the day for the first two days as shown in Figure 3), the ground source is used; otherwise, air-source is used.

For MPC, becasuse the objective is to minimize the operating cost, it results in more frequent swiches between air- and ground-source than RBC. It is to ensure the system is always running at a higher efficiency betweening using the available air- and ground-source. The system tries its best to avoid using any electricity during the peak period. Different from RBC, during the off-peak period, the system operated with MPC sometimes discharged TES for better cost-saving performance. This is a consequence of the system overall COP estimation according to predicted disturbances.

The intelligence of MPC is demonstrated in Figure 5, which shows the heat pump power consumption and the corresponding coefficient of performance (COP). It can be observed that the MPC frequently switches the heat pump source to keep the COP of the system as high as 3.5. However, for RBC, when the air source is used, the COP can be as low as 2.6. And a continuous using of GSHP would result in a lower COP according to g-function. Besides, Figure 5 indicates that the MPC turns on the heat pump less often than the RBC. It indicates another advantage of MPC—an intelligent TES charging strategy, which is illustrated in Figure 6.

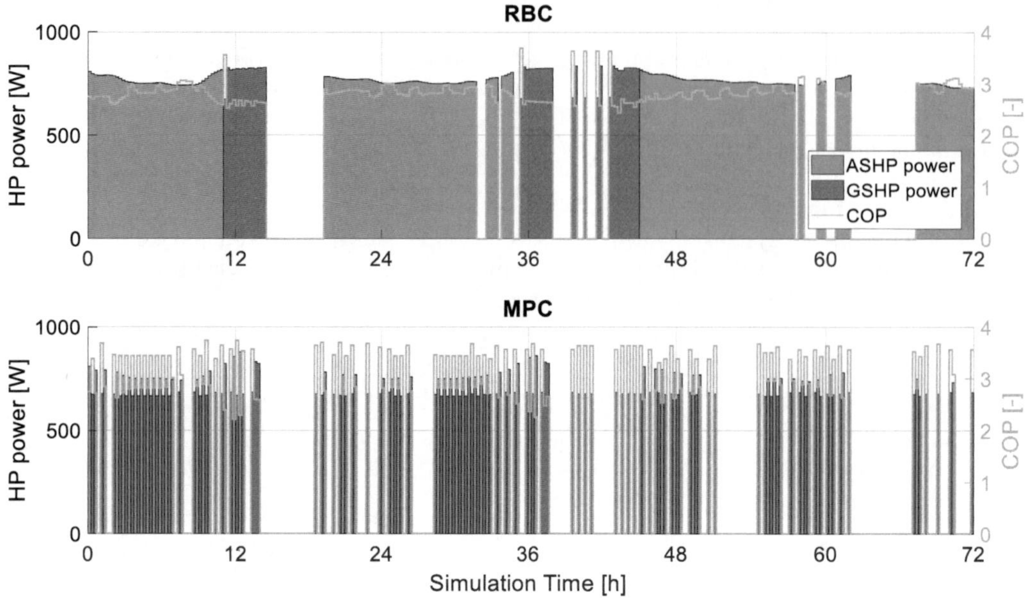

Figure 5: Heat pump power and COP profile

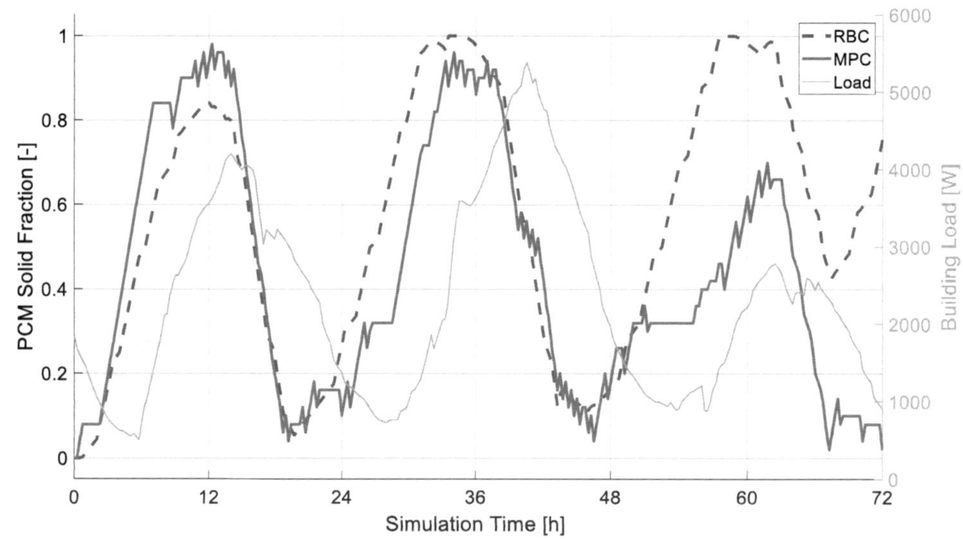

Figure 6: PCM total solid fraction profile

Figure 6 compares the PCM solid fraction profile resulting from the two control strategies. It can be observed that RBC always follows the pre-defined rule (i.e., charge the TES until it is full during the off-peak period). The TES is not fully charged for the first day because the TES is completely discharged at the begning and the charging time is limited. The disadvantage of a high charging ratio is that it leads to noticeable thermal loss to the outer tank, which results in a low overall thermal efficiency of the system. However, for the MPC, it can choose the charging level according to the next day's on-peak period total cooling demand. It is noted that the third-day on-peak cooling demand is much less than the first two days, so the MPC only charges the TES to 70% before the on-peak period on the third day. This operation showed the advantage of MPC over RBC since it is capable of adjusting the charging operation based on the predited needs for stored cooling energy, which can reduce heat loss during standby and improve the overall efficiency of the integrated system during a given time period.

Table 2 summarizes the overall energy consumption and operation cost during the 3-day period. The MPC results savings in both the energy consumption and operating cost compared with the RBC. With minimizing cost as the objective, the MPC saves 38.5% operating cost and reduces 36.9% energy consumption compared with the RBC. It should be noted that the 3-day simulation period may be too short to reflect the real difference between the two controls. The cost reduction rate may not be as high as that in this 3-day period when the system is operated for a cooling season. However, the results during the 3-day period indicate that the MPC is more inteligent than RBC by taking advantages of the forecasted cooling demands and the performance of the integrated system.

Table 2: Summary of total energy consumption and operating cost during a 3-day perod

	Energy Consumption [kWh]	**Energy Reduction [%]**	**Operating Cost [$]**	**Cost Reduction [%]**
RBC	42.6	-	2.39	-
MPC	26.9	-36.9	1.47	-38.5

5. CONCLUSIONS AND FUTURE WORK

An MPC for a novel integrated TES and heat pump system for residential buildings has been developed and evaluated by comparing its performance with a typical RBC through a case study. Simulation results indicate that, operated with the MPC, the integrated TES and heat pump system meets the cooling demand of a typical single family home with high efficiency and shifts electric demand from on-peak to off-peak hours of the electric grid on a daily basis in response to a ToU electricity tariff.

The MPC is developed with a set of simplified component models and a customized Dynamic Programming solver. The MPC can find the optimal control sequence for operating the integrated heat pump and TES system at real time based on forecasts of the cooling demand of the building and the performance of the system. The MPC takes 1.5 min to find the optimal operation mode (e.g., charging or discharging TES, and slecing air- or ground-source for the DSHP) at every 15-min time step over a 24-hour period.

By taking advantage of predicted disturbances, the MPC performs more intelligently than the RBC by (1) adjusting TES charging based on the needed cooling energy during the on-peak period and (2) swiching between air- and ground-source to ensure the DSHP always run at a higher efficiency. A case study shows that, for a 3-day operation, the MPC saves 36.9% energy and reduces 38.5% operating cost compared with a typical RBC.

There are some limitations in the current work that will be solved in the future:

1. The current dynamic system model used in MPC did not consider the detailed building envelope, and the building thermal load was given as input. A resistance-capacitace (RC) model or a data-driven model will be used in the MPC to predict the thermal load of a given building.
2. The error of the disturbances (e.g., the predicted thermal loads) was not considered. A self-adaptive feasture for the building model will be implemented to improve the accuracy of the predicted thermal load.
3. The error of the dynamic system model was not considered. The system model will be calibrated with experimental data of an integrated heat pump and TES system to improve its accuracy. In addition, an algorithm for estimating the error of the model prediction will be implemented in the MPC to fine-tune the model predictions at realtime.

NOMENCLATURE

C	capacitance	(J/°C)
CAP	latent capacity	(J)

c_p	specific heat	(J/kg/°C)
d	disturbance	(-)
k	thermal conductivity	(W/m/°C)
m	mass	(kg)
mFR	mass flow rate	(kg/s)
N	number	(-)
P	power	(W)
$\dot{Q}$/Q/q	heat transfer rate	(W)
s	scaling factor	
T	temperature	(°C)
t	time	(s)
UA	overall heat transfer coefficient	(W/°C)
u	control variable	(-)
x	state variable	(-)
λ	solid fraction	(-)
Subscript		
avg	average	
ent	entering	
FC	fan coil	
i/k/n	time step number	
loa	heat pump load side	
ref	reference	
sou	heat pump source side	
ug	undisturbed ground	
w1	DPUTB inner tank water	
w2	DPUTB outer tank water	

REFERENCES

Baniasadi, A., Habibi, D., Al-Saedi, W., & Masoum, M. A. (2019). A Cost-effective Thermal and Electrical Energy Storage Management Strategy for Smart Buildings. In *2019 IEEE PES Innovative Smart Grid Technologies Europe (ISGT-Europe)* (pp. 1-5). IEEE.

D'Ettorre, F., De Rosa, M., Conti, P., Testi, D., & Finn, D. (2019). Mapping the energy flexibility potential of single buildings equipped with optimally-controlled heat pump, gas boilers and thermal storage. *Sustainable Cities and Society*, *50*, 101689.

DOE (U.S. Department of Energy). *Building Energy Codes Program: Residential Prototype Building Models*. Last modified October 7, 2020, accessed February 20, 2021.

Drgoňa, J., Arroyo, J., Figueroa, I. C., Blum, D., Arendt, K., Kim, D., Ollé, E.P., Oravec, J., Wetter, M., Vrabie, D.L., & Helsen, L. (2020). All you need to know about model predictive control for buildings. Annual Reviews in Control, 50, 190-232.

Eskilson, P., & Claesson, J. (1988). Simulation model for thermally interacting heat extraction boreholes. *Numerical heat transfer*, *13*(2), 149-165.

Golmohamadi, H., Larsen, K. G., Jensen, P. G., & Hasrat, I. R. (2021). Optimization of power-to-heat flexibility for residential buildings in response to day-ahead electricity price. *Energy and Buildings*, *232*, 110665.

Neukomm, M., Nubbe, V., & Fares, R. (2019). Grid-interactive efficient buildings technical report series: Overview of research challenges and gaps.

Schwartz, L., Wei, M., Morrow, W., Deason, J., Schiller, S.R., Leventis, G., Smith, S., Leow, W.L., Levin, T., Plotkin, S., & Zhou, Y. (2017). Electricity end uses, energy efficiency, and distributed energy resources baseline.

Shi, L., Liu, X., Qu, M., Zhang, M., & Wang, L. (2021). *Numerical Modeling and Parametric Study of a Dual Purpose Underground Thermal Battery* (No. ORNL/TM-2020/1796). Oak Ridge National Lab. (ORNL), Oak Ridge, TN (United States).

University of California, Davis, 2021. Low-Cost, Large-Diameter Shallow Ground Loops for Ground-Coupled Heat Pumps.

Wolisz, H., Kull, T. M., Müller, D., & Kurnitski, J. (2020). Self-learning model predictive control for dynamic activation of structural thermal mass in residential buildings. *Energy and Buildings*, *207*, 109542.

Modeling Air Handling Units to Create a Diverse Fault Dataset for FDD Innovation: Lessons Learned and Recommendations

Armando CASILLAS[1]*,Yimin CHEN[1],Guanjing LIN[1], Jessica GRANDERSON[1], Sen HUANG[2], Zhelun CHEN[3]

[1]Lawrence Berkeley National Laboratory, Building Technology and Urban Systems Division, Berkeley, CA, USA

acasillas@lbl.gov (*Corresponding Author), yiminchen@lbl.gov, gjlin@lbl.gov, jgranderson@lbl.gov

[2]Pacific Northwest National Laboratory, Energy and Environment, Richland, WA, USA

huang875@pnnl.gov

[3]Drexel University, Department of Civil, Architectural and Environmental Engineering, Philadelphia, PA, USA

zc98@dragons.drexel.edu

* Corresponding Author

ABSTRACT

As energy management and information systems (e.g., automated fault detection and diagnostics [AFDD] tools) become more prevalent in the commercial building stock, it is important to determine the effectiveness of these technologies by benchmarking their performance. The authors have been working to develop the largest publicly available dataset of HVAC fault datasets for performance benchmarking applications, covering the most common HVAC systems and designs including chiller plants, rooftop packaged units, dual duct air handling unit and single duct air handling units. This study covers the development, modeling, and validation of a synthetic fault dataset for the air handling unit (AHU), one of the most common HVAC configurations found in the commercial building stock. Despite this being a common system, real-world time series data are scarce and usually do not span a wide range of weather conditions. Due to this limitation, two detailed AHU models, which included the single duct AHU and dual duct AHU developed in the Modelica language and HVACSIM+ were employed to carry out annual simulations of numerous common sensor faults, mechanical faults, and control sequence faults. The fault inclusive data were then validated by comparing fault effects on system performance to expected symptoms. We summarize the nature of each fault and their impacts under different weather and operation conditions. We report some lessons learnt during the efforts of validating the high volumes of the FDD data sets. Finally, we highlight considerations for FDD developers that may want to use this dataset to assess their algorithms' performance and their improvement over time.

1. INTRODUCTION

As building data becomes more readily available, and as the budding field of data science and analytics comes to buildings, fault detection and diagnostics (FDD) is of increasing relevance to the research and product development communities. A primary method of improving building controls and operational efficiency is through algorithms developed to perform automated fault detection and diagnostics (AFDD), which use building data to identify the presence of faults and potentially isolate root causes. Estimated energy use reduction from these improvements have been estimated at an average of 29%, which accounts for approximately 5% of overall national energy consumption (Fernandez 2017). Practically, building owners and operators have already leveraged the benefits of AFDD technology, using it to enable median whole-building portfolio savings of 9% (Kramer *et al.* 2020).

Development of FDD for air distribution subsystems, including hydronic air handling units (AHU) system, for example, are presented in a number of studies dating back decades (House, Vaezi-Nejad, and Whitcomb 2001) (Bushby *et al.* 2001) (Bushby and Schein 2006). Since then, a diversity of techniques have been developed for FDD in AHU systems, spanning analytical-based physical or gray box models, data-driven approaches, and knowledge-based heuristic approaches (Yu, Woradechjumroen, and Yu 2014), and developers continuously strive to develop improved algorithms. A persistent challenge, however, has been the lack of common datasets and test methods to benchmark the performance accuracy of FDD methods, and gauge improvement of these tools over time. Granderson (2018) most recently developed a test and benchmarking framework for FDD algorithm performance, demonstrating a growing need for HVAC fault datasets that can be used to further determine the accuracy and effectiveness of FDD algorithms. HVAC performance datasets have been developed before in the form of ASHRAE's RP1312 fault dataset. ASHRAE Project 1312-RP data (Li *et al.* 2010a, Li *et al.* 2010b) is the resulting dataset from a series of experiments that were performed on two multi-zone VAV AHUs (AHU-A and AHU-B) with the same configuration running simultaneously. Further work was initiated to fill this gap with the introduction of an open sourced dataset for FDD evaluation purposes (Granderson *et al.* 2020), which introduced a first of its kind public dataset with ground-truth data on the presence and absence of faults for multiple HVAC systems, including a simulated SDAHU system. This paper will dive into more specifically the expansion of the AHU fault dataset, which is considered one of the most typical HVAC system designs in commercial buildings. Specifically we will be covering two configurations of this system in single duct and dual duct AHU. The data set consists of high resolution, simulated time series HVAC operational data (e.g. temperatures, pressures, control signals, component status, etc.) under a diversity of operating and weather conditions, combined with information on the presence and absence of faults and their associated intensity. Furthermore, the paper applies our previously established data validation and ground truth assessment protocol for the successful development of the AHU FDD test dataset (Casillas *et al.* 2020).

2. METHODS

Two types of AHUs, i.e., the single duct AHU (SDAHU) and the dual duct AHU (DDAHU) were simulated based on the various simulation software tools such as Modelica, EnergyPlus and HVACSIM+. For each type of the fault, the fault was injected to the system and the simulation was performed to output one-year (i.e., 365 days) fault inclusive data. Consequently, the fault inclusive data cover all operating conditions of the AHUs under one year time scope. In this section, we illustrate the simulation method for each type of equipment.

2.1 Single-duct AHU (SD-AHU)

The SDAHU model was developed in the Modelica language by developers at PNNL, based on model components available in open-source Modelica libraries such as the Modelica Buildings, IBPSA libraries. Modelica is an equation-based, objective-oriented modeling language for complex dynamic systems. In order to capture the building's thermal response a reference commercial building model from EnergyPlus (Deru et al. 2011) was integrated. The data exchange between the EnergyPlus IDF model and the Modelica system model was handled by a co-simulation framework, exporting the IDF file as a functional mockup unit, analogous to the methods in Huang et. al (2021). In addition to calculating the thermal loads of the space, the IDF file also stores pertinent weather information that is fed into the modelica model, which allows for annual modeling of a building based on a historical weather data set. For this study's purposes, the climate data modeled was that of Chicago, Il. The major components of the modeled for the SDAHU, as shown in Figure 1, are supply air fan with a variable frequency drive (VFD), return fan with a VFD, cooling coil, cooling control valves, outdoor air (OA) and return air (RA) dampers. The measurement points made available for end users are related to the aforementioned components, including temperature, airflow and static pressure readings, temperature and pressure setpoints, control signal and positions for actuated components, speed and power output for motorized components. The AHU's baseline control sequence is applied from engineering standard best practices (e.g. ASHRAE 90.1). These control parameters and sequences are programmed in the modelica language with control and logic components.

The control loops are mostly concerned with three different components:

- Fan speed control determined by occupancy state and static pressure setpoints
- Cooling coil valve determined by occupancy state and supply air temperature setpoint
- Damper positions determined by occupancy state, outdoor air temperature and mixed air temperature setpoint

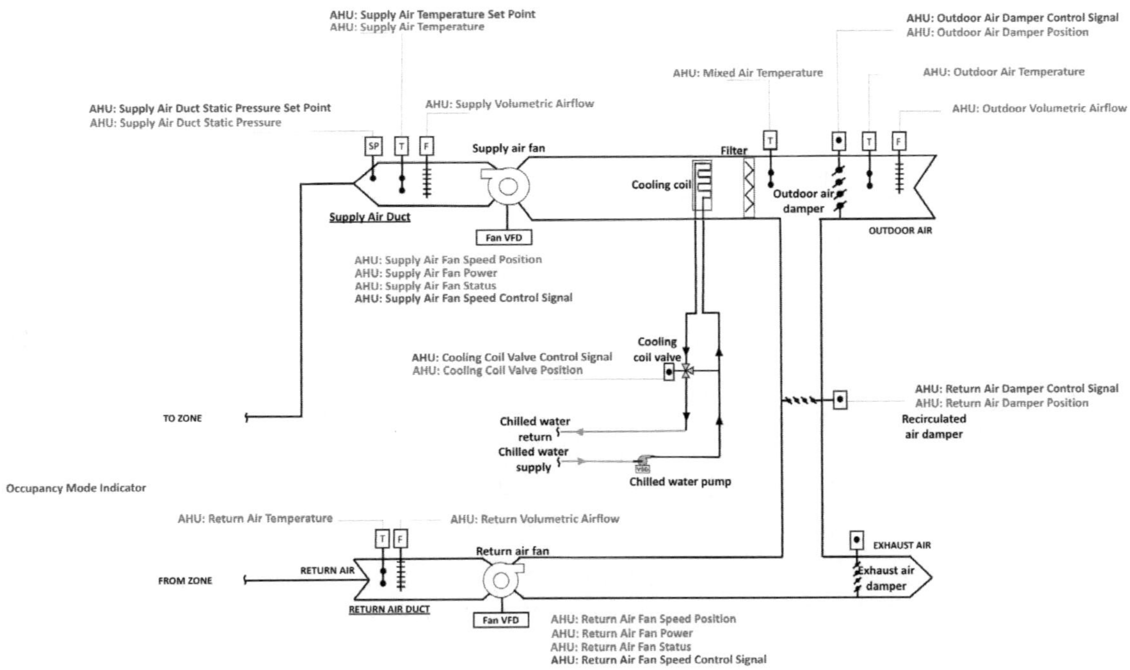

Figure 1: SDAHU diagram with all measurement points denoted

2.2 Dual-duct AHU (DDAHU) system

The DDAHU system equips two separate supply air ducts as hot duct and cold duct, and two supply air fans in each duct to provide desired air circulation and thermal comfort to different zones. In this system, both the heating and cooling coils can operate at the same time. The hot air and the cold air will be mixed with dampers in VAV terminal units at each zone. In this study, the DD-AHU system was developed by HVACSIM+ software tool which was developed by the National Institute of Standards and Technology (Clark & May, 1985). The system contains a dual-duct AHU and four associated VAV terminal units which serve four zones as shown in Figure 2. In the HVACSIM+ simulation platform, various elements such as parts, components and control sequences in a HVAC system can be modeled to create various instance blocks referred to as different "TYPE" components. Various components can be grouped into "superblocks" for simultaneous solution. Consequently, each superblock is a numerically independent subsystem of the overall simulation. Each superblock can independently handle its time evolution and internal solution during the simulation process. In addition, the time step in a superblock is a variable that is automatically and continuously adjusted by a solver subroutine to maintain numerical stability.

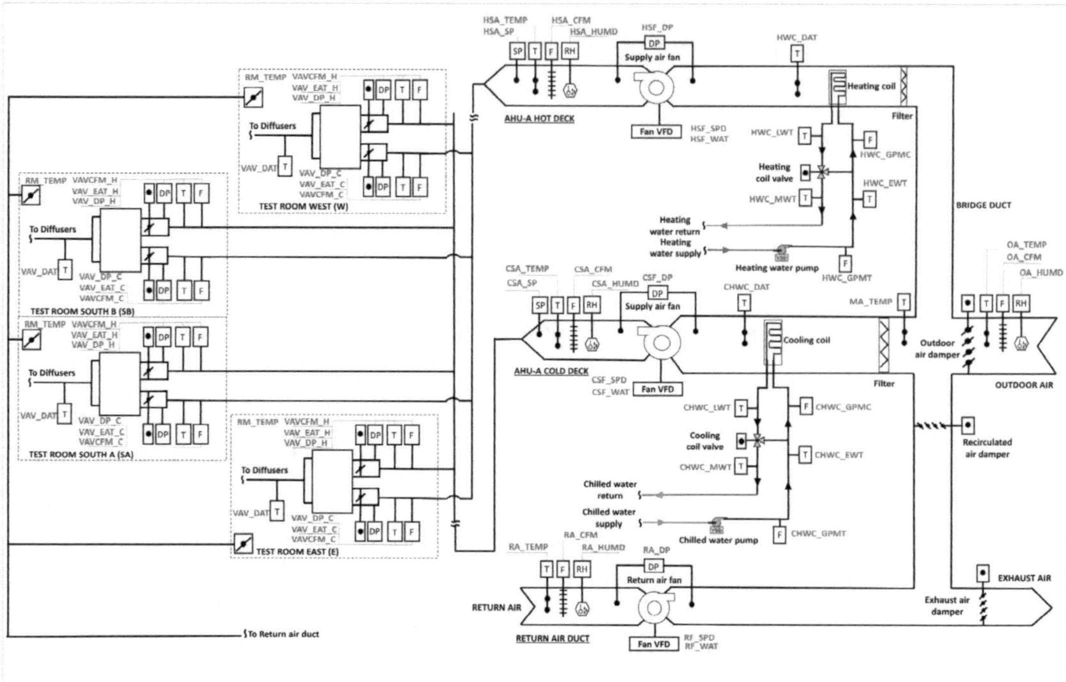

Figure 2: DDAHU diagram with all measurement points denoted

Before imposing faults on the simulation platform, various simulation settings were determined to ensure the simulation accuracy. The simulation setting includes three parts as 1) control sequence and parameter settings, 2) zone load settings, and 3) environment parameter settings. The control sequence includes the operation mode sequence, and individual component control sequences for fan, dampers, cooling coil valve and heating coil valve. The hourly zone internal load value was set according to (Park *et al.*, 1986). The TMY3 weather data for Des Moines, IA was used as the weather inputs as the system model was developed in Iowa Energy Center.

2.3 Fault Modeling

Three different components were targeted for fault modeling in the SDAHU model: the outdoor air damper, the cooling coil and the temperature sensors. The faults are all implemented by modifying or overriding the baseline control logic of the model. For example, the outdoor air damper stuck fault is implemented by overriding the position of the damper component. The fault imposition methods are summarized in the table below. As an example, for each intensity of the OA damper stuck fault, the fault is imposed by overriding the position of the modeled damper to the predetermined value. The scaled dataset creation is carried out with a parametric simulation Modelica script. This allows for the intensity of each fault to be modeled based on a single value that is passed as a parameter into the fault model component such as "TwoWayValveStuck" for both the cooling coil valve and OA damper.

In the DDAHU study, a total of 15 types of faults which are commonly studied by academic publications and reported by field engineers in the AHU side were imposed to obtain fault inclusive operation data (Chen *et al.* 2021; Roth *et al.* 2004; Schein *et al.* 2006; Wang and Xiao 2004; Zhao, Wen, and Wang 2015; Zhao *et al.* 2017). For hardware faults, the simulations with multiple severity levels were performed for each type of the fault. For the software faults, the simulations with single severity level were performed for each type of the fault. Consequently, a total of 55 fault simulation cases were carried out in this study. Each fault case was simulated to generate one year of operation data so that all system's operational conditions can be covered to fully evaluate the measurement sensitivity under various operational conditions.

Table 1: Overview of HVAC fault modeled and imposition method

Fault	Method of fault imposition
Supply, Outdoor Air Temperature Sensor Bias	Add or subtract constant value from initial sensor reading
OA Damper,Cooling Coil Valve Stuck	Automated override of OA damper position to indicate that OA damper is stuck. Automated override of coil valve position to indicate that cooling coil valve is stuck.
Cooling Coil Valve Leak	Adjusted the minimum coil valve position value when control signal is zero
Cooling/Heating coil fouling fault (air side)	Modified fin and tube heat transfer coefficients in the coil component
Cooling/Heating coil fouling fault (water side)	Modified fin heat transfer coefficient, tube heat transfer coefficient, and the coil fluid flow resistance in the coil components
Cooling/Heating control sequence unstable	Changed the absolute value of the proportional band of the cooling and heating control sequences from a properly tuned value 45.7 to an improper value 4

The symptoms of each fault are detailed below:

The **outdoor air damper stuck fault** is a mechanical fault by nature and will directly affect the AHU's ability to take advantage of outdoor air to maintain supply temperatures while minimizing cooling energy as well its ability to maintain effective supply temperature control. During instances in which the OA damper is stuck above minimum position and supply air is cooler than desired setpoint, excess outdoor air may cause the cooling energy to be minimized while dramatically reducing the supply air temperature of the AHU. The case in which warmer temperatures are seen outdoors, the excess outdoor air will cause more cooling energy to be used, driving the control signal of the OA damper to minimum while maximizing the cooling coil control signal. Higher than normal supply air temperatures may occur.

A **stuck cooling coil valve** directly affects the AHU's ability to maintain effective supply temperature control. During instances in which the supply air is warmer than desired, the control signal will be driven to 100% due to the inability of the system to maintain cool enough air to the zone level. This will cause higher than normal supply temperatures, higher than normal return air temperatures, and higher overall cooling energy consumed. During instances in which the cooling coil is providing too much cooling, or a supply temperature colder than the setpoint, the control signal will eventually be driven to zero due to the inability of the system to maintain supply air temperature set point. This will ultimately lead to lower than desired supply and return air temperatures and higher overall cooling energy consumed.

A **leaking cooling coil valve** affects the AHU's ability to fully close the cooling coil valve. During instances in which the control signal is driven to a level below the leakage level or to 0, the ground truth position of the valve will bottom out at the leakage level. This will cause lower than normal supply temperatures during these instances, and higher overall cooling energy consumed. During instances in which the leakage level is higher than the control signal, the fault will behave more like a stuck valve fault.

A **temperature sensor bias fault in the outdoor temperature sensor** would cause an adverse effect on supply temperature control, mainly the modulation of the outdoor air damper according to the economizer control sequence.

As the bias becomes more positive (4C), the seemingly higher outdoor air temperature would result in less activity in the economizer control signal, resulting in higher overall cooling energy consumption.

A **temperature sensor bias fault in the supply temperature sensor** would cause an adverse effect on supply temperature control, mainly the modulation of the cooling coil valve to meet setpoint. As the bias becomes more positive (4C), the seemingly higher supply temperature would result in higher control signal for added cooling, resulting in higher overall cooling energy consumption, cooler rooms (lower return air temperatures)

Both **the coil fouling air side fault and water-side fault** would cause several typical symptoms such as decreased water flow rate, increased coil valve position to provide desired cooling/heating capacity, and increased or decreased supply air temperature. For example, if the heating coil is severely fouling, the heating water flow will be significantly decreased. In the winter season, this may lead the heating coil valve position to be higher than normal to provide the desired heating. Under some extremely cold weather conditions, the supply air temperature may be lower than normal due to the severe impacts on the heating supply capacity. For the fouling fault types, three fault severity levels (minor, moderate and severe) were imposed by decreasing the fluid flow rate and the decreasing heat transfer rate.

Both **cooling and heating control sequence faults occuring in the valve controllers.** The control sequence fault causes valve position to be unstable and consequently causes supply air temperature to be fluctuating. The fault was imposed by changing the absolute value of the proportional band setting from 45.7 to -4 in the PID controller until the valve operation was unstable.

3. RESULTS

The annual fault cases are modeled in their respective tools, then validated to ensure the fault behaviors are as expected based on our knowledge based approach. An example of the SDAHU fault model for an outdoor air damper stuck at minimum (10%) is presented below in Figure 3. The fault is most easily discernible by the constant position reading of 10% throughout the day. More telling however, is the increase in cooling coil valve signal, which indicates the faulted case consumes more cooling energy due to the lack of ability to take advantage of mild conditions and economize.

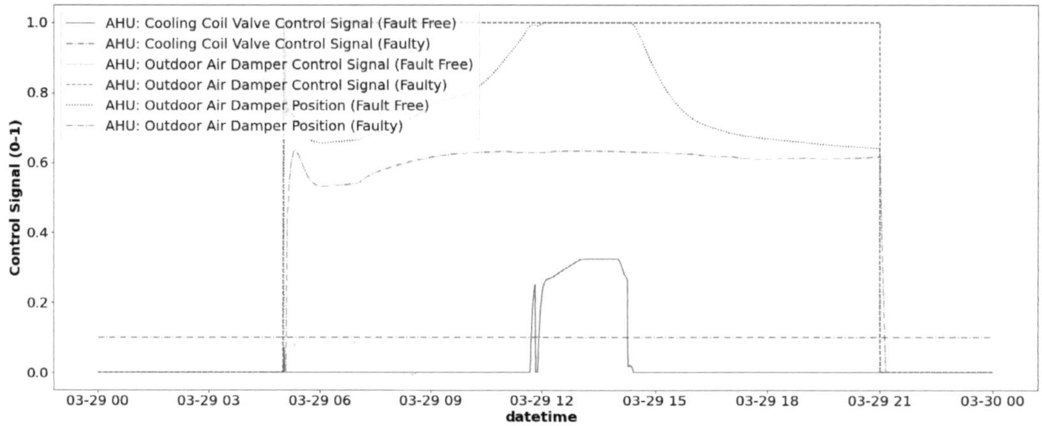

Figure 3: SDAHU fault symptom validation (outdoor damper stuck at minimum [10%])

Figure 5 shows another case example to demonstrate the fault symptoms under the cooling coil valve stuck fault in the DDAHU. When the cooling coil valve is stuck at 80% opening position which is higher than the normal position, the coil valve position signal is frozen at 80% and the valve control signal reaches to 0 to try to offset the effects of the higher coil valve position as shown in the left of the Figure 4. However, because the cooling coil valve is out of control, this fault causes the supply air temperature in the cold deck to be lower than the normal value (i.e., around 10 °C instead of 13 °C under the fault-free operation as given in the right of the Figure 5). In addition, this fault causes the cascading abnormal operation in the downstream VAV terminal units.

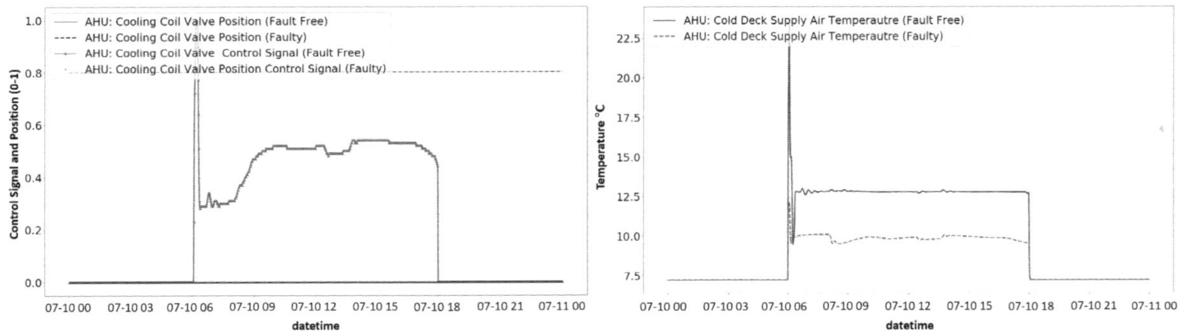

Figure 4: DDAHU fault symptom validation (cooling coil valve stuck at a 80% position fault)
Left: coil control signal and position; Right: cold deck supply air temperature

4. DISCUSSION

A large scale simulation task like this is a multi-year, multi-partner effort that presents challenges along the way. We have compiled these lessons in the following section by system type. These considerations are meant for other model developers so they are able to avoid inefficiencies and pitfalls in their own approaches. The lessons learned are categorized and summarized in Table 2 below.

Table 2: Overview of lessons learned from AHU fault modeling efforts

Lesson	Details	Solutions	System affected
Control sequence refinement	Freeze protection, competing PID loops for economizer	Add preheating, disable cooling coil until OA damper has reached 100% during economizing.	Coil, OA damper
Fault imposition methods	Actuator-related faults (e.g., a damper stuck fault), the changed value should be injected to freeze the actuator	Add position variable to give the control signal the ability to behave according to its PID control loop output	Actuated components (damper, coil valve)
Inappropriate sizing of components	Error of the component model needed better control to enable the acceptable output error of the system model.	Resize components based on observed load given climatic conditions and internal loads of the space conditioned by the system	Supply air temperature, zone level
Modeling low flow conditions	Limitations during low flow conditions. Causes abnormal conditions which manifest in different ways during modeling.	Create virtual bypass values that pass the incoming coil temperatures through the coil without heat transfer calculations applied.	Coil, OA damper, temperature sensor measurements

4.1 DDAHU (HVACSIM+)

Compared with the previous fault model validation process in which a few days were selected and only employed, it is necessary to add control sequences which enable a complete simulation under all operational conditions. For example, in the DDAHU control, a freezing protection functionality was added after encountering an issue during winter operation. In the original DDAHU model, there was no freezing protection control sequence during the unoccupied hours when the system was completely shut off. This caused the SAT to be extremely low (e.g., as low as 10 F) when the OAT was low in the winter season. In the real system, the freezing protection control sequence may be adopted even when the system was switched to the unoccupied mode (i.e., the system does not provide the desired cooling or heating during unoccupied hours). The freezing protection sequence often enables the preheating, or triggers the freezing alarm so that the operators can fix the fault quickly. This issue may only occur when whole year simulation is performed to enable the system control simulation to mimic this operating condition.

The model validation under all operational conditions was also applied to the equipment physical model validation to ensure the equipment model accurately outputs values. For example, in the simulated cooling valve model, the operating position limit should be set to be 100 (i.e., fully opened) as the maximum value even though the cooling supply may not be enough in extremely hot weather. After encountering this issue, all the actuator models were re-examined to add the actuator operating position limit.

Another challenge addressed is the system level fault simulation. In the DDAHU simulation, the system model, which was developed in HVACSIM+ software, consists of various component models (i.e., coils, valves, dampers) in the DDAHU equipment and terminal unit model. Various models were connected with each other to complete the system level simulation. When performing simulation, the output error of the component model needed to be well controlled to enable the acceptable output error of the system model. For example, the heating coil was downsized in the model, this component output error caused the whole system output error, and consequently supply air temperature to be abnormal. This error was eliminated by increasing the heating coil size.

When imposing actuator-related faults (e.g., a damper stuck fault), the changed value should be injected to freeze the actuator action, but not fix the control signal. This is because in the real practice, the actuator stuck was believed to be some mechanical/communication issues (e.g., the linkage between the driver and actuator is broken, or the communication between the DDC and the drive is lost). The DDC should output the proper control signal to compensate/tolerate the fault. Therefore, the control signal should not be frozen. To address this issue, some extra points (i.e., valve position feedback) should be added in the original equipment model to mimic the control behavior and simulate the fault effects.

4.2 SDAHU (Modelica)

A common occurrence for the Modelica based model are indeterminate conditions when fluid flows are near zero. This happens often during overnight scenarios, in which the flow provided to a water coil or the airflow approaches zero due to the lack of demand in the system in unoccupied hours. This low flow causes abnormal conditions which manifest in different ways during modeling. One example is **its effects on supply air temperature during severe damper fault cases.** In reality, when the temperature of the air is too low (less than the temperature of the inlet water), it is possible that it "absorbs" heat from the cooling coil. In simulation, the process for calculating the water temperature in a heat exchanger becomes unreliable when the water flow is approaching 0. Low flow conditions can also cause abnormal temperature deviations and flat profiles given limitations of the sensor modules we use during simulations. These flat temperatures may have effects on their associated control loops, for instance the supply air temperature control loop sends a **cooling coil valve control signal of** 100% due to the flat measurement of the supply air temperature sensor when the supply air flow is zero.

In addition to limitations with modeling low flows, there were some inconsistencies between the control sequence programmed in the model and best practices that we caught during our validation process. The model has two PID control loops, one for the economizer control and one for the cooling coil valve control. The economizer control uses the mixed air temperature as the control variable while the supply air temperature controls the second loop. The problem occurs when economizing is enabled during mild conditions. Both controls loops compete against each other to satisfy their setpoints, which leads to cooling while the outdoor air damper is modulating. Best practice prescribes the cooling coil valve stay disabled until the economizer output is 100%. Modulating the cooling coil valve before this condition results in unnecessary energy consumption and does not allow the economizer to take full advantage of ideal outdoor air conditions. These types of interconnected control loops need to be programmed in sequence so that the cooling coil valve control is disabled in economizing mode until the outdoor air damper reaches 100% open position.

Another consideration is related to co-simulation frameworks such as the one presented here for the SDAHU models. The change of data between the EnergyPlus idf file FMU and the AHU system model is not only limited to building thermal loads but also used to determine when the building is in Occupied or unoccupied mode. The documented occupied times of 6pm to 10pm are scheduled in the FMU and the occupied output is fed to the model as a boolean value. The problem observed during validation is related to the daylight savings time option in the idf file was set to true, which meant that the start of daylight savings time resulted in a 1-hr shift in time reporting from the FMU. This manifests in the model as an occupied time of 5am to 9pm, which is inconsistent with the provided documentation.

5. CONCLUSION

In this study we have established our methods and results of our large-scale fault modeling work. Once validated and fully documented, we plan to publish this dataset in the largest ever publicly available fault database. The study presented here is a short summary of the work we have done for AHU systems. The lessons learned presented in the discussion are common issues found in all modeling approaches and are intended to help developers avoid inefficiencies in their own approach.

REFERENCES

Bharathan B., Bhattacharya, A., Fierro, G., Gao, J., Gluck, J, Whitehouse, K. (2016). Brick: Towards a Unified Metadata Schema For Buildings. In Proceedings of the 3rd ACM International Conference on Systems for Energy-Efficient Built Environments (BuildSys '16). Association for Computing Machinery, New York, NY, USA, 41–50. DOI://doi.org/10.1145/2993422.2993577

Bushby, S. T., & Park, C. (2001). Using the Virtual Cybernetic Building Testbed and FDD Test Shell for FDD Tool Development NISTIR 6818 Using the Virtual Cybernetic Building Testbed and FDD Test Shell for FDD Tool Development.

Casillas, A. Lin, G. Im, P. Granderson, J. (2020). Curation of Ground-Truth Validated Benchmarking Datasets for Fault Detection & Diagnostics Tools. Proceedings of the 2020 ACEEE Summer Study on Energy Efficiency in Buildings.

Chen,Y., Lin,G., Crowe,E., & Granderson, J. (2021). "Development of a Unified Taxonomy for HVAC System Faults." Energies 14 (17): 5581. https://doi.org/10.3390/en14175581.

Clark, D. R., & May, W. B. (1985). "HVACSIM+ Building Systems and Equipment Simulation Program - User's Guide." PB-86-130614/XAB; NBSIR-85/3243. Washington, DC: National Bureau of Standards, Building Equipment Division. https://www.osti.gov/biblio/6127383.

Deru, M., Field, K., Studer, D., Benne, K., Griffith, B., Torcellini, P., … Studer, D. (2011). U . S . Department of Energy Commercial Reference Building Models of the National Building Stock U . S . Department of Energy Commercial Reference Building Models of the National Building Stock. (February 2011).

Fernandez, N, et al. (2017). Impacts on commercial building controls on energy savings and peak load reduction. Pacific Northwest National Laboratory. PNNL Report Number PNNL-25985

Granderson, J., Lin, G., Singla, R., Mayhorn, E., Ehrlich, P., & Vrabie, D. (2018). Commercial Fault Detection and Diagnostics Tools: What They Offer, How They Differ, and What's Still Needed. Lawrence Berkeley National Laboratory Recent Work, 1–12. https://doi.org/10.20357/B7V88H

Granderson, J., Lin, G., Harding, A., Im, P., & Chen, Y. (2020). Building fault detection data to aid diagnostic algorithm creation and performance testing. Scientific Data, 7(1), 1–14. doi.org/10.1038/s41597-020-0398-6

House, J. M., Vaezi-Nejad, H., & Whitcomb, J. M. (2001). An Expert Rule Set for Fault Detection in Air-Handling Units. ASHRAE Winter Meeting CD, Technical and Symposium Papers, 1005–1018.

Huang, S., Zuo, W., Vrabie, D., & Xu, R. (2021). Modelica-based system modeling for studying control-related faults in chiller plants and boiler plants serving large office buildings. Journal of Building Engineering, 44(May), 102654.doi.org/10.1016/j.jobe.2021.102654

Kramer, H., Lin, G., Curtin, C., & Granderson, J. (2020). Proving the Business Case for Building Analytics Energy Technologies Area October , (2020). https://doi.org/10.20357/B7G022

Li, S. & Wen, J., "Development and Validation of a Dynamic Air Handling Unit Model - Part I (RP 1312) ", ASHRAE Transactions, Vol. 116, Pt. 1, pp. 45 - 56, 2010.

Li, S., Wen J., Zhou, X., & Klaassen, C. J. (2010). "Development and Validation of a Dynamic Air Handling Unit Model - Part II (RP 1312) ", ASHRAE Transactions, Vol. 116, Pt 1, pp. 57 - 73,.

Liao, H., Cai, W., Cheng, F., Dubey, S., & Rajesh, P. B. (2021). An Online Data-Driven Fault Diagnosis Method for Air Handling Units by Rule and Convolutional Neural Networks.

Montazeri, A., & Mohamad, S. (2020). Fault detection and diagnosis in air handling using data-driven methods. Journal of Building Engineering, 31(March), 101388. doi.org/10.1016/j.jobe.2020.101388

Park, C,Clark D.R., & Kelly G.E. (1986). "HVACSIM+ Building Systems and Equipment Simulation Program: Building-Loads Calculation." PB-86-189909/XAB; NBSIR-86/3331. Washington, DC: National Bureau of Standards.
Roth, K., Westphalen, D., Patricia, L., & Feng, M. (2004). "The Energy Impact of Faults in U.S. Commercial Buildings." In International Refrigeration and Air Conditioning Conference. West Lafayette, IN: Purdue University.
Schein, J., & Bushby, S. T. (2006). A Hierarchical Rule-Based Fault Detection and Diagnostic Method for HVAC Systems. HVAC&R Research, 12(1), 10–12.
Wang, Shengwei, and Fu Xiao. (2004). "AHU Sensor Fault Diagnosis Using Principal Component Analysis Method." Energy and Buildings 36 (2): 147–60. https://doi.org/10.1016/j.enbuild.2003.10.002.
Wu, S., & Sun, J. (2012). A physics-based linear parametric model of room temperature in of fi ce buildings. Building and Environment, 50, 1–9. doi.org/10.1016/j.buildenv.2011.10.005
Yan, R., Ma, Z., Zhao, Y., & Kokogiannakis, G. (2016). A decision tree based data-driven diagnostic strategy for air handling units. Energy & Buildings, 133, 37–45. doi.org/10.1016/j.enbuild.2016.09.039
Yu, Y., Woradechjumroen, D., & Yu, D. (2014). A review of fault detection and diagnosis methodologies on air-handling units. Energy & Buildings, 82, 550–562. doi.org/10.1016/j.enbuild.2014.06.042
Yun, W., Hong, W., & Seo, H. (2021). A data-driven fault detection and diagnosis scheme for air handling units in building HVAC systems considering undefined states. Journal of Building Engineering, 35(December 2020), 102111. doi.org/10.1016/j.jobe.2020.102111
Zhao, Y., Wen J., & Wang, S. (2015). "Diagnostic Bayesian Networks for Diagnosing Air Handling Units Faults – Part II: Faults in Coils and Sensors." Applied Thermal Engineering 90 (November): 145–57. https://doi.org/10.1016/j.applthermaleng.2015.07.001.
Zhao, Y., Wen, J., Xiao,F.,Yang, X., & Wang, S. (2017). "Diagnostic Bayesian Networks for Diagnosing Air Handling Units Faults – Part I: Faults in Dampers, Fans, Filters and Sensors." Applied Thermal Engineering 111 (January): 1272–86. https://doi.org/10.1016/j.applthermaleng.2015.09.121.
Zhong, C., Yan, K., Dai, Y., Jin, N., & Lou, B. (2019). Energy Efficiency Solutions for Buildings: Automated Fault Diagnosis of Air Handling Units Using Generative Adversarial Networks. Energies. 1–11. doi.org/10.3390/en12030527

ACKNOWLEDGEMENT

This work was supported by the Assistant Secretary for Energy Efficiency and Renewable Energy, Building Technologies Office, of the U.S. Department of Energy under Contract No. DE-AC02-05CH11231. We also recognize each of the fault detection and diagnostic tool developers who participated in this survey. We would also like to thank Erika Gupta and the Building Technologies Office as well as our data contributors.

Evaluation of a Graphene Enhanced Photocatalytic Oxidation Device

Paul HOERTZ[1*], Stephen KUJAK[2], Kathleen OWEN[3]

Trane Technologies
[1]Davidson, NC USA & [2]La Crosse, WI USA
[3]Owen Air Filtration Consulting
Cary, NC, USA

[1]410-456-1441, paul.hoertz@tranetechnologies.com
[2]608-386-8067, skujak@tranetechnologies.com
[3]919-656-5295, kathleenown@att.net

*Corresponding Author

ABSTRACT

The novel Coronavirus disease (COVID-19) pandemic has driven innovation for the development of new air cleaning technologies with renewed interest in determining effectiveness of both new as well as existing cleaning technologies. Standardized test methods for the evaluation of reactive air-cleaning technologies for both volatile organic compounds (VOCs) and microbiological agents do not exist. Since air movement in a building is very dynamic and situational specific, the lack of standardized methods may lead to contradictory results or sub-optimal evaluation of a device's performance in a laboratory setting.

In this paper, a graphene enhanced photocatalytic oxidization (GPCO) device is presented and evaluated for general effectiveness against some challenge VOCs. The authors describe the VOC testing methodology used as well as variables that could impact the reproducibility and repeatability of the results. In addition, measurements of various by-products that could have been generated by the device, e.g., ozone, are presented. The results demonstrated that the tested device reduced concentrations of formaldehyde and toluene over time.

1. INTRODUCTION

The COVID-19 pandemic has spurred a growing body of research around indoor environmental quality (IEQ) factors, with added focus on indoor air quality (IAQ) due to the role it plays in mitigating the risk of transmission of certain airborne diseases. Indoor environmental quality is an occupant's perceived indoor experience of the built environment that includes aspects of thermal comfort, lighting, acoustics and IAQ. Standards to measure, maintain and optimize IEQ for energy consumption, thermal comfort, lighting, and acoustics are well defined, but standards for measuring IAQ are still under development at the time of this publication. IAQ contaminants of concern fall primarily into three classes: particles, gases (including volatile organic compounds, VOCs) and microorganisms. Distinct categorization of IAQ contaminants are not mutually exclusive and similar technology approaches can be used for mitigation. For example, both microorganisms and dust are particles whose total concentration can be reduced by particle removal technologies. IAQ standards to measure and optimize the quality of indoor air have been slow to develop in part because IAQ contaminants of concern are both difficult to control and measure within the built environment. One objective of this paper is to describe a testing methodology that could be used as input to develop a test standard to evaluate the performance of IAQ technologies against VOCs.

There are generally four traditional approaches to removing or reducing various IAQ contaminants. The first two approaches, dilution and exhaust, are usually used in combination to remove contaminants and reduce exposure to IAQ contaminants including particles, VOCs, and microorganisms. Increasing outdoor air ventilation to bring in more fresh air that has lower concentrations of IAQ contaminants will dilute or reduce the concentration of the contaminants in the occupied space. Increasing the supply of fresh air will also increase the exhaust airflow in order to maintain an adequate building pressure. Increasing exhaust airflow increases the rate at which the IAQ contaminants are removed from the occupied space. Dilution and exhaust are effective methodologies and are primary approaches in managing a building's IAQ through the HVAC systems. Of course, there are many building designs or environmental variables that have an impact on the IAQ contaminant removal or reduction effectiveness through the dilute and exhaust methodology; these will not be discussed here.

The third approach is to reduce the concentration of IAQ contaminants of concern through control of indoor humidity. HVACR systems are used to control humidity to improve indoor comfort. Generally, managing indoor humidity can support the mitigation of the growth of airborne and surface bound microorganisms like bacteria and fungi. ASHRAE® recommends 40%-60% relative humidity (RH) to maximize human comfort and to reduce microbial growth. Humidity can also have an impact on particle size and number of water-based aerosols of microorganisms after emission by a host source.

The fourth and last approach is passively cleaning the air through various means. One common approach is to capture the IAQ contaminant through filtration so it cannot reach occupants. Air is circulated through a filter and the contaminant can be captured through various mechanisms, but primarily through impingement or adsorption onto a filtration media. ASHRAE developed Standard 52.2 -2019, Method of Testing General Ventilation Air-Cleaning Devices for Removal Efficiency by Particle Size, to determine the Minimum Efficiency Reporting Value (MERV) for air filters. This method is limited to media filters since it measures the pressure drop as particle loading of the filter increases. Devices, like electrostatic precipitators, which do not have increasing pressure drop with particle loading, cannot be given a MERV rating. Using a filter with a higher MERV rating is an effective tool in removing particles but may cause increased pressure drop in installations with little flexibility in filter choice. Increased pressure drop will increase energy consumption, but this increase in energy use can be more effective in reducing indoor contaminants than that associated with increased outdoor air ventilation or use of in-room air cleaners (Pistochini, 2021).

Another approach to cleaning is to break down or inactivate the IAQ contaminant in the air and/or on surfaces. For example, application-specific ultraviolet germicidal irradiation (UVGI) systems are now a widely accepted technology for controlling microorganisms and may be installed in the HVAC duct system or inside an HVAC unit to irradiate a surface like a coil or drain pan. Another approach that has been used as far back as the 1920s is to bring UVGI into the room (occupied spaces) and either treat a portion of the room's air as it circulates (e.g., upper room UVGI) or treat surfaces directly (typically unoccupied spaces). A wide range of additional air cleaning technologies have also been developed and are currently available on the market and the need for standard methods of testing these devices and verifying performance has become a clear priority. ASHRAE Standard 185.1 is a test standard available for assessing the single pass UV-C light efficacy toward bioaerosols but does not test the combination of the UV-C light in a specific air handler. ASHRAE Standard 185.2 is a standard developed for determining the efficacy of UV-C light irradiance of coils and duct surface inactivation. Standards have not been created for evaluating the efficacy of UVGI in-room inactivation. A safety standard exists for the generation of secondary contaminates such as ozone for air cleaning devices. ASHRAE Standards 62.1-2019, Ventilation for Acceptable Indoor Air Quality, requires that air-cleaning devices comply with UL 2998 which limits ozone to 5 parts per billion (ppb) or less (UL 2998, ED3). ASHRAE Standard 145.2 provides a test method for assessing the performance of gas-phase air cleaning devices, sorbent media, and gas/VOC-removing filters but applies to in-duct devices and media and not for in-room devices.

Recently, Trane Technologies accelerated the development and refinement of a set of HVAC-focused testing methodologies for measuring the safety (e.g., generation of by-products such as ozone), applicability, and efficacy of air cleaning devices and incorporated various facets from existing standards. Our approach was to develop methodologies and procedures that made sense within the framework of accepted test standards while considering the needs of air cleaners that are not yet covered by currently recognized test standards. Current test standards for in-duct devices only allow single-pass testing (ASHRAE 2017, ASHRAE 2016). For in-room air cleaners, AHAM AC-1 (2013), which gives the Clean Air Delivery Rate (CADR), measures reduction of particles but does not address gas-phase contaminants or duct mounted devices. The CADR for gas-phase contaminants can be determined experimentally during chamber experiments by measuring the decay kinetics of the contaminant in the presence (k_{observed}) and absence (i.e., natural decay, $k_{\text{observed, natural decay}}$) of the air cleaning device using the following equation (Ye 2021):

$$\text{CADR} = (k_{\text{observed}} - k_{\text{observed, natural decay}}) \text{ x Chamber Volume} \quad \text{(Equation 1)}$$

More specifically, the CADR is the amount of clean air the air cleaner produces such that a 100 cfm airflow device at 75% single-pass efficiency has a CADR of 75 cfm. Importantly, the chamber decay methodology of AHAM AC-1 allows longer exposure and multiple passes through the air cleaner. Thus, this method serves as the basis of the chamber work discussed here. We use the term CADR in this paper in the sense it is often used colloquially to mean a clean air delivery level but not to indicate the values are from an AC-1 test as is apparent from the descriptions of the testing.

One technology that showed promising efficacy towards VOCs was graphene-enhanced photocatalytic oxidation (GPCO), which employs a composite of graphene with the ubiquitous photocatalyst, titanium dioxide (TiO_2). Photocatalytic oxidation (PCO) has been studied for several decades and its application towards the cleaning of air streams remains a maturing technology (Jacoby, 1996; Tsang, 2019, Noguchi, 1998; Kormann, 1988; Schneider, 2014; Perry, 2011). The most widely used photocatalytic material in PCOs is TiO_2 which has a bandgap of 3.2 eV (corresponding to 388 nm). Bandgap excitation produces both a potent oxidizing agent (i.e., a conduction band hole) and a potent reducing agent (i.e., a conduction band electron), both of which lead to the formation of reactive oxygen species such as hydroxyl radicals ($^{\cdot}OH$), superoxide ($O_2^{\cdot-}$), hydroperoxide radical ($^{\cdot}OOH$), and hydrogen peroxide (H_2O_2) (Howe, 1987; Nosaka, 2017). A review of the literature shows that the following elementary mechanistic steps are the likely key contributors toward the production of reactive oxygen species (ROS) that can cause the subsequent degradation/oxidation of organic species like VOCs: (1) oxidation of surface-bound water molecules to produce hydroxyl radicals, (2) reduction of oxygen by conduction band electrons to produce superoxide followed by hydrogen peroxide, and (3) production of hydrogen peroxide from two adjacent surface-bound hydroxyl radicals (Howe, 1987; Nosaka, 2017; Schneider, 2014; Ishibashi, 1998). The ROS that are formed on the surface of TiO_2 during photocatalysis are often depicted as remaining on the surface until they decay or are utilized in the surface-based oxidation of incoming or nearby organic species Competing with ROS creation is the recombination of the conduction band electrons with valence band holes. This process diminishes the overall yield of ROS that can be created by absorbed photons. The quantum yield for producing ROS by TiO_2 and is reported to fall in the range 1-5% (Nosaka, 2017).

The GPCO device (Figure 1) comprises a photocatalytic structure that combines graphene with TiO_2. The composite photocatalyst becomes activated by nearby 390-395 nm UV-A light-emitting diodes which excites electrons across the bandgap of TiO_2 (ThermoKing 2021). In the transit bus application, the in-duct GPCO device is placed upstream from the HVAC unit to allow for cleaning of the airstream (Figure 1). Importantly, the GPCO device exhibits very low pressure drop.

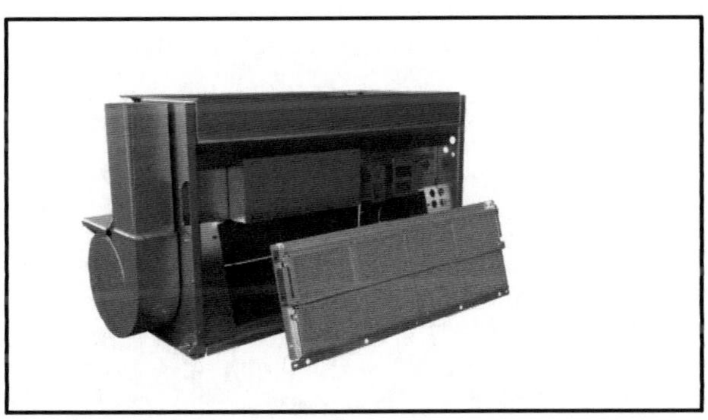

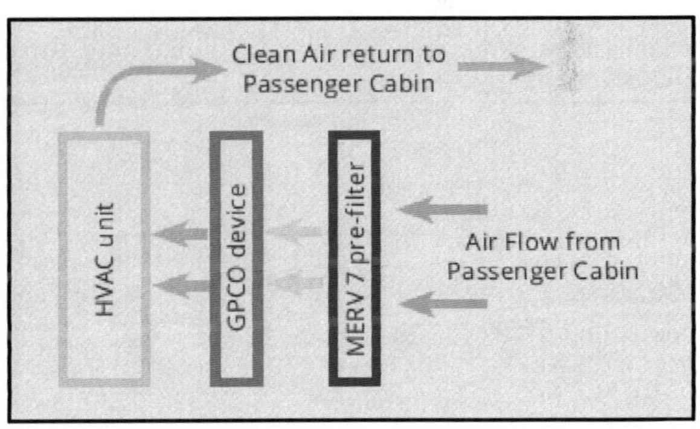

Figure 1. ***Left.*** Image of the Graphene-Enhanced PCO device used in bus applications. ***Right.*** In-duct placement of the Graphene-Enhanced PCO device in the bus application air flow.

The combination of TiO_2 with graphene is expected to have several implications on the resulting photocatalytic performance (Ng, 2010). For example, graphene offers an extremely high specific surface area (typically > 1000 m^2/g). As a carbon-based support with high surface area, this could facilitate the adsorption of gases like VOCs adjacent to TiO_2 photocatalysts that are then available and within the ROS generation field to allow for facile oxidation. The graphene could also play a role in improving charge separation and ROS yields by diminishing the rate of charge recombination (Ng, 2010). Given the potential that graphene could enhance the performance of PCOs, a series of experiments were performed to investigate the performance of the GPCO technology.

In this paper, emphasis will be made on the HVAC-focused testing methodologies that were developed for measuring the safety, applicability, and efficacy of the GPCO device toward select airborne VOCs. It should be noted that the GPCO devices were not manufactured by Trane Technologies.

2. EXPERIMENTAL

2.1 Test Chambers Overview

Many air cleaners require a longer period of time to operate and clean the air within a space; as a result, these air cleaners are not adequately covered by current standard test methods. Because of this, a test chamber similar to the AHAM AC-1 standard for testing in-room air cleaners for particle removal was a good starting point. However, some of these air cleaners are duct-mounted, so a simple chamber is not sufficient, and a combination chamber with a removable side duct apparatus was chosen as the basis for these tests. A removable side duct allowed the same chamber to be used for both in-room and duct-mounted devices. The two test chambers used for these experiments were 1007 ft^3 and 4096 ft^3 in volume, with both having an optional side duct with approximately 80 ft^3 of volume with an available 24"x24" device mounting section for mounting HVAC-mounted devices. Figure 2 provides an overview of the test chambers. Flexible 12-inch HVAC tubing was used to connect the side/by-pass duct with the chamber; for the 4096 ft^3 chamber, an additional ~30 ft of tubing was needed.

The individual sections of this side duct may be exchanged with other sections to provide more space, different access, or simply different sealing options. The smaller chamber size is equivalent to a typical office or bedroom and, as such, is quite appropriate for testing small-to-medium sized in-room air cleaners. It is not, however, the size that a full HVAC-mounted device would service. For example, a 2000 cubic foot per minute (cfm) unit would result in two air exchanges in 1 minute, or 120 air exchanges per hour (ACH) for this chamber, whereas in a real-world situation the same unit might serve 20 similarly sized rooms in a building, providing a more typical 6 ACH. The size of the chamber and of the duct as well as the intended airflow rate of the air cleaner must be considered when reviewing the test data. The chamber also allowed for air cleaners and sampling devices to be positioned in different locations. The internal fan could be run continuously to guarantee good mixing during the injection of contaminants to ensure good initial mixing. Alternatively, the internal fans could not be used at all to test a zero-to-low air flow condition. Three ceiling fans and two box fans at opposite ends of the chambers were used for both chambers. A variable frequency drive (VFD) was operated to adjust the test system blower to the desired test airflow rate in the by-pass recirculating duct. The chamber was purged with high-efficiency particulate air (HEPA) filters, carbon filters, and catalyst filters until stable baseline levels were reached. UV lights were also used to decontaminate the chamber after microbial tests. The

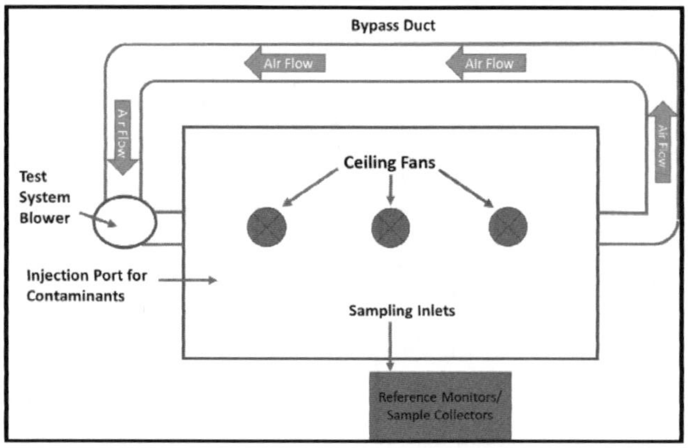

Figure 2. ***Left.*** Environmental test chamber schematic depicting placement of various sensors, injection location, and sampling locations. ***Right.*** Image of the chamber with the bypass recirculating duct.

purging processes lasted 30 to 60 minutes depending on the test contaminate and chamber size. Importantly, the experiments were carried out under closed chamber conditions (i.e., essentially zero outside air). Humidity was adjusted prior to the beginning of any tests to avoid the introduction of particles during the test.

2.2. Instrumentation

Multiple measurements were made during the environmental chamber experiments during each evaluation. Ozone was measured using an ACOEM Ecotech Serinus 10 Ozone Analyzer, with a measurement range of 0-20,000 ppb and a lower detection limit (LDL) of 0.5 ppb. Toluene was measured using a RAE Systems UltraRAE 3000 with a measurement range of 1 ppb – 10,000 parts per million (ppm). Background and generated formaldehyde were measured with an Aerodyne QC-TILDAS Formaldehyde Monitor with measurement range of 0-15 ppm. The potential for ions being produced by an air cleaner was monitored using an AlphaLab Air Ion Counter Model AIC2; two separate devices were used to measure positive and negative ions. Particle sizes and distributions over the size range 15-600 nanometers (nm) and concentration ranges of 1 to 10^7 particles/cm^3 were captured with a TSI Scanning Mobility Particle Sizer Spectrometer 3938. Temperature and relative humidity (RH) within the chamber were also monitored, using a TSI VelociCalc 9545, with typical ranges of 71-74°F and 48-53%, respectively. Power consumption by the devices was also measured. When relevant, the airflow through an in-room device, due to its own fan, was determined, while for duct-mounted devices, the actual airflow rate through the duct was measured.

2.3 Introduction of Formaldehyde and Toluene into the Chamber

Aqueous formaldehyde (Ladd Research) was introduced into the test chamber using a custom-made flask evaporator with dry compressed air as the carrier gas. The toluene challenge was generated from anhydrous toluene (Sigma Aldrich). The analyzers for both formaldehyde and toluene were located outside the chamber. PTFE tubing was run from the inside the chamber to deliver air samples to the analyzers. Averages, changes over time, and changes relative to the timing of the experiment were calculated. For this report, the timescales were converted to time from the end of VOC injection and stabilization to the needed initial concentration to allow easier comparison across the tests. VOC injection times were typically 10-15 min.

2.4 Test Sequences (Induct versus In-Room Devices)

Two types of chamber decay tests, similar to AHAM AC-1, were derived using an environmental chamber, one for in-room devices and one for in-duct or duct-mounted devices. The GPCO devices were run for a certain length of time (typically several hours to overnight) prior to VOC challenge testing. This step was performed to ensure an activated state of the GPCO photocatalysts and removal of any residual organic compounds that may have been present on the photocatalytic surfaces.

In-Room Devices

1. Install air cleaner in the 1007 ft^3 or 4096 ft^3 chamber
2. Purge the chamber with ventilation system
3. Check for background levels of VOCs, ozone, particles
4. Inject VOC challenge into the chamber
5. Mix the chamber well with the fan
6. Collect air samples over time to determine the natural decay rate; measure other by-products such as ozone, ions, particles, and formaldehyde as needed.
7. Purge the chamber with the ventilation system
8. Repeat steps 2-7 but this time with the air cleaner on prior to or following the injection of the VOC
9. Calculate concentrations and decay rate.

In-Duct Devices

1. Set up the chamber with the bypass loop attached (1007 ft^3 or 4096 ft^3 chamber)
2. Install air cleaner in the bypass duct loop attached
3. Purge the chamber with ventilation system
4. Check for background levels of VOCs, ozone, particles
5. Turn on the air flow through the device and bypass loop
6. Inject VOC challenge into the chamber
7. Collect air samples over time to determine the natural decay rate; measure other by-products such as ozone, ions, particles, and formaldehyde as needed.
8. Purge the chamber with ventilation system
9. Repeat steps 3-8 but this time with the air cleaner on prior to or following the injection of the VOC
10. Calculate concentrations and decay rate.

3. RESULTS AND DISCUSSION

Organic chemical compounds are ubiquitous in both indoor and outdoor environments because they are essential ingredients in many products and materials and are generated by living and decomposing matter. In indoor environments, VOCs primarily come from the use and presence of products, materials, and objects (humans as well) which contain and emit VOCs into the air. The main concern in indoor environments is the potential for VOCs to adversely impact the health or comfort of people that are exposed to specific VOCs. The two target VOCs used in this evaluation were formaldehyde and toluene.

Formaldehyde exposure indoors is a problem in many parts of the world and is of concern in the United States (Kaden 2010). Formaldehyde is known to be difficult to generate, sample, and analyze. The Aerodyne QC-TILDAS formaldehyde monitor used in these tests provides the ability to selectively analyze for formaldehyde as both a challenge and a byproduct (either degradation product from larger molecule VOCs or off-gassing by the materials used within the air cleaner or chamber).

Toluene is one of the most commonly detected VOCs in indoor environments, along with benzene, ethanol, xylenes, and D-Limonene. In this study, toluene was chosen since it is the required challenge in ASHRAE 145.2 standard for the VOC category and because it requires multiple oxidative steps before it can be broken down into carbon dioxide (CO_2) and water. As such, this allows toluene to also be utilized to study potential breakdown byproducts for reactive air cleaners such as photocatalytic oxidation devices.

3.1 GPCO Performance Towards VOCs

The concentrations of specific VOCs (i.e., either formaldehyde or toluene) within the environmental chamber were measured as a function of time with and without the GPCO in operation. Plots of $\ln(C/C_o)$ versus time, where C_o is the initial concentration and C is the concentration at a given time point, were used to extract observed first order rate constants (in min^{-1}) for both the absence of the GPCO in the duct (i.e., natural decay) and GPCO device in the duct. These observed rate constants were used to calculate the CADR according to the following equation (Ye 2021):

$$\text{CADR} = (k_{\text{observed, GPCO}} - k_{\text{observed, natural decay}}) \text{ x Chamber Volume} \quad \text{(Equation 2)}$$

For all test cases involving formaldehyde, an initial concentration of 100 ppb was chosen, corresponding to 13% of the OSHA permissible exposure limit for formaldehyde in the workplace measured as an 8-hour time averaged value. Figure 3a provides representative data observed for a formaldehyde challenge test that is representative of air flows and air exchange rates for a residential or commercial building. In this instance, the CADR was determined to be 41 CFM which corresponds to the amount of clean air that the device delivers per minute. Table 1 summarizes the data collected for formaldehyde and provides data for both 1007 and 4096 ft^3 sized chambers as well as two different air exchange rates, specifically 6 and 20 ACH. All of the CADR data shown here falls within the range of 15 to 41. For the three cases of 101 CFM within a 1007 ft^3 chamber, the average CADR is 24 ± 15 CFM, demonstrating good reproducibility across different days of data collection.

Table 1. Summary of GPCO Performance toward Formaldehyde as a Function of Chamber Size and Air Flow

Chamber Size (ft^3)	Air Flow (CFM)	ACH	$k_{\text{observed, GPCO}} - k_{\text{observed natural decay}}$ /min^{-1}	CADR (CFM)
1007	101	6	4.1×10^{-2}	41
1007	101	6	1.5×10^{-2}	15
4096	410	6	9.0×10^{-3}	37
1007	101	6	1.6×10^{-2}	16
1007	336	20	3.4×10^{-2}	34

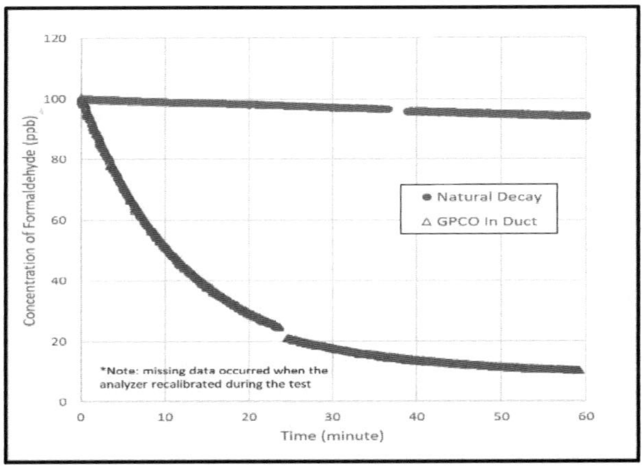

Figure 3a. GPCO performance towards formaldehyde when device is placed in-duct for a 1007 ft^3 chamber at 101 CFM (6 ACH).

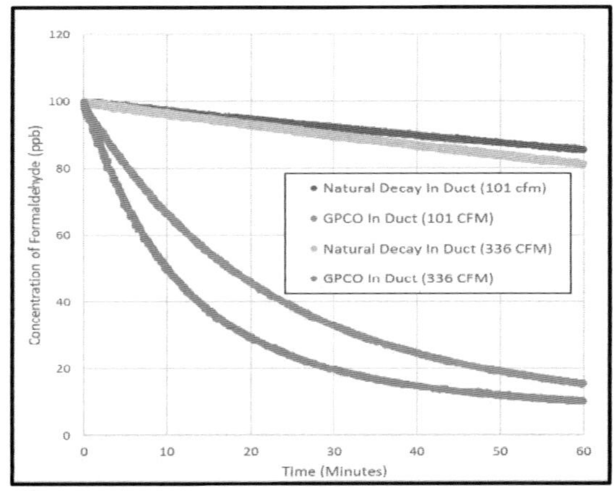

Figure 3b. GPCO performance towards formaldehyde when device is placed in-duct for a 1007 ft^3 chamber at 101 CFM (6 ACH) and 336 CFM (20 ACH).

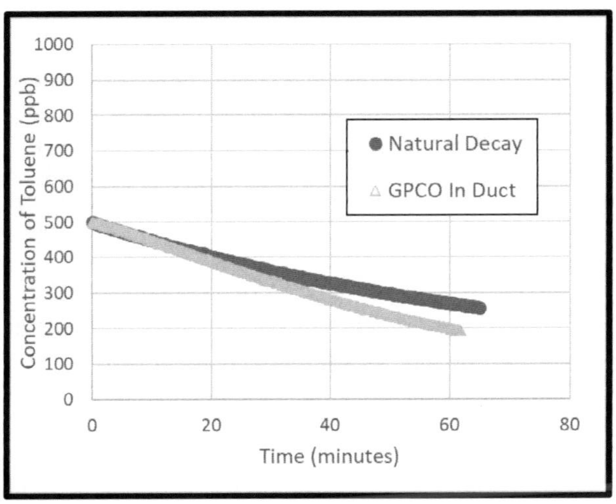

Figure 3c. GPCO performance (50 ACH) towards toluene when the device is placed in-duct for a 1007 ft^3 chamber

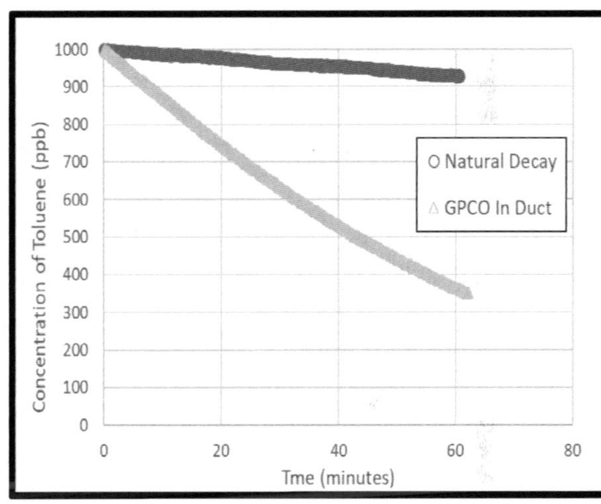

Figure 3d. GPCO performance (6 ACH) towards toluene when GPCO device with 254 nm light output is placed in-duct for a 1007 ft^3 chamber

Decay data was collected for toluene evaluations collected at two different air flows (101 and 840 CFM), corresponding to typical air exchange rates for a residential/commercial building and a bus application (6 and 50 ACH, respectively). Figure 3c shows the data taken at 840 CFM and 50 ACH to represent a typical bus application. The observed rate constants and CADR data collected for toluene are summarized in Table 2. Toluene decays were also compared at 101 CFM for two different initial toluene concentrations (not shown). As can be seen from the data, the CADR for the 50 ACH case increased by a factor of ~3 relative to the 6 ACH scenario. The higher air flow presumably increased the CADR by allowing more passes over the active photocatalyst. The CADR remained constant at both 500 and 1000 ppb initial toluene concentrations, which is consistent with these concentrations falling within the saturation regime of surface adsorption of toluene to the photocatalyst surfaces (Noguchi 1998).

The photoionization detector (PID) used to monitor the toluene concentrations does not differentiate between the different VOCs that are measured. As a result, both toluene and its breakdown products likely contributed to the signal. If this is the case, the kinetics of toluene removal may actually be faster than what is shown here. Data with speciation of the VOCs present is needed to understand this effect. To this end, proton transfer reactive mass spectrometry (PTRMS) will be used in future experiments to focus in on individual parent VOCs, as well as their byproducts (Ye 2021).

Table 2. Summary of GPCO Performance toward Toluene, 1007 ft^3 Chamber

Initial Toluene Concentration (ppb)	Air Flow (CFM)	ACH	$k_{observed, GPCO}$ /min^{-1}	$k_{observed, natural\ decay}$ /min^{-1}	CADR (CFM)
500	101	6	4.2 x 10^{-3}	2.0 x 10^{-3}	2.2
500	840	50	1.6 x 10^{-2}	1.0 x 10^{-2}	6.0
1000	101	6	4.6 x 10^{-3}	2.3 x 10^{-3}	2.3

In a different version of the GPCO device, the 390-395 nm UV-A lamps were replaced with UV-C lamps with 254 nm light output (Figure 3d). The utilization of higher energy photons resulted in a significant improvement in the decay kinetics of toluene corresponding to a CADR of 16 CFM. The larger CADR value may in part be due to a larger absorptance of 254 nm incident light by TiO_2 than in the case of the 390-395 nm LED.

3.2 Potential for By-Products Generation

Ozone is the byproduct that is most often mentioned for air cleaning devices because many electronic air cleaners, including both portable and duct-mounted devices, can generate high amounts of ozone. As a result, air cleaning devices, especially those sold in California, US, are tested to UL 867 and/or UL 2998 environmental standards to show low or no ozone produced. However, this test does not determine whether a device is functional, i.e., that the device neutralizes or inactivates contaminants, as claimed. A check of ozone during a contaminant removal test gives important information about the function of the device when it is working. Ozone was measured during the test sequence and the ozone levels measured during natural decay and during 60 min operation of the GPCO were both ~0.3-1.1 ppb, demonstrating that negligible ozone was generated by the air cleaner.

Ions outputs from air cleaners were measured systematically during our standard test methods because many technologies claim that ion production is an essential feature toward the achievement of high efficacies for microorganism inactivation. The ion counts were similar for both natural decay and VOC challenge tests remaining under 3000 ions/cm^3 for both.

Reactive air cleaning devices can form secondary organic aerosols (SOA) upon reaction with certain VOCs such as terpenes. In addition, certain air cleaning devices are intended to remove particles from the air by inducing particle agglomeration thereby creating larger particles that can more easily be captured by media filters. Thus, the concentration and sizes of the particles in the chamber are significant to understanding the performance of the device and by-products such as SOAs. During this study, cumulative particle counts of smaller particles in the 16-600 nm size range were measured using a Scanning Mobility Particle Sizer Spectrometer (SMPS). For toluene and formaldehyde, the particle concentrations measured during 60 minutes of operation of the GPCO device (~4000-12000 particles/cm^3) increased only slightly compared to natural decay (<2000 particles/cm^3) values, consistent with minimal SOA formation.

4. CONCLUSIONS

In this paper, we have introduced the HVAC-focused testing methodologies that Trane Technologies has recently adapted and developed for measuring the safety (i.e., generation of by-products), applicability, and efficacy of IAQ air cleaning technologies. We have focused here on the development and results of VOC challenge tests for a device based on GPCO and have focused on two VOCs, formaldehyde and toluene, commonly found in indoor spaces. The GPCO displayed a higher CADR for formaldehyde (CADR = 15-41 CFM) than for toluene (CADR = 2-6 CFM). This data is consistent with a slower photocatalytic process and breakdown for toluene, although the lower CADRs for toluene could be artificially low due to some limiting factors with the PID detector that was used since it cannot distinguish between specific VOCs. As a result, both toluene and its breakdown products likely contributed to the signal. In follow up studies, we will be overcome this limitation by using PTRMS to track toluene and other VOC degradation kinetics in the chamber. The results have provided baseline VOC efficacy data for the GPCO that can be applied toward device optimization for specific VOCs, different sized indoor spaces, and various applications. From a product standpoint and beyond its VOC removal capabilities, the GPCO is attractive for the following reasons: (1) it does not emit ozone, (2) it operates at variable air flows and works better at higher air flows, (3) the device can be sized to fit applications where space is a premium, (4) it has a very low pressure drop, and (5) the use of UV-A LEDs avoids harmful deeper UV rays and allows for safer integration into applications that are proximate to occupants of indoor spaces.

A key conclusion of the study is that it is possible to collect useful information about devices like the GPCO in relatively straight-forward chamber tests that utilize a bypass duct. The methodology presented here provides a foundation that can be built upon to more fully understand and benchmark the performance, safety (i.e., the generation of by-products), and applicability of various air cleaning technologies. We believe this methodology can be leveraged in the development of improved standards for testing of air cleaners such as PCOs and assist with improving their acceptance in the market by customers and consumers.

ACKNOWLEDGMENTS

Kevin Kwong and Al Vatine from LMS Technologies, Inc and Chris Hsieh of Trane Technologies.

REFERENCES

ASHRAE. 2017. Standard 52.2-2017, Method of Testing General Ventilation Air-Cleaning Devices for Removal Efficiency by Particle Size, ASHRAE, Inc., Atlanta, GA 30329.

ASHRAE Standard 62.2-2019, Ventilation for Acceptable Indoor Air Quality, ASHRAE, Inc., Atlanta, GA 30329.

AHAM AC-1-2013. 2013. Method for Measuring Performance of Portable Household Electric Room Air Cleaners, Association of Home Appliance Manufacturers.

ASHRAE. 2016. Standard 145.2 "Laboratory Test Method for Assessing the Performance of Gas-Phase Air Cleaning Systems: Air Cleaning Devices," American Society of Heating, Refrigerating, and Air-Conditioning Engineers, Inc.

ASHRAE. 2021. "Equivalent Outdoor Air Calculator." ASHRAE Epidemic Task Force. https://docs.google.com/spreadsheets/d/1GUCcjAyhzrTATHD8SQvNcF7JnuWKpadSVT6LA_8SUIl/edit#gid=0

Burkhead, Bob. 2018. "Chamber Testing" Status Report for ASHRAE SSPC 52.2, presented at the ASHRAE SSPC 52.2 committee meeting, January 20.

Foarde, K.K., E.A. Myers, J.T. Hanley, D.S. Ensor, and P.F. Roessler. 1999. Methodology to perform clean air delivery rate type determinations with microbiological aerosols. *Aerosol Science and Technology* 30:235–245.

Howe, R.F., Gratzel, M. EPR Study and Hydrated Anatase Under UV Irradiation. 1987. *J. Phys. Chem.*, 1987, 91, 3906-3909.

Indoor Air Quality for Transit Buses: A Guide to Selecting the Right Air-Cleaning Solution. ThermoKing White Paper, 2021.

Ishibashi, K., Nosaka, Y., Hashimoto, K., Fujishima, A. 1998. Time-Dependent Behavior of Active Oxygen Species Formed on Photoirradiated TiO_2 Films. *J. Phys. Chem. B*, 102, 2117-2120.

Jacoby, W.A., Blake, D.M., Fennell, J.A., Boulter, J.E., Vargo, L.M., George, M.C., Dolberg, S.K. 1996. Heterogeneous Photocatalysis for Control of Volatile Organic Compounds in Indoor Air. *Journal of the Air and Waste Management Association*, 46, 891-898.

Kaden D.A, Mandin C., Nielsen G.D., et al. Formaldehyde. 2010. WHO Guidelines for Indoor Air Quality: Selected Pollutants. Geneva: World Health Organization; 2010. 3.

Kormann, C., Bahnemann, D.W., Hoffmann, M.R. 1988. Photocatalytic Production of H_2O_2 and Organic Peroxides in Aqueous Suspensions of TiO_2, ZnO, and Desert Sand. *Environ. Sci. Tech*, 22, 798-806.

Macher, J. M. 1989. Positive-Hole Correction of Multiple-Jet Impactors for Collecting Viable Microorganisms, *Am. Ind. Hyg. Assoc.* J. 50:561–568.

Ng, Y.H., Lightcap, I.V., Goodwin, K., Matsumura, M. Kamat, P.V. 2010. To What Extent Do Graphene Scaffolds Improve the Photovoltaic and Photocatalytic Response of TiO_2 Nanostructured Films? *J. Phys. Chem. Lett.*, 1, 2222–2227.

Noguchi, T., Fujishima, A., Sawunyama, P., Hashimoto, K. 1998. Photocatalytic Degradation of Gaseous Formaldehyde Using TiO_2 Film. *Env. Sci. Tech.*, 32, 3831-3833.

Nosaka, Y., Nosaka, A.Y. 2017. Generation and Detection of Reactive Oxygen Species in Photocatalysis. *Chemical Reviews*, 117, 11302-11336.

NRCC (National Resource Council of Canada). 2011. Method for Testing Portable Air Cleaners. NRCC-54013. Prepared for: Government of Canada, Clean Air Agenda, Indoor Air Initiative. Evaluation of IAQ Solutions in Support of Industry Innovation, pp. 1-43, March.

Pistochini, T., et al, 2022. Energy and Long-range Airborne Disease Transmission Impacts of Ventilation and Filtration Methods in Classrooms. *Journal of Building Engineering*, submitted paper.

Professional Standard of the Republic of China: Test of Pollutant Cleaning Performance of Air Cleaners. 2010. Issued by the Ministry of Housing and Urban-Rural Development of the People's Republic of China. December 20.

Perry, J.L., Frederick, K.R., Scott, J.P., Reinemann, D.N. 2011. A Comparison of Photocatalytic Oxidation Reactor Performance for Spacecraft Cabin Trace Contaminant Control Applications. 41st International Conference on Environmental Systems, 1-8.

Schneider, J., Matsuoka, M., Takeuchi, M., Zhang, J., Horiuchi, Y., Anpo, M., Bahnemann, D.W. 2014. Understanding TiO_2 Photocatalysis: Mechanisms and Materials. *Chem. Rev.*, 114, 9919–9986.

Tsang, C.H.A., Li, K., Zeng, Y., Zhao, W., Zhang, T., Zhan, Y., Xie, R., Leung, D.Y.C., Huang, H. 2019. Titanium Oxide-Based Photocatalytic Materials Development and Their Role in Air Pollutants Degradation. *Environmental International*, 125, 200-228

UL Standard 2998: ED 3., Environmental Claim Validation Procedure (ECVP) for Zero Ozone Emissions for Air Cleaners, Underwriters Laboratories, Northbrook, IL

Ye, Q., Krechmer, J.E., Shutter, J.D., Barber, V.P., Li, Y., Helstrom, E., Franco, L.J., Cox, J.L., Hrdina, A.I.H., Goss, M.B., Tahsini, N., Canagaratna, M., Keutsch, F.N., Kroll, J.H. 2021. *Environ. Sci. Technol. Lett.,* 8(12), 1020–1025.

Evaluation of a Gas-Phase Hydrogen Peroxide Generating Device

Paul HOERTZ[1*], Stephen KUJAK[2], Kathleen OWEN[3]

Trane Technologies
[1]Davidson, NC USA & [2]La Crosse, WI USA
[3]Owen Air Filtration Consulting
Cary, NC USA

[1]410-456-1441, paul.hoertz@tranetechnologies.com
[2]608-787-3766, skujak@tranetechnologies.com
[3]919-656-5295, kathleenowen@att.net

*Corresponding Author

ABSTRACT

The novel Coronavirus disease (COVID-19) pandemic has driven innovation of new air cleaning technologies with renewed interest in determining effectiveness of both new and existing air cleaning technologies. A standardized test method for the evaluation of reactive air-cleaning technologies for both volatile organic compounds (VOCs) and microbiological agents does not exist at present. Since air movement in a building is very dynamic and situational, the lack of standardized test methods often leads to contradictory results or sub-optimal evaluation of a device's performance in a laboratory setting.

In this paper, an air cleaning device that generates gas-phase hydrogen peroxide (H_2O_2) is evaluated for general effectiveness against bacteriophage MS2 (*Emesvirus zinderi*). The authors also describe the testing methodology used and discuss variables that could impact the reproducibility and repeatability of the results. In addition, various by-products were measured that could be generated by the device, specifically ozone, ions, and formaldehyde. The results of the study demonstrate that the gas-phase H_2O_2 generator is effective in reducing both airborne MS2 and MS2 on surfaces. Possible implications are discussed.

1. INTRODUCTION

The COVID-19 pandemic has spurred a growing body of research around indoor environmental quality (IEQ) factors, with added focus on indoor air quality (IAQ) due to the role it plays in mitigating the risk of transmission of certain airborne diseases. Indoor environmental quality is an occupant's perceived indoor experience of the built environment that includes aspects of thermal comfort, lighting, acoustics and IAQ. Standards to measure, maintain and optimize IEQ for energy consumption, thermal comfort, lighting, and acoustics are well defined, but standards for measuring IAQ are still under development at the time of this publication. IAQ contaminants of concern fall primarily into three classes: particles, gases (including volatile organic compounds, VOCs) and microorganisms. Distinct categorization of IAQ contaminants are not mutually exclusive and similar technology approaches can be used for mitigation. For example, both microorganisms and dust are particles whose total concentration can be reduced by particle removal technologies. IAQ standards to measure and optimize the quality of indoor air have been slow to develop in part because IAQ contaminants of concern are both difficult to control and measure within the built environment. One objective of this paper is to describe a testing methodology that could be used as input to develop a test standard to evaluate the performance of IAQ technologies against microorganisms like MS2.

There are generally four traditional approaches to removing or reducing various IAQ contaminants. The first two approaches, dilution and exhaust, are usually used in combination to remove contaminants and reduce exposure to IAQ contaminants including particles, VOCs, and microorganisms. Increasing outdoor air ventilation to bring in more fresh air that has lower concentrations of IAQ contaminants will dilute or reduce the concentration of the contaminants in the occupied space. Increasing the supply of fresh air will also increase the exhaust airflow in order to maintain an adequate building pressure. Increasing exhaust airflow increases the rate at which the IAQ contaminants are removed from the occupied space. Dilution and exhaust are effective methodologies and are primary approaches in managing a building's IAQ through the HVAC systems. Of course, there are many building designs or environmental variables that have an impact on the IAQ contaminant removal or reduction effectiveness through the dilute and exhaust methodology; these will not be discussed here.

The third approach is to reduce the concentration of IAQ contaminants of concern through control of indoor humidity. HVACR systems are used to control humidity to improve indoor comfort. Generally, managing indoor humidity can support the mitigation of the growth of airborne and surface bound microorganisms like bacteria and fungi. ASHRAE® recommends 40%-60% relative humidity (RH) to maximize human comfort and to reduce microbial growth. Humidity can also have an impact on particle size and number of water-based aerosols of microorganisms after emission by a host source.

The fourth and last approach is passively cleaning the air through various means. One common approach is to capture the IAQ contaminant through filtration so it cannot reach occupants. Air is circulated through a filter and the contaminant can be captured through various mechanisms, but primarily through impingement or adsorption onto a filtration media. ASHRAE developed Standard 52.2 -2019, Method of Testing General Ventilation Air-Cleaning Devices for Removal Efficiency by Particle Size, to determine the Minimum Efficiency Reporting Value (MERV) for air filters. This method is limited to media filters since it measures the pressure drop as particle loading of the filter increases. Devices, like electrostatic precipitators, which do not have increasing pressure drop with particle loading, cannot be given a MERV rating. Using a filter with a higher MERV rating is an effective tool in removing particles but may cause increased pressure drop in installations with little flexibility in filter choice. Increased pressure drop will increase energy consumption, but this increase in energy use can be more effective in reducing indoor contaminants than that associated with increased outdoor air ventilation or use of in-room air cleaners (Pistochini, 2021).

Another approach to cleaning is to break down or inactivate the IAQ contaminant in the air and/or on surfaces. For example, application-specific ultraviolet germicidal irradiation (UVGI) systems are now a widely accepted technology for controlling microorganisms and may be installed in the HVAC duct system or inside an HVAC unit to irradiate a surface like a coil or drain pan. Another approach that has been used as far back as the 1920s is to bring UVGI into the room (occupied spaces) and either treat a portion of the room's air as it circulates (e.g., upper room UVGI) or treat surfaces directly (typically unoccupied spaces). A wide range of additional air cleaning technologies have also been developed and are currently available on the market and the need for standard methods of testing these devices and verifying performance has become a clear priority. ASHRAE Standard 185.1 is available for single pass in duct inactivation and ASHRAE Standard 185.2 for irradiance of coils and duct surface inactivation, but standards are not available for in-room inactivation. A safety standard exists for the generation of secondary contaminates such as ozone for air cleaning devices. ASHRAE Standards 62.1-2019, Ventilation for Acceptable Indoor Air Quality, requires that air-cleaning devices comply with UL 2998 which limits ozone to 5 parts per billion (ppb) or less (UL 2998, ED3). Current test standards for in-duct devices only allow single-pass testing (ASHRAE 2017, ASHRAE 2016). For in-room air cleaners, AHAM AC-1 (2013), which defines the Clean Air Delivery Rate (CADR), measures reduction of particles but does not address gas-phase contaminants or duct-mounted devices. Still, the chamber decay methodology of AHAM AC-1 allows longer exposure and multiple passes through the air cleaner. Thus, this method serves as the basis of the chamber work discussed here. Methodologies similar to AHAM AC-1 are also recommended in the NRCC version of the chamber test (NRCC 2011).

The COVID-19 pandemic has accelerated the development and testing, and development of various IAQ technologies by the HVAC industry that could facilitate the removal and inactivation of airborne pathogens such as the SARS-CoV-2 virus in various applications. As a result, Trane Technologies accelerated the development and refinement of a set of HVAC-focused testing methodologies for measuring the safety, applicability, and efficacy of various air cleaning technologies and incorporated various facets from existing standards.

One technology that showed promising efficacy towards airborne pathogens is a gas-phase hydrogen peroxide (H_2O_2) generator, which utilizes titanium dioxide photocatalysts to convert atmospheric water into gas-phase H_2O_2 through a photocatalytic oxidation (PCO) process. The new technology releases H_2O_2 into room air at very low levels (<25 ppb) to inactivate microorganisms. H_2O_2 is used commonly to clean surfaces by direct surface application as a liquid formulation or by aerosolization of vapor-phase H_2O_2 into the air to clean entire rooms (e.g., disinfecting hospital rooms). Like other technologies, there is a balance between effectiveness and safe use of the chemistry. The direct products of the PCO process are hydroxyl radicals ($^{\cdot}OH$) and superoxide ($O_2^{\cdot-}$) which can lead to the formation of H_2O_2. Hydroxyl radicals are extremely reactive/unstable and have sub-microsecond lifetimes. As a result, these radicals are not sufficiently long-lived enough to leave the device and inactivate microorganisms in the occupied space. The technology combines unstable hydroxyl radicals generated through the PCO process into more stable H_2O_2 molecules that can then diffuse throughout the space and remotely inactivate microorganisms over time. H_2O_2 is highly hydroscopic and will likely associate with water vapor in the air as it is generated by the device.

Photocatalytic oxidation (PCO) has been studied for several decades and its application towards the cleaning of air streams remains a maturing technology (Jacoby, 1996; Tsang, 2019, Noguchi, 1998; Kormann, 1988; Schneider, 2014; Perry, 2011). The most widely used photocatalytic material in PCOs is titanium dioxide (TiO_2) which has a bandgap of 3.2 eV (corresponding to 388 nm). Bandgap excitation produces both a potent oxidizing agent, i.e., a conduction band hole, and a potent reducing agent, i.e., a conduction band electron, both of which lead to the formation of reactive oxygen species such as hydroxyl radicals ($^{\cdot}OH$), superoxide ($O_2^{\cdot-}$), hydroperoxide radical ($^{\cdot}OOH$), and hydrogen peroxide (H_2O_2) (Howe, 1987; Nosaka, 2017). A review of the literature shows that the following elementary mechanistic steps are the likely key contributors toward the production of reactive oxygen species (ROS) that can cause the subsequent degradation/oxidation of organic species such as microorganism like MS2: (1) oxidation of surface-bound water molecules to produce hydroxyl radicals, (2) reduction of oxygen by conduction band electrons to produce superoxide followed by formation of hydrogen peroxide, and (3) production of hydrogen peroxide from two adjacent surface-bound hydroxyl radicals (Howe, 1987; Nosaka, 2017; Schneider, 2014). Competing with ROS creation is the recombination of the conduction band electrons with valence band holes. This process diminishes the overall yield of ROS that can be created by absorbed photons. The quantum yield for producing ROS by TiO_2 and is reported to fall in the range 1-5% (Nosaka, 2017).

The ROS that are formed on the surface of TiO_2 during photocatalysis are often depicted as remaining on the surface until they naturally decay or are utilized in the surface-based oxidation of incoming or nearby organic species such as nearby microorganisms. However, several literature examples exist that have experimentally demonstrated that these oxidizing agents can enter the gas-phase and be detected down-stream from the photocatalytic structure and utilized in the oxidation of nearby organic species that are located a certain distance away from the photocatalysts (Murakami, 2006; Tatsuma, 2001). These experiments were performed under zero to low air flow conditions. The maximum diffusion length of photocatalytically-produced OH radicals generated by TiO_2 that reach the gas-phase has been determined to be approximately 2 mm under specific experimental setup and conditions (Murakami, 2006). This short diffusion length is consistent with the sub-microsecond lifetimes often reported for airborne OH radicals (Stone, 2012; Nosaka, 2017). Superoxide has been observed to form photocatalytically on TiO_2 surfaces and has been assigned as an intermediate in hydrogen peroxide and hydroperoxide radical formation (Howe, 1987; Ishibashi, 1998). Hydroperoxide radicals, like OH radicals, are short-lived species with sub-second lifetimes. In contrast, the lifetime of gas-phase H_2O_2 is significantly longer (minutes to 0.5 hour) and is highly dependent on environmental conditions such as temperature and humidity (Steris, 2006). Importantly, hydrogen peroxide, hydroperoxide radicals, and OH radicals are all known to exist and form within the Earth's atmosphere with well-documented lifetimes (Zheng, 2003; Das, 1994; Stone 2012). Taken together, it is theoretically possible that H_2O_2, OH radicals, and hydroperoxide radicals can enter an air stream that passes through an operating photocatalytic TiO_2 structure. From an indoor air space standpoint, only H_2O_2 will survive long enough to be detected at distances greater than about 1 cm from the photocatalyst. Over time, the H_2O_2 that has entered the room will either react with organic species within the indoor space or decay naturally into the benign products, water and oxygen.

The gas-phase H_2O_2 generating device comprises a TiO_2-based photocatalytic structure through which airstreams can flow (Figure 1). The photocatalyst becomes activated by light given off by a nearby UV-A bulb which excites electrons across the bandgap of TiO_2. Both in room and in duct device designs are available on the market, the latter of which can be placed wherever it is needed in the duct (e.g., either far away from or adjacent to the diffuser that allows air to re-enter a room). A series of evaluations were conducted to gain further insights into the performance and behavior of gas-phase H_2O_2 generators.

This paper presents initial results of the performance of the gas-phase H_2O_2 generating technology against the MS2 virus. Emphasis is made on the HVAC-focused testing methodologies that were developed for measuring the safety, applicability, and efficacy of gas-phase H_2O_2 generators that are placed within a room (in-room) or placed within the

HVAC ductwork (in-duct). In addition, data is presented that demonstrates the inactivation of both surface-bound and airborne/aerosolized MS2 viruses.

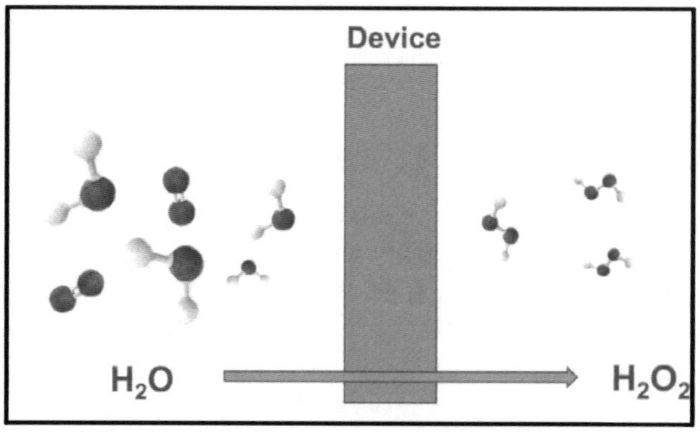

Figure 1. Schematic of the gas-phase H_2O_2 generator showing the utilization of atmospheric water and conversion into H_2O_2 via the photocatalytic oxidation process.

2. EXPERIMENTAL

As COVID-19 focused the industry's attention on air cleaners and the need for air cleaning, Trane Technologies began a project to test several air cleaning devices that were not yet covered by test standards. We worked with LMS Technologies Inc. to determine which methodologies and procedures made sense within the framework of accepted test standards and the needs of air cleaners that are not yet covered by test standards. It should be noted that these air cleaning technologies were not manufactured by Trane Technologies.

2.1 Test Chamber Overview

Many air cleaners require a longer period of time to operate and clean the air within a space; as a result, these air cleaners are not adequately covered by current standard test methods. Because of this, a test chamber similar to the AHAM AC-1 standard for testing in-room air cleaners for particle removal was a good starting point. However, some of these air cleaners are duct-mounted, so a simple chamber is not sufficient, and a combination chamber with a removable side duct apparatus was chosen as the basis for these tests. A removable side duct allowed the same chamber to be used for both in-room and duct-mounted devices. The two test chambers used for these experiments were 1007 ft^3 and 4096 ft^3 in volume, with both having an optional side duct with approximately 80 ft^3 of volume with an available 24"x24" device mounting section for mounting HVAC-mounted devices. Figure 2 provides an overview of the test chambers. Flexible 12-inch HVAC tubing was used to connect the side/by-pass duct with the chamber; for the 4096 ft^3 chamber, an additional ~30 ft of tubing was needed.

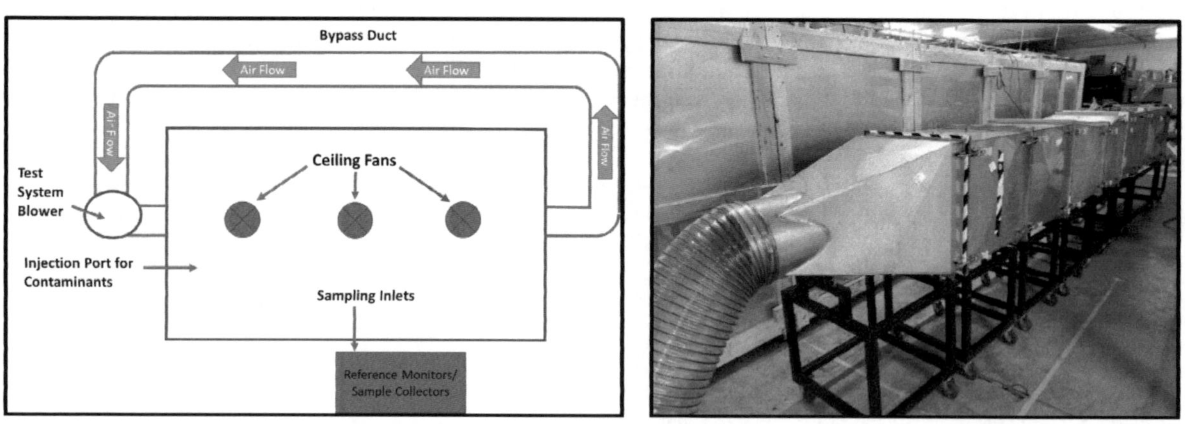

Figure 2. ***Left.*** Environmental test chamber schematic depicting placement of various sensors, injection location, and sampling. ***Right.*** Image of the chamber with the bypass recirculating duct.

The individual sections of this side duct may be exchanged with other sections to provide more space, different access, or simply different sealing options. The smaller chamber size is equivalent to a typical office or bedroom and, as such, is quite appropriate for testing small-to-medium sized in-room air cleaners. It is not, however, the size that a full HVAC-mounted device would service. For example, a 2000 cubic foot per minute (cfm) unit would result in two air exchanges in 1 minute, or 120 air exchanges per hour (ACH) for this chamber, whereas in a real-world situation the same unit might serve 20 similarly sized rooms in a building, providing a more typical 6 ACH. The size of the chamber and of the duct as well as the intended airflow rate of the air cleaner must be considered when reviewing the test data. The chamber also allowed for air cleaners and sampling devices to be positioned in different locations. The internal fan could be run continuously to guarantee good mixing during the injection of contaminants to ensure good initial mixing. Alternatively, the internal fans could not be used at all to test a zero-to-low air flow condition. Three ceiling fans and two box fans at opposite ends of the chambers were used for both chambers. A variable frequency drive (VFD) was operated to adjust the test system blower to the desired test airflow rate in the by-pass recirculating duct. The chamber was purged with high-efficiency particulate air (HEPA) filters, carbon filters, and catalyst filters until stable baseline levels were reached. UV lights were also used to decontaminate the chamber after microbial tests. The purging processes lasted 30 to 60 minutes depending on the test contaminate and chamber size. Importantly, the experiments were carried out under closed chamber conditions (i.e., essentially zero outside air). Humidity was adjusted prior to the beginning of any tests to avoid the introduction of particles during the test.

2.2 Instrumentation

Multiple measurements were made during the environmental chamber experiments during each evaluation. Ozone was measured using an ACOEM Ecotech Serinus 10 Ozone Analyzer, with a measurement range of 0-20,000 ppb and a lower detection limit (LDL) of 0.5 ppb. Background and generated formaldehyde were measured with an Aerodyne QC-TILDAS Formaldehyde Monitor with measurement range of 0-15 ppm. The potential for ions being produced by an air cleaner was monitored using an AlphaLab Air Ion Counter Model AIC2; two separate devices were used to measure positive and negative ions. Particle sizes and distributions over the size range 15-600 nanometers (nm) and concentration ranges of 1 to 10^7 particles/cm^3 were captured with a TSI Scanning Mobility Particle Sizer Spectrometer 3938. Temperature and relative humidity (RH) within the chamber were also monitored, using a TSI VelociCalc 9545, with typical ranges of 71-74°F and 48-53%, respectively. Power consumption by the devices was also measured. When relevant, the airflow through an in-room device, due to its own fan, was determined, while for duct-mounted devices, the actual airflow rate through the duct was measured.

2.3 Generation and Sampling of MS2

MS2 bacteriophage (ATCC 15597-B1) was propagated, made into a solution, and titrated to 3×10^8 plaque forming units per milliliter (PFU/mL) to produce the MS2 challenge. The resulting MS2 suspension from propagation was typically diluted by half with 1% saline to achieve the desired MS2 concentration for the chamber experiments. This suspension of MS2 was aerosolized into the test chamber using a TSI 9302 Single Jet Atomizer with a regulated pressure of 20 pounds per square inch (psi). For data analysis, time zero was set as the end of the injection period and injections typically lasted ~10-14 min. MS2 was sampled from the air within the chamber using impactors located in the center of the room. Three SKC QuickTake 30 high flow pumps equipped with SKC BioStage 400-hole single-stage impactors collected the bioaerosol samples. Air samples were collected in triplicate from the chosen location onto agar plates, usually the center of the test chamber. The QuickTake 30 pumps were calibrated for a flow rate of 1 cfm (28 Lpm), and the collection period was 1 minute.

The sampled MS-2 bacteriophage was grown on double-layer tryptone yeast extract agar with *E. coli* (ATCC 15597) added to the top-layer as the bacteriophage host. The recovered organisms were enumerated after 30 hours of incubation. MS2 plates were quantified by counting the visible plaque forming units (PFU) on the sampling plates. To account for the potential that more than one bacteriophage could have landed in the same location on the plate (i.e., having gone through the same hole in the impactor), a positive hole correction was applied to these numbers before further data analysis (Macher 1989). The efficacy towards MS2 on surfaces was separately evaluated in the following manner. Petri dishes equipped with gaskets under the lid were treated with 20 µL MS2 diluted suspensions and allowed to air dry for 1-1.5 h and then placed at several locations about the chamber. At given time points, the petri dishes were tightly sealed using a pneumatic arm to protect them from further exposure to generated H_2O_2 or to collect natural decays at the same time point.

2.4 Test Sequences (In-duct versus In-Room Devices)

Two types of chamber decay tests, similar to AHAM AC-1, were derived using an environmental chamber, one for in-room devices and one for in-duct or duct-mounted devices.

In-Room Devices

1. Install air cleaner in the 1007 ft^3 or 4096 ft^3 chamber
2. Purge the chamber with ventilation system
3. Check for background levels of VOCs, ozone, particles
4. Inject microorganism challenge into the chamber
5. Mix the chamber well with the fan
6. Collect air samples over time to determine the natural decay rate; measure other by-products such as ozone, ions, particles, and formaldehyde as needed.
7. Purge the chamber with the ventilation system
8. Repeat steps 2-7 but this time with the air cleaner on prior to or following the injection of the microorganism
9. Calculate concentrations and decay rate.

In-Duct Devices

1. Set up the chamber with the bypass loop attached (1007 ft^3 or 4096 ft^3 chamber)
2. Install air cleaner in the bypass duct loop attached
3. Purge the chamber with ventilation system
4. Check for background levels of VOCs, ozone, particles
5. Turn on the air flow through the device and bypass loop
6. Inject microorganism challenge into the chamber
7. Collect air samples over time to determine the natural decay rate; measure other by-products such as ozone, ions, particles, and formaldehyde as needed.
8. Purge the chamber with ventilation system
9. Repeat steps 3-8 but this time with the air cleaner on prior to or following the injection of the microorganism
10. Calculate concentrations and decay rate.

3. RESULTS AND DISCUSSION

3.1 By-products: Ozone, Ions and Formaldehyde

Ozone is the byproduct that is most often mentioned for air cleaning devices because many electronic air cleaners, including both portable and duct-mounted devices, can generate high amounts of ozone. As a result, air cleaning devices, especially those sold in California, US, must qualify to comply with UL 867 and/or UL 2998 environmental standards to show low or no ozone produced. However, qualifying for zero ozone emission does not determine whether a device is functional, i.e., that the device neutralizes or inactivates contaminants, as claimed. A check of ozone during a contaminant removal test gives important information about the function of the device when it is working. Ozone was measured during the test sequence. The ozone levels measured during natural decay and during 60 min operation of the gas-phase H_2O_2 generator were both ~0.4-0.6 ppb, demonstrating that negligible ozone was generated by the air cleaner.

Ions output from air cleaners were measured systematically during our standard test methods because many technologies claim that ion production is an essential feature toward the achievement of high efficacies for microorganism inactivation. The ion counts remained under 2000 ions/cm^3 which is due to the charging inherent in the spraying process used to introduce MS2.

Formaldehyde exposure indoors is a problem in many parts of the world including the United States and is a potential breakdown product of the oxidation of organic species including microorganisms. (Kaden 2010). Formaldehyde levels remained under 7 ppb throughout the 60-min. test, demonstrating that formaldehyde levels remain very low with the gas-phase H_2O_2 generating device responding to a viral bioaerosol challenge.

3.2 Particle Measurements

As described above, an aqueous suspension of MS2 was sprayed into an environmental chamber producing an aerosol consisting of both MS2 and non-MS2 particles. Smaller particles in the 16-600 nm size range were measured using a Scanning Mobility Particle Sizer Spectrometer (SMPS). The average particle size remained <100 nm during the 60-min tests. The diameter of MS2 virus by itself is ~70-100 nm indicating that it is not present in the majority of the generated particles. Calculations based on the concentrations of virus injected into chamber and the SMPS particle concentration data show that about 1 in every 10000 particles that were measured likely contained a virus.

3.3 Microbial Inactivation by Gas-Phase H_2O_2 Generators

An infectious-to-humans microorganism presents aerosolization, sampling, and handling issues. Even some Biosafety Level 1 organisms can present health risks when aerosolized at room or test duct quantities even though they are essentially harmless in liquid suspension. Thus, testing is conventionally accomplished by using surrogate organisms, usually ones with similar characteristics to the desired microorganism. The common viral surrogate MS2, which is a BSL1 organism, was used for these tests. It should be noted that the MS2 bacteriophage virus is a small, non-enveloped virus. Non-enveloped viruses are expected to be more difficult to inactivate than an enveloped virus (Firquet 2015). MS2 is often used as a surrogate for SARS-CoV-2 because SARS-CoV-2 is an enveloped virus and is expected to be inactivated faster than MS2.

It is more complicated to work with viruses than bacteria, since viruses need to replicate in a host to show viability. Thus, growth plates need to be inoculated with *Escherichia coli* (*E. Coli*) first and then used to sample MS2. After growth, the lack of colony growth for *E. coli* indicates viable MS2. The results are quantified as plaque forming units (PFU); the data in this paper are normalized and presented as percent reduction of MS2. The experimental details and subtleties of preparing the MS2 injection, releasing, and analyzing air samples are provided above.

Both in-duct and in-room gas-phase H_2O_2 generators were tested for MS2 inactivation efficacy in a 1007 ft^3 chamber. Figure 3 provides the MS2 challenge response curves for the in-duct device with an air flow of 101 CFM and corresponding to an air exchange rate of 6 ACH. Figure 3 also provides the MS2 challenge response curves for the in-duct device with an air flow of 336 CFM and corresponding to an air exchange rate of 20 ACH. The error bars in all graphs are based on a single standard deviation of the triplicate samples for each test at each time point. Chamber air sampling was performed with impactors at specific time intervals with time = 0 being set as the time-point immediately following the end of the MS2 injection. As can be seen from the data, MS2 in the absence of the air cleaner exhibited a natural decay rate with approximately 70% of the initial active viral concentration becoming inactivated at 60 minutes. When the gas-phase H_2O_2 generator was positioned in the duct adjacent to the diffuser as close to the room as possible, the device gives a faster rate of inactivation than natural decay resulting in a larger overall percent reduction of MS2 (~95%). The data shows that 70% reduction in MS2 was achieved at 20 minutes for the in-room device compared to 60 minutes for natural decay. Importantly, the repositioning of the gas-phase H_2O_2 generator deeper into the duct and further away from air re-entry to the room resulted in no significant inactivation compared with the natural decay. Similar results were also observed for the in-duct unit adjacent to the diffuser when the air flow was increase to 336 CFM (20 ACH), albeit with a lower final efficacy towards MS2 (~88%). Using the data between 0 and 20 minutes for Figure 3, it is estimated that the in-duct device reduces airborne MS2 at a rate that is 1.5-2.0 times faster than natural decay. The requirement of the placement near the diffuser suggests that a gas-phase H_2O_2 generating device needs to be placed either as near as possible to a duct outlet or placed within a space.

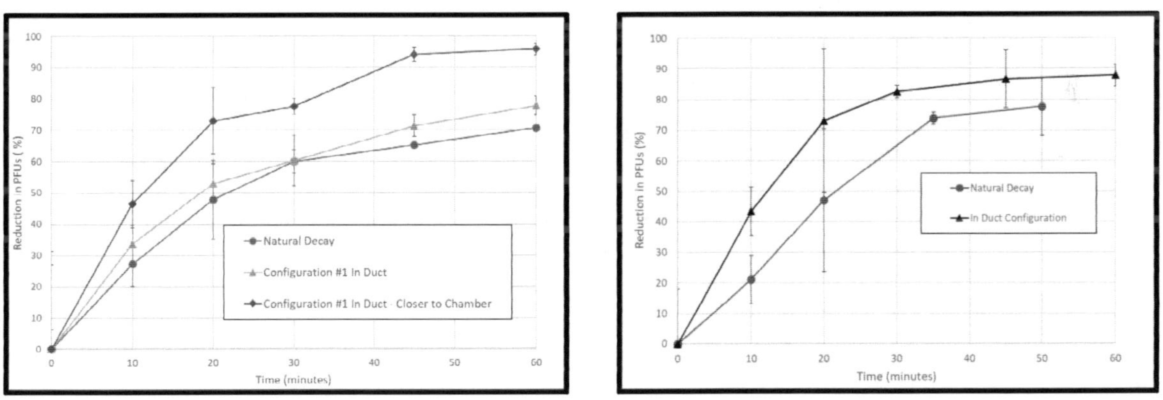

Figure 3. ***Left.*** MS2 challenge data for an in-duct gas-phase H_2O_2 generator at 101 CFM 6 ACH, 1007 ft^3 chamber. ***Right.*** MS2 challenge data for an in-duct gas-phase H_2O_2 generator at 336 CFM, 20 ACH, 1007 ft^3 chamber

The next evaluation was conducted to quantify the efficacy towards MS2 for the in-room gas-phase H_2O_2 generating device in a 4096 ft^3 testing chamber. Figure 4 provides the MS2 challenge response curves for an in-room device with an air flow of 410 CFM, corresponding to an air exchange rate of 6 ACH for the larger chamber. MS2 in the absence of the air cleaner exhibited a natural decay with ~72% of the initial viral concentration after 60 minutes. When the gas-phase H_2O_2 generator was positioned in the room, the device gives a faster rate of inactivation than natural decay resulting in a larger overall percent reduction of MS2 (~88%). The data shows that 70% reduction in MS2 is achieved at 20 minutes for the in-room device compared to 50-60 minutes for natural decay. These results were independent of whether or not the air cleaner was working in the room for 1 hour prior to MS2 injection.

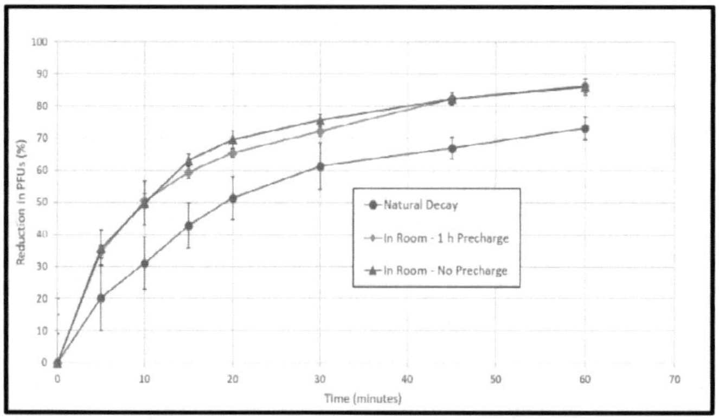

Figure 4. MS2 challenge data for an in-room H_2O_2 generator at 410 CFM, 6 ACH, 4096 ft^3 chamber.

The in-duct device data presented above demonstrates that the position of the unit is critically important. This behavior can be explained by the relatively high reactivity of H_2O_2 which can diminish its effective lifetime. For instance, H_2O_2 is known to react with metal surfaces such as those provided by the metal ductwork in the bypass duct. As the pathlength between the device and the room becomes longer, the H_2O_2 concentration could possibly become diminished (due to reactions with the metal ducting) to a point where MS2 inactivation is minimal or no longer even occurs. The primary reported effectiveness of gas-phase H_2O_2 generating device is their surface cleaning capability for various microorganisms including viruses, bacteria, and fungi. Evaluations were conducted in which closable petri dishes containing MS2 were placed throughout the environmental chamber to measure the decay data for MS2 with and without exposure to gas-phase H_2O_2 generating devices. Figure 5 provides the performance for an in-room device. The data shows that surface deactivation of MS2 by gas-phase H_2O_2 generating devices is highly prevalent at one hour of exposure to H_2O_2.

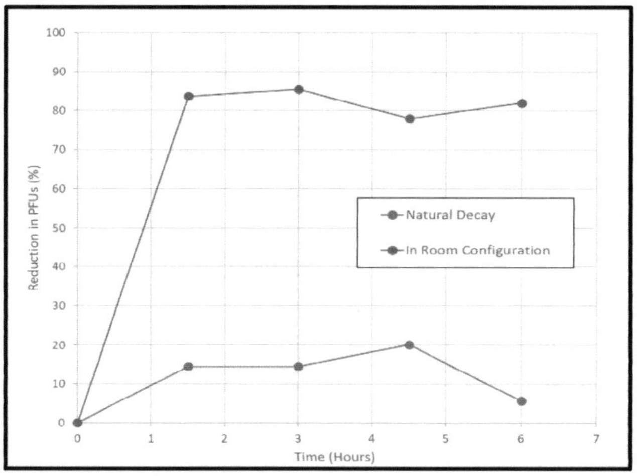

Figure 5. Surface MS2 deactivation data for an in-room H_2O_2 generator.

4. CONCLUSIONS

A potential standard testing methodology for microbiological challenges was provided for evaluating both in room and in duct air cleaning devices to determine their safety, applicability, and efficacy. In-duct and in-room gas-phase

H_2O_2 generating devices were evaluated using this testing method and the devices were shown to exhibit various levels of effectiveness versus natural decay of the MS2 challenge. The efficacy of the gas-phase H_2O_2 generators toward airborne challenges, subtracting out the natural decay of the virus, was observed to be up to25% after 60 min for both in duct and in room devices. At 101 CFM, the in room and in duct devices reach 70% MS2 reduction three times faster than natural decay with rates that are estimated to be 1.5-2.0 times faster. In addition, we observed that an in-room device shows very good performance toward inactivating MS2 virus on contaminated surfaces. Since the MS2 virus is an enveloped virus, it could be expected that these devices would show greater efficacy against non-enveloped viruses, such as SARS-CoV-2. The results shown here provide baseline microorganism efficacy data for gas-phase H_2O_2 generators that can be applied toward device optimization for specific microorganisms and different sizes of indoor spaces.

This work demonstrates that it is possible to collect useful information about devices, like gas-phase H_2O_2 generators, in relatively straight-forward chamber testing that utilize a bypass duct. The methodology presented here provides a foundation that can be built upon to understand and benchmark the performance, safety, and applicability of various air cleaning technologies more comprehensively. The proposed testing methodology can be used as a potential starting point in the development of improved standards of testing for air cleaners and better understanding of safety and performance of various IAQ cleaning technologies

ACKNOWLEDGMENTS

Kevin Kwong and Al Vatine from LMS Technologies, Inc and Chris Hsieh of Trane Technologies.

REFERENCES

AHAM AC-1-2013. 2013. Method for Measuring Performance of Portable Household Electric Room Air Cleaners, Association of Home Appliance Manufacturers.

ASHRAE Standard 52.2-2017, Method of Testing General Ventilation Air-Cleaning Devices for Removal Efficiency by Particle Size, ASHRAE, Inc., Atlanta, GA 30329.

ASHRAE Standard 62.2-2019, Ventilation for Acceptable Indoor Air Quality, ASHRAE, Inc., Atlanta, GA 30329.

ASHRAE. 2016. Standard 145.2 "Laboratory Test Method for Assessing the Performance of Gas-Phase Air Cleaning Systems: Air Cleaning Devices," Atlanta, GA

ASHRAE. 2021. "Equivalent Outdoor Air Calculator." ASHRAE Epidemic Task Force. https://docs.google.com/spreadsheets/d/1GUCcjAyhzrTATHD8SQvNcF7JnuWKpadSVT6LA_8SUII/edit#gid=0

Burkhead, Bob. 2018. "Chamber Testing," Status Report for ASHRAE SSPC 52.2, presented at the ASHRAE SSPC 52.2 committee meeting, January 20.

Das, M, Aneja, V.P. 1994. Analysis of Gaseous Hydrogen Peroxide Concentrations in Raleigh, North Carolina. *Journal of Air and Waste Management Association*, 44, 176-180.

Firquet, S., Beaujard, S., Lobert, P.E., Sané, F., Caloone, D., Izard, D., Hober, D. 2015. Survival of Enveloped and Non-Enveloped Viruses on Inanimate Surfaces. *Microbes Environ.*, 30, 2, 140-144.

Foarde, K.K., E.A. Myers, J.T. Hanley, D.S. Ensor, and P.F. Roessler. 1999. Methodology to perform clean air delivery rate type determinations with microbiological aerosols. *Aerosol Science and Technology* 30:235–245.

Half-Life of Vaporized Hydrogen Peroxide (VHP) During Remediation of the U.S. Department of State SA-32 Diplomatic Pouch and Mail Facility. STERIS Corporation - June 22, 2006.

Howe, R.F., Gratzel, M. EPR Study and Hydrated Anatase Under UV Irradiation. 1987. *J. Phys. Chem.*, 1987, 91, 3906-3909.

Ishibashi, K., Nosaka, Y., Hashimoto, K., Fujishima, A. 1998. Time-Dependent Behavior of Active Oxygen Species Formed on Photoirradiated TiO_2 Films. *J. Phys. Chem. B*, 102, 2117-2120.

Jacoby, W.A., et. al. 1996. Heterogeneous Photocatalysis for Control of Volatile Organic Compounds in Indoor Air. *Journal of the Air and Waste Management Association*, 46, 891-898.

Kaden, D.A, Mandin C., Nielsen G.D., et al. Formaldehyde. 2010. WHO Guidelines for Indoor Air Quality: Selected Pollutants. Geneva: World Health Organization; 2010. 3.

Kormann, C., Bahnemann, D.W., Hoffmann, M.R. 1988. Photocatalytic Production of H_2O_2 and Organic Peroxides in Aqueous Suspensions of TiO_2, ZnO, and Desert Sand. *Environ. Sci. Tech*, 22, 798-806.

Macher, J. M. 1989. Positive-Hole Correction of Multiple-Jet Impactors for Collecting Viable Microorganisms, *Am. Ind. Hyg. Assoc. J.* 50:561–568.

Murakami, Y., et al. 2006. Direct Detection of OH Radicals Diffused to the Gas Phase from the UV-Irradiated Photocatalytic TiO_2 Surfaces by Means of Laser-Induced Fluorescence Spectroscopy. *J. Phys. Chem. B*, 110, 16808-16811.

Noguchi, T., Fujishima, A., Sawunyama, P., Hashimoto, K. 1998. Photocatalytic Degradation of Gaseous Formaldehyde Using TiO_2 Film. *Env. Sci. Tech.*, 32, 3831-3833.

Nosaka, Y., Nosaka, A.Y. 2017. Generation and Detection of Reactive Oxygen Species in Photocatalysis. *Chemical Reviews*, 117, 11302-11336.

NRCC. 2011. Method for Testing Portable Air Cleaners. NRCC-54013. Prepared for: Government of Canada, Clean Air Agenda, Indoor Air Initiative. Evaluation of IAQ Solutions in Support of Industry Innovation, pp. 1-43, March.

Perry, J.L., et al. 2011. A Comparison of Photocatalytic Oxidation Reactor Performance for Spacecraft Cabin Trace Contaminant Control Applications. 41st International Conference on Environmental Systems, 1-8.

Pistochini, T., et al, 2022. Energy and Long-range Airborne Disease Transmission Impacts of Ventilation and Filtration Methods in Classrooms. *Journal of Building Engineering*, submitted paper

Professional Standard of the Republic of China: Test of Pollutant Cleaning Performance of Air Cleaners. 2010. Issued by the Ministry of Housing and Urban-Rural Development of the People's Republic of China. December 20.

Schneider, J., Matsuoka, M., Takeuchi, M., Zhang, J., Horiuchi, Y., Anpo, M., Bahnemann, D.W. 2014. Understanding TiO_2 Photocatalysis: Mechanisms and Materials. *Chem. Rev.*, 114, 9919–9986.

Stone, D., Whalley, L.K., Heard, D.E. 2012. Tropospheric OH and HOO Radicals: Field Measurements and Model Comparisons. *Chem. Soc. Rev.,* 41, 6348–6404.

Tatsuma, T., Tachibana, S., Fujishima, A. 2001. Remote Oxidation of Organic Compounds by UV-Irradiated TiO_2 via the Gas Phase. *J. Phys. Chem. B*, 105, 6987-6992.

Tatsuma, T., Tachibana, S., Miwa, T., Tryk, D.A., Fujishima, A. 1999. Remote Bleaching of Methylene Blue by UV-Irradiated TiO_2 in the Gas Phase. *J. Phys. Chem. B*, 103, 8033-8035.

Tsang, C.H.A., et al., 2019. Titanium Oxide-Based Photocatalytic Materials Development and Their Role in Air Pollutants Degradation. *Environmental International*, 125, 200-228.

UL Standard 2998: ED 3., Environmental Claim Validation Procedure (ECVP) for Zero Ozone Emissions for Air Cleaners, Underwriters Laboratories, Northbrook, IL.

UL Standard 867: ED5. The Standard for Safety for Electrostatic Air Cleaners, Underwriters Laboratories, Northbrook, IL.

Zheng, J., Springston, S.R., Weinstein-Lloyd, J. 2003. Quantitative Analysis of Hydroperoxyl Radical Using Flow Injection Analysis with Chemiluminescence Detection. *Analytical Chemistry*, 75, 4696-4700.

A Luminance-based Approach for Inferring Personal Daylight Preferences Using a New Composite Similarity Index

Dongjun Mah[1,2], Jie Xiong[3], Athanasios Tzempelikos[1,2]*

[1]Purdue University, Lyles School of Civil Engineering,
550 Stadium Mall Dr., West Lafayette, IN, 47907, USA
Contact Information (dmah@purdue.edu)

[2]Ray W. Herrick Laboratories, Center of High Performance Buildings,
140 S. Martin Jischke Dr., West Lafayette, IN, 47907, USA
Contact Information (ttzempel@purdue.edu)

[3]National Renewable Energy Laboratory,
15033 Denver West Parkway, GOLDEN, CO 80401, USA
Contact Information (jie.xiong@nrel.gov)

* Corresponding Author

ABSTRACT

The concept of extracting similarity features between two luminance maps is presented in this study for classifying lighting preferences of office workers. A composite similarity index is proposed to quantify the similarity between different lighting conditions, using processed luminance maps and luminance contrast maps, captured from High Dynamic Range (HDR) images in real offices together with occupant visual preference data. 7 existing lighting parameters were selected for evaluating the performance of the similarity index. Logistic regression models were trained using each parameter separately to classify the occupant's visual preference. A zoning approach of the generated similarity maps was developed for a fair comparison and to prevent information loss during averaging. Similarity indices in each zone of both maps were then averaged and used as input parameters of a logistic regression model. The results showed that: (i) positive and negative signs of similarity indices were necessary for describing the occupant's visual preference; (ii) dividing the similarity maps into 3 zones showed better classification performance; (iii) both luminance similarity and luminance contrast similarity maps are important for classifying the occupant's visual preference; and (iv) similarity index maps could show more stable classification results than existing parameters. This composite similarity index scheme can be used in more complex deep learning techniques to predict and learn occupant's visual preferences.

1. INTRODUCTION

The effect of daylighting is a commonly considered important portion in high-performance building design for occupants' health, comfort, and productivity. Efficiently controlled daylighting can influence significantly to reduce energy consumption of electric lighting, which is at least 17% of entire energy consumption in the US commercial buildings (EIA, 2012). However, inefficient daylighting control can lead to glare and adverse effects (e.g., reduction of daylight access and outside views). Therefore, despite of the benefits, it is necessary to carefully design and control daylighting systems.

Luminance monitoring is useful for assessing visual preferences, and real-time FOV luminance monitoring is required in actual buildings. This is achieved using High Dynamic Range Imaging (HDRI) sensors combined with wide-angle fisheye lens (Cauwerts, 2018). The images are then processed by software like Evalglare (Wienold, J., & Andersen, 2016) for further analysis. HDRI measurements have been used to correlate scene luminance characteristics, measured from HDR images to subjective visual comfort responses (Hirning, Isoardi, & Cowling, 2014; Hirning, Isoardi, & Garcia-hansen, 2017; Konis, 2014). However, detecting visual discomfort scenarios using HDRI sensors is not equivalent to learning preferred visual conditions. Instead, learning lighting preferences without considering discomfort scenarios (Amir, Awalgaonkar, Karava, & Bilionis, 2017; Lindelöf, Morel, & Lindelo, 2008; Xiong et al., 2020; Xiong, Tzempelikos, Bilionis, & Awalgaonkar, 2018; Xiong et al., 2019) is preferred. Selecting the proper variables while learning lighting preferences is not easy. Using simple variables such as horizontal or vertical illuminance, average luminance or simple contrast ratios from captured luminance maps is maybe sufficient for specific glare assessment cases (Wymelenberg & Inanici, 2015; Wymelenberg, Inanici, & Johnson, 2010) but not for inferring preferred conditions, partly due to averaging pixel information and neglecting information in different parts of the luminance field. There is no agreement on which extractable features can better represent visual preferences of occupants in typical daylighting settings.

Instead of trying to find specific parameters that can better explain occupants' visual preferences using surveys and numeric scales, in this study, the pixel-wise luminance in every part of the occupant's field of view was considered under several different visual conditions. Luminance maps from HDR images and respective occupant preference votes were used for generating a comparative visual preference dataset. A new similarity index concept is introduced for inferring user's visual preferences. Reference HDR images were converted to luminance maps and used for generating comparative visual preference data. 7 reference lighting parameters were selected for comparison to check the performance of the similarity index. Classification models were generated individually using these parameters for estimating the visual preferences of three office occupants. For a fair performance comparison, a zoning logic was applied to compute the average similarity index values in each zone according to relative luminance levels. The proposed similarity index with the zoning logic was then applied and a classification model was trained using logistic regression for inferring the user's visual preferences.

2. METHODOLOGY

2.1 Pixelwise Similarity Index

2.1.1 Background

In this study, a pixelwise similarity index is proposed to quantify a degree of similarity between two luminance maps. The origin of this similarity index concept comes from structural similarity index (SSIM), which is widely used for degradation assessment of image qualities between an original image and test image (Wang et al., 2004). The local SSIM is computed using a selected size of small windows (regions), which are stated as **x** and **y** from original image (X) and test image (Y) in Equation (1):

$$SSIM(\boldsymbol{x},\boldsymbol{y}) = \left(1 - \frac{(\mu_x - \mu_y)^2}{\mu_x^2 + \mu_y^2 + c_1}\right) * \left(1 - \frac{(\sigma_x - \sigma_y)^2}{\sigma_x^2 + \sigma_y^2 + c_2}\right) * \left(\frac{\sigma_{xy} + c_3}{\sigma_x \sigma_y + c_3}\right) \tag{1}$$

Where μ_x and μ_y are mean pixel intensities of selected windows from original image and test image respectively. Similarly, σ_x and σ_y are the standard deviation of selected windows, and σ_{xy} is the correlation between two selected regions (**x** and **y**) at same location and c_1, c_2, and c_3 are constants. The size of window (region) for local SSIM can be determined by a user and a typical size of the window is 11x11 pixels. Then, the selected size of small window slides through the entire original and test images and SSIM values at each location can be computed. By averaging the entire local SSIM values, it is possible to simply check a degree of similarity between two images. In addition to the original type of SSIM, several studies have tried to modify SSIM for different applications by changing a form of the equation slightly such as using different type of normalization factor, which was difference between maximum and minimum values of entire pixels in two images (Liu, Soo, Beresford, & Fleisher, 2020; Palubinskas, 2017).

2.1.2 Pixelwise similarity index

The pixelwise similarity index was designed to apply the similarity index concept to comparing two luminance maps. One of the most different parts of the designed similarity index compared to the original one is an existence of small region for comparison. In contrast to the original SSIM or other previous similarity indices, the designed similarity index does not use a small sliding window for computing mean, standard deviation, and correlation values, instead, it compares pixels at the same location from two luminance maps directly (it is possible to directly compare important visual information in a small region of two luminance maps without data loss). Also, it is possible to mitigate a difference between two luminance maps by averaging even though we used small sliding window size in the case of using previous types of similarity indices. The equation of designed similarity index is shown in Equation (2):

$$Luminance\ Sim(x_c, x_p) = [sign(x_c - x_p)] * (1 - \frac{|x_c - x_p|}{\max(x) - \min(x)}) \tag{2}$$

where x is a luminance value at each pixel and the subscription c and p stand for pixels of current and previous luminance maps from one comparative luminance map data respectively. Also, $\max(x)$ and $\min(x)$ compute the maximum and minimum luminance values from entire two luminance maps. Since the shape of the scene was circular, the outer part of the pixels was excluded during computing maximum and minimum luminance values. To consider the direction of luminance change between the previous and current luminance maps, the sign of luminance difference was applied to the final similarity index. In the Equation (2), $sign(x_c - x_p)$ outputs 1 if the difference between x_c and x_p is 0 or positive value and -1 if the difference is negative value. Since there is no window slide for comparing small regions between two luminance maps in the newly proposed similarity index, in contrast to the typical type of similarity indices, there is only one term for directly comparing two pixels at the same location in two luminance maps.

A similarity of 1 means that luminance intensity values from current and previous luminance maps at same pixel location have exactly same values. Positive sign denotes an increase in luminance from previous to current conditions and vice versa. The degree of similarity will be decreased if the absolute value gets closer to 0.

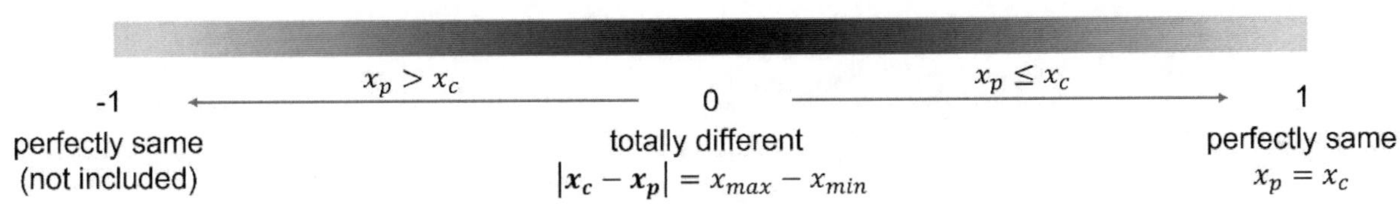

Figure 1: Meaning of pixelwise similarity index degree

2.1.3 Pixelwise Luminance & Luminance Contrast Similarity maps

In this study, two types of similarity maps were generated from the collected HDR images for capturing the occupant's visual information in detail. One was a luminance similarity map, which was generated by directly comparing two luminance maps, and the other was a luminance contrast similarity map extracted by comparing two luminance contrast maps, which were computed by comparing the working area and outer area in the scene. The first step of computing luminance similarity or luminance contrast similarity maps was selecting a task area from the scene. The shape and size of the task area were selected following the recommendation of (Pierson, 2018), which proposed an opening angle of 60 degree with circular shape. The center point coordinate of the task area was fixed to the location of 20 degree downward from the center coordinate of the scene.

A representative example of generating luminance similarity maps is shown in Figure 2. After RGB color data of two HDR images converted to luminance maps, using the Equation (3) described by (Inanici, 2006), the luminance of each pixel was used to calculate the luminance similarity index the current and previous luminance maps.

$$L = 179 * (R * 0.2126 + G * 0.7152 + B * 0.0722) \tag{3}$$

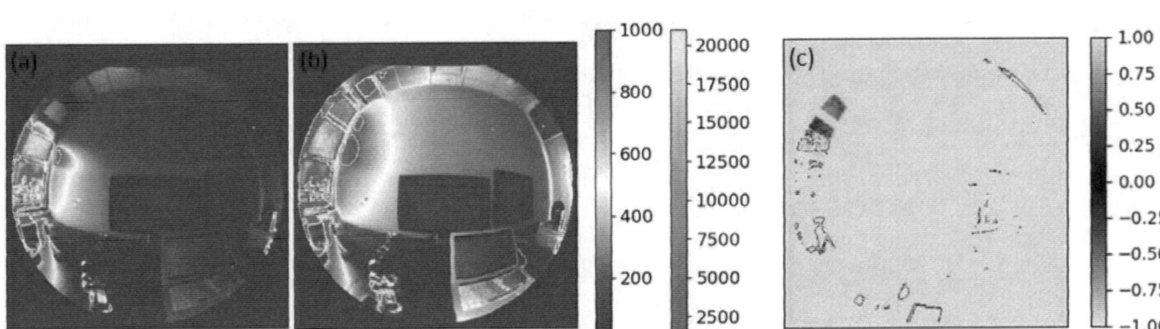

Figure 2: Procedure of generating luminance similarity map
(a: previous luminance map, b: current luminance map, c: luminance similarity map)

In addition to comparing raw luminance values from current and previous luminance maps, luminance contrast values were also compared to capture additional visual information for inferring the user's visual preference. Two methods for generating luminance contrast similarity maps were studied. The luminance contrast map was firstly generated by computing the ratio between each luminance pixel outer to the task area and mean luminance of task area (Figure 3). The luminance contrast values from two luminance contrast maps were compared using the Equation (4):

$$Lum\ Contrast\ Sim(C_c, C_p) = [sign(C_c - C_p)] * (1 - \frac{|C_c - C_p|}{\max(C) - \min(C)}) \quad (4)$$

where C is a luminance contrast value at each pixel and subscripts c and p mean pixels from current and previous luminance contrast maps respectively. There are positive and negative signs in the luminance contrast similarity index and the difference of maximum and minimum values from two luminance contrast maps was used as a normalization factor.

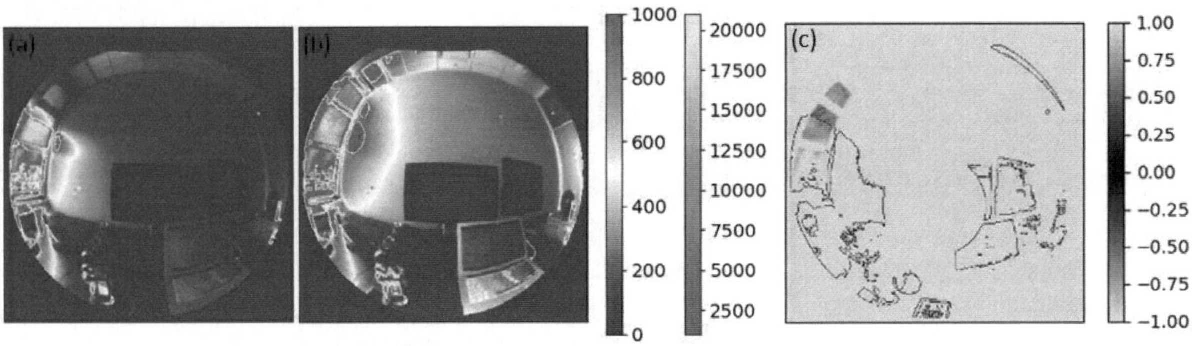

Figure 3: Procedure of generating the first type of luminance contrast similarity map (simple type) (a: previous luminance map, b: current luminance map, c: luminance contrast similarity map)

The other type of luminance contrast map was generated by considering the luminance distribution in the task area, illustrated in Figure 4. The similarity index of mean and std luminance contrast values from two luminance contrast maps were computed using Equation (5) and (6),

$$Mean\ Lum\ Contrast\ Sim(C_{\pi c}, C_{\pi p}) = [sign(C_{\pi c} - C_{\pi p})] * (1 - \frac{|C_{\pi c} - C_{\pi p}|}{\max(C_\pi) - \min(C_\pi)}) \quad (5)$$

$$Std\ Lum\ Contrast\ Sim(C_{\sigma c}, C_{\sigma p}) = 1 - \frac{|C_{\sigma c} - C_{\sigma p}|}{\max(C_\sigma) - \min(C_\sigma)} \quad (6)$$

Where C is a luminance contrast value at each pixel and subscription c and p mean pixels from current and previous luminance contrast maps respectively from one comparative luminance contrast map data. Also, the subscription π means mean luminance contrast map and σ means std luminance contrast map. In contrast to the other similarity index equations in this study, only the equation of std luminance contrast similarity does not have the sign term.

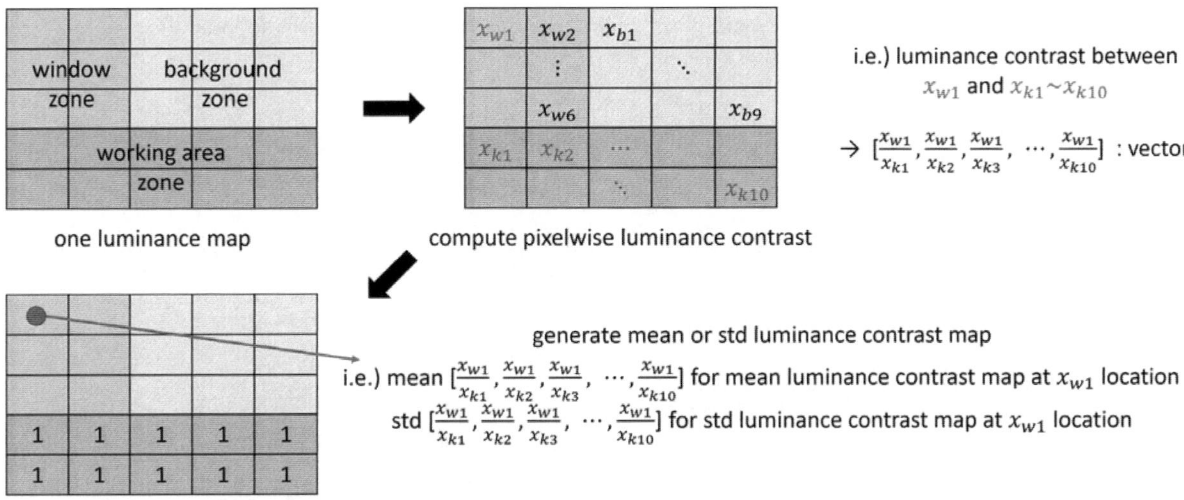

Figure 4: Schematic procedure of generating a luminance contrast map considering each pixel in working area

2.2 Evaluation of Similarity Index

2.2.1 Generate Comparative Luminance Map Data

In order to collect the information of occupant's visual conditions in a private office, HDR images at the occupant's eye level were captured by (Xiong et al., 2018). The measurement was carried out using Raspberry Pi 3 B+ (Raspi) with calibrated Raspi Camera Module V2 and fisheye lens in a mock-up private office in both sunny and cloudy weather conditions. The collected HDR images were converted to luminance maps and directly compared pixel-by-pixel in this study for capturing important visual information of the occupants. These converted luminance maps were further processed to be comparative luminance maps data by grouping two luminance maps collected from a specific time point and previous time point following the collected time information from the (Xiong et al., 2018). In addition, each luminance map from the paired map data was further converted to be a luminance contrast map for capturing additional visual information from raw luminance map data. The specific time point was named current situation and the other time point was named previous situation.

2.2.2 Selected Reference Parameters

To assess the performance of similarity index, 7 parameters, which were considered as the best parameters for estimating the occupant's visual comfort in each reference (Konis, 2014; Wymelenberg & Inanici, 2015; Wymelenberg et al., 2010), were selected for comparison. The percentage of pixels exceeding 2000 cd/m^2, average luminance of entire scene, and DGP were selected from (Wymelenberg et al., 2010). Standard deviation of window luminance and average luminance of 40° horizontal band were selected from (Wymelenberg & Inanici, 2015). In addition, maximum luminance in window and average luminance in window divided by 200 cd/m^2 respectively were selected from (Konis, 2014). The DGP, average luminance of entire scene, and average luminance of 40° horizontal band were computed by Evalglare, and the other parameters were computed manually by masking window part in the scenes.

2.2.3 Zoning luminance Maps

The selected parameters in previous section were used as input parameters of visual preference classification models in the following sections. Although the best way of dealing with similarity index maps for input data was using them directly, it was not fair comparison with the other selected parameter cases because they were constant values extracted from each luminance map. In addition, each scene was divided into specific zones according to relative luminance difference and average similarity index values were computed. Zoning logic of the comparative luminance map data was designed to divide the scene into 2 or 3 parts. The schematic procedure of dividing the pixels in the scene was shown in Figure 5.

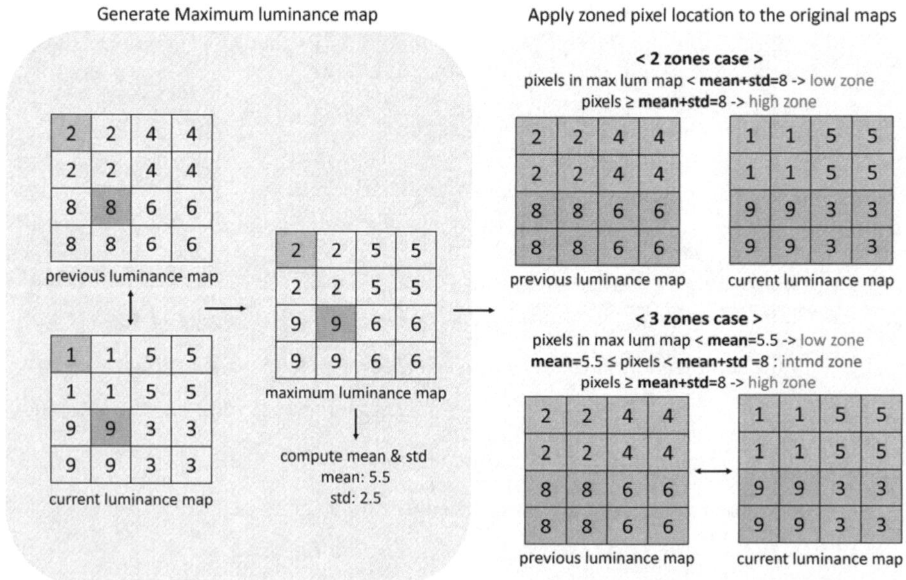

Figure 5: Schematic procedure of zoning pixels according to luminance levels

The first step is computing maximum values between previous and current luminance map pixel by pixel at the same location. Then, mean and standard deviation values of the entire maximum luminance map were computed to determine criteria for dividing pixels. For the case of zoning 2 parts from the scene, pixels of maximum luminance map greater than (mean + 1 std) values were classified as pixel location of high luminance zone. The pixel locations classified as high luminance zone were applied to both previous and current luminance maps. Then, the remain pixel locations were classified as low luminance zone. For the 3-zoning case, the same logic was used but, in addition, another zoning criteria was applied: if the intensity of maximum luminance map were less than mean value of entire maximum luminance map, those pixel locations were classified as a low luminance zone. The remaining pixel locations were classified as an intermediate luminance zone. Then, all the divided pixel location information was applied to the original current and previous luminance maps. Representative masks of each zoning case from one comparative luminance map data were shown in Figure 6.

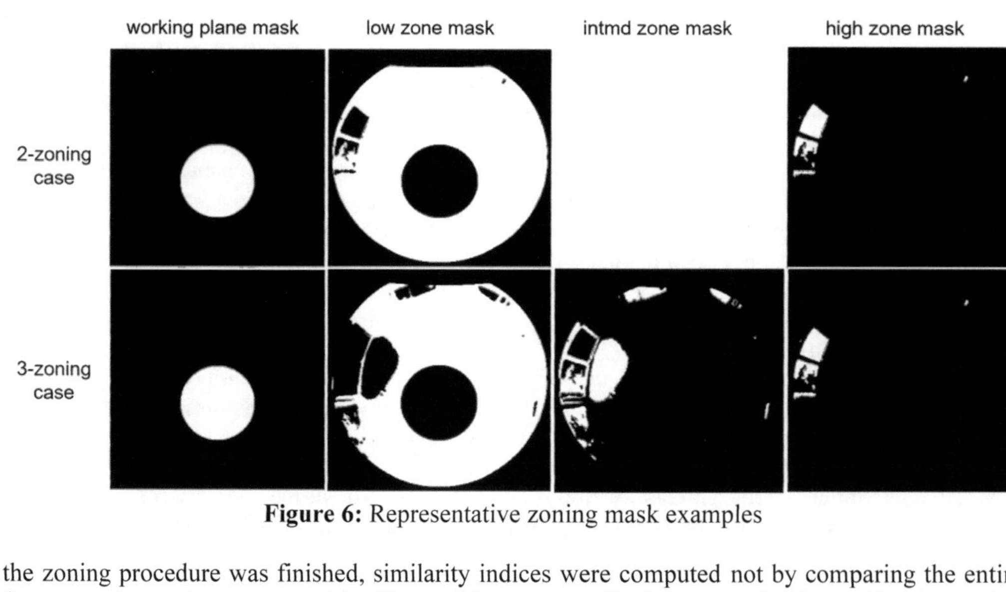

Figure 6: Representative zoning mask examples

After the zoning procedure was finished, similarity indices were computed not by comparing the entire scene at once but by comparing each zone separately. The working area was firstly extracted and considered as an independent area regardless of the original divided zone to capture similarity pattern as much in detail as possible. Therefore, the normalization factor was computed by using only the pixels in the task area. Then, the similarity values were computed and input to the pixel locations of low and high luminance zone in the case of 2 zones case and the pixel locations of low, intermediate, and high luminance zones for the 3 zones case.

2.2.4 Generate Classification Model to Infer Visual Preference

A logistic regression model was trained to classify a user's visual preference using a designed similarity concept with zoning logic and the selected reference parameters respectively. After all the luminance and luminance contrast similarity maps with 2-zoning or 3-zoning logics were generated, average similarity index values in each zone were computed (Figure 7), for input parameters of the classification model using logistic regression. All the input parameters were normalized separately using min-max normalization. During averaging similarity pixels, positive and negative similarity pixels were divided and computed separately. Since the occupants answered to questions asking their preferences between previous and current visual situation every 10 minutes, the collected data was used for test data to evaluate the classification performance of the trained model. For the training data, the automatically saved HDR images with 2 minutes interval were used assuming that the user would have same visual preference with corresponding answered data.

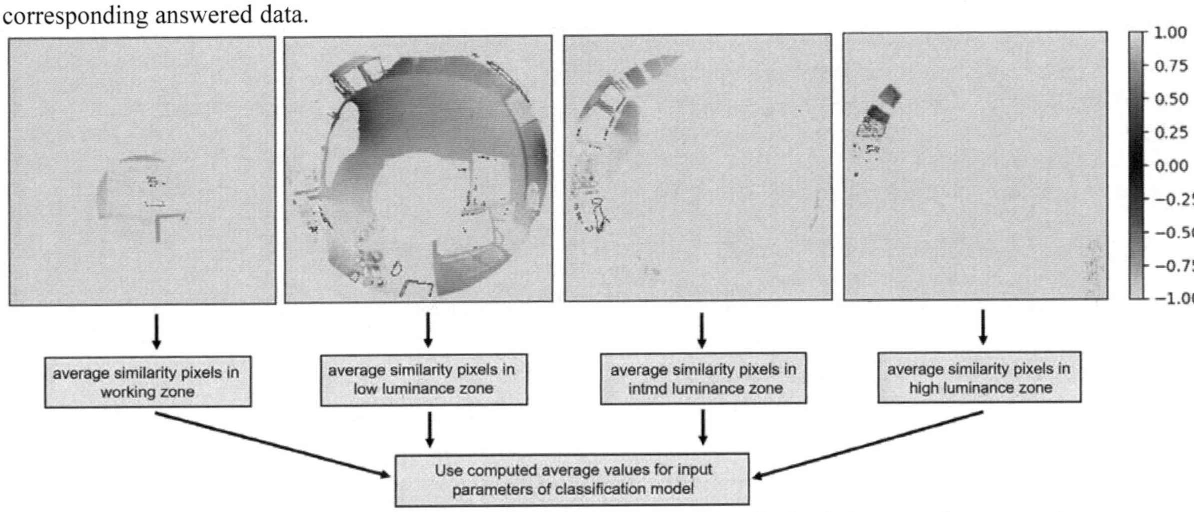

Figure 7: Schematic example of generating input parameters for logistic regression parameters

Several factors were studied in terms of their influence on the classification performance. First, the influence of the sign in similarity index was studied by comparing the test result of classification models trained with average values of absolute similarity indices for input parameters to the result of using average values of original similarity indices. In addition, 2 and 3-zoning cases were studied. At last, several combinations of luminance and luminance contrast similarity maps were tested to estimate the influence of map types on the classification performance. For this assessment, several logistic regression models were trained using different combination of maps such as using either luminance similarity map or luminance contrast similarity map, using both 2 maps separately, and combining 2 maps into one map by pixelwise multiplication. The classification accuracy was tested by computing the ratio between the number of correctly classified test data and the number of entire test data.

3. Results

The results of classification performance using test data are listed in Tables 1, 2, and 3, for three different occupants. Using similarity indices without sign showed worse performance compared to the case of using positive and negative signs in the similarity indices. This trend came from the fact that the users showed their preference to the change of luminance distribution either of increasing or decreasing direction.

From Table 1, dividing the luminance or luminance contrast maps into 3 zones showed slightly greater performance compared to the 2-zoning case. Since logistic regression model got average values of each divided luminance zone as input parameters, 3-zoning case could capture more visual information especially in the intermediate and low luminance zones by adding parameters. In addition, using luminance similarity maps showed slightly greater classification performance compared to using any types of luminance contrast similarity maps. This was because the user #1 showed very clear preference to the brighter situation so the luminance value itself was more helpful than luminance contrast value. In all the cases, using two maps separately showed better classification result compared to the case of using each map individually because of adding user's visual information.

The classification results of other two users, listed in Table 2 and 3, showed lower performance that user #1, and also did not show better performance compared to the case of using other parameters consistently. On the other hand, the case of using both luminance and luminance contrast similarity maps showed relatively better performance consistently. However, the overall classification performance was poor considering the objective of this model, which was classifying 2 classes (prefer previous or current), mainly because the occupant feedback in these cases was less clear. In these cases, it will be necessary to try to capture specific visual information in small areas but this was impossible with the logistic regression model (this model only used average similarity values in each divided zone as input parameters). Therefore, more complicated models such as deep learning methods will be necessary to classify more complex preference data like the case of these two users in the future.

Table 1: result of user #1 (train: 754, test: 31)

Lsim: Luminance similarity map
$Csim_{comp}$: Luminance contrast similarity map with more complex computation
$Csim_{simple}$: Luminance contrast similarity map with simpler computation
Lsim & $Csim_{comp}$: Used Lsim and $Csim_{comp}$ separately
Lsim & $Csim_{simple}$: Used Lsim and $Csim_{simple}$ separately
Abs: similarity index without sign
sign: similarity index with positive and negative signs

P_{2000}: percentage of pixels exceeding 2000 cd/m^2
avlum: average luminance of entire scene
std_{window}: standard deviation of window luminance
$avlum_{40}$: average luminance of 40° horizontal band
R_{max}: maximum luminance in window / 200 cd/m^2
R_{CPU}: average luminance in window / 200 cd/m^2

	Lsim		**$Csim_{comp}$**		**$Csim_{simple}$**		**Lsim & $Csim_{comp}$**		**Lsim & $Csim_{simple}$**	
existence of sign	Abs	sign	Abs	sign	Abs	sign	Abs	sign	Abs	sign
2 zones	0.613	0.903	0.613	0.903	0.613	0.935	0.613	0.903	0.613	0.903
3 zones	0.613	0.935	0.613	0.903	0.613	0.968	0.613	0.968	0.613	0.935

other parameters	**P_{2000}**	**avlum**	**DGP**	**std_{window}**	**$avlum_{40}$**	**R_{max}**	**R_{CPU}**
-	0.935	0.935	0.903	0.903	0.935	0.839	0.968

Table 2: result of user #2 (train: 1472, test: 74)

	Lsim		**$Csim_{comp}$**		**$Csim_{simple}$**		**Lsim & $Csim_{comp}$**		**Lsim & $Csim_{simple}$**	
existence of sign	Abs	sign	Abs	sign	Abs	sign	Abs	sign	Abs	sign
2 zones	0.595	0.676	0.595	0.635	0.595	0.703	0.595	0.676	0.595	0.730
3 zones	0.622	0.676	0.595	0.689	0.595	0.689	0.608	0.689	0.608	0.716

other parameters	**P_{2000}**	**avlum**	**DGP**	**std_{window}**	**$avlum_{40}$**	**R_{max}**	**R_{CPU}**
-	0.689	0.730	0.595	0.662	0.662	0.608	0.716

Table 3: result of user #3 (train: 828, test: 41)

	Lsim		**$Csim_{comp}$**		**$Csim_{simple}$**		**Lsim & $Csim_{comp}$**		**Lsim & $Csim_{simple}$**	
existence of sign	Abs	sign	Abs	sign	Abs	sign	Abs	sign	Abs	sign
2 zones	0.585	0.634	0.585	0.683	0.585	0.659	0.585	0.732	0.585	0.707
3 zones	0.585	0.659	0.585	0.683	0.585	0.561	0.561	0.707	0.585	0.659

other parameters	**P_{2000}**	**avlum**	**DGP**	**std_{window}**	**$avlum_{40}$**	**R_{max}**	**R_{CPU}**
-	0.537	0.610	0.610	0.585	0.683	0.659	0.585

4. CONCLUSIONS

The objective of this study was to propose a method for assessing similarity degree of two luminance maps or luminance contrast maps generated from different visual situations. Reference raw HDR image data, which was collected by Raspi camera with a fisheye lens, was used. In addition to the new similarity concept, zoning logic according to the relative luminance level in each comparative luminance map data was applied to divide the captured scene into several zones to capture the occupant's visual information more clearly. To assess the feasibility of using proposed similarity concepts for explaining the user's visual preference, a classification model was generated using logistic regression with the generated luminance similarity and luminance contrast similarity maps and was compared to the model generated using selected reference parameters. For considering both fair comparison and the objective of similarity index maps, a zoning logic was applied to each comparative luminance maps. Average values in each divided zones of luminance and luminance contrast similarity maps were used for input parameters of the classification model. The result showed that the sign of the similarity index was important to classify a user's visual preference. Dividing each scene into 2 or 3 zones according to the relative luminance level was helpful to increase the classification performance. In addition, it was necessary to use both luminance and luminance contrast similarity maps separately not combining them into one map for maintaining important visual information. At last, in contrast to using existing parameters from references, the classification performance was more stable. Although it was possible to check a possibility of using new composite similarity maps with zoning logic to classify a user's visual preference, more complex model was necessary for using similarity index maps directly to capture the occupant's visual information in detail.

REFERENCES

Amir, S., Awalgaonkar, N. M., Karava, P., & Bilionis, I. (2017). A Bayesian modeling approach of human interactions with shading and electric lighting systems in private offices. *Energy & Buildings*, *134*, 185–201. https://doi.org/10.1016/j.enbuild.2016.10.046

Cauwerts, C. M. B. P. (2018). Application of High-Dynamic Range Imaging Techniques in Architecture : A Step toward High-Quality Daylit Interiors ? *Journal of Imaging*, *4*. https://doi.org/10.3390/jimaging4010019

EIA. (2012). 2012 Commercial Buildings Energy Consumption Survey: Energy Usage Summary.

Hirning, M. B., Isoardi, G. L., & Cowling, I. (2014). Discomfort glare in open plan green buildings. *Energy & Buildings*, *70*, 427–440. https://doi.org/10.1016/j.enbuild.2013.11.053

Hirning, M. B., Isoardi, G. L., & Garcia-hansen, V. R. (2017). Prediction of discomfort glare from windows under tropical skies. *Building and Environment*, *113*, 107–120. https://doi.org/10.1016/j.buildenv.2016.08.005

Inanici, M. N. (2006). Evaluation of high dynamic range photography as a luminance data acquisition system. *Lighting Research and Technology*, *38*(2), 123–136. https://doi.org/10.1191/1365782806li164oa

Konis, K. (2014). Predicting visual comfort in side-lit open-plan core zones : Results of a field study pairing high dynamic range images with subjective responses. *Energy & Buildings*, *77*, 67–79. https://doi.org/10.1016/j.enbuild.2014.03.035

Lindelöf, D., Morel, N., & Lindelo, D. (2008). Bayesian estimation of visual discomfort Bayesian estimation of visual discomfort. *Building Research & Information*, *3218*. https://doi.org/10.1080/09613210701544061

Liu, Y., Soo, K., Beresford, R. M., & Fleisher, D. H. (2020). Ecological Informatics A generic composite measure of similarity between geospatial variables. *Ecological Informatics*, *60*(October), 101–169. https://doi.org/10.1016/j.ecoinf.2020.101169

Palubinskas, G. (2017). Image similarity / distance measures : what is really behind MSE and SSIM ? *International Journal of Image and Data Fusion*, *9832*(December), 32–53. https://doi.org/10.1080/19479832.2016.1273259

Pierson, C. (2018). Daylight Discomfort Glare Evaluation with Evalglare : Influence of Parameters and Methods on the Accuracy of Discomfort Glare Prediction. *Buildings*, (July). https://doi.org/10.3390/buildings8080094

Wang, Z., Bovik, A. C., Sheikh, H. R., Member, S., Simoncelli, E. P., & Member, S. (2004). Image Quality Assessment : From Error Visibility to Structural Similarity. *IEEE Transactions on Image Processing*, *13*(4), 1–14.

Wymelenberg, K. Van Den, & Inanici, M. (2015). The journal of the Illuminating Engineering Society of North America Evaluating a New Suite of Luminance-Based Design Metrics for Predicting Human Visual Comfort in Offices with Daylight Evaluating a New Suite of Luminance-Based Design Metrics for Predict. *LEUKOS*,

2724(September). https://doi.org/10.1080/15502724.2015.1062392
Wymelenberg, K. Van Den, Inanici, M., & Johnson, P. (2010). The Effect of Luminance Distribution Patterns on Occupant Preference in a Daylit Office Environment. *LEUKOS*, *7*, 103–122. https://doi.org/10.1582/LEUKOS.2010.07.02003
Xiong, J., Awalgaonkar, N. M., Tzempelikos, A., Bilionis, I., Performance, H., Laboratories, R. W. H., … Lafayette, W. (2020). Efficient learning of personalized visual preferences in daylit offices : An online elicitation framework. *Building and Environment*, *181*(March), 107013. https://doi.org/10.1016/j.buildenv.2020.107013
Xiong, J., Tzempelikos, A., Bilionis, I., & Awalgaonkar, N. M. (2018). Inferring personalized visual satisfaction pro fi les in daylit o ffi ces from comparative preferences using a Bayesian approach. *Building and Environment*, *138*(April), 74–88. https://doi.org/10.1016/j.buildenv.2018.04.022
Xiong, J., Tzempelikos, A., Bilionis, I., Karava, P., Performance, H., Laboratories, R. W. H., … Lafayette, W. (2019). Energy & Buildings A personalized daylighting control approach to dynamically optimize visual satisfaction and lighting energy use. *Energy & Buildings*, *193*, 111–126. https://doi.org/10.1016/j.enbuild.2019.03.046

Hybrid modeling approach for better identification of building thermal network model and improved prediction

Sang woo Ham[1*], Donghun Kim[2]

[1] Building Technology & Urban System Division, Lawrence Berkeley National Laboratory, Berkeley, CA, United States of America
sham@lbl.gov

[2] Building Technology & Urban System Division, Lawrence Berkeley National Laboratory, Berkeley, CA, United States of America
donghunkim@lbl.gov

* Corresponding Author

ABSTRACT

The gray-box modeling approach (i.e., semi-physical thermal network model) has been widely used for prediction applications for buildings such as a model predictive control (MPC). However, applying the modeling approach for practical buildings is still challenging due to unmeasured disturbances such as occupants, lighting, appliances, and in/exfiltration loads. To overcome this problem, several system identification approaches have been proposed by considering the dynamics of unmeasured disturbance. However, the performance of long-term (e.g., one day) zone temperature or load predictions could still be very poor, and this is an important research topic for enabling grid-interactive buildings. In this study, we propose a hybrid modeling approach to improve long-term temperature or load predictions. Several system identification approaches for gray-box models are compared using simulations to understand the limitations. A neural network model that accounts for unmeasured disturbance is developed by considering the limitation of the gray-box model and is combined with the gray-box model. This hybrid model approach shows 0.24°C root mean squared error (RMSE) for 1-day ahead temperature prediction on average, while the conventional gray-box model shows 1.1°C RMSE on average.

1. INTRODUCTION

The gray-box modeling approach (i.e., semi-physical thermal network model) has been widely used for prediction applications for buildings such as model predictive control (MPC) because it has a flexible structure representing a complex building with a low order R-C (thermal resistance and capacitance) network model from measured data while keeping its physical principles. However, in real buildings, some information for the gray-box model such as internal heat gains from occupants, lighting, appliances, and in/exfiltration is generally not available due to the lack of sensors, and the retrofit of these sensors for the unmeasured disturbance is costly prohibited especially in small and medium-sized commercial buildings (SMCBs).

Accordingly, a conventional system identification for a real-building, which minimizes the sum of squared error between n-step ahead simulation and measurement (Braun and Chaturvedi, 2002), often results in significant bias in parameters, especially the unmeasured disturbance is not negligible (Kim et al., 2016). It could yield a critical problem in predictive application. For example, the unmeasured disturbance could be represented as unrealistic large thermal capacitance and solar heat gains in the conventional system identification, so the predictive controller's performance (e.g., load shift) may be wrongly executed.

Several system identification approaches (Coffman and Barooah, 2018; Kim et al., 2016) have been proposed to correctly estimate model parameters of a gray-box model by handling the internal heat gains as a lumped unmeasured disturbance. Although they showed better identification performance, the success of the identification highly depends on the quality of data (i.e., uncorrelated control inputs and measured/unmeasured disturbances)(Kim et al., 2018). Furthermore, the unmeasured disturbance information is not given for the future prediction, so the predictive performance is poor regardless of the accuracy of identified system parameters.

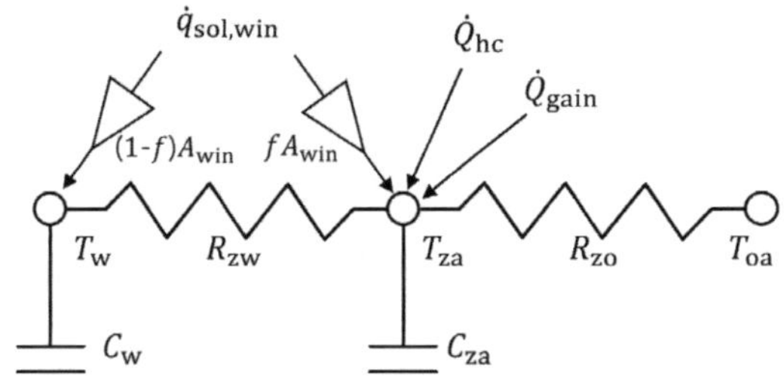

Figure 1: A simple RC network model for a case study building.

In building simulation, the unmeasured disturbances (i.e., occupant, appliance, and infiltration heat gains) are generated by time-based scheduled values in deterministic (ASHRAE, 2017) or stochastic ways (Hong et al., 2020). This approach is valid in a predictive application for white-box simulation because the goal is to estimate typical energy consumption based on typical unmeasured disturbance schedules for building design. However, the unmeasured disturbance needs to be estimated from the data in a real building in operation. While it can be easily obtained via a Kalman (Coffman and Barooah, 2018; O'Neill et al., 2010) or a particle filter (Ham et al., 2021) from the measured data, these methods do not provide a future prediction. Some studies (Dong et al., 2016; Lee and Zhang, 2021) developed a black-box model to predict non-HVAC energy consumption from historical data and use it to quantify unmeasured disturbances. One study models unmeasured disturbances as a function of time-related features (i.e., time of the day, day of the week, and day of the year) via feed-forward neural network (Ellis, 2021). The model is trained after estimating the disturbance via a Kalman filter or together with a building gray-box model. Though previous studies show good prediction results on their data, it is necessary to investigate more in-depth investigation on the impact of the gray-box model quality and the proper structure of the black-box model because the gray-box model quality is not guaranteed in the real building due to the lack of sensors and low data quality.

In this research, we propose a hybrid approach that integrates a neural network-based black-box model for unmeasured disturbances with a gray-box model for a predictive application. First, we provide a simulation case study to show how the conventional identification algorithm poorly behaves under unmeasured disturbances and propose and compare alternative identification algorithms to overcome this issue. After investigating the limitation of alternative identification algorithms in a predictive application, we propose a neural network model by considering the gray-box model structure. Finally, the impact of input features and model types on the performance is investigated with actual prediction performances.

2. COMPARISONS OF IDENTIFICATION APPROACHES UNDER UNMEASURED DISTURBANCE AND EFFECT ON PREDICTION: SIMULATION CASE STUDY

For real buildings, the various heat gain information (e.g., occupants, lighting, appliances, and in/exfiltration loads) is often not available. and it often leads to failure of the conventional system identification approach (CONV) (Braun and Chaturvedi, 2002), which minimizes a norm of errors between simulation and measurement, despite its popularity in the literature (Kim et al., 2016; Kim et al., 2018). Different identification approaches are needed to overcome this problem. We provide a simulation case study to show how the conventional identification algorithm poorly behaves under unmeasured disturbances and propose and compare alternative identification algorithms to overcome this issue. The proposed algorithms model not only the thermal dynamics of a building envelope system but also the dynamics of unmeasured disturbances in the form of either a lumped input disturbance (ID) (Coffman and Barooah, 2018) or output disturbance (OD) (Kim et al., 2016; Kim et al., 2018).

2.1 True system description and data generation

It is very difficult to compare the performance of identification approaches for a real building since we do not know the true system dynamics or parameters. Instead, we created a hypothetical building envelope model and treated it as the *True model* for this study. The True model was tuned with a data set that was generated from a laboratory environment (FLEXLAB) (Lawrence Berkeley National Laboratory, 2021). The laboratory represents a single-zone office space. The True model has the RC network shown in Fig. 1 and the state-space form of Eq. 1 (state transition process and measurement process).

$$\underbrace{\begin{bmatrix}\dot{T}_{\mathrm{w}}\\ \dot{T}_{\mathrm{za}}\end{bmatrix}}_{\dot{\mathbf{x}}}=\underbrace{\begin{bmatrix}\frac{-1}{C_{\mathrm{w}}R_{\mathrm{zw}}} & \frac{1}{C_{\mathrm{w}}R_{\mathrm{zw}}}\\ \frac{-1}{C_{\mathrm{za}}R_{\mathrm{zw}}} & \frac{-1}{C_{\mathrm{za}}R_{\mathrm{zw}}}+\frac{-1}{C_{\mathrm{za}}R_{\mathrm{zo}}}\end{bmatrix}}_{\mathbf{A}}\underbrace{\begin{bmatrix}T_{\mathrm{w}}\\ T_{\mathrm{za}}\end{bmatrix}}_{\mathbf{x}}+\underbrace{\begin{bmatrix}0 & \frac{(1-f)A_{\mathrm{win}}}{C_{\mathrm{w}}} & 0 & 0\\ \frac{1}{C_{\mathrm{za}}R_{\mathrm{zo}}} & \frac{fA_{\mathrm{win}}}{C_{\mathrm{za}}} & \frac{1}{C_{\mathrm{za}}} & \frac{1}{C_{\mathrm{za}}}\end{bmatrix}}_{\mathbf{B}}\underbrace{\begin{bmatrix}T_{\mathrm{oa}}\\ \dot{q}_{\mathrm{sol,win}}\\ \dot{Q}_{\mathrm{hc}}\\ \dot{Q}_{\mathrm{gain}}\end{bmatrix}}_{\mathbf{u}} \tag{1}$$

$$y_{\mathrm{za}}=\underbrace{\begin{bmatrix}0 & 1\end{bmatrix}}_{\mathbf{C}}\mathbf{x}$$

where T_{w}, T_{za}, and T_{oa} are temperature nodes of large thermal mass (w), zone air (za), and outdoor air (oa), R_{zw} and R_{zo} are thermal resistances between temperature nodes [K/kW], C_{w} and C_{za} are thermal capacitances [kWh/K], $\dot{q}_{\mathrm{sol,win}}$ is incident solar radiation per area on windows [kW/m^2], A_{win} is effective window area [m^2], f is convective fraction of the incident solar radiation, $\dot{Q}_{\mathrm{hc}}$ is heating/cooling rate from maximum heating/cooling rate ($\dot{Q}_{\mathrm{heat,max}}/\dot{Q}_{\mathrm{cool,max}}$) times signal ($i_{\mathrm{heat}}/i_{\mathrm{cool}}$) [kW], $\dot{Q}_{\mathrm{gain}}$ is internal heat gain (e.g., summation of plug, lighting, and occupancy loads) [kW], and y_{za} is measured zone air temperature [°C].

The parameters to be estimated through the system identification are $\theta = [C_{\mathrm{w}}, C_{\mathrm{za}}, R_{\mathrm{zw}}, R_{\mathrm{zo}}, f, A_{\mathrm{win}}]$ and were tuned with the complete measurements including $\dot{Q}_{\mathrm{gain}}$. The values of the True model were set to C_{w}: 4.0 kWh/K, C_{za}: 1.0kWh/K, R_{zw}: 1.2 K/kW, R_{zo}: 9 K/kW, f: 0.3, and A_{win}: 3.0 m^2. The maximum heating and cooling rates ($\dot{Q}_{\mathrm{heat,max}}$ and $\dot{Q}_{\mathrm{cool,max}}$) are set to 6kW and −6kW, respectively. Many SMCBs are controlled by a on/off thermostat controller with its own thermostat setting such as minimum cycling time. To simulate this behavior, an ideal on-off heating/cooling controller that sending heating and cooling operation signal (i_{heat} and i_{cool}, respectively) with minimum cycling time of 5-minute is implemented. Then, the data is resampled 15-minute interval through centered moving average.

To test the performance of the identification algorithms of the CONV, ID and OD under a practical scenario, $\dot{Q}_{\mathrm{gain}}$ is assumed to be unknown. Instead, it is assumed that we can design experiments and actively control indoor temperature setpoint (within an acceptable room air temperature band) for the purpose of better system identification. Two weeks of data were generated from the True model with Oakland, CA, weather data. During the weekdays, cooling and heating setpoints are set to 25°C and 21°C for occupied times (6 – 18 hour) and 30°C and 16°C for the unoccupied time. However, the setpoint of the first weekend (2 days) was perturbed according to a pseudo-binary random signal (PRBS) with 2-hour of time-scale and 4th order. The binary signal is mapped to sampled setpoints between 18 – 26°C. The generated data set is visualized in Fig. 2.

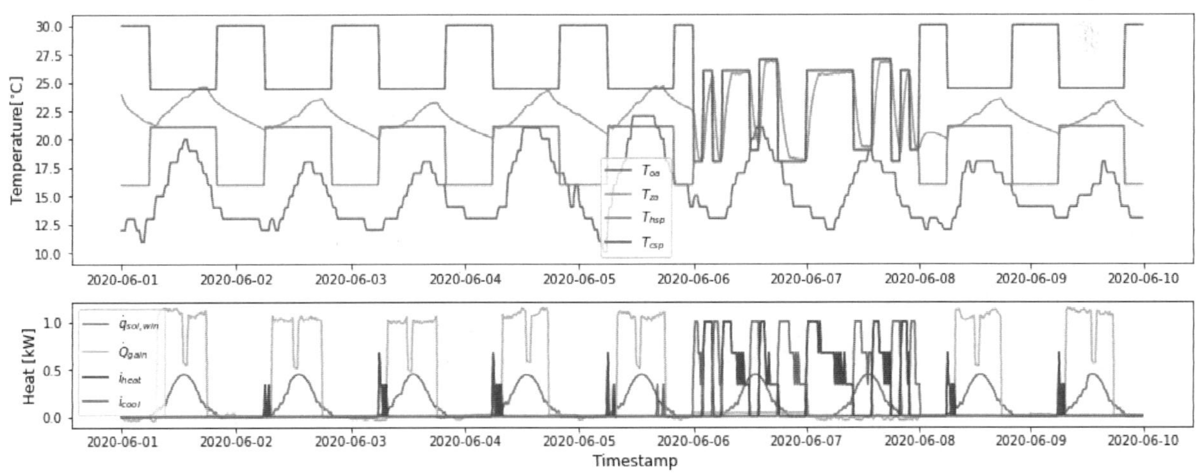

Figure 2: Temperatures and measured/unmeasured disturbances/resampled control signals of simulated data.

2.2 Summary of system identification approaches

Eq. 1 was discretized in a 15-min sampling time by using the zero-order hold (Rouchier et al., 2019). Three different system identification approaches (CONV, ID, and OD) were tested with the test data set aiming at identifying the True model. They have the same model structure for the building envelope dynamics, although their disturbance model structures differ. Note that the unmeasured disturbance ($\dot{Q}_{\text{gain}}$) was not provided to the identification algorithms. In other words, the identified model G maps only the measured disturbances (T_{oa}, $\dot{q}_{\text{sol,win}}$) and control input ($\dot{Q}_{\text{hc}}$) to the indoor air temperature (y_{za}). In the following paragraphs, we will introduce the basic concept of each identification approach because of the lack of space.

Conventional simulation error minimization approach: The CONV approach (Braun and Chaturvedi, 2002) assumes that all unmeasured disturbances can be expressed as white noise (e_{conv}) in the measurement process. The discretized system can be written as Eq. (2);

$$\begin{aligned} \mathbf{x}(k+1) &= \mathbf{A}_{\text{d}}\mathbf{x}(k) + \mathbf{B}_{\text{d}}\mathbf{u}(k) \\ y(k) &= \mathbf{C}_{\text{d}}\mathbf{x}(k) + e_{\text{conv}}(k) \end{aligned} \tag{2}$$

where $\mathbf{A}_{\text{d}}$, $\mathbf{B}_{\text{d}}$, and $\mathbf{C}_{\text{d}}$ from $\mathbf{A}$, $\mathbf{B}$, and $\mathbf{C}$ in Eq. (1).

The set of parameters (θ^*_{conv}) is estimated by minimizing the sum of squared errors between simulation ($\hat{y}(k;\theta) = \mathbf{C}_{\text{d}}\hat{\mathbf{x}}(k;\theta)$) and measurement ($y(k)$) via the nonlinear optimization (Eq. 3);

$$\begin{aligned} \hat{\mathbf{x}}(k+1;\theta) &= \mathbf{A}_{\text{d}}(\theta)\hat{\mathbf{x}}(k;\theta) + \mathbf{B}_{\text{d}}(\theta)\mathbf{u}(k) \\ y(k) &= \mathbf{C}_{\text{d}}\hat{\mathbf{x}}(k;\theta) + \varepsilon_{\text{conv}}(k;\theta) \\ \theta^*_{\text{conv}} &= \arg\min_{\theta} \sum_{k=1}^{N} \left(\varepsilon_{\text{conv}}(k;\theta)\right)^2, \end{aligned} \tag{3}$$

where subscript d indicates a discretized system and, k is a discretized time index.

Input disturbance identification approach: The ID approach assumes that unmeasured disturbances come from the input channel, i.e., the heat gain term, and treats the input disturbance as an additional dynamic state. This can be written as an augmented state space form (Eq. 4) (Coffman and Barooah, 2018).

$$\begin{aligned} \underbrace{\begin{bmatrix} \dot{\mathbf{x}} \\ \dot{\zeta}_{\text{ID}} \end{bmatrix}}_{\dot{\mathbf{x}}_{\text{ID}}} &= \underbrace{\begin{bmatrix} \mathbf{A} & \mathbf{A}_{\zeta_{\text{ID}}} \\ \mathbf{0} & \mathbf{0} \end{bmatrix}}_{\mathbf{A}_{\text{ID}}} \underbrace{\begin{bmatrix} \mathbf{x} \\ \zeta_{\text{ID}} \end{bmatrix}}_{\mathbf{x}_{\text{ID}}} + \mathbf{B}\mathbf{u} + \mathbf{w}_{\text{ID}}, \text{ and } \mathbf{A}_{\zeta_{\text{ID}}} = \begin{bmatrix} \mathbf{0} \\ \frac{1}{C_{za}} \end{bmatrix} \\ y_{\text{za}} &= \underbrace{\begin{bmatrix} \mathbf{C} & 0 \end{bmatrix}}_{\mathbf{C}_{\text{ID}}} \mathbf{x}_{\text{ID}} + e_{\text{ID}} \end{aligned} \tag{4}$$

where $\mathbf{w}_{\text{ID}}$ and e_{ID} are state and measurement noises. ζ_{ID} represents the lumped input disturbance term.

The key idea in this approach is to treat ζ_{ID} as Wiener process, so therefore , it behaves as the Brownian motion after discretization according to its noise level (i.e., $\zeta_{\text{ID}}(k+1) = \zeta_{\text{ID}}(k) + w_{\zeta_{\text{ID}}}(k)$). For the system identification, the ID approach calculates one-step ahead prediction errors after filtering states via Kalman filter at each time. More details can be found in Coffman and Barooah, 2018.

Output disturbance identification approach: The OD approach (Kim et al., 2016; Kim et al., 2018) models the effect of unmeasured heat gain on the output (i.e., room air temperature) because the output more smoothly changes compared to the input heat gain. The aggregated contribution of the unknown heat sources to the output is called the output disturbance. The OD approach models the output disturbance as a filtered process of white noise ($e_{\text{OD}}(k)$), which is called output disturbance ($v_{\text{OD}}(k)$ in Eq. 5). The output disturbance dynamics can be modeled with two more parameters, ρ_1 and ρ_2 (Eq. 5).

$$\begin{aligned} \mathbf{x}(k+1) &= \mathbf{A}_{\text{d}}\mathbf{x}(k) + \mathbf{B}_{\text{d}}\mathbf{u}(k) \\ y(k) &= \mathbf{C}_{\text{d}}\mathbf{x}(k) + v_{\text{OD}}(k) \\ \zeta_{\text{OD}}(k+1) &= \left[\rho_1\right]\zeta_{\text{OD}}(k) + \left[\rho_2\right]e_{\text{OD}}(k) \\ v_{\text{OD}}(k) &= \zeta_{\text{OD}}(k) + e_{\text{OD}}(k) \end{aligned} \tag{5}$$

The OD approach estimates a set of parameters by minimizing the square sum of one-step ahead prediction error. More details for the identification algorithm can be found in Kim et al., 2016.

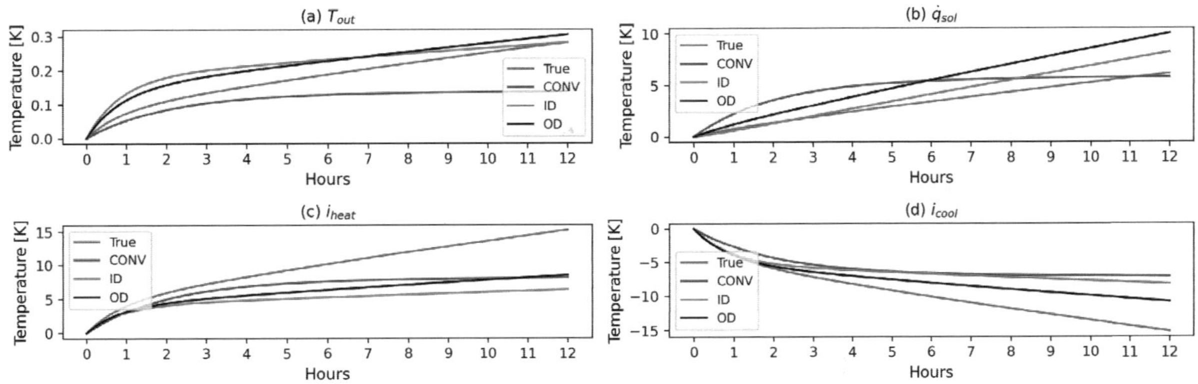

Figure 3: Comparison of step response of measured disturbances (T_{oa} and $\dot{q}_{\mathrm{sol,win}}$) and control input ($\dot{Q}_{\mathrm{hc}}$) for each method.

2.3 System identification results and discussion

For each identification algorithm, the optimization problem is non-convex, and hence to find a better optimal solution, we randomly sampled initial starting points and repeatedly solved the optimization problems 50 times. The estimation results of CONV, ID, OD identification approaches are summarized in Table. 1.

Table 1: Comparison of estimated parameters from system identification approaches

	C_{w} $[\frac{\mathrm{kWh}}{\mathrm{K}}]$	C_{za} $[\frac{\mathrm{kWh}}{\mathrm{K}}]$	R_{zw} $[\frac{\mathrm{K}}{\mathrm{kW}}]$	R_{zo} $[\frac{\mathrm{K}}{\mathrm{kW}}]$	f [-]	A_{win} $[\mathrm{m}^2]$	$\dot{Q}_{\mathrm{heat,max}}$ [kW]	$\dot{Q}_{\mathrm{cool,max}}$ [kW]
True	4.0	1.0	1.2	9.0	0.3	3.0	6.0	-6.0
CONV	40	0.38	6.17	40.0	1.0	1.06	1.48	-1.3
ID	20.63	1.37	0.81	3.37	0.03	20.0	6.7	-8.7
OD	10.58	1.33	0.84	4.14	0.15	12.37	6.72	-8.6

Overall, the estimated R, C, and $\dot{Q}$ values of ID and OD have realistic scale compared to CONV, but all methods wrongly estimate solar-related parameters (i.e., f and A_{win}). A building thermal dynamics is the combination of all the inputs and current state information, so some of wrong parameters are affecting the long-term thermal dynamics of the building. Fig. 3 shows the comparison of step response of measured disturbances and control inputs for each method. While all three methods fail to capture long-term responses, ID and OD show better performance overall, especially for the short-term period. For example, ID and OD capture a short-term dynamics of cooling control input (i_{cool}) for 2 hours.

2.4 Challenges in prediction application

In a predictive application such as model predictive control (MPC), the performance of prediction is critically important to correctly respond to the current grid's status (e.g., price signal). However, it is difficult for a building when the unmeasured disturbances are significant, even if we have a perfect gray-box model. For example, the predicted temperature profiles for daytime are much lower than the measurement if there is no credible internal heat gain prediction. This results in a wrong prediction for the cooling period. This section provides a more detailed simulation study that shows this issue and motivates developing an additional model, namely the hybrid modeling approach described in the following section, to overcome this problem.

In Fig. 4, the 1-day ahead, room air temperature predictions of the True, CONV, ID, and OD models are compared with the measurement. Like before, the heat gain ($\dot{Q}_{\mathrm{gain}}$, marked with the red line in the bottom figure) information was not given to all models. Not surprisingly, all methods show large deviations from the measurement in future prediction, even for the True model. This is because the True model did not receive the heat gains while the physical building did. This wrong prediction is problematic when the predictive application targets daily demand management, such as load shift. Therefore, in addition to the correct parameter estimation, it is also essential to forecast the unmeasured disturbances for the predictive application.

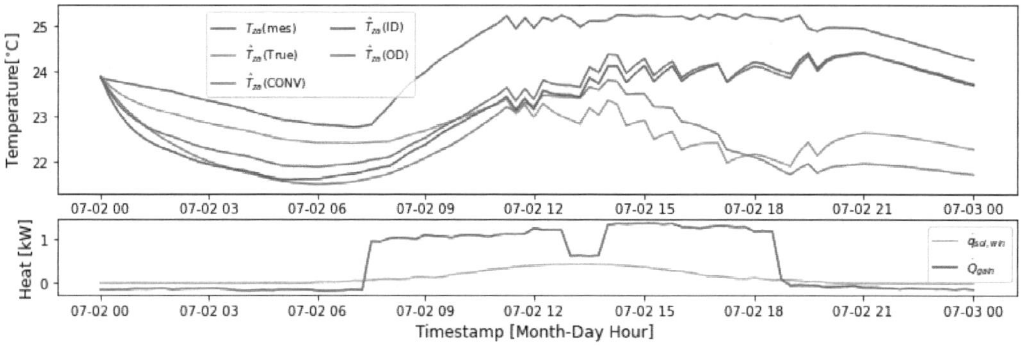

Figure 4: 1-day ahead temperature predictions of different approaches without unmeasured disturbances.

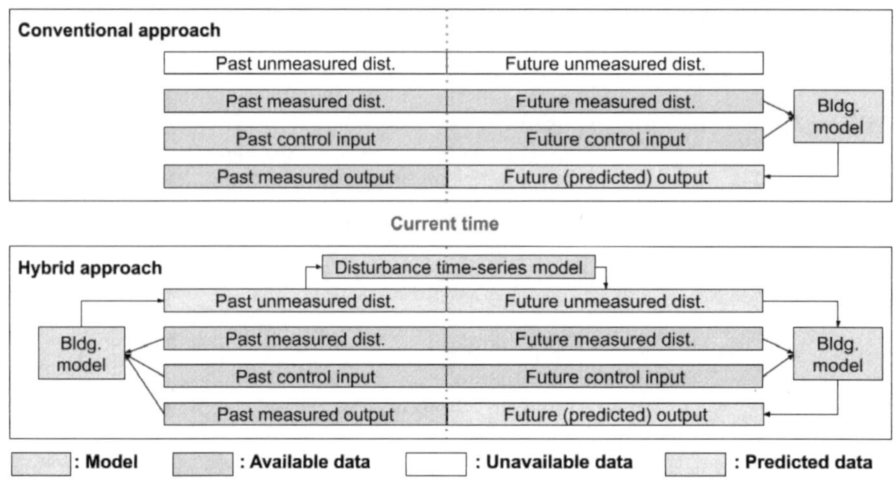

Figure 5: Comparison between Conventional and Hybrid approaches.

3. HYBRID MODELING APPROACH FOR PREDICTIVE APPLICATION

3.1 Overview of Hybrid modeling approach

In this section, we introduce a hybrid modeling approach that combines the OD system identification with a method of forecasting unmeasured disturbances for predictive applications. Fig. 5 shows the schematic diagram of the proposed Hybrid approach compared to the Conventional approach. As shown in Figures 5 and 6 (a), a building thermal model is a transfer function between the inputs (control inputs, measured disturbances, and unmeasured disturbances) and the output (temperature).

In Conventional approach, the building model is used to predict future output (i.e., temperature) by using the future measured disturbance (i.e., forecast of T_{oa} and $\dot{q}_{\mathrm{sol,win}}$) and future control inputs ($\dot{Q}_{\mathrm{hc}}$). However, as discussed in section 2.3 (see Fig. 3), this approach might not provide reliable prediction when the absence of unmeasured disturbances is not negligible.

On the other hand, the Hybrid approach attempts to include future unmeasured disturbance information through a disturbance model. The key steps are as follows. First, a gray-box model is trained through OD system identification. Then, we estimate the past unmeasured input disturbance ($\hat{\zeta}_{\mathrm{ID}}$) at each sampling time by utilizing the ID model (Eq. 5). Once the unmeasured input disturbance is estimated, then we perform the time series modeling explicitly, e.g., via machine learning techniques, to forecast the unmeasured disturbances. In this case, we use unmeasured input disturbance because the prediction errors on the input disturbance are filtered through the gray-box model on temperature prediction, while the prediction errors of the unmeasured output disturbance directly go to the temperature prediction.

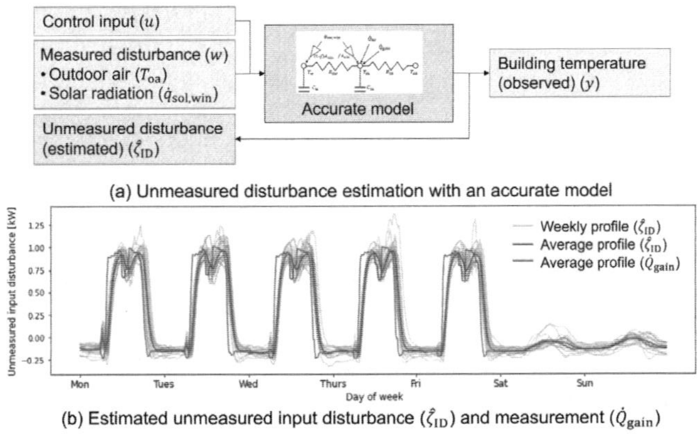

Figure 6: Unmeasured disturbance estimation with the accurate model.

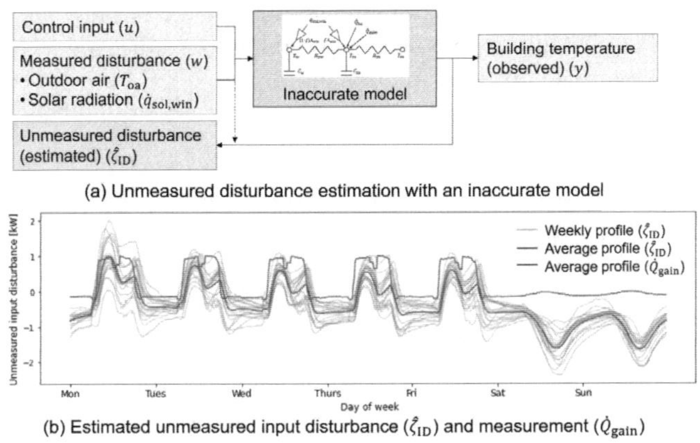

Figure 7: Unmeasured disturbance estimation with the inaccurate model.

3.2 Design of Hybrid model structure

The unmeasured disturbances are mainly composed of internal heat gains from occupants, appliances, and plug loads, and it is typically modeled as a function of daily/weekly schedules (Ellis, 2021; Hong et al., 2020). It is a valid approach when the gray-box model is accurate.

Fig. 6 shows the schematic diagram of unmeasured disturbance estimation ($\hat{\zeta}_{ID}$) with an accurate (True) model (a), and the average of estimated weekly profiles are compared with the average of measured values ($\dot{Q}_{gain}$) (b). Since the model is the True model, the estimation shows good agreement with the true profile.

On the other hand, Fig. 7 shows the schematic diagram of unmeasured disturbance estimation with an inaccurate model (a), and the average of estimated weekly profiles are compared with the average of the measured values ($\dot{Q}_{gain}$) (b). Although the unmeasured disturbance with an inaccurate model has a weekly pattern, it deviates from the measured values. In addition, each week's estimation has more variations compared to the accurate model case (Fig. 6 (a)), meaning it has some more dynamics in addition to the weekly pattern. As shown in Fig. 7 (a), the estimated unmeasured disturbance could be affected by control input and measured disturbance because the system model is not accurate.

From this investigation, a design matrix of input features and model types is established as shown in 8. Based on the design matrix, we develop several neural network models and evaluate the performance. Input features are composed of three components: time, pattern, and measured disturbance. Time (e.g., hour of week, etc.) and pattern features (past n days of $\hat{\zeta}_{ID}$) are typically used in the time-series modeling. It explains the time-specific and autoregressive pattern characteristics of time-series data. In addition, we also include measured disturbance (w) as shown in Fig. 7.

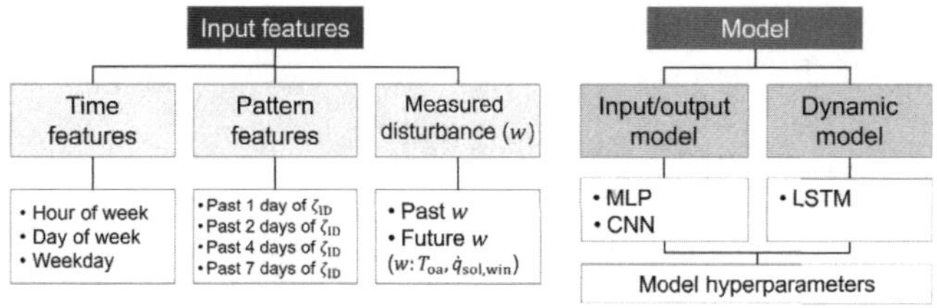

Figure 8: Design matrix of input features and model types for the unmeasured disturbance model.

Three types of models are used in this study: Multi-Layer Perceptron (MLP), Convolution Neural Network (CNN), and Long short-term memory (LSTM). MLP and CNN have input-output structure that maps input time-series vector ($\mathbf{x} \in \mathbb{R}^{(n_x n_{k,x})}$) to output time-series vector ($\mathbf{y} \in \mathbb{R}^{n_{k,y}}$) (i.e., feed-forward network). Input time-series vector is 1-dimensional vector concatenated by input features for a given input time-period ($n_{k,x}$), and output time-series vector is 1-dimensional vector of estimated unmeasured disturbance for the prediction time-horizon ($n_{k,y}$). On the other hand, LSTM is a dynamical model that maps input features ($\mathbf{x}_k \in \mathbb{R}^{n_x}$) to output ($\mathbf{y}_k \in \mathbb{R}^1$) in each time-step ($k$) (i.e., feed-back network). All the three models are commonly used in deep-learning, so we simply summarize the structures below due to the lack of space. MLP is a fully connected feed-forward neural network (Murphy, 2022). The output time-series vector ($\mathbf{y}$) is modeled from input time-series vector ($\mathbf{x}$) through n_{layer} numbers of hidden linear layers with activation function (φ) and dropout layer (Eq. 6).

$$\begin{aligned} &\mathbf{z}_0 = \text{dropout}(\varphi(\mathbf{W}_0\mathbf{x} + \mathbf{b}_0)) \\ &\text{for } i \text{ in } 1 : n_{\text{layer}} : \\ &\quad \mathbf{z}_i = \text{dropout}(\varphi(\mathbf{W}_i\mathbf{z}_{i-1} + \mathbf{b}_i)) \\ &\mathbf{y} = \mathbf{W}_y\mathbf{z}_{n_{\text{layer}}} + \mathbf{b}_y \end{aligned} \tag{6}$$

where $\mathbf{W}_i$ and $\mathbf{b}_i$ are weights and bias parameters that maps hidden layers from $\mathbf{z}_{i-1}$ to $\mathbf{z}_i$ is hidden. φ is activation function, dropout is a dropout layer, which is commonly used to prevent from over-fitting (Murphy, 2022).

CNN is one type of neural network that maps input-output data and is widely used for image classification problems because it extracts meaningful features from raw data through the convolution kernel (Murphy, 2022). The structure is similar to MLP, but the convolution filter and max-pooling layer are used instead of the linear layer, and one more linear layer is for the last convolution layer (Eq. 7).

$$\begin{aligned} &\mathbf{z}_0 = \varphi(\text{maxpool}(\text{conv}(\mathbf{x}))) \\ &\text{for } i \text{ in } 1 : n_{\text{layer}} : \\ &\quad \mathbf{z}_i = \varphi(\text{maxpool}(\text{conv}(\mathbf{z}_{i-1}))) \\ &\mathbf{y} = \mathbf{W}_y(\text{dropout}(\varphi(\mathbf{W}_{n_{\text{layer}}+1}\mathbf{z}_{n_{\text{layer}}} + \mathbf{b}_{n_{\text{layer}}+1}))) + \mathbf{b}_y \end{aligned} \tag{7}$$

where conv is convolution filter and maxpool is max-pooling layer.

LSTM is one type of recurrent neural network (RNN) using LSTM cell that maps input ($\mathbf{x}_k \in \mathbb{R}^{n_x}$) to output ($\mathbf{y}_k \in \mathbb{R}^1$) in each time ($k$) (Eq. 8) (Murphy, 2022). There are several types of cells for RNN, but LSTM, which includes both short- and long-term memory states, shows a good performance for several time-series tasks.

$$\begin{aligned} &\text{for } i \text{ in } 1 : (n_{k,x} + n_{k,y}) : \\ &\quad \mathbf{h}_{i+1}, \mathbf{c}_{i+1}, = \text{LSTM}(\mathbf{h}_i, \mathbf{c}_i, \mathbf{x}_i) \\ &\text{for } i \text{ in } 1 : (n_{k,y}) : \\ &\quad \mathbf{y}_i = \mathbf{W}_y(\text{dropout}(\varphi(\mathbf{W}_h\mathbf{h}_i + \mathbf{b}_h))) + \mathbf{b}_y \end{aligned} \tag{8}$$

where $\mathbf{h}$ and $\mathbf{c}$ are hidden and cell state vector that holds short- and long-term memory, respectively. $\mathbf{x}_i$ is subset of $\left(\text{time}_i, w_i, \hat{\zeta}_{\text{ID},i}\right)$, and $\hat{y}_{(n_{k,x}+1):(n_{k,x}+n_{k,y})}$ are used to for $\hat{\zeta}_{(n_{k,x}+1):(n_{k,x}+n_{k,y})}$.

For all three models, there are several hyperparameters such as types of activation function, sizes and number of hidden layers, sizes of convolution filter and pooling layers, etc. Model training is conducted on various combinations

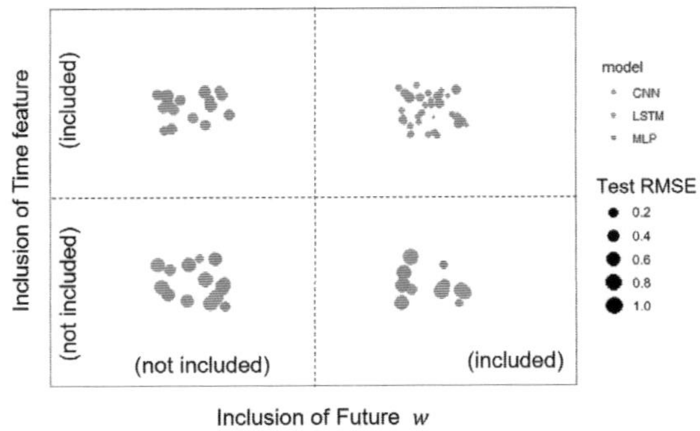

(a) Test RMSE vs. model and input features

Test RMSE=β_0+$\beta_1$time+$\beta_2$future w+β_3CNN+β_4LSTM

Variable	β	p-value
Intercept	0.532	2e-16***
time	-0.243	2e-16***
future w	-0.091	1e-4***
CNN	-0.037	0.128
LSTM	0.0067	0.807

(b) Regression of test RMSE vs. model and input features

Figure 9: Impact of time feature and future unmeasured disturbance on test RMSE (test RMSE data is scattered in each box for readability without any meaning).

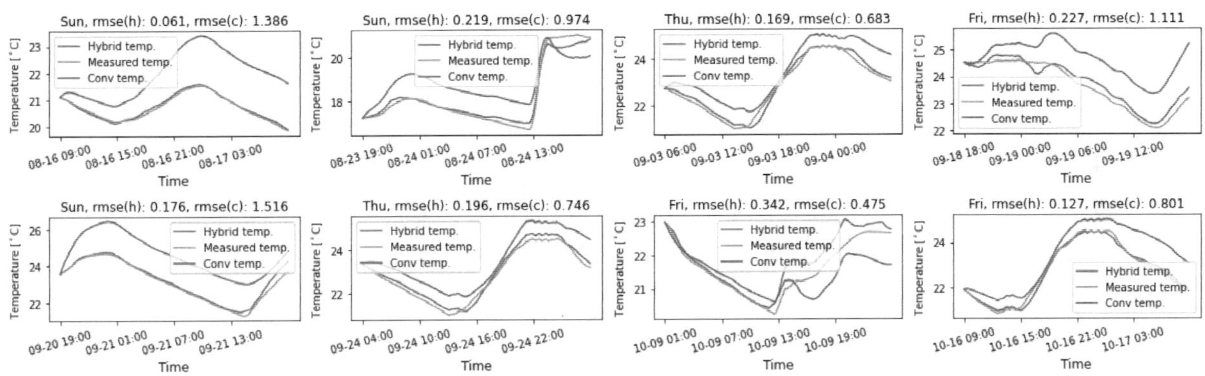

Figure 10: One-day ahead temperature prediction of Hybrid and Conventional approaches on test data (h: Hybrid, c: Conventional).

of hyperparameters, and the best hyper-parameter sets that shows the minimum RMSE on test data is chosen. The details are omitted due to the lack of the space. All the models are developed using PyTorch (Paszke et al., 2019). One month of data (June, 2020) is used for training, and the following four months of data (July-Oct, 2020) is used for the test. All the data is resampled in hourly interval and then scaled by standardization (i.e., z-score normalization).

4. RESULT AND CONCLUSION

One of the main contributions of this study is to propose a Hybrid modeling approach in consideration of the limitation of the gray-box system identification. To validate the design approach proposed in section 3.2, we compare the test RMSE values of various models of input features and model structures. Specifically, the impact of time feature, future measured disturbance (w), and model type on test RMSE is tested using linear regression (Fig. 9 (b)) and visualized in Fig. 9 (a). The inclusion of any time features and future w clearly shows the better performance, and it is statistically significant (p-value<0.05). However, the model types do not show the statistical differences.

Since there are no significant differences between models that include time feature and future w, we choose one specific model (LSTM with weekday time feature, 2 days historic ζ_{ID}, and both past and future w) that shows consistent performances over different data and practical issues (e.g., the required amount of historic data, number of parameters, etc.). The one-day ahead predictions of the Hybrid approach on test data are randomly selected and visualized in Fig. 10. The Hybrid (h) approach shows better prediction than Conventional (c) approach, and the average RMSE of Hybrid approach is 0.24°C while Conventional approach shows 1.1°C.

In this study, we propose the Hybrid approach that overcomes the gray-box model in a predictive application for real buildings. Throughout the investigation of the various system identification approaches and their limitations, we design a neural network model that predicts unmeasured disturbance. The Hybrid approach shows better performance in temperature prediction on the test data. In the future study, we will investigate the performance of the proposed model in various real data with hyperparameter sensitivity analysis. Also, it is necessary to be discussed how to make this approach in scale over spaces (i.e., a limited number of data) and time (i.e., different seasons). One practical solution is to provide a guideline to deploy the model, but it is worth testing the applicability of Bayesian model update or Transfer learning approach.

REFERENCES

ASHRAE. (2017). 2017 ASHRAE handbook: Fundamentals, chapter 18 SI: Nonresidential cooling and heating load calculations.

Braun, J., & Chaturvedi, N. (2002). An inverse gray-box model for transient building load prediction. *HVAC&R Research*, *8*(1), 73–99. https://doi.org/10.1080/10789669.2002.10391290

Coffman, A. R., & Barooah, P. (2018). Simultaneous identification of dynamic model and occupant-induced disturbance for commercial buildings. *Building and Environment*, *128*, 153–160. https://doi.org/10.1016/j.buildenv.2017.10.020

Dong, B., Li, Z., Rahman, S. M., & Vega, R. (2016). A hybrid model approach for forecasting future residential electricity consumption. *Energy and Buildings*, *117*, 341–351. https://doi.org/10.1016/j.enbuild.2015.09.033

Ellis, M. J. (2021). Machine learning enhanced grey-box modeling for building thermal modeling. *2021 American Control Conference (ACC)*, 3927–3932. https://doi.org/10.23919/ACC50511.2021.9482715

Ham, S. W., Karava, P., Bilionis, I., & Braun, J. (2021). Real-time model for unit-level heating and cooling energy prediction in multi-family residential housing. *Journal of Building Performance Simulation*, *14*(4), 420–445. https://doi.org/10.1080/19401493.2021.1968495

Hong, T., Macumber, D., Li, H., Fleming, K., & Wang, Z. (2020). Generation and representation of synthetic smart meter data. *Building Simulation*, *13*(6), 1205–1220. https://doi.org/10.1007/s12273-020-0661-y

Kim, D., Cai, J., Ariyur, K. B., & Braun, J. E. (2016). System identification for building thermal systems under the presence of unmeasured disturbances in closed loop operation: Lumped disturbance modeling approach. *Building and Environment*, *107*, 169–180. https://doi.org/10.1016/j.buildenv.2016.07.007

Kim, D., Cai, J., Braun, J. E., & Ariyur, K. B. (2018). System identification for building thermal systems under the presence of unmeasured disturbances in closed loop operation: Theoretical analysis and application. *Energy and Buildings*, *167*, 359–369. https://doi.org/10.1016/j.enbuild.2017.12.007

Lawrence Berkeley National Laboratory. (2021). *FLEXLAB the world's most advanced integrated building and grid technologies testbed.* Retrieved December 24, 2021, from https://flexlab.lbl.gov/

Lee, Z. E., & Zhang, K. M. (2021). Scalable identification and control of residential heat pumps: A minimal hardware approach. *Applied Energy*, *286*, 116544. https://doi.org/10.1016/j.apenergy.2021.116544

Murphy, K. P. (2022). *Probabilistic machine learning: An introduction.* MIT Press. probml.ai

O'Neill, Z., Narayanan, S., & Brahme, R. (2010). Model-based thermal load estimation in buildings. *Proceedings of SimBuild*, *4*(1), 474–481.

Paszke, A., Gross, S., Massa, F., Lerer, A., Bradbury, J., Chanan, G., Killeen, T., Lin, Z., Gimelshein, N., Antiga, L., Desmaison, A., Kopf, A., Yang, E., DeVito, Z., Raison, M., Tejani, A., Chilamkurthy, S., Steiner, B., Fang, L., … Chintala, S. (2019). PyTorch: An imperative style, high-performance deep learning library. In H. Wallach, H. Larochelle, A. Beygelzimer, F. d. Alché-Buc, E. Fox, & R. Garnett (Eds.), *Advances in neural information processing systems*. Curran Associates, Inc. https://proceedings.neurips.cc/paper/2019/file/bdbca288fee7f92f2bfa9f7012727740-Paper.pdf

Rouchier, S., Jiménez, M. J., & Castaño, S. (2019). Sequential monte carlo for on-line parameter estimation of a lumped building energy model. *Energy and Buildings*, *187*, 86–94. https://doi.org/10.1016/j.enbuild.2019.01.045

ACKNOWLEDGMENT

The work was supported by the California Energy Commission-sponsored project entitled "HP-Flex: Next Generation Heat Pump Load Flexibility" and DOE BTO sponsored CRADA project entitled "A Low-cost, Scalable Control Solution for Grid-Interactive Small and Medium Sized Commercial Buildings".

Fault Detection and Diagnostic Method Based on Evolving Data-driven Model for Radiant Floor Heating and Cooling Systems

Sujit DAHAL[1]*, Liping WANG[1], James BRAUN[2]

[1]University of Wyoming, Department of Civil and Architectural Engineering,
Laramie, WY, USA

[2]Purdue University, School of Mechanical Engineering,
West Lafayette, IN, USA

* Corresponding Author: email: sdahal@uwyo.edu, Tel: (307) 223-6025

ABSTRACT

Faults in components (valves, sensors, etc.) of radiant floor heating and cooling systems affect the efficiency, cooling and heating capacity as well as the reliability of the system. While various fault detection and diagnostic (FDD) methods have been developed and tested for building systems, FDD algorithms for radiant heating and cooling systems have previously not been available. This paper presents an evolving learning-based FDD approach for a radiant floor heating and cooling system based on growing Gaussian mixture regression (GGMR). The experimental space was controlled with a building automation system (BAS) in which the operating conditions can be monitored, and control parameters can be overridden to desired values. Trend data for normal operation and faulty operation were collected. A total of six fault types with different severities in a radiant floor system were emulated through overriding control parameters.

An FDD model based on the GGMR approach was developed with training data and the performance of the model was tested for "known" faults that were including in the training and new "unknown" faults that were implemented in the fault testing. The prediction accuracy for each known fault was extremely high with the lowest prediction accuracy of 98% for one of the faults. The algorithm was successful in detecting the new fault as an unknown state before evolving the model and in diagnosing it as a new fault after evolving the model.

1. INTRODUCTION

Hydronic radiant systems have been adopted as an alternative to conventional HVAC systems for achieving higher thermal comfort and energy savings (Fabrizio et al., 2012; Joe & Karava, 2019). Heating or cooling in a radiant system is provided by circulating water through pipes installed within the structure (floors, ceilings, beams, *etc.*). Hydronic radiant heating and cooling systems consist of several components, including hot and cold-water sources, valves controlling water flow, pumps, temperature sensors, flow sensors, pressure sensors, thermostats, *etc.* Faults in any of these components, errors in the sensors and instruments, and degradation due to normal wear or lack of maintenance deteriorates the performance of the system. Identifying aberrations from normal operation plays an important role in reducing energy use, maintaining a healthy indoor environment, and prolonging the equipment service life (Chakraborty & Elzarka, 2019). Although a significant amount of data is available from sensors installed in buildings which contains information about faults, it is difficult and time-consuming to analyze and detect faults manually (West et al., 2011). Automated fault detection and diagnostics (AFDD) procedures present the potential to reduce energy waste (Wall et al., 2011).

Methods for fault detection can be categorized into data-driven or process history-based methods, quantitative physical model-based, and qualitative rule and physics-based methods (Katipamula & Brambley, 2005). Qualitative rule and physics-based methods use certain if-then-else rules which can be obtained from first principles or expert knowledge. Rule-based methods are inefficient in real-time computation and in detecting and diagnosing sudden faults if the system requires a long time to reach a steady-state (Yu et al., 2014). Quantitative physical model-based methods are based on physical laws that regulate the system's behavior. The system behavior is predicted for a given set of measured input and model parameters which is compared to measured outputs. The key advantage of a quantitative (data-driven) model over other FDD methods is that it makes predictions using historical data, does not require detailed building information, and has lower complexity (Katipamula & Brambley, 2005).

The limitation of existing data-driven models is that the models are static. Although data-driven models have been extensively used in the field of fault detection and diagnosis of the HVAC system, the FDD method cannot properly handle the unknown faults in the system. Most of the existing FDD methods can diagnose the faults which are used during the training. When new faults occur in the system, identifying this as an unknown state and integrating its information into the model is crucial to prevent the fault from being misdiagnosed. This limitation can be handled by using adaptive learning which is capable of processing new measurements and performing runtime updates of model parameters in response to the changes in the building system (Karami & Wang, 2018).

Although extensive research is available on the FDD of HVAC systems and components, FDD for radiant heating and cooling system is still lacking. Also, the methods of evolving the FDD algorithm and model in response to the detection of an unknown fault in the system are yet to be explored in the radiant heating and cooling systems and such studies are very rare in the HVAC system as well. In this study, learning based Gaussian mixture model (GMR) model is used for fault detection and diagnosis for the radiant floor heating and cooling systems.

2. METHODOLOGY

To develop the proposed FDD model, experiments were conducted to collect normal operation data that was used to train a baseline model and FDD model under fault-free conditions and faulty data through the implementation of individual faults in the system. A GMR model was developed as a baseline model for predicting radiant slab (RS) flow and heating and cooling capacity. Next, using a feature selection algorithm the features which showed promising characteristics in detecting and diagnosing faults were selected from a list of features measured during the experiments. The features were used to develop an FDD model based on GGMR trained with the normal operation and faulty data collected during the experiments. The GGMR model was tested with known and unknown fault data and if any unknown state was detected and was verified as a new fault, the FDD model was evolved to incorporate the information of the new fault.

2.1. System description

The Living Lab 1 at the Herrick Laboratories, Purdue University was used as the case study to evaluate the proposed learning based FDD method. Living Lab 1 is equipped with a hydronic radiant floor system that provides both cooling and heating. The system diagram of the radiant slab system is shown in Figure 1. The concrete slabs and heat exchangers have RTD sensors to measure the slab temperature and the supply and return water temperatures. The main supply water pipe is divided into 10 channels with individual control valves which supply water to the 10 slabs (A-K) in parallel. The 10 valve positions are determined using a PI feedback loop with space temperature as the controlled variable with a setpoint temperature of 73°F (22.78 °C). The maximum valve position of all the valves is set at 100% and the minimum valve position is set at 35%. The water flow is maintained in the loops by a VFD pump. The supply water temperature is controlled through a heat exchanger connected to a district cool water supply and hot water supply. The cool water flow and hot water flow are controlled using cool water and hot water valves on the supply side of the heat exchanger. Both the cool water valve position and hot water valve position are determined using individual PID loops with supply water temperature as the controlled variable and the supply water temperature setpoint. If the space temperature is above 73 °F (22.78 °C), the system will be in the cooling mode and if it is lower than 73 °F (22.78 °C), the system will be in heating mode. During the cooling mode, the maximum of either the space dew point temperature plus 3 °F (1.7 °C) or 58 °F (14.4 °C) is used as the supply water setpoint temperature to calculate the cool water valve position. During the heating mode, 85 °F (29.44 °C) is used as the supply water setpoint temperature to calculate the hot water valve position. The supply water pump is enabled if either the cool water loop or hot water loop is enabled. The supply water pump position is also controlled based on the output of a PI loop with differential pressure in the supply pipe as the controlled variable with a setpoint pressure of 10 psi (68947.6 pa). The maximum pump output is set at 100% and the minimum is set at 0%.

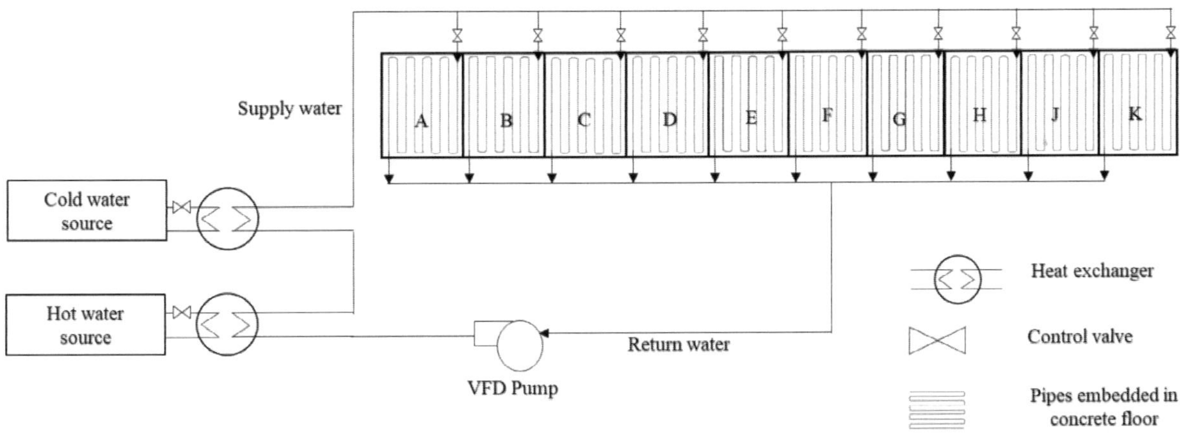

Figure 1: System diagram of the Living Lab 1 radiant floor

2.2. Gaussian Mixture Regression (GMR)

Gaussian mixture regression (GMR) was used to create a baseline model for flow through the radiant slabs, a baseline model heating capacity and cooling capacity of the radiant slab system, and an FDD model. Gaussian Mixture Model (GMM) is a model-based clustering method where the data is identified as a population with K different components. Each component of the GMM is approximated by a Gaussian distribution characterized by a mean as the center of the cluster and a variance. The Gaussian mixture probability density function is a weighted sum of Gaussian probability density functions (pdfs) and is given by Equation 1.

$$f(x, y|\pi, \mu, \delta) = \sum_{j=1}^{k} \pi_j N(x, y|\mu_j, \delta_j) \tag{1}$$

where $N(x, y|\mu_j, \delta_j)$ is a multivariate Gaussian density function, k is the number of Gaussian mixtures, μ_j is the mean vector, δ_j is the covariance matrix of j-th Gaussian component and π_j is the weight of the j-th component.

GMM is trained using a training dataset to find the unknown parameters of the mixture model (π_j, μ_j, δ_j) using the Expectation-Maximization (EM) algorithm in which the maximum likelihood estimates of the GMM parameters are predicted. A k-means clustering algorithm is used in the EM process to get the initial values for the parameters. When the new input (x) is provided, Gaussian Mixture Regression (GMR) is used for the prediction of the dependent variable Y which can be obtained by computing expectation over $[Y|X = x]$ as equation 2. $\hat{y}(x)$ is the prediction of the dependent variable Y.

$$\hat{y}(x) = E[Y|X = x] = \sum_{j=1}^{k} w_j(x) m_j(x) \tag{2}$$

where m_j is the regression function derived from the joint mixture Gaussian density and w_j is the weight of each regression model which is extracted using the Bayes rule. The detailed equations and working mechanism of the Gaussian mixture model and Gaussian mixture regression are presented in (Sung, 2004).

2.3. Evolving Mechanism

The GMR model is trained with the baseline data and when there is new data available, the existing GMR is updated. The evolving mechanism can be described using three steps.

Step 1: Calculate the probability density of the new sample data for each Gaussian component, apply the distance criterion and determine the best-matched component using equation 3.

$$P_j = \begin{cases} \pi_j N(\mathbf{d}(t)|\boldsymbol{\mu}_j(t), \delta_j(t)), & \frac{|\mathbf{d}(t) - \boldsymbol{\mu}_j(t)|}{|\delta_j(t)|} < \frac{1}{k}\sum_{j=1}^{k}\frac{|\mathbf{d}(t) - \boldsymbol{\mu}_j(t)|}{|\delta_j(t)|} \\ 0 \quad , & \text{otherwise} \end{cases} \quad (3)$$

Step 2: If a match is found ($P_j \neq 0$), the parameters (π_j, μ_j, δ_j) of the matching Gaussian are updated.

Step 3: If no existing Gaussian mixture component matches the new sample data ($P_j = 0$) and the number of existing components is less than a maximum threshold provided, a new Gaussian component is generated and the parameters prior, mean, and covariance matrix of the newly generated component are initialized using equation 4.

$$(\pi, c, \mu, \delta) = (\alpha, 1, x_i, \delta_0) \quad (4)$$

As shown in equation (5), after implementing all the steps of the evolving learning approach, the priors of the new Gaussian mixture model are normalized to satisfy the condition of forming a standard Gaussian mixture model which requires the sum of components' priors to be one.

$$\pi_j = \frac{\pi_j}{\sum_{j=1}^{k'} \pi_j} \quad (\text{for } j = 1 \ldots k') \quad (5)$$

k' is the number of components in the updated Gaussian mixture model. The evolving process is described briefly in section 2.6 and in detail in the research work by (Bouchachia & Vanaret, 2011).

2.4. Selection of number of Gaussian components

In GMM it is crucial to determine the number of Gaussian components (k) to group the data into an optimal number of clusters. In this study, Akaike Information Criterion (AIC) was used as a model selection criterion. The data was fitted to a Gaussian mixture model and the model was tested for several k values. The model with the smallest AIC and its corresponding k value was chosen as the number of parameters in the Gaussian model. The equation for computing AIC for parameters in the fitted model (k) is given by equation 6 (Akaike, 1998). Here, L is the maximum value of the likelihood function. This criterion was used to select the number of Gaussian components for the flow prediction, heating and cooling capacity prediction, and the FDD model.

$$AIC = -2\ln L + 2k \quad (6)$$

2.5. Fault selection and implementation

The experimental space is controlled with a building automation system (BAS) in which the operating conditions can be monitored, and control parameters can be overridden to desired values. Normal operation and faulty data were collected through experiments performed using an RS system in a real open-plan office setting. A total of 6 fault types with different severities were emulated by overriding control parameters which were selected considering the likeliness of their occurrence in a radiant floor system and the feasibility of implementation using the BAS. The fault types include stuck water flow valves for the radiant floor, hot water supply, cool water supply; sensor offsets for the space thermostat and supply water temperature sensors; and a VFD pump that is stuck at a fixed speed. The implemented faults with varying severity levels along with the dates of implementation are listed in Table 1.

Table 1: Implemented faults

Fault ID	Description (Implementation dates)	Fault ID	Description (Implementation dates)
F1	Space temperature sensor offset of +5 °F (2.78 °C) (12/10-12/2021)	F8	Hot water valve stuck at 0% open (12/6-8/2021)
F2	Space temperature sensor offset of -5 °F (2.78 °C) (12/4-6/2021)	F9	Hot water valve stuck at 50% open (12/8-10/2021)
F3	RS valves stuck at 0% open (12/14-16/2021)	F10	Hot water valve stuck at 100% open (12/12-14/2021)
F4	RS valves stuck at 50% open (12/16-18/2021)	F11	Cold water valve stuck at 0% open (12/25-27/2021)
F5	RS valves stuck at 100% open (12/18-20/2021)	F12	Cold water valve stuck at 50% open (1/7-9/2022)
F6	Supply water temperature sensor offset of +9 °F (5 °C) (12/20-22/2021)	F13	Cold water valve stuck at 100% open (1/9-12/2022)
F7	Supply water temperature sensor offset of -9 °F (5 °C) (12/22-25/2021)	F14	VFD pump stuck at 50% of its maximum speed 1/12-14/2022)

2.6. Fault detection and diagnosis method

The fault detection and diagnosis process along with evolving mechanism is shown in Figure *2* and proceeds through steps 1 to 8.

1. The baseline GMR model for energy use is trained with baseline data.
2. A GGMR FDD model is trained with faulty and normal operation training data.
3. The testing data containing known faults, normal operation, and the unknown fault is run through the Baseline GMR model, and the difference between measured and predicted RS heating and cooling capacity use is compared with the mean baseline difference.
4. If the error is less than the threshold, it is regarded as normal operation and the existing baseline GMR model is evolved with normal operation data. If the error is larger than the threshold, it is deemed as faulty data.
5. The testing data is then run through the GGMR FDD model to get the prediction of the fault label.
6. If the testing data has a predicted fault id that belongs to one of the labeled data, the fault id is reported. If the testing data is predicted as 0 or other labels not included in the training data, it is deemed as an unknown state.
7. The unknown state data is analyzed by an expert and if it is evaluated as a new fault, a new label for the new fault is provided and the GGMR FDD model is evolved to incorporate the information of the new fault.
8. The performance of the GGMR FDD model is tested both before and after evolving the model with new fault data.

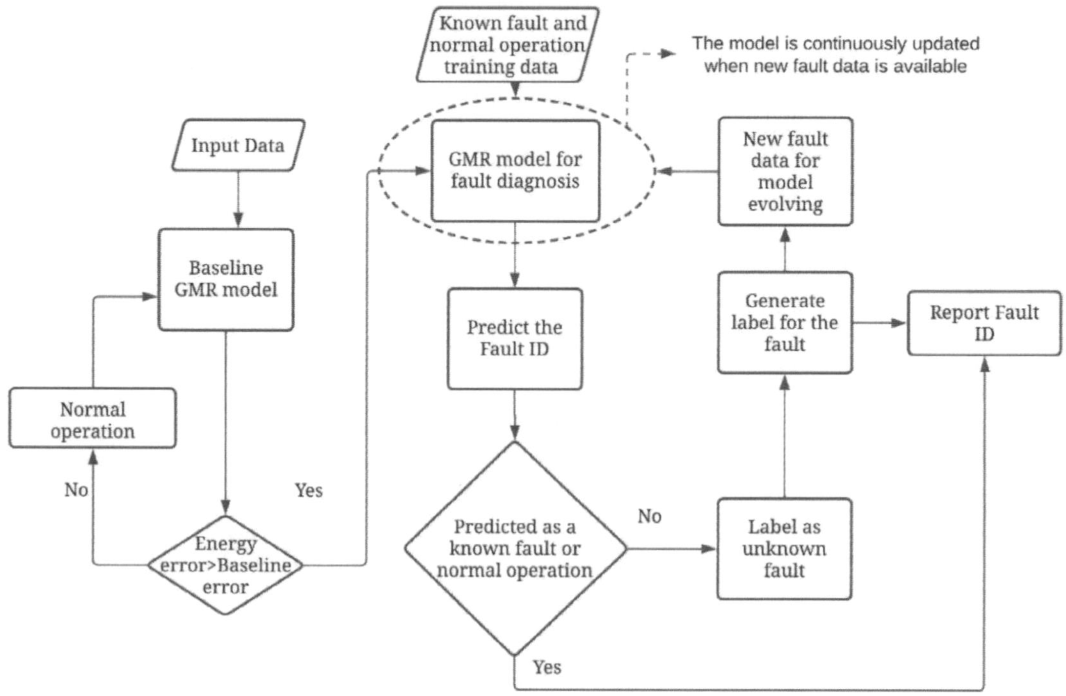

Figure 2: GGMR model generation, evolving, and Fault detection and diagnosis process diagram.

3. RADIANT SLAB SYSTEM HEATING AND COOLING CAPACITY PREDICTION MODEL

The RS heating and cooling capacity was calculated based on the water flow rate, supply water, and return water temperature using equation 7.

$$q = 0.063 \times \frac{5}{9} \times M \times C_p \times \Delta T \tag{7}$$

where q is the calculated heating or cooling capacity (KW), M is the mass flow rate of supply water (gal/min), $C_p$is the specific heat capacity of water = 4.2 $\frac{KJ}{Kg.K}$, and ΔT is the difference between supply and return water temperature. 0.063 is the factor to convert water flow rate (gal/min) to mass flow rate (kg/s), and $\frac{5}{9}$ is the factor to convert °F to °C. Negative values of q represent cooling capacity and positive values represent heating capacity. If the calculated q was positive but the supply water temperature was less than the average slab temperature, the RS heating capacity was set as 0. Similarly, if the calculated q was negative but the supply water temperature was greater than the average slab temperature, the RS cooling capacity was set as 0.

RS heating and cooling capacity are predicted using the Gaussian mixture regression model. The input data used in the heating and cooling capacity prediction model were identified using a correlation coefficient. GMR with evolving was used to develop the model. Normal operation data were divided into training data (11/30 to 12/2 2021) which was used to train the GMR model and testing data (1/14 to 2/3 2022) which was first used to test the model and later evolve the model. The evolved model is then used as the baseline heating and cooling capacity prediction model. When faulty data was collected, this model was used to get the heating and cooling capacity of the RS system under normal operating conditions. Along with the heating and cooling capacity, the water flow rate through the RS system was also predicted using GMR. The predicted water flow rate is one of the inputs for predicting the RS heating and cooling capacity. The input parameters selected for the prediction of water flow rate are the space temperature, RS valve position, and differential pressure. The input parameters selected for the prediction of RS heating and cooling capacity are the space temperature, RS valve position, average slab temperature, cold water valve position, hot water valve position, and the GMR predicted water flow rate. Ten Gaussian components were selected for both the RS water flow rate prediction model and the heating and cooling capacity prediction model. Both of those values were selected based on the AIC discussed in the previous section. The selected values for learning rate and closeness threshold are 0.001 and 10, respectively. Comparisons between measured and predicted RS flows and measured and predicted heating and cooling capacities during model testing are shown in Figure 3. The coefficient of determination R^2 between predicted and measured RS flow for the training data was 0.95 and between predicted and measured RS heating and cooling capacity was 0.98, which shows excellent prediction accuracies. For model testing data, the CV-RMSE for the predicted RS flow was 6% and was 15.6% for the predicted heating and cooling capacities, which is within the ASHRAE recommended threshold of 30% for the baseline model.

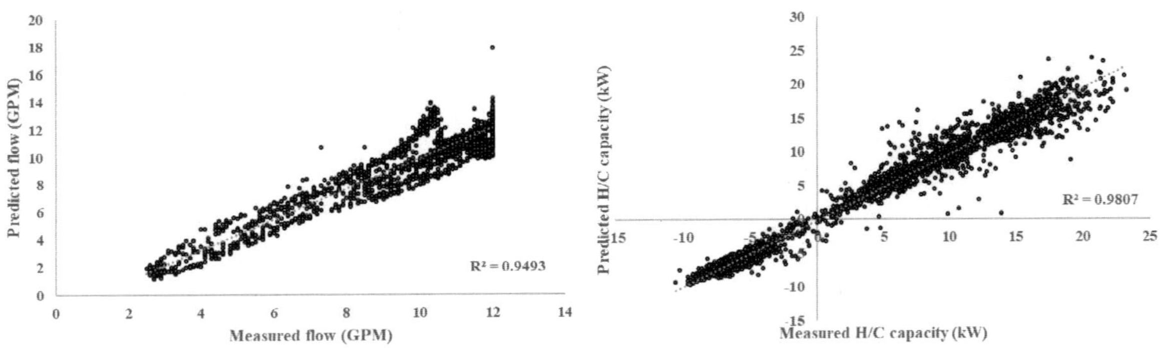

Figure 3: Measured vs Predicted RS flow (left) and Measured vs Predicted RS heating and cooling capacity (right) during model testing

4. FEATURE SELECTION FOR FDD MODEL

The features (input parameters) required for fault detection by GGMR were selected using the feature selection method presented in the research work of (Fijany & Vatan, 2006). The features selected contain the required ability to distinguish between all faults. In the first step, training data is used to recognize the symptoms associated with each fault. In the second step, an optimization algorithm is applied to find the minimum number of features required to diagnose the faults. The occurrence of each fault will have some effect on features that are correlated to the fault. This effect on features could be used as a symptom to identify the faults by comparing the feature values during faulty conditions with feature values during the fault-free condition. Table 2 lists the features used for the case study.

Table 2: Initial set of features

Initial features	
1. Space temperature difference from the setpoint (ΔT)	10. 12-hour cumulative sum of ΔT
2. Differential pressure difference from setpoint (ΔP)	11. 12-hour cumulative sum of ΔP
3. Measured-predicted flow (ΔF)	12. 12-hour cumulative sum of ΔF
4. Measured – predicted Heating and Cooling capacity (ΔC)	13. 12-hour cumulative sum of ΔC
5. Slab temperature difference from setpoint ($T_{slab-sp}$)	14. 12-hour cumulative sum of $T_{slab-sp}$
6. Supply water temperature-return water temperature (T_{sw-rw})	15. 12-hour cumulative sum of T_{sw-rw}
7. Change of RS valve signal (ΔV)	16. 12-hour cumulative sum of ΔV
8. The ratio of pump speed to maximum speed (P_s)	17. 12-hour cumulative sum of P_s
9. Flow change rate for the 24 hours (Δf)	18. 12-hour cumulative sum of Δf

Each data point for each feature was converted into a feature matrix by assigning 2 for symptoms resembling normal operation with the consideration of uncertainties, 1 for symptoms of a faulty condition where the measured values exceed the baseline upper limit, and -1 for symptoms of a faulty condition where the measured values are lower than the baseline lower limit to form a signature matrix. The feature matrix for one of the faults (space temperature positive sensor offset fault) is shown in Table 3 in which each column is an individual feature, and the rows represent each data point obtained with the faulty condition. Only a few features are shown in the table but this process was followed for all the features and all the faults and normal operation data.

Table 3: Feature matrix for space temperature positive sensor offset fault

ΔT	ΔP	ΔF	ΔC	ΔT(slab-sp)	ΔT(sw-rw)	ΔV	Ps	Δf
2	2	2	2	-1	2	2	2	2
2	2	2	2	-1	2	2	2	2
2	2	2	2	-1	2	2	2	2
2	2	2	2	-1	2	2	2	2
2	2	2	2	-1	2	2	2	2

In the next step, key signatures for each fault are identified. If a feature column has the same number in each row for a fault, the symptom is retained (2, 1, or -1), or else the symptom is represented by 0 meaning the feature does not provide a clear symptom required to diagnose the fault. Following this process for all the faults, a sensor matrix (S) is developed which has a key signature for each fault. Now, assuming S(x) is a vector containing the number of features in each row of the sensor matrix S, the optimal sensor placement problem can be formulated as a constrained optimization problem illustrated in equation 8 (Fijany & Vatan, 2006).

$$\text{minimize} \sum_{i=1}^{m} S(x_i) \qquad \text{Constraint:} \quad \widetilde{M}x \geq 1 \;, x_i = 0 \text{ or } 1 \tag{8}$$

The function in equation 8 minimizes the number of the sensors while the constraint assures that each row of sensor matrix is non-zero and unique. A linear integer programming routine in MATLAB was used to obtain the solution to the optimization problem.

Five different baseline scenarios as listed in Table 4 were developed with varying upper and lower limits to find the best performing baseline threshold. Here, Max and Min represent the maximum and minimum value of the feature respectively and SD is the standard deviation of the feature during the baseline period. The feature selection was done for all five cases with 18 initial features. The selected features for FDD are features 9, 11, 12, 13, 14, 15, and 17 from Table 2.

Table 4: Cases for baseline establishment and features selected for each case

Cases	Upper limit	Lower limit
1	Max	Min
2	Max+0.5*SD	Min-0.5*SD
3	Max+1*SD	Min-1*SD
4	Max+2*SD	Min-2*SD
5	Max+3*SD	Min-3*SD

5. FDD RESULTS

A total of 14 faults were implemented among which 13 known faults along with normal operation period were used to train the FDD model. One of the faults (VFD pump stuck at 50% of its maximum speed) was treated as a new fault. Five different GGMR FDD models with corresponding features and the number of Gaussian components were trained using the criteria for the five cases discussed in

Table **4**. The selection of the best case was done by evaluating the model training accuracy, false alarm rate, and fault misdiagnosis. All five cases had excellent model training accuracy as all the faults were trained with 100% accuracy.

All the faults except the ones presented in Table 5 did not have any misdiagnosed faults for all the cases. There were no false alarms when normal operation data was presented to the FDD model for any of the five cases. The space temperature negative sensor offset fault was misdiagnosed as supply water positive sensor offset fault (0.7%) and cool water valve stuck at 100% open fault (3.5%) in Case 1. It was misdiagnosed as the space temperature positive sensor offset fault (7.7%) in Case 5. The cool water valve stuck at 0% open was misdiagnosed as the hot water valve stuck at 100% open fault (5.6%) in Case 2 and Case 3. The cool water valve stuck at 50% open fault was also misdiagnosed in Case 2 and Case 3 as the cool water valve stuck at 0% open fault (20%). The cool water valve stuck at 100% open fault was misdiagnosed as a different fault in all 5 cases where the lowest misdiagnosis percent was for Case 1 and Case 4 (1.8%). Overall, Case 4 performed the best with the lowest misdiagnosis rates of the known faults. Hence, this baseline case was adopted to test the model in the presence of an unknown fault.

Table 5: Misdiagnosis of known faults during model testing

Faults	Case 1	Case 2	Case 3	Case 4	Case 5
F2	4.22	0	0	0	7.74
F11	0	5.6	5.6	0	0
F12	0	20	20	0	0
F13	1.796407	55.68862	55.68862	1.796407	16.16766

The performance of the FDD model determined using the criteria for Case 4 from

Table **4** was tested in presence of an unknown fault. All the unknown fault data was predicted as 0 which was not a label for any fault. As there was no information regarding the data presented to the model, the prediction was 0 for all data points. All the other faults had a testing accuracy of 100% except the cool water valve stuck at 100% open fault, which was predicted as normal operation (1.8%). Upon recognition of the type of the new fault by an expert, the data associated with the new fault was labeled (15) and fed to the evolving engine to update the Gaussian mixture regression model in the diagnosis section. Values for learning rate (0.01) and the closeness threshold (10) that resulted in the lowest prediction error were selected while evolving the GMR model. In the evolved diagnosis model, two new components were created to learn the information about the new fault. Before evolving the model, the new fault data fell into one of the 16 existing Gaussian components (Gaussian component 8) which was allocated to Fault 7 and Fault 8. After evolving the model, two new Gaussian components (17 and 18) were added to the model which had the information about the new fault.

The performance of the evolved FDD model was evaluated using the same dataset used to test the model before evolving. The results showed that the FDD model remains stable after being evolved since the accuracy of the true diagnosis lies in the range of 98% to 100% for known faults. Only 1.8% of the cool water valve stuck at 100% open fault was misdiagnosed as a normal operation. The evolved model also exhibits excellent performance in the diagnosis of the new fault as the accuracy of true diagnosis for the new fault was 100%.

6. CONCLUSIONS

The focus of this study was to develop a learning-based FDD model for radiant floor heating and cooling systems. An experimental study was conducted to investigate the performance of the learning-based GGMR model for fault detection and diagnostics of faults in an radiant slab system. A total of 6 common faults of different categories and severities (14 total faults) were implemented and were divided into known and new faults. The performance of the AFDD algorithm in the diagnosis of experimental fault data significantly improved after implementing the evolving process. A GGMR model was developed and trained with normal operation and 13 known faults data and was evolved using a new fault data. The performance of the model was tested using both known and new faults before and after evolving the model. The prediction accuracy for each fault using test data was extremely high with the lowest prediction accuracy of 98%. The algorithm was successful in detecting the new fault as an unknown state before evolving the model and in diagnosing it as a new fault after evolving the model. Further research can be conducted to evolve the model with the data from more than one new fault in the system. Another addition to this study would be to modify and test the model if simultaneous faults occur in the system. Further study can be conducted to test the performance of the model in presence of the data from the real buildings. This will also help to know if the model can be transferred to a real building setting without compromising the fault detection accuracy.

ACKNOWLEDGEMENT

This study was supported by the National Science Foundation EPSCoR Research Infrastructure program under Grant No. 1929209. Any opinions, findings, and conclusions, or recommendations expressed in this material are those of the authors and do not necessarily reflect the views of the National Science Foundation.

REFERENCES

Akaike, H. (1998). Information theory and an extension of the maximum likelihood principle. In *Selected papers of* 199-213). Springer.

Bouchachia, A., & Vanaret, C. (2011). Incremental learning based on growing gaussian mixture models. 2011 10th International Conference on Machine Learning and Applications and Workshops,

Chakraborty, D., & Elzarka, H. (2019). Early detection of faults in HVAC systems using an XGBoost model with a dynamic threshold. *Energy and Buildings*, *185*, 326-344.

Fabrizio, E., Corgnati, S. P., Causone, F., & Filippi, M. (2012). Numerical comparison between energy and comfort performances of radiant heating and cooling systems versus air systems. *HVAC&R Research*, *18*(4), 692-708.

Fijany, A., & Vatan, F. (2006). A new efficient algorithm for analyzing and optimizing the system of sensors. 2006 IEEE Aerospace Conference,

Joe, J., & Karava, P. (2019). A model predictive control strategy to optimize the performance of radiant floor heating and cooling systems in office buildings. *Applied Energy*, *245*, 65-77.

Karami, M., & Wang, L. (2018). Fault detection and diagnosis for nonlinear systems: A new adaptive Gaussian mixture modeling approach. *Energy and Buildings*, *166*, 477-488.

Katipamula, S., & Brambley, M. R. (2005). Methods for fault detection, diagnostics, and prognostics for building systems—a review, part I. *Hvac&R Research*, *11*(1), 3-25.

Sung, H. G. (2004). *Gaussian mixture regression and classification.* Rice University.

West, S. R., Guo, Y., Wang, X. R., & Wall, J. (2011). Automated fault detection and diagnosis of HVAC subsystems using statistical machine learning. 12th International Conference of the International Building Performance Simulation Association.

Wall, J., Guo, Y., Li, J., & West, S. (2011). A Dynamic Machine Learning based Technique for Automated Fault Detection in HVAC Systems. *Ashrae Transactions*, *117*(2).

Yu, Y., Woradechjumroen, D., & Yu, D. (2014). A review of fault detection and diagnosis methodologies on air-handling units. *Energy and Buildings*, *82*, 550-562.

Outdoor Comfort in Public Spaces, a Critical Review

Mingliang Li[1], Hongxi Yin[1]*, Ming Qu[2], Ian Trivers[1],

[1]Sam Fox School of Design & Visual Arts, Washington University in St. Louis,
St. Louis, MO 63130
Tel: (314)935-6200, E-mail: mingliang.li@wustl.edu; hongxi.yin@wustl.edu; itrivers@wustl.edu

[2] Lyles School of Civil Engineering, Purdue University,
West Lafayette, IN 47906
Tel: (765)494-9125, E-mail: mqu@purdue.edu

* Corresponding Author

ABSTRACT

A comfortable outdoor microclimate encourages outdoor physical activity and enhances user health and wellbeing. This study reviews the literature on outdoor comfort. The literature review focuses on the following four critical aspects of outdoor comfort: 1) the types of comfort that play significant roles in human outdoor activities, 2) the parameters (and/or values) used as indicators to evaluate outdoor comfort, 3) the critical urban design features that improve outdoor comfort, and 4) urban design interventions (guidelines) to enhance outdoor comfort in public spaces. This paper provides a comprehensive understanding of the physical, psychological, and design factors that influence the outdoor comfort of public spaces. It concludes with a discussion of future research directions in outdoor comfort and public space design.

1. INTRODUCTION

Outdoor comfort has become the primary goal to assess the quality of public space design. However, controversially, there is no universally accepted standard for outdoor comfort. Existing studies on outdoor comfort have focused on the definition of thermal comfort and its influencing factors, including climatic and physical factors (Elnabawi & Hamza, 2020; Lai, 2017; Mackey et al., 2017; Rakha et al., 2017; Shooshtarian et al., 2020). Except for the general concept of thermal comfort, there is no strict definition of outdoor thermal comfort. Thermal comfort is generally defined for indoor conditions as a psychological state that expresses satisfaction with the thermal environment and is assessed through subjective evaluations (ASHRAE, 2020). It is usually expressed using a thermal comfort index determined by measuring the heat balance of the human body and the surrounding environment. The thermal comfort indices were developed in the specific steady-state controlled environment. They are not suitable for outdoor environments, which are dynamic with constantly changing temperature, wind speed, radiation, lighting, landscape, and other environmental factors. The dynamic outdoor environments pose significant challenges to maintaining the homeostasis of the microclimate associated with the human body, so the thermal equilibrium between the human body and the environment is in a state of change that is difficult to stabilize. In addition to a physiological indicator, the thermal comfort index is a psychological state that is influenced by environmental changes. Outdoor thermal comfort studies are not a substitute for studies of real-world outdoor comfort (Peng et al., 2019a). The assessment of outdoor comfort is complex. It should consider the psychological aspects of the user's interaction with the design of the outdoor public space (e.g., the atmosphere, climate, noise, and lighting) and the socio-demographic background of the user (gender, age, education, height, and income) in addition to the physiological experience (Chen & Ng, 2012; Mackey et al., 2017; Reiter, 2004; Reiter & De Herde, 2003; Shooshtarian et al., 2020). Assessing outdoor comfort involves complex evaluation parameters based on outdoor thermal comfort and some psychological/human factors (MH Elnabawi, N Hamza, 2020; Pardeep Kumar & Amit Sharma, 2020).

As of March 2022, a search for studies on outdoor comfort and human behavior in public spaces by using sciencedirect and google scholar reveals that no more than ten research teams have investigated the topic. Most of these have studied outdoor thermal comfort by examining people's comfort and behavior in public spaces. The earliest research on outdoor thermal comfort was conducted by Reiter & Herde (2003) two decades ago. They used both quantitative and qualitative analyses to study outdoor comfort. Quantitative research focuses on physiological thermal comfort, including climatic factors such as air temperature, wind speed, radiation, and relative humidity, and human factors such as activity and clothing (Elnabawi & Hamza, 2020; Lai, 2017; Mackey et al., 2017; Rakha et al., 2017; Shooshtarian et al., 2020). The qualitative analysis focuses on design aspects such as the atmosphere of the outdoor environment, whether the climate is diverse, whether people believe they can potentially control the environment, and the natural attributes of the site. Except for Reiter & Herde (2003), studies related to outdoor comfort and human behavior in space are all from the past five years and examine either psychological or design factors that affect human outdoor comfort. Therefore, research on outdoor comfort and human behavior in public spaces must be strengthened and clarified. It is also necessary to review and summarize the current state of the art. Defining which comfort levels and factors influence outdoor thermal comfort and how outdoor comfort can help future public space design assessments and promote sustainable urban development.

2. METHODOLOGY

A search using the keywords of public space, human behavior, outdoor thermal comfort, and outdoor comfort identified forty-four articles from 2001 to 2021, with most studies published in the past five years (figure 1). Research fields covered are outdoor thermal comfort, outdoor comfort, human behavior, urban design factors, and psychological factors (figure 2). The research methods include case study calibration, case study analysis and comparison, onsite observation and analysis, validation experiments, comfort mapping, critical review, creating new assessment levels/frameworks, online surveys, theoretical analysis, or a combination of these methods. As figure 3 shows, nearly half of the researchers use simulation and comparison methods to do the research. 20% of studies used observation, field measurement, questionnaire, and analysis methods. The simulation and field observation methods count for 48% of total research (of the total). Other methods are data analysis (23% of the total), critical review (11% of the total), and mapping (5% of the total). As shown in figure 4, ten out of forty-four studies (22% of the total) do not explicitly state the research's season. Twenty-eight studies (63% of the total) were conducted in the summer. Sixteen studies (36% of the total) were conducted in winter. Eleven studies (25%) were conducted in spring, and five studies were conducted in the fall, 11%. Nineteen studies were conducted in more than one season, 43% of the total.

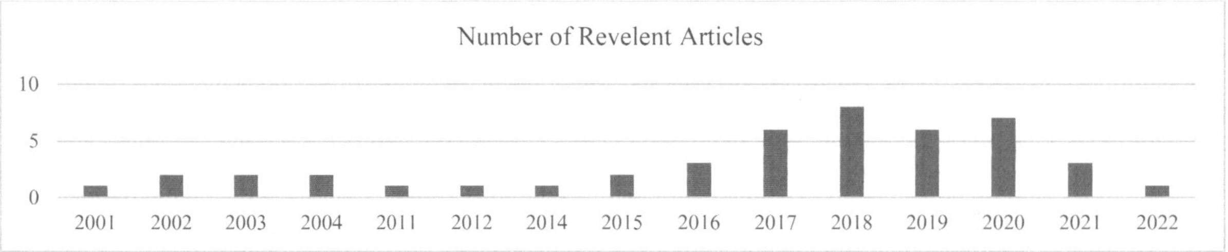

Figure 1: Number of articles related to public space, human behavior, outdoor thermal comfort, and outdoor comfort

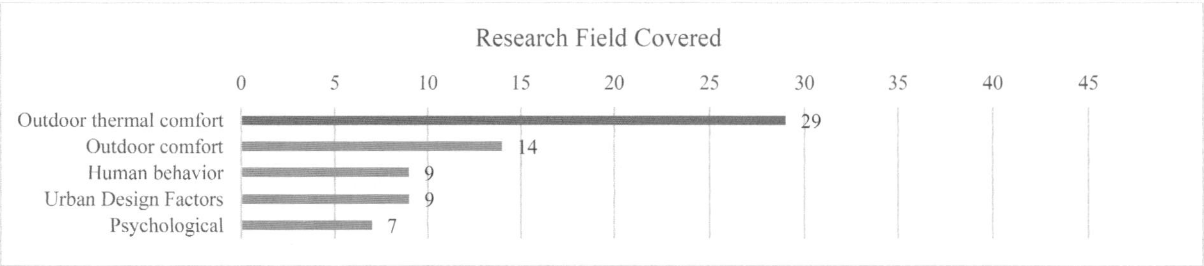

Figure 2: Research areas of the reviewed papers

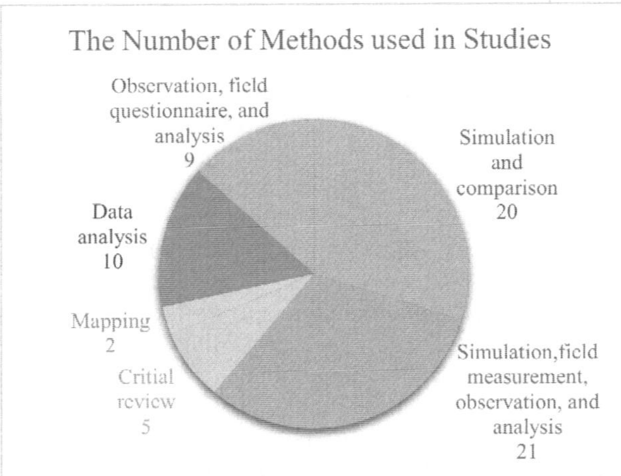

Figure 3: Number of methods used in 44 studies

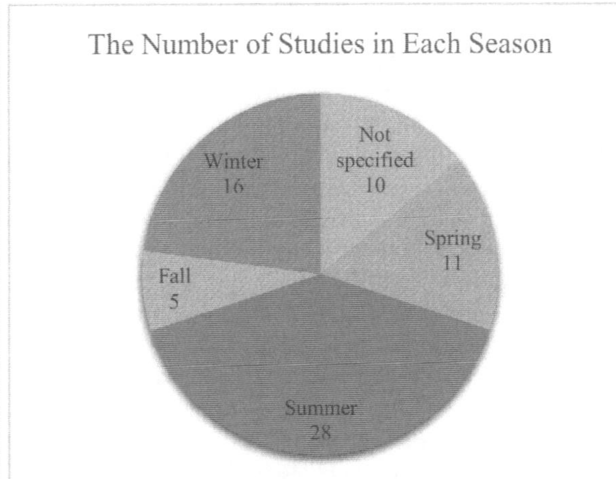

Figure 4: Number of studies in each season

Most studies have three steps for outdoor comfort in public spaces: selecting the site and research interest, simulation, and collecting data on outdoor comfort or thermal comfort. Most of the studies used the following methods: simulation and comparison, field measurements, observations, interviews, questionnaires, mapping, data analysis, and structural equation modeling. Some studies directly use thermal comfort index without valibarting or adjusting. For example, Sodoudi et al. (2018) used thermal comfort index and focused on simulating and analyzing several different outdoor design scenarios. Farajzadeh & Matzarakis (2012) used thermal comfort index to find the best months for tourism in Ourmieh Lake, Iran. Tumini (2016) used microliacte simulate software and thermal comfort index to test the different microclimate condition and thermal comfort perception for best urban renovation strategies. Naboni et al.(2018) only compared and analyzed the difficulty of using different climate simulation software for designers, but their study do not involve verification. They did not focus on outdoor comfort assessment or thermal comfort indicators. There is no universally accepted detailed index representing outdoor thermal comfort. Most of the case studies directly use some existing thermal comfort indices such as predicted mean vote (PMV), percentage dissatisfaction index (PPD), physiologically equivalent temperature (PET), standard effective temperature (SET), and universal thermal climate index (UTCI) to evaluate outdoor thermal comfort or outdoor comfort (Du et al., 2020; Fang et al., 2017; Lai, 2017; Salata et al., 2016).

Other studies have validated the calculated thermal comfort against empirical studies, often producing inconsistent results. Therefore, some scholars focused on improving outdoor comfort and outdoor thermal comfort threshold (Du et al., 2020; Fang et al., 2017; Salata et al., 2016). Some studies used questionnaires and field measurements to adjust the optimal thermal comfort interval calculated using a thermal comfort index (Farshid Aram, 2019; Sharifi et al., 2016; Ehsan Sharifi & John Boland, 2018; Jiawei Yao et al., 2018). Some studies directly measured the physical factors in the field and observed the environmental factors that affect human activities outdoors. Some studies include questionnaires for calibrating the outdoor thermal comfort index. Researchers then performed data analysis to determine the outdoor comfort zone and the factors that affect human outdoor comfort. However, the optimal thermal comfort interval for human outdoor activities is often difficult to predict. Some authors argued that the problem arises because rational thermal comfort metrics alone are insufficient to reflect the full range of human thermal comfort (Nikolopoulou & Steemers, 2003; Peng et al., 2021). After all, the underlying assumption of rational indicators is that thermal equilibrium between the human body and its surroundings is equivalent to thermal comfort (Peng et al., 2021). Thermal equilibrium reflects the physiological perception of an objective world as hot or cold.

On the other hand, thermal comfort is a sensation, an emotional experience relative to expectations (Ehsan Sharifi & John Boland, 2018, 2020; ASHRAE, 2020; JLM Hensen, 1990; Peng You et al., 2019). The difference between thermal equilibrium and thermal comfort was confirmed by experimental studies long ago (AP Gagge, 1969; Peng You et al., 2019). Research that goes beyond the purely physiological outdoor thermal comfort threshold should consider the influence of human emotions, expectations, and perceptions to fall into the broader category of outdoor comfort studies (Peng et al., 2019a, 2019b, 2020; Reiter, 2004; Reiter & De Herde, 2003).

3. RESULTS

Thermal Comfort Factor

According to figure 2, twenty-nine of forty-four studies on thermal comfort, accounting for 66%, reveals that thermal comfort is a significant factor in outdoor comfort and influences human activity in public spaces (table 1). Sharifi et al.'s (2016, 2020) research studied the relationship between outdoor thermal comfort and human behavior. They have conducted studies in several Australian cities such as Adelaide, Sydney, and Melbourne using case studies, field measurements, observations, and online questionnaires. They classified human behavior in public spaces as necessary, optional, and social. According to the changes of people engaged in the three behaviors in public areas at different air temperatures in summer, they derived the degree of influence of thermal comfort on human behavior and estimated the outdoor thermal neutrality threshold and thermal adaptation limit of public spaces and using it to modify the outdoor thermal comfort index. The results showed that thermal comfort is the dominant factor in outdoor comfort. Nikolopoulou (2001) concluded that the thermal environment is the primary factor affecting human use of these spaces by observing people in squares, streets, and gardens in Cambridge, UK, during four seasons. Farajzadeh and Matzarakis (2012) also considered thermal comfort an essential factor influencing travel. They conducted a simulation study in Ourmieh, Iran, using outdoor thermal comfort simulation software to explore this possibility. They concluded that June through September is the most comfortable time for tourism, sports, and recreational activities in Ourmieh.

Table 1: Thermal comfort factors

Author-year	Sub-field	Thermal Comfort Factors	Type	Season	Methods
Sharifi et al. (2016)	Outdoor thermal comfort & human behavior & outdoor comfort	Outdoor thermal comfort measurement (humidty, wind, sunshine), thermal environment, heat stress	Online survey	Not specified	Data analysis, survey
Ehsan Sharifi & John Boland, (2020)	Outdoor thermal comfort & human behavior	Outdoor thermal comfort measurement, thermal comfort index, thermal adaptation	Case study and field observation	Summer	Observation, analysis, passive activity observation
Marialena Nikolopoulou (2001)	Outdoor comfort, outdoor thermal comfort, Urban design, Human Parameter	Comfort condition outdoor, thermal comfort measurement, human parameter	Case study analysis	Spring, summer, winter	Data analysis
Farajzadeh et al. (2011)	Outdoor thermal comfort	Outdoor thermal comfort measurement, thermal comfort index	Case study simulation	Summer	Simulation and comparison
Shooshtarian et al. (2020)	Outdoor thermal comfort & human behavior	Thermal comfort assessment, human behavior, thermal comfort index	Summarize and analysis review (critical)	Not specified	Literature review
Liang Chen, Edward Ng (2012)	Human behavioral aspects of outdoor thermal comfort	Thermal comfort assessment, human behavior, thermal comfort index, assessment level (physical, psychological, social and behavioral)	Critical review	Not specified	Create a new outdoor thermal comfort assessment, survey, observation
MH Elnabawi and N Hamza. (2020)	Outdoor thermal comfort & human behavior	Thermal comfort assessment, human behavior, thermal comfort index	Critial review	Not specified	Create new assessment level, objective based and subjective assessment

One individual study and two critical literature reviews examining thermal comfort and human behavior have concluded that thermal comfort and human activity in public spaces are highly correlated. Shooshtarian (2017) reviewed Australian studies related to thermal comfort assessment in a literature review. Most of these studies used a combination of subjective thermal comfort assessment and field observations. They employed different thermal comfort indicators to determine the thermal neutrality of the study site and the choice of outdoor activities under other meteorological conditions in summer. Chen and Ng (2012) created an innovative framework for evaluating outdoor thermal comfort based on aspects of human performance after reviewing research in the past decade on outdoor thermal comfort and outdoor activities. This framework is divided into four levels: physical, physiological, psychological, and social/behavioral. The first two levels are obtained by measuring, simulating, and monitoring objective influences related to thermal comforts, such as sun, temperature, wind, and energy balance. The latter two levels are obtained by surveying, interviewing, and observing the main factors related to human behavior, such as expectations, past experiences, and preferences. A decade later, Elnabawi and Hamza (2020) refined Chen and Edward's outdoor thermal comfort evaluation framework to create a more comprehensive framework for designers and planners. The framework requires that the outdoor thermal comfort index be corrected by onsite measurement and thermal sensation questionnaires before the index is used. The thermal sensation is linked to the usage pattern of space; for example, the perception of outdoor thermal comfort is assessed in behavior.

Environmental factors that influence thermal comfort

Table 2: Environmental factors that influence thermal comfort

Author-year	Sub-field	Environmental Factor	Type	Season	Methods
Christopher Mackey (2017)	Universal thermal climate index (UTCI) Mapping, microclimate maps	Wind speed, sun, surface temperature, heat island	Case study calibration	Not specified	Simulation and comparison
Rakha et al. (2017)	Thermal comfort mapping	Mean radiant temperature	Case study simulation	Spring, summer, and winter	Simulation, MRT mapping
Pamela Smith, Cristián Henríquez (2018)	Outdoor comfort & recommendations for analysis and design; urban design	Morphoclimatic parameters of selected public spaces: (Land) surfaces, sky view factor, H/W, shadow, and radiation; land use, the proportion of built space, thermic quality of materials, orientation and slope of the surface, vegetal cover, ground humidity	Case study simulation & experiment	summer	Field measurement, observation, analysis
Katzschner et al. (2003)	Bioclimatic comfort mapping	Morphology (surface, tree, vegetation), meteorology (solar radiation, wind speed), sky view factor	Comfort mapping	Not specified	Analysis & mapping
Jason D. Wark et al. (2020)	Thermal comfort in the zoo	Material, shade	Case study analysis	Spring, summer, fall	Observation and analysis

Environmental stimuli influence human thermal comfort. Therefore, researchers have also studied environmental factors (table 2). For example, Sharifi and Boland (2020) categorized human activity outdoors as necessary, optional, and social and examined it through passive activity observation in Australia. Mackey (2017) argues that the sun brings heat exchange, and wind (rather than surface temperature or heat island) is the leading cause of the inability to predict outdoor thermal comfort accurately. Rakha et al. (2017) argued that MRT, a physical quantity representing the radiation around the human body, is a key factor influencing outdoor thermal comfort assessment. They proposed a simulation method to improve MRT measurements by conducting experiments in an outdoor space in Syracuse, NY, USA. In Chillán City, Chile, Smith, and Henríquez (2018) considered relative humidity a key factor influencing outdoor thermal comfort. They found that the most uncomfortable users occurred along the Parque Estero Las Toscas

River, where the air was unexpectedly dry. Katzschner et al. (2003) argued that design, morphology (geometry of buildings, surfaces, trees/vegetation), time, and meteorology (solar radiation and wind speed) are the fundamental structures of thermal comfort zoning mapping. Wark et al. (2020) found that habitat material and shading are the main factors influencing outdoor thermal comfort by studying zoo animals' activities. In the literature mentioned above, the parameters and values used to evaluate the thermal comfort aspects of outdoor comfort are air temperature, mean radiation temperature, wind, radiation/sun/solar radiation/shading, relative humidity, activity, clothing, morphology, and time.

Table 3: Psychological factors that influence thermal comfort

Author-year	**Sub-field**	**Psychological factors**	**Type**	**Season**	**Methods**
Tian et al. (2022)	Thermal perception (comfort; design aspect)	Thermal sensation acceptability, factors influencing thermal perception (physical, individual, social, psychological)	Field experiment and analysis	Spring, summer, and winter	Observations, simulation, questionnaire, analysis
Elham Zabetiana and Reza Kheyroddin (2019)	Thermal adaptation	Feeling and perception of it (psychological), thermal comfort felt in the two selected urban spaces, seven levels of sense of place	Case study experiment	Summer and winter / cold and warm	Spatial analysis, questionnaire, simulation and comparison
Ehsan Sharifi & John Boland (2018)	Outdoor thermal comfort & thermal adaptation & human behavior & outdoor comfort	Local and seasonal climate expectations, comfort perceptions, demographic specifications, activity choices, and socio-cultural norms	Field observation and analysis	Winter	Observations, questionnaire, and analysis
Peng et al. (2019)	Outdoor comfort	Need satisfaction of outdoor activity (expectation of thermal and wind condition), acceptability of outdoor activity (emotional status, preference of wind and sunlight)	Field measurement and analysis	Spring	Survey and analysis
Peng et al., (2019)	Outdoor comfort	Environment sensation, urban setting perception, expectation, preference, emotion	Case study simulation, experiment, and analysis	winter/ spring	Data analysis (qualitative); simulation and comparison, field measurement
Peng et al., (2021)	Outdoor comfort	Acceptability/need of outdoor activity,	Case study simulation, experiment and analysis	All four seasons	Case study, data analysis (qualitative), survey, simulation and comparison, field measurement

Psychological Factor

In recent years, researchers have slowly begun to highlight the role of psychological factors in evaluating outdoor comfort (table 3). Tian et al. (2022) studied the thermal comfort of open spaces in Xi'an, China. Using meteorological measurements and questionnaires, they concluded that psychological and individual factors and physical factors are the main determinants of the comfort of outdoor spaces in spring and summer. Physical and social factors are the primary influences on human activities in outdoor spaces in winter. The psychological factors in this study include the length of residence, time of exposure, frequency of visits, overall satisfaction, differences in climatic region, and the purpose of visit. Zabetiana and Kheyroddin (2019) argued that the different sense of place in urban spaces is directly related to thermal comfort, i.e., the psychological states influenced by contextual variables and thermal comfort are correlated. A study of two German squares in winter and summer found that the respondents' mood and sense of place (including the diversity of the venue's activities and the attractiveness) correlated with thermal comfort (Zabetiana and Kheyroddin, 2019). However, age, gender, and clothing had low correlations with thermal comfort. From his survey of several Australian cities, Sharifi and Boland (2018) concluded that outdoor climate expectations, comfort perceptions, and socio-cultural norms are psychologically biased factors that impact outdoor comfort. Peng et al. (2019a, 2019b, 2021) argued that current thermal comfort research is not an adequate substitute for comfort research. They sought to create a new modeling framework to extend outdoor comfort assessment by adding the following human factors: socio-demographic, psychological, and behavior. The socio-demographic characteristics are age, sex, education, and body mass index. The psychological factors include environmental sensation, urban setting perception, and expectation. The parameters and values used as the indicators to evaluate the psychological aspects of outdoor comfort are a sense of place/environment sensation/urban setting perception, length of residence, frequency of visits, time of exposure, the respondent's emotion, overall satisfaction, the purpose of the visit, differences of climatic region, expectation, preference, diversity of site activities, and attractiveness of the activities.

Urban Design Factor

In addition to thermal comfort and psychological factors, public space design also influences outdoor comfort (table 4). The impact of the design of urban or public spaces on human comfort outdoors can be broadly divided into two categories: the effects of the background design of public spaces and the impact of the ambient design of public spaces. For example, most studies emphasized the materials and layouts within public areas that can affect thermal comfort. Through onsite measurements and simulation analysis of the campus of Changwon National University in South Korea, Dain et al. (2015) concluded that pavements of different materials have different impacts on thermal comfort. Perini et al. (2017) used two simulation programs to simulate urban form and vegetation to determine their importance for thermal comfort. Through field measurements and simulations, Aram (2019) concluded that Large-Scale Urban Parks positively affect the thermal comfort of the surrounding environment. Nouri et al. (2018) presented four Measure Review Frameworks to refine the analysis of greenery, shading, vertical façade surface materials, and fountain sizes to determine how different combinations mitigate thermal comfort in public spaces. Smith and Henríquez (2018) observed and measured five public spaces in central and southern Chile, South America. They evaluated the effect of impervious (land) surfaces, sky view factor, H/W, land use, the proportion of built space, the thermic qualities of materials, vegetal cover, orientation, and slope of the surface on human quality of life. In a study on three typical weather days in Gulou Square, Nanjing, China, Xu et al. (2019) found that outdoor thermal perception, sunshine perception, and visual perception have an essential impact on human quality of life and human activities in public spaces.

On the other hand, the atmosphere of public space also impacts human activities. Reiter & Herde (2003, 2004) have suggested the importance of various qualitative factors in evaluating outdoor comfort, including the following: "identification of an atmosphere, relation with the context, continuity of the environmental conditions, diversity of the environmental conditions, variability of the environmental conditions, perception of potential control, the capacity of adaptation, naturality of the place, meaning of the place, globality of the comfort feeling." In research on German squares, Zabetiana and Kheyroddin (2019) proposed a seven-level sense of place emphasizing the significance of the ambiance of public places. The literature mentioned above shows that researchers use the following design-related parameters and values as indicators to evaluate outdoor comfort: greenery/tree/vegetation/park/vegetal cover, different materials of artificial pavement/land surface, thermic qualities of materials, the surface materials of vertical façade, water, shading/sunshine perception/sky view factor, height-to-width ratio, the proportion of built space, orientation, the slope of the surface, urban form, land use, urban atmosphere (as mentioned by Reiter & Herde, (2003)), and sense of place (as mentioned by Zabetiana and Kheyroddin (2019)).

Table 4: Urban design factors that influence thermal comfort

Author-year	Sub-field	Design factors	Type	Season	Methods
Dain et al. (2015)	Outdoor thermal comfort	Paving material, tree, building type	Validation experiment	Summer	Simulation and comparison
Perini et al. (2017)	Outdoor comfort, built environment	Urban form, vegetation and canyon proportion affecting urban microclimate	Case study analysis and comparison	Summer	Experiment, simulation and field, measurement comparison
Farshid Aram (2019)	Outdoor thermal comfort	Size of the urban park, distance from the park	Case study experiment	Summer	Data analysis, simulation, survey
Nouri et al. (2018)	Outdoor thermal comfort & Public Place Design	The different techniques, and measures are reviewed and framed into four measure review frameworks (green, sun, surface, water)	Critical review	Summer	Simulation, and comparison
Xu et al. (2019)	Outdoor comfort & human behavior	Location of public facilities (such as seats, pergola, entrances and exits) and extension of vision	Case study simulation and experiment	Spring, summer, winter	Observation, simulation, comparison and data analysis

Many design factors influence outdoor human comfort. However, similar factors can be grouped into categories to reduce the number of indicators. For example, land surface and vegetation cover are artificial pavement but in different materials; these indicators can be summarized as natural properties. The urban design features that can enhance outdoor comfort and attract more people can be summarized into the following categories: natural properties, visual effect, urban form, function, variety, control, variability, activity, and the sense of place. More specific interventions include greening rate, sunlight, shading, sky view factor, fountains, pools, the thermal quality of different artificial pavement and façade materials, type and number of events, movable components, and the possibility of versatile use (localization change).

4. CONCLUSIONS

This paper reviews the research on the human impact of outdoor comfort in public spaces over the past two decades through a literature review. It clarified the differences between outdoor and thermal comfort and categorized the numerous factors that may affect outdoor comfort. The most important contribution of this literature review paper is the elucidation of three important categories of factors that affect human outdoor comfort: physical, psychological, and urban design factors. This article distinguishes outdoor comfort and outdoor thermal comfort research at the present stage. The concept of outdoor comfort responds to the narrowness of the more prevalent vision of outdoor thermal comfort.

As research on outdoor comfort reveals, thermal comfort remains a major factor influencing human outdoor activity, but psychological and design factors also influence human perceptions of the comfort of the surrounding environment. Psychological changes follow the influence of external conditions. In particular, the environment's design affects the perception of outdoor comfort through sight and sensation. Good design provides a variety of microclimatic conditions for the site. These microclimatic conditions are linked to the site's thermal comfort and psychological aspects, influencing the evaluation of outdoor comfort. Therefore, future outdoor comfort research in public spaces should focus on design scenarios. Improving outdoor comfort through scene design will attract more people outdoors while providing them better thermal comfort, ultimately increasing human outdoor activities.

REFERENCES

Aram, F., Solgi, E., Higueras García, E., Mosavi, A., & R Várkonyi-Kóczy, A. (2019). The cooling effect of large-scale urban parks on surrounding area thermal comfort. *Energies*, *12*(20), 3904.

Chen, L., & Ng, E. (2012). Outdoor thermal comfort and outdoor activities: A review of research in the past decade. *Cities*, *29*(2), 118–125.

Coccolo, S., Pearlmutter, D., Kaempf, J., & Scartezzini, J.-L. (2018). Thermal comfort maps to estimate the impact of urban greening on the outdoor human comfort. *Urban Forestry & Urban Greening*, *35*, 91–105.

Dain, J., Park, K., Song, B., Kim, G., Choi, C., & Moon, B. (2015). Validation of ENVI-met PMV values by in-situ measurements. *Proceedings of the 9th International Conference on Urban Climate Jointly with 12th Symposium on the Urban Environment, Toulouse, France*, 20–24.

Elnabawi, M. H., & Hamza, N. (2019). Behavioural perspectives of outdoor thermal comfort in urban areas: A critical review. *Atmosphere*, *11*(1), 51.

Elnabawi, M. H., & Hamza, N. (2020). A behavioural analysis of outdoor thermal comfort: A comparative analysis between formal and informal shading practices in urban sites. *Sustainability*, *12*(21), 9032.

Fan, Q., Du, F., Li, H., & Zhang, C. (2021). Thermal-comfort evaluation of and plan for public space of Maling Village, Henan, China. *PloS One*, *16*(9), e0256439. https://doi.org/10.1371/journal.pone.0256439

Farajzadeh, H., & Matzarakis, A. (2012). Evaluation of thermal comfort conditions in Ourmieh Lake, Iran. *Theoretical and Applied Climatology*, *107*(3), 451–459.

Kastner, P., & Dogan, T. (2019). Towards high-resolution annual outdoor thermal comfort mapping in urban design. *Proceedings of the 16th International IBPSA Conference, Rome, Italy*, 2–4.

Katzschner, L. (2004). Open space design strategies based on thermal comfort analysis. *Proc. PLEA*, *1*, 47–52.

Katzschner, L., Bosch, U., & Röttgen, M. (2003). A methodology for bioclimatic microscale mapping of open spaces. *Fifth International Conference on Urban Climate*.

Kumar, P., & Sharma, A. (2020). Study on importance, procedure, and scope of outdoor thermal comfort–A review. *Sustainable Cities and Society*, *61*, 102297.

Lai, D. (2017). *Modeling Thermal Comfort in Outdoor Environments* [PhD Thesis]. Purdue University.

Lee, H., & Mayer, H. (2018). Maximum extent of human heat stress reduction on building areas due to urban greening. *Urban Forestry & Urban Greening*, *32*, 154–167.

Mackey, C., Galanos, T., Norford, L., Roudsari, M. S., & Architects, P. (2017, August). Wind, sun, surface temperature, and heat island: Critical variables for high-resolution outdoor thermal comfort. In *Proceedings of the 15th international conference of building performance simulation association. San Francisco, USA*.

Naboni, E., Coccolo, S., Meloni, M., & Scartezzini, J.-L. (2018). Outdoor comfort simulation of complex architectural designs: A review of simulation tools from the designer perspective. *2018 Building Performance Analysis Conference and SimBuild Co-Organized by ASHRAE and IBPSA-USA, CONF*.

Nikolopoulou, M., Baker, N., & Steemers, K. (2001). Thermal comfort in outdoor urban spaces: Understanding the human parameter. *Solar Energy*, *70*(3), 227–235. https://doi.org/10.1016/S0038-092X(00)00093-1

Peng, Y., Feng, T., & Timmermans, H. (2019a). A path analysis of outdoor comfort in urban public spaces. *Building and Environment*, *148*, 459–467.

Peng, Y., Feng, T., & Timmermans, H. J. P. (2019b). Expanded comfort assessment in outdoor urban public spaces using Box-Cox transformation. *Landscape and Urban Planning*, *190*, 103594. https://doi.org/10.1016/j.landurbplan.2019.103594

Peng, Y., Feng, T., & Timmermans, H. J. P. (2021). Heterogeneity in outdoor comfort assessment in urban public spaces. *Science of The Total Environment*, *790*, 147941. https://doi.org/10.1016/j.scitotenv.2021.147941

Peng, Y., Peng, Z., Feng, T., Zhong, C., & Wang, W. (2021). Assessing Comfort in Urban Public Spaces: A Structural Equation Model Involving Environmental Attitude and Perception. *International Journal of Environmental Research and Public Health*, *18*(3), 1287. https://doi.org/10.3390/ijerph18031287

Perini, K., Chokhachian, A., Dong, S., & Auer, T. (2017). Modeling and simulating urban outdoor comfort: Coupling ENVI-Met and TRNSYS by grasshopper. *Energy and Buildings*, *152*, 373–384.

Rakha, T., Zhand, P., & Reinhart, C. (2017). A Framework for Outdoor Mean Radiant Temperature Simulation: Towards Spatially Resolved Thermal Comfort Mapping in Urban Spaces. *Proceedings of the 15th IBPSA*, 2414–2420.

Reiter, S. (2004). Correspondences between the conception principles of sustainable public spaces and the criteria of outdoor comfort. *The 21th Conference on Passive and Low Energy Architecture*.

Reiter, S., & Herde, A. D. (2003). Qualitative and quantitative criteria for comfortable urban public spaces. In *Research in Building Physics*. CRC Press.
Salata, F., Golasi, I., de Lieto Vollaro, R., & de Lieto Vollaro, A. (2016). Urban microclimate and outdoor thermal comfort. A proper procedure to fit ENVI-met simulation outputs to experimental data. *Sustainable Cities and Society*, *26*, 318–343.
Santos Nouri, A., Costa, J. P., Santamouris, M., & Matzarakis, A. (2018). Approaches to Outdoor Thermal Comfort Thresholds through Public Space Design: A Review. *Atmosphere*, *9*(3), 108. https://doi.org/10.3390/atmos9030108
Setaih, K., Hamza, N., Mohammed, M., Dudek, S., & Townshend, T. (2014). CFD modeling as a tool for assessing outdoor thermal comfort conditions in urban settings in hot arid climates. *Journal of Information Technology in Construction (ITCon)*. https://eprints.ncl.ac.uk
Sharifi, E., & Boland, J. (2018). Limits of thermal adaptation in cities: Outdoor heat-activity dynamics in Sydney, Melbourne and Adelaide. *Architectural Science Review*, *61*(4), 191–201.
Sharifi, E., & Boland, J. (2020). Passive activity observation (PAO) method to estimate outdoor thermal adaptation in public space: Case studies in Australian cities. *International Journal of Biometeorology*, *64*(2), 231–242.
Sharifi, E., Sivam, A., & Boland, J. (2016). Resilience to heat in public space: A case study of Adelaide, South Australia. *Journal of Environmental Planning and Management*, *59*(10), 1833–1854.
Sharifi, E., Sivam, A., & Boland, J. (2017). Spatial and activity preferences during heat stress conditions in Adelaide: Towards increased adaptation capacity of the built environment. *Procedia Engineering*, *180*, 955–965.
Sharmin, T., Steemers, K., & Matzarakis, A. (2017). Microclimatic modelling in assessing the impact of urban geometry on urban thermal environment. *Sustainable Cities and Society*, *34*, 293–308.
Shooshtarian, S., Lam, C. K. C., & Kenawy, I. (2020). Outdoor thermal comfort assessment: A review on thermal comfort research in Australia. *Building and Environment*, *177*, 106917. https://doi.org/10.1016/j.buildenv.2020.106917
Shooshtarian, S., & Ridley, I. (2017). The effect of physical and psychological environments on the users thermal perceptions of educational urban precincts. *Building and Environment*, *115*, 182–198. https://doi.org/10.1016/j.buildenv.2016.12.022
Smith, P., & Henriquez, C. (2018). Microclimate metrics linked to the use and perception of public spaces: The case of Chillán city, Chile. *Atmosphere*. https://doi.org/10.3390/atmos9050186
Sodoudi, S., Zhang, H., Chi, X., Müller, F., & Li, H. (2018). The influence of spatial configuration of green areas on microclimate and thermal comfort. *Urban Forestry & Urban Greening*, *34*, 85–96.
Tian, Y., Hong, B., Zhang, Z., Wu, S., & Yuan, T. (2022). Factors influencing resident and tourist outdoor thermal comfort: A comparative study in China's cold region. *Science of The Total Environment*, *808*, 152079.
Tumini, I., Higueras García, E., & Baereswyl Rada, S. (2016). Urban microclimate and thermal comfort modelling: Strategies for urban renovation. *International Journal of Sustainable Building Technology and Urban Development*, *7*(1), 22–37. https://doi.org/10.1080/2093761X.2016.1152204
Wang, C., Zhan, W., Liu, Z., Li, J., Li, L., Fu, P., Huang, F., Lai, J., Chen, J., & Hong, F. (2020). Satellite-based mapping of the Universal Thermal Climate Index over the Yangtze River Delta urban agglomeration. *Journal of Cleaner Production*, *277*, 123830.
Wark, J. D., Wierzal, N. K., & Cronin, K. A. (2020). Mapping shade availability and use in zoo environments: A tool for evaluating thermal comfort. *Animals*, *10*(7), 1189.
Xu, T., Tong, Z., & Xu, S. (2019). Integration of Microclimate into the Multi-Agent System Simulation in Urban Public Space. *Smart Cities*, *2*(3), 421–432. https://doi.org/10.3390/smartcities2030026
Yao, J., Yang, F., Zhuang, Z., Shao, Y., & Yuan, P. F. (2018). The effect of personal and microclimatic variables on outdoor thermal comfort: A field study in a cold season in Lujiazui CBD, Shanghai. *Sustainable Cities and Society*, *39*, 181–188. https://doi.org/10.1016/j.scs.2018.02.025
Zabetian, E., & Kheyroddin, R. (2019). Comparative evaluation of relationship between psychological adaptations in order to reach thermal comfort and sense of place in urban spaces. *Urban Climate*, *29*, 100483.

Performance Modeling and Analysis of a Thermoelectric Building Envelope for Space Heating

Xiaoli LIU[1], Ming QU[1]*, Kazuaki YAZAWA[2], Jorge KOHANOFF[3], Piotr CHUDZINSKI[4], Lorenzo STELLA[4], Brian NORTON[5], Niall HOLMES[5], Ruchita JANI[5], Hongxi YIN[6], Conrad JOHNSTON[7]

[1]Lyles School of Civil Engineering, Purdue University
West Lafayette, Indiana, USA
mqu@purdue.edu

[2]Birck Nanotechnology Center, Purdue University
West Lafayette, Indiana USA
kyazawa@purdue.edu

[3]Instituto de Fusion Nuclear "Guillermo Velarde", Universidad Politecnica de Madrid
Madrid, Spain
j.kohanoff@upm.es

[4]School of Mathematics and Physics, Queen's University Belfast
Belfast, United Kingdom
l.stella@qub.ac.uk

[5]School of Civil and Structural Engineering, Technological University Dublin,
Dublin, Ireland
brian.norton@tyndall.ie

[6]Sam Fox School of Design and Visual Arts, Washington University in St Louis, St Louis, Missouri, USA
hongxi.yin@wustl.edu

[7]Pacific Northwestern National Laboratory, Richland, WA 99354, USA
conrad.s.johnston@googlemail.com

* Corresponding Author

ABSTRACT

To provide energy-efficient space heating and cooling, a thermoelectric building envelope (TBE) embeds thermoelectric devices in building walls. The thermoelectric device in the building envelope can provide active heating and cooling without requiring refrigerant use and energy transport among subsystems. Thus, the TBE system is energy and environmentally friendly. A few studies experimentally investigated the TBE under limited operating conditions, and only simplified models for the commercial thermoelectric module (TEM) were developed to quantify its performance. A holistic approach to optimum system performance is needed for the optimal system design and operation. The study developed a holistic TBE-building system model in Modelica for system simulation and performance analysis. A theoretical model for a single TEM was first established based on energy conversion and thermoelectric principles. Subsequently, a TBE prototype model combining the TEM model was constructed. The prototype model employing a feedback controller was used in a whole building system simulation for a residential house. The system model computed the overall building energy efficiency and dynamic indoor conditions under varying operating conditions. Simulation results indicate the studied TBE system can meet a heating demand to maintain the desired room temperature at 20 °C when the lowest outdoor temperature is at -26.3 °C, with a seasonal heating COP near 1.1, demonstrating a better heating performance than electric heaters. It suggests a potential energy-efficient alternative to the traditional natural gas furnaces and electric heaters for space heating.

1. INTRODUCTION

Thermoelectric technology uses thermoelectric materials to realize the mutual conversion between temperature differences and electricity (Luo et al., 2020). Researchers have integrated thermoelectric materials and devices into building walls for space heating and cooling (Ahmad Gondal, 2019; Aksamija et al., 2019; Ibañez-Puy et al., 2018; Irshad et al., 2019; Martín-Gómez et al., 2021; Zuazua-Ros et al., 2018, 2019). As shown in Figure 1(a), the TBE-building system is a highly integrated thermal-activated heating and cooling without the requirements of transporting energy and synergy among subsystems. Meanwhile, the TBE system requires no refrigerant and no release of chemicals, making it environmentally friendly. The TBE-building system has great flexibility due to the ultra-low-profile thermoelectric devices (e.g., TEM) and expandable modulus TBE components. A typical TBE component, as displayed in Figure 1(b), has three primary parts, including (1) TEMs, (2) heat dissipation (e.g., heat sinks and circulation fans), and (3) associated wall design.

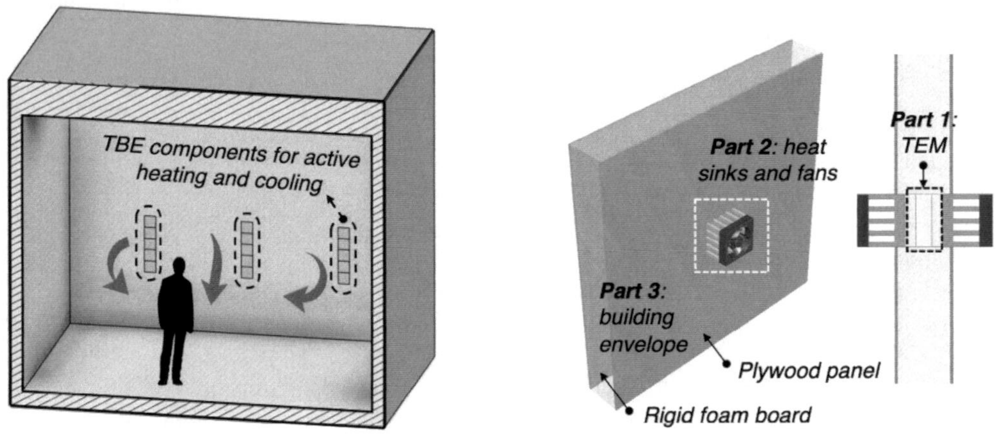

Figure 1: The schematic diagrams of (a) a TBE-building system and (b) a ventilation-type TBE prototype.

However, the research on the TBE system is still at an early stage. Due to physical and technical challenges, only a few experimental studies have been carried out under limited weather conditions. Martin *et al.* reported that the TBE's heating COP including and excluding fan power at an ambient condition were 0.82–1.01 and 1.02–1.91, respectively (Martín-Gómez et al., 2021). Researchers seek approaches to improve the energy conversion efficiency of the TBE system and to investigate the adaptivity of the TBE system in different climates. Therefore, models of a complete TBE-Building system are highly needed to comprehensively guide the system design and operation and the adoption of thermoelectric technology in buildings.

Most models developed to evaluate a TBE device focused on TEMs used in the device and only reflected the steady-state performance without considering varying outdoor conditions and system control. Then in 2020, Luo *et al.* developed a state-space model of a photovoltaic-thermoelectric wall system demonstrating a net-zero heat flux through the building envelope via a model predictive control (Luo et al., 2020). This study pointed out a limitation in whole building simulation and optimization, which is rather critical for building energy system analysis. However, limited studies integrated the TBE model within building energy modeling for seasonal performance analysis. Wang *et al.* conducted the annual energy analysis of a TBE system utilizing renewable energy for domestic heating by only comparing the heating demand of a house with TBE's heating supply assuming a constant COP and a given renewable power supply (Wang et al., 2017). The model's result indicated a possible energy-saving of 64% and a reduction of CO_2 emission of 4305 kg/year for the proposed system. However, this model ignored the impact of varying outdoor temperatures and controls of power on the TBE's annual performance. To study the thermoelectric technology for room temperature control, Hagenkamp *et al.* established a steady-state model for thermoelectric elements with optimized power input to meet the fixed heating demand (Hagenkamp et al., 2022). This model indicated an efficient heating performance of thermoelectricity elements by using an advanced control system. Since the TBE system is still a new system for space air conditioning, current modeling methods cannot reflect a comprehensive performance of a TBE system. It is necessary to develop a holistic approach toward the optimum performance of the TBE system. Hence, our study aims to establish a TBE-building system model to guide the optimal system design and operation.

In this paper, a transient TBE system model for space heating was investigated by using Modelica language. The system model integrated a verified TBE prototype model employing a Proportional-Integral-Differential (PID) power controller within a residential house. The model was used to predict the transient heat transfer and energy efficiency of a TBE system integrated into buildings with feedback control to maintain the desired indoor thermal comfort. The seasonal indoor air temperature, current input, heat deliverable, and COPs of TEM and TBE under historical weather conditions were computed and analyzed. Key contributions include the whole system simulation, transient heat transfer analysis, and seasonal energy analysis. Finally, conclusions and suggestions were provided to guide the design, sizing, and control of a TBE system to facilitate the deployment of thermoelectric applications in buildings.

2. METHODOLOGY

2.1 Methodology Overview

As displayed in Figure 2, the heat and electric transfer models at three different hierarchical levels have been investigated in this study. First, at the module level (level 1), an analytic model for a single TEM was established based on the energy conversion and thermoelectric principles, including the Peltier effect, Joule heating, heat conduction, and heat storage, as described in Section 2.2. Subsequently, at the prototype level (level 2), a TBE model was developed by combining the TEM at level 1 with the envelope design, as presented in Section 2.3. Experimental data have validated the prototype model. At the system level (level 3), the verified prototype model (at level 2) employing a PID controller was plugged into a residential house to construct the TBE-Building system model, as illustrated in Section 2.4. Three models were developed using the Modelica language and solved by an ordinary-differential-equation solver. Finally, the system model computed the overall building energy consumption under different operating conditions, including the outdoor temperature, power input to TBE, etc.

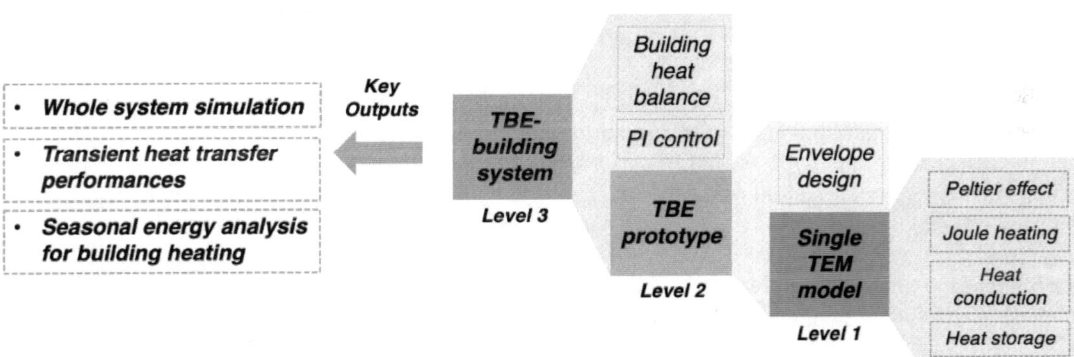

Figure 2: The hierarchy of three Modelica models developed in this study

2.2 Thermoelectric Module Model

A simplified theoretical model of thermoelectric elements has often been used for analyzing thermal and electrical power because it can be easily integrated with other sub-models and reduces computation time, making it suitable for simulating a complex system. The model holds several assumptions:

a) The material properties are independent of temperature.
b) Thomson effect and the surface heat losses of thermoelectric elements are negligible.
c) Joule heating contributes equally to the extremes of thermoelectric elements.
d) The model only considers one-dimensional heat transfer for TEMs.

This first-principal model is based on energy conversion and thermoelectric fundamentals. The heating power ($\dot{Q}_h$) at the hot surface and the cooling power ($\dot{Q}_c$) at the cold surface of a thermoelectric element are given by

$$\dot{Q}_{h,te} = SIT_{h,te} - 2K_{te}(T_{h,te} - T_{m,te}) + 0.5I^2R_{te} \tag{1}$$

$$\dot{Q}_{c,te} = SIT_{c,te} - 2K_{te}(T_{m,te} - T_{c,te}) - 0.5I^2R_{te} \tag{2}$$

where S is the Seebeck coefficient; I is the current input; K_{te} is the thermal conductance, kA/L, where k is the thermal conductivity, A is the cross-sectional area, and L is the leg length of the thermoelectric element; R_{te} is the electrical resistance; $T_{m.te}$, $T_{h,te}$, and $T_{c,te}$ are middle-point, hot surface, and cold surface temperatures of thermoelectric elements, respectively. As can be seen from equation (1) and equation (2), the heat transfer model of a thermoelectric element contains three primary effects, including the Peltier effect (a heating power of $SIT_{h,te}$ at the hot surface and cooling power of $SIT_{c,te}$ at the cold surface), heat conduction ($K_{te}(T_{h,te} - T_{m,te})$), and Joule heating ($I^2R_{te}$). Thermal storage in a thermal mass is considered to model the transient behavior of the component. For thermoelectric elements, the heat balance in the two-resistance-one-capacitance (2R1C) model can be written as

$$m_{te}c_{p,te}\frac{d(T_{m,te})}{dt} = 2K_{te}(T_{h,te} - T_{m,te}) - 2K_{te}(T_{m,te} - T_{c,te}) \tag{3}$$

A TEM is a thermoelectric device containing a total number of N n-type and N p-type thermoelectric elements, N metal interconnects, and two ceramic substrates. Metal interconnects link thermoelectric elements electrically in series to expand the power output. Ceramic plates are thermally conductive but electrically insulative for insulation protection. For metal interconnects and ceramic substrates, the heat transfer equations modeled by a 2R1C thermal network are expressed as:

$$m_{me}c_{p,me}\frac{d(T_{m,me})}{dt} = 2K_{me}(T_{1,me} - T_{m,me}) - 2K_{me}(T_{m,me} - T_{2,me}) + I^2R_{me} \tag{4}$$

$$m_{ce}c_{p,ce}\frac{d(T_{m,ce})}{dt} = 2K_{ce}(T_{1,ce} - T_{m,ce}) - 2K_{me}(T_{m,ce} - T_{2,ce}) \tag{5}$$

where the subscripts 1 and 2 represent the left and right sides of material along the direction of heat flow. Since the ceramic-metal interface and the metal-thermoelectric interface are ideal surfaces without considering contact resistance, then the heat transferred to and out of the interfaces is balanced. Thus, the boundary conditions for interfaces include, $\dot{Q}_{2,me} = 2\dot{Q}_{c,te}$ and $N\dot{Q}_{1,me} = \dot{Q}_{2,ce}$ for the cold side, and $\dot{Q}_{1,me} = 2\dot{Q}_{h,te}$ and $N\dot{Q}_{2,me} = \dot{Q}_{1,ce}$ for the hot side of a TEM.

Modelica model can reduce such a physical problem to a thermal and electrical resistance network. The input of the TEM model is the electrical current signal (I), and the outputs of the TEM model are the temperature (T_{ce}) and heat flux ($\dot{Q}_{ce}$) at two ends of the TEM. An electric circuit is generated with an electric source, and resistances of metal interconnects in the electrical field and n-type and p-type thermoelectric elements. Once the current signal is given, the Joule heating due to electrical resistance is computed and transferred to the thermal field. The heat generated due to the Peltier effect can also be computed by combining the current signal with material properties and the computed surface temperature. In the thermal field, each thermoelectric element, metal interconnects, and ceramic substrates are modeled by 2R1C models. Joule heating and the Peltier effect are considered heat sources applied to the capacitance.

2.3 TBE Prototype Model

The TBE model contains a TEM model, an envelope model, and boundary conditions (i.e., heat sinks and fans). The TBE prototype studied in this paper has three TEMs connected both electrically and thermally in series. The detailed design of the TBE prototype can be found in the literature (Liu et al., 2022). The 3D rendering image of a prototype is shown in Figure 1. Two TEM surfaces are attached with heatsinks and fans to help dissipate heating and cooling power generated from the surface. The overall thermal resistance of the heatsink and fan in operation are obtained from manufacturers' datasheets. Thus, the heat balances for heat sink at TEM's hot and cold sides are expressed by:

$$m_{hs}c_{p,hs}\frac{dT_{h,hs}}{dt} = \dot{Q}_{h,ce} - h_{hs}A_{hs}(T_{h,hs} - T_{h,air}) \tag{6}$$

$$m_{hs}c_{p,hs}\frac{dT_{c,hs}}{dt} = \dot{Q}_{c,ce} - h_{hs}A_{hs}(T_{h,hs} - T_{c,air}) \tag{7}$$

Moreover, the model employed a three-dimensional thermal network to capture the spatial heat transfer of the envelope associated with the prototype design. The TBE prototype model developed in Modelica combined the equations above by referring to the physical construction of the TBE prototype.

2.4 TBE-Building System Model

As shown in Figure 3, the TBE-building system model has four parts: a TBE prototype model, a heat balance model for a residential house, the weather data as an input, and a PID controller to control the current provided to the TBE prototype for room temperature regulation.

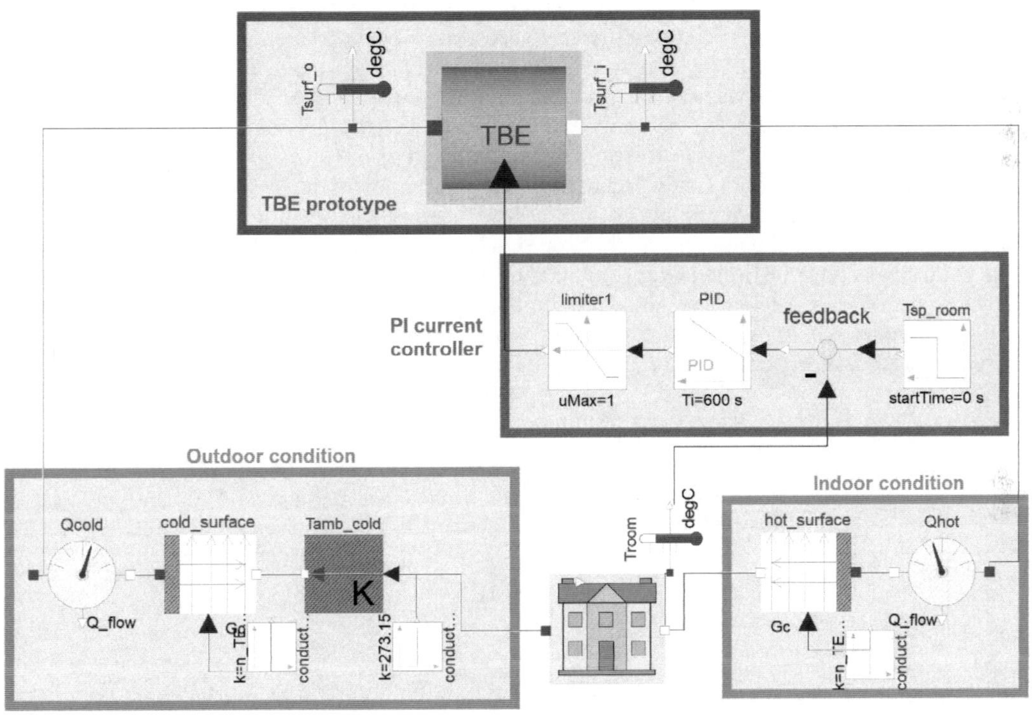

Figure 3: Thermal network of a TBE-Building system model using Modelica language

In this paper, the heating performance of the TBE-building system was investigated during the heating season. Given the outdoor and indoor air temperatures, the controller regulates the power input to the TBE prototype based on the temperature difference between the indoor temperature and the setpoint. Meanwhile, the TBE prototype dissipates heat to the indoor space to meet the heating demand required by the house and adjusts the air temperature to reach a set point of 20 °C in this study. The residential home has a volume of 3.05 × 3.05 × 2.44 m^3 (10 × 10 × 8 ft^3) with 0.102 m (4-in) thick foam boards covered on four exterior walls and floor with a thermal resistance of R-20 and a 0.305 m (12-in) thick foam board covered on the roof with a thermal resistance of R-60. The house model includes the envelope heat losses, 18 CFM infiltration, and 80 W internal heat gain in the heat balance. Both daily and seasonal performance were investigated under different outdoor conditions: three heating days and a heating season from January to March in Chicago.

2.5 Solver and Performance Indicators

The overall model is a combination of ordinary differential equations (ODEs) with respect to time only, which is also named the differential-algebraic equation (DAE) system. The Modelica model can be transformed into an ODE representation to perform simulation by using numerical integration methods. The solver is DASSL to solve the ODEs using the implicit and backward differentiation formula. The model can compute transient profiles of the air and surface temperatures and heat flux for desired materials and intersections.

The performance analysis of this TBE-building system includes the study of room air temperature, heat deliverable to the room, electricity consumption, and heating COP of TEMs. In this study, the COP of the TEM ($COP_{h,tem}$) is the ratio of heat flux at hot surfaces ($\dot{Q}_{h,tem}$) to the electrical power consumption of the module (W_{tem}), given by

$$COP_{h,tem} = \frac{\dot{Q}_{h,tem}}{W_{tem}} \tag{8}$$

3. MODEL VALIDATION

A convective TBE prototype designed by the research team was tested in Herrick Lab at Purdue. The test setup mainly contains a TBE prototype for heat generation installed between a pair of psychrometric chambers to simulate different indoor and outdoor conditions, a DC power supply unit for controlling power input, and thermocouples and DAQ for temperature measurement. Three TEMs used are from the market (model number: TEM-199-1.4-0.8, TEM-127-1.4-2.5, and TEM-199-1.4-0.8) and connected thermally and electrically in series. A tall heat sink and DC fan were used to dissipate the heat to the surroundings effectively. The experimental data (Liu et al., 2022) was used to validate the prototype model. Two groups of tests in heating scenarios were selected, with the indoor temperature at 22 °C and the outdoor temperature at 12 °C and current inputs of 0.5 A or 1 A. Simulation results from the TBE prototype model were compared with the measured data. The comparison of TEM's surface temperatures versus time is presented in Figure 4. The solid lines show TBE modeling results, and they yield good agreement with the measurement data represented by the dashed lines. The most significant discrepancy between the measured data and the modeling results is about 0.47 °C in winter scenarios, 3%. In conclusion, the Modelica model can accurately simulate the heat and electric transfer in a TBE prototype and can be used to predict the heating behavior in building system simulation.

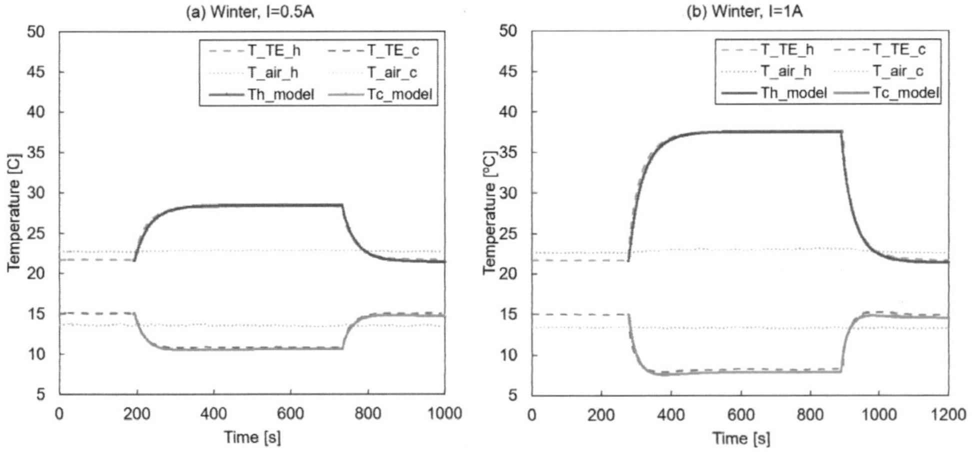

Figure 4: Comparison of test results and model-predicted temperatures for a TBE with (a) I=0.5A and (b) I=1A in winter.

4. SIMULATION RESULT

The developed TBE-building system model is used to predict the performance of the TBE prototype using TEMs (HP-127-1.4-2.5) for heating a residential house with a volume of 22.6 m^3 (800 ft^3). The number of TEMs was selected for two reasons. One must provide sufficient heating and maintain the indoor air temperature at 20 °C. The other reason is to lower the indoor surface temperature under 46.7 °C for safety operations. Two cases were investigated in this study, including the operation of the TBE-building system at an outdoor air temperature fluctuating within -2.5–2.5 °C and a historical weather condition in Chicago from January to March.

4.1 System Performance in Three Heating Days

Figure 5 displays the simulation result of using TBE with 22 TEMs for heating a residential house for three days. As shown in Figures 5(a) and (d), a cosine wave is used to simulate the change in outdoor air temperature within -2.5–2.5 °C. With different outdoor air temperatures, the heating demand of the space to reach a setpoint of 20 °C is different. A PID controller controls the current input and the heat rate deliverable to the space changes from 389–513 W accordingly (Figures 5(b) and (e)). As a result, the indoor air temperature can be maintained at 19.86–20.14 °C. The system has a fast response and can accurately regulate indoor air temperature. The studied system demonstrates a heating COP of TEMs ($COP_{h,tem}$) at 1.00–1.14 and 1.05–1.22, for TEMs with boundary thermal resistance of 0.9 K/W and 0.75 K/W, respectively, excluding the system startup.

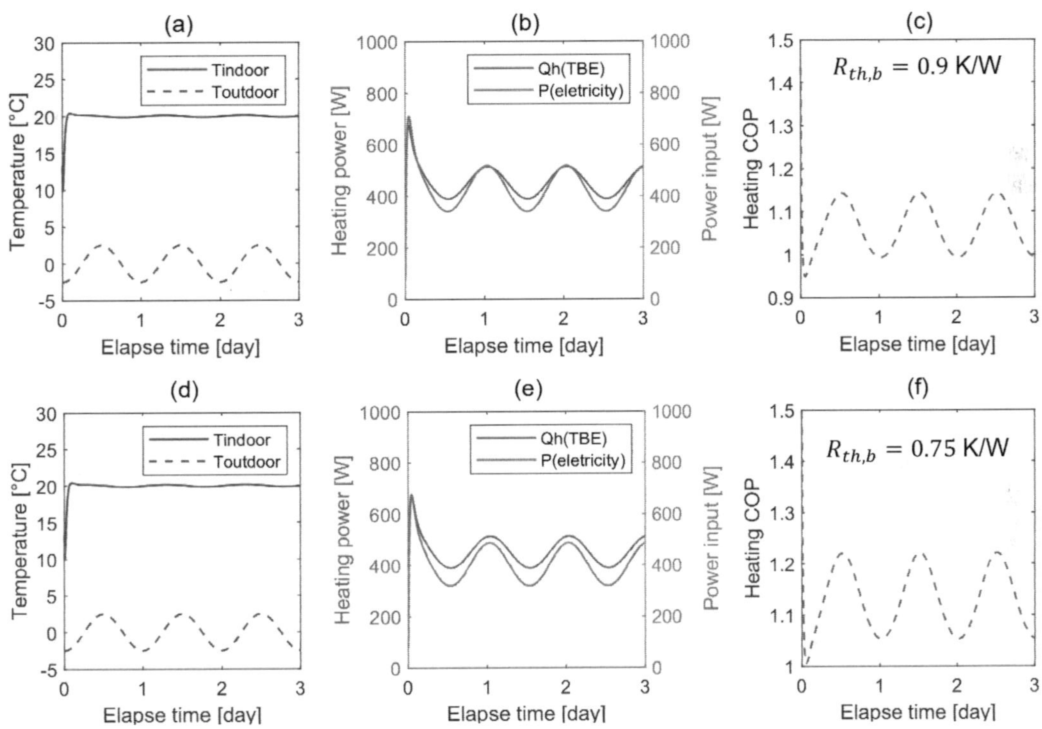

Figure 5: Simulation results of TBE-building system on three heating days: (a) indoor and outdoor temperatures, (b) heating power and electricity, and (c) heating COP with thermal boundary resistance of 0.9 K/W; (d) indoor and outdoor temperatures, (e) heating power and electricity, and (f) heating COP with thermal boundary resistance of 0.75 K/W.

4.2 System Performance in a Heating Season

Figure 6 presents the simulation result of heating a residential house from January to March in Chicago using a TBE system with 35 TEMs. As shown in Figure 6(a), the outdoor air temperature (T_o) fluctuates from -26.3 °C to +21.5 °C. The average value of these 90-day temperatures is -2.36 °C. The TBE system is the only source of heating power for the space. The indoor air temperature (T_i) can be kept at 20 °C, and the heating demand can be fulfilled by TBE even under extremely cold outdoor conditions. When T_o is -26.3 °C, a large current is needed to provide a heating power of 1217 W to the space (Figure 6(b)). The TBE system demonstrates a high ability in air temperature elevation. In this operating condition, the heating COP of TEM ($COP_{h,tem}$) is lower than one due to the significant heat loss, as can be seen in Figure 6(c). The COP of TBE varies with different outdoor air temperatures. A higher T_o leads to a higher heating COP. When T_o is -5 °C, $COP_{h,tem}$ is around 1. When T_o is +5 °C, $COP_{h,tem}$ is around 1.34. The seasonal heating COP of TEMs is about 1.07, which is higher than the efficiency of an electric heater.

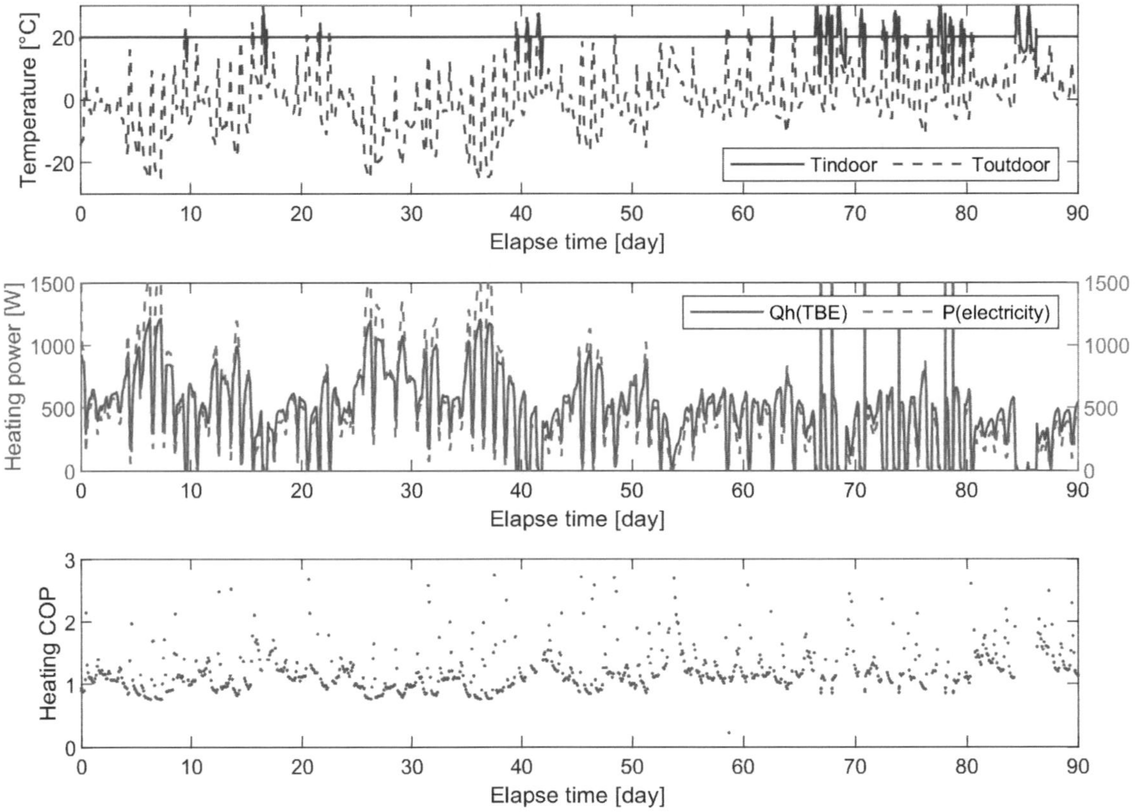

Figure 6 Simulation results of the TBE-Building system model in a heating season: (a) indoor and outdoor temperatures, (b) current input and heat deliverable, and (c) COP.

5. DISCUSSION

The TBE system integrates the building envelope, heating generation system, and heating delivery system into one system to provide accurate temperature control and reduce the heat loss from traditional HVAC subsystems. Thus, the performance of the TBE system is sensitive to outdoor conditions. This study combined a building envelope model, a control model, and a TBE model into one TBE-Building system model to comprehensively predict the system's transient behavior under varying operating conditions. The critical parameters influencing the COP include the design of TBE (e.g., the geometry of TEM, the heat dissipation technology, and the heat loss), the number of TEMs used, and the temperature difference between indoor and outdoor air, etc.

Energy-efficient technology for heat dissipation is the key to improving a TBE's heat deliverable and COP in the TBE design. In the study, every TEM is attached to heat sinks and a fan with thermal resistance at around 0.75–0.9 K/W on hot and cold surfaces. COP can be further improved by reducing the fan's power consumption or reducing the thermal resistance at TEM's boundary. Reducing heat loss is also critical in the TBE design. The heat loss might come from the heat conduction across the TEM, which can be reduced by optimizing thermoelectric elements' design (i.e., dimension) and physical properties (i.e., thermal conductivity) for lower heat flux through conduction. The leakage of the envelope also needs to be avoided to lower the heat loss through infiltration.

In terms of system sizing, the selection of the number of modules and the corresponding current input also affects the system performance. The larger current allows fewer TEMs but leads to a risk of high surface temperature exceeding the highest limit of a TEM and associated with more significant heat loss through conduction due to a large temperature difference. Therefore, the sizing of the system depends on the maximum operating current, voltage, temperature of a TEM, the outdoor temperatures (or the climate zone), the thermal load of buildings, the cost, etc.

A TBE's COP is a function of indoor and outdoor air temperatures. The greater the temperature difference, the more significant the thermoelectric material's surface temperature difference needs to be overcome. As a result, there is a greater loss of thermal surface (indoor surface) due to heat transfer. The model developed in this study indicates that the heating COP can be near 1.1 in winter operation in Chicago, suggesting that the TBE system is an energy-efficient alternative to the traditional electric heater. Some strategies to cope with extreme weather can be referred to. Firstly, heat recovery technology can be used to reduce the temperature difference by increasing the air temperature on the cold side, thus improving the TEM performance. In addition, using high-performance TE materials that can hold a larger temperature difference to reduce heat loss can also improve the overall TBE performance.

6. CONCLUSION

This study developed a TBE-Building system model for space heating using Modelica language. The model integrated models of the building's heat balance, a PI controller, and the TBE prototype to predict the transient performance of air temperatures, heat fluxes, and energy efficiency under dynamic operating conditions.

The daily simulation result demonstrates that when the outdoor temperature fluctuates from -2.5 to +2.5 °C, the TBE with 22 TEMs can deliver heating power of 389–513 W to the space to maintain a desired indoor temperature, with heating COP of TEMs ($COP_{h,tem}$) at 1–1.34, depending on outdoor conditions and boundary thermal resistances. Seasonal simulation in a colder climate with 35 TEMs presents the seasonal heating COP ($COP_{h,tem}$) of 1.07.

Key parameters influencing the heating COP include temperature difference between indoor and outdoor air, the electricity input to the TEM and the number of TEMs used, the design of the TEM construction, and the heat transfer on the wall. In addition to the thermoelectric materials. Strategic items to improve the performance might be energy-efficient heat removal, high-performance (larger ZT) thermoelectric materials, optimum TEM design, and minimization of heat leak in TBE devices.

In summary, the TBE-building system developed in this paper provides critical guidance to the design and operation at a system application level. The TBE system demonstrates a better heating performance than a backup electric heater for the state-of-the-art heat pump, which has a lower COP under a cold climate. It also suggests potential energy-efficient replacement options for the traditional natural gas furnaces and electric heaters for space heating.

NOMENCLATURE

Symbol

c	specific heat	(J-kg^{-1}-K^{-1})
I	current	(A)
K	heat conductance	(W-K^{-1})
$\dot{Q}$	heat flux	(W)
R	electrical resistance	(ohm)
S	Seebeck coefficient	(V-K^{-1})
T	temperature	(°C)
t	time	(sec)
V	volume	(m^3)
ρ	density	(kg-m^{-3})

Abbreviation

DC	direct current
PI	proportional-integral
TBE	thermoelectric building envelope
TEM	thermoelectric module
2R1C	two-resistance-one-capacitance

Subscript

c	cold side
ce	ceramic substrate
h	hot side
hs	heat sink
me	metal interconnect
te	thermoelectric element
tem	thermoelectric module

REFERENCES

Ahmad Gondal, I. (2019). Design and experimental analysis of solar thermoelectric heating, ventilation, and air conditioning system as an integral element of a building envelope. *Building Services Engineering Research and Technology*, *40*(2), 220–236. https://doi.org/10.1177/0143624418814067

Aksamija, A., Aksamija, Z., Counihan, C., Brown, D., & Upadhyaya, M. (2019). Experimental Study of Operating Conditions and Integration of Thermoelectric Materials in Facade Systems. *Frontiers in Energy Research*, *7*(FEB), 6. https://doi.org/10.3389/fenrg.2019.00006

Hagenkamp, M., Blanke, T., & Döring, B. (2022). luo *International Journal of Energy and Environmental Engineering*, *13*(1), 241–254. https://doi.org/10.1007/S40095-021-00424-X/TABLES/3

Ibañez-Puy, M., Martín-Gómez, C., Bermejo-Busto, J., Sacristán, J. A., & Ibañez-Puy, E. (2018). Ventilated Active Thermoelectric Envelope (VATE): Analysis of its energy performance when integrated in a building. *Energy and Buildings*, *158*, 1586–1592. https://doi.org/10.1016/J.ENBUILD.2017.11.037

Irshad, K., Habib, K., Saidur, R., Kareem, M. W., & Saha, B. B. (2019). Study of thermoelectric and photovoltaic facade system for energy efficient building development: A review. *Journal of Cleaner Production*, *209*, 1376–1395. https://doi.org/10.1016/J.JCLEPRO.2018.09.245

Liu, X., Qu, M., & Yazawa, K. (2022). Performance Assessment of Integrated Building Envelopes Using Thermoelectric Modules for Temperature Regulation. *2022 ASHRAE Winter Conference*.

Luo, Y., Zhang, L., Liu, Z., Yu, J., Xu, X., & Su, X. (2020). Towards net zero energy building: The application potential and adaptability of photovoltaic-thermoelectric-battery wall system. *Applied Energy*, *258*, 114066. https://doi.org/10.1016/J.APENERGY.2019.114066

Martín-Gómez, C., Zuazua-Ros, A., del Valle de Lersundi, K., Sánchez Saiz-Ezquerra, B., & Ibáñez-Puy, M. (2021). Integration development of a Ventilated Active Thermoelectric Envelope (VATE): Constructive optimization and thermal performance. *Energy and Buildings*, *231*, 110593. https://doi.org/10.1016/j.enbuild.2020.110593

Wang, C., Calderón, C., & Wang, Y. (2017). An experimental study of a thermoelectric heat exchange module for domestic space heating. *Energy and Buildings*, *145*, 1–21. https://doi.org/10.1016/J.ENBUILD.2017.03.050

Zuazua-Ros, A., Martín-Gómez, C., Ibañez-Puy, E., Vidaurre-Arbizu, M., & Gelbstein, Y. (2019). Investigation of the thermoelectric potential for heating, cooling, and ventilation in buildings: Characterization options and applications. *Renewable Energy*, *131*, 229–239. https://doi.org/10.1016/J.RENENE.2018.07.027

Zuazua-Ros, A., Martín-Gómez, C., Ibáñez-Puy, E., Vidaurre-Arbizu, M., & Ibáñez-Puy, M. (2018). Design, assembly and energy performance of a ventilated active thermoelectric envelope module for heating. *Energy and Buildings*, *176*, 371–379. https://doi.org/10.1016/j.enbuild.2018.07.062

ACKNOWLEDGEMENT

This material is based upon work supported by the National Science Foundation under Grant No. CBET-1805818.

A Framework for Analyzing Widespread Grid Intervening Technologies: A Case Study of Heat Pump-Integrated Thermal Energy Storage Systems in Buildings

Jason Hirschey[1]*, Richard A. Simmons[1], Tim J. LaClair[2], Kyle R. Gluesenkamp[2], Samuel Graham[1]

[1]George W. Woodruff School of Mechanical Engineering, Georgia Institute of Technology
North Ave NW, Atlanta, GA, USA

[2]Oak Ridge National Laboratory
1 Bethel Valley Rd, Oak Ridge, TN, USA

* Corresponding Author (jhirschey13@gatech.edu)

ABSTRACT

Heating and cooling demands of buildings require an immense amount of energy from the electric grid. Extreme temperatures or other weather events may stress the grid further by increasing the heating or cooling demands of buildings to maintain a livable indoor environment. When spread across the grid, this increase in energy demand requires grid operators to increase energy supply, often by using so-called peaker power plants. These plants are generally combustion turbines burning either natural gas or diesel fuel. As such, the emissions associated with these plants are higher than most baseload grid energy generation emissions. A building's peak energy demand from the grid may be reduced by employing energy storage systems that activate during these periods of high building energy demand to replace or supplement energy from the grid. Heating and cooling can benefit from thermal energy storage (TES). TES integrated into a heat pump (HP) system may reduce a building's peak energy demands during extreme temperature conditions. When deployed on a large-scale, the cumulative effect of HP-TES may reduce grid demand during critical times and abate the need for peaker power plants. This study establishes a framework around which the effect on grid emissions (CO_2, NO_x, and SO_x) of intervening technologies, such as TES, may be examined on a large scale. As a case study, residential HP-TES is analyzed for select households in the Southern US.

Keywords: *thermal energy storage, heat pump, residential buildings, emissions, intervening technologies*

1. INTRODUCTION

Energy storage has been deemed a critical component for the complete transition to a renewable energy-based energy grid (Gallo *et al.*, 2016). As such, energy storage is a rapidly advancing field of study with a large body of work dedicated to thermal energy storage (TES) systems. Generally, TES systems operate by storing and releasing heat in a narrow temperature range. Near-ambient temperature TES systems are well-suited for applications where low-grade heat is the primary end use: refrigeration, domestic hot water, and indoor building thermal conditioning.

Residential buildings occupy a large share of the US national energy consumption, upwards of 22% (*Monthly Energy Review*, 2021). Much of this energy consumption is spent on thermally conditioning the indoor spaces. This work presents a framework for analyzing the scaled impacts of residential heat pump-integrated TES (HP-TES) systems for indoor thermal space conditioning. As a case study, this framework is used to conjecture on the effect of grid-level emissions produced within the Southern Company Balancing Authority territory in 2018 attributed to a hypothetical widespread adoption of residential HP-TES for select residential building types.

Figure 1 illustrates the operating principle of such HP-TES systems for both heating and cooling. An HP-TES system acts as an intervention technology on building energy demand during peak hours by providing a more efficient HP operation to thermally condition the indoor space. TES provides an isothermal source or sink for the HP to couple. In doing so, the HP-TES may be acting with a more favorable temperature gradient to thermally condition the indoor space than if the HP is coupled to the ambient. During this discharging mode, the HP may use less electrical energy than the conventional HP system. This relative reduction in energy may be timed to coincide with the daily extreme ambient temperature, the peak hours. Much like an electrical battery, the TES needs to be recharged. This is done by coupling the HP to the TES and the ambient, ideally by choosing the time in which the ambient temperature is most favorable for the TES recharging, the off-peak hours. By taking advantage of naturally occurring diurnal ambient temperature changes, HP-TES systems may shift building energy consumption from peak hours to off-peak.

In this current work, the same TES system is used for both indoor heating and cooling applications. For this model, the TES temperature is chosen as 20°C. It is assumed that no temperature gradients exist within the TES, and it is well-oversized such that its capacity to condition the building in any 24-hour period is never deplete. Charging is still necessary to return the TES to a full state of charge.

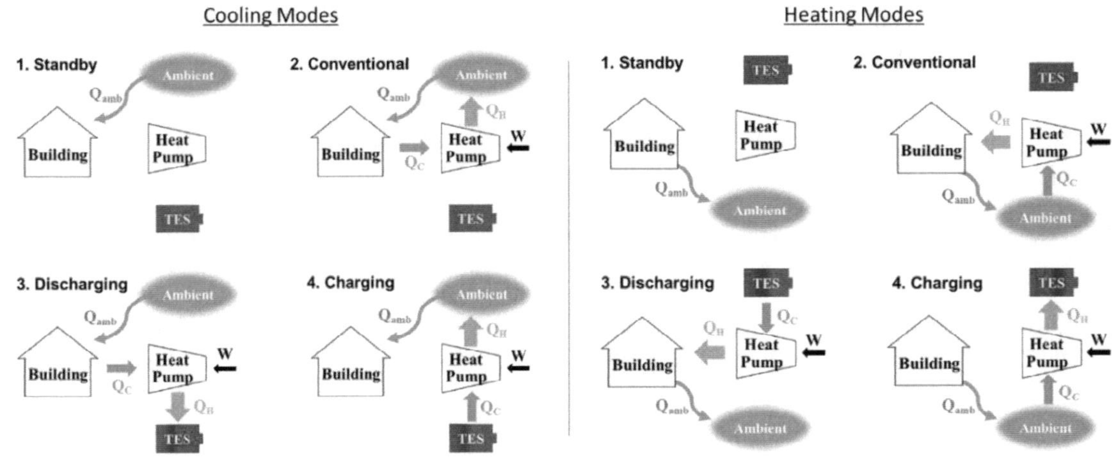

Figure 1. Operating modes of HP-TES system for heating and cooling

This present work builds a framework to analyze large-scale impacts of HP-TES systems but may be generalized for other intervening technologies. This framework is loosely based on the analysis done by Siler-Evans *et al.* (2012) as a method of determining marginal technologies and emissions. Compared to that work, this framework is more localized and specific; evaluating emissions of a single year and region, and speculating the impact a specific intervening technology may have. There are other tools to estimate emissions impacts, such as the Environmental Protection Agency (EPA) AVoided Emissions and geneRation Tool (AVERT) and some consumer-facing products.

The framework developed here is employed in a cast study evaluating the widespread effect that HP-TES systems may have on grid energy generation and emissions if these systems replaced conventional HP & air conditioning units in select residential homes within the Southern Company (SOCO) Balancing Authority (BA) territory in 2018. This paper outlines the framework in the context of the case study: discussing how the building model, the HP model, TES model, and the emissions model coalesce to estimate the large-scale effects of HP-TES systems.

2. MODEL OVERVIEW

2.1 Case Study Scope

This case study is limited to select residential households in the 2018 SOCO BA territory. However, the framework built will enable future examination of other building types, sizes, regions, times, generation portfolios, and what-if scenarios for TES and other intervening technologies. The results discussed herein are not widely applicable to other regions, HP-TES systems generally, or other energy storage technologies; care must be taken to understand the assumptions that led to the presented results.

The overall framework is composed of several component-level models that coalesce into one unified model. Each component-level model is discussed in Sections 2.3 – 2.5. The methodology for scaling the component-level models across the entire region is discussed in Section 2.6.

2.2 Weather Data

The weather data was collected from National Oceanic and Atmospheric Administration (NOAA) Local Climatological Data (LCD) Stations located within the territory of interest. This data provides hourly data presented in METeorological Aerodrome Reports (METAR) FM-15 or FM-16 format, historically reported for aviation purposes. The data includes dry bulb temperature, relative humidity, wind speed, any overhead sky conditions including cloud altitude. For the simplicity of this case study analysis, only dry bulb temperature is considered. Future versions of this model may correlate sky conditions and time of day and year to solar irradiance on a building or seek to find historical data on geospatial solar irradiance. This model does not consider relative humidity, though this is important for dehumidification purposes. These are outside the scope of this current report.

The NOAA LCD data was collected for 119 weather stations across SOCO BA territory in 2018. However, only a subset of 13 locations is used in this current analysis, and the data are extrapolated and scaled across the entire territory. Future versions of this model may apply more advanced regression and interpolation between weather stations to better characterize the effects of weather to the energy usage of households throughout the territory. Figure 2 shows the map of selected weather stations with the whole territory outlined.

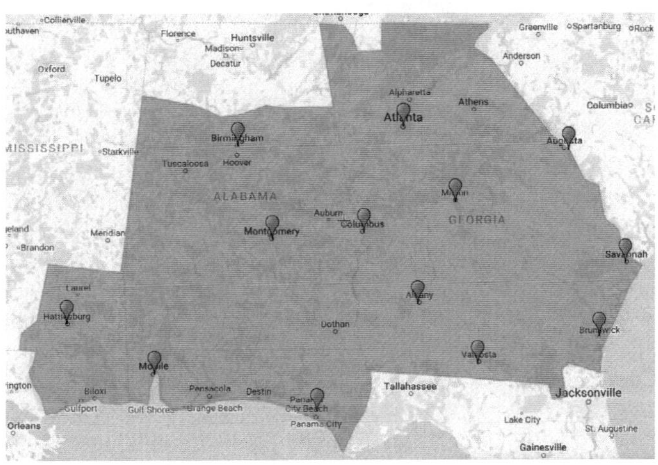

Figure 2. Map of selected weather stations with 2018 SOCO BA territory outlined

Alternatively, Typical Meteorological Year (TMY) data from the National Renewable Energy Lab (NREL) was considered. However, historical weather of the same year as power plant generation was deemed necessary for this framework; a correlation exists between local weather conditions and energy demand, and this was not captured properly with TMY data.

2.3 Vapor Compression Heat Pump and Thermostat Model

An R410A vapor compression heat pump (VCHP) model was written in MATLAB using open source CoolProp refrigerant data (Bell *et al.*, 2014). The condenser and evaporator temperatures were selectively tied to the indoor building, outdoor ambient, or TES temperatures. By assuming constant approach temperatures, superheating, and subcooling, the model iteratively solved for the heat removed from the evaporator side and pumped to the condenser side. The VCHP model includes a compressor operating with a constant volumetric displacement rate which was sized to provide approximately 4 refrigeration tons of cooling.

Figure 3 shows an example operation of the HP-TES model for a single day (Jan 12, 2018) as it switches between different operating modes and compares it to a conventional HP system. The operating mode of the HP is displayed on the upper graph. The electrical energy to operate the HP is shown on the lower graph. The conventional system is only capable of switching between conventional operation and standby. The HP-TES system operates in discharging mode (utilizing the TES for indoor thermal heating) at the beginning of the day, and its associated electrical energy use is less during these hours. During midday, however, the HP-TES enters charging mode and switches between conventional heating mode to heat the indoor space and charging mode to heat the TES. Its average energy consumption is higher than the conventional system during these hours.

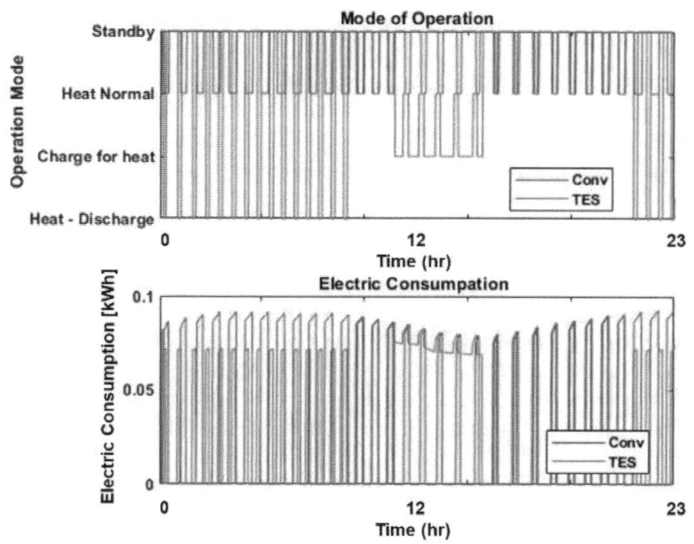

Figure 3. Operating mode and energy consumption of conventional HP system and a HP-TES

The VCHP uses a day-ahead weather forecasting to predict the hours of the next day's ambient temperature extremes (high and low) using the NOAA LCD data. This is highly idealized as the data contains perfect information. However, the level of detail necessary for this control algorithm could likely rely on real weather forecasting reports without issue.

This thermostat model records information on the thermal conditioning inertia of the building by tracking the total time the building is engaged in either heating or cooling for the past 72 hours. This inertia then dictates the preferred mode for TES-assisted on-peak conditioning and ensures the HP will not overcorrect itself. For example, in the winter months where heating is more likely to be required, the system will prime the TES for heating modes. If an unseasonably warm day occurs and building cooling is required, the HP will couple the building to the TES so that the building cooling recharges the TES for a future heating operation as the inertia would indicate as the more likely mode. Furthermore, if the HP overshoots the building setpoint temperature, this thermal conditioning inertia will prevent the system from switching to the opposite mode to overcorrect itself and will instead be put on standby. Implementation of this is elementary in conventional HP systems, but this dual mode HP-TES model adds some complexity that required this additional control algorithm. Future versions of the model may improve upon this.

Determined in part by the thermal conditioning inertia of the thermostat system, the thermostat bounds charging and discharging around the extreme temperature hours accordingly. For example, discharging the TES to cool the indoor space is done only during the five-hour window bounding the hour of the day's ambient high temperature, and charging to cool the TES is done only during the window bounding the hour of the day's ambient low temperature. No HP-TES decision making was influenced by emissions, utility rates, etc. Future iterations of this model may include such factors for advanced emissions mitigation or to maximize consumer monetary savings.

2.4 Building Model

The building type analyzed is estimated to be a 2000-3000 square foot, single-family detached home with central air conditioning as the primary source of cooling and central heat pump heating as the primary source of heating.

The building model was built based on the results found in Cui *et al.* (2018). The building model is a 4R4C thermal circuit. The RC constants were adjusted slightly from the source, but retain the same order of magnitude as presented there. The building model calculates the indoor space temperature from heat gains or losses from the ambient, and the heat from the VCHP. Dehumidification of the indoor space is not considered in the present report, but it is understood that dehumidification is an important use of HPs operating in this region. Future iterations of the model will include dehumidification of indoor spaces. The RC constants were validated by comparing the building's modeled energy consumption for heating and cooling to responses from the 2015 Energy Information Administration (EIA) Residential Energy Consumption Survey (RECS) for similarly featured buildings in the South and South East Central regions (*Residential Energy Consumption Survey (RECS)*, 2015).

Figure 4 shows the total hourly HP energy consumption for a residential building modeled in Macon, GA for January 2018. The energy consumption for the conventional system is highest in the early morning hours, which roughly corresponds to the lowest average hourly temperature. The HP-TES system has the lowest energy consumption during these hours due to the TES intervention. However, the HP-TES system uses more energy during midday when the TES is recharged. For this example, the HP-TES energy consumption to maintain a thermally conditioned space is nearly inverted compared to the conventional HP system.

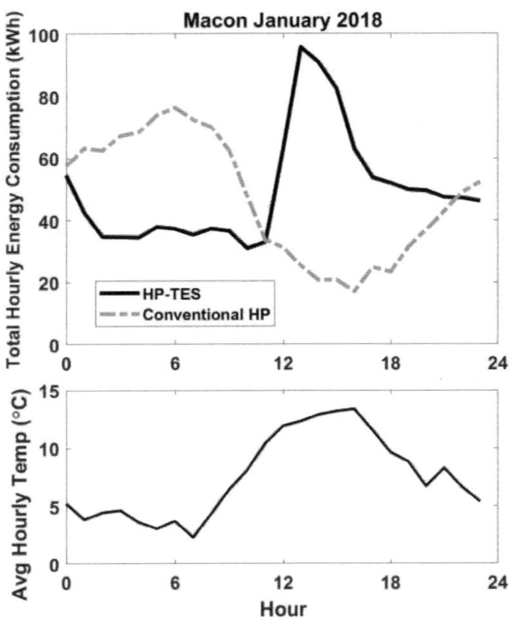

Figure 4. Total hourly HP-TES energy consumption modeled for Macon, GA in January 2018

2.5 Emissions and Grid Generation Model

Power plant emissions data were gathered from the EPA Continuous Emissions Monitoring (CEM) program, regulated by the US Code of Federal Regulations, Title 40, Chapter I, Subchapter C, Part 75, Continuous Emissions Monitoring. The data were accessed through the EPA Field Audit Checklist Tool (FACT). Data were collected for 2018 across all power plants operating in the Southern Company Balancing Authority territory, approximately 63 generating plants and 316 reporting generating units or monitoring locations. The CEM data collected contains only the fossil fuel power generation and do not include solar, wind, hydro, or nuclear energy sources. It is assumed that the non-fossil fuel and nuclear generation is unaffected by intervening technologies; these technologies are typically non-dispatchable or baseload. Thus, the analysis proceeds treating only fossil fuel generation as marginal technologies that respond to demand changes across the grid.

In the CEM data, the emissions from different generating technologies can be determined. The four main fossil fuel technologies in this region are natural gas combustion turbines (NGCT), natural gas combined cycle (NGCC), natural gas boilers (NGB), and coal. Oil or diesel combustion turbines comprise a minuscule portion of the energy generation in this region and are typically only called upon during transient periods to ramp up or complement the other four technologies. Figure 5 shows the relationship between change in generation of these technologies and the associated change in emissions. The slope of the line indicates the change of tons of CO_2 emitted by the change in MWhe of generation from that technology. The outlying data points in coal may be due to startup or shutdown procedures, or may be undercounting or over counting if the CEM monitoring location is temporarily moved or the reporting is out of sync. Future work is necessary to investigate this further and identify the source of outliers. However, these outlying data do not affect the linear trend observed. As expected, coal is the most CO_2 polluting technology (at about 0.98 t/MWh) followed by NGCT (0.62 t/MWh), NGB (0.53 t/MWh), and NGCC (0.42 t/MWh), and all agree with (Bruckner *et al.*, 2014). A similar regression was performed for SO_2, NO_x, and Hg emissions data collected through CEM, however, these did not yield a good linear fit (not shown here). Future work will investigate this further.

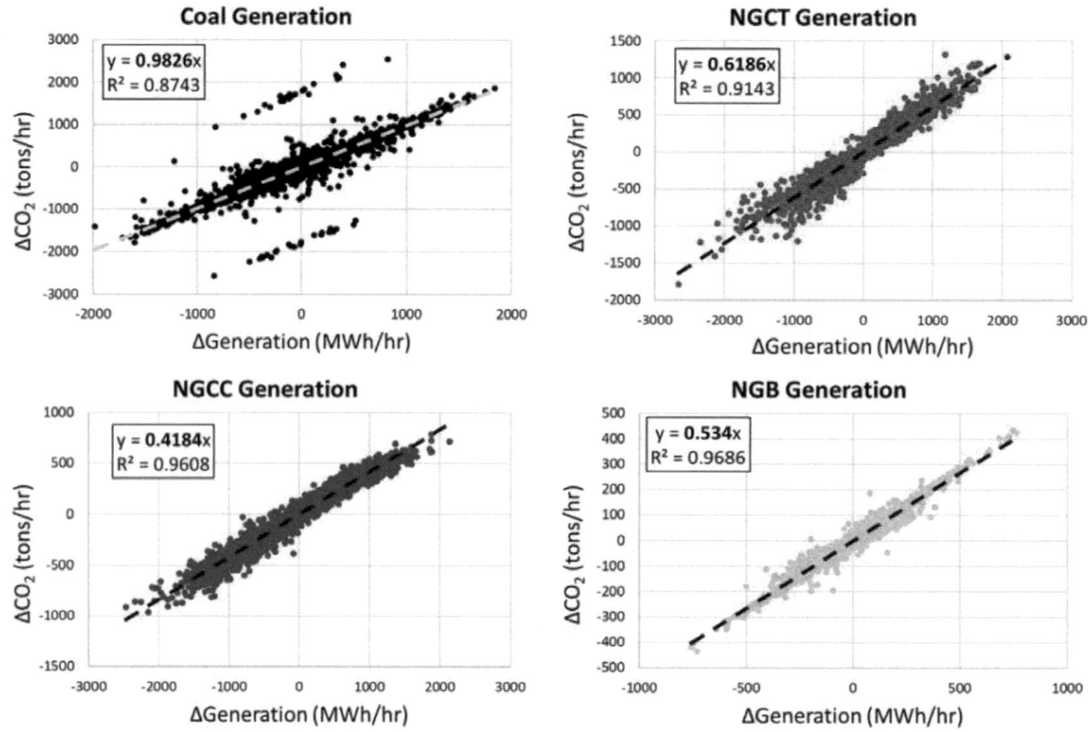

Figure 5. Emission factors (EF) for main fossil fuel technologies from 2018 CEM data

As the demand on the grid changes, grid operators respond by dispatching different power plants and different generation technologies. Deciding which plants and technologies are dispatched is dependent on the current total generation, the capacity factor of online units, operating costs, demand forecasts, and more. Those generating units and technologies that are dispatched to meet demand are on the margin. The model developed to determine the marginal technology, and thus associated emissions, is loosely based on the methodology outlined in Siler-Evans et al. (2012).

For every day of each month, the hourly generation and emissions of each technology are summed. This creates a profile containing all energy generation and emissions for a month in a 24-hour format. For example, all data corresponding to 2pm are summed across each day of the month. To determine a typical hourly marginal technology dispatch for that month, the change in hour-to-hour generation of each technology is compared to the hour-to-hour change of all fossil fuel generation. This results in a normalized marginal factor (MF) which weights the magnitude of each technology to respond to changes in generation across the whole system.

To illustrate the MF, Figure 6 shows the results for July 2018. In this figure when the graph is negative, the total system generation is decreasing and when the graph is greater than zero, the total system generation is increasing. As an example, in hour 13, the MF for NGCT is ~0.33. This indicates that as the system generation increases, NGCT account for ~33% of this new generation; NGCT are turning on to meet demand for this hour. In hour 22, the MF for is ~-0.44. This indicates that as the system generation decreases, 44% of this decrease are NCGT shutting down or reducing generation. In hour 1, the MF for NGCT is ~0.05; any change in system demand is not significantly the result of NGCT. This illustrates NGCT used as a marginal technology during this month and for specific hours. Furthermore, the MF from Figure 6 allow speculation as to how the system generation portfolio might respond if demand changes for a specific hour as the result of an intervening technology. These MF are used to determine the effect of intervention as if it happened proportionately across all technologies.

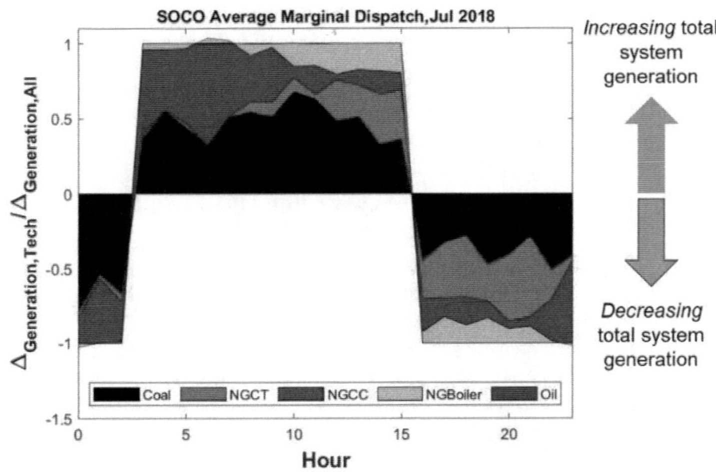

Figure 6. Marginal factors (MF) of different generation technologies for July 2018

It may appear that coal is a dominant technology on the margin. However, coal does not respond to changes in demand in the same way NGCT is dispatched; operators generally run coal plants on a predetermined schedule spanning several days for forecasted baseload. This schedule is based on historical data and demand predictions made by the system operator hours or days in advance. Coal is not load-following and is not well equipped to respond to short-term hourly demand fluctuations. By evaluating these MF data over a month, this fact is obscured by day-to-day variations.

2.6 Model Scalability

To estimate the number of affected households, data from EIA RECS was used. Filtering for the South and South East Central regions, the approximate number of households matching the building model analyzed was found. However, the EIA regions do not coincide with the SOCO region under analysis. Therefore, county-level population data from the US Census Bureau was used to determine the approximate number of persons in the SOCO territory relative to the EIA regions. This provided a weight that was applied to the number of households from the EIA RECS microdata. This estimation resulted in approximately 330,000 households within in the SOCO region that match the housing profile modeled here.

The county-level US Census Bureau population data is also used to weight the building model results around each of the 13 NOAA weather stations. It is assumed that all 330,000 households utilize an identical HP-TES system and use the same decision-making operating logic. All households surrounding one of the weather stations are responding to its weather data only. Differences arise across the territory due to slightly different ambient conditions recorded at different weather stations, and because this region spans two time zones. All aggregated data is converted to EST, but households will operate in accordance with their local time and weather conditions.

3. RESULTS & DISCUSSION

3.1 Aggregated Building Energy Consumption

The building model integrated with and without HP-TES model were run for the 13 NOAA weather stations. The models without TES are assumed to be the baseline energy consumption portfolio of the building. The models with HP-TES systems show the energy consumption portfolio if this intervention technology were implemented. The difference in energy consumption between the two models quantifies the effect of the HP-TES system. These results are then scaled and aggregated according to the procedure outlined in Section 2.6.

Figure 7 shows these aggregated results for the total change in energy consumption of all modeled buildings across the region. It is evident that there is a large increase in energy consumption in the early morning hours of the summertime months. This corresponds to the recharging of the TES system for the next day's indoor cooling service.

However, there is not an equivalent reduction in energy consumption during the midday hours of the summertime months. This indicates that the energy consumption to charge the TES for cooling, even when the ambient temperature is lowest, is greater than the energy savings achieved by employing the TES for indoor cooling. The only instance in which there is a strong inversion of energy usage is during January. It is noted that January and February 2018 were colder than normal for this region.

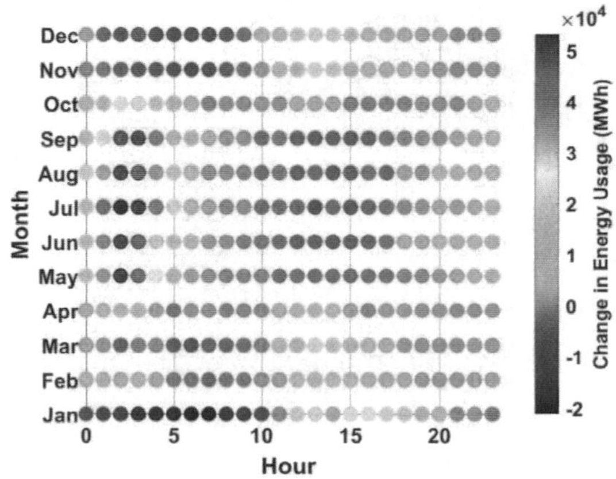

Figure 7. Aggregated results for total change in energy consumption if all eligible buildings operated HP-TES systems

3.2 Effect of Total Grid Generation

If the results of Figure 7 are assumed to be intervening on the grid-level fossil fuel generation, there may be potential to smooth the energy demand curve. Figure 8 shows the effect that this HP-TES system may have across the entire region for the month of January. The differences are small, but there is an increase in generation in the midday hours, and a small reduction of generation in the early morning hours.

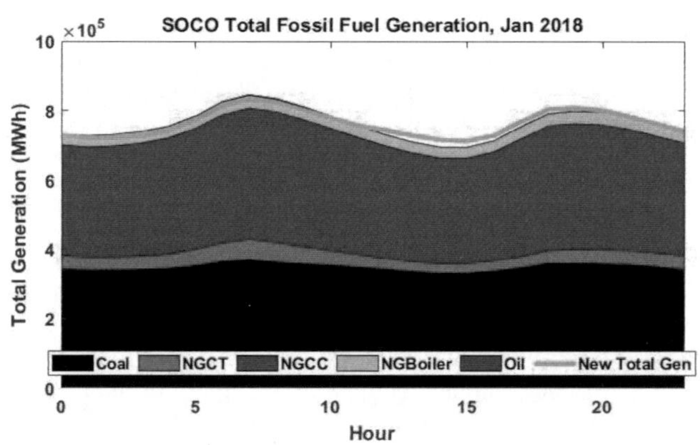

Figure 8. Estimated change in total fossil fuel generation for January 2018 if all eligible buildings operated HP-TES systems

3.3 Effect of Total Grid Emissions

It is difficult to know what a system grid operator may do if an intervening technology at scale significantly changes the system demand. This decision ultimately affects the emissions produced by grid-level energy generation. How the grid operator would change the generation portfolio is assumed here in two ways: 1) weighted favorably to reduce CO_2 emissions, and 2) weighted according to the MF for each technology (Figure 6). Then using the emission rates shown in Figure 5 for each technology, the total effect that this HP-TES system may have on grid-level emissions can be determined

By weighting favorably to reduce CO_2 emissions, it is assumed that any decrease in total generation due to the intervening HP-TES system reduces the use of NGCT. And any increase in total generation increases NGCC. The rationale for this is that NGCT is a predominant marginal technology and NGCC is considered more baseload.

Effectively, this means that each incremental MWh has a net emissions reduction of 0.20 tCO_2 (i.e., equivalent to the difference between the respective emissions rates for NGCT-NGCC, or 0.62-0.42). It would be more favorable to reduce coal, but this is considered baseload in this region and time, and thus unlikely to be affected by intervening technologies.

By weighting according to the MF (Figure 6), any increase or decrease in total system generation increases or decreases all technologies proportionately to their MF for that hour and month. The rationale for this weighting method is that if all households did install this technology and operated it in this manner, the system grid operator would be able to predict and plan for this in their baseload schedules, and thereby mimic historical dispatch protocols and observed trends.

The results for CO_2 emissions are shown in Figure 9. Here, only in the month of January under the favorable CO_2 weighting does the system decrease its total CO_2 emissions. All other months and using the MF weighting, CO_2 emissions increase overall. This is attributed to the net increase in energy of operating the HP-TES system as outlined in Section 3.1. For the favorable CO_2 weighting, the net increase in NGCC generation offsets any reduction of NGCT, despite NGCC having a lower CO_2 emission rate than NGCT. And for the MF weighting, the increase in baseload generation, particularly coal, increases the CO_2 emissions across the whole year.

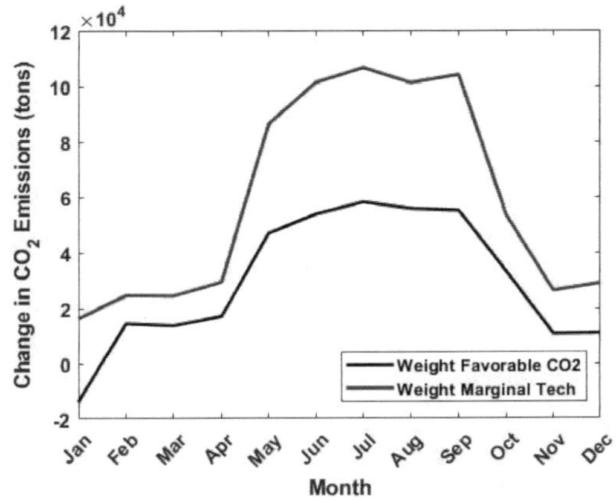

Figure 9. Effect of grid-produced CO_2 emissions if all eligible buildings operated HP-TES systems

Effects on SO_2 and NO_x emissions can be determined using the data from CEM, and an identical process to the CO_2 methodology. The only instance in this case study where emissions decrease is NO_x under the favorable CO_2 scenario.

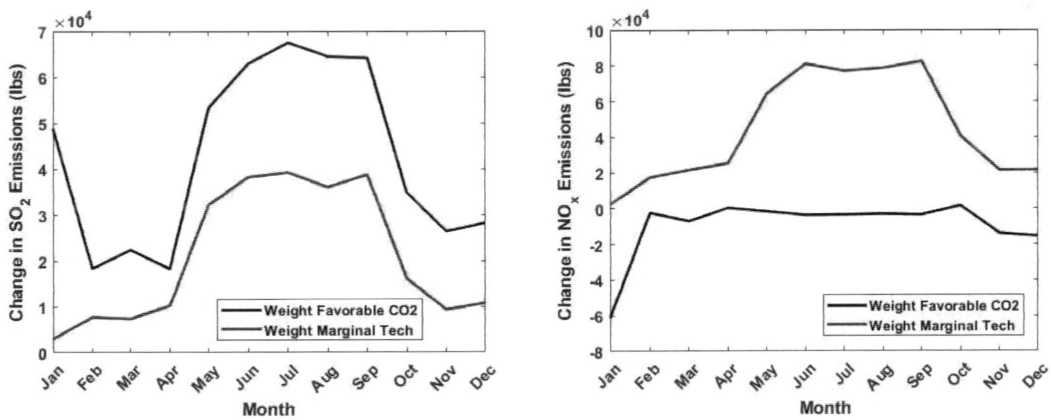

Figure 10. Effect of grid-produced SO_2 and NO_x emissions if all eligible buildings operated HP-TES systems

4. CONCLUSION

This study sought to build a framework around which the grid-level emissions impact of intervening technologies can be estimated. As a case study, the framework is used to estimate the effect of wide-spread adoption of residential HP-

TES systems in the 2018 Southern Company Balancing Authority region. 2000-3000 square foot single-family, detached homes with central air conditioning cooling and primary heat pump heating are included in this case study. The TES stores heat at 20°C without losses. Buildings in 13 select locations were modeled using NOAA LCD weather data, a 4R4C building thermal model, and a VCHP as a heating and cooling source, and the results scaled across the region according to US Census Bureau county-level population estimates and EIA RECS household numbers. Emissions data were collected from EPA CEM data for all fossil fuel generation plants within the SOCO region.

This case study indicates that installing this HP-TES system and operating under the strict assumptions made in this model, a total increase in energy generation might be expected. Some leveling of the demand curve is observed, however. Constrained by the 2018 SOCO generating portfolio, a net increase in power plant emitted CO_2, NO_x, and SO_2 is shown. This region was not equipped with low-carbon marginal technologies in 2018 and thus the HP-TES systems could not exploit diurnal load-shifting.

The results from this case study are not definitive, or widely applicable to HP-TES systems or other intervening technologies. Many assumptions and simplifications are necessary for a single unified model of this scope, and there is much room for improvement around this framework. Future steps include refining the building model, exploring more building types and sizes, using real HP-TES performance data, modeling other TES temperatures, including higher resolution weather data, interpolating between weather stations, using advanced regression of grid emissions, exploring other energy grids and regions, and investigating low-carbon generation portfolios including curtailed renewables or predicted future grid mixtures.

5. REFERNCES

Bell, I. H., Wronski, J., Quoilin, S., & Lemort, V. (2014). Pure and Pseudo-pure Fluid Thermophysical Property Evaluation and the Open-Source Thermophysical Property Library CoolProp. *Industrial & Engineering Chemistry Research, 53*(6), 2498-2508. doi:10.1021/ie4033999

Bruckner, T., Fulton, L., Hertwich, E., McKinnon, A., Perczyk, D., Roy, J., Schaeffer, R., Schlömer, S., Sims, R., & Smith, P. (2014). Technology-specific cost and performance parameters [annex III]. In *Climate Change 2014: Mitigation of Climate Change* (pp. 1329-1356): Cambridge University Press.

Cui, B., Dong, J., Munk, J., Mao, N., & Kuruganti, T. (2018). *A simplified regression building thermal modelling method for detached twofloor house in U.S.* Paper presented at the International High Performance Buildings Conference at Purdue, Purdue, Indiana.

Gallo, A. B., Simões-Moreira, J. R., Costa, H. K. M., Santos, M. M., & Moutinho dos Santos, E. (2016). Energy storage in the energy transition context: A technology review. *Renewable and Sustainable Energy Reviews, 65*, 800-822. doi:https://doi.org/10.1016/j.rser.2016.07.028

Monthly Energy Review. (2021). Retrieved from https://www.eia.gov/totalenergy/data/monthly/archive/00352104.pdf

Residential Energy Consumption Survey (RECS). (2015).

Siler-Evans, K., Azevedo, I. L., & Morgan, M. G. (2012). Marginal Emissions Factors for the U.S. Electricity System. *Environmental Science & Technology, 46*(9), 4742-4748. doi:10.1021/es300145v

ACKNOWLEDGEMENT

This work was sponsored by the U. S. Department of Energy's Building Technologies Office under Contract No. DE-AC05-00OR22725 with UT-Battelle, LLC. The authors would like to acknowledge Mr. Sven Mumme, Technology Manager – Building Envelope, U.S. Department of Energy Building Technologies Office.

Demonstration of Thermal Energy Storage System with Salt Hydrate Phase Change Material Composite

Jason Hirschey[1]*, Kyle R. Gluesenkamp[2], Samuel Graham[1]
[1]George W. Woodruff School of Mechanical Engineering, Georgia Institute of Technology
North Ave NW, Atlanta, GA, USA
jhirschey13@gatech.edu

[2]Oak Ridge National Laboratory
1 Bethel Valley Rd, Oak Ridge, TN, USA

* Corresponding Author (jhirschey13@gatech.edu)

ABSTRACT

Thermal energy storage (TES) decouples heat generation from use, providing a crucial tool to mitigate fluctuating thermal loads. TES systems may often contain a phase change material (PCM) which stores heat isothermally through the enthalpy of phase change, and thus the TES as a whole operates at a near constant temperature. Integrated into a heat pump (HP) or heat transfer fluid (HTF) circulation loop, the TES will behave as an isothermal heat exchanger (HX), absorbing or releasing heat into the HTF. For near-ambient TES (-15-85°C), solid-liquid PCMs provide the highest energy storage density. However, the PCMs available in this temperature range often have low thermal conductivity that hinder the power capacity of a TES. As such, modifications such as metallic extended surfaces or fins are made to the TES to increase its power capacity. This work instead uses an enhanced PCM composite material in a simple and scalable shell-and-tube design. In this work, two prototype-scale TES units are demonstrated: 1) a benchtop scale unit as a proof-of-concept with targeted specifications 50 W, 100 Wh, and 2) an intermediate scale with target specifications 200 W, 0.8 kWh. The PCM composite is based on sodium sulfate decahydrate (SSD) and high thermal conductivity expanded graphite (EG). Both TES units containing this material met or exceeded design performance. The design, production method, and performance results are discussed.

Keywords: *Thermal energy storage, heat exchanger, salt hydrate, graphite, phase change material*

1. INTRODUCTION

Thermal energy storage (TES) is a powerful energy storage mechanism that is perfectly suited for applications where the end use is heat. TES may be especially useful for low-grade thermal energy near ambient temperatures which traditionally rely on a near-instantaneous conversion of one energy form into heat. For example, heat pumps and refrigeration cycles typically utilize electrical energy to move thermal energy for heating or cooling the target application, or resistive heating elements that convert electrical energy or furnaces that convert chemical energy into thermal energy for heating applications. TES provides a tool that may enable an application to decouple this conversion, and also may offer a solution for capturing and reusing waste heat.

Two important characteristics of any TES system are the quantity of heat storable within, and the rate at which this heat can be dispatched or moved in and out of the TES. The energy storage capacity of a TES is almost entirely dictated by the materials used. Including a phase change material (PCM) into the TES sharply increases its energy storage capacity; through the latent heat of phase change, the PCM can isothermally absorb and release quantities of heat much greater than most sensible thermal storage. Furthermore, the PCM allows the TES to operate nearly isothermally, a feature which can improve reliability and target TES for specific applications.

---

[1] DISCLAIMER: This manuscript has been authored by UT-Battelle, LLC under Contract No. DE-AC05-00OR22725 with the U.S. Department of Energy. The United States Government retains and the publisher, by accepting the article for publication, acknowledges that the United States Government retains a non-exclusive, paid-up, irrevocable, world-wide license to publish or reproduce the published form of this manuscript, or allow others to do so, for United States Government purposes. The Department of Energy will provide public access to these results of federally sponsored research in accordance with the DOE Public Access Plan (http://energy.gov/downloads/doe-public-access-plan).

The thermal power of a TES is dictated by the materials and application, but also largely dependent on the geometry of the system. Material thermal conductivity and thermal effusivity characterize the ability of these materials to move heat through themselves. But heat moving in and out of the TES relies on its geometric interfacing with external heat transfer surfaces. In this sense, a TES system may be considered an isothermal heat exchanger (HX) that couples in some way to its application. The TES could be integrated into a heat pump (HP) refrigeration loop or some other heat transfer fluid (HTF) circulation loop. Often, extended metallic fins or complex serpentine coils are used to increase the heat transfer contact area thereby increases TES thermal power (Abdulateef *et al.*, 2019; Asgari *et al.*, 2021; Herbinger & Groulx, 2022; Pakalka *et al.*, 2020). However, these features displace PCM lowering the energy storage capacity and may not be cost-effective at large scales. To this end, this work presents a TES PCM-HX design with a PCM composite material of enhanced thermal conductivity and a simple HX design to demonstrate that high TES thermal performance can be achieved in a scalable design.

In this work, the design and performance of two TES units are presented. The two TES discussed are 1) a small benchtop unit with target specifications nominally 50W of thermal power and 100Wh of thermal energy, called TES-α, and 2) a larger unit with target specifications nominally 200W of thermal power and 800Wh of thermal energy, called TES-β. The PCM is a composite salt hydrate and expanded graphite. The addition of the graphite increases the bulk thermal conductivity of the PCM, negating the need for fins or extended heat exchanger surfaces in the HX. This work evaluates the viability of this design and PCM-graphite composites materials for TES systems.

2. MATERIALS

The TES are based on a shell-and-tube design. The shell is sourced from standard schedule 40 PVC pipe, and the tube is standard ¼ in. outer diameter copper pipe. PVC knockout caps are used to encapsulate the PCM in the shells. The caps are epoxied to the shells using a two-part steel enforced epoxy.

The PCM is a composite of consisting of mainly salt hydrate and graphite. The salt hydrate is sodium sulfate decahydrate, $Na_2SO_4 \cdot 10H_2O$, (SSD). The salt was sourced in its anhydrous form from Alfa Aesar. The graphite is expanded graphite (EG) which has been commonly studied for use in PCM composites (Fang *et al.*, 2019; Mao *et al.*, 2017; Ye *et al.*, 2017). The SSD, EG, and other materials are mixed and compressed into the annulus of the HX shells.

For TES-α, borax, $B_4O_7Na_2 \cdot 10H_2O$, was added as a nucleating agent and sodium polyacrylate (SPA) added as a thickening agent. Borax is historically a nucleating agent for SSD due to similar crystal characteristics (Ruben *et al.*, 1961; Telkes, 1952). The borax and SPA were sourced from Alfa Aesar. For TES-β, borax was again added as a nucleating agent. Dextran sodium sulfate (DSS) was added as a thickening agent. Both sourced from Alfa Aesar. The change in thickening agent is a result of research conducted between the fabrication and testing of the two TES units analyzed here; the DSS showed greater stability in repeated thermal cycling tests (Li *et al.*, 2022).

The composition of the PCM composites of both TES units are reported by mass percent in Figure 1. Both units were composed of several individual modules that were assembled containing PCM in these same ratios, although some variations exist between modules for each. The total PCM composite mass for TES-α was 3.2 kg with a packing density of 0.84 kg/m^3. The total PCM composite mass for TES-β was 27.537 kg with a packing density of 0.92 kg/m^3. The additional material and their ratios relative to the SSD were not arbitrary; they were found to be high performing material compositions based on other research being conducted congruently to this work. The details of why these materials are chosen and in what ratios are outside the scope of this present work.

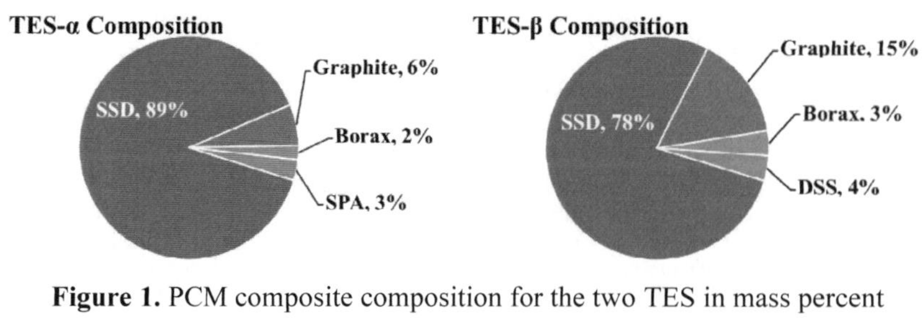

Figure 1. PCM composite composition for the two TES in mass percent

3. DESIGN

The design of these TES units is a pseudo shell-and-tube design. Figures 2 & 3 depict the general geometry of these two designs. The HTF passes through an axially centered copper tube; the total metallic content composes less than 5% of the total TES volume. The PCM is packed into the annulus of the shell. The shell is PVC plastic material which is low thermal conductivity and acts as a relative insulator. The TES modules are wrapped in insulating foam during testing. Thermocouples are placed in the HTF stream at the inlet and outlet of all modules. A mass flow meter measures the HTF flow rate.

The dimensions of each design were determined by an iterative process. With certain target performance metrics (power, energy, time), known PCM composite and HTF thermophysical properties (phase change temperature, enthalpy, density, specific heat), and capabilities and standard dimensions of the instrumentation available (pump flow rate, hot water and cold water bath temperature limits, standard copper tubing dimensions, stand PVC dimensions), a series of equations were found to relate these constraints to the unknown quantities: overall TES length, number of parallel TES modules, inlet and outlet target temperatures, and desired flowrate.

Figure 2 illustrates the iterative design process to achieve the defined TES performance metrics. There are two classes of design variables: selectable and customizable. Selectable variables include standard sized, off-the-shelf components: schedule 40 PVC shell size, copper tube size, PCM properties. Customizable variables include parts that are easily adjusted around the selectable variables, such as HTF flow rate and temperature, and number of modules in parallel or series. The first step in the iterative design process is arbitrarily choosing some selectable variables. Then the customizable variables are tuned to meet certain criteria for the TES, including the performance metrics, but also more tangible metrics like size and weight, and some less tangible metrics like if the design can be produced in the lab with the tools available. A series of equations were written containing heat transfer, thermodynamic, and fluid dynamics relationships to check for consistency in power and energy between the HTF and the TES. These relationships capture most of the selectable and customizable variables, and most are reversible so that the process could be done if other metrics are important considerations, if necessary.

As an example of the design process, the thermal energy performance metric is used to determine the quantity of PCM composite necessary in the annulus of the shell-and-tube design based on the phase change enthalpy and density of the PCM composite. Once the diameter of schedule 40 PVC shell and copper tube size are selected, the quantity of necessary PCM informs the overall length of the TES. This length influences the necessary HTF flow rate, and inlet and outlet temperatures to achieve the desired thermal power metric. Fluid dynamics equations would be used to compute the HTF heat transfer rate, and this compared to heat transfer equations on the PCM composite under these temperature conditions to ensure thermodynamic consistency. On the last step, the feasibility of production was considered for the determined size of each TES module. If infeasible, one selectable variable would be changed and the process repeated until a feasible design met all considerations.

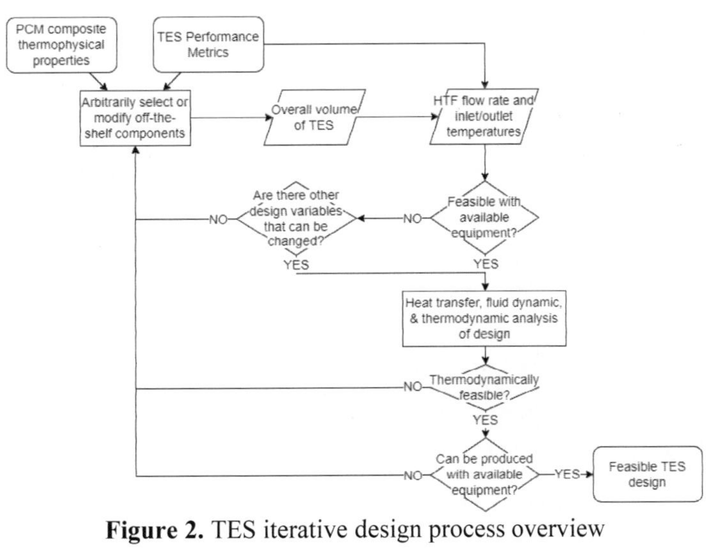

Figure 2. TES iterative design process overview

For each TES system, the PCM composite material was synthesized in large quantities. The graphite-PCM material was then compressed into the annulus of the shell-and-tube design with the tube in place, shown in Figure 3: A) The material was first poured into the shell (A) until nearly full (B). A piston/plunger added atop the material (C). Handheld bar clamps used to apply even force, compressing the material (D). Several piston/plungers of varying sizes were used to compress deep into the shell. Processes A-D repeated until the shell is filled before being capped (E). This process resulted in good thermal contact between the tube and the PCM composite material. Producing several identical modules increases the scalability of this design; several modules may be attached to achieve certain performance metrics by following the process outlined in Figure 2.

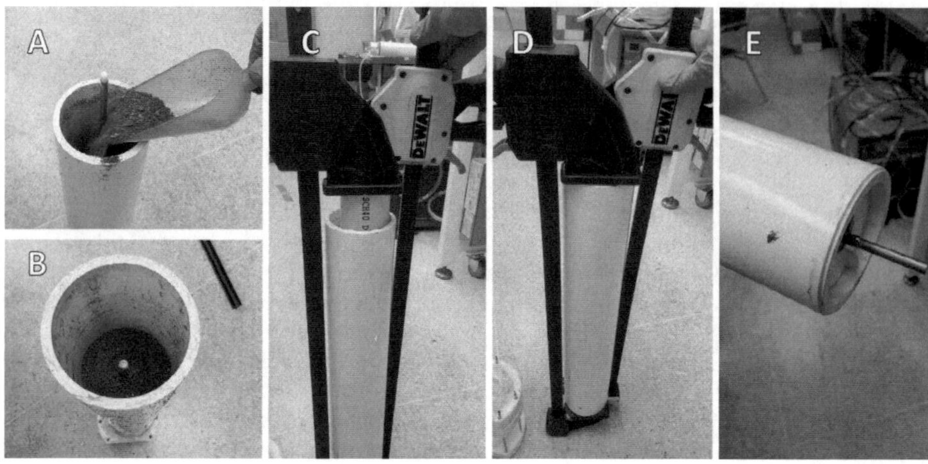

Figure 3. Process of compressing PCM composite into TES shell

For TES-α, two shell-and-tube modules were synthesized. These were arranged in series with each other. Figure 4 shows the dimensions of this design and the design installed on the test bed before insulation foam was added to the exterior of the shells. A peristaltic pump was needed to throttle the HTF flow rate from the hot water bath and the chillder for this small TES size.

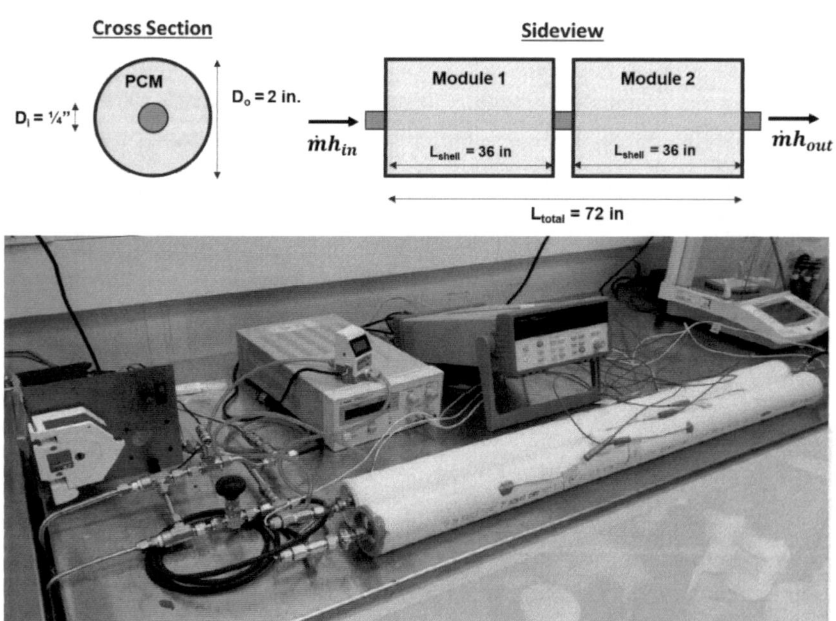

Figure 4. TES-α dimensions and test bed setup

For TES-β, fourteen shell-and-tube modules were constructed. They were arranged in two banks in series composed each of seven modules in parallel. Figure 5 shows this arrangement and the TES on the testbed with insulating foam blanketing the two banks.

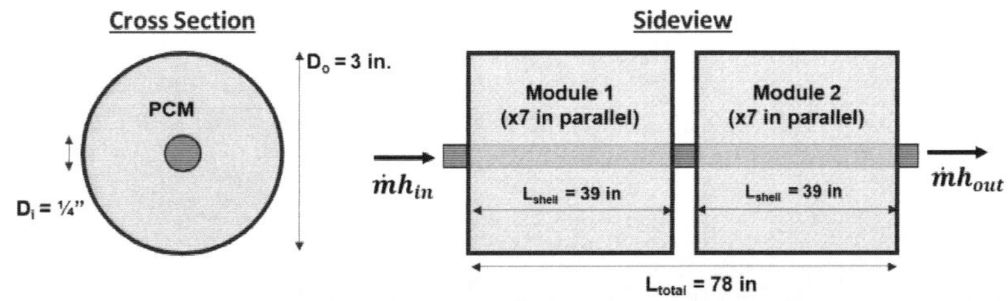

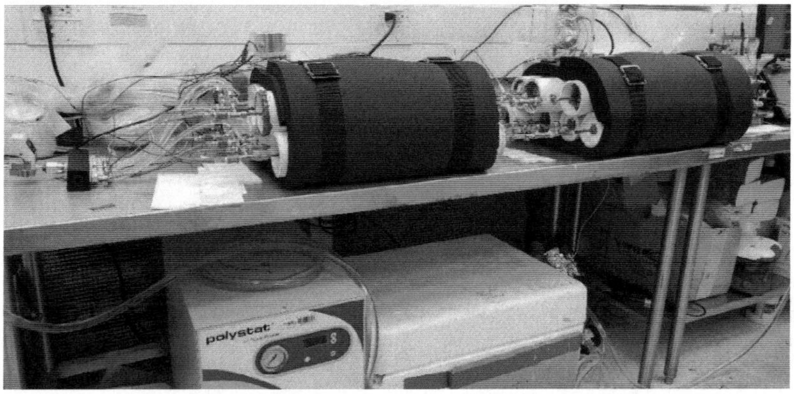

Figure 5. TES-β dimensions and test bed setup

4. OPERATION & POWER CALCULATION

The HTF used in these TES units is water which has well-characterized thermophysical properties for ease in analysis. Two water baths, one hot and one cold, are separately controlled with PID temperature controls. The cold bath is a Cole-Parmer Polystat chiller with integrated water pump. The hot bath is a custom-built system with an external water pump and enclosed in an insulating chamber. Details of this hot water bath are omitted for brevity; any commercially available or custom-built water bath capable of maintaining a constant temperature and pumping water with steady flow rate will suffice in replicating this experiment.

The baths are connected to the TES units with three-way valves that enable the operator to switch between heating and cooling modes. Figure 4 depicts the orientation of the water baths on TES-β.

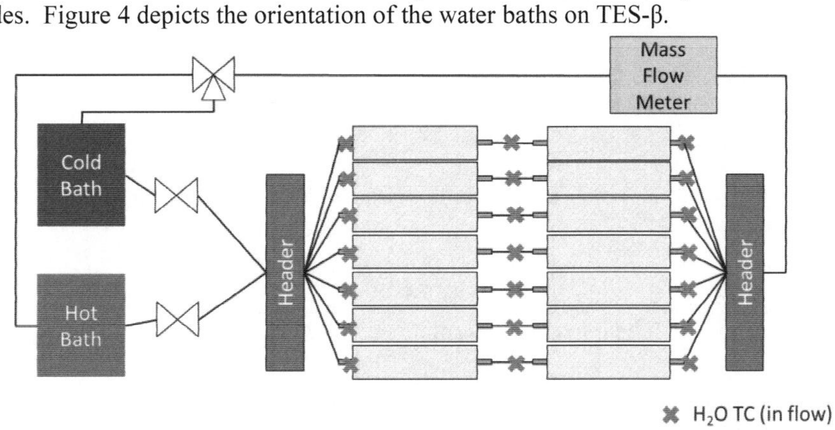

Figure 4. Integration of TES-β with hot- and cold-water baths

Thermocouples are embedded in the HTF stream, and the thermocouple measurement location are placed approximately at the inlet and outlet of each TES module. For TES-β, a header splits the flow into the seven parallel shells. It is assumed that the flow is split approximately evenly between the seven parallel pathways.

The calculation of thermal power *P(t)* is done by measuring the HTF temperature difference at the inlet and the outlet of the TES modules multiplied by the mass flow rate, $\dot{m}(t)$, and the specific heat of the HTF, $c_{p,HTF}$, Eq. 1. However, the HTF takes some time to traverse from the inlet to the outlet. Thus, rather than measuring an instantaneous spatial HTF temperature difference, Eq. 1 is modified to measure a temporal HTF temperature difference. This is accomplished by offsetting the time reading of inlet temperature by τ which is the calculated time the water travels from inlet to outlet. τ is calculated by using the flow rate measurement, density of the HTF, and geometry of the tube in the TES unit.

$$P(t) = \dot{m}(t) * c_{p,HTF} * (T_{HX,in}(t - \tau) - T_{HX,out}(t)) \quad (1)$$

The total energy *E(t)* is determined by integrating the power over time of operation, Eq. 2.

$$E(t) = \int_{t=0}^{t=end} P(t)\, dt \quad (2)$$

5. RESULTS

The thermal power and energy results of TES-α are shown in Figures 5 & 6 for a single heating cycle. The green region of Figure 5 shows a time of approximately 1 hour wherein TES-α maintained a thermal power greater than the target of 50W. This small TES required the peristaltic pump to be adjusted mid-experiment which explains the jumps in the power curve prior to this window and at the end of the experimental run. Note how the first shell in series is responsible for nearly twice as much power as the second shell during the first three hours of the run, then its performance drops below the second shell drastically. This is the effect of phase change within the TES-α. Once the phase change is complete on the first shell, it is incapable of storing heat isothermally and thus its temperature rises to match the HTF, significantly reducing the driving temperature and lowering its effect on overall thermal power. Real systems utilizing this design will need to account for this drop in power.

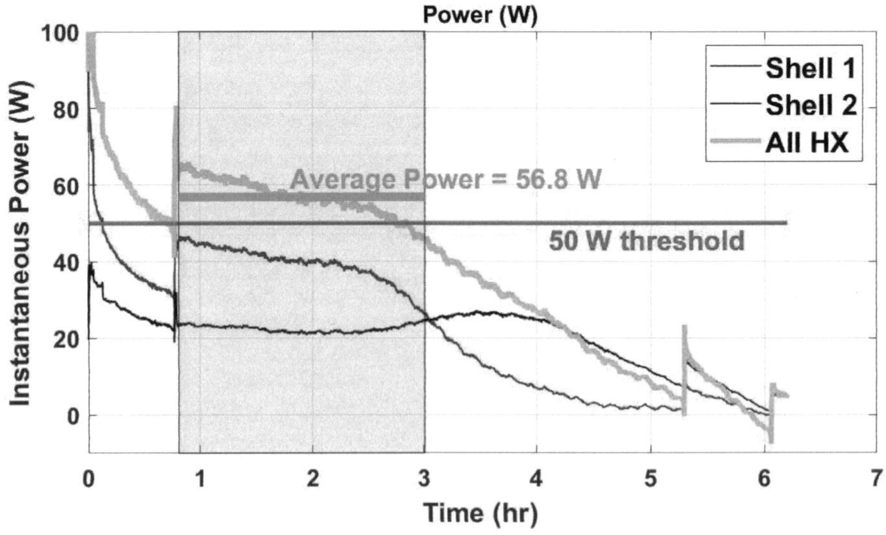

Figure 5. Thermal power of TES-α

Figure 6 shows the cumulative energy stored within the TES-α during this same heating cycle. The target for this TES-α was 100 Wh of storage which was exceeded by ~20% in the green highlighted window, within added the factor of safety. Even if some energy stored is accounted for by sensible storage, the majority is assumed to be latent storage of the PCM due to the plateaued power curve of Figure 5.

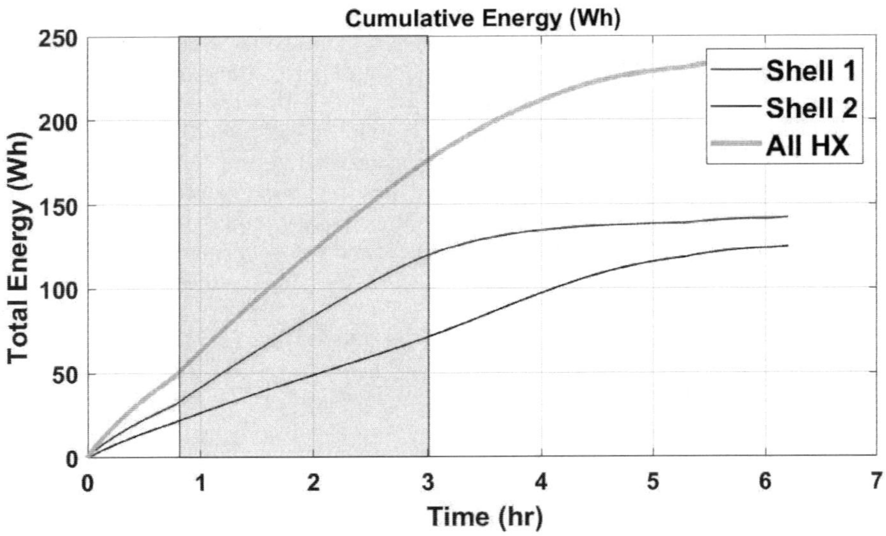

Figure 6. Thermal energy storage of TES-α

The thermal power results of TES-β are shown in Figure 7. A large spike of thermal power at the beginning of the test is due to sensible heating from the relatively cold steady state to the sudden influx of relatively hot HTF. However, once the power settles from this initial spike, its output is nearly 2-3 times the expected value. It was later understood that the HTF flow rate and the hot water bath temperature were both higher than was planned, an oversight by the operator. As such, the power for each bank in this TES achieves > 200 W of thermal power for nearly two hours.

Due to the flowrate mishap, there is no clear distinction between the first and second bank of tubes like was observed between the first and second modules in TES-α. It is speculated that the higher-than-expected flow rate moved the HTF though the TES more quickly than the initial design calculations predicted. As such, the HTF temperature change was small for each bank of the TES and thus each bank saw approximately the same thermal forcing temperature gradient. Paired with slight variations in the 14 TES modules, both the first and second bank acted approximately identically. In a real system, this may be ideal; there may not be a performance drop until all TES capacity is depleted. But this is erroneously done here.

After operation for 2 hours, the hot water bath water pump developed a leak. An attempt to patch the leak mid-test was unsuccessful and the test had to be aborted after 2.5 hours.

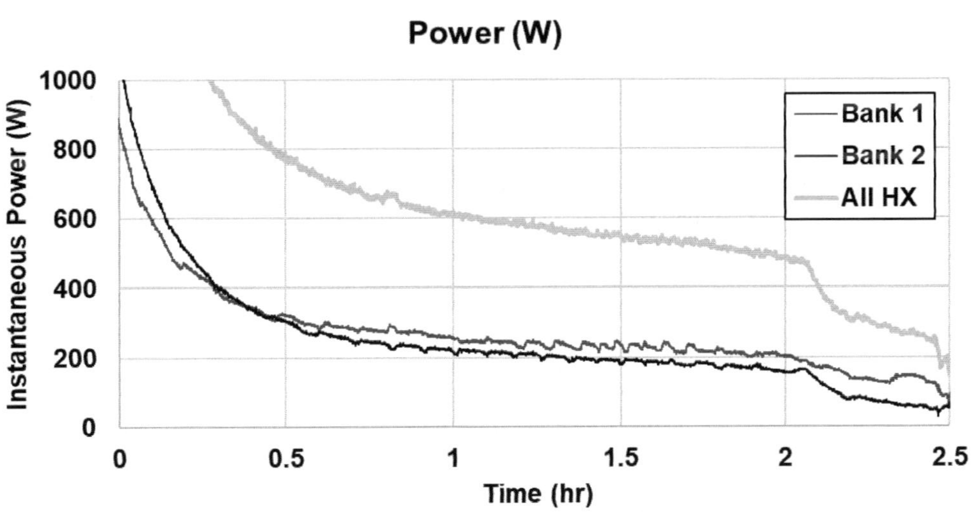

Figure 7. Thermal power of TES-β

Despite the issues encountered during the testing of TES-β, its energy storage far exceeded the performance metric, as shown by Figure 8. Even if the initial spike at the start of the test is ignored, and the period between 0.5 and 2.5 hours is analyzed, this TES stored over 1 kWh of thermal energy. This higher-than-designed energy storage is not surprising as the updated PCM composite mixture was seen to have much higher performance than the PCM composite mixture in TES-α, with higher enthalpy and more stability, but this was not updated in the TES design process. The energy storage values are reasonable and within the designed factor of safety once accounting for the updated thermophysical properties. However, because the test was aborted early, the heat leakage rate was unable to be quantified. As such, some of this energy may be lost to the ambient and not truly stored in the TES.

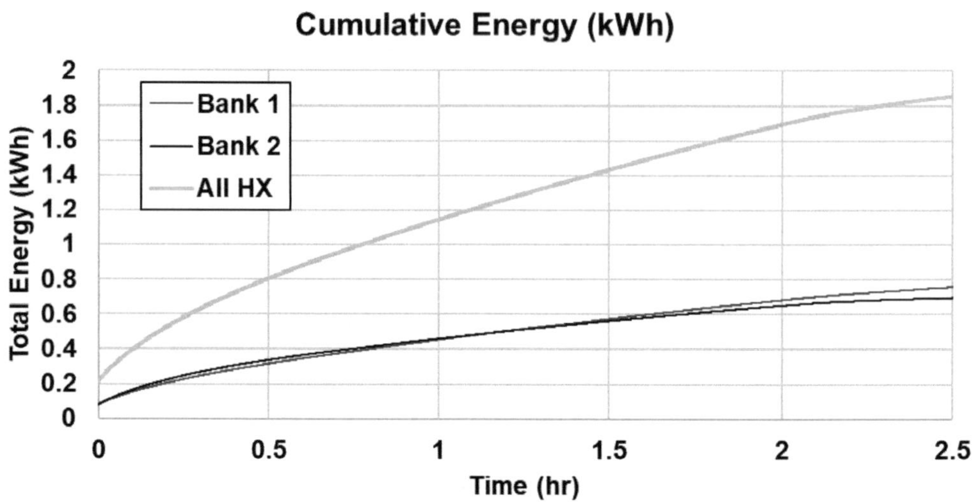

Figure 8. Thermal energy storage of TES-β

Furthermore, this TES-β showed that the design was capable of high thermal power performance. This exemplifies the potential for TES systems containing an enhanced PCM-graphite composite material in a simple and scalable shell-and-tube HX setup; no extended metallic fins, heat exchange surfaces, or complex designs were necessary to achieve high performance.

6. CONCLUSION

Thermal energy storage (TES) is an ideal mechanism for recouping waste heat that may be used in applications where low-grade heat is the end use: refrigeration, indoor heating and cooling, domestic hot water. TES system are often characterized by its energy storage capacity and its thermal power which will vary between applications.

Implementing a phase change material (PCM) into TES can drastically improve the energy storage capacity by leveraging the latent heat of phase change as an isothermal heat source or sink. In regard to increasing the thermal power, the TES with integrated PCM may be better described as a PCM heat exchanger (HX). These PCM HX have been the subject of much research, but typically rely on some metallic extended finned surface or complex geometry to aid in moving heat in and out of the TES and to the target application.

This work demonstrated a high performing TES using a simple shell-and-tube geometry and an enhanced high thermal conductivity PCM composite material made of an inorganic salt hydrate and expanded graphite. Two TES units were demonstrated: a 50W, 100Wh proof-of-concept (TES-α), and a 200W, 800Wh intermediary to a full-scale TES system (TES-β). The design was based on an iterative method that factored in the constraints of the equipment available, commercially available products, and a laboratory-based manufacturing process without heavy equipment.

TES-α performed as designed, averaging 56.8W of thermal power for 1.2 hours, and ~120Wh of thermal energy in this time. TES-β encountered issues during test, but still surpassed the target performance metrics. This system experienced over 400W of thermal power for over 1.5 hours and stored over 1kWh of thermal energy. The success of the TES-β is attributed to adjustments to the experimental test bed as a result of lessons learned from the TES-α, and an improved PCM composite recipe with higher enthalpy and more stability.

These tests show that high performing, high thermal conductivity PCM composite materials can be effectively utilized in simple TES system geometries without the need for extended heat transfer surfaces. Future work will explore these TES designs for stability, adjustable power modes, and heat pump integration.

NOMENCLATURE

TES	Thermal Energy Storage
PCM	Phase Change Material
HP	Heat Pump
HTF	Heat Transfer Fluid
HX	Heat Exchanger
TES-α	Benchtop demonstration unit with targeted specifications 50 W, 100 Wh
TES-β	Larger demonstration unit with targeted specifications 200 W, 0.8 kWh

REFERNCES

Abdulateef, A. M., Abdulateef, J., Sopian, K., Mat, S., & Ibrahim, A. (2019). Optimal fin parameters used for enhancing the melting and solidification of phase-change material in a heat exchanger unite. *Case Studies in Thermal Engineering*, 100487. doi:https://doi.org/10.1016/j.csite.2019.100487

Asgari, M., Javidan, M., Nozari, M., Asgari, A., & Ganji, D. D. (2021). Simulation of solidification process of phase change materials in a heat exchanger using branch-shaped fins. *Case Studies in Thermal Engineering, 25*, 100835. doi:https://doi.org/10.1016/j.csite.2020.100835

Fang, Y., Ding, Y., Tang, Y., Liang, X., Jin, C., Wang, S., Gao, X., & Zhang, Z. (2019). Thermal properties enhancement and application of a novel sodium acetate trihydrate-formamide/expanded graphite shape-stabilized composite phase change material for electric radiant floor heating. *Applied Thermal Engineering, 150*, 1177-1185. doi:https://doi.org/10.1016/j.applthermaleng.2019.01.069

Herbinger, F., & Groulx, D. (2022). Experimental Comparative Analysis of Finned-Tube PCM-Heat Exchangers' Performance. *Applied Thermal Engineering*, 118532. doi:https://doi.org/10.1016/j.applthermaleng.2022.118532

Li, Y., Kumar, N., Hirschey, J., Akamo, D. O., Li, K., Tugba, T., Goswami, M., Orlando, R., LaClair, T. J., Graham, S., & Gluesenkamp, K. R. (2022). Stable salt hydrate-based thermal energy storage materials. *Composites Part B: Engineering, 233*, 109621. doi:https://doi.org/10.1016/j.compositesb.2022.109621

Mao, J., Hou, P., Liu, R., Chen, F., & Dong, X. (2017). Preparation and thermal properties of SAT-CMC-DSP/EG composite as phase change material. *Applied Thermal Engineering, 119*, 585-592. doi:https://doi.org/10.1016/j.applthermaleng.2017.03.097

Pakalka, S., Valančius, K., & Streckienė, G. (2020). Experimental comparison of the operation of PCM-based copper heat exchangers with different configurations. *Applied Thermal Engineering, 172*, 115138. doi:https://doi.org/10.1016/j.applthermaleng.2020.115138

Ruben, H. W., Templeton, D. H., Rosenstein, R. D., & Olovsson, I. (1961). Crystal Structure and Entropy of Sodium Sulfate Decahydrate. *Journal of the American Chemical Society, 83*(4), 820-824. doi:10.1021/ja01465a019

Telkes, M. (1952). Nucleation of Supersaturated Inorganic Salt Solutions. *Industrial & Engineering Chemistry, 44*(6), 1308-1310. doi:10.1021/ie50510a036

Ye, R., Lin, W., Yuan, K., Fang, X., & Zhang, Z. (2017). Experimental and numerical investigations on the thermal performance of building plane containing $CaCl_2 \cdot 6H_2O$/expanded graphite composite phase change material. *Applied Energy, 193*, 325-335. doi:https://doi.org/10.1016/j.apenergy.2017.02.049

ACKNOWLEDGEMENT

This work was sponsored by the U. S. Department of Energy's Building Technologies Office under Contract No. DE-AC05-00OR22725 with UT-Battelle, LLC. The authors would like to acknowledge Mr. Sven Mumme, Technology Manager – Building Envelope, U.S. Department of Energy Building Technologies Office.

A Review on Carbon Emissions of Ultra-High-Performance Fiber Reinforced Concrete as a Building Construction Material

Syed Muhammad AQIB[1]*, Z. John MA[1]

[1]University of Tennessee, Civil and Environmental Engineering
Knoxville, TN, USA
saqib@vols.utk.edu

*Corresponding Author

ABSTRACT

Cement, as the most important component used in concrete, contributes 5-8% of global CO_2 emissions due to its highly energy intensive manufacturing process. Traditional reinforced concrete (RC), which is commonly used building material contains approximately 15% of the cement. But due to increasing demand and bulk design volume, RC requires high quantity of cement. In 2021, the global cement production was 4.4 billion metric tons. Ultra-High-Performance Fiber Reinforced Concrete (UHPFRC) is an emerging high-tech building material with an increased strength and durability compared to conventional concrete. This is achieved by low water-to-binder ratio ($w/b<0.25$), steel fibers and a very dense particle packing using finer sand and silica fume. UHPFRC requires significantly less material and less maintenance for similar mechanical performance and provides higher lifespan. But it also consists higher cement content contributing to high carbon footprint. The mechanical properties of UHPFRC are being widely researched and improved but the environmental impacts are rarely considered. So, there is a need to review and compare the long-term environmental impacts of UHPFRC and conventional RC. In this study, a review of the life cycle environmental impact analysis is presented which compares the Global Warming Potential (GWP) of conventional concrete to UHPFRC over a construction phase and cradle-to-grave phase. It is observed that the GWP of UHPFRC is at least 60% greater than conventional concrete but when the cradle-to-grave phase is considered, the GWP of UHPFRC outperforms the conventional concrete.

1. INTRODUCTION

Buildings and construction sector account for 40% of global energy consumption and concrete has largely been used as a building material. The most significant environmental impact of concrete is cement, which reportedly contributed 36% to the carbon dioxide emissions of 7.7 gigatons (Bajželj et al., 2013). Hence, building construction materials present a huge potential for decarbonization. Considering that new concrete constructions account for 8% of global greenhouse gas emissions (Favier et al., 2019), it is imperative to reduce their global warming potential. Many researchers have demonstrated that the type of cement used and concrete design mix are major factors influencing the environmental impacts (Bertola et al., 2021).The use of alternative binders to replace cement such as silica fume, fly ash, and slag, have been developed as solutions for reducing concrete's environmental impact.

Ultra-High-Performance Concrete (UHPFRC) has garnered increased interest in the past three decades for building applications because of its excellent durability (Alkaysi et al., 2016; Graybeal & Tanesi, 2008; Magureanu et al., 2012) and high compressive strength compared to conventional concrete (ACI 239R, 2018; de Larrard & Sedran, 1994). The specialized concrete intended for high performance like ultra-high-performance concrete (UHPFRC) often includes steel fibers as a part of the concrete granular structure to improve the mechanical performance of concrete. The optimization of specialized concrete for both environmental performances along with the mechanical performance is crucial to maintain the benefits of these novel construction practices.

Ultra-high-performance concrete (UHPFRC) can be defined as fiber-reinforced cement-based composite material with characteristic strain hardening properties, compared with conventional concrete where once concrete cracks, it loses its tensile capacity very sharply. Finely graded, homogeneous matrix and high fiber content allow UHPFRC to achieve very high mechanical properties (e.g., flexural strength higher than 2,000 psi comparing with about 500 psi for conventional concrete). Figure 1 provides a comparison of the mechanical properties of conventional concrete and UHPFRC.

Table 1: Mechanical properties of UHPFRC and conventional concrete (ACI 239R, 2018)

Material characteristics	Conventional Concrete	Ultra-High-Performance Concrete
Compressive strength	3000 to 6000 psi (20 to 40 MPa)	22,000 to 36,000 psi (150 to 250 MPa)
Direct tensile strength	150 to 440 psi (1 to 3 MPa)	900 to 1700 psi (6 to 12 MPa)
Elastic modulus (ASTM C469/C469M)	3,600,000 to 4,400,000 psi (25 to 30 GPa)	6,000,000 to 7,200,000 psi (40 to 50 GPa)

UHPFRC is being widely researched but not yet fully commercialized due to its high commercial cost (Administration, 2013; Kay Wille & Boisvert-Cotulio, 2015). A major portion of the material cost of UHPFRC can be attributed to the two constituents of the mix design: cementitious materials and steel fibers. Cementitious materials used in UHPFRC usually consist of cement and silica fume and they can contribute as much as 20% and steel fibers can contribute as much as 40% of the overall cost of UHPFRC (Administration, 2013; Kay Wille & Boisvert-Cotulio, 2015). Figure 2 provides the constituents of a typical UHPFRC mixture design and their contribution to the overall cost.

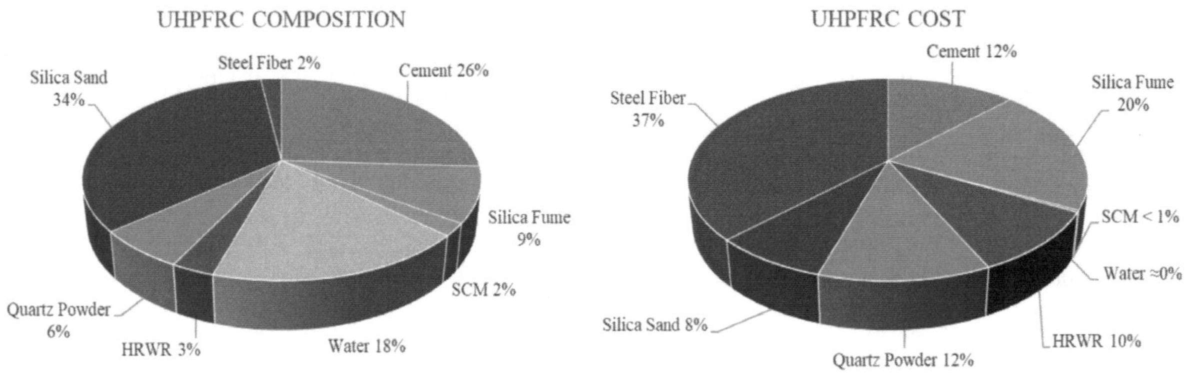

Figure 1: UHPFRC components and their contribution toward UHPFRC cost (Abellán-García et al., 2021)

UHPFRC exhibits a brittle failure mode and fibers are added to provide enhanced ductility (ACI 239R, 2018; Larsen & Thorstensen, 2020). Steel fibers are the most commonly used fibers in UHPFRC and their performance is very well documented (Christ et al., 2019; Larsen & Thorstensen, 2020; Shehab El-Din et al., 2016; K. Wille et al., 2014). Steel fibers make it possible for UHPFRC to exhibit superior flexural properties and strain-hardening behavior after the UHPFRC matrix cracks initially to exhibit superior ductility (K. Wille et al., 2014). The most common obstacle of using UHPFRC has been the steel fiber cost.

Fibers other than steel (glass and PVA fibers) can be used in UHPFRC (Kay Wille & Boisvert-Cotulio, 2015). Pourjahanshahi et. al used a combination of steel fibers with barchip, kortta, glass, carbon, and polypropylene fibers to investigate a UHPFRC mix design with hybrid fibers (Pourjahanshahi & Madani, 2021a). Among non-metallic fibers, glass fibers have become a popular alternative to steel fibers especially where steel corrosion is a concern. Glass fibers have relatively low cost than steel fibers (Ghosh et al., 2021; Kay Wille & Boisvert-Cotulio, 2015) and good durability properties (Gooranorimi & Nanni, 2017; He et al., 2021; Rigaud et al., 2012).

Non-proprietary cost-effective UHPFRC materials can be developed for building construction. The mixture designs can be optimized in their efficiency considering workability, mechanical performance, and cost-effectiveness, which will significantly impact the current construction practices.

A comprehensive literature review is conducted to analyze various mixture designs of UHPFRC with different fibers. It is noticeable that most work on UHPFRC focuses on investigating the mechanical performance. Very few researchers have analyzed the cost and emissions. A handful of articles have conducted a life cycle analysis and reported global warming potential of concrete and UHPFRC, both during construction process and maintenance. Furthermore, all those papers are based on the bridge construction as a case study. We couldn't find any work that investigates cost-saving and decarbonization potential of UHPFRC for buildings' construction.

This paper provides a review on UHPFRC to establish that it can be a promising building construction material to decarbonize concrete construction industry. The literature review in section 2 summarizes the advancements in incorporating alternate fibers. Section 3 and 4 provide an analysis on the cost and carbon emissions of conventional concrete versus UHPFRC.

2. LITERATURE REVIEW ON UHPFRC

As mentioned above, the high cost of UHPFRC compared to conventional concrete is a hurdle in the widespread adoption of UHPFRC and hence, the development of cost-effective UHPFRC has been a major research focus. It has been shown that this can be achieved to variable extents in different ways such as by using locally available materials, substituting Portland cement by cheaper and sustainable binder alternatives, and replacing steel fibers by cost effective glass, carbon, and synthetic fibers.

Numerous research studies have investigated these cost-cutting measures (Abellán-García et al., 2021; Christ et al., 2019; Gesoglu et al., 2016; He et al., 2021; Holubová et al., 2017; Liu et al., 2019; Madhkhan & Saeidian, 2021; Mohammed et al., 2021; Pourjahanshahi & Madani, 2021b; Kay Wille & Boisvert-Cotulio, 2015; Yan et al., 2021) but the primary focus has been the mechanical properties and the impact on environment and cost is seldom reported. Because of that, it is difficult to comprehend what environmental and cost benefits these studies bring.

Gooranirimi et. al investigated the performance of glass fiber reinforced polymer (GFRP) bars that were exposed to concrete alkalinity conditions after 15 years of service and did not find any deterioration in the microstructures and change in chemical composition (Gooranorimi & Nanni, 2017). He et.al used glass fiber (GF) and high-performance polypropylene (HPP) fiber in UHPFRC and claimed that glass fibers outperformed the high-performance polypropylene fibers in compressive, tensile and bending properties (He et al., 2021). Riguad et. al developed a self-placing, ductile, ultra-high performance concrete containing glass fibers for thin structural elements. This UHPFRC maintained ductility even after accelerated aging tests (Rigaud et al., 2012).

Ghosh et. al used glass bars in High-Early-Strength concrete and achieved a compressive strength of 22.1 MPa at 6 hours (Ghosh et al., 2021). UHPFRC is characterized by a minimum compressive strength of 150 MPa at 28 days as defined in ACI 239R-18. According to the cost estimation by Ghosh et. al, these fibers are 33% cheaper compared to steel fibers and therefore, they have the potential to reduce the cost of UHPFRC.

Table 1 below summarizes a comprehensive literature review where attempts are made to find suitable alternatives to steel fibers that exhibit similar mechanical properties. Different types of fibers include glass, polypropylene, carbon, basalt, and PVA fibers. These fibers are used as either partial or complete replacements of steel fibers and their mechanical properties (compressive strength, tensile strength, ductility etc), fracture properties, and durability are evaluated.

Table 2: UHPFRC research with various fibers

Author	Fiber	Cost reported	Emissions reported	Key Findings
Madhkhan et.al 2021	Glass fibers	No	No	Brittle after accelerated aging
He et. al 2021	Glass and High-performance Polypropylene fibers	No	No	2% fiber provides better performance. GF is better for compressive and tensile strength. HPP is better for toughness
Liu et. al 2019	GF,	No	No	Hybrid fibers have best

	Polypropylene, and Hybrid			performance. Polypropylene has better performance than GF
Chanvillard et. al 2012	GF	No	No	Durability and ductility maintained after aging tests
Christ et. al 2019	Hybrid steel and polypropylene fibers (50-100% steel, 0-50% PP)	No	No	80% steel and 20% PP gave the best performance in lowering cement content while maintaining compressive strength, and toughness
Pourjahanshahi et. al 2021	Steel, glass, carbon, and hybrid	No	No	Some fibers and combinations performed better in various scenarios of strength and durability
Yan et. al 2021	Basalt fiber, glass fiber, and polypropylene fibers	No	No	Enhanced compressive, flexural, modulus of rupture, toughness index with the addition of fibers compared to no fibers
Wille et. al 2015	Steel, glass, and PVA	Yes	No	Proposed cost-effective mix designs in different regions of US
Mohammed et. al 2021	Micro Glass fibers	No	No	Performance and ductility increased with addition of fibers
Gesoglu et. al 2016	Micro steel, and glass fibers	No	No	Properties enhanced with increase in fiber percentage
Holubova 2017	Alkali-resistant glass fibers	No	No	Durability was not affected, and fibers did not deteriorate even after 1 year
Pernicova 2015	Glass fibers	No	No	Type of glass fibers did not impact the compressive, flexural, and modulus of elasticity
Abellán-García 2021	Steel fibers	Reviewed	No	1% fiber was enough to achieve strain hardening. 2% of deformed fiber produced more strain-hardening

It is evident from these studies that a lot of emphasis is placed on the mechanical properties and durability but the attempt to quantify and compare the cost and environmental impact against conventional concrete is lacking. Mechanical strength and durability are very important for a construction material, and while numerous studies have investigated the performance and reported promising results, it is important to study the environmental impacts. Upcoming sections attempt to provide an overview of the life cycle analysis of UHPFRC to determine its global warming potential in comparison to conventional concrete.

3. EMISSIONS ASSOCIATED WITH CONVENTIONAL REINFORCED CONCRETE

The construction of new buildings is a significant source of greenhouse emissions. The emissions are produced at various stages when building materials are processed, manufactured, and transported for construction. Concrete is a widely used building construction material and while some residential buildings are made of sustainable materials like wood and glass, concrete structures still dominate the commercial and industrial sectors. Particularly, concrete and steel production contribute significantly to global greenhouse gas emissions (Amiri et al., 2020; Davis et al., 2018).

Concrete structures are not only resilient and durable but are easy to build and relatively cheaper than other building materials. That's why reinforced concrete is commonly used for buildings ("Concrete Needs to Lose Its Colossal Carbon Footprint," 2021). About 70% of the global population lives in reinforced concrete buildings. Each year, 30 billion tonnes of concrete are used around the world (Monteiro et al., 2017). In the United States, more than 50% of low-rise buildings are constructed from concrete (*Buildings & Structures*, 2016).

A typical concrete is composed of hydraulic cement, water, fine and coarse aggregates, and chemical admixtures in a properly proportioned mixture. Concrete is categorized based on compressive strength and normal or conventional concrete has a compressive strength ranging from 20 to 40 MPa.

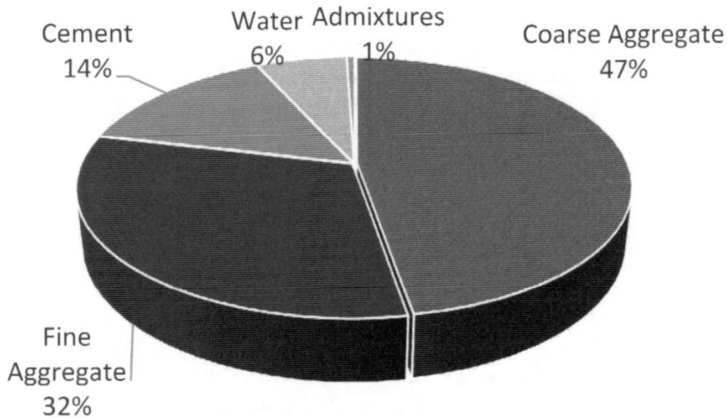

Figure 2: Composition of conventional reinforced concrete (RC) (Behravan et al., 2022)

Concrete is a composite material composed of 60-75% aggregates, 5-10% water and 10-15% cement. An important part of the cement making process entails heating the limestone to temperatures of 1500 °C, for a long period of time, producing high amount of carbon emissions (U.S. Geological Survey, 2020). 80% of the CO_2 generated during the concrete production, comes from this process. 822 kg of CO_2 is added to the environment for 1 ton of cement produced (Andrew, 2022).

From 1928 to 2018, cement production has generated an estimated 38 billion tonnes of total global greenhouse emissions. Of that amount, 71% was generated after 1990. In 2019 alone, 4.1 billion tonnes of cement was produced globally (Kuijpers, 2020). As of today, cement production generates the largest emissions of CO2 in the industrial sector, accounting for 8% of global CO2 emissions (2.8 Gtons/yr), and 2-3% of energy usage for fuel combustion (Ellis et al., 2020).

It's not surprising that concrete is increasingly being used for climate-resilient construction, as it is durable and long-lasting, but it comes at an environmental cost of a colossal carbon footprint. Cement production is projected to increase annually by 50% by 2050. The anticipated increase in greenhouse gas emissions and energy demand over the next 33 years would be additional 85–105 Gt of CO2 and energy demand of 420–505 TJ (equivalent to the combined global emissions from 2009 and 2010, and the world's primary energy supply in 2005) (Miller et al., 2016; Monteiro et al., 2017).

The sustainability standards and regulations are usually based on buildings' energy consumption and performance, causing the construction and maintenance pollution go undetected. More than half of a building's lifetime carbon emissions come from the embodied carbon, i.e., the CO_2 emitted during building construction. The embodied carbon is almost never taken into the account and needs to be addressed by shifting towards alternate building materials and UHPC.

4. REVIEW ON LIFE CYCLE ENVIRONMENTAL IMPACT ANALYSIS

Previous research work on UHPC focuses mostly on the mix design and mechanical properties. Very few works have been done to evaluate the environmental impact of conventional concrete and UHPFRC. As explained in the previous section, enormous amount of carbon emissions is generated during the concrete mixing, and cement production. The final built structure requires maintenance during all its lifetime. The emissions analysis of the maintenance is completely different and has been considered by some researchers while determining the global warming potential of conventional concrete and UHPC.

A life-cycle assessment (LCA) is an evaluation of the potential environmental impacts of a system during their entire life cycle. This helps understand how a process impacts the environment and is very beneficial for researchers, policy makers, and end users (Dong, 2018). In this paper, the life cycle of concrete is defined in two phases: construction and maintenance. The construction phase includes the production of individual components of concrete such as production of cement, aggregates, and chemical admixtures. After the production, the individual components are transported to the concrete mixing plant where they are mixed to produce concrete and this concrete is transported to the construction site where it is placed and cured. The curing duration of concrete is typically 28 days. The maintenance phase comprises of routinely monitoring the surface conditions, cracks, deformations, and signs of degradation of the concrete structure. If needed, repairs are performed to maintain the serviceability or if that is not possible, the structure is demolished, and the waste is recycled (Sameer et al., 2019).

In this section, a review is presented on the Life-cycle assessment of conventional concrete and UHPFRC. There can be several parameters to determine the environmental impacts of any materials such as global warming potential (GWP), acidification potential, eutrophication potential, and ozone depletion among others (Dong, 2018; Sameer et al., 2019). Since GWP is the most reported parameter, it is quantitively compared for conventional concrete and UHPFRC.

4.1 Global Warming Potential (GWP)

The global warming potential is chosen as the environmental indicator herein. The cumulative environmental impact in terms of CO_2 emissions of the infrastructure over its service life is expressed as global warming potential. It is measured in $kgCO_{2\,eq}$ which is the equivalent weight of CO_2 produced.

4.2 Construction Stage

The Global Warming Potential of UHPFRC and conventional concrete is compared quantitatively (Bertola et al., 2021; Dong, 2018; Rangelov et al., 2018; Sameer et al., 2019). It can be observed that in this stage, the GWP of UHPFRC is higher than conventional concrete.

Table 3: GWP comparison of UHPFRC and conventional RC in construction stage

Reference	**Application**	**Global Warming Potential ($kgCO_{2\,eq}$)**		
		Conventional RC	**UHPFRC**	**Percentage difference**
Dong 2018	Bridges	348	877	60%
Bertola et. al 2021	Bridges	7217	10914	66%
Sameer et. al 2019	Bridges	390	1700	77%
Rangelov et. al 2020	Bridges	434	1930	78%

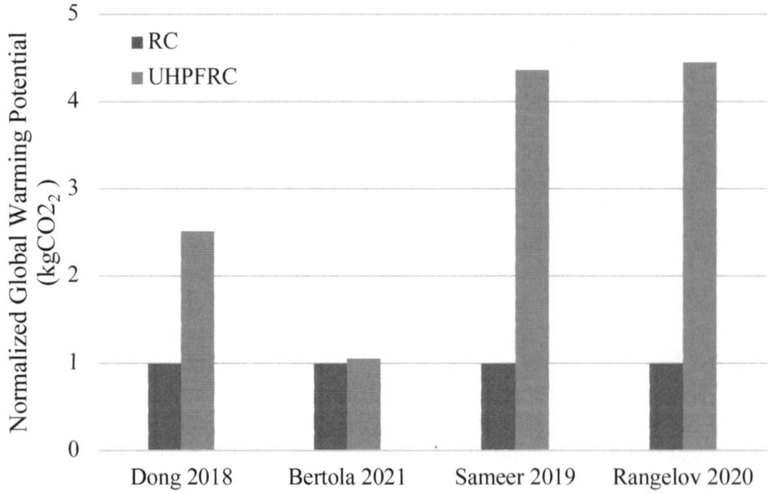

Figure 3: Normalized GWP comparison of UHPFRC and reinforced RC in construction stage

4.3 Cradle-to-Grave

In this stage, the maintenance phase of the life cycle is added which can span decades after construction depending on the specific application. It can be observed that the GWP of UHPFRC is considerably lower than conventional concrete in this phase. In the initial construction phase, the GWP of UHPFRC is higher than conventional concrete. This can be attributed to the increased cement usage in UHPFRC and the presence of steel fibers. As the cement content is much lower in conventional concrete, the GWP is also lower.

But as time goes on, maintenance issues begin to arise. Owing to its superior durability, the maintenance needs of UHPFRC are very low as compared to conventional concrete and hence, the life cycle GWP of UHPFRC is very low as compared to conventional concrete.

Table 4: GWP comparison of UHPFRC and conventional RC in cradle-to-grave stage

Reference	**Application**	**Global Warming Potential ($kgCO_{2\,eq}$)**		
		Conventional RC	**UHPFRC**	**Percentage reduction**
Dong 2018	Bridges	430,370	290,960	32%
Bertola et. al 2021	Bridges	15,878	11,236	29%
Sameer et. al 2019	Bridges	280,000	250,000	11%
Rangelov et. al 2020	Bridges	2,800,000	651,000	77%

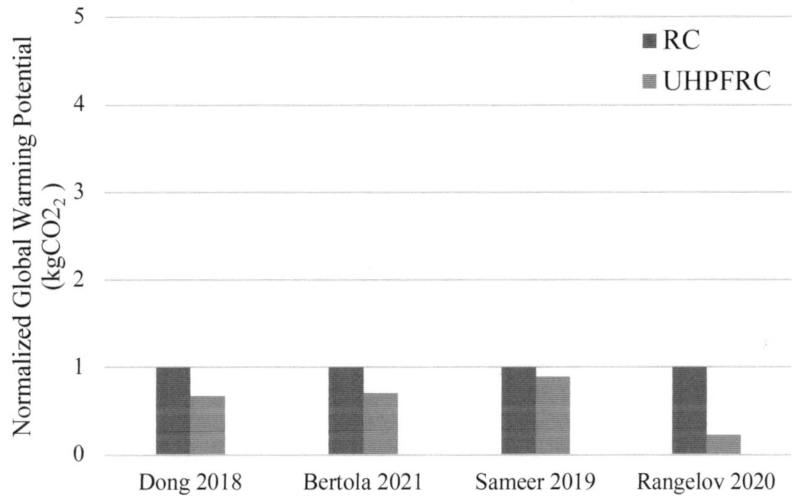

Figure 4: Normalized GWP comparison of UHPFRC and conventional RC in cradle-to-grave stage

Figure 3 gives a comparison of the life cycle GWP between UHPFRC and conventional concrete (Bertola et al., 2021). The GWP of UHPFRC hardly rises even after decades whereas the conventional concrete structure has undergone several maintenance routines in the same time period. This is because the conventional concrete is highly susceptible to damaging durability mechanisms such as alkali-silica reaction, carbonation, chloride penetration, and cracking among others. Because of the tight microstructure and dense particle packing, UHPFRC performs much better in terms of durability.

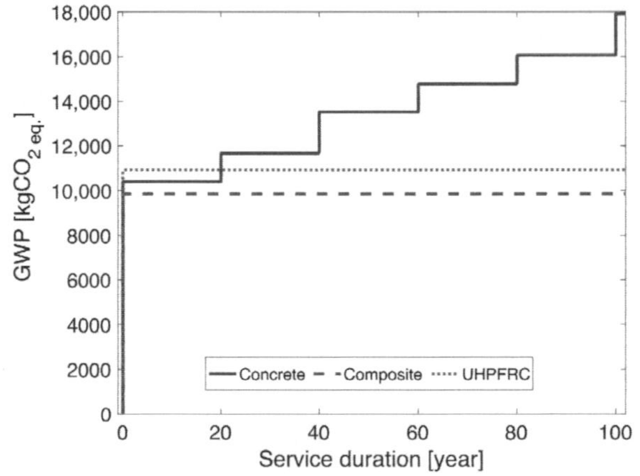

Figure 5: Influence of service duration on the environmental impact (Bertola et al., 2021)

5. CONCLUSIONS

The mechanical properties of UHPFRC are being widely researched with improvements being made in compressive, tensile, and durability properties. However, the environmental impacts are rarely considered. In this study, a review of the life cycle environmental impact analysis is presented which compares the Global Warming Potential (GWP) of conventional concrete to UHPFRC over a construction phase and cradle-to-grave phase. Based on the comparative life-cycle assessment, it can be concluded that although initial GWP of UHPFRC is higher than conventional concrete, UHPFRC requires significantly less maintenance and provides better durability in one lifespan which is from raw material acquisition to demolition. Although the initial GWP of UHPFRC is 60-80% higher than conventional RC, the reduction in GWP over a lifespan in UHPFRC ranges from 11-77% based on application and material design. As the service duration of concrete structure increases, the GWP of conventional concrete exceeds the GWP of composite of UHPFRC of similar capacity. Thus, UHPFRC is estimated to have lower GWP compared to conventional concrete. This comparative analysis is very relevant to current construction industry as a huge portion of USA infrastructure nears the end of its life-expectancy and require more frequent maintenances. Inclusion of UHPFRC in construction projects, despite its initial cost and high GWP, can prove beneficial and economical in the long term.

NOMENCLATURE

UHPFRC	Ultra High-Performance Fiber Reinforced Concrete	
RC	Reinforced Concrete	
GF	Glass Fibers	
GWP	Global Warming Potential	(kg $CO_{2\,eq}$)
PVA	Polyvinyl Alcohol	

REFERENCES

Abellán-García, J., Fernández-Gómez, J. A., Torres-Castellanos, N., & Núñez-López, A. M. (2021). Tensile behavior of normal-strength steel-fiber green ultra-high-performance fiber-reinforced concrete. *ACI Materials Journal*, *118*(1), 127–138. https://doi.org/10.14359/51725992

ACI 239R. (2018). Ultra-high-performance concrete: An emerging technology report. *American Concrete Institute ACI 239.*

Administration, F. H. (2013). *Development of Non-Proprietary Ultra-High Performance Concrete for Use in the Highway Bridge Sector.*

Alkaysi, M., El-Tawil, S., Liu, Z., & Hansen, W. (2016). Effects of silica powder and cement type on durability of ultra high performance concrete (UHPC). *Cement and Concrete Composites*, *66*, 47–56. https://doi.org/10.1016/j.cemconcomp.2015.11.005

Amiri, A., Ottelin, J., Sorvari, J., & Junnila, S. (2020). Cities as carbon sinks - Classification of wooden buildings.

Environmental Research Letters, *15*(9). https://doi.org/10.1088/1748-9326/aba134
Andrew, R. (2022). *Global CO2 emissions from cement production*. Zenodo. https://doi.org/10.5281/zenodo.6553090
Bajželj, B., Allwood, J. M., & Cullen, J. M. (2013). Designing climate change mitigation plans that add up. *Environmental Science and Technology*, *47*(14), 8062–8069. https://doi.org/10.1021/es400399h
Behravan, A., Aqib, S. M., Delatte, N. J., Ley, M. T., & Rywelski, A. (2022). Performance Evaluation of Silane in Concrete Bridge Decks Using Transmission X-ray Microscopy. *Applied Sciences*, *12*(5). https://doi.org/10.3390/app12052557
Bertola, N., Küpfer, C., Kälin, E., & Brühwiler, E. (2021). Assessment of the environmental impacts of bridge designs involving UHPFRC. *Sustainability (Switzerland)*, *13*(22), 1–19. https://doi.org/10.3390/su132212399
Buildings & Structures. (2016). Portland Cement Association. http://www.tefugen.com/index.php/analysis/cfd-analysis/ buildings-and-structural
Christ, R., Pacheco, F., Ehrenbring, H., Quinino, U., Mancio, M., & Tutikian, B. F. (2019). Study of mechanical behavior of ultra - high performance concrete (UHPC) reinforced with hybrid fibers and with reduced cement consumption. *Revista Ingeniería de Construcción*, *34*(2), 159–168. https://doi.org/10.4067/s0718-50732019000200159
Concrete needs to lose its colossal carbon footprint. (2021). *Nature*, *597*(7878), 593–594. https://doi.org/10.1038/d41586-021-02612-5
Davis, S. J., Lewis, N. S., Shaner, M., Aggarwal, S., Arent, D., Azevedo, I. L., Benson, S. M., Bradley, T., Brouwer, J., Chiang, Y. M., Clack, C. T. M., Cohen, A., Doig, S., Edmonds, J., Fennell, P., Field, C. B., Hannegan, B., Hodge, B. M., Hoffert, M. I., … Caldeira, K. (2018). Net-zero emissions energy systems. *Science*, *360*(6396). https://doi.org/10.1126/science.aas9793
de Larrard, F., & Sedran, T. (1994). Optimization of ultra-high-performance concrete by the use of a packing model. *Cement and Concrete Research*, *24*(6), 997–1009. https://doi.org/10.1016/0008-8846(94)90022-1
Dong, Y. (2018). Performance assessment and design of ultra-high performance concrete (UHPC) structures incorporating life-cycle cost and environmental impacts. *Construction and Building Materials*, *167*, 414–425. https://doi.org/10.1016/j.conbuildmat.2018.02.037
Ellis, L. D., Badel, A. F., Chiang, M. L., Park, R. J. Y., & Chiang, Y. M. (2020). Toward electrochemical synthesis of cement—An electrolyzer-based process for decarbonating CaCO3 while producing useful gas streams. *Proceedings of the National Academy of Sciences of the United States of America*, *117*(23), 12584–12591. https://doi.org/10.1073/pnas.1821673116
Favier, A., Scrivener, K., & Habert, G. (2019). Decarbonizing the cement and concrete sector: Integration of the full value chain to reach net zero emissions in Europe. *IOP Conference Series: Earth and Environmental Science*, *225*(1). https://doi.org/10.1088/1755-1315/225/1/012009
Gesoglu, M., Güneyisi, E., Muhyaddin, G. F., & Asaad, D. S. (2016). Strain hardening ultra-high performance fiber reinforced cementitious composites: Effect of fiber type and concentration. *Composites Part B: Engineering*, *103*, 74–83. https://doi.org/10.1016/j.compositesb.2016.08.004
Ghosh, D., Abd-Elssamd, A., Ma, Z. J., & Hun, D. (2021). Development of high-early-strength fiber-reinforced self-compacting concrete. *Construction and Building Materials*, *266*, 121051. https://doi.org/10.1016/j.conbuildmat.2020.121051
Gooranorimi, O., & Nanni, A. (2017). GFRP Reinforcement in Concrete after 15 Years of Service. *Journal of Composites for Construction*, *21*(5), 04017024. https://doi.org/10.1061/(asce)cc.1943-5614.0000806
Graybeal, B., & Tanesi, J. (2008). A cementitious long-life wearing course to reduce frequency of maintenance works on high-traffic roads. *Transport Research Arena Europe 2008*, *1561*(February), 454–461. https://doi.org/10.1061/(ASCE)0899-1561(2007)19
He, J., Chen, W., Zhang, B., Yu, J., & Liu, H. (2021). The mechanical properties and damage evolution of UHPC reinforced with glass fibers and high-performance polypropylene fibers. *Materials*, *14*(9). https://doi.org/10.3390/ma14092455
Holubová, B., Hradecká, H., Netušilová, M., Gavenda, T., & Helebrant, A. (2017). Corrosion of glass fibres in ultra high performance concrete and normal strength concrete. *Ceramics - Silikaty*, *61*(4), 319–326. https://doi.org/10.13168/cs.2017.0031
Kuijpers, M. (2020). Concrete is one of the most polluting materials in the world. Here's how we can make it sustainable. *The Correspondent*. https://thecorrespondent.com/678/concrete-is-one-of-the-most-polluting-materials-in-the-world-heres-how-we-can-make-it-sustainable/844886973084-cc8f94da
Larsen, I. L., & Thorstensen, R. T. (2020). The influence of steel fibres on compressive and tensile strength of ultra high performance concrete: A review. *Construction and Building Materials*, *256*, 119459.

https://doi.org/10.1016/j.conbuildmat.2020.119459
Liu, J., Jia, Y., & Wang, J. (2019). Experimental Study on Mechanical and Durability Properties of Glass and Polypropylene Fiber Reinforced Concrete. *Fibers and Polymers*, *20*(9), 1900–1908. https://doi.org/10.1007/s12221-019-1028-9

Madhkhan, M., & Saeidian, P. (2021). Mechanical properties of ultra-high performance concrete reinforced by glass fibers under accelerated aging. *International Journal of Engineering, Transactions B: Applications*, *34*(5), 1074–1084. https://doi.org/10.5829/ije.2021.34.05b.01

Magureanu, C., Sosa, I., Negrutiu, C., & Heghes, B. (2012). Mechanical properties and durability of ultra-high-performance concrete. *ACI Materials Journal*, *109*(2), 177–184. https://doi.org/10.14359/51683704

Miller, S. A., Horvath, A., & Monteiro, P. J. M. (2016). Readily implementable techniques can cut annual CO2 emissions from the production of concrete by over 20%. *Environmental Research Letters*, *11*(7). https://doi.org/10.1088/1748-9326/11/7/074029

Mohammed, B. H., Sherwani, A. F. H., Faraj, R. H., Qadir, H. H., & Younis, K. H. (2021). Mechanical properties and ductility behavior of ultra-high performance fiber reinforced concretes: Effect of low water-to-binder ratios and micro glass fibers. *Ain Shams Engineering Journal*, *12*(2), 1557–1567. https://doi.org/10.1016/j.asej.2020.11.008

Monteiro, P. J. M., Miller, S. A., & Horvath, A. (2017). Towards sustainable concrete. *Nature Materials*, *16*(7), 698–699. https://doi.org/10.1038/nmat4930

Pourjahanshahi, A., & Madani, H. (2021a). Chloride diffusivity and mechanical performance of UHPC with hybrid fibers under heat treatment regime. *Materials Today Communications*, *26*(January), 102146. https://doi.org/10.1016/j.mtcomm.2021.102146

Pourjahanshahi, A., & Madani, H. (2021b). Chloride diffusivity and mechanical performance of UHPC with hybrid fibers under heat treatment regime. *Materials Today Communications*, *26*(January), 102146. https://doi.org/10.1016/j.mtcomm.2021.102146

Rangelov, M., Spragg, R. P., Haber, Z. B., & H, D. (2018). *Life-cycle assessment of ultra-high performance concrete bridge deck overlays* (1st Editio).

Rigaud, S., Chanvillard, G., & Chen, J. (2012). Characterization of bending and tensile behavior of ultra-high performance concrete containing glass fibers. *RILEM Bookseries*, *2*(January 2012), 373–380. https://doi.org/10.1007/978-94-007-2436-5_45

Sameer, H., Weber, V., Mostert, C., Bringezu, S., Fehling, E., & Wetzel, A. (2019). Environmental assessment of ultra-high-performance concrete using carbon, material, and water footprint. *Materials*, *16*(6). https://doi.org/10.3390/ma12060851

Shehab El-Din, H. K., Mohamed, H. A., Khater, M. A. E.-H., & Ahmed, S. (2016). *Effect of Steel Fibers on Behavior of Ultra High Performance Concrete. 2004*, 1–10. https://doi.org/10.21838/uhpc.2016.11

U.S. Geological Survey. (2020). Mineral commodity summaries 2020. In *U.S Departtment OF The Interior, U.S Geological Survey* (Issue 703). https://pubs.usgs.gov/periodicals/mcs2020/mcs2020.pdf

Wille, K., El-Tawil, S., & Naaman, A. E. (2014). Properties of strain hardening ultra high performance fiber reinforced concrete (UHP-FRC) under direct tensile loading. *Cement and Concrete Composites*, *48*, 53–66. https://doi.org/10.1016/j.cemconcomp.2013.12.015

Wille, Kay, & Boisvert-Cotulio, C. (2015). Material efficiency in the design of ultra-high performance concrete. *Construction and Building Materials*, *86*, 33–43. https://doi.org/10.1016/j.conbuildmat.2015.03.087

Yan, P., Chen, B., Afgan, S., Aminul Haque, M., Wu, M., & Han, J. (2021). Experimental research on ductility enhancement of ultra-high performance concrete incorporation with basalt fibre, polypropylene fibre and glass fibre. *Construction and Building Materials*, *279*, 122489. https://doi.org/10.1016/j.conbuildmat.2021.122489

PCM Selection for Heat Pump Integrated with Thermal Energy Storage for Demand Response in Residential Buildings

Sara Sultan[1]*, Tugba Turnaoglu[3], Damilola Akamo[1], Jason Hirschey[2], Xiaobing Liu[3], Tim Laclair[3], Kyle R. Gluesenkamp[3]

[1]University of Tennessee, Bredesen Center of Interdisciplinary Education
Knoxville, TN, USA
ssultan1@vols.utk.edu

[2]Georgia Institute of Technology
Atlanta, GA, USA

[3]Oak Ridge National Laboratory
Oak Ridge, TN, USA
gluesenkampk@ornl.gov

* Corresponding Author

ABSTRACT

Phase Change Materials (PCM) based Thermal energy storage (TES) is a widespread solution to shift buildings' peak energy demand and add stability to the grid. PCMs can be used for space heating and cooling applications in residential buildings by integrating into the heat pump equipment or building envelope via several possible configurations. The heat pump integrated active PCM storage can provide significant energy savings and reduce peak demand, but the benefits vary substantially depending on the system configuration, storage capacity, and the temperature range of the PCM used. It is critical to provide a review of the materials and their suitability in heat pump integrated PCM systems. This paper will present an analysis of the PCM material selection for heat pump integrated TES in various configurations. The comprehensive review will compare the phase change materials with different melting temperatures, and configurations, corresponding to the energy savings and demand impact reported in the literature. This work is significant to guide the design of high-performance heat pump integrated PCM systems.

Keywords: *Thermal Energy Storage, Phase Change Materials, Heat Pumps, Demand Response, Residential Buildings*

1. INTRODUCTION

Building's energy consumption and peak demand is often attributed to air conditioning applications. More than half of the energy consumed in the residential buildings is used for air conditioning and space heating. According to the DOE report, 54% of the site energy consumption is accounted for space heating and cooling combined, driving major portion of the residential energy demand (U.S. Energy Information Administration EIA, 2015, 2019). The residential buildings can play a major role to reduce energy demand and increase the building flexibility while maintaining the temperature in a comfortable range, by designing a load shifting strategy (Arteconi et al., 2019; Y. Chen et al., 2018; Jensen et al., 2017; Junker et al., 2018). The major electric load from the residential buildings comprises 80% of the total electricity use in the U.S., and 78% of the peak period use during 2-8 PM (Goetzler et al., 2019).

About 80% of American homes are detached single-family houses of an average area of about 2600 ft^2, with an average occupancy of 2.5 (U.S. EIA, 2020). Furthermore, 90% of homes in the U.S. use air conditioning as of 2019, and 75%

of those air-conditioned homes use central units. In 2019, the average annual electricity consumption for a U.S. residential utility customer was about 877 kWh per month. The consumers end up paying additional peak demand charges due to stress on the utility grid.

Though, the energy consumption and demand in the U.S. declined by 14% in response to the COVID-19 pandemic, however, the residential buildings energy consumption rose by 8% due to the increased residential demand shift as a result of working from home (*U.S. Energy Information Administration - EIA - Independent Statistics and Analysis*, n.d.). Power generation grids around the world are undergoing major changes to cope with the increasing energy demand and maintain physical and economic stability in response to the growing need for renewable energy generation (Blum, 2016). The technological advances and flexibility in the design and operation of HVAC systems can allow them to help the grid overcome these new load patterns challenges if properly managed.

Typically, in US residential buildings, there is no mechanical outdoor air exchange, or cross ventilation using natural resources, therefore, heat pumps are important to regulate the humidity inside the house as well as the indoor temperature. It is often cost-prohibitive to rebuild houses or update the conventional methods, but the heat pump systems can be made smart to respond to the peak demand changes. Today's high-efficiency HVAC and appliance products can provide energy savings, but do not generally help with demand management and grid stress (Sultan et al., 2021).

Thermal energy storage (TES), which has capability to decouple the energy generation from its demand and usage, can be a crucial tool to shift the thermal loads. TES systems can be easily incorporated in small to large single-family or multi-family residential buildings. They are designed to trim the building's electrical demand profile by reducing the energy usage and high demand costs arising from summer cooling and winter heating needs (Cui et al., 2014). When properly controlled, thermal storage can be useful in an economic or behavioral curtailment scenario for demand response. Major factors influencing an appropriate TES system's choice include the energy storage period and duration, cost and economic feasibility, and operational conditions. The building's thermal load is also dependent on the weather conditions and the thermal quality of the building envelope.

Phase Change Materials (PCMs) are widely used commercial thermal storage materials that could be used for space heating and cooling applications in various configurations. PCM-TES can store and release energy through charging and discharging process, respectively. By utilizing TES to store the energy for later use during peak hours, demand management systems can be designed to take advantage of lower utility rates during off-peak hours.

Researchers have used various configurations to employ the economic and grid benefits of PCM-TES for heating and cooling applications in buildings. The most common configurations are to install the material in building envelope via passive storage where the charging and discharging process is the function of ambient temperature, hence the grid benefits can't be maximized. Active storage configurations allow real-time control of TES and are proven to be more effective than passive configurations. Active TES systems mechanically transfer the heat and hence can control when to charge or discharge the TES. This allows the TES systems to achieve grid flexibility, making them more suitable for load shifting and demand management. Another benefit of active configuration is the flexibility to be installed in an existing building infrastructure without needing a significant construction. The TES may be separate and removed from the space.

PCM-TES can behave as a standalone isothermal heat exchanger when integrated with heat pump (HP), and can offer advantages by shifting the peak demand while maintaining an adequate internal temperature in comfort range, without excessively using heat pump equipment. HP-TES in an active configuration, can enhance these benefits with a better performance if heat transfer is controlled by the heat pump (Dong et al., 2019).

The key to using TES to provide economic and grid benefits is the development of materials and packaging in optimized configuration and temperature range. Low temperature salt hydrate PCMs with melting point closer to the room temperature have garnered interest for near ambient TES systems because of their high storage density. However, the PCMs in this range often have low thermal conductivity that impacts the storage capacity. As such, modifications are needed to enhance the stability and thermal properties of PCM. Researchers have used single material in passive configurations for both high and low temperature applications but for active configuration, the materials with different melting temperature are selected for each application. Melting temperature for cooling applications is below 10°C, while for heating is above 40°C.

The benefits also vary largely due to other factors such as PCM material and temperature, configuration, and climate zone. Figure 1 below shows the demand impact and load shifting benefits reported in literature for Heat pump coupled passive and active TES. Active TES dominate mostly because the high level of control, allows designing and incorporating effective load shifting strategy.

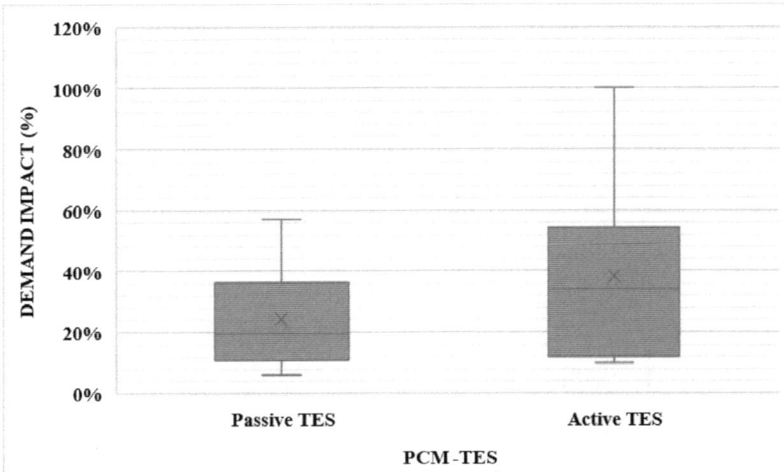

Figure 1: Comparison of load shifting benefits for Active and Passive TES

The focus of this paper is to review the phase change materials used in active heat pump integrated thermal storage systems. Literature review is conducted to evaluate the properties of PCM used in active HP-TES, as well as to compare the different configurations.

Section 2 will discuss the salt hydrate PCMs, focusing on melting temperature and storage capacity. Section 3 explains how PCMs are integrated in heat pump via different configurations. The demand impact is also reported and compared against the PCM types, heat pump configurations, and TES application.

2. PCM FOR HP-TES

Phase change materials (PCM) perform solid-liquid phase changes at temperatures in the operational range of the designated thermal application. When a phase transformation from a solid to a liquid occurs, the energy is absorbed by the phase transitioning material from the surroundings, though maintaining almost the same temperature (Fleischer, 2015). There are generally three classifications of PCMs; organic PCMs, inorganic PCMs, and eutectics. Organic PCMs are attributed to low heat conductivities and are flammable, while inorganic PCMs are relatively inexpensive, easily synthesized, nonflammable and possess high thermal conductivity and high storage capacity (Sarbu et al., 2018).

Any energy storage system should at least possess the following components:

- A PCM with the desired temperature range
- A compatible heat exchanger

This section reviews important factors and parameters for PCM selection to be integrated in heat pump componeents or coupled to the vapor compression system. The salt hydrate PCMs integrated with heat pump for space heating or cooling applications are reviewed.

2.1 PCM Selection

PCM temperature and storage capacity are considerably important to design the TES that can meet the buildings' load and provide necessary comfort. Various approaches can be found in literature to choose the melting point of PCMs to maximize their performance for space conditioning in buildings. Free cooling for continental climates was studied by Medved and Arkar (Arkar et al., 2007), and they found the PCM with a melting point equal to the average temperature of hottest month, most appropriate choice. The melting temperature range 20 and 22 °C was found suitable for passive cooling. Waqas and Kumar (Waqas et al., 2011) demonstrated an optimum PCM storage unit performance by selecting a PCM with the melting point equal to the comfort temperature of hottest month for free cooling.

The PCM temperature optimization studies (Alam et al., 2014; Arıcı et al., 2020; Berardi et al., 2019; Evola et al., 2013; Lei et al., 2016; Markarian et al., 2019; Piselli et al., 2020; Prabhakar et al., 2020; Saffari et al., 2017; Zhu et al., 2021) have commonly utilized passive cooling with organic PCMs. Paraffins are the frequently investigated materials with 20-30 °C melting temperature range. Furthermore, for a given TES, different PCM melting temperatures have been studied for both heating and cooling applications.

Table 1 below reviews salt hydrate PCMs integrated with different types of heat pumps via heat exchangers for space heating and cooling.

Table 1: PCMs for HP-TES

Author	PCM material	Application	Heat Exchanger	Thermophysical Properties	Performance
(Alkhwildi et al., 2020)	S8, S17, S27 (8 °C, 17 °C, 27 °C)	Geothermal heat pump for heating and peak load shifting	Single U-Tube ground heat exchanger	S27 showed highest savings and had thermal conductivity of 0.540 W/m.K	COP increased from 3.8 to 5.5
(Xin Jin et al., 2021)	Composite modified SAT (47.8°C)	Coupled to air source heat pump for heating	Shell and tube heat exchanger	Latent heat 219.8 kJ/kg (presented longer cycling lifetime)	Heat pump COP increased by 5.1%, 10 year payback period
(Han et al., 2008)	$CaCl_2.6H_2O$ (29.9 °C)	Solar assisted ground source heat pump	U-Tube soil exchanger	PCM thermal conductivity 1.09 W/m.K	COP increased from 3 to 5.95
(Yamaguchi et al., 1995)	$Na_2SO_4.10H_2O$ (32 °C)	Coupled to floor heating system for heating	Tube heat exchanger	PCM thermal conductivity 1.09 W/m.K	COP reported 3.0
(Kaygusuz et al., 1999)	$CaCl_2.6H_2O$ (28°C)	Coupled to solar and heat pump for heating	Fin tube heat exchanger	PCM thermal conductivity 1.09 W/m.K	COP increased to 4.0
(Benli, 2011)	$CaCl_2.6H_2O$ (28°C)	Coupled to ground source heat pump for green house heating	Ground heat exchanger	PCM thermal conductivity 1.09 W/m.K	COP increased to 4.2
(Kelly et al., 2014)	Commercial salt hydrate (48 °C)	Coupled to heat pump for heating	PCM enhanced buffer tank	Thermal conductivity 0.45 W/m.K	COP reduced from 3.0 to 2.4
(Xu et al., 2021)	Sodium Acetate Trihydrate SAT (58°C)	Coupled to condenser for heating	Plate heat exchanger	Thermal conductivity 0.57 W/m.K	COP reported 4.5
(Kumirai et al., 2019)	SP24E, RT27, RT25 (24 °C - 25 °C)	Coupled to ventilation system for cooling	Plate heat exchanger	Thermal conductivity 0.60 W/m.K	Salt hydrate PCM had highest storage capacity
(Alam et al., 2019)	FlatICE PCM (15 °C)	Coupled to chiller for space cooling and peak load shifting	Plate heat exchanger	Thermal conductivity 0.43 W/m.K	Chiller load reduced by 37%. PCM utilized 15% of its storage capacity

Most of the active TES studies reported in literature, as shown in the table above, have selected a melting temperature closer to the typical room temperature. A few of these works have investigated more than one type of PCM and have reported that the salt hydrate PCMs performed better than the organic PCMs in terms of storage capacity, and economic benefits.

3. INTEGRATION IN HEAT PUMP COMPONENTS

Active PCM-TES can be coupled to heat pump in various configurations. Table 1 in previous section described different types of heat exchangers and heat pumps. This section will particularly focus on PCM location within the heat pump and charging operation for the configuration. Researchers have reported energy and cost savings, and demand impact benefits for off-peak heating and cooling. Besides the PCM type and melting temperature, the savings and benefits largely depend on how and when PCM has been charged and discharged.

3.1 Heat Pump Integration

PCM-TES can be installed in building envelope, air distribution loops, or the vapor compression system itself. For an active PCM-TES to be installed in the building envelope, a heat exchanger is used and coupled to the air handling unit (Kondo et al., 2006). In the vapor compression system, either the PCM is directly installed in the condenser or evaporator, or coupled via a stand alone storage tank (Bruno et al., 2014; Real et al., 2014). Air distribution system has been the most commonly used configuration where the PCM is installed in the supply or return air ducts (Chaiyat, 2015; Hlanze et al., 2021).

3.2 Application and Configuration

Active HP-TES systems have been investigated for the demand response and load shifting potential. The literature reported in Table 2 and 3 show that the cooling applications are widely explored and peak cooling demand is shifted to off peak hours using different control strategies.

A paraffin PCM was installed in a return air duct coupled with evaporator to enhance the cooling efficiency of air conditioning system (Chaiyat, 2015). The PCM was solidified during nighttime by the cold air from the heat pump. During the daytime, return air from the room was passed through the PCM, liquifying it and reducing the cooling load at the evaporator. The time-of-use TOU tarrif was used and 3 hour load was shifted to off-peak hours. 9% daily energy and cost savings were reported.

Bruno et al. (Bruno et al., 2014) used ice and two salt hydrates PCMs installed in a heat exchanger coupled to chillder. The cooling energy is provided through chiller to either the PCM for solidification, or to the building during off-peak hours. The charging time and duration was optimized depending on the outside temperature to minimize energy usage and the PCM with 10°C melting temperature could save 13% energy.

Table 2 and 3 below compare the energy savings and demand impact for Active TES systems integrated with heat pump via different configurations. PCM melting point and charging operations are reported. Table 2 summarizes the salt hydrate PCMs while Table 3 reviews other PCMs used for HP-TES.

Table 2: Active HP-TES Configurations with Salt hydrate PCMS

Author	PCM material	Application	Configuration & HP integration	Charging Operation	Key Results
(Real et al., 2014)	S10 (10 °C) S27 (27 °C)	Both heating and cooling	In two storage tanks coupled to HVAC	Cooling energy is stored in cold tank at night using HVAC	18.97% energy saved
(Xu et al., 2021)	Sodium Acetate Trihydrate (SAT) (58 °C)	Space Heating	In storage tank coupled to condenser & electric heater	Heat pump condenser is used to charge PCM during off-peak hours	6 hours of peak load shifted
(Farah et al., 2019)	Salt Hydrate PCM 18T (18°C)	Space Cooling	In storage tank coupled to an air source heat pump	Return air from room passes through PCM before	Peak demand reduced by 32%

				entering the evaporator to increase COP	
(Mosaffa et al., 2016)	S27 (27°C) S25 (25°C)	Space Cooling	In storage tank coupled to vapor type compressor of heat pump	PCM is solidified using air-conditioner during off-peak hours	7-9% energy savings
(Bruno et al., 2014)	S4, S10 (4, 10 °C)	Space Cooling	In heat exchanger coupled to chiller	Chiller provides cooling to PCM during off-peak hours	Peak electric consumption reduced by 36.6%

Table 3: Active HP-TES Configurations with other PCMS

Author	PCM material	Application	Configuration & HP integration	Charging Operation	Key Results
(Takeda et al., 2004)	35% paraffin GR25 composite (25°C)	Space Cooling	In supply air duct of a ventilation system	PCM is solidified during night using the cool air from AHU	Ventilation load reduced by 62.8%
(Kondo et al., 2006)	Micro capsulate (25°C)	Space Cooling	In ceiling board coupled with AHU	PCM is charged overnight using AHU	9.4% load reduction
(Chaiyat, 2015)	RT20 Paraffin (20 °C)	Space Cooling	In return air duct coupled with the evaporator	Return air from room passes through PCM before entering the evaporator to increase COP	9% energy saved
(X. Chen et al., 2020)	RT11HC Paraffin (10-12°C)	Space Cooling	In storage tank coupled to heat pump	Heat pump directs cold water to solidify PCM	11.28% peak demand charges reduced
(Hlanze et al., 2021)	PureTemp 15T (15 °C)	Space Cooling	In supply air duct of air conditioning system	Cold supply air solidifies PCM during off-peak hours	Peak energy consumption reduced by 20-25%
(Hu et al., 2021)	ATS 30 (28 °C–33 °C)	Space Heating	In supply air system of heat pump	PCM is charged by heat pump at lower utility rates	6.7% energy consumption reduced

4. RESULTS AND DISCUSSIONS

The demand impact reported in the tables 2 and 3 corresponds to peak load shifted, peak electricity consumption reduced, or heat pump cooling and heating loads reduced. This section would discuss the PCM melting temperature and heat pump TES configuration for maximum demand impact. There are still number of factors to be considered such as the ambient climate conditions, size of TES, and heat transfer coefficient. The scope of this paper is limited to PCM material, melting temperature, and heat pump configuration.

4.1 Melting Temperature

Figure 2 shows the range of PCM temperature corresponding to demand impact for space heating and cooling. The PCM materials with the melting temperature ranging from 15 to 27 °C showed the most savings and demand benefits. Majority of the HP-TES systems reported in literature for demand response have investigated a cooling system which shows that cooling demand dominates in the centrally air conditioned residential homes. The PCM melting temperatures reported lie within the narrow range closer to room temperature except the Sodium Acetate Trihydrate (58 °C) which is used for space heating.

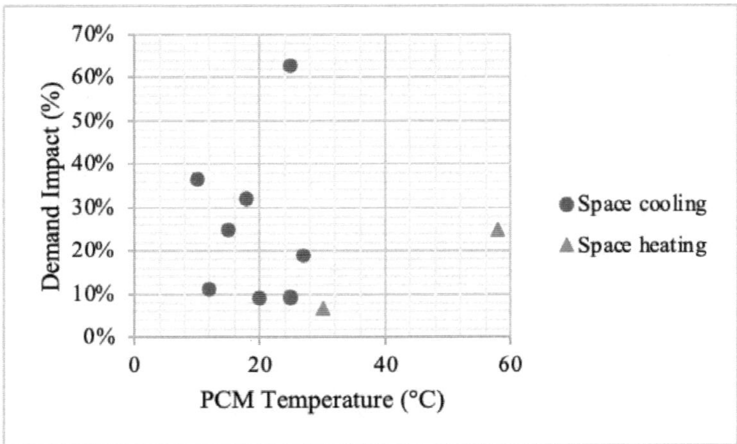

Figure 2: Demand Impact and PCM Temperature

4.2 Configuration

PCM-TES has been integrated with heat pump in various different configurations. Heat exchanger or storage tank containing PCM is coupled to either building envelope or heat pump components. Heat pump components can be on the air-side or the refrigirent-side of the heat pump. The refrigirent-side is referred as vapor compression system components (condenser, evaporator, and compressor), and the air-side corresponds to the air distribution system (supply air duct and return loop). Figure 3 shows that the air-side configurations are the ones most commonly investigated and report maximum demand impact and energy savings.

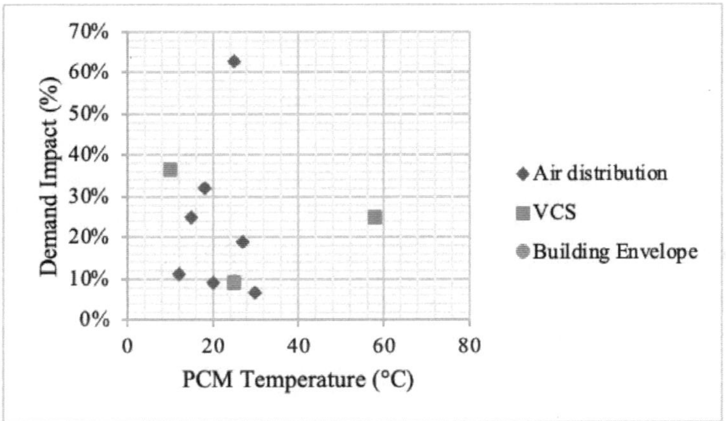

Figure 3: Demand Impact and heat pump configurations

5. CONCLUSIONS

Phase Change Material (PCM) based Thermal energy storage (TES) can provide economic and environmental benefits for grid-interactive residential buildings. The value and performance of TES vary greatly depending on the temperature range and storage capacity of the material, as well as the system configuration and ambient temperature. Active storage configurations allow real-time control of TES and are proven to be more effective than passive configurations, however, the active TES studies found in literature have utilized different materials for both heating and cooling.

This paper reviews the PCM materials, melting temperatures, and heat pump configurations to compare the energy savings and demand benefits of heat-pump integrated TES. The optimum demand impact and energy savings are reported for salt hydrate PCM with melting temperature between 15 and 27 °C integrated via air distribution system of the heat pump.

NOMENCLATURE

PCM	Phase Change Material	
TES	Thermal Energy Storage	
HP-TES	Heat Pump integrated Thermal Energy Storage	
AHU	Air Handling Unit	
TOU	Time of Use	
HVAC	Heating, Ventilation, and Air Conditioning	
HX	Heat Exchanger	
COP	Coefficient of Performance	
VCS	Vapor Compression System	
T	Temperature	(°C)

REFERENCES

Alam, M., Jamil, H., Sanjayan, J., & Wilson, J. (2014). Energy saving potential of phase change materials in major Australian cities. *Energy and Buildings*, *78*, 192–201. https://doi.org/10.1016/j.enbuild.2014.04.027

Alam, M., Zou, P. X. W., Sanjayan, J., & Ramakrishnan, S. (2019). Energy saving performance assessment and lessons learned from the operation of an active phase change materials system in a multi-storey building in Melbourne. *Applied Energy*, *238*, 1582–1595. https://doi.org/10.1016/j.apenergy.2019.01.116

Alkhwildi, A., Elhashmi, R., & Chiasson, A. (2020). Parametric modeling and simulation of Low temperature energy storage for cold-climate multi-family residences using a geothermal heat pump system with integrated phase change material storage tank. *Geothermics*, *86*, 101864. https://doi.org/10.1016/j.geothermics.2020.101864

Arıcı, M., Bilgin, F., Nižetić, S., & Karabay, H. (2020). PCM integrated to external building walls: An optimization study on maximum activation of latent heat. *Applied Thermal Engineering*, *165*, 114560. https://doi.org/10.1016/j.applthermaleng.2019.114560

Arkar, C., & Medved, S. (2007). Free cooling of a building using PCM heat storage integrated into the ventilation system. *Solar Energy*, *81*(9), 1078–1087. https://doi.org/10.1016/j.solener.2007.01.010

Arteconi, A., Mugnini, A., & Polonara, F. (2019). Energy flexible buildings: A methodology for rating the flexibility performance of buildings with electric heating and cooling systems. *Applied Energy*, *251*(May), 113387. https://doi.org/10.1016/j.apenergy.2019.113387

Benli, H. (2011). Energetic performance analysis of a ground-source heat pump system with latent heat storage for a greenhouse heating. *Energy Conversion and Management*, *52*(1), 581–589. https://doi.org/10.1016/j.enconman.2010.07.033

Berardi, U., & Soudian, S. (2019). Experimental investigation of latent heat thermal energy storage using PCMs with different melting temperatures for building retrofit. *Energy and Buildings*, *185*, 180–195. https://doi.org/10.1016/j.enbuild.2018.12.016

Blum, D. H. (2016). *Improving the Use of Commercial Building HVAC Systems for Electric Grid Ancillary Services*.

Bruno, F., Tay, N. H. S., & Belusko, M. (2014b). Minimising energy usage for domestic cooling with off-peak PCM storage. *Energy and Buildings*, *76*, 347–353. https://doi.org/10.1016/j.enbuild.2014.02.069

Chaiyat, N. (2015a). Energy and economic analysis of a building air-conditioner with a phase change material (PCM). *Energy Conversion and Management*, *94*, 150–158. https://doi.org/10.1016/j.enconman.2015.01.068

Chen, Y., Xu, P., Gu, J., Schmidt, F., & Li, W. (2018). Measures to improve energy demand flexibility in buildings for demand response (DR): A review. In *Energy and Buildings* (Vol. 177, pp. 125–139). Elsevier Ltd. https://doi.org/10.1016/j.enbuild.2018.08.003

Cui, B., Wang, S., & Sun, Y. (2014). Life-cycle cost benefit analysis and optimal design of small scale active storage system for building demand limiting. *Energy*, *73*, 787–800. https://doi.org/10.1016/j.energy.2014.06.084

Dong, J., Shen, B., Munk, J., Gluesenkamp, K. R., Laclair, T., & Kuruganti, T. (2019). Novel PCM Integration with Electrical Heat Pump for Demand Response. *IEEE Power and Energy Society General Meeting*. https://doi.org/10.1109/PESGM40551.2019.8973936

Evola, G., Marletta, L., & Sicurella, F. (2013). A methodology for investigating the effectiveness of PCM wallboards for summer thermal comfort in buildings. *Building and Environment*, *59*, 517–527. https://doi.org/10.1016/j.buildenv.2012.09.021

Farah, S., Liu, M., & Saman, W. (2019). Numerical investigation of phase change material thermal storage for space cooling. *Applied Energy*, *239*, 526–535. https://doi.org/10.1016/j.apenergy.2019.01.197

Fleischer, A. S. (2015). *Thermal Energy Storage Using Phase Change Materials*. Springer International Publishing. https://doi.org/10.1007/978-3-319-20922-7

Goetzler, B., Guernsey, M., & Kassuga, T. (2019). *Grid-interactive Efficient Buildings Technical Report Series: Heating, Ventilation, and Air Conditioning (HVAC); Water Heating; Appliances; and Refrigeration*.

Han, Z., Zheng, M., Kong, F., Wang, F., Li, Z., & Bai, T. (2008). Numerical simulation of solar assisted ground-source heat pump heating system with latent heat energy storage in severely cold area. *Applied Thermal Engineering*, *28*(11–12), 1427–1436. https://doi.org/10.1016/j.applthermaleng.2007.09.013

Hlanze, P., Elhefny, A., Jiang, Z., Cai, J., & Shabgard, H. (2021). *Integration of Phase Change Material-Based Storage in Air Distribution Systems to Increase Building Power Flexibility*. https://docs.lib.purdue.edu/cgi/viewcontent.cgi?article=1365&context=ihpbc

Hu, Y., Heiselberg, P. K., Drivsholm, C., Søvsø, A. S., Vogler-Finck, P. J. C., & Kronby, K. (2021). Experimental and numerical study of PCM storage integrated with HVAC system for energy flexibility. *Energy and Buildings*, 111651. https://doi.org/10.1016/j.enbuild.2021.111651

Jensen, S. Ø., Marszal-Pomianowska, A., Lollini, R., Pasut, W., Knotzer, A., Engelmann, P., Stafford, A., & Reynders, G. (2017). IEA EBC Annex 67 Energy Flexible Buildings. *Energy and Buildings*, *155*, 25–34. https://doi.org/10.1016/j.enbuild.2017.08.044

Jin, X., Wu, F., Xu, T., Huang, G., Wu, H., Zhou, X., Wang, D., Liu, Y., & Lai, A. C. (2021). Experimental investigation of the novel melting point modified Phase–Change material for heat pump latent heat thermal energy storage application. *Energy*, *216*, 119191. https://doi.org/10.1016/j.energy.2020.119191

Junker, R. G., Azar, A. G., Lopes, R. A., Lindberg, K. B., Reynders, G., Relan, R., & Madsen, H. (2018). Characterizing the energy flexibility of buildings and districts. *Applied Energy*, *225*(April), 175–182. https://doi.org/10.1016/j.apenergy.2018.05.037

Kaygusuz, K., & Ayhan, T. (1999). Experimental and theoretical investigation of combined solar heat pump system for residential heating. *Energy Conversion and Management*. https://doi.org/10.1016/S0196-8904(99)00026-6

Kelly, N. J., Tuohy, P. G., & Hawkes, A. D. (2014). Performance assessment of tariff-based air source heat pump load shifting in a UK detached dwelling featuring phase change-enhanced buffering. *Applied Thermal Engineering*, *71*(2), 809–820. https://doi.org/10.1016/j.applthermaleng.2013.12.019

Kondo, T., & Ibamoto, T. (2006). Research on thermal storage using rock wool phase-change material ceiling board. *ASHRAE Transactions*, *112 PART 1*, 526–531.

Kumirai, T., Dirker, J., & Meyer, J. (2019). Experimental analysis for thermal storage performance of three types of plate encapsulated phase change materials in air heat exchangers for ventilation applications. *Journal of Building Engineering*, *22*, 75–89. https://doi.org/10.1016/J.JOBE.2018.11.016

Lei, J., Yang, J., & Yang, E. H. (2016). Energy performance of building envelopes integrated with phase change materials for cooling load reduction in tropical Singapore. *Applied Energy*, *162*, 207–217. https://doi.org/10.1016/j.apenergy.2015.10.031

Markarian, E., & Fazelpour, F. (2019). Multi-objective optimization of energy performance of a building considering different configurations and types of PCM. *Solar Energy*, *191*, 481–496. https://doi.org/10.1016/j.solener.2019.09.003

Mosaffa, A. H., & Garousi Farshi, L. (2016). Exergoeconomic and environmental analyses of an air conditioning system using thermal energy storage. *Applied Energy*, *162*, 515–526. https://doi.org/10.1016/j.apenergy.2015.10.122

Piselli, C., Prabhakar, M., de Gracia, A., Saffari, M., Pisello, A. L., & Cabeza, L. F. (2020). Optimal control of natural ventilation as passive cooling strategy for improving the energy performance of building envelope with PCM integration. *Renewable Energy*, *162*, 171–181. https://doi.org/10.1016/j.renene.2020.07.043

Prabhakar, M., Saffari, M., de Gracia, A., & Cabeza, L. F. (2020). Improving the energy efficiency of passive PCM system using controlled natural ventilation. *Energy and Buildings*, *228*, 110483. https://doi.org/10.1016/j.enbuild.2020.110483

Real, A., García, V., Domenech, L., Renau, J., Montés, N., & Sánchez, F. (2014a). Improvement of a heat pump based HVAC system with PCM thermal storage for cold accumulation and heat dissipation. *Energy and Buildings*, *83*, 108–116. https://doi.org/10.1016/j.enbuild.2014.04.029

Saffari, M., de Gracia, A., Fernández, C., & Cabeza, L. F. (2017). Simulation-based optimization of PCM melting temperature to improve the energy performance in buildings. *Applied Energy*, *202*, 420–434. https://doi.org/10.1016/j.apenergy.2017.05.107

Sarbu, I., & Sebarchievici, C. (2018). A comprehensive review of thermal energy storage. *Sustainability (Switzerland)*, *10*(1). https://doi.org/10.3390/su10010191

Sultan, S., Gluesenkamp, K. R., & States, U. (2021). The State of Art of Heat-Pump integrated Thermal Energy Storage for Demand Response. *IEA Heat Pump Technologies Magazine*, *40*(2), 27–30. https://doi.org/https://doi.org/10.23697/62tr-nt79

Takeda, S., Nagano, K., Mochida, T., & Shimakura, K. (2004). Development of a ventilation system utilizing thermal energy storage for granules containing phase change material. *Solar Energy*, *77*(3), 329–338. https://doi.org/10.1016/j.solener.2004.04.014

U.S. EIA. (2020). *Annual Energy Outlook 2020*. https://www.eia.gov/outlooks/aeo/

U.S. Energy Information Administration - EIA - Independent Statistics and Analysis. (n.d.). Retrieved April 25, 2022, from https://www.eia.gov/todayinenergy/detail.php?id=44556

U.S. Energy Information Administration EIA. (2015). *Residential Energy Consumption Survey (RECS) 2015 Consumption and Expenditures Technical Documentation Summary*. https://www.eia.gov/consumption/residential/reports/2015/methodology_c&e/

U.S. Energy Information Administration EIA. (2019). *"Use of Energy in the United States Explained." US Energy Information Administration.* EIA. https://www.eia.gov/energyexplained/index.php?page=us_energy_homes

Waqas, A., & Kumar, S. (2011). Utilization of latent heat storage unit for comfort ventilation of buildings in hot and dry climates. *International Journal of Green Energy*, *8*(1), 1–24. https://doi.org/10.1080/15435075.2010.529406

Xu, T., Humire, E. N., Chiu, J. N., & Sawalha, S. (2021). Latent heat storage integration into heat pump based heating systems for energy-efficient load shifting. *Energy Conversion and Management*, *236*, 114042. https://doi.org/10.1016/j.enconman.2021.114042

Yamaguchi, M., Sayama, S., Yoneda, H., Iwamoto, K. ya, Harada, M., Watanabe, S., & Fukai, K. (1995). Heat Storage-Type Floor Heating System with Heat Pump Driven by Discount Electricity Supplied at Night. *KAGAKU KOGAKU RONBUNSHU*, *21*(5), 853–858. https://doi.org/10.1252/kakoronbunshu.21.853

Zhu, N., Deng, R., Hu, P., Lei, F., Xu, L., & Jiang, Z. (2021). Coupling optimization study of key influencing factors on PCM trombe wall for year thermal management. *Energy*, *236*, 121470. https://doi.org/10.1016/j.energy.2021.121470

ACKNOWLEDGEMENT

This material is based upon work supported by the U. S. Department of Energy's Building Technologies Office under Contract No. DE-AC05-00OR22725 with UT-Battelle, LLC. The authors would like to acknowledge Mr. Sven Mumme, Technology Manager, U.S. Department of Energy Building Technologies Office.

Multiple Packaged Rooftop Unit Coordination for Peak Demand Reduction

Joshua DEVENEZIA[1], Todd ROSSI[2], Michael R. MULLER[3]
[1,2,3] Rutgers University, Mechanical and Aerospace Engineering
Piscataway, NJ, USA
caes.rutgers@gmail.com
848-445-3651

ABSTRACT

More than half of commercial buildings in the United States are served by packaged rooftop units (Schantz, 2015). Rooftop units are often controlled by dedicated thermostats acting on fixed heuristic setpoint schedules; thus, it is common for multiple units to operate as independent systems. As a result, many buildings are not taking advantage of cooperative energy management opportunities such as peak power reduction. Enabling grid responsiveness and curtailing lofty demand charges, at scale, adopting integrated unitary HVAC control represents an area of improvement that is mutually beneficial to the power producer and consumer. Putta et al. (2015) proposed a dynamic programming-based model predictive control framework for scheduling rooftop unit operations, requiring only onsite thermostats. Kim and Braun (2016) built on this research by demonstrating a practical integrated control layer for coordinating rooftop units. This research adapts the optimization concept and control methodology to develop a cloud-based application that limits cooling demand spikes. The controller was tested at a pilot location in Northern California for seven months. This work leveraged feedback from the testbed to gain practical insights for implementing a cloud-based thermostat controller. The application successfully achieved instances of up to 33% peak cooling demand reduction. The start-to-finish methodology used to develop and deploy a unit coordination application is explained in the sections of this paper, results and lessons learned are discussed.

1. INTRODUCTION

The Energy Information Administration (Eia.gov) Energy Outlook reports that commercial buildings used 153 billion kWh of electricity for cooling in 2021, which is about 12% of total commercial consumption and four percent of overall U.S. power usage. In aggregate, even marginal improvements to regulating commercial demand could greatly benefit power grid entities, reducing end use load variability and peak demand. Steadier, more responsive end use consumption can simplify balancing the grid, which decreases technical losses, excess/standby generation, and electricity price volatility, while extending the economic lifecycle of grid infrastructure. Consequently, utilities often charge sizable monthly premiums known as demand charges based on a facility's monthly peak power demand. Eia.gov estimates that demand charges comprise 30 to 70 percent of monthly commercial electricity costs. Although rates vary by utility, location, season, and business type, in general, better management of peak demand can significantly reduce operational costs for many commercial/industrial facilities.

Contemporary research on this topic has laid a twenty-year foundation for Model Predictive Control (MPC) in building energy management (BEM), with the latest studies incorporating utility price signaling and demand response timers (Kim & Braun, 2022). However, despite increasing need for demand flexibility, commercial adoption has been slow. This research seeks to highlight the accessibility of unitary HVAC MPC in commercial buildings by demonstrating a straightforward approach to implementing a peak demand limiting MPC application. The controller is developed from fundamentals laid out by Kim and Braun (2018); it is designed to proactively coordinate flexible rooftop unit (RTU) loads such that unnecessary overlap in operation is avoided. Section 2 describes the methodology used to develop the optimization algorithm, coined Unit Coordination (UC); the UC controller uses predictive resistor-capacitor (RC) thermal models to rank simulated outcomes by favorability. Section 3 explains how to configure and implement the UC application, and details a data driven way to evaluate the controllers' performance, uncover problems, and make refinements. Section 3 also provides an overview of using a generic cloud server to interface with IoT thermostats, in which the server executes API push/pull requests to get thermal zone data and update onsite thermostats. The controller does not require additional hardware; the thermostats' programming interface is all that is required to configure and deploy the UC application, an important nod to this solution's low-overhead and scalability. As IoT and API enablement becomes part of the standard thermostat package, unit coordination comes into focus as an increasingly viable BEM option with minimal barriers to entry.

2. METHODOLOGY

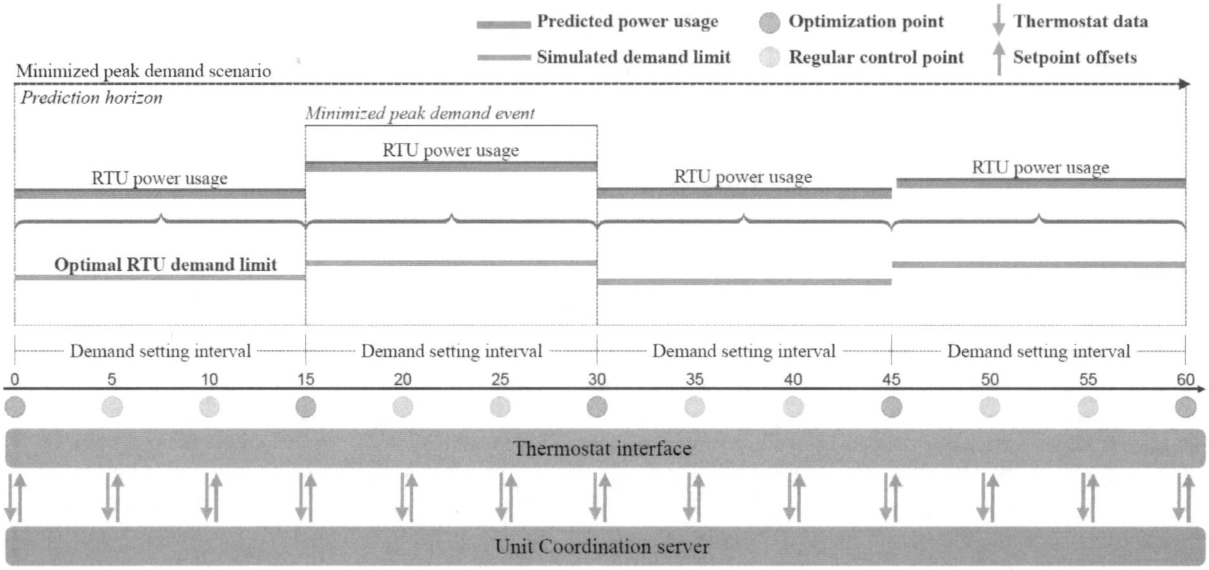

Figure 1: Unit Coordination application

Peak demand management is intuitive because a main advantage of an integrated controller is the ability to address unitary HVAC operation as a whole, rather than independent systems. The driving concept is to attenuate cooling demand by imposing an optimized throttle, or upper bound, a facility's power usage. The optimal upper bound is is identified through simulation as the least-cost comfort-keeping RTU demand limit, which the controller then enforces by alternating rooftop unit runtimes, thus avoiding coincident cooling loads as much as possible. Predictive RC thermal zone models are used every 15 minutes to play forward all possible control decisions contained in a one-hour prediction horizon, simulating the thermal outcomes and resulting peak demand of each scenario. These simulated scenarios are then assessed on the criteria of comfort and demand minimization to select a best case RTU demand limit. If there is no previously higher peak demand, the modeled scenario that predicts to maintain comfort with the lowest peak demand is selected as the optimal control scenario. However, a facility's peak demand is ratcheted throughout the billing cycle, so the demand limiting selection criteria relaxes to any previously established peak event in the billing cycle. Although the optimizer considers a one-hour horizon, the RTU demand limit is re-optimized every 15 minutes to account for unmodeled changes, denoted by every third dot in Figure 1. On a five-minute basis, the controller reallocates the allotted RTU power usage according to each zone's need for cooling, denoted by the *regular control points* in Figure 1. Control decisions are communicated to a facility's onsite thermostats as fixed 2°F offsets, which are used to turn units on or off (within the bounds of the offset). In doing so, the risk of a total controller failure is capped at all zones being two degrees above their desired temperature setpoint or the current zone temperature if the zone is already uncomfortable.

In adherence to common practice by the utility, the optimizer aggregates simulated average power use into four consecutive 15-minute demand intervals (one-hour horizon) and uses the highest of these four demand values to grade the scenario's demand limiting performance. All scenarios are evaluated (exhaustive search), and the optimum is found by ranking the simulated outcomes according to a tiered list of priorities; comfort, peak demand, and RTU demand limit. An exhaustive search is usable because the controller's fixed decision points and the discrete operating states of the RTUs only allow for a finite set of producible outcomes to exist. For example, two equal two stage RTUs can only have five combined distinct power usage levels. If the units' states are {off, stage 1, stage 2} or {0, 1, 2}, then the set of combined power use levels reduces as follows; {(0,0) (0,1) (1,0) (1,1) (0,2) (2,0) (1,2) (2,1) (2,2)} → {0, 1, 1, 2, 2, 2, 3, 3, 4} → {0, 1, 2, 3, 4}. The optimization algorithm simulates the effect of enforcing these RTU power levels (referred to as the demand limits) in every permutation over the proceeding four demand intervals. In the example, there are 625 (5^4) simulations to search for the optimal scenario. It is important to note that the number of RTUs and stages combinatorically increases the number of RTU power levels, L_k, that need to be simulated. At a point, the number of modeled scenarios may become computationally excessive. In such cases, computation time can be reduced by simulating a subset of RTU demand limits (e.g. 10 of 20).

The following table summarizes variables used for optimization, the RC model parameters are represented implicitly by the predicted zone temperature ($T_{z,j,i}$) and RTU power usage (P).

Table 1: Optimization variables

$P_{z,j,i}$ (variable)	$T_{z,j,i}$ (predicted)	$C_{z,i}$ (given)	$\epsilon_{comfort}$ (static)	L_k	$\boldsymbol{D}$
Power use of unit z for i^{th} timestep of jth demand interval	Max temperature of zone z in i^{th} timestep of jth demand interval	Desired temperature of zone z at i^{th} timestep	Discomfort tolerance threshold	Set of k testable demand limits (combine power levels)	The current peak demand established for the billing cycle
$i \in N^+: \{0 \to 3\}, j \in N^+: \{1 \to 4\}, z \in N^+: \{1 \to n_z\}, C_z \in N^+, \epsilon \in \mathbb{R}^+, L_k \in \mathbb{R}^+: \{0, L_1, \ldots, L_n\}, \boldsymbol{D} \in \mathbb{R}^+$					

The manipulated variable is the RTU demand limit, L_k, which is applied at every 5-minute increment. The objective function is a least-cost demand minimization, subject to comfort constraints defined by the thermostats' programmed cooling setpoint. A scenarios "comfort" is measured by the max error between the predicted and desired zone temperatures, which is the reference value used in testing the optimizer's comfort criteria. There may be several qualifying solutions that all predict to produce the same minimum peak demand over the forecasted hour. The choice among these options is discretionary and tunable; this controller opts to impose the most lenient RTU demand limit because it presents the least risk to building comfort. The optimizer first finds all scenarios where the zone temperature is within a reasonable tolerance of the desired setpoint (i.e. comfortable). If such scenarios are present, all other scenarios are rejected. If no comfortable scenarios exist, all scenarios within 10% of the *least uncomfortable* scenario are treated as options to prevent the algorithm from selecting an insignificantly more comfortable option with a significantly higher demand penalty. Expressed as a dynamic program, the control problem is;

$$\min\left\{\left[\max_{i\in N^+, j\in N^+, z\in N^+, C_z\in N^+, \epsilon\in\mathbb{R}^+}\left\{\left[\frac{1}{3}\sum_{i=0}^{3}\sum_{z=1}^{n_z} P_{z,j,i}(L_{k,j}),\ \forall j \in \{1{:}4\},\ L_k \in \{0{:}L_n\}\right]\right\} \geq \boldsymbol{D}\right]\right\} \tag{1}$$

s.t. (hierarchical)

1. If comfortable scenarios exist, any sequence with a setpoint error greater than epsilon is not a solution.

$$\boldsymbol{T}_{z,j}(L_{k,j}) - C_{z,i} < \epsilon \Leftrightarrow \exists\left(\boldsymbol{T}_{z,j} - C_{z,i} < \epsilon\right),\ \ \forall\, i, j \in N^+,\ \boldsymbol{T}_{z,j} \in \mathbb{R},\ L_k\text{: } \{0, L_1, \ldots, L_n\} \tag{1a}$$

2. Otherwise, a sequence within 10% of the least uncomfortable scenario is treated as a solution.

$$\frac{\boldsymbol{T}_{z,j}(L_{k,j}) - C_{z,i}}{\min\{\boldsymbol{T}_{z,j}(L_{k,j}) - C_{z,i}\}} < 0.1,\ \ \forall\, i, j \in N^+,\ \boldsymbol{T}_{z,j} \in \mathbb{R},\ L_k\text{: } \{0, L_1, \ldots, L_n\} \tag{1b}$$

The case study uncovered a common peak setting situation, which happens when the building moves into its occupied period and cooling setpoints drop by a large amount causing units to run at full load. This situation is handled naturally by the optimization algorithm, as seen in the following figures.

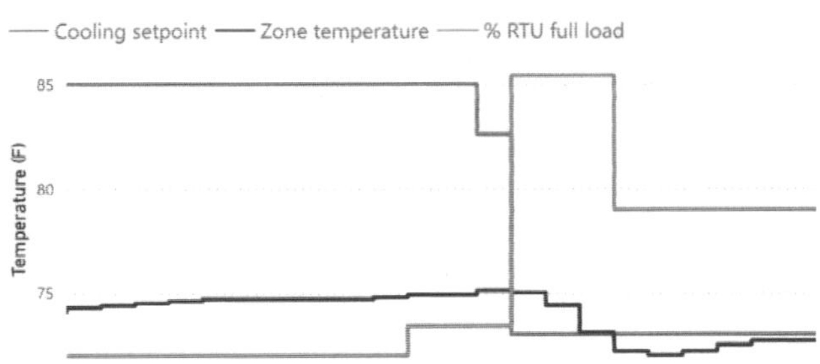

Figure 2: Normal transition to occupied

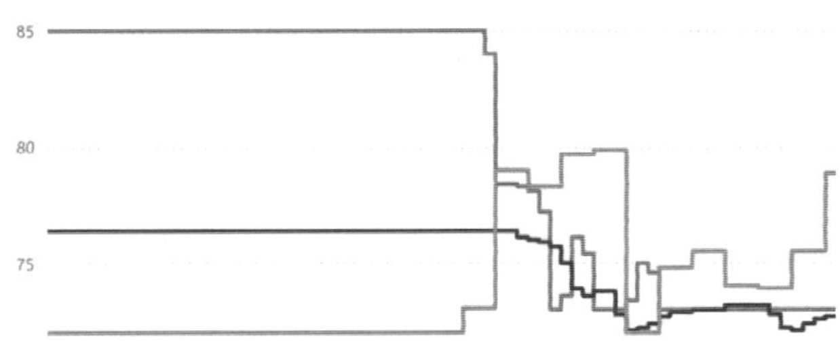

Figure 3: Optimized transition to occupied

Figure 2 shows how moving into the occupied time period normally creates a demand spike as the zones are aggressively cooled to comply with the occupied temperature setpoint and Figure 3 shows how Unit Coordination walks zone temperatures down gradually, reducing the demand spike caused by this transition by about 50%.

3. SETUP AND IMPLEMENTATION

3.1 Site Selection

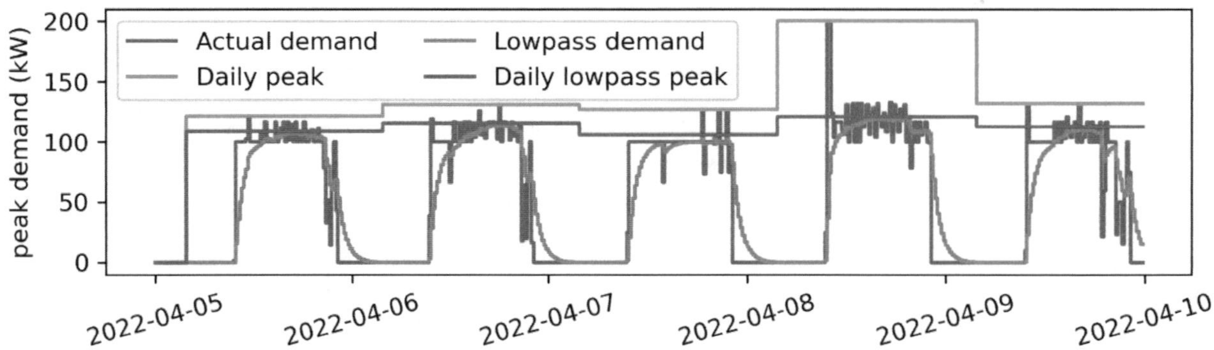

Figure 4: RTU cooling demand profile with lowpass filter comparison

Buildings can be modeled like thermal storage systems that charge and discharge with behavior close to an RC network. As such, a first order lowpass filter can be applied over a buildings' cooling (or total) demand profile as a preliminary check for demand savings potential. The lowpass filter approximates the effect of using a building's thermal mass to distribute cooling within a time constant related to a building's actionable thermal storage capacity (e.g. 1 hour). The time constant can be tuned in reference to best fit RC model parameters for a more accurate representation of demand savings potential. In Figure 4, demand savings are estimated by the difference between the actual peak demand and peak demand after running the buildings cooling load profile through a lowpass filter with a 1-hr time constant. Figure 4 shows how the lowpass curve redistributes and smooths demand spikes, see Apr 8th for effect. The lowpass filter suggested that the case study could realize a peak cooling demand reduction of 30% during shoulder months, which is consistent with measured results.

3.2 Modeling

The various zones of a building may require different RC thermal circuit configurations, so a variety of model forms are tested for a best fit. The form that best fits (least-squared error) historical thermostat data is used for modelling scenarious in the MPC algorithm. A simple RC model explanation (Figure 5), model best fit plots of tested configurations (Figure 6), and RC modelling best practices are provided in this section.

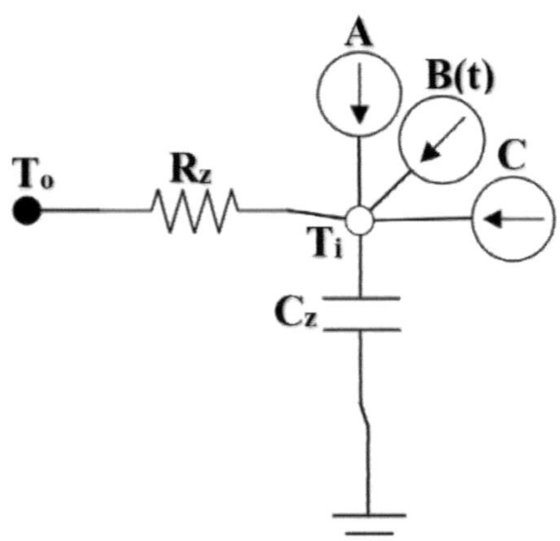

Figure 5: Thermal circuit example

- **T_o**: time varying outdoor temperature (input)
- **C** : time varying percent cooling runtime (input, parameter)
- **T_i** : time varying zone temperature (output)
- **A** : time invariant constant internal gains (parameter)
- **B** : time invariant occupied internal gains (parameter)
    - **B(t)** : *time varying* occupied internal gains
- **R_z**: time invariant thermal resistance (parameter)
- **C_z**: time invariant thermal capacitance (parameter)

Note: the testbed's thermal response lagged cooling signals by at least five minutes, so it is important to evaluate model performance under the correct delay conditions.

Although it is often necessary to account for thermal coupling in building zones, this research found simple uncoupled RC models to be sufficient for the case study, which is a two-zone restaurant with 10-ton RTUs serving a kitchen and dining area. One practical way to check for coupling is to find or create examples of a single zone being cooled/heated and observe its influence on adjacent zones.

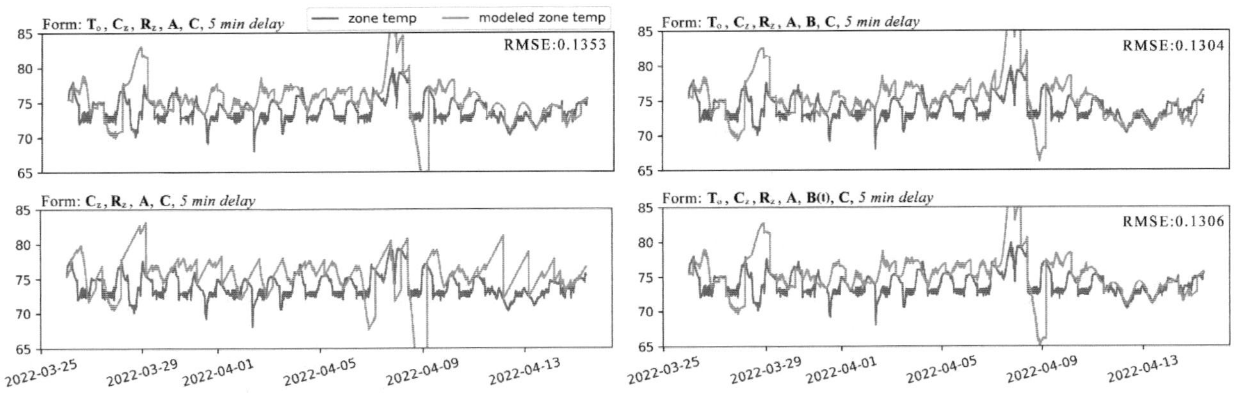

Figure 6: Fitted thermal circuit model forms

Figure 6 shows a sample of tested RC circuit configurations where the predicted thermal outcomes from each is compared against real data collected from the thermostats' historians. Visual inspection of model fits can help to identify problems in the training data when unaccounted for thermal influences may be causing undue modelling error. Testing model performance against the testbed's historical data uncovered useful best practices for RC modelling;

- Model occupied internal gains as its own piecewise current source with appropriate time varying profiles
- Check for lag between the thermostat signal and the HVAC units' response
- Use least squared error and zero bias initial conditions to measure model accuracy
- Test model accuracy within its scope of application (i.e. one hour prediction horizon)

Basic current analysis is used to generate first order ordinary differential equations (ODEs) which model versions of the circuit in Figure 5. The differential forms are discretized, and model parameters are optimized using a least-squares error numerical solving method. The studied zones exhibited time varying internal gains during occupied hours. A concave parabola symmetric over the occupied period was used to characterize the time dependent behavior. By identifying a characteristic shape for the time varying thermal influence, the solving methodology is relatively unaffected because the quadratic shape can be conveyed numerically as a series of multipliers tracing a parabolic shape, $-\left(\frac{t}{a}\right)^2 + 1,\ t \in [-a, a]$, over the occupied timeframe $[-a, a]$. A scaling factor, $\boldsymbol{Q}_{ig}$, is then optimized to develop the occupied internal gains profile, $\boldsymbol{Q}(t) = \boldsymbol{Q}_{ig}\left(-\left(\frac{t}{a}\right)^2 + 1\right)$ (2).

3.3 Simulation testing

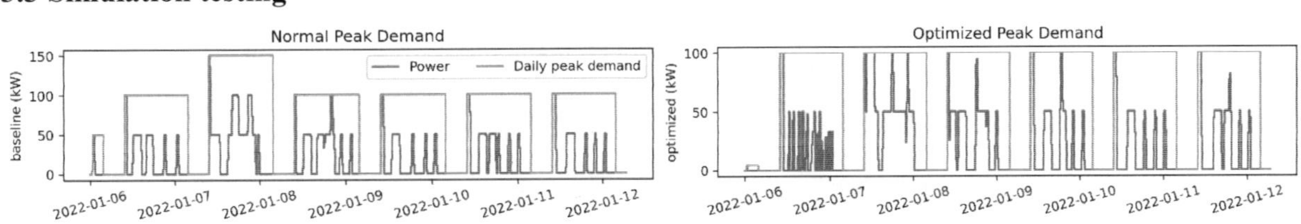

Figure 7: Simulated UC comparison to baseline performance

A final preparatory step before implementing the UC application is to simulate the controller on a portion of historical data and compare optimized operation to normal operation. In a simulated environment, the algorithm does not suffer from uncertainties like model mismatch, unknown perturbations, and inaccurate or corrupted measurements, so it is useful for detecting problems with the cost function and overall optimization framework. For example, the UC algorithm did not initially check the thermostat schedule for upcoming setpoint changes, which resulted in erroneously stringent RTU demand limits. As in Figure 7, this step can also serve as a secondary savings estimation, where the historical dataset acts as a performance baseline and the UC application is evaluated under the same historical conditions. If the savings estimate from simulation agrees closely with the lowpass results it verifies that the UC controller is successfully coordinating cooling runtimes, in so far as models are accurate.

3.4 Implementation

IoT enabled smart thermostats are becoming ubiquitous, most large commercial thermostat vendors offer thermostats with remote interfacing and API access. The UC application makes use of one such interface to put the controller into action, where setpoint updates made via API request serve as indirect ON/OFF control signals. Thermostat APIs are internet communication protocols to pull thermostat data such as; outdoor/indoor temperature, relative humidity, operating unit stage, setpoint schedules, etc. and push setpoint changes to the thermostat. API requests are made using HTTP GET or POST URL methods, depending on the request type. Retrieving information is accomplished with a GET request whereas updating information is achieved with a POST request. In this work, a virtual machine hosting a Python-executable Apache web server is used to communicate with Ecobee smart thermostats installed at the testbed. The UC application is split into an API-specific "interpreter" layer and an optimization layer to keep case-by-case aspects of implementation separate from the optimization algorithm. The interpreter creates an thermostat API client, which is an HTTP channel for issuing push/pull requests, information is passed between the server and thermostat interface in JSON format. Zone thermal data is pulled and reformatted by the interpreter as a table of state variables containing outdoor/zone temperatures, setpoints and RTU runtimes for each zone, the table is then read in by the optimization layer to initialize the simulation procedure described in Section 2. The returned control decisions are sent back to the interpreter and relayed to thermostats as setpoint offsets using the appropriate POST URL request.

4. RESULTS AND DISCUSSION

A day-on-day-off strategy is used to compare normal operation against Unit Coordination performance. RTU power usage is calculated using the thermostats' stage and runtime historical record, Figure 8 demonstrates how the UC application consistently brings down daily RTU peak cooling demand by ~33%, where the highlighted regions of Figure 8 denote days when UC is active.

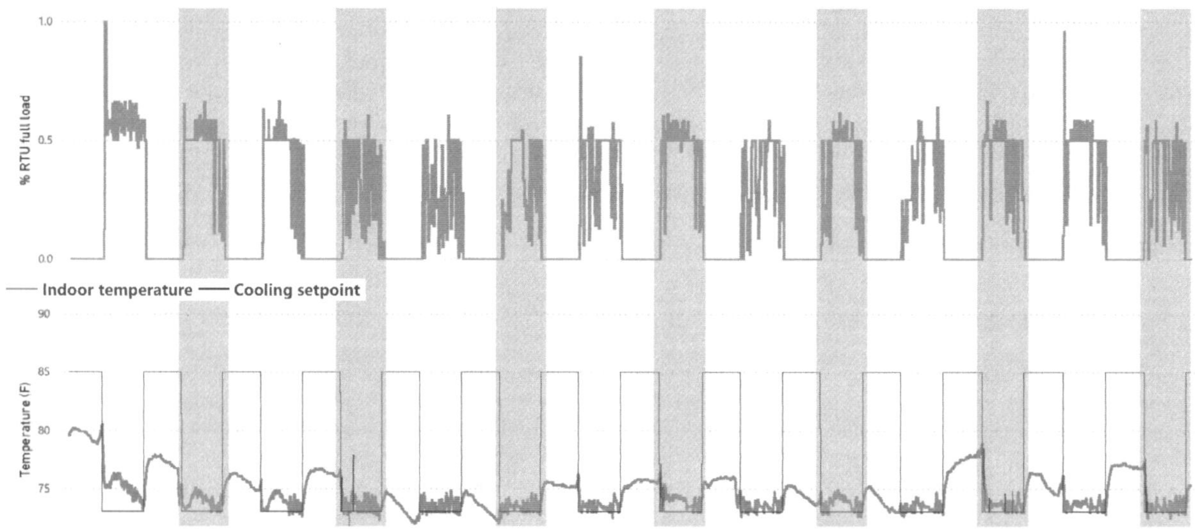

Figure 8: Day-on-day-off Unit Coordination

Figure 8 shows a day-to-day comparison of optimized and normal operation and captures several instances where the application avoids a peak setting event adjusting thermostat setpoints. Large demand spikes are observable on days when UC is turned off, these spikes are significantly reduced on adjacent days, when the application is on. Cooling setpoints are relaxed by 2°F to turn units off in compliance with the controllers determined optimal RTU demand limit. Figure 9 vertically aligns RTU power use and optimized cooling setpoints for the two thermostats present at the testbed during a day with active unit coordination. Figure 9 demonstrates how the controller strategically alternates setpoint offsets to actuate RTUs asynchronously. As intended, RTU power usage is clearly staggered, when the power use of RTU 1 increases, the power use of RTU 2 generally decreases, implying that the RTU demand limit is causing the units to tradeoff operation.

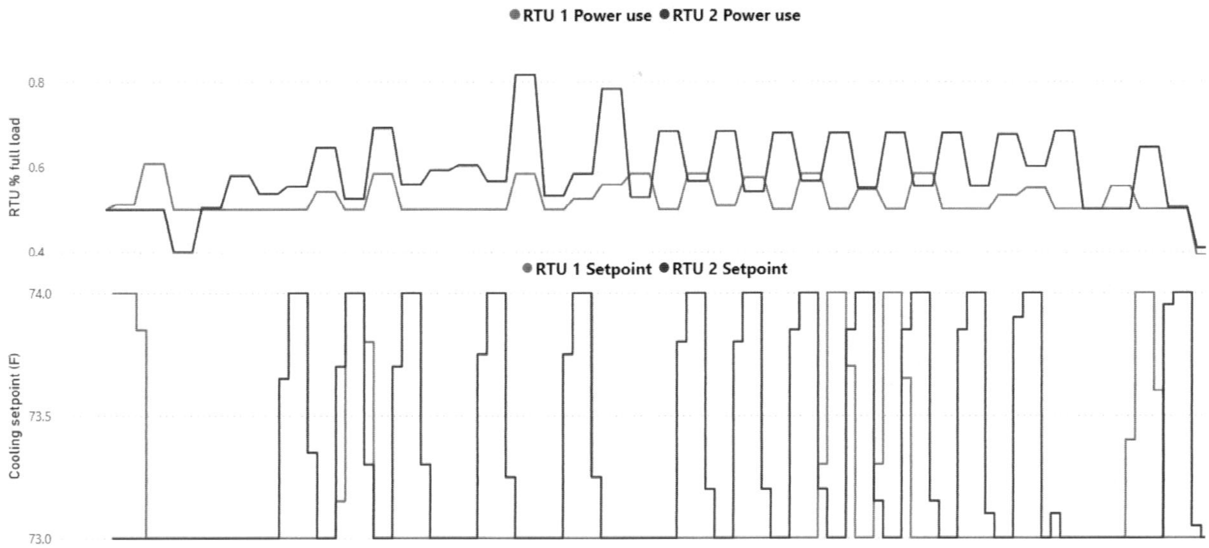

Figure 9: Coordinated setpoint changes

It is possible to tune the comfort tolerance, ϵ, seen in section 2 in order to make the unit coordination effect more pronounced, in exchange for higher transient discomfort. In this application, the UC controller was tuned to accept an ϵ of 0°F, which means the predicted zone temperature cannot exceed the desired zone temperature at any time, when possible. In practice, zone thermal comfort depends heavily on model accuracy, so it is best to err on the side of caution and use a model that makes conservative zone cooling predictions because models that overstate the RTUs cooling influence are likely to cause undue discomfort. It is important to regularly update the learned RC model parameters to account for changes at the building. The RC models used in this research performed adequately, but could be improved by representing the thermal mass using more than one capacitor.

Although there is no definitive way to know the precise percent cooling demand savings based on the day-on-day-off demand trend, such as in Fig. 8, the UC controller provided a roughly 33% decrease in peak cooling demand. The applications demand limiting performance was evaluated using 15-minute power consumption intervals over the buildings' demand setting window. It is important to note that adding more RTUs (degrees of freedom) to the control problem will emphasize Unit Coordination's effectiveness because demand flexibility increases when more rooftop units are introduced. The lowpass filter forecasts potential demand savings in the range of 15% to 30% from May to October. Demand savings is generally expected to drop for summer months because many RTUs are sized to run at full load during hot summer days, and so RTU peaks may not be avoidable.

5. CONCLUSION

The UC application developed reliably produced control decisions that maintain zone comfort while avoiding cooling demand spikes. The controller helped realize a roughly 33% drop in peak cooling demand over a studied month. The application continues to run and feedback data is still being collected, so it may be iteratively improved using better fit RC models (currently uncoupled 1R1C thermal circuits) and modifying the control algorithm to address special or extreme cases. The work performed verifies that thermostat APIs offer sufficient HVAC controllability to enable Unit Coordination. This research highlights that integrated control over multiple rooftop units is an accessible demand response resource that is well positioned for high levels of adoption soon. The MPC framework is very generalizable and lends itself to many BEM situations, including the integration of energy producing and storing assets, which will be increasingly relevant as distributed energy resources popularize.

ACKNOWLEDGEMENT

Special thanks to Dr. Jim Braun and Dr. Donghun Kim at Purdue's Herrick Labs for their contributions to this research field. Thanks to Dr. Robert M. Muller, Head of Rutgers' Center for Advanced Energy Systems for supporting this work. And finally, thanks to mCloud Corp for providing the case study building used in the forgoing research.

REFERENCES

Kim, D., Braun, J. E., Cai, J., & Fugate, D. L. (2015). Development and experimental demonstration of a plug-and-play multiple RTU coordination control algorithm for small/medium commercial buildings. *Energy and Buildings*, *107*, 279–293. https://doi.org/10.1016/j.enbuild.2015.08.025

Kim, D., & Braun, J. E. (2020). Model predictive control for supervising multiple rooftop unit economizers to fully leverage free cooling energy resource. *Applied Energy*, *275*(May), 115324. https://doi.org/10.1016/j.apenergy.2020.115324

Kim, D., & Braun, J. E. (2018). Development, implementation and performance of a model predictive controller for packaged air conditioners in small and medium-sized commercial building applications. *Energy and Buildings*, *178*, 49–60. https://doi.org/10.1016/j.enbuild.2018.08.019

Kim, D., & Braun, J. E. (2022). MPC solution for optimal load shifting for buildings with ON/OFF staged packaged units: Experimental demonstration, and lessons learned. Energy and Buildings, 266, 112118. https://doi.org/10.1016/j.enbuild.2022.112118

Putta, V. K., Kim, D., Cai, J., Hu, J., & Braun, J. E. (2015). Dynamic programming based approaches to optimal rooftop unit coordination. *Science and Technology for the Built Environment*, *21*(6), 752–760. https://doi.org/10.1080/23744731.2015.1059153

Schantz, M. (2015, November 10). What's on your roof? rooftop unit (RTU) efficiency advice and guidance from the Advanced RTU Campaign. Energy.gov. Retrieved April 14, 2022, from https://www.energy.gov/eere/buildings/articles/whats-your-roof-rooftop-unit-rtu-efficiency-advice-and-guidance-advanced

Assessment of short-term aquifer thermal energy storage for energy management in greenhouse horticulture: modeling and optimization

Queralt ALTES-BUCH[1*], Tanguy ROBERT[2], Sylvain QUOILIN[1], Vincent LEMORT[1]

[1] Energy Systems Research Unit, Faculty of Applied Sciences, University of Liege, Liege, Belgium
{qaltes, squoilin, vincent.lemort}@uliege.be
[2] Urban and Environmental Engineering Research Unit, Faculty of Applied Sciences, University of Liege, Liege, Belgium
tanguy.robert@uliege.be

* Corresponding Author

ABSTRACT

In this work, the use of the thermal energy storage in alluvial aquifers is proposed as a sustainable solution for supplying the energy demand of greenhouses. Air-conditioning greenhouses during summer allows for storing their energy surplus in an underground seasonal buffer that can be used, in winter, for heating via a heat pump. The evaluation of the benefits arising from using aquifer thermal energy storage (ATES) for supplying the energy demand in greenhouse applications requires the modeling of a greenhouse climate model, heating distribution systems (i.e., heating pipes and coils), a heat pump and the ATES system. The models are sized to a case study, which is based on a 100 m thick aquifer in the Cretaceous chalk located in Wallonia (Belgium), and are simulated for two years. The first year is only devoted to storage and emulates the creation of the cold and warm wells. In the second year, the thermal energy stored in the aquifer is recovered according to the needs of the greenhouse (the cold well is used during air-conditioning and the warm well during heating). Results show that shallow alluvial aquifers can be very valuable and can provide a sustainable solution when heating and cooling greenhouses. However, the greenhouse climate controller must be correctly calibrated in order to maintain a balanced ATES system. More specifically, greenhouses in the latitude of Belgium are forced to increase their cooling consumption, which is possible thanks to the flexibility offered by greenhouses. Otherwise, the system may not be sustainable due to the higher needs of heating.

1. INTRODUCTION

Glass greenhouses are a well established technology used to cultivate crops. By heating and increasing the exposure to light via lamps, they allow colder regions for cultivating breeds of flowers and vegetables that naturally grow in temperate or tropical climates. One of the many benefits of technologically equipped greenhouses is the potential they offer for optimizing crop production. On the one hand, the focus on climate control has enabled maximizing crop production both quantitatively and qualitatively. The latter has lead countries such as the Netherlands to be the largest exporter of tomatoes in Europe, even above Spain (Commission, 2021). On the other hand, climate control comes at a high energy cost. To maintain the optimal temperature, light and CO_2 levels that maximize the crop yield, greenhouses may consume (prohibitively) substantial amounts of energy for heating, electricity, CO_2 and cooling.

Even in the Atlantic and Central European regions, cooling is needed because greenhouses have a surplus of thermal energy in summer that must be removed to avoid yield reduction and crop diseases (Stanghellini, Ooster, and Heuvelink, 2018). The most established cooling method in these regions is natural ventilation. However, when ventilating the thermal energy surplus is lost to the environment. In addition, if the target temperature is lower than the outside (dry or wetbulb) temperature, non-passive cooling is required. Most of the existing mechanical cooling methods for greenhouses have been developed for hot, dry climate regions whose primary energy consumption is cooling. Based on evaporative cooling, these systems are not suited to the Atlantic and Central European climates. With an increased need for active cooling during summer in regions with mainly heating requirements, sustainable technologies that allow for seasonal energy recovery in greenhouses have become a relevant scope of research.

In a greenhouse with reduced ventilation, active cooling in summer allows for storing the thermal energy surplus in an underground seasonal buffer that can later be used for heating in winter (Stanghellini, Ooster, and Heuvelink, 2018). Among the available underground storage technologies, aquifer thermal energy storage (ATES) is the most suitable for that purpose. The high percentage of heat recovery makes ATES useful for demand-side management applications such as greenhouses (De Schepper et al., 2019).

The main goal of this work is to assess the energy recovery potential of greenhouses with thermal energy storage in shallow alluvial aquifers in the Atlantic climate. To that end, this work models and simulates a case study based on a greenhouse coupled to a 100 m thick synthetic chalk aquifer mimicking a real Cretaceous chalk aquifer in the Mons area in Wallonia (Belgium). The aquifer is characterized by a slow ambient groundwater flow. The flows are at 10-15°C for heating and at 5-10°C for cooling. While heating requires the use of a heat pump, cooling is possible directly at the aquifer temperature. The greenhouse is used for growing tomato crop and is equipped with lighting and a CO_2 enrichment system. The aquifer is assumed to supply the thermal demand at all times.

The greenhouse and the thermal distribution and generation systems are modeled by means of the Greenhouses Library (Altes-Buch, Quoilin, and Lemort, 2019), an open-source Modelica library for the dynamic simulation of greenhouse climate and energy systems. The Greenhouses library also provides control strategies for the heating, ventilation and other appliances (e.g. CO_2 enrichment and lighting), which determine the combined demand of the greenhouse during the simulation period. The aquifer is modeled by a deterministic 3D ground water flow and heat transport numerical model in the presented case study.

The case study is simulated for a period of two years. Results show that shallow alluvial aquifers can be very valuable and can provide a sustainable solution when heating and cooling greenhouses. However, to achieve so, the greenhouse climate controller must be correctly calibrated. In fact, to maintain a balanced system, greenhouses in the latitude of Belgium are forced to increase their cooling consumption. Otherwise, the system is imbalanced due to the higher needs of heating.

2. MODELING

The evaluation of the benefits arising from using ATES for supplying the energy demand in greenhouse applications requires to model several thermal systems, namely: a greenhouse climate model, heating distribution systems (i.e. heating pipes, heating and cooling coils), generation systems (i.e. heat pump) as well as the ATES system.

2.1 Greenhouse climate

Greenhouse climate models are useful to quantify the energy performance and the crop production given a greenhouse design, outdoor climate and a specifc control. Similar to building energy modeling, greenhouse climate models compute the indoor climate of a greenhouse by describing the flows between its components and solving their respective energy balances. Since in greenhouses the indoor climate is described by three variables (i.e. the temperature, the humidity and the CO_2 concentration in the air), the greenhouse model also includes moisture and CO_2 mass balances on the air.

The sensible energy balances are written in the form of Equation (1) and are applied to all components of the greenhouse. As shown in Figure 1, the greenhouse model consists of the air, the canopy, the envelope (i.e. the cover and the floor), the heating pipes and the thermal screen. The thermal screen is an horizontally movable membrane used to reduce the far-infrared radiative losses to the cover and to the sky. When the screen is deployed, the air of the greenhouse is divided in two zones, i.e. below and above the screen. These zones are modeled separately and their respective climate is assumed to be homogeneous (Altes-Buch, Quoilin, and Lemort, 2022).

$$\rho c_p V \frac{dT}{dt} = \sum \dot{Q} + \sum \dot{H}_{sens} \quad (1)$$

Apart from the energy flows exchanged within the forementioned greenhouse components, the model takes into account the moisture exchanged by natural ventilation with the outside air, the condensation at the cover and thermal screen, as well as the transpiration and photosynthesis (CO_2 consumption) from the canopy.

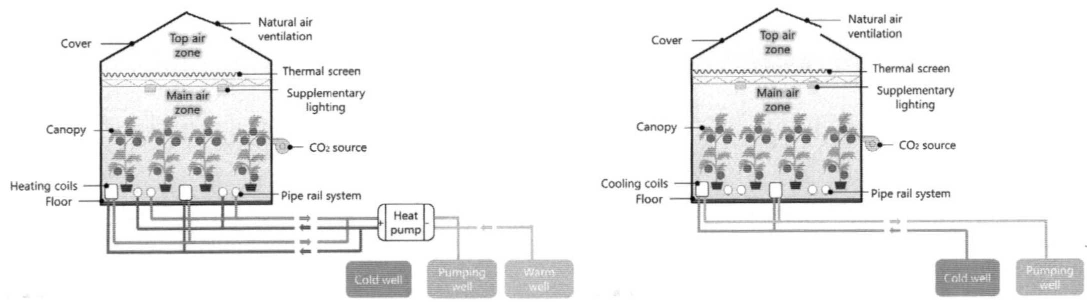

(a) Heating **(b)** Cooling

Figure 1: Sketch of the system

2.2 Heat distribution systems

In order to have an optimal thermal energy recovery, the thermal energy distribution systems inside the greenhouse must be selected by taking into account potential constraints from the supply source, such as the supply temperature, and from the crop (Urban and Urban, 2010). For instance, in order to keep crop yield at an optimal level, it has been proven that, the temperature of the canopy is of utmost importance, not the temperature of the air that surrounds it. To that end, radiative systems (e.g. water pipes) are preferred over purely convective systems, since not only they heat the air through convection, but also they heat the canopy through long-wave radiation. As an illustration, in (Altes-Buch, Quoilin, and Lemort, 2022) it is demonstrated that in winter the canopy absorbs as much radiative heat flow from the pipes than convective heat flow from the air.

In this work, water pipes are selected as the primary heating distribution system. Water pipes are the most established technology for distributing heat in regions with high heating requirements. They usually cover the entire greenhouse floor through parallel loops that go along the crop rows. Therefore, despite their high investment cost, they allow for an horizontal temperature distribution of the highest quality. In order to be used as transport rails for the harvest, they are plain pipes (i.e. without fins) placed some centimeters above the ground. The installed capacity (diameter and length) follows greenhouse crop rows setup standards. Depending on the heat generation system that supplies them, they work between 45°C and 90°C.

Contrary to the case of heating, radiative systems are less adequate for cooling greenhouses. The plain nature of the water pipes implies a low heat transfer coefficient to the air. For the given aquifer supply temperature and the standard piping installed capacity (restricted to increase due to the crops setup), the water pipes are not able to meet the peak cooling demand. Additionally, the radiative exchange decreases the canopy temperature below the optimal range, which implies substantial yield reduction. It is therefore not desirable to lower supply temperature by means of a chiller to meet the peak demand. In order to be able to meet the peak demand without production losses, purely convective systems are a preference. Out of the available convective distribution systems, only water-to-air units are suitable for ATES applications. This work proposes the use of coils.

Given that the standard installed piping capacity is insufficient to supply the peak heating demand in winter (cfc Section 3.3), the coils are also used for heating distribution. They are used during peak heating demand hours as an accessory to the pipe system. This combination allows maintaining the canopy temperature at an optimal level both when heating and cooling to avoid yield reduction.

2.2.1 Heating pipes

The pipes heat the greenhouse air through hindered convection, but also the canopy, the floor and the greenhouse cover through long-wave radiation. Because of their length, a constant heat transfer coefficient cannot be assumed along the pipe. As a consequence, the heating pipes are modeled using a finite volume approach by means of a discretized model that divides the pipes into several cells, each one connected in series by a node (Quoilin et al., 2014). In each cell, the flow is described with enthalpy as a state variable. The dynamic energy balance and static mass and momentum balances are applied in each cell. The model assumes uniform speed through the cross section as well as constant pressure. Axial thermal energy transfer is neglected. The heat flow is computed by an ideal heat transfer model with constant heat transfer coefficient. The energy balance on the fluid for a cell i is described by:

$$V_i \rho_i \frac{dh_i}{dt} + \dot{M}(h_{ex,i} - h_{su,i}) = A_i \dot{q}_i, \quad i = 1, 2 ... N_{cell} \tag{2}$$

where h_i is the fluid specific enthalpy at cell i, and h_{ex} and h_{su} are the enthalpy at the cell's outlet and inlet nodes, respectively.

2.2.2 Heating and cooling coils

Coils can work in dry, partially wet or wet regimes. In most cases, cooling and drying are done simultaneously with the coil working in a wet regime. In a partially wet regime, condensation appears in a point where the contact temperature is lower than the dew point temperature of the air. In this case, a part of the coil works in dry regime, while the rest works in a wet regime.

Over the past decades, water cooling coil models have been the subject of a substantial literature. Many models using different approaches have been developed and validated, the most common ones being:

- Single zone epsilon-NTU models (Lebrun et al., 1990; Morisot, Marchio, and Stabat, 2002),
- Variable boundary models (Brandemuehl, 1993), and
- Finite elements (Gang, Mingsheng, and Claridge, 2007) and lumped models.

In this work, the model proposed in Lebrun et al., 1990 is implemented because of its increased robustness for similar levels of accuracy in comparison to the rest of the models, as reviewed in Gendebien, Bertagnolio, and Lemort, 2010. The model is characterized by three main hypothesis. The first one is that it assumes the cooling coil either completely dry or completely wet. In reality, both options underestimate the heat transfer rate. On the one hand, if the coil is assumed completely dry, the latent heat transfer is neglected. On the other hand, if the coil is assumed completely wet, the model predicts air moisture in the dry part of the coil, which means it considers a negative latent heat transfer rate. Given that both descriptions underestimate the total exchange, the model sets the cooling power to the one with the highest exchanged power:

$$\dot{Q}_{ccoil} = \max\left(\dot{Q}_{dry}, \dot{Q}_{wet}\right) \tag{3}$$

The heating coil is modeled in a simpler manner since only the dry regime must be described.

The second hypothesis states that in the energy balance in wet regime (which takes into account the latent heat), the air can be subtituted by a ficticious fluid the enthalpy of which is completely defined by the wet temperature of the fluid. Therefore, the energy balance described by Equation (4) is equivalent to Equation (5).

$$\dot{Q}_{wet} = \dot{M}_{air} \cdot (h_{air,su} - h_{air,ex,wet}) - \dot{M}_{air} \cdot (w_{air,su} - w_{air,ex,wet}) \cdot c_w \cdot T_{air,ex,wet} \tag{4}$$

$$\dot{Q}_{wet} = \dot{M}_{air} \cdot c_{p,airf} \cdot (T_{wb,su} - T_{wb,ex,wet}) \tag{5}$$

Finally, in order to compute the state in which the air exits the coil, the model defines a semi-isothermal exchanger that takes the air as one fluid and the external surface of the coil as the other, as described by:

$$h_{air,su} - h_{air,ex,wet} = \varepsilon_{c,wet} \cdot (h_{air,su} - h_{c,wet}) \tag{6}$$

$$w_{air,su} - w_{air,ex,wet} = \varepsilon_{c,wet} \cdot (w_{air,su} - w_{c,wet}) \tag{7}$$

$$\varepsilon_{c,wet} = 1 - exp(-NTU_{c,wet}) \tag{8}$$

$$NTU_{c,wet} = \frac{1}{R_{air} \cdot \dot{C}_{air}} \tag{9}$$

2.3 Water-to-water heat pump

The heat pump model implemented in this work predicts the performance at both full- and partial-load operation for different working conditions by means of three polynomial laws fitted through manufacturing data (Bolther et al., 1999). The first law describes the coefficient of performance (COP) at full load as a function of the temperatures at the evaporator inlet and at the condenser outlet:

$$EIRFT = \frac{COP_n}{COP_{fl}} = \alpha_0 + \alpha_1 \cdot \Delta T + \alpha_2 \cdot \Delta T^2 \tag{10}$$

where

$$\Delta T = \frac{T_{su,ev}}{T_{ex,cd}} - \frac{T_{su,ev,n}}{T_{ex,cd,n}} \tag{11}$$

with the temperatures values being in Kelvins.

The second law describes the heating power at full load as a function of the temperatures at the evaporator inlet and at the condenser outlet:

$$CAPFT = \frac{\dot{Q}_{cd,fl}}{\dot{Q}_{cd,n}} = \beta_0 + \beta_1 \cdot (T_{su,ev} - T_{su,ev,n}) + \beta_2 \cdot (T_{ex,cd} - T_{ex,cd,n}) \tag{12}$$

Finally, the third law describes the performance at partial load:

$$EIRFPLR = \frac{\dot{W}_{pl}}{\dot{W}_{fl}} = \delta_0 + \delta_1 \cdot PLR_{cd} + \delta_2 \cdot PLR_{cd}^2 \tag{13}$$

where the part load ratio is described by:

$$PLR_{cd} = \frac{\dot{Q}_{cd,pl}}{\dot{Q}_{cd,fl}} \tag{14}$$

and the electrical power required at the compressor at full load operation is described by:

$$\dot{W}_{fl} = \frac{\dot{Q}_{cd,fl}}{COP_{fl}} \tag{15}$$

2.4 Aquifer Thermal Energy Storage

The synthetic model used to simulate groundwater flow and heat transport represents a 100 m thick aquifer typical of the Cretaceous chalk area located in Mons, in Belgium. These aquifers are very productive and are good targets for ATES systems. As an example, pumping rates can reach up to 300 m^3/h. This model was built with the MODFLOW (Harbaugh et al., 2000) and MT3DMS (Zheng and Wang, 1999) suites. For the latter code simulating only solute transport, we took advantage of the analogy between the equations of heat transport and solute transport, as previously shown and validated in Hecht-Méndez et al., 2010.

The modeled ATES system consists of three wells, the so-called triplet. The first well (the pumping well) always supplies water at 10°C. The other two wells are used to store (1) cold water (at about 5°C) when

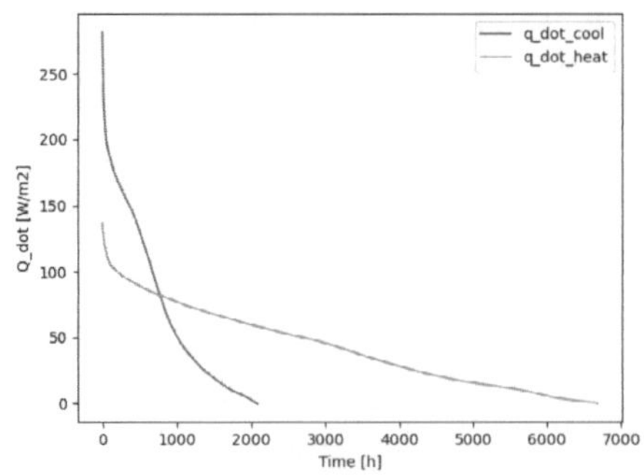

Figure 2: Yearly load duration curves of cooling and heating

the greenhouse is heated and (2) hot water (at about 15°C) when the greenhouse is air-conditioned. These two wells are called the cold well and the warm well respectively. The inputs of the model depend on the direction of pumping. When pumping from the pumping well, the inputs are the pumping rate in the pumping well and the injection rates at the cold and warm well with their associated temperatures. When pumping from the warm and cold wells, the input is still the pumping rates at the different wells (which are defined by the greenhouse demand). However, the water temperature of the pumped water from the cold and warm wells is now an output of the model.

The hydraulic conductivity is the main parameter for groundwater flow simulation. The hydraulic conductivity is set to 10^{-4} m s^{-1}, which is a mean value for this kind of aquifer. As for the thermal parameters that do not vary in great proportion in aquifers, we took the same parameters that can be found in Hecht-Méndez et al., 2010.

The ATES system was designed according to Dutch best practice. This means for example that the distance between the different wells is sufficiently large to avoid any thermal interaction between the thermal plumes stored in the aquifer (Bloemendal, Olsthoorn, and Boons, 2014; Bloemendal, Jaxa-Rozen, and Olsthoorn, 2018).

3. SIZING OF THE COMPONENTS

The simulation of the case study requires the sizing of the modeled components. This section summarizes the sizing of all of the models presented in Section 2.

3.1 Greenhouse sizing

The greenhouse is sized to the geological site. The alluvial plain from this case study allows for a construction of a 1 ha greenhouse. In order to size the greenhouse energy systems, the greenhouse model described in Section 2.1 has been simulated for a period of one year. The goal is to obtain the peak heating and cooling powers required by the greenhouse. The climate set-points are computed as proposed in Altes-Buch and Lemort, 2018 and are suited to the crop needs. The obtained yearly load duration curves for heating and cooling are shown in Figure 2.

Results show that energy for heating purposes is demanded about three quarters of the year, with a peak demand of 137 W m^{-2}. Cooling is required only in summer and has a higher peak demand at 278 W m^{-2}. This is, as expected, due to the greenhouse effect on sunny warm summer days.

3.2 Heating pipes sizing

As previously mentioned, water pipes are placed along the crop rows to keep a homogeneous climate on an horizontal axis and to serve as rails for harvest transport. The installed capacity of the pipes is therefore defined by the setup of the crop rows over the greenhouse floor area. In a classic Venlo type greenhouse, crop rows are separated by 1.6 m (De Zwart, 1996). Therefore, the required installed capacity for a greenhouse of 1 ha, assuming 200 m over 50 m, is 125 loops of 100 m.

Given the considerable length of each pipe loop, two conditions must be respected in order to have a homogeneous climate on an horizontal axis. First, the power output is adjusted by temperature control. The mass flow rate is therefore constant to the nominal value. Second, small temperature differences from the supply to the exhaust are desired. Typical controls aim for a temperature difference of about 5 to 7 degrees.

Assuming a loop completion time of 20 min (De Zwart, 1996) and a standard pipe diameter of 51 mm (Urban and Urban, 2010), the nominal mass flow rate is computed to be 0.14 kg s^{-1}. Considering a 6 degrees temperature difference supply-exhaust, the maximum power output of the pipes is around 46 W m^{-2}, which is 42% of the the peak demand (assuming the peak at 110 W m^{-2} instead of 137 W m^{-2}, given the low frequency of the points above 110 W m^{-2}). As a reminder, the rest of the heat is supplied by the coils.

3.3 Coils sizing

The coils are used for both heating and cooling. Given that the cooling demand is higher than the heating demand, the coils have been sized for cooling. When observing the steepness of the load duration curve, one may conclude that it is not cost-efficent to size the system according to the peak demand. Indeed, demand values above 200 W m^{-2} are only required few hours per year. However, the installed capacity must be high enough to allow for maintaining the air temperature below a certain threshold defined by the crop. The main reason is that the crop has temperature and humidity thresholds that, if exceeded, lead to crop yield reduction and potentially to crop diseases. For instance, many diseases are favored by temperatures between 24°C and 35°C (Urban and Urban, 2010). Additionnaly, the harvest rate at a maximum day temperature of 30°C, 35°C and 40°C is 91%, 79% and 52% of the one at 25°C (Vanthoor, 2011). In this work, the maximum acceptable air temperature has been established at 32°C. To comply with the later, the cooling coils must be able to supply 250 W m^{-2}. Given the characteristics of the modeled coils, the greenhouse must be equipped with 192 units (i.e. a share of 52 m^2 of greenhouse floor per unit).

As previously mentioned, coils are also used as an accessory to the heating pipes in case the heating demand is higher than the pipes installed capacity. Considering a heating peak demand of 110 W m^{-2} and the pipes maximum output of 46 W m^{-2}, the heating coils must be able to supply 64 W m^{-2}. Therefore, only a quarter of the installation (sized according to cooling needs) is required to provide the heating demand.

3.4 Heat pump sizing

Manufacturer data are necessary to fit the performance laws of the heat pump model. The heat pump must be able to provide the total heat demand, which is translated to a power of 1.13 MW at the condenser. The peak power can either be achieved by a single unit or by several smaller units working in parallel or series (e.g. 4 units of 283 kW). Manufacturer data that fits the size range useful for our case study could only be obtained for a water-to-water chiller. The characteristics at nominal load are displayed in Table 1. The coefficient of performance (COP) of a machine working in heat pump mode can be assumed as its energy efficiency ratio (EER) plus one:

$$COP = EER + 1 \tag{16}$$

Given the specifications and together with Equation (15), we compute an equivalent nominal heating power of 445 kW, which allows for supplying the peak demand by three units in parallel. The nominal cooling capacity and the power input are based on $\Delta T = 5$K entering/leaving condenser water temperature. The chiller's working temperature range is:

- Evaporator outlet: from 4°C to 9°C, with 7°C as nominal operation.
- Condenser inlet: from 25°C to 45°C, with 30°C as nominal operation.

The chiller is a water-cooled machine equipped with 2 single-screw compressors and working with HFC-134a. Each compressor, integrated in 2 separate refrigerant circuits, can modulate continuously its capacity down to 25% by means of capacity slides, offering a cooling capacity reduction of the chiller down to 12.5%. Evaporator and condenser are shell-and-tube heat exchangers.

Equations (10) and (12) were fit with the data at full load operation and the working $\Delta T = 5$K. Equation (13) was fit with the data at part load operation (shown in Figure 3), which is given for a cooling power of

Table 1: Characteristics of chiller *ECOPLUS XE - ST 100.2*

Specifications	Values	Units
$\dot{Q}_{ev,n}$	366	kW
EER_n	4.62	-
$T_{ev,ex,n}$	7	°C
$T_{ev,su,n}$	12	°C
$T_{cd,su,n}$	30	°C
$T_{cd,ex,n}$	35	°C

324.2 kW and a compressor input of 84.3 kW. As it can be observed, the chiller performance are higher at part load operation, with a peak at 40% of load. The obtained calibration parameters are displayed in Table 2.

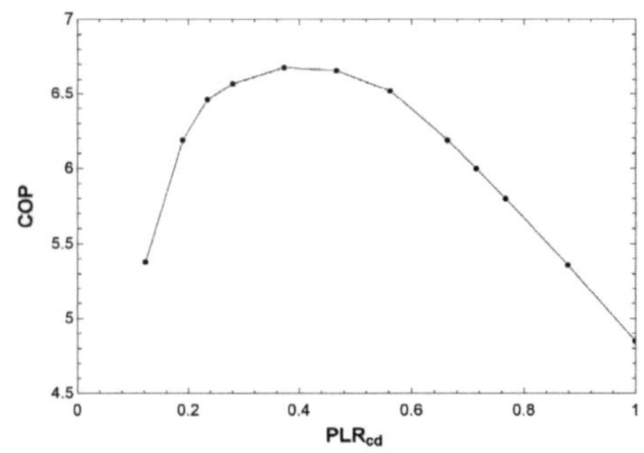

Figure 3: Performance at part load as a function of the part load ratio at the condenser

Table 2: Coefficients of the heat pump model

Coefficient	Value
α_0	0.99777
α_1	-8.9186
α_2	39.826
β_0	1.00145
β_1	0.026347
β_2	-0.0026717
δ_0	0.0898214
δ_1	0.186482
δ_2	0.710761

3.5 ATES sizing

The ATES wells must be sized to the extraction volumes required by the greenhouse. These are the volumes needed at the heat pump evaporator and the cooling coils to provide the greenhouse demand. The volumes are computed by means of Simulation 1 in the experimental protocol. The simulation details are described in Section 4.1 and the results are later discussed in Section 5.

4. CASE STUDY

In order to quantify the energy recovery potential of the study site, the modeled system is simulated for two years:

- First year: the goal is to recover energy from the greenhouse and store it underground, thus forming a warm and a cold plume in the aquifer. To that end, the greenhouse demand for both heating and cooling is provided by flows extracted from a pumping well at 10°C (i.e. the temperature of the site). The flows to the greenhouse are controlled such that the return temperature is ±5K with respect to the supply temperature. The greenhouse heating and cooling returns are injected in two separate locations, the goal being to create a warm plume at 15°C around the warm well and inversely, a cold plume at 5°C around the cold well.

- Second year: the goal is to exploit the energy stored in the warm and cold wells as shown in Figure 1. To that end, the greenhouse demand flows are extracted from the warm and cold wells for heating and cooling (respectively). Both returns are injected to the pumping well.

Because of crop constraints, the start of the simulation must be equivalent to the start of the growing year. Since the studied crop is tomato, the simulation starts on December 10th, as proposed in Grisey and Brajeul, 2007. In the greenhouse model, a typical meteorological year of Brussels is used to represent the outdoor climate.

4.1 Experimental protocol

The overground models (i.e. the greenhouse climate model, the heat pump model, the water pipes model and the coils model) cannot be simulated simultaneously with the ATES model because they are based on two different simulation software. Therefore, the simulation of the global system requires iterations between the two platforms. To tackle the iteration process, the following experimental protocol is followed:

- **Simulation 1**: simulate the overground models over the first year

  Goal: compute the greenhouse demand at the ATES level (i.e. the heat pump and coils flow rates and their return temperature)

  To that end, the models have been interconnected as follows. The greenhouse climate model, in which the water pipes and coils models are integrated, is coupled to three heat pump models connected in parallel. The aggregated condensers' outlet/inlet are connected to the inlet/outlet of the water pipes and heating coils. The evaporators' and cooling coils' inlets are connected to a supply source model representing the pumping well of the ATES. The flows of the source model are regulated by the climate controller and are assumed at 10°C (i.e. the ATES extraction temperature). Since the demand must be met at all times, the supply source has no flow restriction. Given an ATES target temperature of 15°C and 5°C, the models have been calibrated to ensure that the return temperature from the evaporator and the cooling coils is at ±5K from the supply.

- **Simulation 2**: simulate the ATES model over the first year

  Goal: compute the cold and warm plume formations and their status after the one year of injection.

  In the first year, the ATES model is made of one extraction well and two injection wells. The injection flows and temperatures are imposed to the profiles computed in step one. For the cold well, these are the heat pump evaporator flows and return temperature. For the warm well, these are the cooling coils flows and their return temperature. The extraction flow is equivalent to the addition of the latter.

- **Simulation 3**: simulate the ATES model over the second year

  Goal: compute the temperature evolution of the warm and cold wells over the year as a function of extracted volume.

  In the second year, the ATES model is made of two extraction wells and one injection well. The model initialization conditions are set to the status at the end of one year of injection (i.e. the simulation from step two). Given that the greenhouse is assumed to have the same consumption, the same flow profiles are used. Over the second year, these are not injection but extraction flows: the cooling coils flows are the cold well extraction, the heat pump evaporator flows are the warm well extraction. The extraction temperature profile evolves due to the storage level and the losses to the environment. The temperature profile as a function of the extracted volume is the main result of the simulation.

- **Simulation 4**: simulate the overground models over the second year

  Goal: compute the greenhouse demand with a supply from the cold and warm wells. Assess the energy efficiency of the global system in the ex-post.

  The interconnected overground models are simulated for one year. The evaporator and cooling coils inlets are connected to two supply source models representing the warm and cold wells of the ATES. The temperature of the sources evolve as a function of the extracted volume given the results from step three.

5. RESULTS AND DISCUSSION

The main results of the overground models simulations (i.e. simulations 1 and 4) are summarized in Table 3. As it can be observed, for both simulations the climate controller succeeds in maintaining the air temperature within one degree of difference from the set-point during 82-83% of the time. The air temperature is outside the acceptable temperature range for the crop less than 1% of the time. This can be seen in Figure 4, which displays quarter-hourly data of the air temperature as a function of the set-point together with the thresholds of acceptable temperatures defined by the crop. The colors indicate whether the greenhouse was being heated, cooled or none of the above. It is observable that most of the points outside a 2 degrees range from the set-point value are on the upper side. This is because as explained in Section 3.3, the cooling system is not sized to supply the peak cooling demand but to respect the maximum acceptable threshold.

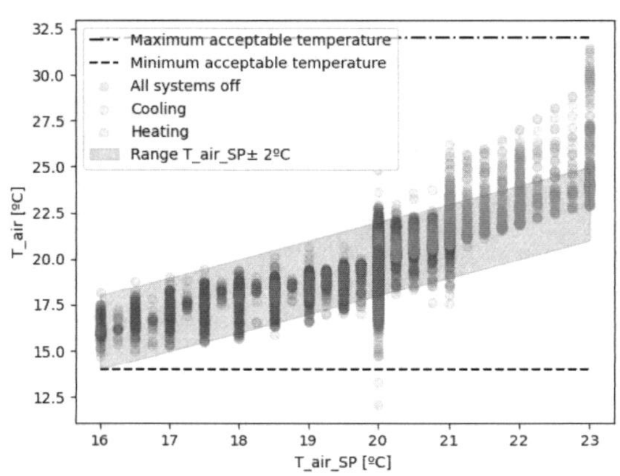

Figure 4: Comparison of the air temperature with respect to its set-point (quarter-hourly data points of simulation 1)

During the first simulation year, the greenhouse recovers 2338 MWh at the evaporator and 1546 MWh at the cooling coils. The accumulated volumes used in these exchangers are 403 dam^3 and 398 dam^3, respectively. The evolution of the accumulated volumes over the year is shown in Figure 5 (blue curves). The maximum extraction volumetric flow rate, which used to size the ATES, is 180 $m^3\ h^{-1}$. The return temperatures of the heat pumps and coils circuits are well controlled at ± 5 degrees with respect to the supply. After one year of simulation (Simulation 2), the warm and cold wells temperature is stable at 14.5°C and 5.7°C, respectively.

In Simulation 3, the recovered temperatures over a year as a function of the extraction volumes are obtained (Figure 5, red curves). As a reminder, Simulation 3 assumes that the greenhouse has the same demand profile as Simulation 1. As expected, the amount of energy stored decreases proportionally to the increase in extraction. In the warm well, the recovered temperature decreases over the year by a linear trend that follows the more or less linear demand profile. In the cold well, on the contrary, the absence of cooling demand at the begining of the year allows for the temperature to be impacted by diffusion losses to the underground flows. When the cooling demand becomes stable throughout summer, the recovered temperature increases in a linear fashion.

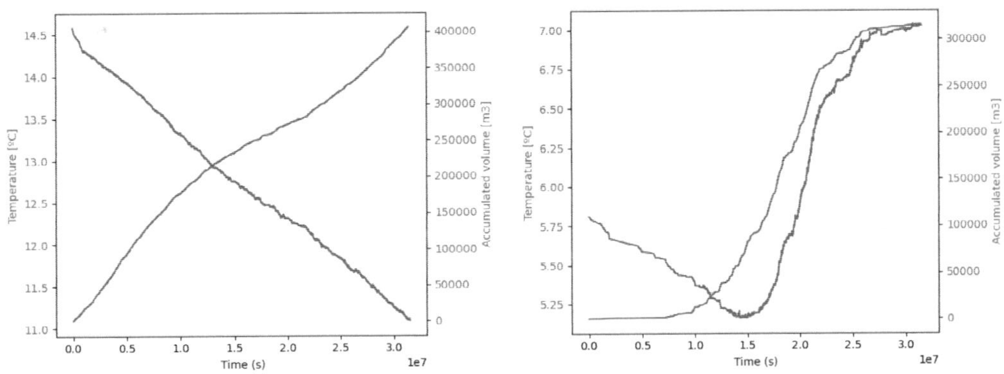

(a) Warm well **(b)** Cold well

Figure 5: Evolution of the recovered temperature at the wells over one extraction year

In Simulation 4, the second simulation year in which the greenhouse demand is produced by the warm and cold wells (assumed at the temperature profiles from Simulation 3), the greenhouse recovers 3127 MWh at the evaporator and 1599 MWh at the cooling coils. On the one hand, and as expected since the climate set-points remain unchanged, these values are very similar to the ones from Simulation 1. On the other hand, given that the supply temperatures at the evaporator and the cooling coils are different, the required volumes for producing a heating or cooling capacity differ from the ones considered in Simulation 1. The volumes consumed at the evaporator and the cooling coils are now 437 dam^3 and 216 dam^3, respectively. This means that the heat volume consumed from the warm well in the second year (437 dam^3) is higher than the volume stored in the first year (308 dam^3). Similarly, the cooling volume consumed from the cold well in the second year (216 dam^3) is lower than the volume stored in the first year (403 dam^3). Over the years, this may lead to a system imbalance, mainly in the cold well side, since more energy is stored than consumed.

A way to avoid this imbalance is through the flexibility offered by greenhouses. Among the available measures that would help balancing the system, the ones that generate the most impact are:

i) Reducing ventilation for temperature control: The greenhouse climate controller of Simulations 1 and 4 is programmed to use natural ventilation to cool the greenhouse when the outside temperature is lower than the targetted set-point. The greenhouse cooling consumption could therefore be increased by not opening the vents and using active cooling instead.

ii) Decreasing the air temperature set-point to reduce heating in winter: As previously mentioned in this work, extreme temperatures are not optimal for crop yield and must be avoided. However, there is a range of near-optimal temperatures that allow for similar values of crop production rate. For instance, the harvest rate at a mean temperature of 14°C is 18% of the harvest rate at a mean temperature of 22°C. At the same time, the harvest rate at a mean temperature of 18°C is 87% of the one at 22°C (Vanthoor, 2011).

iii) Increasing the air relative humidity ventilation threshold: natural ventilation is used when the air relative humidity increases above a threshold mainly to reduce the risk of crop diseases. Higher levels of humidity do not necessarily imply that the crop is affected by diseases, but because the risk on such diseases increases as humidity becomes higher, growers try to avoid such high moisture contents.

In order to illustrate the impact of these measures, Simulation 4 has been repeated twice producing variants 4b and 4c. The first, Simulation 4b, applies all three methods in the following fashion. The ventilation threshold for temperature control is increased from 1.5 to 2 degrees when heating is used, and from 1 to 2 degrees when no heating is used (i). The temperature set-point is modified in such way that the average temperature set-point is decreased by 2 degrees with respect to the original simuation, but the lowest temperature set-point remains unchanged (ii). The threshold of ventilation for humidity control is increased from 85% to 90% from sunset until midnight (iii). The second, Simulation 4c, applies measures (ii) and (iii), and forbids ventilation for temperature control. That means, the vents can only open for humidity control. Measures (ii) and (iii) are applied as in Simulation 4b.

The obtained results, displayed in Table 3 and Figure 6, indicate that the measures can have a substantial impact on the greenhouse energy consumption. The modified set-points (measures ii and iii) decrease the heating needs down to 2502 MWh and increase the cooling demand up to 1995 MWh. In other words, measures (ii) and (iii) allow for increasing the cold well extraction by 40% (304 dam^3) and decreasing the warm well extraction by 29% (311 dam^3). A reduction on the ventilation for temperature control (measure i) as an additional measure allows for increasing the cold well extraction by 72% (371 dam^3) and decreasing the warm well extraction by 45% (241 dam^3), with respect to Simulation 4. Therefore, both methods decrease the system imbalance considerably, with measure (i) being the one with the most impact. However, as shown in Figure 7, when using measure (i) (Simulation 4c) the air relative humidity is above 90% during long periods, amounting for 30% of the year. The higher risk of diseases, although not quantifiable by the model, must be kept in mind and considered a drawback of this type of measure. In Simulation 4c, the integrated cooling rate (2561 MWh) is higher than the integrated heating rate (2012 MWh). This implies that, even in the Atlantic climate, the greenhouse can become a major cooling consumer.

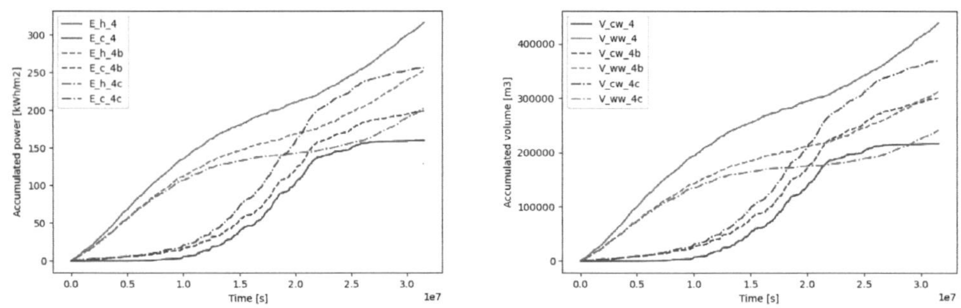

(a) Heat transfer rate

(b) Volume

Figure 6: Accumulated heat transfer rate and volumes for simulations 4 and 4b

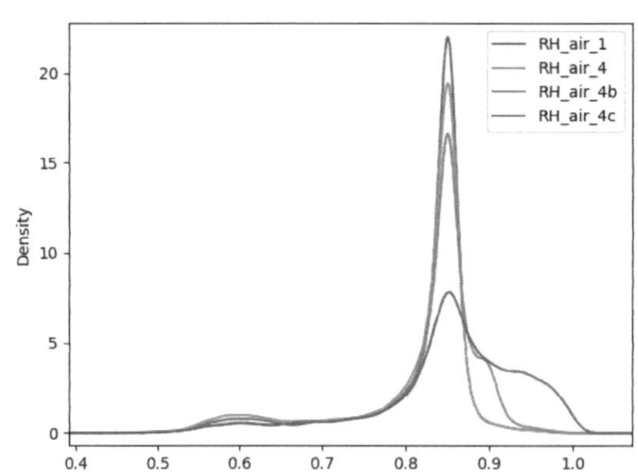

Figure 7: Density distribution of the air relative humidity over one year

Table 3: Simulation results

Variable	Unit	Sim 1	Sim 4	Sim 4b	Sim 4c
Simulation year	-	First	Second	Second	Second
Time T_air is within T_air_sp ± 1°C	%	82.12	83.70	84.57	86.84
Time T_air is within [14, 32]°C	%	100	100	99.40	99.45
Time RH_air is below 90 %	%	98.27	98.32	94.11	70.38
Average air temperature	%	20.04	20.02	18.36	18.39
Average air relative humidity	%	81.59	80.85	82.72	85.29
E_heat_total	MWh	2885	3127	2502	2012
E_heat_pipes	MWh	2117	2236	1979	1630
E_heat_coils	MWh	768.2	891.3	523.2	381.8
E_cool_total	MWh	1546	1599	1995	2561
E_el_total	MWh	3341	2802	2674	2589
E_el_lamps	MWh	2809	2265	2267	2272
E_el_hp	MWh	532.6	536.5	406.9	316.8
E_ev	MWh	2338	2595	2068	1661
V_ev	dam^3	403.3	436.9	310.8	241.6
V_cc	dam^3	307.8	216.4	304.1	371.3
V_ww	dam^3	307.8	436.9	310.8	241.6
V_cw	dam^3	403.3	216.4	304.1	371.3
Direction	-	Injection	Extraction	Extraction	Extraction
V_pw	dam^3	711.1	653.3	614.9	612.9
Direction	-	Extraction	Injection	Injection	Injection

6. CONCLUSIONS

In this work, the use of the thermal energy storage in shallow alluvial aquifers is proposed as a sustainable solution for supplying the energy demand of greenhouses. Air-conditioning greenhouses in summer allows for storing their energy surplus in an underground seasonal buffer that can later be used, in winter, for heating via a heat pump. The evaluation of the benefits arising from using ATES for supplying the energy demand in greenhouse applications requires the modeling of a greenhouse climate model, heating/cooling distribution and emission systems (i.e. heating pipes and coils), a heat pump and the ATES system. The models have been sized to the presented case study, which is based on a 100 m thick aquifer in the Cretaceous chalk, located in Wallonia (Belgium).

The case study has been simulated for two years. Due to the nature of the models, the experimental protocol is iterative and consists of 4 simulations. Results show that the recovered energy and volumes for heating and cooling are not in equilibrium. Greenhouses located in the Belgian latitude can be primarily considered as consumers of heating energy. Although energy is recovered during 3 months per year from cooling the greenhouse, the total energy consumed for heating is much higher. This results in a system imbalance on the cold well, since more energy is stored than consumed.

Despite this fact, a sustainable system can be achieved if the greenhouse climate controller is correctly calibrated. On the one hand, the great potential of flexibility offered by greenhouses allowed for improving the results in two extra simulations in which the greenhouse is forced to increase its cooling consumption. On the other hand, we observed a higher risk for crop diseases since a reduction of the ventilation to the outside air lead to a higher moisture content in the greenhouse. Not authorizing the opening of the vents for temperature control is therefore not desirable unless the greenhouse is equipped with a dehumidifier.

At this stage, the coupling between the two models (the aquifer and greenhouse ones) is not perfect. The results of one model are simply injected as inputs into the second and vice versa. This prevents us from going beyond the second year or simulating more complex (but probably more efficient) strategies. Therefore, a more tight coupling between the two models would be the natural continuation of this work for tackling optimisation strategies that lead to a sustainable, self-sufficient system.

NOMENCLATURE

α	polynomial coefficient	(–)
β	polynomial coefficient	(–)
δ	polynomial coefficient	(–)
ε	efficiency	(–)
ρ	density	(kg m^{-3})
A	surface	(m^2)
c	specific heat capacity	(J kg^{-1} K^{-1})
COP	coefficient of performance	(–)
E	accumulated energy	(MWh)
EER	energy efficiency ratio	(–)
h	specific enthalpy	(J kg^{-1})
$\dot{H}$	heat flow associated to a mass transfer	(W)
$\dot{M}$	mass flow	(kg s^{-1})
$\dot{Q}$	heat flow	(W)
$\dot{q}$	specific heat flow	(W m^{-2})
RH	relative humidity	(–)
T	temperature	(°C)
t	time	(s)
V	volume	(dam^3)
$\dot{W}$	electrical power	(W)
w	absolute humidity	(kg kg^{-1})

Subscript

air	air
c	contact
cc	cooling coils
cd	condenser
cool	cooling
cw	cold well
dry	dry
el	electricity
ev	evaporator
ex	exhaust
f	ficticious
fl	full-load
heat	heating
hp	heat pump
n	nominal
p	pressure
pl	part-load
pw	pumping well
su	supply
sens	sensible heat
sp	set-point
total	total
w	water
wb	wet-bulb
wet	wet
ww	warm well

REFERENCES

Altes-Buch, Q. and V. Lemort (Dec. 2018). "Modeling framework for the simulation and control of greenhouse climate". In: *Proceedings of the 10th International Conference on System Simulation in Buildings*. Liege.

Altes-Buch, Q., S. Quoilin, and V. Lemort (May 2022). "A modeling framework for the integration of electrical and thermal energy systems in greenhouses". en. In: *Building Simulation* 15.5, pp. 779–797. ISSN: 1996-8744. DOI: 10.1007/s12273-021-0851-2.

Altes-Buch, Queralt, Sylvain Quoilin, and Vincent Lemort (Mar. 2019). "Greenhouses: A Modelica Library for the Simulation of Greenhouse Climate and Energy Systems". English. In: DOI: 10.3384/ecp19157533.

Bloemendal, Martin, Marc Jaxa-Rozen, and Theo Olsthoorn (Apr. 2018). "Methods for planning of ATES systems". en. In: *Applied Energy* 216, pp. 534–557. ISSN: 0306-2619. DOI: 10.1016/j.apenergy.2018.02.068.

Bloemendal, Martin, Theo Olsthoorn, and Frank Boons (Mar. 2014). "How to achieve optimal and sustainable use of the subsurface for Aquifer Thermal Energy Storage". en. In: *Energy Policy* 66, pp. 104–114. ISSN: 0301-4215. DOI: 10.1016/j.enpol.2013.11.034.

Bolther, A. et al. (1999). *Méthode de calcul des consommations d'énergie des bâtiments climatisés ConsoClim*. Tech. rep. CSTB ENEA/CVA-99.176R. Paris: Ecole des Mines.

Brandemuehl, Michael J. (1993). *HVAC 2 Toolkit: A Toolkit for Secondary HVAC System Energy Calculations*. en. Google-Books-ID: Bq7fAAAACAAJ. ASHRAE. ISBN: 978-0-910110-98-3.

Commission, European (2021). *The tomato market in the EU: Vol. 3a: Trade for fresh products*. Tech. rep. AGRI.G2 - F&V - 2021, p. 29.

De Schepper, Guillaume et al. (May 2019). "Assessment of short-term aquifer thermal energy storage for demand-side management perspectives: Experimental and numerical developments". en. In: *Applied Energy* 242, pp. 534–546. ISSN: 0306-2619. DOI: 10.1016/j.apenergy.2019.03.103.

De Zwart, H.F. (1996). "Analyzing energy-saving options in greenhouse cultivation using a simulation model". PhD Thesis. Wageningen University.

Gang, Wang, Liu Mingsheng, and David E. Claridge (Aug. 2007). "Decoupled modeling of chilled-water cooling coils". English (US). In: *ASHRAE Transactions*. Vol. 113 PART 1. ISSN: 0001-2505 Journal Abbreviation: 2018 ASHRAE Annual Conference. ASHRAE, pp. 484–493.

Gendebien, Samuel, Stephane Bertagnolio, and Vincent Lemort (2010). "Comparative and empirical validation of three water cooling coil models". en. In: p. 8.

Grisey, A. and E. Brajeul (2007). *Serres chauffées: réduire ses dépenses énergétiques*. 1st. Editions Centre technique interprofessionnel des fruits et légumes (CTIFL).

Harbaugh, A.W. et al. (Jan. 2000). "MODFLOW-2000, the U.S. geological survey modular ground-water flow model-User guide to modularization concepts and the ground-water flow process". In: *U.S. Geological Survey Open-File Report 00-92*.

Hecht-Méndez, Jozsef et al. (2010). "Evaluating MT3DMS for Heat Transport Simulation of Closed Geothermal Systems". en. In: *Groundwater* 48.5, pp. 741–756. ISSN: 1745-6584. DOI: 10.1111/j.1745-6584.2010.00678.x.

Lebrun, J. et al. (1990). "Cooling Coil Models to be used in Transient and/or Wet Regimes". In: *Proceedings of the International Conference on System Simulation in Buildings 1990*, pp. 405–411.

Morisot, O., D. Marchio, and P. Stabat (Apr. 2002). "Simplified Model for the Operation of Chilled Water Cooling Coils Under Nonnominal Conditions". In: *HVAC&R Research* 8.2. Publisher: Taylor & Francis, pp. 135–158. ISSN: 1078-9669. DOI: 10.1080/10789669.2002.10391433.

Quoilin, S. et al. (2014). "ThermoCycle: A Modelica library for the simulation of thermodynamic systems". en. In: *Proceedings of the 10th International Modelica Conference*.

Stanghellini, Cecilia, Bert Van 't Ooster, and Ep Heuvelink (Sept. 2018). *Greenhouse horticulture*. Wageningen Academic Publishers. ISBN: 978-90-8686-329-7. DOI: 10.3920/978-90-8686-879-7.

Urban, L. and I. Urban (2010). *La production sous serre: La gestion du climat*. French. 2nd. Vol. 1. Lavoisier.

Vanthoor, B. H. E. (2011). "A model-based greenhouse design method". dut. PhD Thesis. Wageningen University.

Zheng, Chunmiao and Pu Wang (Jan. 1999). "MT3DMS: A Modular Three-Dimensional Multispecies Transport Model for Simulation of Advection, Dispersion, and Chemical Reactions of Contaminants in Groundwater Systems; Documentation and User's Guide". In.

ACKNOWLEDGMENT

The authors would like to thank the Walloon Region of Belgium for funding this research in the context of the EcoSystemePass project (convention 1510610).

Optimal Load Shifting for Multiple ON/OFF Air Conditioning Units: How to Avoid Unnecessary Peak for Precooling?

Donghun Kim[1*], James E. Braun[2]

[1] Lawrence Berkeley National Laboratory, Building Technology & Urban Systems Deiision,
Berkeley, CA, USA
Contact Information (donghunkim@lbl.gov)
* Corresponding Author

[2] Purdue University, Mechanical Engineering,
West Lafayette, IN, USA
Contact Information (jbraun@purdue.edu)

ABSTRACT

Small and medium-sized commercial buildings (SMCB) are significant demand response resources, and it is important to develop grid-responsive control algorithms that exploit those resources and create financial benefits for building owners and HVAC service providers. Furthermore, unlike large-sized commercial buildings, there is an opportunity to have universally applicable control solutions for many SMCBs since those buildings have a consistent HVAC system configuration: SMCBs are commonly served by multiple-staged air conditioning units controlled by their own thermostats. Despite the demand response potential and scalability, however, very few control solutions are available for SMCBs. Typical model predictive control (MPC) and heuristic control approaches for cooling load shifting that lower thermostat setpoints before an electric price jump are suitable mainly for large-sized commercial buildings where a continuous capacity modulation is possible, e.g., via dampers in variable air volume terminal units. However, those approaches can cause undesired, high peaks for SMCBs due to the nature of ON/OFF unit staging and narrow thermostat deadbands. This could discourage the use of advanced grid-responsive controls for SMCBs due to the concern of high demand charges, and has to be resolved. This paper presents a MPC solution that overcomes this challenge. It has a hierarchical MPC structure where an upper level MPC is responsible for electrical load shifting in response to an electric price signal while a lower level MPC is responsible for coordinating compressor stages to eliminate unnecessary peaks and follows the setpoints determined by the upper level MPC. A comprehensive, long-term laboratory test has been carried out to demonstrate load shifting and evaluate cost savings of the algorithm.

1. INTRODUCTION

Since buildings account for over 70% of U.S. total electricity use, and contribute significantly to electric power demand peaks and carbon emission, integrating building controls with the grid can improve grid reliability, resiliency, and efficiency, and help in achieving states' and government's carbon neutrality goals. Buildings have inherent energy storage, i.e., the thermal mass associated with building structure, which can be used for shifting and shedding significant thermal loads. The use of thermal mass in cooling seasons can involve pre-cooling the building prior to an event (e.g., an ON peak price period) in order to lower cooling load and electric peak demand during the event. One of the most popular and widely studied control approaches for managing thermal mass is model predictive control (MPC), and a large number of previous studies have demonstrated that it can provide grid services while presenting economic benefits to building owners without compromising thermal comfort.

Small-sized ($< 464\ m^2$, $< 5{,}000\ ft^2$) and medium-sized ($< 4{,}640\ m^2$, $< 50{,}000\ ft^2$) commercial buildings (SMCB) account for about 95% of U.S. commercial buildings, 50% (3.6 billion m^2) of the total commercial indoor floor space and 60% of energy usage in the commercial building sector. Thus, SMCBs are undoubtedly tremendous, readily-available demand response resources. Furthermore, unlike large-sized commercial buildings, SMCBs have a fairly consistent HVAC and control configuration: commonly served by multiple staged air conditioning units, e.g., rooftop units (RTU), for space heating and cooling where each unit stage is controlled by its own thermostat. Examples are banks, retail stores, restaurants, schools and factories. This implies that there is a potential to have a universal and

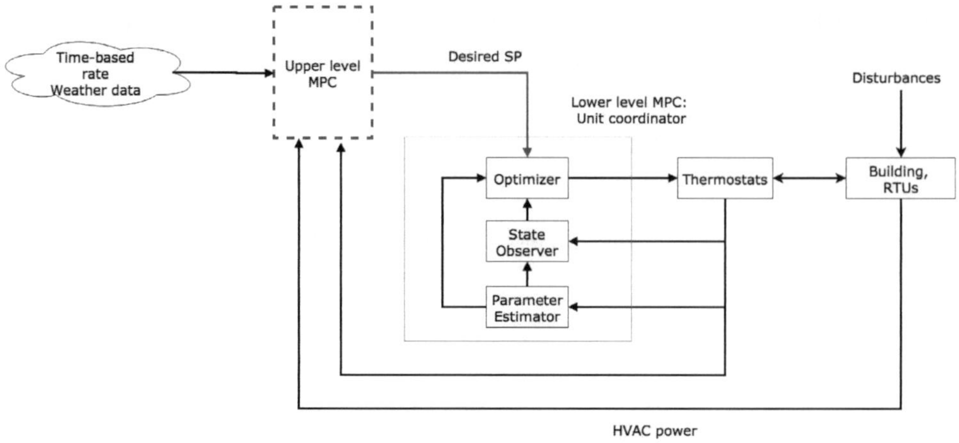

Figure 1: Conceptual diagram of a two-layer MPC for optimal load shifting and further peak demand reduction

scalable control solution which can be applied to many SMCBs and could rapidly transition them to Grid-interactive Efficient Buildings (GEB).

A limited number of MPC papers are found in the literature for SMCBs. Nutaro et al. (2014) developed a MPC algorithm and implemented it at a gymnasium having four identical 10-ton RTUs. This coordinator combines a simplified ARX type input-output model with heuristics and optimization to limit electric peak demand. The reduction of peak power consumption was about 15% with respect to a conventional thermostat control for the building. Kim et al. (2015) developed a MPC algorithm for minimizing energy consumption and electric peak demand for multiple RTUs especially for open-spaced buildings. The control approach was designed to minimize sensor and configuration requirements in order to enable a more cost effective control implementation for small/medium commercial buildings. It was implemented at a laboratory and muliple field sites, and reported overall up to 10% building HVAC energy savings and 20% peak demand reduction (Kim and Braun, 2018). Biyik et al. (2015) developed a RTU coordinator to minimize peak power. The goal was achieved by assigning different time varying penalties to different RTUs in order to prevent all units from operating simultaneously. It is reported that the peak power reduction with the optimal control strategy was about 20-40% for existing buildings. Putta et al. (2015) applied dynamic programming to solve an optimization problem for minimizing energy consumption and reducing compressor short cycling. In the case study, about 10% energy savings were estimated for a building served by 4 RTUs. Zhang et al. (2017) developed a RTU coordination control algorithm that shaves power usage within a limit during a demand response event for 90 minutes when triggered by a utility company, while minimizing comfort impacts.

However, those studies focus mainly on peak demand reduction and energy efficiency not load flexibility. SMCBs have a unique technical difficulty in load shifting. Typical MPC approaches in the literature, for variable air volume systems for example, which lower thermostat setpoints to shift load would not work for the type of buildings. This is because when the setpoints are lowered by, e.g., a couple of degree Celsius, full compressor stages for all units would be simultaneously activated creating a high electric peak due to the nature of ON/OFF stage of the units and narrow thermostat deadbands (usually $< 1^oC$). This is a very significant issue, since the undesired high peak and hence potentially high demand charge to building owners makes no economic sense for load shifting. To fully exploit SMCBs' demand response resources, this problem has to be solved.

This paper develops and demonstrates a MPC solution for SMCBs to overcome this problem. The solution incorporates a hierarchical MPC approach. Long-term experiments were carried out to demonstrate the load shifting and demand reduction capabilities. The MPC behaviors under two different time of use (TOU) scenarios were investigated, and one of the results for a typical TOU case is presented in this paper. Interesting trade-offs between energy efficiency and load flexibility were observed and are discussed.

2. METHODOLOGY: HIERARCHICAL MPC DESIGN

The proposed MPC for optimal load shifting for SMCBs has a hierarchical control structure as shown in Fig. 1 in which the upper level MPC (UMPC) marked with the blue dashed box optimizes desired setpoints while the lower level MPC (LMPC) marked with the red box acts as a slave controller. The UMPC is responsible for the optimal load shifting in response to utility price and weather signals by looking at long-term building thermal dynamics, while the LMPC optimally coordinates multiple RTUs or other unit packages to track the optimal setpoints determined by the UMPC while preventing unnecessary peaks discussed in the previous section.

2.1 Design of Lower MPC

In this section, mathematical descriptions of the LMPC are presented. It should be noted that the prediction horizon of the LMPC (N_p^L) is relatively short, e.g. 30 min to 1 hour. The control problem at a current time step k is:

$$\begin{aligned} \min \quad & \sum_{j=1}^{N_p^L} \sum_{i=1}^{p} P_i u_i(k+j-1) + \omega_d \delta + \omega_l \Gamma_l + \omega_u \Gamma_u \\ s.t. \quad & T_l^L - \Gamma_l \leq E(y(k+j)|\mathcal{G}_k) \leq T_u^L + \Gamma_u \\ & \sum_{i=1}^{p} P_i u_i(k+j-1) \leq \delta \;\; (\forall j \in \{1, \cdots, N_p^L\}), \end{aligned} \tag{1}$$

where j represents the time step for control changes measured from the current value of k. $u_i(k+j)$ and $y_i(k+j)$ represent unit stage and thermostat temperature for the ith unit at the $k+j^{th}$ timestep. P_i is the rated power for the ith unit. $E(y(k+j)|\mathcal{G}_k)$ is the optimal j-step prediction of thermostat temperatures given data $\mathcal{G}_k = \{y(k-1), y(k-2), \cdots, u(k+j-1), u(k+j-2), \cdots\}$. The prediction is calculated using the identified model, G_u (See Section **??**). $T_u^U (\in \mathbb{R}^n)$ are the desired setpoints that will be specified by the upper level MPC, and $T_u^L (\in \mathbb{R}^n)$ are temperature lower bounds. $\omega_l, \omega_u (\in \mathbb{R}^+)$ and $\omega_d (\in \mathbb{R}^+)$ are weights on variables of $\Gamma_l, \Gamma_u (\in \mathbb{R}^+)$ and $\delta (\in \mathbb{R}^+)$. Γ_l and Γ_u can be seen as comfort violations from the first constraint of (1) that are introduced to ensure a feasible solution. The variables to be optimized are the unit stages for all units over the prediction horizon and $\delta, \Gamma_l, \Gamma_u$, which form a mixed integer linear programming problem.

From the last constraint, it can be seen that δ is an upper bound on the electric demand for each time interval over the prediction horizon. Therefore, minimizing δ will naturally lower the electric peak demand over the prediction horizon. The coordination of unit stages to minimize the electric demand is the key component of the LMPC to eliminate high peaks for load shifting. For the capability of peak regulation and reliability of the LMPC, we refer to Kim and Braun (2018) which reports long-term field site MPC results.

2.2 Description of Upper MPC (UMPC)

In this section, mathematical descriptions of the UMPC are presented. To provide the optimal upper bounds (or desired cooling temperature setpoints) to the LMPC, i.e., T_u^L, it is natural to optimize setpoints by predicting a cost associated with the setpoints. However, it is challenging to obtain a dynamic model which maps setpoints to cooling load or power, because of the nonlinearity (hysteresis) of thermostats. Instead, an alternative approach is used: input trajectories are optimized first and desired setpoint trajectories are back calculated from the optimal inputs and G_u. With this in mind, the control problem of the UMPC at a current time step k is:

$$\min \quad \sum_{j=1}^{N_p^U} \sum_{i=1}^{p} R(k+j-1) P_i(k+j-1) \bar{u}_i(k+j-1) + \omega_d \delta + \omega_l \Gamma_l + \omega_u \Gamma_u \tag{2}$$

$$\begin{aligned} s.t. \quad & T_l^U - \Gamma_l \leq E(\bar{y}(k+j)|\mathcal{G}_k) \leq T_u^U + \Gamma_u \\ & \sum_{i=1}^{p} P_i(k+j-1) \bar{u}_i(k+j-1) \leq \delta \\ & 0 \leq \bar{u}_i(k+j-1) \leq 1 \;\; (\forall j \in \{1, \cdots, N_p^U\}), \end{aligned} \tag{3}$$

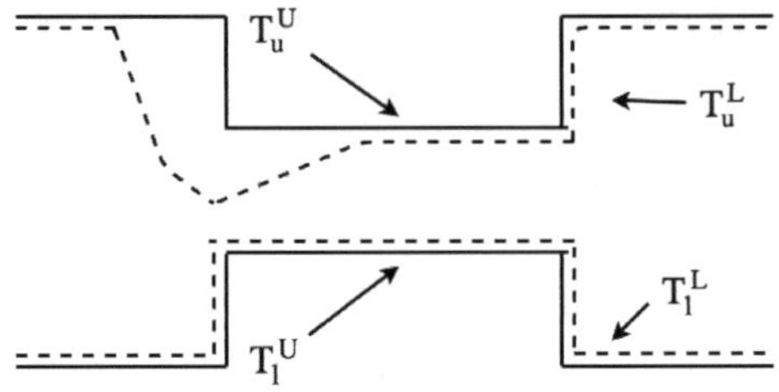

Figure 2: Example of upper and lower bounds for the UMPC and LMPC

where $\bar{u}$ and $\bar{y}$ are the moving averaged unit stages and thermostat temperatures, respectively, over a window (a 30-minute interval is used for this study). Because of averaging, $\bar{u}_i$ for each sampling time is a real number bounded by 0 and 1 (for a single stage unit): $\bar{u}$ can be viewed as unit runtime fractions (RTF) over the averaging window. A sequence of $\bar{u}(\cdot)$ is the optimization variable of (2). The desired setpoint trajectories are calculated with the optimal RTFs and an input-output model.

Differences compared with the LMPC formulation in (1) are explained as follows. N_p^U is the prediction horizon for the UMPC (> 6 hours), which is longer than N_p^L of the LMPC (around 1 hour). The control implementation time for the UMPC is also longer than that of the LMPC: That is, the UMPC and LMPC are implemented for every 30 min and 5 min, respectively. R which is the electricity rate is provided only to the UMPC for load shifting. The feedback and future prediction data set, i.e. $\mathcal{G}_k$, includes forecast of outdoor air temperature as follows. $\mathcal{G}_k = \{\bar{y}(k-1), \bar{y}(k-2), \cdots, \bar{u}(k+N_p^U-1), \bar{u}(k+N_p^U-2), \cdots, \bar{T}_o(k+N_p^U-1), \bar{T}_o(k+N_p^U-2), \cdots\}$ where $\bar{T}_o$ is the moving averaged signal with the same window size for $\bar{u}, \bar{y}$.

The upper and lower temperature bounds, i.e. T_l^U, T_u^U indicate an acceptable temperature range specified by users and are different from the bounds for the LMPC. The bounds for the LMPC could be narrower than that of the UMPC and are determined by the UMPC as follows.

$$T_{u,i}^L(k) = \begin{cases} \bar{y}_i^*(k), & if\, \bar{u}_i^*(k) > 0 \\ T_{u,i}^U(k), & if\, \bar{u}_i^*(k) = 0 \end{cases} \tag{4}$$
$$T_{l,i}^L(k) = T_{l,i}^U(k), \quad \forall i \in \{1, 2, \cdots, p\}$$

where $\bar{u}_i^*(k)$ indicates the optimal RTF over the averaging window for the ith unit at the kth timestep of the UMPC and $\bar{y}_i^*(k)$ is the corresponding desired temperature. The first equation of (4) means the desired temperature profiles are used for the upper bound for the LMPC. However, it is modified with the following heuristic strategy: if the UMPC does not use the ith unit, i.e. $\bar{u}_i^*(k) = 0$, put the upper bound of the LMPC for the ith unit as high as possible in order to turn off the unit.

Example profiles of the upper and lower bounds for the UMPC and LMPC are shown in Fig. 2. The bounds for the UMPC, solid lines in the figure, are based on a predefined temperature band while the bounds of the LMPC, dashed lines, follow a decision of the UMPC according to (4).

2.3 Modeling: System Identification

To predict zone air temperature profiles with a forecast for the outdoor air temperature and a candidate profiles of control inputs, we utilized a methodology termed the Lumped Disturbance modeling approach (LD) (Kim et al., 2016). We refer to Kim et al. (2018) for the mathematical description and background of the algorithm, and validation for the case study building of this paper.

3. LABORATORY TEST RESULTS

3.0.1 Site Description A cooling system for a conference room (about 15 m long, 7 m wide and 3.5 m high) in the Ray W. Herrick Laboratories at Purdue University, IN, U.S. (see Fig. 3) was retrofitted to test the overall MPC approach. Two packaged air conditioners (termed U1 and U2) were installed and the air duct system was reconfigured accordingly.

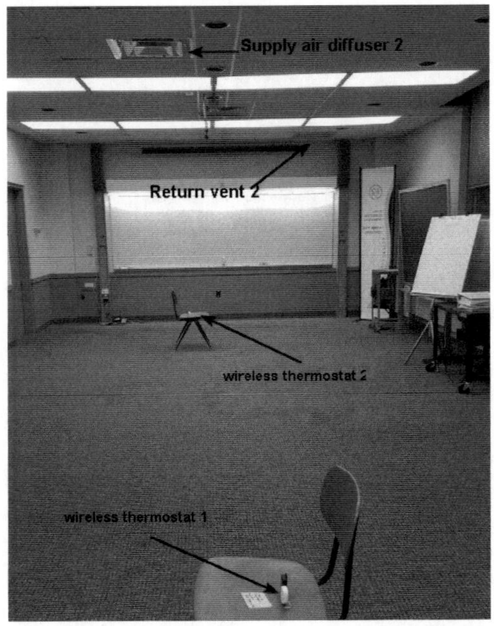

Figure 3: Case study building (supply air diffuser 1 and return vent 1 are not shown)

U1 is a 1-ton single stage unit with an energy efficiency ratio (EER) 9 and U2 is a 2-ton single stage unit with 10 EER. Thermostats, supply and return vents associated with the two units are shown in Fig. 3 and Fig. 4. Note that there is a strong coupling between the two sub-zones (or thermostats), and hence the operation of one unit can influence both thermostat temperatures. Unmeasured heat sources are lighting gains, loads from electric appliances (a small freezer and one laptop computer), infiltration/exfiltration, solar gains through windows, and occupancy gains. The lights were turned ON and OFF by occupant random behavior.

It should be noted that the conference room is not very representative of many SMCBs in the sense that 1) it has a carpet and mat (see Fig. 3) and 2) it has a high U-value for the wall and window (See the external view of Fig. 4): the external wall consists of pure bricks (without insulation) and single pane windows covering around 1/3 of the wall area. These system properties restrict the MPC savings potential which will be discussed in detail.

Conventional Control Set Up: A conventional thermostat control (Conv) turns on and off the unit stages based on thermostat deadband(s) and tracking error between the measured thermostat temperature and setpoint. The Conv fol-

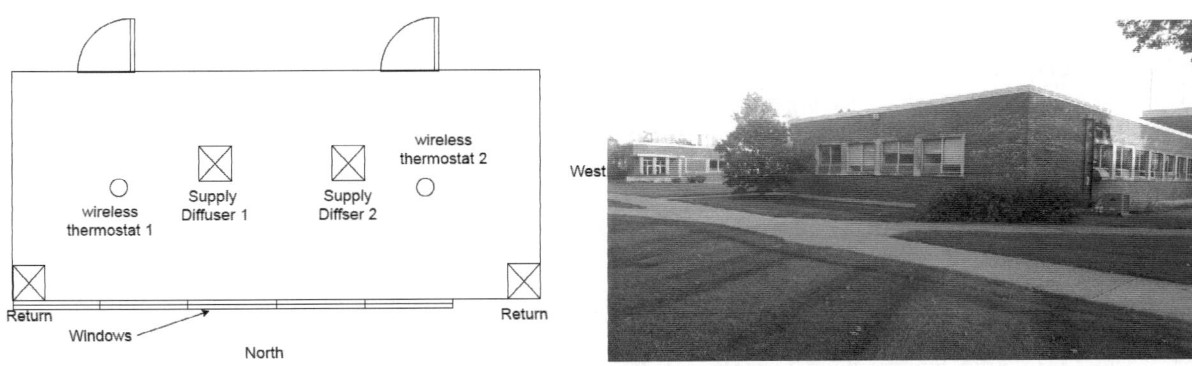

(a) (b)

Figure 4: (a) Floor plan with locations of thermostats, supply diffusers and return vents and (b) external view of the demonstration site (Purdue Herrick Laboratory)

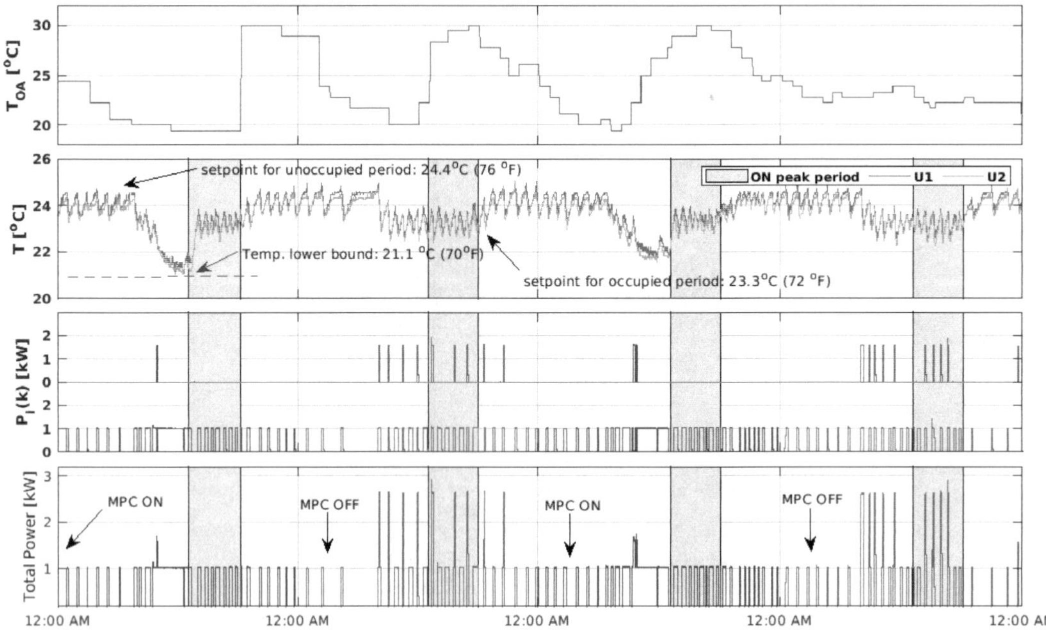

Figure 5: Sample profiles of outdoor air temperature, zone air temperature, setpoints, instantaneous unit powers, and total HVAC power under conventional control and MPC

lows a night setup schedule, 23.3^oC (74^oF) during the occupied period (9:00 AM to 6:00 PM) and 24.4^oC (76^oF) during the unoccupied period.

Hierarchical MPC Set Up: For the lower level MPC (LMPC), the sampling time for control implementation was 5 minutes and the prediction horizon was set to 40 minutes. For the upper level MPC (UMPC), the control implementation time interval and averaging window were set to 30 minutes with a 12-hour prediction horizon. The temperature upper limit (T_u^U) followed the same night setup schedule. That is, 23.3^oC (74^oF) during the occupied period and 24.4^oC (76^oF during the unoccupied period. The lower temperature limits for both LMPC and UMPC (T_l^U and T_l^L) were set to 21.1^oC (70^oF). The NOAA API was used to retrieve weather forecasts (using the Purdue airport weather station, which is around 1.5 miles away from the lab) for the UMPC's 12-hr ahead prediction and the forecast was updated every 30 min, which is the UMPC implementation time. Once a decision was made by the UMPC, it was applied to the LMPC according to Eqn. (4). Then, a stage decision was made by the LMPC, and was transmitted to the thermostats every 5 minutes, which is the control implementation time interval for the LMPC.

Utility Cost Setup: The utility cost is a key parameter that decides the behavior and energy cost savings of the MPC. A few different electricity rates were investigated, and in this paper the results for the following rate structure are presented.

- 0.20 \$/kWh from 1:00 PM to 6:00 PM (the ON peak period), and 0.10 \$/kWh at other times.

3.1 Experimental Test Results (ON-to-OFF peak price ratio of 2)

3.1.1 Experimental Results: Short-Term Comparison: Fig. 5 shows sample test results for 4 days for the conventional control and MPC on days in August under the ON-to-OFF peak price ratio of 2 and an on-peak period from 1:00 PM to 6:00 PM. As shown at the bottom of the figure, the two control algorithms were switched back and forth on a daily basis. Thermostat temperatures (blue line: U1, brown line: U2) are shown in the second subfigure. The unit powers are displayed in the third figure and the total HVAC power (a 15-min moving average) is shown in the last (red: conventional controller days, blue: the MPC days).

The temperatures in the second figure show that the MPC clearly responds to the on-off peak price signal (0.2 \$/kWh

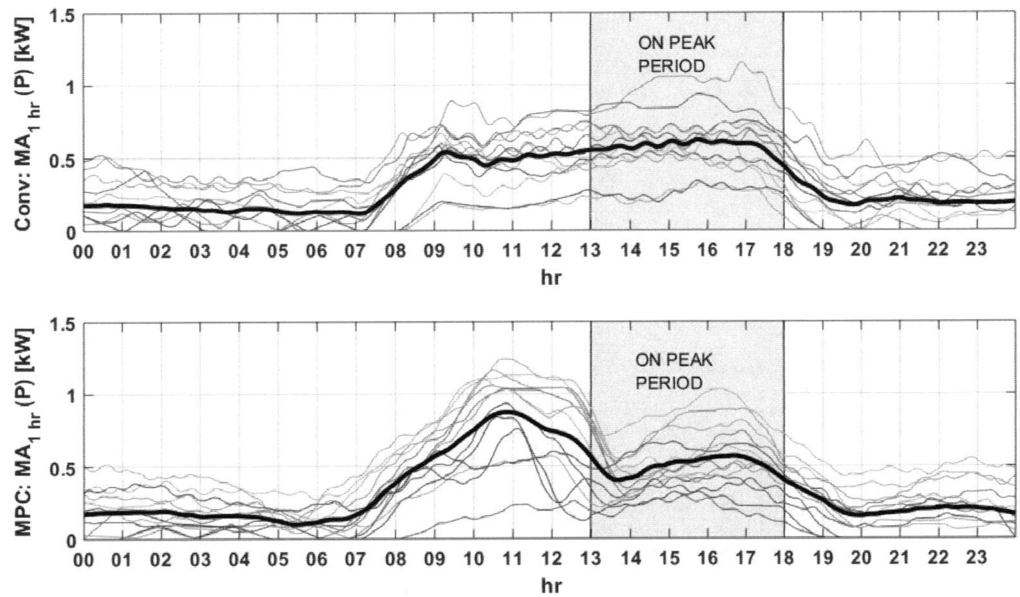

Figure 6: Distributions of daily profiles of total 1-hr moving average HVAC power for the conventional control and MPC. For each figure, different colors represent moving-averaged power measurements for different days, and the thick black line indicates the mean trajectory.

from 1:00 PM- 6:00 PM, 0.1 $/kWh elsewhere). The MPC precools the space down to the lower temperature limit (21.1^oC, 70^oF) when the electricity price is low, and turns off the units for a period after the price jumps. The precooling started at around 2 to 4 hours ahead of the price jump and the temperatures were smoothly lowered down. During the precooling periods, the peaks are under regulation as shown in the bottom of the figure.

3.1.2 Experimental Results: Long Term Comparison: Fig. 6 shows sets of daily HVAC power profiles (1-hr moving averaged) for the conventional and MPC. Fig. 7 shows comparisons of the two representative profiles between the two controllers. To ensure that the comparison of sampled powers is meaningful, daily temperature distributions are also provided in Fig. 8. Clearly, the MPC regulated thermostat temperatures within the specified temperature range (21.1 to 23.3^oC, 70 to 74^oF during the occupied period of 9:00 AM to 6:00 PM, and 21.1 to 24.4^oC, 70 to 76^oF during the unoccupied period). As shown in the last subfigure of Fig. 8, the MPC precooled the space to around 22.2^oC (72^oF) although the lower limit was set to 21.1^oC (70 ^oF). This is due to the two conflicting driving potentials of the MPC, i.e., the utility cost rate structure and increment of loads from the outdoor air for precooling.

Fig. 7 demonstrates the electrical load shifting capability of the MPC. It shifts electrical loads for about 3 to 4 hours. The MPC gradually increases the charging until the price jump and it discharges after that. The discharging capacity is high at the beginning but drops exponentially. The discharging process is effective for around 4 hrs (1:00 PM – 5:00 PM). The increase of energy consumption associated with the precooling compared with that of the baseline is 9.77%.

The energy cost savings for this experiment was only 3.61%. As mentioned, the small energy cost savings are due to a significant thermal resistance between the zone air and building structure and poor insulation. Fig. 9 shows 15-min moving averaged total power profiles for the two controllers to compare peak demand charges. The MPC's anytime demand savings are 36% (compare the black dashed lines showing the maximum values for the two controllers). For the case of on-peak demand charge, savings are 60% (see the red dashed lines showing the maximum peaks over the on-peak period). The greater on-peak demand savings are the combined effect of reduced cooling loads due to precooling and the narrower on-peak period where the discharging is effective in addition to the LMPC's coordination effort.

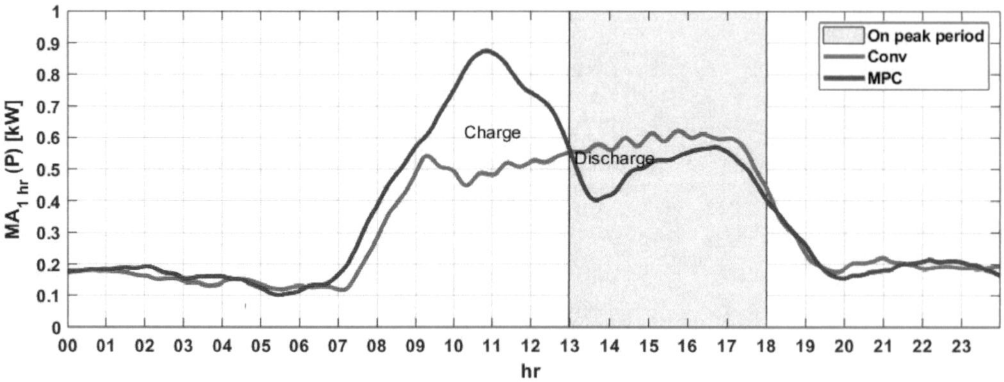

Figure 7: Comparison of representative daily profiles of total 1-hr moving average HVAC power between the conventional control and MPC

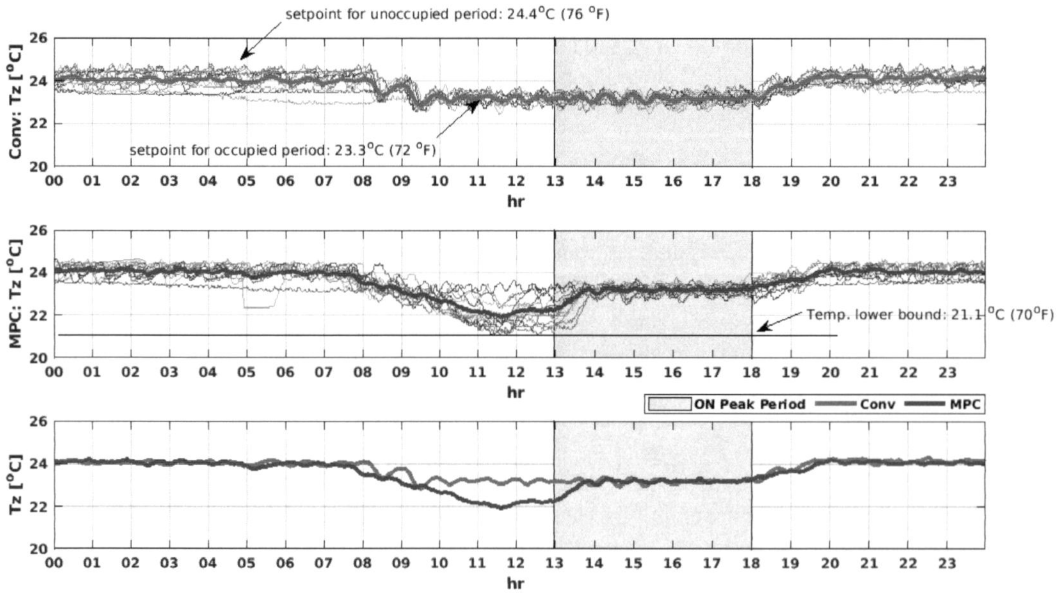

Figure 8: Distributions of daily temperature profiles between the conventional control and MPC. For each figure, different colors represent thermostat temperature measurements for different days, and the thick black line indicates the mean trajectory.

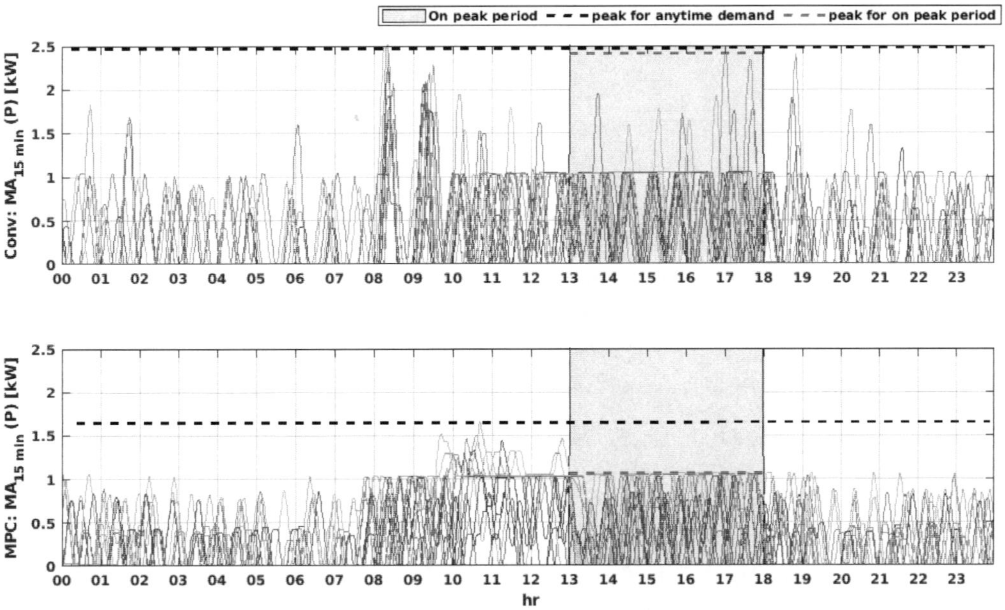

Figure 9: Distributions of daily profiles of total 15-min moving average HVAC power for the conventional control and MPC. For each figure, different colors represent moving-averaged power measurements for different days.

4. CONCLUSIONS

Despite a long history of studying MPCs applied to buildings and increasing needs for load flexibility in small and medium sized commercial buildings (SMCB), few advanced control algorithms have been developed for SMCBs. This paper discussed the unique MPC challenge of high peak demand occurring when shifting HVAC load for SMCBs served by multiple staged packaged air conditioners, and presented a hierarchical MPC to overcome this issue. The load shifting capability of the MPC while regulating the unnecessary peaks during precooling has been tested for a laboratory space over two months for two different time-of-use (TOU) scenarios. The MPC successfully shifted electrical loads in response to TOU rates, with remarkable HVAC demand cost savings, anytime demand cost savings over 30% and on-peak demand cost savings over 40%, but with only modest energy cost savings. The virtual battery, i.e. the building thermal mass, was effective for about 4-5 hours with a substantial load reduction during the ON peak period. The energy cost savings were less than 10%. One of the reasons for the low energy cost savings is the fact that the site has a carpet, mat and a poor envelope with single pane windows which results in inefficient charging and discharging of the building mass. It is expected that savings should be significantly greater for more typical commercial buildings.

REFERENCES

Biyik, E., Brooks, J. D., Sehgal, H., Shah, J., and Gency, S. (2015). Cloud-based model predictive building thermostatic controls of commercial buildings: Algorithm and implementation. In *American Control Conference (ACC), 2015*, pages 1683–1688. IEEE.

Kim, D., Braun, J., Cai, J., and Fugate, D. (2015). Development and experimental demonstration of a plug-and-play multiple rtu coordination control algorithm for small/medium commercial buildings. *Energy and Buildings*, 107:279–293.

Kim, D. and Braun, J. E. (2018). Development, implementation and performance of a model predictive controller for packaged air conditioners in small and medium-sized commercial building applications. *Energy and Buildings*, 178:49–60.

Kim, D., Cai, J., Ariyur, K. B., and Braun, J. E. (2016). System identification for building thermal systems under the presence of unmeasured disturbances in closed loop operation: Lumped disturbance modeling approach. *Building and Environment*, 107:169–180.

Kim, D., Cai, J., Braun, J. E., and Ariyur, K. B. (2018). System identification for building thermal systems under the presence of unmeasured disturbances in closed loop operation: Theoretical analysis and application. *Energy and Buildings*, 167:359 – 369.

Nutaro, J. J., Fugate, D. L., Kuruganti, T., and Starke, M. R. (2014). An inexpensive retrofit technology for reducing peak power demand in small and medium commercial buildings. Number 3494. Conference: 3rd International High Performance Buildings Conference at Purdue.

Putta, V., Kim, D., Cai, J., Hu, J., and Braun, J. E. (2015). A switched dynamic programming approach towards optimal control of multiple rooftop units. In *American Control Conference (ACC), 2015*, pages 281–287. IEEE.

Zhang, X., Pipattanasomporn, M., and Rahman, S. (2017). A self-learning algorithm for coordinated control of rooftop units in small-and medium-sized commercial buildings. *Applied Energy*, 205:1034–1049.

ACKNOWLEDGEMENT

The work was supported by the Center for High Performance Building (CHPB) at Purdue University, and completed with the support of the Assistant Secretary for Energy Efficiency and Renewable Energy, Building Technologies Office of the U.S. Department of Energy under Contract No. DE-AC02-05CH11231.

Assessing EMIS Benefits: A new field evaluation protocol offers rigor and flexibility

Guanjing Lin[1]*, Eliot Crowe[1], Valerie Nibler[1], Jessica Granderson[1]

[1] Lawrence Berkeley National Laboratory,
Berkeley, CA, USA
Contact Information (gjlin@lbl.gov; Tel.: +01-510-486-5979)

* Corresponding Author

ABSTRACT

Energy management and information systems (EMIS) are combined hardware and software products that comprise a broad range of analytics and/or optimal control functionality to manage commercial building energy use, covering three main types of functionality: energy information system focus on interval meter data analysis, fault detection and diagnostics focus on system or equipment-level data analysis, and automated system optimization focus on optimal control. More and more researchers and building owners seek to field test and pilot EMIS. However, the studies were conducted differently with disparate design and metrics. The field validation of EMIS also facing the challenges of qualification of non-energy benefits and linkage between information analysis and energy savings.

In response, this paper presents a standardized field testing protocol for EMIS assessment based on literature review and stakeholder interviews. The protocol provides a standardized approach for assessing EMIS' energy and non-energy benefits. It balances rigor and flexibility by specifying a set of required and optional evaluation metrics that can be tailored to the needs of technology evaluators, including energy savings, demand reduction, cost effectiveness, operation and maintenance improvement, capability to enable efficiency, etc. With this protocol, consistent evaluation results can be generalized across multiple studies to support the adoption and further advancement of EMIS.

1. INTRODUCTION

Energy management and information systems (EMIS) are combined hardware and software products that comprise a broad range of analytics and/or optimal control functionalities to uncover energy savings potential and to realize cost savings through improved energy management based on building data (Granderson et al. 2021). The data integrated into EMIS primarily comes from smart meters and/or building automation systems (BAS). Other building data sources such as weather data, internet of things sensors, or occupancy data, may be integrated into EMIS. EMIS may offer one or a combination of three main functionalities:

- Energy information systems (EIS): EIS analytics focus on meter-level monitoring, is used to store, analyze, and display energy meter data. Example analytics include time-series load profiling, energy benchmarking, peak load analysis, and energy anomaly detection.
- Fault detection and diagnostics (FDD): FDD analytics automate the process of detecting faults and suboptimal performance of building systems and help to diagnose potential causes. FDD obtains data from the BAS and focuses on system or equipment-level monitoring, analysis, and charting.
- Automated system optimization (ASO): ASO analytics continuously analyze, determine and write optimal schedule or control setpoints (e.g. supply air temperature setpoint, chilled water leaving temperature setpoint) back to Heating, Ventilation and Air Conditioning (HVAC) system through building automation system (BAS) to optimize energy use while maintaining occupant comfort.

The digitalization in the building sector drives the increasing interest in EMIS. There is a growing body of EMIS field evaluation projects conducted by researchers, utility programs, and building owners to assess the performance of a specific EMIS technology in real buildings. However, there has not been a standardized protocol for EMIS assessment. Evaluation studies are often conducted differently, with disparate design and metrics, and are difficult to generalize for any stakeholder to support the adoption and further advancement of EMIS. This paper presents a standardized field evaluation protocol for assessing EMIS' energy and non-energy benefits. It illustrates the steps in the evaluation process (Section 3), the key areas covered in a field evaluation plan (Section 4), and the minimum and optional evaluation parameters and approaches that can be tailored to the needs of technology evaluators (Section 5).

2. METHODOLOGY

The authors reviewed a set of EMIS field evaluation literatures from utility emerging technologies programs, Federal emerging technology demonstration programs (e.g., the Department of Energy High Impact Technology Program, the General Services Administration Proving Ground Program, the Department of Defense Environmental Security Technology Certification Program), and other R&D organizations. There were four literatures evaluating EMIS with EIS functionality (Henderson and Waltner 2013, Mercado and Elliott 2012, Owen et al. 2010, Russell J et al. 2011), seven literatures evaluating EMIS with FDD functionality (Ferretti et al. 2017, Gorbounov et al. 2016, Katipamula et al. 2003, Lane and Epperson 2015, Summers et al. 2012, Wall and Gao 2018, Loftness et al. 2016), and five literatures evaluating EMIS with ASO functionality (Granderson et al. 2018, Hail et al. 2016, Howell et al. 2015, Parthasarathy 2016, SDG&E 2015). The common and unique elements in the areas of field test design and evaluation parameters and approaches were identified from literature review. In addition, interviews were conducted with potential users of EMIS field evaluation results, such as researchers, building owners, utility administrators, and building energy codes and standard developers, to understand the priority decision-making and data needs that could be met through a validation protocol. Development of the protocol drew upon the literature review on a wide range of EMIS assessments, interviews with key stakeholders, and past EMIS evaluation projects led by Berkeley Lab.

3. OVERVIEW OF EMIS FIELD EVALUATION

Evaluating an EMIS can take considerable time and effort, from six months to a couple of years, so it is important to take a methodical approach to maximize the value of the eventual results. Table 1 illustrates the key steps in the EMIS evaluation process, which are designed to ensure that roles are clearly understood, data and risks are managed effectively, and that final results are accurate and meet the project sponsor's ultimate objectives.

Table 1: Key steps in the EMIS evaluation process

Step	Description
1: Select the EMIS test site.	An information-gathering screening form is developed based on technology requirements and evaluation objectives. The form lists the required and preferred site characteristics, such as building size, type and accessibility of building automation system (BAS), HVAC system configurations, control baseline, and metering conditions. The key screening considerations include the satisfaction of the required site and system characteristics, the availability of baseline data, the changes in occupancy, and if any major energy efficiency project happened in the baseline period or will happen in the post-installation period.
2: Develop an evaluation plan.	The EMIS field evaluation plan presents the technology and site information and also defines the performance objectives, metrics, analysis approaches, and schedule.
3: Collect baseline data and information.	Baseline data and information are collected at the beginning of the evaluation and the defined baseline period. Depending on the selected performance objectives, the baseline data may include energy use, weather data, utility tariffs, space conditions, existing operation and maintenance process, and more.
4: Track the technology installation / commissioning.	To evaluate the effort needed for installation and commissioning, information is gathered to document the items like the activities implemented during the process, the responsibilities of different stakeholders for each activity, the lead time of this stage, challenges, and best practices.
5: Collect and analyze post-installation data and information.	For each EMIS performance parameter being evaluated, data will be gathered after the EMIS has been operational for the required amount of time. Performance data may be a combination of quantitative and qualitative data. In the case of quantitative data, it is recommended to review the data shortly after EMIS installation to ensure that data will be of sufficient quality.
6: Produce an evaluation report.	An evaluation report is the final deliverable in the field evaluation.

4. EMIS FIELD EVALUATION PLAN

Once an EMIS has been chosen for evaluation and a test site has been selected, the next step in the EMIS assessment process is to develop a field evaluation plan. Two key areas that should be emphasized in the EMIS field evaluation plan, including:

- **Description of the technology and the field evaluation site**: Documenting the technology and the site characteristics in the field evaluation Plan (and reproducing them when documenting assessment results) helps any reader understand that context. EMIS analytics (i.e. EIS, FDD, ASO) and data points integrated (e.g. meter data, BAS data, weather data, and other pertinent data) should be documented in the technology description. Building type, BAS model and make, building systems or equipment covered by EMIS are the essential things needs to be included in the site description. If the ASO is evaluated, the existing control sequence of the optimized control setpoints should be described, e.g., "The chilled water setpoint is reset between 42°F and 48°F based on the maximum chilled water valve position from each air handling unit."
- **Evaluation objectives and approaches**: The EMIS field evaluation plan should document the objectives of the evaluation and reference the key stakeholders targeted by the EMIS implementation (those people driving the objectives and/or expected to be affected by the EMIS). Example objectives may include energy savings, perhaps in line with organizational sustainability targets or a strategic plan; load reduction during peak periods, when electricity costs are highest; improvements to occupant comfort satisfaction (e.g., reducing hot/cold calls from occupants); and improvements in operations and maintenance (O&M) practices. Defining evaluation objectives helps when selecting the key performance parameters that will be included in the design. Analysis approaches for the selected parameters also need to be presented in the evaluation plan. The required or optional parameters and their analysis approaches are discussed in Section 5.

5. FIELD EVALUATION PARAMETERS AND APPROACHES

At the core of the EMIS validation protocol is a set of performance evaluation parameters that will allow for a consistent comparison between EMIS tools (see Table 2), along with associated methods for determining those parameters. The assessment approaches taken may be based on quantitative data analysis, surveys, or a combination of both. These evaluation parameters were chosen based on literature review and stakeholder interviews to determine the highest value core metrics, along with those that may apply in some, but not all, circumstances. At a minimum, energy savings, energy cost savings, EMIS cost, simple payback, and capability to enable energy efficiency are required for a basic EMIS assessment. Additional optional parameters fall under the following four categories:

- Energy and utility cost: In addition to annual energy savings and cost savings, stakeholders may want to assess peak demand reductions, particularly in regions where utilities apply high demand charges. Where utilities offer demand response (DR) programs, stakeholders may also want to assess the ability of EMIS to drive temporary load reductions during DR time periods.
- Non-energy benefits: Non-energy benefits can provide significant value to building owners. An occupant comfort metric allows for quantification of improvements in indoor environmental quality, and an O&M metric can verify EMIS impact on internal operational practices.
- Cost-effectiveness: While simple payback is relatively easy to calculate and understand, some organizations employ more sophisticated methods to calculate long term return on investment. Net present value (NPV) and savings-to-investment ratio (SIR) are two common examples included as optional metrics under this protocol.
- Operational capabilities: Capturing the overall impact of an EMIS is critical to most stakeholders, but for many it is also important to validate specific performance claims. It is essential to understand how effective an EMIS is at enabling energy efficient operational practices, and how the tools contribute to the energy savings. It also is necessary to provide instructions for integrating the tools into the energy management process with a "standard operating procedure." The protocol offers another two optional metrics, to address how easy an EMIS is to install and commission, and how accurately an EMIS can identify operational faults and make appropriate recommendations.

Among the optional parameters listed in Table 2, 'Occupant comfort satisfaction' is highly recommended for the evaluation of ASO, as optimizing existing controls should not adversely affect comfort. Also 'Accuracy of issues/opportunities identified by the FDD' applies only for the evaluation of FDD. As noted earlier, this EMIS validation protocol is not intended to address every possible evaluation parameter that could be applied to any situation. The key objective is to define a clear set of core parameters that will align with most stakeholders' objectives.

Table 2: EMIS field evaluation parameters summary

Performance Evaluation Parameter	Required or Optional	Approach
Energy and utility cost		
Annual energy savings (kBtu per ft^2, percent reduction)	Required	Data analysis
Annual energy cost savings ($)	Required	Data analysis
Annual greenhouse gas emission reduction (pounds of carbon dioxide equivalent, $lbCO_2e$)	Optional	Data analysis
Monthly peak demand reduction (kW)	Optional	Data analysis
Demand flexibility (W/ft^2, kW, percent)	Optional	Data analysis
Non-energy benefits		
Occupant comfort satisfaction	Optional (recommended for ASO)	Data analysis and/or survey
Operations and maintenance	Optional	Data analysis and/or survey
EMIS cost		
EMIS cost ($, $ per ft^2)	Required	Survey
Cost-effectiveness		
Simple payback (years)	Required	Data analysis
Net present value ($)	Optional	Data analysis
Savings-to-investment ratio	Optional	Data analysis
Operational capabilities		
Effort of the installation and commissioning process	Optional	Survey
Capability to enable energy efficiency	Required	Survey
Accuracy of issues/opportunities identified by the FDD	Optional (applicable to FDD only)	Data analysis

The following subsections introduce how to access energy savings, cost effectiveness, capability to enable energy efficiency in an EMIS field evaluation.

5.1 Energy savings

Defining EMIS energy savings is one of the most challenging aspects of EMIS validation, and it faces three major challenges:

- EMIS is not a widget technology. The use of EMIS leads to multiple energy efficiency measures. The energy savings should capture the impacts of all the measures.
- The building energy consumption is affected by various factors. The savings estimation needs to consider the changes in these variables, such as weather conditions and building occupancy.
- From a practical standpoint, an EMIS validation project may have time and/or budget constraints that affect the level of rigor that can be applied to an EMIS validation project.

The International Performance Measurement and Verification Protocol (IPMVP) (EVO, 2012) defines four generic M&V approaches for determining energy savings: Option A - Retrofit Isolation with Key Parameter Measurement, Option B - Retrofit Isolation with All Parameter Measurement, Option C - Whole Building Utility Data Analysis, and Option D - Calibrated Computer Simulation. Under this protocol, the recommended savings estimation method for determining annual energy savings is to follow Option C or Option B, which determines savings impacts based on actual metered data using Equation 1.

$$\text{Savings} = \text{Baseline Projected Energy} - \text{Post Installation Energy} \pm \text{Adjustments} \quad (1)$$

With Option B or C, a mathematical baseline model is created from meter data when the technology is not operating. The baseline model is then forward projected into the measure post-installation period to determine what the energy use would have been in the absence of the technology. The difference between this baseline projected energy use, and the metered post installation energy use is taken as the energy savings. The Adjustments term is used to capture the effects of variables not included in the baseline model, and not associated with the technology, such as increased internal loads, or changes to equipment or building occupancy. Linear, change-point linear, and polynomial regression models are often used to create the mathematical baseline model for IPMVP Option B or C applications. The primary independent variables used for the model include weather conditions (usually outside air temperature), building operation schedule, and building occupancy. For the regression energy model of a chiller plant or chilled water system, the cooling load is the key independent variable. If there is no Btu meter installed for measuring the cooling load, the cooling load can be estimated using outside air temperature, outside air relative humidity (or outside air wet-bulb temperature), and day of the week. The day of the week is often best included as a categorical value (e.g., Sunday, Monday, Tuesday) and not as a numerical value. The savings analysis based on the metered data provides an accurate means of verifying the impact of the multiple energy efficiency measures enabled by the EMIS. ASHRAE Guideline 14 (ASHRAE 2014), BPA Verification by Energy Modeling Protocol (BPA 2012a), and BPA Regression for M&V reference guide (BPA 2012b) provide detailed guidance on the application of meter-based Option B and Option C approaches; for example, regression energy model types, development of the energy model, and software tools to assist with energy modeling. As a general rule of thumb, IPMVP Option C using monthly data requires expected savings > 10 percent of the whole building energy savings, and > 5 percent if using hourly data.

The engineering calculation in IPMVP Option A is usually used for estimating savings of an individual efficiency measure, and therefore is only acceptable as a backup if options B or C are not possible (e.g., due to insurmountable issues with obtaining meter data or project delays resulting in lack of time to gather reporting period meter data). IPMVP Option D uses simulation software (e.g., EnergyPlus, OpenStudio) to model energy performance of a whole building. Models must be calibrated with actual hourly or monthly billing data from the facility. After the model has been calibrated, savings are determined by comparing a simulation of the baseline with either a simulation of the performance period or actual utility data. Option D is acceptable as a backup where a baseline does not exist (e.g., new construction or major building modification in the baseline period).

5.2 Cost-effectiveness

Determining the cost-effectiveness of EMIS implementation is not straightforward since EMIS with the EIS or FDD functionality is an enabling tool—installation of the software does not directly create savings. Rather, savings are achieved by acting upon the information the technology provides (i.e., the improvement opportunities that are identified). The only functionality of EMIS that achieves direct savings is ASO, since the control optimization is performed directly by the ASO functionality. Simple payback period (SPP) is the most widely used financial metric for energy efficiency projects. It is the number of years required to recover the initial investment through project savings. The simple payback period of EMIS can be established using Equation 2:

Simple payback period = (EMIS Implementation Costs + ECM Costs)/(Annual Energy Cost Savings - Annual EMIS Operating Costs) (2)

Where energy conservation measure (ECM) Costs are the costs incurred for implementing the ECMs found by the EMIS (e.g., adjusting system schedules, fixing leaking valves, improving setpoint reset)[1]. The elements included in EMIS implementation costs and annual EMIS operating costs are explained in Table 3. With Equation 2, SPP captures not only EMIS implementation costs, but also the costs for implementing ECMs discovered through the use of the EMIS. The denominator in Equation 2 may be considered the net annual cost savings, based on subtracting annual EMIS operating costs (EMIS software subscription, third party support, internal labor) from the annual energy cost savings.

Table 3: Key elements of EMIS costs

<table>
<tr><th colspan="3">Cost Items</th><th>Description</th></tr>
<tr><td rowspan="3">EMIS Implementation Costs</td><td rowspan="2">A: Base costs for EMIS technology</td><td>A.1: Hardware costs</td><td>Costs for hardware installation and upgrade (e.g., adding meters and sensors during the project for EMIS monitoring purposes, installing gateways for communication, getting data servers for data storage)</td></tr>
<tr><td>A.2: Software costs</td><td>Costs for the EMIS software installation and configuration to bring in all the data points, alteration of the existing BAS to expose legacy data points, and training to site staff, including EMIS vendor and service provider costs</td></tr>
<tr><td colspan="2">B: In-house labor costs for EMIS installation and commissioning</td><td>Approximate total labor costs spent by in-house staff to support installation and configuration of the EMIS</td></tr>
<tr><td rowspan="3">Ongoing Annual EMIS Operating Costs</td><td rowspan="2">C: Ongoing annual costs for EMIS technology</td><td>C.1: Annual EMIS costs</td><td>The recurring annual cost for a software license, software-as-a-service fees, or hardware (e.g. occupancy counters)</td></tr>
<tr><td>C.2: Annual third-party consultant costs</td><td>The average annual cost paid to a third-party consultant for support in analyzing and implementing EMIS findings</td></tr>
<tr><td colspan="2">D: Ongoing annual in-house labor costs for EMIS use</td><td>Approximate labor costs spent by in-house staff reviewing EMIS reports, identifying opportunities for improvement, and implementing measures (may be based on average hours spent per month)</td></tr>
</table>

5.3 Capability to enable energy efficiency

Capability to enable energy efficiency can well explain the linkage between information analysis and energy savings. In this protocol, the capability to enable energy efficiency has different meanings for ASO and EIS/FDD analytics. For ASO, it means the targeted control setpoints can be successfully changed by the ASO. For EIS/FDD, it means the

[1] This cost category is not applicable for ASO, as it directly makes the efficiency changes in its system. Costs for internal staff to implement ECMs does not need to be accounted for.

ability to generate actionable information that leads to the actual efficiency measures. Evaluating the capabilities for identifying efficiency opportunities and supporting the implementation of the efficiency measures will help potential adopters understand how the tools contribute to the energy savings. It can also provide support for successful integration of an EMIS into a building's energy management process.

To evaluate the capability to enable energy efficiency of ASO, the data trends of the targeted control setpoints and measurements should be compared in the baseline and optimizer (post-installation) periods. An assessment can validate if the setpoints change and if the measurements follow the optimized setpoints via BAS data trend analysis. For example, Figure 1 shows the ASO successfully reduces the air handler unit (AHU) static pressure setpoint (SP) by 0.5 pounds per square inch (psi) compared with the baseline. The static pressure SP and static pressure in the optimizer period are shown in blue lines, and the static pressure SP and static pressure in the baseline period are shown in red lines.

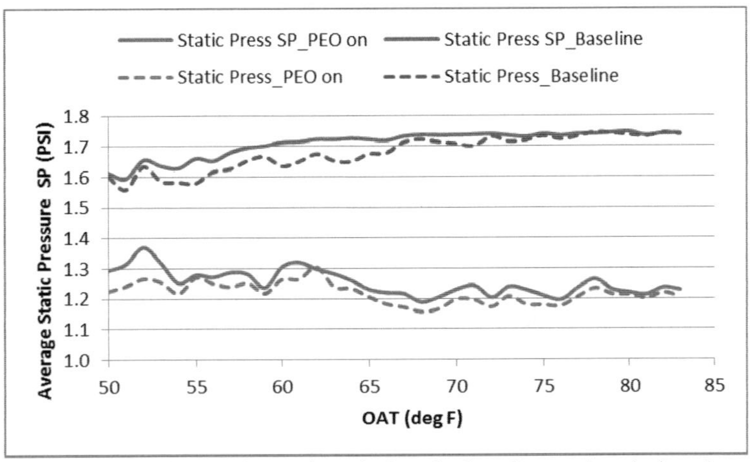

Figure 1: Example of comparison of control setpoints during the baseline and optimizer (post-installation) periods

To evaluate the capability to enable energy efficiency of EMIS with EIS and FDD analytics, the following items should be documented through building operator interviews and the results shown on the EMIS.

- *A record of the implemented efficiency measures based on EIS/FDD analytics.* An example summary is provided in Table 4.

Table 4: Summary of the identified faults in an EMIS with FDD analytics and the implemented efficiency measures

System/Equipment	Identified Faults	Implemented Efficiency Measures
AHU 1-1 AHU 1-2	AHU on all the time	Enable calendar control
VAV 1-1, VAV 1-2, VAV 1-3, VAV 1-4	Zones are outside an acceptable comfort temperature range	Reset the automatic setpoint, tune VAV supply air flow
AHU 2-1	Incorrect economizer control	Reset the minimum outside air intake ratio
AHU 2-2	Valve cycling	Change control logic proportional–integral–derivative loop
Outdoor lighting	Outdoor lighting on a fixed schedule	Introduce daylight harvesting control

- *The workflow of identifying, prioritizing, and taking actions on the issues or opportunities identified by the EIS or FDD analytics.* Faults are prioritized using criteria like impact on energy, comfort, or existence of known issues. Determine which departments or business units are involved, and who is responsible for responding to the finding. Prioritize and assign a list of faults for inspection, inspect the faults, and implement the efficiency measures. Sometimes actions such as equipment scheduling can be addressed by site-level operational staff. In other cases, further investigation may be required, and control and mechanical subcontractors need to be involved. The documentation of workflow leads to a "standard operating procedure" which is easily repeated in the future application.

6. CONCLUSIONS

This paper presents a standardized field testing protocol for EMIS assessment based on literature review, stakeholder interviews, and previous field evaluation project experience. It describes the steps in the evaluation process and the key areas covered in a field evaluation plan. It also balances rigor and flexibility by specifying a set of required and optional performance evaluation parameters that can be tailored to the needs of technology evaluators, including energy savings, demand reduction, cost effectiveness, operation and maintenance improvement, capability to enable efficiency, etc. With this protocol, consistent evaluation results can be generalized across multiple studies to support the adoption and further advancement of EMIS.

REFERENCES

ASHRAE. ASHRAE Guideline 14-2014: Measurement of Energy, Demand and Water Savings. 2014.

Bonneville Power Administration (BPA). 2012a. Verification by Energy Modeling Protocol.

Bonneville Power Administration (BPA). 2012b. Regression for M&V Reference Guide.

Efficiency Valuation Organization (EVO). 2012. International Performance Measurement and Verification Protocol: Concepts and Options for Determining Energy and Water Savings Volume I. EVO-10000-1.

Ferretti, N., Galler, M., Bushby, S., 2017. Performance Monitoring of Chilled-Water Distribution Systems Using HVAC-Cx. ASHRAE Transactions 123, 53–63.

Gorbounov, M., Wang, J., Yasar, M., Reeve, H., 2016. Field testing of diagnostics for state-of-the-art RTUs. Consortium for Building Energy Innovation.

Granderson J, Lin G, Singla R, Fernandes S, Touzani S. Field evaluation of performance of HVAC optimization system in commercial buildings. Energy and Buildings. 2018 (173):577-86.

Granderson, J., Lin, G., Kramer, Hannah, Nibler, V., 2021. A Primer on Organizational Use of Energy Management and In-formation Systems. Better Buildings, U.S. Department of Energy.

Hannah, K., Lin, G., Curtin, C., Crowe, E., Granderson, J., 2020. Proving the Business Case for Building Analytics, Better Buildings, U.S. Department of Energy.

Hail, J., Hatley, D., Underhill, R., 2016. Optimization of Variable Speed Chiller Plants: Frank M. Johnson Jr. Federal Building and U.S. Courthouse, Montgomery, Alabama. United States General Services Administration.

Henderson, P., Waltner, M., 2013. Real-Time Energy Management: A Case Study of Three Large Commercial Buildings in Washington, D.C (No. CS:13-07-A). Natural Resources Defense Council.

Howett, D., Rizy, C., Wolfe, A., 2015. Socially Driven HVAC Optimization Federal Building and US Courthouse Phoenix, Arizona. United States General Services Administration.

Katipamula, S., Bauman, N., Pratt, R.G., Brambley, M.R., 2003. Demonstration of the Whole-Building Diagnostician in a Single-Building Operator Environment (No. PNNL-14239). Pacific Northwest National Laboratory.

Lane, K., Epperson, L., 2015. Enterprise Plug-and-Play Diagnostics and Optimization for Smart Buildings (No. CEC-500-2015-084). California Energy Commission.

Loftness, V., Aziz, A., Lasternas, B., 2016. Building Performance Optimization while Empowering Occupants Toward Environmentally Sustainable Behavior through Continuous Monitoring and Diagnostics (No. EW-201406). the Department of Defense Environmental Security Technology Certification Program.

Mercado, A., Elliott, J., 2012. Energy Performance Platform: Revealing and Maintaining Efficiency With a Customized Energy Information System. Presented at the ACEEE Summer Study on Energy Efficiency in Buildings.

Parthasarathy, G., 2016. Central Plant Optimization for Waste Energy Reduction (No. EW-201349). the Department of Defense Environmental Security Technology Certification Program.

Russell J., Salanti S., Mitchell J., 2011. Assessment of an Energy Information System for the Grocery Sector (No. ET10PGE1031). Pacific Gas and Electric Company.

SDG&E’ s Emerging Technologies Program., 2015. Model-Based Predictive HVAC Control Enhancement Software (No. DR13SDGE0006). San Diego Gas & Electric.

Summers, H., Hilger, C., 2012. Fault Detection and Diagnostic Software (No. ET11PGE3131). Pacific Gas and Electric Company.

Owen, T., Pape-Salmon, A., McMurchy, B., 2010. Employee Engagement and Energy Information Software Supporting Carbon Neutrality. Presented at the ACEEE Summer Study on Energy Efficiency in Buildings.

Wall, J., Guo, Y., 2018. Evaluation of Next-Generation Automated Fault Detection & Diagnostics, Diagnostics (FDD) Tools for Commercial Building Energy Efficiency – Final Report Part I: Case Studies in Australia, FDD Case Studies in Australia (No. RP1026). Low Carbon Living CRC.

ACKNOWLEDGEMENT

This work was supported by the Assistant Secretary for Energy Efficiency and Renewable Energy, Building Technologies Office, of the U.S. Department of Energy under Contract No. DE-AC02-05CH11231. The authors wish to acknowledge Hannah Debelius and Harry Bergmann for their guidance and support of the research.

AUTHOR INDEX

Ray W. Herrick Laboratories
School of Mechanical Engineering
Purdue University
177 S. Russell Street
West Lafayette, IN 47907-2099, USA.

ISBN 978-1-7138-5930-7

Annual Conference and Expo of the Association of Avian Veterinarians (AAV 2022)

Held at ExoticsCon 2022

Denver, Colorado, USA
14 – 18 August 2022